METHODS IN CELL BIOLOGY

VOLUME 36
Xenopus laevis: Practical Uses in Cell and Molecular Biology

Series Editor

LESLIE WILSON

*Department of Biological Sciences
University of California, Santa Barbara
Santa Barbara, California*

METHODS IN CELL BIOLOGY

Prepared under the Auspices of the American Society for Cell Biology

VOLUME 36
Xenopus laevis: Practical Uses in Cell and Molecular Biology

Edited by

BRIAN K. KAY

DEPARTMENT OF BIOLOGY
THE UNIVERSITY OF NORTH CAROLINA AT CHAPEL HILL
CHAPEL HILL, NORTH CAROLINA

H. BENJAMIN PENG

DEPARTMENT OF CELL BIOLOGY AND ANATOMY
THE UNIVERSITY OF NORTH CAROLINA AT CHAPEL HILL
CHAPEL HILL, NORTH CAROLINA

ACADEMIC PRESS, INC.
Harcourt Brace Jovanovich, Publishers

San Diego   New York   Boston   London   Sydney   Tokyo   Toronto

This book is printed on acid-free paper. ∞

Copyright © 1991 BY ACADEMIC PRESS, INC.
All Rights Reserved.
No part of this publication may be reproduced or transmitted in any form or by any means, electronic or mechanical, including photocopy, recording, or any information storage and retrieval system, without permission in writing from the publisher.

Academic Press, Inc.
San Diego, California 92101

United Kingdom Edition published by
ACADEMIC PRESS LIMITED
24-28 Oval Road, London NW1 7DX

Library of Congress Catalog Card Number: 64-14220

ISBN 0-12-564136-2 (alk. paper)

PRINTED IN THE UNITED STATES OF AMERICA
91 92 93 94   9 8 7 6 5 4 3 2 1

CONTENTS

Contributors — xiii
Foreword — xvii
Preface — xix

PART I. GENERAL INFORMATION

1. *Raising Xenopus in the Laboratory*
   M. Wu and J. Gerhart

I.	Introduction	3
II.	Equipment	5
III.	Infectious Diseases	11
IV.	Raising Tadpoles from Ovulated Eggs	13
V.	Raising and Maintaining Frogs	15
	References	17

2. *Genetics of Xenopus laevis*
   Jean-Daniel Graf and Hans Rudolf Kobel

I.	Introduction	19
II.	Genome Organization	21
III.	Estimation of Proportion of Conserved Duplications	24
IV.	Gene Mapping	27
V.	Conclusion	30
	References	31

3. *Experimentally Induced Homozygosity in Xenopus laevis*
   Robert Tompkins and Dana Reinschmidt

I.	Introduction	35
II.	Methods	36
III.	Conclusions	43
	References	44

4. *Oogenesis and Oocyte Isolation*
   L. Dennis Smith, Weilong Xu, and Robert L. Varnold

I.	Introduction	45
II.	Oogenesis: An Overview	46

III.	Isolation of Individual Oocytes from the Ovary	47
IV.	Protein Synthesis during Oogenesis: Practical Hints	51
V.	Labeling of Oocytes	54
VI.	Oocyte Maturation	56
VII.	Concluding Comments	57
	References	58

5. *Early Embryonic Development of Xenopus laevis*
   Ray Keller

I.	Introduction	62
II.	Origins of the Gastrula: Cleavage and Blastula Stages	62
III.	Gastrulation and Neurulation	68
IV.	Investigating Cell Behaviors	82
V.	Experimental Design and the Context-Dependent Function of Cells and Tissues	101
VI.	Microsurgical Methods, Tools, and Manipulations	102
VII.	Microscopy, Image Processing, and Recording: Epi-illumination and Low-Light Fluorescence Microscopy	108
	References	109

PART II. OOCYTES

6. *Vitellogenin Uptake and in Vitro Culture of Oocytes*
   Lee K. Opresko

I.	Introduction and History	117
II.	Methods	119
III.	Discussion	130
	References	131

7. *Biochemical Fractionation of Oocytes*
   Janice P. Evans and Brian K. Kay

I.	Introduction	133
II.	Methods	135
III.	Discussion	146
	References	147

8. *Lampbrush Chromosomes*
   Joseph G. Gall, Harold G. Callan, Zheng'an Wu, and Christine Murphy

I.	Introduction	150
II.	Protocol for Lampbrush Chromosomes of *Xenopus*	151
III.	Composition of Isolation and Dispersal Media	157

IV.	Alternatives If Germinal Vesicle Contents Fail to Disperse	159
V.	Use of Lampbrush Chromosome Preparations	159
VI.	Solutions	162
VII.	Materials	164
	References	165

9. Preparation of Synthetic mRNAs and Analyses of Translational Efficiency in Microinjected Xenopus Oocytes
Mike Wormington

I.	Introduction	167
II.	Methods	171
	References	182

10. Use of Oligonucleotides for Antisense Experiments in Xenopus laevis Oocytes
Carol Prives and Diana Foukal

I.	Introduction	185
II.	Oligonucleotides Reveal Aspects of snRNA Structure and Function in Oocytes	196
	References	208

PART III. EMBRYOS

11. Fertilization of Cultured Xenopus Oocytes and Use in Studies of Maternally Inherited Molecules
J. Heasman, S. Holwill, and C. C. Wylie

I.	Introduction	214
II.	Protocol for Oocyte Fertilization Using the Host Transfer Technique	216
III.	Protocol for Oocyte Fertilization Using *in Vitro* Fertilization Method	219
IV.	Choice of Method	222
V.	Applications of the Techniques	222
VI.	Summary	227
	References	228

12. Isolation of Extracellular Matrix Structures from Xenopus laevis Oocytes, Eggs, and Embryos
Jerry L. Hedrick and Daniel M. Hardy

I.	Introduction	232
II.	Methods	233
III.	Discussion	243
	References	246

13. *Analysis of Cellular Signaling Events, the Cytoskeleton, and Spatial Organization of Macromolecules during Early Xenopus Development*
    David G. Capco and William M. Bement

I.	Introduction	249
II.	Methodology	252
III.	Conclusions	268
	References	269

14. *Generation of Body Plan Phenotypes in Early Embryogenesis*
    Ken Kao and Mike Danilchik

I.	Introduction	272
II.	Axis Perturbations in Early Development	272
III.	Scoring of Phenotypes	277
IV.	Methods	278
V.	Summary	283
	References	283

15. *Fluorescent Dextran Clonal Markers*
    Robert L. Gimlich

I.	Introduction	285
II.	Methods	288
III.	Practical Considerations	293
	References	296

16. *Nuclear Transplantation in Xenopus*
    J. B. Gurdon

I.	Introduction and Historical Background	299
II.	Results Attainable with Nuclei Transplanted to Eggs	301
III.	Methodology	301
	References	309

17. *Mesoderm Induction*
    Igor B. Dawid

I.	Introduction	311
II.	Mesoderm Induction Assay	312
III.	Assay End Points: Markers of Mesoderm Induction	316
IV.	Mesoderm-Inducing Factors	319
V.	Specification of Cell Fate, Organization of Tissues, and Axis Determination	322

	VI.	Outlook	324
		References	325

18. *Neural Induction*
    Carey R. Phillips

	I.	Neural Induction: Historical Perspective	329
	II.	Recent Experimental Approaches to Neural Induction	331
	III.	Signals Initiating Neural Induction	335
	IV.	Methods	338
	V.	Summary	342
		References	344

19. *Analysis of Class II Gene Regulation*
    Thomas D. Sargent and Peter H. Mathers

	I.	Introduction	347
	II.	Brief Review of Class II Gene Transfer into Frog Embryos	348
	III.	Gene Selection	349
	IV.	Initial DNA Injection Experiments	350
	V.	Preliminary Mapping by Polymerase Chain Reaction Footprinting Analysis	353
	VI.	Regulatory Element Functional Studies	363
		References	364

20. *Assays for Gene Function in Developing Xenopus Embryos*
    Peter D. Vize, Ali Hemmati-Brivanlou, Richard M. Harland,
    and Douglas A. Melton

	I.	Introduction	368
	II.	Methods	368
	III.	Results	372
	IV.	Conclusions	383
		References	384

21. *Histological Preparation of Xenopus laevis Oocytes and Embryos*
    Gregory M. Kelly, Douglas W. Eib, and Randall T. Moon

	I.	Introduction	389
	II.	Tissue Processing for Light Microscopy	391
	III.	Light Microscope Level Immunohistochemistry	401
	IV.	Tissue Preparation for Electron Microscopy	404
	V.	Immunoelectron Microscopy	407
	VI.	Photomicrography	412
		References	414

22. *Whole-Mount Staining of Xenopus and Other Vertebrates*
    Michael W. Klymkowsky and James Hanken

    I. Introduction   420
    II. Methods   429
    III. Formulations   437
    IV. Conclusion   438
    References   439

23. *In Situ Hyridization*
    Heather Perry O'Keefe, Douglas A. Melton, Beatrice Ferreiro, and Chris Kintner

    I. Introduction   444
    II. *In Situ* Hybridization to Sectioned Tissue   445
    III. *In Situ* Hybridization to Whole-Mount Tissue   456
    IV. Evaluation of Method   459
    References   462

PART IV. MODEL SYSTEMS USING OOCYTES, EGGS, AND EMBRYOS

24. *DNA Recombination and Repair in Oocytes, Eggs, and Extracts*
    Dana Carroll and Chris W. Lehman

    I. Introduction   467
    II. Methods   469
    III. Results and Discussion   476
    References   484

25. *Expression of Ion Channels by Injection of mRNA into Xenopus Oocytes*
    Alan L. Goldin

    I. Introduction   487
    II. Studies Using Oocytes to Express Ion Channels   488
    III. Cautionary Notes   490
    IV. Preparation of RNA for Injection   493
    V. Preparation and Injection of Oocytes   498
    VI. Biochemical Analysis of Expression   499
    VII. Electrophysiological Analysis   502
    VIII. Conclusions   507
    References   508

26. *Tissue Culture of Xenopus Neurons and Muscle Cells as a Model for Studying Synaptic Induction*
    H. Benjamin Peng, Lauren P. Baker, and Qiming Chen

    I. Introduction   511

II.	Preparation of Neuron and Myotomal Muscle Cultures	512
III.	Characteristics of *Xenopus* Cultures and Development of Neuromuscular Junctions *in Vitro*	519
IV.	Preparation of Myotube Cultures	523
V.	Discussion	525
	References	526

27. The Xenopus Embryo as a Model System for the Study of Cell–Extracellular Matrix Interactions
Douglas W. DeSimone and Kurt E. Johnson

I.	Introduction	527
II.	Methods	529
III.	Concluding Remarks	537
	References	539

28. Chromatin Assembly
Alan P. Wolffe and Caroline Schild

I.	Introduction	542
II.	Preparation of Extracts	543
III.	Assays for Chromatin Assembly	548
IV.	Unresolved Problems	553
	References	557

29. DNA Replication in Cell-Free Extracts from Xenopus laevis
Gregory H. Leno and Ronald A. Laskey

I.	Introduction	561
II.	Preparation of Egg and Oocyte Extracts	562
III.	Analysis of DNA Replication in *Xenopus* Cell-Free Systems	570
IV.	Selecting an Appropriate System	575
	References	578

30. Cell Cycle Extracts
Andrew W. Murray

I.	Introduction	581
II.	Reagents and Equipment	584
III.	Methods	587
IV.	Conclusion and Prospects	603
	References	604

31. Egg Extracts for Nuclear Import and Nuclear Assembly Reactions
Donald D. Newmeyer and Katherine L. Wilson

I.	Introduction	608
II.	Preparatory Methods	616

	III.	Nuclear Assembly Reactions and Assays	624
	IV.	Nuclear Transport Substrates, Reactions, and Assays	625
		References	631

32. Xenopus Cell Lines
J. C. Smith and J. R. Tata

	I.	Introduction	635
	II.	Uses of *Xenopus* Cell Lines	636
	III.	Solutions	637
	IV.	Preparation of *Xenopus* Cell Cultures	638
	V.	Culture Methods	640
	VI.	Transfection of *Xenopus* Cell Lines	644
	VII.	XTC Mesoderm-Inducing Factor	649
	VIII.	Conclusions	652
		References	653

PART V. APPENDIXES

Appendix A. *Solutions and Protocols*   657
H. Benjamin Peng

Appendix B. *Injections of Oocytes and Embryos*   663
Brian K. Kay

Appendix C. *Mutants of Xenopus laevis*   671
Anne Droin

Appendix D. *Codon Usage for Xenopus laevis*   675
J. Michael Cherry

Appendix E. *Pictorial Collage of Embryonic Stages*   679
Mike Danilchik, H. Benjamin Peng, and Brian K. Kay

Appendix F. *Xenopus Suppliers in the United States*   683

Appendix G. *In Situ Hydridization: An Improved Whole Mount Method for Xenopus Embryos*   685
Richard M. Harland

INDEX   697

CONTRIBUTORS

Numbers in parentheses indicate the pages on which the authors' contributions begin.

LAUREN P. BAKER, Department of Cell Biology and Anatomy, University of North Carolina at Chapel Hill, Chapel Hill, North Carolina 27599 (511)

WILLIAM M. BEMENT, Department of Zoology, Arizona State University, Tempe, Arizona 85287 (249)

HAROLD G. CALLAN, Gatty Marine Laboratory, University of Saint Andrews, Saint Andrews, Scotland (149)

DAVID G. CAPCO, Department of Zoology, Arizona State University, Tempe, Arizona 85287 (249)

DANA CARROLL, Department of Biochemistry, University of Utah School of Medicine, Salt Lake City, Utah 84132 (467)

QIMING CHEN, Department of Cell Biology and Anatomy, University of North Carolina at Chapel Hill, Chapel Hill, North Carolina 27599 (511)

J. MICHAEL CHERRY, Department of Molecular Biology, Massachusetts General Hospital, Boston, Massachusetts 02114 (675)

MIKE DANILCHIK, Department of Biology, Hall-Atwater and Shanklin Laboratories, Wesleyan University, Middletown, Connecticut 06457 (271)

IGOR B. DAWID, NIH/NICHD, Laboratory of Molecular Genetics, Bethesda, Maryland 20892 (311)

DOUGLAS W. DESIMONE, Department of Anatomy and Cell Biology, University of Virginia Health Sciences Center, Charlottesville, Virginia 22908 (527)

ANNE DROIN, Département de Zoologie et Biologie Animale, Université de Genève, 1224 Chêne-Bougeries, Switzerland (671)

DOUGLAS W. EIB, Department of Pharmacology, University of Washington Medical Center, Seattle, Washington 98195 (389)

JANICE P. EVANS, Department of Molecular Biology, Scripps Clinic and Research Foundation, La Jolla, California 92037 (133)

BEATRICE FERREIRO, Department of Biology, University of California, San Diego, La Jolla, California 93128 (443)

DIANA FOUKAL, Department of Biological Sciences, Sherman Fairchild Center, Columbia University, New York, New York 10027 (185)

JOSEPH G. GALL, Department of Embryology, Carnegie Institute of Washington, Baltimore, Maryland 21210 (149)

J. GERHART, Department of Molecular and Cell Biology, University of California, Berkeley, Berkeley, California 94720 (3)

ROBERT L. GIMLICH, Department of Anatomy and Cellular Biology, Harvard Medical School, Boston, Massachusetts 02115 (285)

ALAN L. GOLDIN, Department of Microbiology and Molecular Genetics, University of California at Irvine, Irvine, California 92717 (487)

JEAN-DANIEL GRAF, Laboratoire d'Examens Biologiques, Hôpital Cantonal Universitaire, 1211 Genève 4, Switzerland (19)

J. B. GURDON, Wellcome/CRC Institute, Molecular Embryology Group, University of Cambridge, Cambridge CB2 3EJ, England (299)

JAMES HANKEN, Environmental Population and Organismic Biology, University of Colorado, Boulder, Colorado 80309 (419)

DANIEL M. HARDY, Howard Hughes Medical Institute, University of Texas Southwestern Medical Center, Dallas, Texas 75235 (231)

RICHARD M. HARLAND, Department of Molecular and Cell Biology, Division of Biochemistry and Molecular Biology, University of California at Berkeley, Berkeley, California 94720 (367)

J. HEASMAN, Department of Zoology and Wellcome/CRC Institute, University of Cambridge, Cambridge CB2 3EJ, England (213)

JERRY L. HEDRICK, Department of Biochemistry and Biophysics, University of California, Davis, Davis, California 95616 (231)

ALI HEMMATI-BRIVANLOU, Department of Biochemistry and Cell Biology, Harvard University, Cambridge, Massachusetts 02138 (367)

S. HOLWILL, Department of Zoology and Wellcome/CRC Institute, University of Cambridge, Cambridge CB2 3EJ, England (213)

KURT E. JOHNSON, Department of Anatomy, George Washington University Medical Center, Washington, D.C. 20037 (527)

KEN KAO, CRC Molecular Embryology Group, Department of Zoology, University of Cambridge, Cambridge CB2 3EJ, England (271)

BRIAN K. KAY, Department of Biology, University of North Carolina at Chapel Hill, Chapel Hill, North Carolina 27599 (133)

RAY KELLER, Department of Molecular and Cell Biology, University of California, Berkeley, Berkeley, California 94720 (61)

GREGORY M. KELLY, Department of Pharmacology, University of Washington School of Medicine, Seattle, Washington 98195 (389)

CHRIS KINTNER, Molecular Neurobiology Laboratory, Salk Institute, La Jolla, California 93128 (443)

MICHAEL W. KLYMKOWSKY, Department of Molecular, Cellular and Developmental Biology, University of Colorado, Boulder, Boulder, Colorado 80309 (419)

HANS RUDOLF KOBEL, Departement de Zoologie et Biologie Animale, Université de Genève, 1224 Chêne-Borgeries, Geneva, Switzerland (19)

RONALD A. LASKEY, Wellcome Trust & Cancer Research Campaign, Institute of Cancer and Developmental Biology, University of Cambridge, Cambridge CB2 1QR, England (561)

CHRIS W. LEHMAN, Department of Biochemistry, University of Utah School of Medicine, Salt Lake City, Utah 84132 (467)

GREGORY H. LENO, Wellcome Trust & Cancer Research Campaign, Institute of Cancer and Developmental Biology, University of Cambridge, Cambridge CB2 1QR, England (561)

PETER H. MATHERS, Laboratory of Molecular Genetics, National Institute of Child Health and Human Development, National Institutes of Health, Bethesda, Maryland 20892 (347)

DOUGLAS A. MELTON, Department of Biochemistry and Molecular Biology, Harvard University, Cambridge, Massachusetts 02138 (367, 443)

RANDALL T. MOON, Department of Pharmacology, University of Washington Medical Center, Seattle, Washington 98195 (389)

CHRISTINE MURPHY, Department of Embryology, Carnegie Institute of Washington, Baltimore, Maryland 21210 (149)

ANDREW W. MURRAY, Department of Physiology, UCSF Medical Center, University of California at San Francisco, San Francisco, California 94143 (581)

DONALD D. NEWMEYER, La Jolla Cancer Research Center, La Jolla, California 92037 (607)

HEATHER PERRY O'KEEFE, Department of Biochemistry and Molecular Biology, Harvard University, Cambridge, Massachusetts 02138 (443)

LEE K. OPRESKO, Department of Pathology, School of Medicine, University of Utah, Salt Lake City, Utah 84108 (117)

H. BENJAMIN PENG, Department of Cell Biology and Anatomy, University of North Carolina at Chapel Hill, Chapel Hill, North Carolina 27599 (511)

CAREY R. PHILLIPS, Department of Biology, Bowdoin College, Brunswick, Maine 04011 (329)

CAROL PRIVES, Department of Biological Sciences, Sherman Fairchild Center, Columbia University, New York, New York 10027 (185)

DANA REINSCHMIDT, Department of Cell and Molecular Biology, Tulane University, New Orleans, Louisiana 70118 (35)

THOMAS D. SARGENT, Laboratory of Molecular Genetics, National Institute of Child Health and Human Development, National Institutes of Health, Bethesda, Maryland 20892 (347)

CAROLINE SCHILD, Institut de Biologie Animale, Université de Lausanne, CH-1015 Lausanne, Switzerland (541)

J. C. SMITH, Laboratory of Developmental Biology, National Institute for Medical Research, London NW7 1AA, England (635)

L. DENNIS SMITH, University of California, Irvine, Irvine, California 92717 (45)

J. R. TATA, Laboratory of Developmental Biochemistry, National Institute for Medical Research, London NW7 1AA, England (635)

ROBERT TOMPKINS, Department of Cell and Molecular Biology, Tulane University, New Orleans, Louisiana 70118 (35)

ROBERT L. VARNOLD, Developmental Biology Center, University of California, Irvine, Irvine, California 92717 (45)

PETER D. VIZE, Department of Biochemistry and Molecular Biology, Harvard University, Cambridge, Massachusetts 02138 (367)

KATHERINE L. WILSON, Department of Cell Biology and Anatomy, The Johns Hopkins University School of Anatomy, Baltimore, Maryland 21205 (607)

ALAN P. WOLFFE, Laboratory of Molecular Embryology, National Institute of Child Health and Human Development, National Institutes of Health, Bethesda, Maryland 20892 (541)

MIKE WORMINGTON, Department of Biology, University of Virginia, Charlottesville, Virginia 22901 (167)

M. WU, Department of Molecular and Cell Biology, University of California, Berkeley, Berkeley, California 94720 (3)

ZHENG'AN WU, Institute of Developmental Biology, Academia Sinica, Beijing, People's Republic of China (149)

C. C. WYLIE, Department of Zoology and Wellcome/CRC Institute, University of Cambridge, Cambridge CB2 3EJ, England (213)

WEILONG XU, Developmental Biology Center, University of California, Irvine, Irvine, California 92717 (45)

FOREWORD

Let me give a short historical survey. *Xenopus laevis,* the South African clawed toad, had been used for research and for educational purposes for many years in South Africa before it received a more worldwide recognition. The reason for this increase in popularity was the discovery that *Xenopus* could be used for human pregnancy tests. *Xenopus,* a primarily aquatic animal, turned out to be a very suitable laboratory animal which could easily be kept and could be induced to spawn with a simple gonadothropic hormone injection. This discovery led to its worldwide distribution.

The publication of Normal Tables of embryonic development formed one of the original aims of the Hubrecht Laboratory as the seat of the "International Institute of Embryology". Dr. Faber and I therefore decided to try to prepare a Normal Table of *Xenopus laevis'* embryological development, covering the entire development from the fertilized egg till the end of metamorphosis. We wanted to make this an example for future similar enterprises. We therefore asked a large number of specialists to contribute to this monograph. It was published in 1956 by North Holland Publ., Amsterdam and had two further editions respectively in 1967 and 1975 (Nieuwkoop, P. D., and Faber, J. (1967). "Normal Table of *Xenopus laevis* (Daudin)." North-Holland Publ., Amsterdam). I think it has fulfilled our expectations and may actually be one of the most widely cited publication on *Xenopus.* Unfortunately, it has been out of print for quite a number of years. It may also fall more and more out of general use, because its text was predicated upon the general knowledge of anatomy, histology and cytology of its readers. Since anatomy, histology and cytology are no longer adequately taught, many young scientists ask now for a pictorial presentation of the internal development supplemental with extensive legends, so that they may recognize the individual structures they meet in serial sections. I think this general wish may soon be fulfilled, at least for *Xenopus'* early development.

As already mentioned, *Xenopus laevis* can be easily reared in the laboratory and, in comparison with other amphibians, requires a relatively short period to reach metamorphosis and ultimately to reach sexual maturity. It is therefore now also widely used in genetical analysis, F_1 and F_2 generations being obtainable in a few years time. The present volume demonstrates very clearly that *Xenopus laevis* is at present extensively used in many different fields of biological research.

As an experimental embryologist, I feel I must make a few cautionary remarks. The title of this volume emphasizes the practical use of *Xenopus* as an experimental tool; the real aim of the research must however be a better understanding of the development and ultimate function of this particular anuran species.

Xenopus laevis actually is a very particular anuran species; it is not only an evolutionarily "old" species, but is also adapted to a special environment. Among the different anuran groups, it shows a specialized embryonic development. The early embryo is clearly double-layered; mesoderm formation occurs fully internally and neural formation likewise, except for its ependymal layer which is formed in the outer layer. This has rather far-reaching consequences for experimental analysis, for example for cinematographic analysis and cell lineage studies as well as for extirpation and transplantation experiments.

I would like to stress that *Xenopus laevis* is a very suitable experimental animal. I must only warn against generalizations on the basis of its development, since it is a specially adapted anuran species. A comparison with other amphibian groups will certainly require a comparative analysis. Of course, any experimental animal shows certain experimental restrictions as well as advantages due to its special biological adaptations. I warmly recommend *Xenopus laevis* as a laboratory animal suitable for many different lines of research, but not for all.

I feel very honored to have been asked to write a short foreword for this volume in which such an extensive survey of its usefulness for modern cell biological and molecular research is given.

P. D. NIEUWKOOP

PREFACE

Over the past twenty years, amphibian species have become popular model animal systems. They are vertebrates, and can be maintained easily and inexpensively in the laboratory. Oocytes at different stages of differentiation can be readily obtained from adult females. Moreover, their embryos are large and heal well after surgery. Among amphibians, *Xenopus* is one of the champion organisms. *Xenopus laevis* is quite hardy and fairly disease resistant. This volume is dedicated to the practical uses of the oocytes and embryo of this animal in cell, molecular, and developmental biology.

This book is timely for a number of reasons. First, the "frog scientific community" is growing. At the 1990 International *Xenopus* meeting in Switzerland, there were over 250 scientists in attendance. Second, the use of *Xenopus* has permeated every aspect of biological research, according to a literature database. Third, the "bible" of *Xenopus* development, written by Pieter Nieuwkoop and J. Faber in 1975, is out of print and is no longer available.

The goals of this book are to provide useful background information and references for specific uses of *Xenopus'* oocytes and embryos in cells, molecular, and developmental biology. Each chapter features working protocols with valuable tips (do's and don'ts). We have attempted to archive as much useful information as possible about this system in one place. We should note though, that due to space limitations, we have addressed only a few issues concerning the adults themselves.

A few points should be made regarding *Xenopus*. There is no taxonomic rule to define *Xenopus* as a frog or toad; *Rana* and *Bufo* are representatives of the frog and the toad, respectively. However, since *Xenopus* is totally aquatic, it may be considered a frog, since frogs are generally aquatic and toads are primarily terrestial. (In addition, for purposes of self-respect, we will refer to *Xenopus* as a frog in the book. People frown at you when you say you work with toads.) As a frog, *Xenopus* would be considered primitive, because it lacks ears and eyelids, and does not have an evolved tongue that catches insect prey. The most popular of *Xenopus* species is *X. laevis*, although *X. borealis* and *X. tropicalis* are used in a limited fashion in laboratories. In all, approximately 20 different species have been identified in this genus.

The genus is endogenous to the southern part of Africa. One of its first applications in biological experiments came in the 1950's, with its use in human pregnancy tests. The urine of a woman would be injected into an adult female frog; when the urine contained placental chorionic gonadotropin, the frog laid eggs. Through this bioassay, *Xenopus* was introduced into a large number of reproductive biology laboratories around the world.

Early experimental uses of *Xenopus* included sex reversal by hormones (Witschi, Gallieu) and embryological work of Balinsky and Nieuwkoop. In the 1950's, the laboratory of Fischberg used *Xenopus;* Fischberg's student, John Gurdon, pioneered the early nuclear transplantation work. During the 1960's, Brown and his colleagues Dawid and Reeder in the United States, and Birnsteil in Europe initiated biochemical work on *Xenopus*. *Xenopus* oocytes became very popular for the identification of receptor components following the 1971 finding of Gurdon and his colleagues that injected messenger RNA is efficiently translated. A number of scientific firsts have been achieved with this animal, such as the first isolation of any eukaryotic gene, the discovery of rDNA gene amplification, the first isolation of 5S and 4S eukaryotic genes, the first preparation of RNA polymerase III transcription extracts, the first expression of ligand-gated ion channels in a heterologous cell type, the isolation of the maturation promoting factor (MPF), the elucidation of the localization signal sequence of nuclear proteins, the development of anti-sense approaches *in vivo* and the identification of growth factors that can induce mesoderm formation. Not only has *Xenopus* become widely distributed among world academic centers, the genus now populates many drainage ditches in Florida and California; it is successful in the North American wild because it has no natural predators.

This volume is organized into four sections. These comprise background information, uses of oocytes, uses of embryos, and special applications with oocytes, eggs or embryos. The volume consists of 32 chapters and represents the efforts of more than sixty authors. A general description of the volume and the Appendix follows below.

In the first five chapters of the book, general information is presented regarding animal husbandry, oogenesis and embryogenesis. It is relatively easy to rear *Xenopus laevis* in the laboratory, and the introductory chapter (#1) gives useful details for cultivating a frog colony. However, due to a lengthy generation time (~1 year) and the fact that *X. laevis* is a pseudotetraploid, use of this animal for traditional genetic experiments is limited. Nevertheless, two chapters (#2, 3) are presented that describe successful approaches in assigning chromosomes locations for genes, based on their segregation patterns, and generating homozygous animals for expression of recessive mutations. As the last part of the introductory section, there are excellent chapters detailing the fundamentals of *X.laevis* oogenesis (#4) and early embryo development (#5).

The second section of the volume is devoted to *X. laevis* oocytes. Chapter #6 explains how to culture oocytes and follow the receptor-mediated endocytosis of vitellogenin from the culture medium. To analyze the oocyte biochemically, several protocols (#7) are presented on how to prepare oocyte nuclei manually and how to extract proteins from oocytes for SDS-PAGE and western blot analysis. A method (#8) is presented on how to isolate the oocyte meiotic chromosomes, termed lampbrush chromosomes, and spread them on glass slides for cytological analysis. The last two chapters of this section describe experimentation involving injection of synthetic mRNA (#9) and oligonucleotides (#10) into oocytes. Injections are an ideal method of modulating processes within living cells. Oocytes are large cells (1.2 mm diameter, ~1 µl volume) that can be injected easily with various macromolecules and with little mortality.

The third section of the volume includes thirteen chapters dealing with eggs and embryos. In one chapter (#11), two methods are presented on how to convert full-grown oocytes into fertilizable eggs. These techniques will permit manipulations on maternal mRNA and protein stores so that their effects can be followed during embryogenesis. In the transition from oocyte to zygote there are a number of significant biochemical changes inside and outside the cell; these changes are described in two chapters (#12, 13). Chapter no. 14 describes various methods to perturb the dorsal-rentral, anterior-posterior axes of the frog embryo. Chapter #15 demonstrates the utility of fate mapping particular embryonic cells by injecting fluorescent dextran into blastomeres. The dextran has a long half-life and is too large to distribute through the gap junction of the embryo freely, hence identifying descendants for a particular injected blastomere. In Chapter #16, nuclear transplantation experiments are outlined. In the Chapters #17 and 18, the processes of mesoderm and neural induction are discussed, respectively. In Chapter #19, injection experiments are outlined using the embryo as an *in vivo* host for molecular dissection of tissue-specific RNA polymerase II promoters. Chapter #20 presents an excellent review of the successes and failures of manipulating gene expression in developing embryos with injection of DNA, mRNA, oligonucleotides, and antibodies. The final three chapters (#21, 22, 23) of this section are concerned with localizing specific proteins and mRNAs *in situ* in the *Xenopus* embryo. Due to the large size of oocytes and embryos and their yolky nature, special considerations are needed for maximal detection of positive cytological signals, with minimal background levels.

The fourth section of the volume is concerned with experimental uses of oocytes, eggs and embryos. For example, it is possible to use oocytes and eggs, in whole cell or extracted form, for studies of DNA recombination (#24). Another very popular use of frog oocytes (#25), taking advantage of the oocytes' translational capabilities, is the expression of ion channel proteins on the oocyte surface after injection of polyadenylated mRNA extracted from neural tissue. It is possible to culture neurons and muscle cells from embryos as an *in vitro* model for synapse formation (#26). The *Xenopus* embryonic system has also been fruitful for the study of cell-extracelluar matrix interactions (#27). Another popular use of the *Xenopus* system is in emulating important biological processes *in vitro,* through the use of egg extracts. Four chapters describe several *in vitro* assays that have been based on cell-free egg extracts, such as chromatin assembly, DNA replication, cell cycle, and nuclear import and assembly; these are described in chapters #28–31. Finally, the last chapter (#32), describes what *X. laevis* cell lines are available for cell and molecular analyses, how they can be used as sources of growth factors, and how they can be transfected for DNA engineering experiments.

At the end of this volume are eight appendixes. Recipes for popular experimental solutions are compiled, followed by a section devoted to injections of oocytes and embryos. A list of known, characterized mutations in *X. laevis* is cataloged in table format. A codon usage table for *X. laevis* proteins is included; this will be valuable in aiding the design of oligonucleotide probes for cDNAs/ genes encoding *X. laevis* proteins of interest. We include a list of commercial suppliers of adult animals in the United States. Finally, an improved *in situ* hybridization protocol based on whole mounts is described in the last appendix.

In closing, we wish to acknowledge Drs. Carroll, Dawid, Gerhart, Gurdon, Melton, and Sargent for their input and encouragement in preparing this volume. We also wish to thank the members of our laboratories for their help in editing the manuscripts.

BRIAN K. KAY
H. BENJAMIN PENG

Part I. General Information

Chapter 1

Raising Xenopus in the Laboratory

M. WU AND J. GERHART

Department of Molecular and Cell Biology
University of California at Berkeley
Berkeley, California 94720

I. Introduction
II. Equipment
   A. Tanks and Water Changing
   B. Water Quality
   C. Salinity of Water
   D. Light and Temperature
   E. Aeration
   F. Feeding
   G. Population Density
   H. Marking Animals
III. Infectious Diseases
   A. Red Leg
   B. Nematodes
   C. Fungal Infections
IV. Raising Tadpoles from Ovulated Eggs
V. Raising and Maintaining Frogs
   A. Commercially Obtained Adults
   B. Laboratory-Raised Frogs
   C. Recycling Frogs after Ovulation
   D. Genetic Variety
   References

I. Introduction

Xenopus laevis is well suited to the laboratory, perhaps because it survives and proliferates in a variety of nonfastidious conditions in the wild. The animal's tolerance, even preference, for brackish water is well established. As described by Nieuwkoop and Faber (1967, p. 13), "In South Africa, *Xenopus laevis* lives in practically any kind and amount of water.... The frogs live by the thousands in silty farm ponds devoid of any higher plant vegetation.... [They] breed both in moor waters with a relatively low pH, and in lakes... with a rather high pH... as high as 9.0, caused by the presence of large quantities of lime in the fertilizing-mixture used in... [fish hatchery] ponds." This species escaped from pregnancy testing laboratories and established itself in Southern California, for example, in the Los Angeles drainage system (hence the California Department of Fish and Game considers it an ecological hazard and requires permits for importation, maintenance, and shipping). It is occasionally collected there (J. Hedrick, personal communication).

Many laboratories have raised *X. laevis*, and the procedures vary greatly, probably reflecting the animal's tolerance. Reports of frog-rearing procedures tend to be anecdotal since no laboratory, to our knowledge, has compared all procedures, and many researchers have developed workable methods independently. We summarize both unpublished (i.e., personal communications) and published procedures from several laboratories, and we emphasize our own procedures for medium-scale production of frogs on the assumption that some readers will want to start their own colonies.

Limitations on the number of animals one can raise in the laboratory seem mainly set by space and the quality of the water and food. For 15 years we have raised and maintained a colony of about 2000 adult females (each used once or twice a year for ovulation) and 5000 juveniles at various stages of maturity in a space of about 600 square feet, and we have provided animals on a recycle basis to several other laboratories in the Berkeley–San Francisco area. The attention of one person for approximately 1 hour per day is required on average to maintain the colony. It costs us far less to raise animals than to buy them commercially (now $10–15 per adult female). Admittedly, many of our costs are hidden (e.g., heating, water, rental of space) compared with those of the commercial producer. We estimate that our main expenses of food, antibiotics, and salt come to about $1 to raise a 3-year-old frog.

Whenever a few researchers in one area expect to use *X. laevis* as a main experimental animal for a period of years, it is worth their making a joint effort to raise them. Tadpoles metamorphose within 2–3 months of egg laying, and frogs reach sexual maturity within a year or two (Brown, 1970; Callen *et al.*, 1980), depending on the temperature and frequency of feeding. Adult females reach optimum egg production (quality and quantity) at 2–3 years of age and sustain this level for several years more. We have not pushed the animals to grow faster (more food, higher temperature), but others (R. Tompkins, personal communication; Nieuwkoop and Faber, 1967, p. 15) report that mature animals can be obtained in 6–8 months.

1. RAISING *Xenopus* IN THE LABORATORY 5

II. Equipment

A. Tanks and Water Changing

1. Large Tanks; Automatic Discontinuous Water Changing

We have used large tanks to minimize the equipment needed to change water automatically and because we believe that animals grow faster and stay healthier in a large volume of water. Inexpensive tanks (about $50 per tank) are constructed as follows: A rectangular wooden frame 14 feet (4.2 m) long and 4 feet (1.2 m) wide is nailed together from redwood boards 12 inches (0.3 m) high and 2 inches (5 cm) thick. The frame is set on a slightly sloped concrete floor. Styrofoam sheets 1 inch (2.5 cm) thick and 2.5 feet (64 cm) wide are cut to 4-foot (1.2 m) lengths and placed side by side on the floor area enclosed by the frame. Then, a sheet of thick polyethylene plastic (10–16 mil thick, or a double layer of thinner plastic) is cut 18 feet (5.5 m) long and 8 feet (2.4 m) wide (a roll may be purchased from a local hardware store; the plastic is sold commercially to protect basements from water seepage) and draped across the frame to form a water-tight container. Folds of plastic accumulate in the corners but do not lead to leaks. Small holes in the plastic seem to self-seal under the weight of the water.

At the downhill end of the tank is placed an electric immersion pump (Little Giant #2P352, $70 per pump, resting on the bottom and connected to a thick rubber hose leading over the edge of the tank to a drain. The pump is operated from a timer (Dayton lawn sprinkler 2E220, $100, serving two tanks) set to activate the pump and drain the tank for 30 minutes each day or at 2-day intervals if the tank is not crowded (see Section II, G). Above water level at the uphill end of the tank is placed a water inlet pipe controlled by a toilet float valve (with a polypropylene bulb, $20 per unit). This valve opens when the water level drops (when the pump starts) and closes as the water level returns to its original depth (after the pump stops). Water is maintained at a depth of 6–10 inches (15–25 cm); the total volume is about 800 liters (800 kg; 1800 lb) per tank. Since adult frogs try to escape from the tanks, we rest a second tier of thinner boards, 10 inches high, on the frame.

Such tanks have been in continual use for 10 years. In our laboratory, a mat of algae forms on the bottom. The plastic is changed every 2–3 years. Animals are raised from eggs to full-grown reproductive adults in a single tank. As many as 1500 full grown adults, 4000 juveniles, or 10,000 tadpoles are kept in a tank. Since adults tend to hide in the shadowy parts of the tank, we darken one-third of the length with styrofoam sheets. Whereas this type of tank is fully suitable for *X. laevis*, it did not meet with the codes of the animal care inspectors because moisture collected under the plastic and algae tended to accumulate in the tank, making it look like a natural pond rather than a laboratory facility.

2. Medium Tanks; Automatic Continuous Water Changing

We have more recently switched to the use of fiberglass tanks 6 feet (1.8 m) long, 2.5 feet (0.75 m) wide, and 2 feet (0.6 m) deep, since these tanks are recommended by animal care inspectors. Upper and lower tanks are mounted in racks to save floor space. Tanks are tipped slightly lengthwise, and a stand pipe is placed in a drain hole at the lower end where feces and food particles accumulate. The stand pipe has an inner and an outer tube; the inner tube, which is 10 inches (25 cm) high, determines the water level. The outer tube is 12 inches (30 cm) high and is pierced by small holes at the level of the tank bottom, providing entry to waste particles. At the upper end of the tank rests a hose from which water drips slowly into the tank at a rate of about one tank volume (200 liters) per day. This turnover rate is recommended by several authors (cf. Nieuwkoop and Faber, 1967). There are reports that adult frogs actually suffer if the water flow is too rapid; nitrogen collects in the dorsal lymph sac and animals float at the surface, eventually dying (J. Hedrick, personal communication).

3. Small Tanks; Nonautomatic Discontinuous, Water Changing

Many laboratories use plastic dishpans (with lids), plastic food-storage boxes, or aquaria and clean them by emptying and replacing the water every 2 or 3 days. However, this method requires a person's full attention if many frogs are kept. A lid or high sides are needed to prevent animals from escaping.

B. Water Quality

We run tap water directly into the tanks since the water in Berkeley is not heavily chlorinated. In many areas, though, tap water is treated with chlorine or chloramine, a mixture of chlorine and ammonia especially toxic to aquatic animals (Fox, 1984; R. Tompkins, personal communication). In the summer it is especially necessary to take precautions, and this makes frog rearing more complicated and expensive. Treatments include the following:

1. Letting water stand for a day before use, allowing chlorine to react with contaminants or evaporate. This method requires large tanks and extra space.

2. Adding 0.1 mM sodium thiosulfate, or commercial aquarium products such as Dechlor or Novaqua. These agents seem adequate in many situations.

3. Passing water through commercial charcoal filters (e.g., Barnstead organic removal cartridge filter). This treatment is effective in removing chlorine and ammonia, although some filters reportedly release noxious materials. Filters must be changed at intervals depending on use.

4. Recycling. Douglas Melton (personal communication) has equipped his facility with a pumping system to recirculate 90% of the water each day through a filter of sand interspersed with denitrifying bacteria (Fox, 1982, 1984), to remove particles and nitrogenous wastes, and through a UV-illuminated trough to kill disease organisms; a 10% volume of fresh water is introduced each day after passing it through a charcoal filter.

Further considerations of water quality and treatments for ammonia products and chlorine can be found in Fox (1982, 1984) and Buttner and Nace (1984).

C. Salinity of Water

It is worth obtaining a water analysis report from the local water district office. Our water has a low ion content (conductivity equal to 0.5 mM NaCl) compared with water in many other areas. Low salinity is said to stress *X. laevis*, which is continuously aquatic and may lose ions across its skin. If a few tadpoles or adults die in a tank, we often add NaCl to a concentration of 0.6% (0.10 M), which approximates frog blood tonicity. Kiln-dried coarse salt (pickling salt, Morton Salt Co., Newark, CA) is adequate for the purpose and less expensive than chemical grades of salt. Many authors recommend salt addition when any health problem arises (Reichenbach-Klinke and Elkan, 1965; Brown, 1970). *Xenopus laevis* can withstand NaCl concentrations up to 0.3 M for prolonged periods without harm (Balinsky, 1981).

D. Light and Temperature

For 15 years we have raised animals under natural light in a converted greenhouse. The glass roof is whitewashed, providing illumination like that of a bright cloudy day. Water temperature falls to 15°–17°C in the winter and rises to 21°–23°C in the summer. *Xenopus laevis* is stressed by prolonged exposure to temperatures below 14°C and above 26°C; although adults survive, the quality of eggs and oocytes decreases. Stress can be followed by atresia (death and resorption) of the oocytes.

Our latitude is about the same as that of South Africa, where *Xenopus* is indigenous. The natural light and temperature cycle has advantages and disadvantages: In spring the increasing light and warmth stimulate feeding, growth of the animal, and oocyte growth. Animals spontaneously mate when the population is thinned. However, especially if the animals do not ovulate in the spring, they enter a period of torpor in the summer, reflected in reduced food intake. Eggs tend to become overripe and more human chlorionic gonadotropin (HCG) is needed to induce ovulation. Many researchers have noted the difficulty of obtaining high quality eggs in late summer and early fall (Gurdon, 1967). The shortening day length of late summer may accentuate this tendency. Egg quality seems to improve again in the fall and winter. Others have kept *Xenopus* in constant light (e.g., 12 hours light, 12 hours dark) and temperature conditions and report successful growth. Constant conditions may smooth out the annual cycle, preventing entry of the frogs into torpor but also preventing the spring growth spurt of the ovary and frog.

E. Aeration

For tadpoles, we immerse sintered glass bubblers, one at each end of a large tank, setting the airflow rate to a steady stream of small bubbles. This seems to accelerate tadpole growth and perhaps counteract the die-off occasionally caused by fouling of the water after overfeeding. We do not bother to aerate the tank water of adult frogs, which come to the surface regularly.

F. Feeding

1. TADPOLES

We use a mixture of nettle powder (*Urtica* sp., Wunderlich-Diez, Hasbrouck Heights, NJ), bakers' yeast (Fleischmann, active dry yeast for baking), and bone meal (obtained at a local plant nursery), in the proportions of 7:2:1. These ingredients can be obtained at health food stores or from commercial distributors. *Xenopus laevis* tadpoles, compared with those of other anurans, are particularly efficient in collecting yeast and other small particles (Wasserzug, 1972). The powder mixture is slurried in a small volume of tap water and then delivered under the surface of the water of the tank to avoid the formation of scum on the top. It is important not to overfeed the tadpoles and thereby form a sludge of unused food on the bottom; tadpoles should clear the food from the water within a few hours. They continue to graze on "infusoria" growing in the light sediment on the bottom.

Other researchers use straight nettle powder with success (Gurdon, 1967), although in our experience tadpoles grow faster on yeast-supplemented food. Bone meal serves as a calcium and phosphate source. Without it we tend to get animals with skeletal deformities. A diet of fresh whole milk and lyophilized yeast has been used successfully (H. R. Kobel, personal communication). Milk is added in small amounts twice a day to cloud the water but not to build up a scum. Others have used liver powder with success (Nieuwkoop and Faber, 1967).

2. Frogs

These carnivores accept a variety of foods. Raw meat, for example, beef or ox liver, heart, or lung, has been used in many laboratories because it can be cut or ground ahead of time and kept frozen until use. R. Tompkins (personal communication) recommends American Beauty sliced beef liver for fastest growth. For a full-size frog, about 8 g per feeding, twice weekly, is a standard ration (Brown, 1970). Live food such as earthworms, blowflies, meal worms, or 12-day chick embroys can also be used. (There is some concern, though, that excess insect chitin can block the intestine.) A drawback of raw meat is that unconsumed pieces must be removed from the tanks a few hours after feeding time, before pieces putrify, and that feces from a meat diet tend to foul the water badly. In addition, they contain enough connective tissue to clog drains and pumps (precluding automatic water changing). Still, many laboratories use meat and report full success in raising animals, especially when fast growth is desired.

In raising large numbers of animals, we have found it much more convenient to feed adults prepared dry foods such as Purina "trout chow" (38% protein, $0.25–0.40/lb) or Nasco frog brittle (Nasco Co., Fort Atkinson, WI; 44% protein, $1.50–3.00/lb), or Rangen "soft moist salmon diet" (Rangen Inc., Buhl, ID; 45% protein, $0.60/lb), the latter being the highest in protein content (Abel, 1988). In fact, switching to prepared food was the most important factor in our development of a large, nearly self-maintaining colony. We prefer trout chow for large adults because the pellets are hard enough to remain intact in the water, whereas salmon diet and Nasco brittle tend to break into crumbs not usable by adults. Trout chow also, is produced in several pellet sizes suitable for frogs of different sizes. Small-size trout chow pellets (#4) sink to the bottom and are suitable for froglets up to 2 inches (5 cm) long. Larger pellets (#5) float and frogs come to the surface to collect them, making it easy to see the number of pellets still uneaten. Salmon chow is also suitable for froglets, which are willing to scavenge the crumbs, and the high protein content may favor faster growth. Feces from these diets break into small particles that pass through the holes of the pump or stand pipe. The water remains cleaner than with meat, and it is unnecessary to change water soon after feeding.

Adults do not adapt to trout chow for approximately 2 weeks if they have previously been fed on meat. However, they eventually do adapt and seem to do so faster if a few adapted frogs are put in a tank to start a feeding frenzy. We feed growing adults two or three times a week, with sufficient pellets that they clear them from the surface in 15–20 minutes. An adult frog will consume 5–10 pellets at a feeding. We have raised frogs for five generations on trout chow with no apparent detrimental effects. We do not bother to supplement with vitamins (the chow contains vitamins) or live food. Supplementation with meal worms or tubifex worms every few weeks, or with vitamin drops, can be included at the choice of the keeper and probably to the liking of the frogs.

G. Population Density

Several researchers stress the importance of keeping the density low to optimize the rate of development, for example, 10 small tadpoles per liter, one large tadpole or one small froglet per liter, and one large frog per 3 liters (H. R. Kobel and R. Tompkins, personal communications). For a large colony, however, space is often limited. We raise tadpoles at a density of 5–10 per liter. Development is somewhat slower and more asynchronous, but, starting with about 10,000 tadpoles, we can successfully obtain 2000–3000 large frogs (half females, half males) per tank of 800 liters. This result is only possible because of the use of trout chow instead of liver, permitting daily automatic water changing.

Researchers wishing to obtain the fastest development should probably use lower densities, more intensive feeding, a richer food source, temperatures close to 23°C, and more frequent water changing. Of course, this method is only feasible for a small number of animals.

H. Marking Animals

Occasionally it is necessary to identify individual animals in numbers too large to allow the use of individually marked tanks.

1. Branding

A number or letter is shaped from 5-mm-diameter brass wire, one end of which is mounted in a wooden handle. Some letters and numbers may not be easily distinguished on frog skin after a few months (e.g., 3 and 5; 8 and 0; E and C). The number or letter is dipped in liquid nitrogen for cooling and lightly pressed into the back or belly skin of an anesthetized animal for 10 seconds. We use two-place number–letter combinations. The wound heals in a few days, with sloughing of skin. The wound should not be so deep as to expose flesh. These marks last at least 2 years, with better visibility on the belly, though it is more convenient to see back marks in a large population.

2. Toenail Clipping

Various researchers have used this method, clipping combinations of nails of each foot to get 2^6 possibilities (Gurdon, 1967).

3. Beads on Thread

Fine plastic thread is run through the skin behind the armpit, and various colors and combinations of small colored plastic beads are run onto the thread, which is then well knotted. This method is said to last for several years (A. J. Brothers, personal communication).

4. Skin Transplants

A patch of pale belly skin is transplanted to a site in the dark back skin. The size, shape, and position of the graft can be used to identify the animal (Chardonnens, 1975).

III. Infectious Diseases

Animals captured in South Africa or Southern California contain a number of parasites (nematodes, trematodes, protozoa), and generally it is advisable not to mix these animals with ones raised in the laboratory, which tend to be parasite free. We do not have extensive experience with *Xenopus* diseases, in part because we have not used wild animals for the past 15 years. Therefore, the following outline combines our experience with three common diseases and their treatments and a summary of recommendations from published works, such as those of Reichenbach-Klinke and Elkan (1965) and Brown (1970). As general precautions against the spread of diseases, R. Tompkins (personal communication) recommends that water never be mixed between tanks and that live aquatic food organisms (e.g., tubifex worms) be avoided since they may carry disease organisms.

A. Red Leg

The common affliction red leg can be recognized by the reddening of the ventral surface, medial thighs, and foot web, but apparently the disease can be well advanced even before these symptoms appear. Seemingly healthy animals can die within a day of showing symptoms or may linger for weeks, ignoring food and losing weight. Entire colonies can be wiped out in a matter of days. It is referred to as "a form of severe generalized sepsis caused by hemolytic bacteria easily transferred from one animal to the next" (Reichenbach-Klinke and Elkan, 1965, p. 234). A common bacterium associated with the disease is *Aeromonas hydrophila* (gram negative), although other bacteria (species of *Mimeae, Citrobacter*, and *Staphylococcus*) may also be at cause (Cosgrove and Jared 1974; Gibbs et al., 1966). In our experience, animals raised from eggs in the laboratory do not contract red leg until after they have been handled roughly (squeezed) or dropped on the floor during ovulations, perhaps after the protective mucous layer of the skin has been broken.

Animals are said to be more susceptible to the disease when stressed (e.g., after overinjection of hormone and excessive handling) or in poor health (e.g., poor food, temperature extremes, waterborne toxic agents). *Aeromonas* is reportedly present even in healthy populations (Cosgrove and Jared, 1974) and presumably invades animals systemically when they are weak. The healthier the animals, the lower their susceptibility (Reichenbach-Klinke and Elkan, 1965).

Gibbs *et al.* (1966) and Gibbs (1973) report successful rescue of *Rana pipiens* in the initial stages of disease by the treatment of individual animals with tetracycline, 5 mg dissolved in 0.2 ml water, introduced into the stomach via a soft plastic tube pushed down the throat, administered twice daily for 5 days. Animals are fed after each treatment. We have not tried this individualized treatment with *X. laevis*, but we have reduced losses by other means. Following the recommendation of several authors, we isolate animals showing symptoms and treat them separately from the remaining animals. If only a minority of animals in a tank are infected, we raise the NaCl concentration to 100 mM for the remaining animals and stop feeding them for 2 or 3 weeks until newly diseased and dying animals no longer appear. The symptomatic animals (or all animals if the colony shows severe infection, such as frequent deaths or many reddened animals) are treated as follows: In addition to increasing salt in the water and withholding food, we add oxytetracycline (Sigma Chemical Co., St. Louis, MO) rather than tetracycline to the tank water (50 μg/ml) for 2 weeks or longer and replace the antibiotic and tap water and salt every 3 days. It is, of course, important to remove dead and severely diseased animals as soon as they appear. On one occasion we continued this treatment for 2 months in a badly diseased population, of which half the frogs finally survived. We cannot say we have saved animals already suffering from an advanced case of red leg, but we think we can stop the spread of the disease and perhaps reverse mild cases or the earliest stages. Some animals may show no reddening but may remain sluggish, lose weight, and be useless for ovulation. We generally dispose of these animals.

B. Nematodes

Capillaria xenopodis or *Pseudocapillariodes xenopi* has been reported to proliferate in the dorsal skin of *X. laevis* and cause "flaky skin disease," a condition in which frogs slough large pieces of skin, become rough and gray on the dorsal surface and sometimes brown or reddish on the ventral surface, and lose weight. Opportunistic invasion by "red leg" bacteria (Cohen *et al.*, 1984; Stephens *et al.*, 1987) may accompany the nematode infection. Treatment of adult animals with the vermacide thiabendazole (Sigma; also known as Mintezol from Merck, Sharpe, Dome, Rahway, NJ), added to the tank at a concentration of 100 μg/ml for 10–15 hours (overnight) reverses the condition effectively (Cohen *et al.*, 1984) We have applied this treatment when the skin becomes slightly gray and rough, especially after frogs have been ovulated, handled, and dropped in the laboratory. Because frogs try to escape from the water in the presence of this agent, as if it irritates them, the tank should be well enclosed. This treatment improves the health of animals considerably and seems to reduce the sensitivity of the frogs to red leg (A. Coleman, personal communication).

C. Fungal Infections

Occasionally froglets contract fungal infections of the foot webbing. They can be successfully treated as follows. We raise the salt concentration in the tank to 100 mM (0.6%) and add 0.3 μg/ml methylene blue and 0.9 μg/ml acriflavin to the tank for a week or two, or alternatively place animals in a solution of 0.02–0.1% potassium permanganate in tap water–0.6% NaCl for 30 minutes to 1 day (Gurdon, 1967).

IV. Raising Tadpoles from Ovulated Eggs

Several adult male and female *X. laevis* are placed in a large tank (see other chapters of this volume, for schedules of HCG injections) in a depth of 6–10 inches of water. Females can be recognized by their larger size, pear shape, and cloacal valves. Males have black "nuptial pads" on the inside of the forearms. For obtaining a large number of eggs, Hedrick and Nishihara (1990) recommend injecting females with pregnant mare serum gonadotropin (PMSG) (30 IU) 3–4 days before HCG injection (500–800 IU). If a shorter or longer interval is allowed between the PMSG and HCG injections, the number of ovulated eggs is less, but even a 1-day interval has beneficial effects.

Injections of PMSG at 10 days as well as at 3 days before HCG injection ensure good ovulation in even the most refractory frogs (Thornton, 1971).

After 6–8 hours, a male clasps a female (amplexus) and the pair moves about the tank leaving a trail of eggs. Ovulation can continue for a day. Thereafter, adults are removed and the tadpoles allowed to hatch and develop. Alternatively, embryos can be collected from experiments in the laboratory provided, of course, that these eggs have not been excessively-dejellied or overcrowded, which would lead to deformities. We often collect eggs at the end of experiments, especially if the eggs have been of high quality, raising the young tadpoles initially in a one-tenth-strength modified Ringer's solution (full-strength modified Ringer's solution is 100 mM NaCl, 1.5 mM KCl, 2 mM $CaCl_2$, 1.0 mM $MgCl_2$), 1 inch deep, about 10 eggs per square inch, switching to tap water after 2 weeks.

We now summarize our procedure for raising tadpoles; details can be found in Sections II and III. Feeding with the nettle powder–yeast–bone meal mixture is begun a week after fertilization. We have kept small tadpoles at population densities as high as 10–15 animals per liter, although faster and more uniform growth is obtained at lower densities, such as 1–2 tadpoles per liter as the tadpoles grow larger. A sintered glass immersion bubbler provides a stream of fine bubbles to aerate the water and stir the food particles. The tank water is changed every 3 or 4 weeks, leaving 10% of the old water to "condition" the new tap water. There are indications that too frequent changing of the water precludes the establishment of denitrifying bacteria that remove ammonia and nitrite, growth-inhibiting waste products of protein breakdown (Fox, 1982, 1984). With this diet there is no need to remove tadpole feces. Many researchers stress the need to avoid overfeeding; tadpoles should be able to clear the cloudy suspension of food within a few hours. Overfeeding causes, among other things, oxygen depletion owing to eutrophication (air bubblers may counteract this problem) and ammonia and nitrite accumulation. We have added 1–5 mM sodium dihydrogen phosphate to reduce pH and thereby to reduce the amount of free ammonia (ammonium ion is less harmful).

Metamorphosis begins within 2–3 months. We generally do not remove metamorphosing animals from the large tanks, and the success of metamorphosis is roughly in the range of 50%. Since animals may drown during metamorphosis (as they switch to more extensive lung breathing; Nieuwkoop and Faber, 1967, p. 15), it helps to remove metamorphosing animals into a dish of shallow tap water [1 inch (3 cm) deep] with antibiotic (gentamicin, 50 μg/ml). When this is done, survival can approach 100%.

We have not tried to sex-reverse animals, although this process is reported to work well in the direction of males to females (Chang and Witschi, 1955; Gallien, 1953, 1956). Estradiol (1 μg/ml) is put in the water for the period of a week at the time the hind limb buds are erupting (stages 52–56; 3–4 weeks), and frogs of male sex chromosome constitution thereafter develop into productive adult females. The opposite direction requires grafting of a fragment of adult testis under the skin of a genetically female tadpole at the time of limb eruption (Mikamo and Witschi, 1963).

V. Raising and Maintaining Frogs

A. Commercially Obtained Adults

Sources of adult frogs in the United States include Xenopus I (Ann Arbor, MI), Nasco Co., (Ft. Atkinson, WI), Charles Sullivan (Nashville, TN), or Scientific Animal Import (Glen Ridge, NJ), which imports animals from South Africa. Sources abroad include Xenopus Limited (England) and S.T.A.C.E.L. (France). In some cases, commercial sources import frogs captured in South Africa and maintain them until sale, whereas others produce animals from breeding colonies. Animals can be imported from the African Reptile Park, P.O. Box 30129, Tokai 7966, South Africa. Up to 40 full grown adult females can be shipped in a 5-gallon (20-liter) plastic food bucket with a tight-fitting lid, containing sphagnum moss or soft plastic foam as a cushioning and dampening agent with a minimum of water but enough to preclude the frogs' drying out. Temperatures above 26°C or below 14°C are to be avoided; animals can survive temperatures several degrees on either side of this range, but oocytes suffer from the stress to the animal. On arrival in the laboratory, it may take 2–3 weeks for animals to become hungry enough to switch to trout chow as a food if they have been fed on raw meat.

B. Laboratory-Raised Frogs

There is a compromise to be made between space per animal, quality of food, and speed of growth. Newly metamorphosed froglets are fed trout chow, particle size #4, or Rangen soft salmon diet (both of which sink) after metamorphosis and are kept at room temperature between 18° and 22°C. As animals reach 2–3 inches (6–9 cm) in length, we switch them to trout chow pellet #5, which floats. Feeding is conducted daily for small frogs and as seldom as once a week for full grown adults not in active use for egg laying. As described earlier, water is changed every 2 days for large adults, and once every several weeks for froglets depending on the population density.

After a year, the fastest growing frogs are 3–4 inches (7–10 cm) long, and we remove 500–1000 of the largest females to a new tank of 800 liters, where the animals grow more rapidly and become good ovulators. A second batch of smaller females is removed from the source tank a year later for special attention before ovulation. Under our conditions, females can be induced to ovulate at 14 months of age, but the number and quality of eggs improve as the animal ages to 3 years. Males can be mated at 8 or 9 months and may improve with further age. Callen et al. (1980) have related the increasing size of oocytes to the increasing size of laboratory-raised frogs, and they report that, in their experience, although 1-year-old animals contain many oocytes at stage IV (600–1000 μm), another year is required to enlarge these oocytes to stage VI (1200 μm). We have the impression that the oocyte and egg cortex tend to be thicker and stronger (reflected in less oocyte and egg breakage) in second- and third-year frogs. However, R. Tompkins (personal communication) reports ovulation of females just 7 months old (especially in the case of albinos) under conditions in which froglets are fed to saturation on meat and the temperature is kept at 23°–25°C, the upper end of the range.

Occasional injections of low doses of follicle-stimulating hormone–related hormones (e.g., PMSG) or luteinizing hormone–related hormones (e.g., HCG) may hasten ovary development. These hormones are reported to accelerate vitellogenin production and uptake (Wallace et al., 1970; Callen et al., 1986).

C. Recycling Frogs after Ovulation

We induce ovulation in adult female frogs every 2–3 months, after intervening periods of extra feeding and sometimes an anti-nematode treatment (see Section III). Wallace et al. (1970) induced ovulation in frogs every week as part of a study to determine the reservoir of full grown oocytes in the ovary. They obtained an average of 2000 eggs per ovulation in weeks 1 and 2, 1000 eggs per ovulation in weeks 3 and 4, and 500 eggs in weeks 5 and 6. Thus, it seems possible to use frogs more often than every 2–3 months, provided they are healthy and have not been ovulated too heavily. Our frogs frequently release 10,000 eggs at one ovulation and probably would not have a significant reserve for use 1 week later. Three months is considered a reasonable time for the frog to bring intermediate oocytes to full size (Keem et al., 1979). In the recovery period, the frogs consume more trout chow pellets, presumably as they replace ovulated occytes by rapidly growing smaller oocytes.

In a breeding colony, it is a significant problem to dispose of male frogs, which are needed in smaller numbers than females. We offered them to a local zoo to feed to large wading birds, but this was declined because *X. laevis* is considered an ecological hazard in California. Some male frogs have been given to undergraduate laboratory classes for dissection.

D. Genetic Variety

We have not given the topic of genetic variety sufficient attention during the eight or nine generations of our colony. To produce a new cohort of 2000–3000 frogs, we have tended to use three or four mating pairs of adults, of which one or two pairs may produce most of the eggs and viable tadpoles. Over time there is a successive reduction of the genetic variety of the colony. The question is, Which adults should be chosen to start the next generation? At first we used the most rapidly growing animals for mating and inadvertently produced a line of big-bodied animals with large fat reserves but rather small ovaries. Another time we bred animals that seemed particularly disease-resistant and later got healthy offspring that, unfortunately, produced easily broken eggs that fertilized poorly. In recent generations we have screened females in the course of experiments for those producing numerous, unbroken, well-fertilized, and regularly cleaving eggs. Males, though, tend to be selected only for health and fertility. It may be advisable to restore the genetic variety of the colony by occasional breeding with wild-captured animals or, in the other direction, to use of one of the inbred lines with established desirable properties (see Chapter 3).

Acknowledgments

We thank the following researchers for information provided for this chapter. Their contributions are cited in the text as personal communications, and they should be consulted for details about their experience in raising animals: Dr. A. Janice Brothers, School of Veterinary Medicine, 5305 Mill St., University of Nevada, Reno, NV 89502; Dr. Alan Coleman, Department of Biochemistry, University of Birmingham, P.O. Box 363, Birmingham B15 2TT, England; Dr. Anthony Durston, Hubrecht Laboratorium, Uppsalalaan 8, 3584 CT Utrecht, The Netherlands; Dr. John Gurdon, CRC Molecular Embryology Research Group, Department of Zoology, Cambridge University, Cambridge CB2 3EJ, England; Dr. Jerry Hedrick, Department of Biochemistry and Biophysics, University of California, Davis, CA 95616; Dr. H. R. Kobel, Departement de Biologie Animale, Universite de Geneve, 154 Route de Malagnou, CH-1224 Chene-Bougeries, Geneva, Switzerland; Dr. Douglas Melton, Department of Biochemistry and Molecular Biology, 7 Divinity Avenue, Harvard University, Cambridge, MA 02138; Dr. Kenneth Souza, NASA Ames Research Center, Moffett Field, CA 94035; Dr. Robert Tompkins, Department of Cell and Molecular Biology, Tulane University, New Orleans, LA 70118; Dr. Robin Wallace, The Whitney Laboratory, 9505 Ocean Shore Blvd., St. Augustine, FL 32086.

References

Abel, D. J. (1988). An economical, balanced diet for *Xenopus*. *Axolotl Newsl.* **17,** 19–20. (Available from the Department of Biology, Indiana University, Bloomington IN 47401.)
Balinsky, J. B. (1981). Adaptation of nitrogen metabolism to hypertonic environments in amphibia. *J. Exp. Zool.* **215,** 335–350.

Brown, A. L. (1970). "The African Clawed Toad, *Xenopus laevis*: A Guide for Laboratory Practical Work," Butterworth, London.

Buttner, J. K., and Nace, G. W. (1984). Labile water quality parameters in amphibian holding systems. *Axolotl Newsl.* **13**, 38–43.

Callen, J.-C., Dennebuoy, N., and Mounolou, J. C. (1980). Kinetic analysis of entire oogenesis in *Xenopus laevis. Dev., Growth Differ.* **22**, 831–840.

Callen, J.-C., Dennebouy, N., and Mounolou, J.-C. (1986). Early onset of a large pool of previtellogenic oocytes and cyclic escape by vitelogenesis: the pattern of ovarian activity of *Xenopus laevis* females and its physiological consequences. *Reprod. Nutr. Dev.* **26**, 13–30.

Chang, C. Y., and Witschi, E. (1955). Breeding of sex-reversed males of *Xenopus laevis* Daudin. *Proc. Soc. Exp. Biol. Med.* **89**, 150–152.

Chardonnens, X. (1975). Skin grafting in *Xenopus. Experentia* **31**, 237–239.

Cohen, N., Effrige, N. J., Parsons, S. C. V., Rollins-Smith, L. A., Nagata, S., and Albright, D. (1984). Identification and treatment of a lethal nematode (*Capillaria xenopodis*) infestation in the South African frog, *Xenopus laevis. Dev, Comp. Immunol.* **8**, 739–741.

Cosgrove, G. E., and Jared, D. W. (1974). Diseases and parasites of *Xenopus*, the clawed toad. *In* "Gulf Coast Regional Symposium on Diseases of Aquatic Animals" (R. L. Amborski, M. A. Hood, and R. R. Miller, eds.), pp. 225–242. Center for Wetlands Research, Louisiana State University, Baton Rouge.

Fox, W. F. (1982). Ammonia, ammonium, nitrite and nitrate: Causes of disease in axolotls and prevention of their toxic effects. *Axolotl Newsl.* **11**, 24–26.

Fox, W. F. (1984). A cautionary note concerning chloramine treated tap water. *Axolotl Newsl.* **13**, 37.

Gallien, L. (1953). Inversion totale du sexe chez *Xenopus laevis* Daudin a la suite d'un traitement gynogene par le benzoate d'oestradiol administre pedent la vie larvaire. *C. R. Hebd. Services Acad. Sci.* **237**, 1565.

Gallien, L. (1956). Inversion experimentale du sexe chez un Anoure inferieur, *Xenopus laevis* (Daudin). Analyse des consequences genetiques. *Bull. Biol. Fr. Belg.* **90**, 163.

Gibbs, E. L. (1973). *Rana pipiens*: Health and disease—how little we know. *Am. Zool.* **13**, 93–96.

Gibbs, E. L., Gibbs, T. J., and Van Dyck, P. C. (1966). *Rana pipiens*: Health and disease. *Lab. Anim. Care* **16**, 142–155.

Gurdon, J. B. (1967). African clawed frogs. *In* "Methods in Developmental Biology" (F. Wilt and N. Wessels, eds.), pp. 75–84. Thomas Y. Crowell Co., New York.

Hedrick, J. L., and Nishihara, T. (1990). Structure and function of the extracellular matrix of anuran eggs. *J. Electron Microsc. Tech.* (*in press*).

Keem, K., Smith, L. D., Wallace, R. A., and Wolf, D. (1979). Growth rate of oocytes in laboratory maintained *Xenopus laevis. Gamete Res.* **2**, 125–135.

Mikamo, K., and Witschi, E. (1963). Functional sex reversal in genetic females of *Xenopus laevis*, induced by implanted testes. *Genetics* **48**, 1411–1421.

Nieuwkoop, P. D., and Faber, J. (1967). "Normal Table of *Xenopus laevis* (Daudin)." North-Holland Publ., Amsterdam.

Reichenbach-Klinke, H., and Elkan, E. (1965). "The Principal Diseases of Lower Vertebrates: Diseases of Amphibians." Academic Press/T.F.H. Publications, Hong Kong (or P.O. Box 427, Neptune, N. J. 07753).

Stephens, L. C., Cromeens, D. M., Robbins, V. W., Stromberg, P. C., and Jardine, J. H. (1987). Epidermal capillarisis in South African clawed frogs (*Xenopus laevis*). *Lab. Anim. Sci.* **37**, 341–344.

Thornton, V. F. (1971). A bioassay for progesterone and gonadotropins based on the meiotic division of *Xenopus* oocytes *in vitro. Gen. Comp. Endocrinol.* **16**, 599–610.

Wallace, R. A., Jared, D. W., and Nelson, B. L. (1970). Protein incorporation by isolated amphibian oocytes. I. Preliminary studies. *J. Exp. Zool.* **175**, 259–270.

Wasserzug, R. (1972). The mechanism of ultraplanktonic entrapment in anuran larvae. *J. Morphol.* **137**, 279–287.

Chapter 2

Genetics of Xenopus laevis

JEAN-DANIEL GRAF

Laboratoire d' Examens Biologiques
Hôpital Cantonal Universitaire
1211 Genève 4, Switzerland

HANS RUDOLF KOBEL

Département de Zoologie et Biologie Animale
Université de Genève
1224 Chêne-Bougeries, Switzerland

I. Introduction
II. Genome Organization
    A. Karyotype, Bivalents, and Replication
    B. Repetitive Sequences
    C. Gene Clusters
III. Estimation of Proportion of Conserved Duplications
    A. Electrophoretic Studies of Isoenzymes
    B. Molecular Analysis of Gene Structure and Expression
IV. Gene Mapping
    A. Chromosome Mapping Using Hyperdiploid Hybrids
    B. Cytological Localization by *in Situ* Hybridization
    C. Linkage Mapping
    D. Gene–Centromere Mapping
V. Conclusion
    References

I. Introduction

Almost all research on *Xenopus laevis* is done with specimens from the Cape Flats around Cape Town, South Africa. The species, however, has a wide distribution, from Nigeria and Sudan to Cape Agulhas. Several subspecies have been recognized. These show a rather important genetic divergence that may provide useful allelic markers since there is no reproductive barrier between the various subspecies in contrast to interspecific *Xenopus* hybrids. The clawed frog or "Platanna" from South Africa belongs to the subspecies *Xenopus laevis laevis* (Daudin).

A peculiarity of the genus *Xenopus* is that all but 1 of its 16 to 20 species are of polyploid origin, forming a polyploid series in the proportions $2:4:8:12$ (Kobel and Du Pasquier, 1986). *Xenopus laevis* belongs to the tetraploid class. Evidence for tetraploidy comes from comparisons of DNA content in various species of the family Pipidae and from the fact that, in *X. laevis*, a number of genes are represented by two copies, generally of less than 10% sequence divergence.

On the other hand, *X. laevis* shows many characteristics of a diploid species: a karyotype of 18 pairs of distinct chromosomes forming bivalents in meiosis (Tymowska and Fischberg, 1973; Müller, 1974), independent disomic inheritance of duplicated genes (Graf, 1989b), and inviability of "haploid" 18-chromosome zygotes (Hamilton, 1963). In addition, there are about 40 recessive mutants known (Appendix C, this volume). Haploid inviability and recessive mutations indicate that a certain part of the genetic information is present in a nonduplicated form, as demonstrated also for a variety of enzymes and other proteins (Graf, 1989b), as well as particular multigenic loci such as nucleolar organizer (Tymowska and Kobel, 1972; Pardue, 1973), major histocompatibility complex (Du Pasquier *et al.*, 1977), and immunoglobulin heavy chain (Schwager *et al.*, 1988).

Thus, *X. laevis* displays all features of an ancient tetraploid species that is now completely diploidized. If polyploidy resulted from interspecific hybridization (allopolyploidy), which presumably is the case in *Xenopus* (Kobel and Du Pasquier, 1986), diploidization (i.e., disomic inheritance) would probably have occurred at once, because interspecific divergence in genome structure could by itself prevent homeologous chromosome pairing in newly created tetraploids.

The tetraploid origin of *X. laevis* and the high proportion of conserved gene duplications complicated its genetic analysis. Nevertheless, several genetic aspects of this species have been studied in great detail. For instance, the complete nucleotide sequence of the mitochondrial genome has been determined: Its 17,553 base pairs (bp) contain genes for the 12 S and 16 S rRNAs, as well as 22 tRNAs and 13 proteins (Roe *et al.*, 1985). In this chapter, we summarize various approaches to the nuclear genome of *X. laevis*; the methods and results discussed represent selected examples rather than an exhaustive review.

II. Genome Organization

A. Karyotype, Bivalents, and Replication

As discussed above, *X. laevis* is a functional diploid, although its genome (Table I) shows several features reminiscent of its allotetraploid origin. The mitotic karyotype (Fig. 1) of 18 chromosome pairs can be divided into seven groups of morphologically more or less distinct mediocentric and acrocentric chromosomes (Tymowska and Kobel, 1972). Chromosome 12 bears a secondary constriction on the short arm, representing the nucleolar organizer (NOR; 18 S–5.8 S–28 S rRNA gene cluster). Banding techniques have so far not succeeded in producing a consistent mammalian-like banding pattern, except for some heterochromatic bands (Schmid *et al.*, 1987). However, *X. laevis* chromosomes do contain blocks of early and late replicating regions that can be resolved by bromodeoxyuridine incorporation and appropriate staining techniques (H. R. Kobel, 1986, unpublished observations).

Replication of the genome, as analyzed in cultured somatic cells (Callan, 1973), proceeds with 1.5×10^4 replicons of about 195 kilobases (kb) or

TABLE I

The Genome of *Xenopus laevis laevis*

Feature	Value			Reference
Number of chromosomes, somatic	36			Tymowska and Kobel (1972)
Number of bivalents, both sexes	18			Müller; Tymowska and Fischberg (1973)
Number of chiasmata				Müller (1974)
Per oocyte	33.4			
Per bivalent	1.86 (range 1.4–2.6)			
Genome size, gametic	3.18 pg, 3.07×10^9 bp			Thiébaud and Fischberg (1977)
Number of replicons, somatic	1.5×10^4			Callan (1973)
Mean length of replicons	195 kb			Callan (1973)
Components of genome	%	Frequency	Complexity	Lewin (1980, p. 966)
Unique to low copy number	64	1–110	2.8×10^6 to 1.6×10^9	
Repetitive sequences	31	2×10^3	4.5×10^5	
	6	3×10^5	6.3×10^2	
	5	1×10^5	Low	

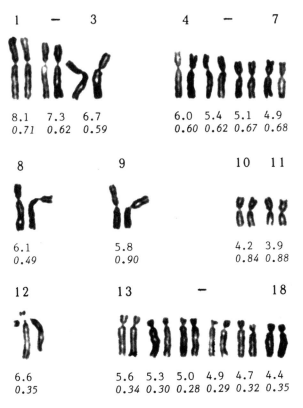

FIG. 1. Mitotic karyotype of *Xenopus laevis laevis*. The relative length (% of the total genome) and the centromere index p/q are indicated below each chromosome pair. (Modified from Tymowska and Kobel, 1972.)

20–100 μm, and its duration is 13 hours with a mean rate of 500 bp per minute. Embryonic cells replicate their genome much faster, probably by using more initiation points in synchrony rather than increasing the replication rate within the replicons (Callan, 1973).

Meiotic cells show 18 bivalents. The morphology of the lampbrush chromosomes allow to identify the bivalents individually (Müller, 1974; Callan *et al.,* 1987). However, it is presently not possible to establish homology between lampbrush and mitotic chromosomes . Cytological analyses of oocytes indicate a mean of 1.86 (range, 1.4–2.6) chiasmata per bivalent (Müller, 1974). Analysis of neither lampbrush nor mitotic chromosomes has allowed identification of sex chromosomes, although it is known (Chang and Witschi, 1956) that sex determination operates through female heterogamety (female *WZ*, male *ZZ*).

B. Repetitive Sequences

From 20 to 30% of the genome of *X. laevis* consists of repetitive sequences (Davidson *et al.*, 1973). More than 20 different elements have been cloned and sequenced, representing together about 20% of all repetitive sequences or 5% of the total genome. The more abundant elements have been studied for chromosomal location (Table II); the remainder mainly belong to the dispersed middle repetitive type. Several of the elements are transcribed but, except for those encoding various ribosomal and transfer RNAs, the function of repetitive sequences remains largely undefined. Nontranscribed sequences may represent constituents of the chromosomes structure, such as telomere and centromere sequences, or may be related to replication, gene regulation, or other functions.

C. Gene Clusters

The fine mapping of the globin gene family in *X. laevis* (Fig. 2) revealed two remarkable features. (1) The α- and β-globin genes are closely linked in the same cluster, whereas in human, mouse, and chicken they lie on different chromosomes. (2) The similarity of architecture of the two clusters found in *X. laevis* supports the tetraploid origin of this species (Hosbach *et al.*, 1983).

TABLE II

CHROMOSOMAL LOCATION OF REPETITIVE GENES AND ELEMENTS

Element	Length	Copies	Chromosome	Reference
rRNA, 18–5.8–28 S	10–17 kb	400–600	12p, secondary constriction	Pardue (1973)
5 S RNA, major oocytic	800 bp	2×10^4	Most chromosomes, telomeric on q	Pardue (1973)
tRNA, 8 genes	3.18 kb	150	One of group 13–18, proximal to 5 S RNA	Fostel *et al.* (1984)
X 132A	77–79 bp	1×10^4	Most, telomeric on p and q	Jamrich *et al.* (1983)
E 1723	6–8 kb	8500	All, dispersed	Kay *et al.* (1984)
Satellite I	741 bp	$2–4 \times 10^4$	About 9 pairs, near the centromere	Jamrich *et al.* (1983)
Rem 1	500 bp	2.5×10^4	Most, telomeric and centromeric	Hummel *et al.* (1984)
Rem 2	487 bp	5000	Most, dispersed	Hummel *et al.* (1984)
Rem 3	463 bp	1200	1q, near the centromere	Hummel *et al.* (1984)

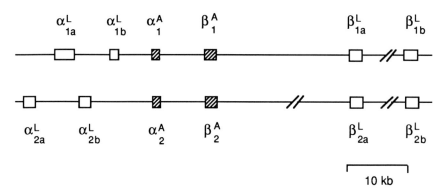

FIG. 2. Putative organization of the *Xenopus laevis* globin gene family. □, Larval genes; ▨, adult genes. (After Hosbach et al., 1983; copyright by Cell Press.)

Similarly, Fritz et al. (1989) identified in *X. laevis* two clusters of three homeobox genes each. Sequence comparisons suggest that the two clusters are duplicate copies of each other. In addition, they appear to be homologous to the human *Hox-2* gene complex.

The histone genes of *X. laevis* are arranged in distinct families of tandemly repeated clusters containing at least one copy of each of the five histone genes. This organization apparently represents an intermediate type between the highly reiterated sea urchin histone gene clusters and the dispersed mammalian histone genes (Perry et al., 1985). Molecular studies of the immunoglobulin heavy chain locus revealed a gene cluster with an overall organization similar to that of mammals (Schwager et al., 1988; Du Pasquier et al., 1989).

III. Estimation of Proportion of Conserved Duplications

A. Electrophoretic Studies of Isoenzymes

The degree of duplicate gene silencing (or elimination) can be estimated on the basis of electrophoretic analyses of biochemical markers. One should keep in mind, however, that this technique is likely to underestimate the number of conserved duplications, since detection of duplicated pairs requires that the two products can be distinguished on the basis of electrophoretic mobility. In *X. laevis*, expression of the two copies of duplicated pairs can be detected in 54% of a sample of 24 proteins (Table III). The identification of duplicated

TABLE III

Single or Duplicated Loci for Enzymes and Blood Proteins in *Xenopus laevis* as Deduced from Electrophoretic Phenotypes and Segregation Analysis in Backcrosses of Heterozygotes

Enzyme or blood protein	EC No.	Single	Duplicated	Reference
Aconitase, s (cytosolic)	4.2.1.3	X	—	Graf (1989b); Bürki (1987)
Aconitase, m (mitochondrial)	4.2.1.3	—	X	Graf (1989b); Bürki (1987)
Albumin (serum)	—	—	X	Graf (1989b); Graf and Fischberg (1986)
Alcohol dehydrogenase	1.1.1.1	—	X	Graf (1989b); Wesolowski and Lyerla (1983)
Creatine kinase (muścle)	2.7.3.2	—	X	Bürki (1985); Robert and Kobel (1988); Wolff and Kobel (1985)
Creatine kinase (brain)	2.7.3.2	X	—	Graf (1989b); Wolff and Kobel (1985)
Fumarate hydratase	4.2.1.2	X	—	Graf (1989b)
Glucose-phosphate isomerase	5.3.1.9	—	X	Graf (1989b)
Glycerol-3-phosphate dehydrogenase	1.1.1.8	—	X	Graf (1989b)
Glyoxalase I	4.4.1.5	—	X	Graf (1989b)
Isocitrate dehydrogenase, s	1.1.1.42	X	—	Graf (1989b); Bürki (1987)
Isocitrate dehydrogenase, m	1.1.1.42	—	X	Graf (1989b); Bürki (1987)
Malate dehydrogenase, s	1.1.1.37	—	X	Kobel and Du Pasquier (1986)
Malate dehydrogenase, m	1.1.1.37	—	X	Kobel and Du Pasquier (1986)
Malic enzyme, s	1.1.1.40	X	—	Graf (1989b); Bürki (1987); Graf (1989a)
Malic enzyme, m	1.1.1.40	X	—	Graf (1989a, b); Bürki (1987)
Mannose-phosphate isomerase	5.3.1.8	—	X	Graf (1989b)
Nucleoside phosphorylase	2.4.1.2	X	—	Graf (1989b)
Phosphogluconate dehydrogenase	1.1.1.44	—	X	Graf (1989b)
Sorbitol dehydrogenase	1.1.1.14	X	—	Graf (1989b)
Superoxide dismutase, s	1.15.1.1	—	X	Graf (1989b); Burki (1987)
Superoxide dismutase, m	1.15.1.1	X	—	Graf (1989b); Bürki (1987)
Transferrin	—	X	—	Graf (1989b)
Xanthine dehydrogenase	1.2.1.37	X	—	Lyerla and Fournier (1983)

loci was based on a comparison with the diploid anuran species *Rana pipiens* (Wright *et al.*, 1983).

Detailed studies of enzyme gene expression in *X. laevis* indicate that the duplicated loci may exhibit differential expression resulting in tissue- or development-specific patterns. For instance, of the two loci encoding glucose-phosphate isomerases (GPI), GPI-1 shows highest activity in heart and skeletal muscle, whereas GPI-2 is mainly active in liver and kidney (Graf, 1989b).

Similarly, the two loci encoding the muscle-specific creatine kinase isozymes show a striking difference in developmental profile (Robert et al., 1990). These observations suggest that those duplicate loci that have not undergone silencing are not simply redundant copies but loci showing newly acquired specialization.

B. Molecular Analysis of Gene Structure and Expression

Comparative molecular studies of sarcomeric actin (Stutz and Spohr, 1986), serum albumin (Westley et al., 1981), and hemoglobin (Jeffreys et al., 1980; Hosbach et al., 1983) indicate that the genome of X. laevis contains twice the number of genes (or gene sets) for these proteins compared with the diploid species Xenopus tropicalis. For albumin and adult globins, it has been shown that both copies of each duplicate pair are expressed (Westley and Weber, 1982; Sandmeier et al., 1986). The expression of the duplicate actin genes has not been established so far, owing to the low divergence between the coding sequences of the duplicated copies (Stutz and Spohr, 1986).

The X. laevis genome appears to retain duplicated genes for a fairly large number of different characters (Table IV). There are some notable exceptions, however. For instance, the constant region of immunoglobin M appears to be encoded in a single locus (Schwager et al., 1988), suggesting that one copy has been lost during the evolutionary history of X. laevis. The vitellogenin gene family illustrates a more complex situation. Xenopus laevis has four vitellogenin genes (A1, A2, B1, and B2), of which three (A1, A2, and B1) show tight genetic linkage with one another (Schubiger and Wahli, 1986). Since the diploid specis X. tropicalis has three vitellogenin genes (two closely related type-A genes and a single type-B gene), the chromosomal arrangement found in X. laevis supports an evolutionary scenario combining gene duplications (A/B, then A1/A2), genome duplication (A1-A2-B1/A1'-A2'-B2), and gene elimination (A1', A2'). Although considerably many genes from X. laevis have been cloned and characterized, this wealth of information does not provide a reliable quantitative estimate of the degree of duplicate gene retention since many studies did not specifically address the question of gene duplication.

The degree of sequence divergence between duplicate copies has been investigated for several gene pairs. For instance, the two genes encoding the adult α-globins (Fig. 2) diverge by 7.6% in the translated nucleotide regions (Knöchel et al., 1985). The divergence amounts to 10.8% for the adult β-globin gene pair, 24.1% for the larval α-globins, and 24.9% for the larval β-globins (Knöchel et al., 1985). Similarly, the duplicate genes encoding serum albumin diverge by about 7% in their coding sequences (Moskaitis et al., 1989), whereas the duplicate XlHbox-2 genes show a 4.5% divergence in the translated region (Fritz et al., 1989).

TABLE IV

Examples of Duplicated Genes in *Xenopus laevis*, as Deduced from Cloned Sequences

Gene	Duplication	Reference
Actin, sarcomeric		
Cardiac	a1-I/a1-II	Stutz and Spohr (1986)
Skeletal	a2-I/a2-II	Stutz and Spohr (1987)
Femoral	a3-I/a3-II	Mohun et al. (1988)
Albumin	68K/74K	Moskaitis et al. (1989)
Calmodulin	I/II	Chien and Dawid (1984)
Cytokeratin, type I (acidic): embryonic	A1B1/A2B2	Miyatani et al. (1986)
Fibroblast growth factor, basic	I/II	Volk et al. (1989)
Hemoglobin cluster (6 gene pairs)	I/II	Hosbach et al. (1983)
Homeobox cluster	6A 7A 2A/6B 7B 2B	Fritz et al. (1989)
Integrins, β subunit	b1/b1*	DeSimone and Hynes (1988)
MyoD	a/b	Harvey (1990)
Neural cell adhesion molecule: 1d-NCAM	I/II	Krieg et al. (1989)
Proopiomelanocortin	A/B	Martens (1986)
Protooncogene		
c-ets-1	a/b	Stiegler et al. (1990)
c-ets-2	a/b	Wolff et al. (1990)
c-myc	I/II	Vriz et al. (1989)
src	1/2	Steele et al. (1989)
Ribosomal proteins	L1a/L1b	Loreni et al. (1986)
	L14a/L14b	Beccari et al. (1986)
	S8a/S8b	Mariottini et al. (1988)
Vimentin	Vim1/Vim4	Herrmann et al. (1989)
Vitellogenin	B1/B2	Schubiger and Wahli (1986)

IV. Gene Mapping

A. Chromosome Mapping Using Hyperdiploid Hybrids

Chromosome mapping using hyperdiploid hybrids, in analogy to the powerful somatic cell hybrid technique used in mammals, makes use of the possibility to generate frogs possessing two chromosome sets of one species plus a random collection of supernumerary chromosomes from another species (Kobel and Du Pasquier, 1979). In *Xenopus*, it is relatively easy to produce allotriploid individuals, either by backcrossing female interspecific hybrids (which often lay endoreduplicated eggs) or by suppressing the second meiotic division in eggs fertilized with sperm of a different species (technique in Müller et al., 1978). Allotriploid females are raised to the adult stage. Meiosis in the oocytes of these triploids usually results in aneuploid gametes. For instance, allotriploid females with two *X. laevis* chromosome sets plus one haploid set from another species (e.g., *Xenopus gilli*) produce aneuploid gametes that contain 18 *X. laevis* chromosomes, plus a random sample of 1 to 18 chromosomes of *X. gilli*. By fertilizing these ova with sperm from *X. laevis*, one obtains hyperdiploid offspring with 36 *X. laevis* chromosomes and several supernumerary *X. gilli* chromosomes (Kobel and Du Pasquier, 1979). A fraction of these animals (from 1 to 3%) reach adulthood, providing an efficient means of identifying syntenic groups and assigning them to specific *X. gilli* chromosomes.

This procedure has been used to map immunological markers in *X. gilli* (Du Pasquier and Kobel, 1979). However, owing to the low number of genetic markers tested, this study has not revealed linkage groups. It is evident that the mapping technique based on hyperdiploid hybrids would bring a wealth of information on *Xenopus* genetics if a large number of loci [e.g., restriction fragment length polymorphisms (RFLPs), isozymes, immunological markers] could be tested simultaneously.

B. Cytological Localization by *in Situ* Hybridization

For obvious technical reasons, chromosomal localization by *in situ* hybridization on mitotic or lampbrush chromosomes has hitherto been restricted to repetitive sequences and repetitive gene clusters (Table II). Recent refinements in the methodology, using much longer cosmid probes along with suppression strategies (Lichter and Ward, 1990), will probably allow the localization of single-copy genes. Since even interphase nuclei provide information on the number of loci present and their physical distance, these hybridization techniques could bring about rapid progress in gene mapping of *Xenopus*.

C. Linkage Mapping

Because *X. laevis* has a long generation time and few characterized genetic markers, it has been considered that the classic mapping procedure based on backcross analysis could not be applied successfully to this organism. However, the high level of genetic differentiation between the various subspecies, along with their interfertility, allow one to produce "hybrids" heterozygous at a large number of scorable genetic loci (isozymes, RFLPs, visible mutations). These highly heterozygous individuals can be backcrossed to one of the parental subspecies, and the offspring can be analyzed for segregation of the genetic markers (Graf, 1989b). The major advantage of this procedure is that every single cross allows a large number of loci to be tested simultaneously, thereby increasing the probability of detecting genetic linkage.

A provisional linkage map of *X. laevis* (Fig. 3) indicates that in some of the linkage groups, the position of the centromere could be determined by

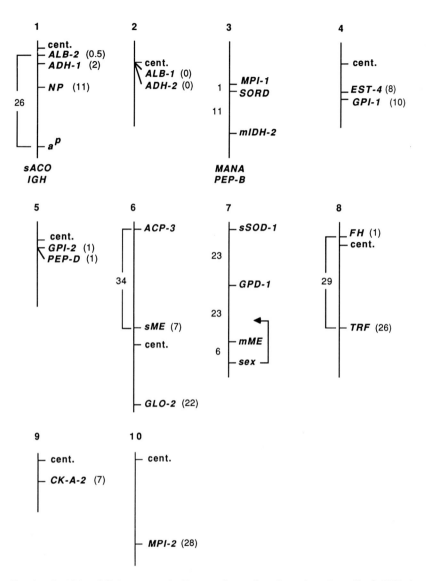

FIG. 3. Provisional linkage map in *Xenopus laevis*, based on data from Graf (1989a,b; 1990, unpublished data). The groups have been numbered from 1 to 10 in an arbitrarily chosen order. Numbers in parentheses indicate the recombination value (%) between a locus and the centromere. Numbers to the left of each group indicate recombination frequencies between loci. Genes listed below linkage groups 1 and 3 segregate with these groups, but their position in the sequence of loci could not be determined. a^p, Periodic albinism; *ACP*, acid phosphatase; *sACO*, cytosolic aconitase; *ALB*, serum albumin; *ADH*, alcohol dehydrogenase; cent., centromere; *CK-A*, creatine kinase (skeletal muscle); *EST*, esterase; *FH*, fumarate hydratase; *GLO*, glyoxalase; *GPI*, glucose-phosphate isomerase; *GPD*, α-glycerophosphate dehydrogenase; *IGH*, immunoglobulin heavy chain; *mIDH*, mitochondrial isocitrate dehydrogenase; *sME*, cytosolic malic enzyme; *mME*, mitochondrial malic enzyme; *MANA*, α-mannosidase; *MPI*, mannose-phosphate isomerase; *NP*, nucleoside phosphorylase; *PEP-B*, peptidase B; *PEP-D*, peptidase D; *sSOD-1*, cytosolic superoxide dismutase; *SORD*, sorbitol dehydrogenase; *TRF*, transferrin.

analyzing gynogenetic offspring (see Section IV,D). The salient features of these preliminary results can be summarized as follows (Graf, 1990, unpublished data). (1) When duplicate copies of a gene could be tested simultaneously, they showed independent segregation (e.g., *ALB-1/ALB-2, ADH-1/ADH-2, MPI-1/MPI-2, GLO-1/GLO-2*). (2) The close linkage uniting *ALB-2* to *ADH-1* is also observed between their duplicate copies (i.e., *ALB-1* and *ADH-2*); moreover, both *ALB–ADH* groups are closely linked to their respective centromeres. (3) Sex is determined by alleles at one locus showing recombination with other loci located in the same linkage group. It is worth mentioning that several syntenies were found to be common to the Xenopus genome and the human genome: *ALB* and *ADH* are syntenic on human chromosome 4; *MPI, SORD, MANA,* and *mIDH* are syntenic on human chromosome 15; *GPI* and *PEP-D* are linked on human chromosome 19; *GLO-1* and *sME* are syntenic on human chromosome 6.

D. Gene–Centromere Mapping

In *Xenopus* and other amphibians, one can easily obtain diploid gynogenetic offspring by suppression of the second meiotic division in eggs submitted to elevated hydrostatic pressure, following activation with UV-irradiated sperm (Tompkins, 1978). Examination of gynogenetic diploids for characters heterozygous in the mother provides information on the map distance between the corresponding loci and their centromere (hemitetrad analysis). This procedure has allowed estimations of gene–centromere distances for about 30 loci of *X. laevis*, regulating or encoding sex (Colombelli *et al.*, 1984), embryo development mutations (Thiébaud *et al.*, 1984; Reinschmidt *et al.*, 1985), and the major histocompatibility complex (Reinschmidt *et al.*, 1985), as well as isoenzymes and blood proteins (Fig. 3).

V. Conclusion

The amount of information on the genetics of *X. laevis* has been growing ever faster over the past 20 years. Undoubtedly, this species is now the most studied of all amphibians. An impressive number of genes have been characterized, 10 linkage groups have been identified, and gene–centromere distances have been estimated. Mitotic and lampbrush karyotypes have been described. The architecture and composition of the nuclear genome as a whole, and of particular genes and gene clusters, have been analyzed in detail. The nucleotide sequence of the mitochondrial genome has been determined.

In addition, a large variety of genes, with emphasis on growth and transcription factors, are presently under investigation for expression patterns and functional aspects, giving *X. laevis* a key position in developmental genetics.

There remain, however, many gaps in our knowledge of *Xenopus* genetics. For instance, a classification of mitotic chromosomes based on banding patterns still needs to be established. The gene map of *X. laevis* is in its infancy, and the linkage groups identified so far still await their assignation to individual chromosomes. Since *X. laevis* is a tetraploid-derived species, most likely of hybrid origin, the genetic basis of certain phenotypic traits might be particular to this species. This often neglected problem could be solved by comparative studies using the diploid species *X. tropicalis*, as well as other tetraploid-derived *Xenopus* (e.g., *X. borealis* and *X. fraseri*).

Another problem is the very limited availability of genetically defined strains (Kobel and Du Pasquier, 1977; Reinschmidt *et al.*, 1979; Chapter 3, this volume). Usually, researchers still rely on wild-caught, genetically undefined specimens whose origin is often uncertain. Since population genetics of *X. laevis* has remained an unexplored field, the level of genetic heterogeneity within and among natural populations is unknown. More attention should be paid to this point in the future.

ACKNOWLEDGMENTS

Research was supported by the Swiss National Science Foundation (Grants 31-9181.87 to J.-D.G. and 3.388-0.78 to H.R.K.) and the Georges and Antoine Claraz Donation.

REFERENCES

Beccari, E., Mazzetti, P., Mileo, A.-M., Bozzoni, I., Pierandrei-Amaldi, P., and Amaldi, F. (1986). Sequences coding for the ribosomal protein L 14 in *Xenopus laevis* and *Xenopus tropicalis*: Homologies in the 5′ untranslated region shared with other r-protein mRNAs. *Nucleic Acids Res.* **14**, 7633–7646.

Bürki, E. (1985). The expression of creatine kinase isozymes in *Xenopus tropicalis*, *Xenopus laevis laevis*, and their viable hybrid. *Biochem. Genet.* **23**, 73–88.

Bürki, E. (1987). Comparative analysis of electrophoretic protein phenotypes in the genus *Xenopus* (Anura: Pipidae). Ph.D. Thesis No. 2244, University of Geneva, Switzerland.

Callan, H. G. (1973). DNA replication in the chromosomes of eukaryotes. *Cold Spring Harbor Symp. Quant. Biol.* **38**, 195–203.

Callan, H. G., Gall, J. G., and Berg, A. (1987). The lampbrush chromosomes of *Xenopus laevis*: Preparation, identification, and distribution of 5S DNA sequences. *Chromosoma* **95**, 236–250.

Chang, C. Y., and Witschi, E. (1956). Genetic control and hormonal reversal of sex differentiation in *Xenopus*. *Proc. Soc. Exp. Biol. Med.* **93**, 140–144.

Chien, Y. H., and Dawid, I. B. (1984). Isolation and characterization of calmodulin genes from *Xenopus laevis*. *Mol. Cell. Biol.* **4**, 507–513.

Colombelli, B., Thiébaud, C. H., and Müller, W. P. (1984). Production of WW superfemales by diploid gynogenesis in *Xenopus laevis*. *Mol. Gen. Genet.* **194**, 57–59.

Davidson, E. H., Hough, B. R., Amenson, C. S., and Britten, R. J. (1973). General interspersion of repetitive with non-repetitive sequence elements in the DNA of *Xenopus*. *J. Mol. Biol.* **77**, 1–23.

DeSimone, D. W., and Hynes, R. O. (1988). *Xenopus laevis* integrins. *J. Biol. Chem.* **263**, 5333–5340.

Du Pasquier, L., and Kobel, H. R. (1979). Histocompatibility antigens and immunoglobulin genes in the clawed toad: Expression and linkage studies in recombinant and hyperdiploid *Xenopus* hybrids. *Immunogenetics* **8**, 299–310.

Du Pasquier, L., Miggiano, V. C., Kobel, H. R., and Fischberg, M. (1977). The genetic control of histocompatibility reactions in natural and laboratory-made polyploid individuals of the clawed toad *Xenopus*. *Immunogenetics* **5**, 129–141.

Du Pasquier, L., Schwager, J., and Flajnik, M. F. (1989). The immune system of *Xenopus*. *Annu. Rev. Immunol.* **7**, 251–275.

Fostel, J., Narayanswami, S., Hamkalo, B., Clarkson, S. G., and Pardue, M. L. (1984). Chromosomal location of a major tRNA gene cluster of *Xenopus laevis*. *Chromosoma* **90**, 254–260.

Fritz, A. F., Cho, K. W. Y., Wright, C. V. E., Jegalian, B. G., and DeRobertis, E. M. (1989). Duplicated homeobox genes in *Xenopus*. *Dev. Biol.* **131**, 584–588.

Graf, J.-D. (1989a). Sex linkage of malic enzyme in *Xenopus laevis*. *Experientia* **45**, 194–196.

Graf, J.-D. (1989b). Genetic mapping in *Xenopus laevis*: Eight linkage groups established. *Genetics* **123**, 389–398.

Graf, J.-D., and Fischberg, M. (1986). Albumin evolution in polyploid species of the genus *Xenopus*. *Biochem. Genet.* **24**, 821–837.

Hamilton, L. (1963). An experimental analysis of the development of the haploid syndrome in embryos of *Xenopus laevis*. *J. Embryol. Exp. Morphol.* **11**, 267–278.

Harvey, R. P. (1990). The *Xenopus MyoD* gene: an unlocalized maternal mRNA predates lineage-restricted expression in the early embryo. *Development (Cambridge, U.K.)* **108**, 669–690.

Herrmann, H., Fouquet, B., and Franke, W. W. (1989). Expression of intermediate filament proteins during development of *Xenopus laevis*. I. cDNA clones encoding different forms of vimentin. *Development (Cambridge, U.K.)* **105**, 279–298.

Hosbach, H. A., Wyler, T., and Weber, R. (1983). The *Xenopus laevis* globin gene family: chromosomal arrangement and gene structure *Cell (Cambridge, Mass.)* **32**, 45–53.

Hummel, S., Meyerhof, W., Korge, E., and Knöchel, W. (1984). Characterization of highly and moderately repetitive 500 bp EcoRI fragments from *Xenopus laevis*. *Nucleic Acids Res.* **12**, 4921–4938.

Jamrich, M., Warrior, R., Steele, R., and Gall, J. G. (1983). Transcription of repetitive sequences on *Xenopus* lampbrush chromosomes. *Proc. Natl. Acad. Sci. U.S.A.* **80**, 3364–3367.

Jeffreys, A. J., Wilson, V., Wood, D., and Simons, J. P. (1980). Linkage of adult α- and β-globin genes in *X. laevis* and gene duplication by tetraploidization. *Cell (Cambridge, Mass.)* **21**, 555–564.

Kay, B. K., Jamrich, M., and Dawid, I. B. (1984). Transcription of a long, interspersed, highly repeated DNA element in *Xenopus laevis*. *Dev. Biol.* **105**, 518–525.

Knöchel, W., Meyerhof, W., Stalder, J., and Weber, R. (1985). Comparative nucleotide sequence analysis of two types of larval β-globin mRNAs of *Xenopus laevis*. *Nucleic Acids Res.* **13**, 7899–7908.

Kobel, H. R., and Du Pasquier, L. (1977). Strains and species of *Xenopus* for immunological research. *In* "Developmental Immunobiology" (J. B. Solomon and J. D. Harton, eds.), pp. 299–306. Elsevier/North-Holland Biomedical Press, Amsterdam.

Kobel, H. R., and Du Pasquier, L. (1979). Hyperdiploid species hybrids for gene mapping in Xenopus. *Nature (London)* **279**, 157–158.

Kobel, H. R., and Du Pasquier, L. (1986). Genetics of polyploid *Xenopus*. *Trends Genet.* **2,** 310–315.

Krieg, P. A., Sakaguchi, D. S., and Kintner, C. R. (1989). Primary structure and developmental expression of a large cytoplasmic domain form of *Xenopus laevis* neural cell adhesion molecule (NCAM). *Nucleic Acids Res.* **17,** 10321–10335.

Lewin, B. (1980). "Gene Expression, Volume 2: Eukaryotic Chromosomes." Wiley, New York.

Lichter, P., and Ward, D. C. (1990). Is non-isotopic *in situ* hybridization finally coming of age? *Nature (London)* **345,** 93–94.

Loreni, F., Ruberti, I., Bozzoni, I., Pierandrei-Amaldi, P., and Amaldi, F. (1986). Nucleotide sequence of the L1 ribosomal protein gene of *Xenopus laevis*: Remarkable sequence homology among introns. *EMBO J.* **4,** 3483–3488.

Lyerla, T. A., and Fournier, P. C. (1983). Xanthine dehydrogenase activity in the clawed frog, *Xenopus laevis*. *Comp. Biochem. Physiol. B.* **76B,** 497–502.

Mariottini, P., Bagni, C., Annesi, F., and Amaldi, F. (1988). Isolation and nucleotide sequences of cDNA for *Xenopus laevis* ribosomal protein S8: Similarities in the 5' and 3' untranslated regions of mRNAs for various r-proteins. *Gene* **67,** 69–74.

Martens, G. J. M. (1986). Expression of two proopiomelanocortin genes in the pituitary gland of *Xenopus laevis*: Complete structure of the two preprohormones. *Nucleic Acids Res.* **14,** 3791–3798.

Miyatani, S., Winkles, J. A., Sargent, T. D., and Dawid, I. B. (1986). Stage-specific keratins in *Xenopus laevis* embryos and tadpoles: the XK81 gene family. *J. Cell Biol.* **103,** 1957–1965.

Mohun, T., Garrett, N., Stutz, F., and Spohr, G. (1988). A third striated muscle actin gene is expressed during early development in the amphibian *Xenopus laevis*. *J. Mol. Biol.* **202,** 67–76.

Moskaitis, J. E., Sargent, T. D., Smith, L. H., Jr., Pastori, R. L., and Schoenberg, D. R. (1989). *Xenopus laevis* serum albumin: Sequence of the cDNA encoding the 68- and 74-Kd peptides and the regulation of albumin gene expression by thyroid hormone during development. *Mol. Endocrinol.* **3,** 464–473.

Müller, W. P. (1974). The lampbrush chromosomes of *Xenopus laevis*. *Chromosoma* **47,** 283–296.

Müller, W. P., Thiébaud, C. H., Ricard, L., and Fischberg, M. (1978). The induction of triploidy by pressure in *Xenopus laevis*. *Rev. Suisse Zool.* **85,** 20–26.

Pardue, M. L. (1973). Localization of repeated DNA sequences in *Xenopus* chromosomes. *Cold Spring Harbor Symp. Quant. Biol.* **38,** 475–482.

Perry, M., Thomsen, G. H., and Roeder, R. G. (1985). Genomic organization and nucleotide sequence of two distinct histone gene clusters from *Xenopus laevis*. *J. Mol. Biol.* **185,** 479–499.

Reinschmidt, D. C., Simon, S. J., Volpe, E. P., and Tompkins, R. (1979). Production of tetraploid and homozygous diploid amphibians by suppression of first cleavage. *J. Exp. Zool.* **210,** 137–143.

Reinschmidt, D. C., Friedman, J., Hauth, J., Ratner, E., Cohen, M., Miller, M., Krotoski, D., and Tompkins, R. (1985). Gene-centromere mapping in *Xenopus laevis*. *J. Hered.* **76,** 345–347.

Robert, J., and Kobel, H. R. (1988). Purification and characterization of cytoplasmic creatine kinase isozymes of *Xenopus laevis*. *Biochem. Genet.* **26,** 543–555.

Robert, J., Wolff, J., Jijakli, H., Graf, J.-D., Karch, F., and Kobel, H. R. (1990). Developmental expression of the creatine kinase isozyme system of *Xenopus*: Maternally derived CK-IV isoform persists far beyond the degradation of its maternal mRNA and into the zygotic expression period. *Development (Cambridge, U.K.)* **108,** 507–514.

Roe, B. A., Ma, D.-P., Wilson, R. K., and Wong, J. F.-H. (1985). The complete nucleotide sequence of the *Xenopus laevis* mitochondrial genome. *J. Biol. Chem.* **260,** 9759–9774.

Sandmeier, E., Gygi, D., Wyler, T., Nyffenegger, U., and Weber, R. (1986). Analysis of globin transition in *Xenopus laevis* and identification of globins by *in vitro* translation of hybrid-selected mRNA. *FEBS Lett.* **205,** 219–222.

Schmid, M., Vitelli, L., and Batistoni, R. (1987). Chromosome banding in amphibia. XI. Constitutive heterochromatin, nucleolus organizers, 18S + 28S and 5S ribosomal RNA genes in Ascaphidae, Pipidae, Discoglossidae and Pelobatidae. *Chromosoma* **95**, 271–284.

Schubiger, J.-L., and Wahli, W. (1986). Linkage arrangement in the vitellogenin gene family of *Xenopus laevis* as revealed by gene segregation analysis. *Nucleic Acids Res.* **14**, 8723–8734.

Schwager, J., Grossberger, D., and Du Pasquier, L. (1988). Organization and rearrangement of immunoglobulin M genes in the amphibian *Xenopus*. *EMBO J.* **7**, 2409–2415.

Steele, R. E., Unger, T. F., Mardis, M. J., and Fero, J. B. (1989). The two *Xenopus laevis SRC* genes are co-expressed and each produces functional pp60^{src}. *J. Biol. Chem.* **264**, 10649–10653.

Stiegler, P., Wolff, C. M., Balzinger, M., Hirzlin, J., Senan, F., Meyer, D., Ghysdael, J., Stéhelin, D., Befort, N., and Remy, P. (1990). Characterization of *Xenopus laevis* cDNA clones of the c-*ets*-1 oncogene. *Nucleic Acids Res.* **18**, 5298.

Stutz, F., and Spohr, G. (1986). Isolation and characterization of sarcomeric actin genes expressed in *Xenopus laevis* embryos. *J. Mol. Biol.* **187**, 349–361.

Stutz, F., and Spohr, G. (1987). A processed gene coding for sarcomeric actin in *Xenopus laevis* and *Xenopus tropicalis*. *EMBO J.* **6**, 1989–1995.

Thiébaud, C. H., and Fischberg, M. (1977). DNA content in the genus *Xenopus*. *Chromosoma* **59**, 253–257.

Thiébaud, C. H., Colombelli, B., and Müller, W. P. (1984). Diploid gynogenesis in *Xenopus laevis* and the localization with respect to the centromere of a gene for periodic albinism a^p. *J. Embryol. Exp. Morphol.* **83**, 33–42.

Tompkins, R. (1978). Triploid and gynogenetic diploid *Xenopus laevis*. *J. Exp. Zool.* **203**, 251–256.

Tymowska, J., and Fischberg, M. (1973). Chromosome complements of the genus *Xenopus*. *Chromosoma* **44**, 335–342.

Tymowska, J., and Kobel, H. R. (1972). Karyotype analysis of *Xenopus muelleri* (Peters) and *Xenopus laevis* (Daudin), Pipidae. *Cytogenetics* **11**, 270–278.

Volk, R., Köster, M., Pöting, A., Hartmann, L., and Knöchel, W. (1989). An antisense transcript from the *Xenopus laevis* bFGF gene coding for an evolutionarily conserved 24 Kd protein. *EMBO J.* **8**, 2983–2988.

Vriz, S., Taylor, M., and Méchali, M. (1989). Differential expression of two *Xenopus* c-*myc* proto-oncogenes during development. *EMBO J.* **8**, 4091–4097.

Wesolowski, M. H., and Lyerla, T. A. (1983). Alcohol dehydrogenase isozymes in the clawed frog, *Xenopus laevis*. *Biochem. Genet.* **21**, 281–290.

Westley, B., and Weber, R. (1982). Divergence of the two albumins of *X. laevis*. Evidence for the glycosylation of the major 74K albumin. *Differentiation (Berlin)* **22**, 227–230.

Westley, B., Wyler, T., Ryffel, G., and Weber, R. (1981). *Xenopus laevis* serum albumins are encoded in two closely related genes. *Nucleic Acids Res.* **9**, 3557–3574.

Wolff, C. H., Stiegler, P., Baltzinger, M., Meyer, D., Ghysdael, J., Stéhelin, D., Befort, N., and Remy, P. (1990). Isolation of two different c-*ets*-2 proto-oncogenes in *Xenopus laevis*. *Nucleic Acids Res.* **18**, 4603.

Wolff, J., and Kobel, H. R. (1985). Creatine kinase isozymes in pipid frogs: their genetic bases, gene expressional differences, and evolutionary implications. *J. Exp. Zool.* **234**, 471–480.

Wright, D. A., Richards, C. M., Frost, J. S., Camozzi, A. M., and Kunz, B. M. (1983). Genetic mapping in amphibians. *Isozymes Curr. Top. Biol. Med. Res.* **10**, 287–311.

Chapter 3

Experimentally Induced Homozygosity in *Xenopus laevis*

ROBERT TOMPKINS AND DANA REINSCHMIDT

Department of Cell and Molecular Biology
Tulane University
New Orleans, Louisiana 70118

I. Introduction
II. Methods
   A. Propagation of Primary Homozygous Diploid Animals
   B. Defining Animals Homozygous at All Loci
III. Conclusions
   References

I. Introduction

Genomic manipulations of amphibians were developed to study nucleocytoplasmic interactions (Fankhauser, 1955), to alter ploidy (Tompkins, 1978; Reinschmidt et al., 1979), to construct cell markers (Szaro et al., 1985), and to study the allometric effects of such changes on cellular morphology (Szaro and Tompkins, 1987). By contrast, genetically defined lines of vertebrates have usually derived from inbreeding strategies (Klein, 1975; Festing, 1979). Several groups have generated inbred lines of *Xenopus laevis*, such as our congenic resistant lines (Tompkins et al., 1980). Difficulties in inbreeding strategies common to all animals include inbreeding depression and fixation of deleterious alleles, demanding that numerous lines be initiated to ensure some will be fertile and healthy enough for routine use. This mass inbreeding process is very costly and beyond the capacity of most individual laboratories using amphibians.

An obvious alternative is to suppress first cleavage division in haploid individuals and to screen for viability and fecundity, eliminating in one generation the bulk of the cost of developing inbred lines and, in addition, producing a product of at least theoretical perfection, that is, diploid animals homozygous at all loci. Although there were earlier indications that this strategy might work (Asher, 1970), proof that diploidization of haploid zygotes had indeed occurred needed to be addressed, along with the problem of reproducing these unique animals as clones so that they might be of long-term use. We have investigated strategies for altering the meiotic and mitotic processes to solve these problems.

Genomic manipulation of amphibians during meiosis or mitosis has been accomplished using primarily physical agents such as cold, heat, and pressure. Chemical methods, such as those involving the use of cytochalasin B, are useful in organisms with small eggs that facilitate removal of the agents by washing but are difficult to control in larger eggs such as those of amphibia. Under the correct conditions the spindle apparatus is arrested or destroyed so that a restitution nucleus containing all sets of chromosomes destined for the meiotic or mitotic products forms in the cell whose division has been aborted. These experimental strategies have typically been limited to second meiotic and first mitotic divisions since their precise timing under controlled conditions fosters reliable manipulations. Other induced or naturally occurring subversions of the meiotic aand mitotic processes have been described by Asher (1970) in a variety of organisms, but these affect processes not readily accessible in amphibians.

Gynogenetic reproduction occurs naturally in several amphibian species, and it has been induced in many species by suppression of second meiotic division in eggs fertilized with irradiated sperm or with other sperm cells that do not contribute nuclear material to the zygote. These results suggest that suppression of first mitotic division in a gynogenetic haploid zygote might be an effective way to short-circuit the laborious production of inbred lines. Defined lines of *X. laevis* have proven useful in immunological research (Kaye et al., 1983), in construction of genomic libraries in which characteristics can be reproduced for molecular biology, and in various applications in toxicology. Genetic demonstration of their qualities should stimulate their fruitful application to new problems.

II. Methods

The pressure method of suppression of first mitotic division was chosen because it is more easily controlled that other physical and chemical methods. Genetically marked females were used to facilitate genetic analysis of the progeny. For the group most completely analyzed (Reinschmidt et al., 1991), namely, that comprising homozygous diploid (HD) animals HD-1 to HD-20 (the summary data from which analysis will be used throughout this chapter), the female and only parent was homozygous for the A allele of the major histocompatibility complex (MHC) (Tompkins et al., 1980) but heterozygous at many minor histocompatibility loci. In addition, the female was heterozygous for periodic albinism (a^p) and sex (ZW), two genes remote from their centromeres. The sperm donor was chosen so that it did not share the MHC allele of the female and was homozygous for the normal allele of a^p and for the Z sex allele. Tester stocks for test crosses to assess MHC, pigmentation, and sex locus genotype were routinely maintained.

The strategy for producing animals homozygous at all loci is to fertilize eggs with sperm cells previously irradiated with short-wave ultraviolet light, which stimulate development without participating genetically. The nonparticipation of the sperm nucleus must be demonstrated in the progeny of this process. Since fertilization occurs prior to metaphase of meiosis II, meiosis must be completed before mitotic divisions can begin. Since occasional females lay a fraction of their eggs that spontaneously fail to undergo meiosis II, the completion of normal meiosis must be demonstrated in the progeny. The haploid zygotes that begin to develop from this process are then subjected to pressure in order to block the first mitotic division. Successful pressure treatments lead to diploids homozygous at all loci. Both direct observations and indirect methods are used to assess the success of these processes.

The female chosen to produce HD animals was stimulated to ovulate by a single injection of 300 IU of chorionic gonadotropin (Sigma Chemical Co., St. Louis, MO) 6–8 hours prior to when eggs were desired. Sperm was obtained from the chosen male by removing the testis from the pithed male, mascerating the testis in 10 ml of 100% Steinberg's solution (Steinberg, 1957), and irradiating each milliliter of sperm suspension in an uncovered 60-mm petri dish with a short-wave ultraviolet light source (Mineralight). The petri dish was placed at the bottom of an ice bucket lined with aluminum foil and containing a 2-inch layer of ice. The dish was exposed to ultraviolet light for 1.5 minutes at a distance of 15 cm from the source, swirled, and finally exposed for another 1.5 minutes.

After removal from ice, eggs were squeezed manually from the ovulating female into the petri dish containing the irradiated sperm. After 1 minute, the dish was flooded gently with dechlorinated tap water to activate the sperm. Control aliquots of eggs thus fertilized were set aside to ensure that all developed into haploid embryos. Control eggs fertilized with unirradiated sperm were set aside to indicate the exact time to first cleavage. It is essential that the MHC genotype of the sperm donor is known and that samples of this genotype are maintained so that the nonparticipation of the male in the zygote genomes can be demonstrated by histoincompatibility of the HD animals with the sperm donor genotype tissue.

Three to four minutes after insemination, the eggs were washed with 10% Steinberg's solution and placed in a pressure bomb. Pressure was applied at 7500 psi for 6 minutes commencing 6 minutes before control cleavage in the previous group. The time of first cleavage depends on temperature and other environmental factors, but it usually occurs between 75 and 95 minutes after insemination at room temperature. After the pressure was realeased, the timing and form of first and second cleavage were recorded for comparison with normal controls. Those zygotes which cleaved first when controls were undergoing second cleavage, and which developed as normal diploids as shown by chromosomal squash analysis, were reserved. After determination of the normal diploid nature of the normally developing embryos, genetic analysis suggested that this method was successful. Complete genetic analysis was carried out at maturity, using the appropriate tester stocks for test crosses.

Mature animals were analyzed for histocompatibilities using the skin-grafting methods outlined in Tompkins et al. (1980). Grafts to AA animals confirmed the homozygosity of the HD animals at the MHC; grafts from the sperm donor stock confirmed the effectiveness of ultraviolet irradiation of the sperm in inactivating the sperm nuclei. Grafts between clones from different HD animals showed a high frequency of graft rejection (73/75), each rejection taking more than 30 and less than 120 days, diagnostic for minor histoincompatibilities. In contrast, all intraclonal grafts were accepted for as long as the animals lived; most animals are still alive and their grafts unrejected after 4 or more years.

The resulting progeny of the above process are referred to as the primary homozygous diploid animals. To be classified as primary HD animals, the following criteria must be met:

1. The first mitotic division of the egg following pressure treatment must be delayed about 30 minutes and occur at a time coincident with second cleavage in unpressed control animals. Animals that cleave immediately on pressure release either die soon thereafter or developed as inviable haploids. The delay of first cleavage is not definitive; our experience with producing tetraploid animals by suppression of first cleavage in diploids has shown that following delayed cleavage many abnormalities may ensue. Tripolar and tetrapolar mitoses are common and lead to abnormal development and death, as expected. More interestingly, normal diploid animals and animals chimeric for diploid and tetraploid tissues are often seen. The occurrence of these animals suggests some orderly ploidy reduction, perhaps based on dropping a round of DNA synthesis, other than those based on use of supernumerary centrioles. Thus, observed suppression of first cleavage is a necessary condition of successful HD production, but it does not guarantee it.

2. It is essential to examine the primary HD animals to ensure they have a normal diploid karyotype. Spontaneous failure of the second meiotic division

coupled with successful mitotic suppression will result in tetraploid animals. Abnormal centriolar divisions may have resulted in a variety of aneuploidy. These conditions usually preclude normal reproduction in any event, but primary HD groups with such abnormalities in viable animals should not be used further since it is obvious that unwanted processes are contaminating the resulting offspring.

3. The genetic structure of the population of primary HD animals must be probed to determine if the desired processes of sperm inactivation, normal meiosis, and first mitotic suppression have occurred and that unwanted processes have not. To complete these analyses it was necessary first to clone the primary HD animals. Methods for genetic analysis are considered below, after the method for cloning.

A. Propagation of Primary Homozygous Diploid Animals

Diploid *Xenopus* homozygous at all loci are necessarily either male (ZZ) or female (WW), and thus propagation of these genotypes requires parasexual or nonsexual methods that will not affect the genomic quality. The simplest method, which we have used extensively, is to induce diploid gynogenesis in WW females via terminal meiotic fusion (Tompkins, 1978). Suppression of second polar body formation by hydrostatic pressure is accomplished as follows. Irradiated sperm cells are prepared as previously described. Eggs are stripped from chorionic gonadotropin-stimulated females into small petri dishes each containing 1 ml of sperm suspension. After 1 minute, dishes are flooded with dechlorinated tap water, then the eggs are carefully transferred to a pressure bomb. Exactly 5 minutes after flooding, a pressure of 4800 psi is applied and held for 6 minutes. After pressure release the eggs are spread in small groups in 10% Steinberg's solution. This method has proved effective with *Xenopus* and internal controls using sperm with dominant color gene markers and/or codominant MHC markers. Since there is no heterozygosity in the primary HD animals, homologous recombination during meiosis I will have no effect on the chromosomal transmission to the progeny, which are thus clones of the primary HD animals. In this way the female primary HD animals have been cloned up to five generations to date.

Further reproduction of HD animals would be facilitated if some members of each clone could be sex reversed and cloning could be done by sexual reproduction. Treatment of ZW and WW animals with androgenic hormones merely masculinizes the female without inducing sperm production. Full sex reversal of these animals is possible only through the arduous and lengthy method of Humphrey (1945), which involves implantation of a small testis in the midlarval tadpole followed by removal of the implant 1 year after metamorphosis. In a small percentage of cases, the frog's own inhibited gonadal tissue will then develop into a sperm-producing testis, and the ZW or WW animal can participate in normal sexual reproduction as a male. On the other hand, full sex reversal from male to female is more easily accomplished hormonally in *Xenopus* (Witschi and Allison, 1950). This strategy is being used to produce primary HD animals, as previously described, from ZW females and to treat all tadpoles with estrogenic steroids. All treated animals develop as functional females and can be cloned by gynogenesis. Test matings are used to discover the sex-reversed ZZ animals. The gynogenetic offspring of the ZZ sex-reversed females can be separated into two groups: one untreated, which will develop as males; and one treated with estrogenic steroids, which will develop as females. Sexual reproduction of these clones can then be established, requiring only that some ZZ clines be sex reversed in each generation.

B. Defining Animals Homozygous at All Loci

Two issues must be addressed in defining animals homozygous at all loci. First, has the procedure for producing a group of animals homozygous at all loci been successful? Second, is there evidence that each individual is a homozygous diploid animal (or, perhaps, was one inadvertent product of some other process included by error in the putative homozygous diploid group)? The first problem is addressed by analysis of the formal genetics of the whole group of sibling animals; the latter is accomplished by comparisons of intraclonal histocompatibility of siblings.

Suppression of first cleavage in a haploid zygote will produce diploid animals homozygous at all loci, but other processes must be eliminated as possible contributors to the genomic structure of these animals, particularly failure of meiosis II division. Gynogenesis arising from suppression of the second meiotic division also may result in homozygosity at many loci and therefore may be confused with the suppression of first mitosis in haploid embryos. To distinguish these two processes genetically, we use markers remote from their centromeres such that the proportion of heterozygous offspring, following gynogenesis by suppression of meiosis II, is large (Nace *et al.*, 1970). The proportion of heterozygosity for a given locus is determined by the genetic distance of the locus from its centromere and the interference levels. Interference has no effect on the occurrence of the first crossover on a chromosome arm between a locus and its centromere. However, interference reduces the frequency of second and subsequent events.

Because each rank order of crossover event proceeds with the unitary increment of the power to which the frequency of the first crossover event is raised, the effect of interference is to increase the relative frequency of heterozygosity caused by odd-number events and to decrease the crossover class contributions to homozygosity due to the even-number class of crossover events (Barratt et al., 1954). Hence the frequency of heterozygosity in the presence of interference increases above the two-thirds expected at infinite genetic distance without interference. Thus, females heterozygous for several loci mapped remote from their centromeres are best chosen for production of homozygous animals. Fortunately, several nonlethal markers known to be remote from their centromeres are available and therefore, in heterozygotes, can generate large numbers of heterozygous offspring by suppression of second meiosis following fertilization with irradiated sperm. These markers include *unresponsive* (*ur*), *periodic albinism* (a^p), and the sex locus (Krotoski et al., 1985; Reinschmidt et al., 1985). Chi-square analysis of the data derived from the aforementioned group of putative HD animals (Tables I and II) demonstrates that the data do not fit the meiosis II suppression gynogenetic hypothesis but do fit the suppression of first cleavage hypothesis.

For individual homozygous diploid animals, as well as their clones, chi-square analysis in the same vein does not supply unequivocal results. For the

TABLE I

CHI-SQUARE ANALYSIS
OF GENETIC MARKER DATA FROM ONE HD GROUP OF ANIMALS
COMPRISING HD-1 TO HD-20, INCLUSIVELY[a]
(Hypothesis: That animals in the HD-20 group
were the gynogenetic progeny of female 140-5
produced by meiosis II suppression)

Genotype	Observed (O)	Expected (E)	$(O - E)^2/E$
a^p/a^p	10	2.41	23.9
+/+	10	2.41	23.9
$a^p/+$	0	15.18	15.18
Total[b]	—	—	62.98
ZZ	11	2.81	23.87
WW	9	2.81	13.63
ZW	0	14.38	14.38
Total[b]	—	—	51.88

[a] The markers are analyzed individually since the values obtained using combined genotypes are larger than those listed in the chi-square tables and would lead to an overestimation of the probability of gynogenesis by meiosis II suppression.

[b] $p < .001$ for the a^p locus and for the sex locus. The combined probabilities of both observed results, given the meiosis II suppression hypothesis, is less than 10^{-6}, and the hypothesis is rejected at the .05 level of acceptance on the basis of either locus or both.

TABLE II

CHI-SQUARE ANALYSIS
OF GENETIC MARKER DATA FROM ONE HD GROUP OF ANIMALS
COMPRISING HD-1 TO HD-20, INCLUSIVELY
(Hypothesis: That this group of animals was produced by
suppression of first mitotic division of gynogenetic haploid zygotes)

Genotype	Observed (O)	Expected (E)	$(O-E)^2/E$
a^p/a^p ZZ	5	5	0
a^p/a^p WW	4	5	1
+/+ ZZ	6	5	1
+/+ WW	5	5	0
Total	—	—	2[a]

[a] The chi-square value 2, with 3 degrees of freedom, corresponds to a value of p between .3 and .5; hence, the hypothesis is accepted given an arbitrary cutoff value of .05.

gynogenetic progeny of meiosis II suppression in the eggs of heterozygous females, the frequency of homozygosity at the periodic albino locus (0.24) multiplied by the frequency of homozygosity at the sex locus (0.30) gives the probability (0.072) of each of the observed animal's genotype at these loci, if one putative primary HD animal were produced in this way. Clearly this is not adequate to exclude this possibility at the .05 level of probability. Hence, skin grafting is used to ensure their individual identities as diploids homozygous at all loci. This method requires estimating the number of minor histocompatibility loci for which the mother of the primary HD animals was heterozygous, and from that value calculating the probability of the observed histocompatibility in the offspring of the primary HD animals. Since the previous analysis showed that the overall results are compatible with HD production, the frequency of skin-graft acceptances (S) (in this case 2/75) between the different clones was used to estimate the number of minor histocompatibility loci (L) for which the original female was heterozygous. For each heterozygous locus of the mother of the HD group, two classes of offspring, each homozygous for one allele, were produced. Therefore, 50% $(1/2)^1$ skin-graft acceptances should occur in random grafts between the different cloned HD animals for one locus, $(1/2)^2$ for two loci, $(1/2)^3$ for three loci, etc. Since for L heterozygous loci $(1/2)^L$ acceptances should be observed, the number of loci can be found by

$$S = (\tfrac{1}{2})^L$$

or

$$L = \frac{\ln S}{\ln(\tfrac{1}{2})} = \frac{\ln(\tfrac{2}{75})}{\ln(\tfrac{1}{2})} = 5.23$$

Hence, the mother of the group of HD animals under discussion was heterozygous for at least five minor histocompatibility loci.

If the average gene–centromere distance for these five minor histocompatibility loci were equal to the average for mapped genes in *X. laevis*, the corresponding fraction of heterozygous offspring from terminal meiotic fusion would be 0.452 (Reinschmidt, *et al.*, 1985). It is necessary to make the conservative assumption, namely, that minor histocompatibility loci are randomly distributed, to complete this simple analysis. More complex nonrandom distribution assumptions entail more complex analyses that do not affect the conclusion. The chance of homozygosity at any one locus is 0.548, and for five loci [$(0.548)^5$] it is 0.049. If any primary HD animal were heterozygous for a minor histocompatibility locus, the frequencies of genotypes in the gynogenetic offspring, determined by the average frequency of heterozygosity, would cause an above average gene–centromere distance. If 10 random grafts are done among the gynogenetic offspring of each primary HD animal, the chances of 10 acceptances [$(0.602)^{10}$] is 0.00039; for larger numbers of graft acceptances and/or heterozygosity at more than one locus, the possibility of heterozygous primary HD animals becomes vanishingly small.

The total probability of error with respect to individual HD identification (identification of an animal as a primary HD when in fact it resulted from meiosis II suppression), then, is the chance (A) of gynogenetically producing homozygotes at (in this case) five minor histocompatibility loci, plus the probability (B) of the alternate pathway to histocompatibility within the progeny of each HD (chance association of histocompatible animals within heterogenous population), all multiplied by the probability (C) of homozygosity at the two marker loci used for each HD animal:

$$(A + B)C = (0.049 + 0.0039)\,0.072 = 1.37 \times 10^{-5}$$

This analysis, along with the overwhelming assurance that the overall results are compatible with HD production and incompatible with meiosis II suppression, ensures that each individual primary HD animal and its gynogenetically cloned offspring are in fact homozygotes at all loci produced by suppression of first cleavage in haploid zygotes.

III. Conclusions

The genetic methods developed give confidence that the procedures we have used for constructing and cloning diploid animals homozygous at all loci are effective and that the quality of individual HD animals can be assured. Unlike inbred strains of animals in which levels of homozygosity rise asymptotically, approaching levels of homozygosity determined by the inbreeding coefficients and the mutation rates, the primary HD animals have no heterozygosity at all. This level of homozygosity is not maintained because of mutations; it will decline asymptotically to a level determined by the inbreeding coefficient and the mutation rate. For applications of HD stocks that utilize them as references, such as genomic libraries, continuing old HD lines will have value. Other applications that exploit their homozygosity per se might well be better served with newly made and genetically tested HD lines.

Acknowledgments

This work was supported in part by Department of Defense Grant No. 2 (Order No. 89-116-88-150) from the Defense Nuclear Agency to Tulane University and administered by the Center for Bioenvironmental Research. We also thank the Coordinated Instrumentation Facility of Tulane University for equipment use and maintenance.

References

Asher, J. H., Jr. (1970). *Genetics* **66**, 369–391.
Barratt, R. W., Newmeyer, D., Perkins, D. D., and Garnjobst, L. (1954), *Genetics* **6**, 1–93.
Fankhauser, G. (1955). *In* "Analysis of Development" (B. H. Willier, P. A. Weiss, and V. Hamburger, eds.), pp. 126–150. Saunders, Philadelphia, Pennsylvania.
Festing, M. F. W. (1979). "Inbred Strains in Biomedical Research." Oxford Univ. Press, New York.
Humphrey, R. R. (1945). *Am. J. Anat.* **76**, 33–66.
Kaye, C., Schermer, J. A., and Tompkins, R. (1983). *Dev. Comp. Immunol.* **7**, 497–506.
Klein, J. (1975). "Biology of the Mouse Histocompatibility-2 Complex." Springer-Verlag, Berlin.
Krotoski, D. M., Reinschmidt, D. C., and Tompkins, R. (1985). *J. Exp. Zool.* **233**, 443–449.
Nace, G. W., Richards, C. M., and Asher, J. H., Jr. (1970). *Genetics* **66**, 349–368.
Reinschmidt, D. C., Simon, S. J., Volpe, E. P., and Tompkins, R. (1979). *J. Exp Zool.* **210**, 137–143.
Reinschmidt, D. C., Friedman, J., Hauth, J., Ratner, E., Cohen, M., Miller, M., Krotoski, D., and Tompkins, R. (1985). *J. Hered.* **76**, 345–347.
Reinschmidt, D. C., *et al.* (1991). In preparation.
Steinberg, M. (1957). (*in* Report by J. D. Ebert) *Year Book—Carnegie Inst. Washington* **56**, 347.
Szaro, B. G., and Tompkins, R. (1987). *J. Comp. Neurol.* **358**, 304–316.
Szaro, B., Ide, C., Kaye, C., and Tompkins, R. (1985). *J. Exp. Zool.* **234**, 117–129.
Tompkins, R. (1978). *J. Exp. Zool.* **203**, 251–256.
Tompkins, R., Reinschmidt, D. C., Wilson, J. M., and Volpe, E. P. (1980). *In* "Phylogeny of Immunological Memory" (M. J. Manning, ed.), pp. 187–195. North Holland Biomedical Press, Amsterdam.
Witschi, E., and Allison, J. (1950). *Anat. Rec.* **108**, 101.

Chapter 4

Oogenesis and Oocyte Isolation

L. DENNIS SMITH, WEILONG XU, AND ROBERT L. VARNOLD

*Department of Developmental and Cell Biology
and Developmental Biology Center
University of California at Irvine
Irvine, California 92717*

I. Introduction
II. Oogenesis: An Overview
III. Isolation of Individual Oocytes from the Ovary
IV. Protein Synthesis during Oogenesis: Practical Hints
V. Labeling of Oocytes
VI. Oocyte Maturation
VII. Concluding Comments
   References

I. Introduction

Use of the amphibian egg as a system to evaluate the function of components injected into it traces back to the pioneering experiments of Briggs and King (1952) on nuclear transplantation. These early studies using eggs from the leopard frog *Rana pipiens* were soon followed by work on eggs of *Xenopus laevis* (Fischberg *et al.*, 1958; Gurdon, 1960). The initial use of oocytes as recipients for microinjected material also involved *R. pipiens* and included studies concerned with protein synthesis prior to fertilization (Ecker and Smith, 1968). Related to this work, the *R. pipiens* oocyte was established *in vitro* as a model to investigate the completion of meiosis, that is, oocyte maturation (Schuetz, 1967; Masui and Markert, 1971; Smith *et al.*, 1968). However, *R. pipiens* oocytes proved to be very difficult material for biochemical experiments, especially those dealing with extraction of nucleic acids. In contrast, a series of experiments during the 1960s by Brown and colleagues and Davidson and colleagues, among others, had established *X. laevis* oocytes as the favorite for biochemical analyses. Such work established the conclusion that full grown oocytes contain large reserves of the components required for protein synthesis during early embryogenesis, including maternal mRNAs, ribosomes, and tRNAs (Davidson, 1986).

Considering the aforementioned background, Gurdon *et al.* (1971) initially reported that rabbit globin mRNA could be translated when microinjected into *Xenopus* oocytes. This observation has led to a literature that now encompasses literally thousands of studies in which *Xenopus* oocytes have been used to study the transcription and translation of a wide variety of messages as well as a test system to identify mRNAs from diverse cells and tissues. Numerous reviews have appeared over the past few years describing the use of *Xenopus* oocytes for the translation of heterologous mRNAs (Lane, 1983; Soreq, 1985; Colman, 1984b; Kawata *et al.*, 1988). In addition, the oocyte has been used to study phosphorylation, glycosylation, acetylation, and cleavage of proteins (Kawata *et al.*, 1988; Smith, 1989, 1991), as well as the processes of protein secretion and protein compartmentalization (Ceriotti and Colman, 1989, 1990). Furthermore, studies on the synthesis and assembly of membrane receptors can be carried out conveniently in the oocyte (Yu *et al.*, 1989; Froehner *et al.*, 1990). However, even considering all of this work, there have been few attempts to use oocytes under conditions that have been optimized for the purpose at hand.

This Chapter describes methodology used routinely in our laboratory for biochemical studies on oocytes, not only with respect to endogenous events that occur during oogenesis, but especially with respect to the use of oocytes as an assay system for a variety of injected molecules. In so doing, we present some practical suggestions for obtaining material, culturing oocytes, and microinjecting them. These topics are also discussed in other chapters in this volume. In addition, we present information obtained over the years on the synthetic capabilities of oocytes, which, it is hoped, will allow design of experiments maximized for synthetic activity after microinjection.

II. Oogenesis: An Overview

Oogenesis in adult *Xenopus laevis*, maintained in the laboratory, is asynchronous, meaning that all stages of growth are usually found in the ovary at any given time. Dumont (1972) has classified the various stages largely according to size, the smallest designated as stage I and the largest as stage VI (Fig. 1). Overall, the process of oogenesis, namely, growth from stage I

FIG. 1. *Xenopus* oocytes of different stages defolliculated manually in OR2 medium. The smallest transparent oocyte is at stage I (Dumont, 1972), while oocytes of increasing size represent stages II–VI, respectively. Bar represents 500 μm.

FIG. 3. Defolliculation of *Xenopus* oocytes. (A) One pair of forceps grabs the ovary near the stalk where the follicle is attached (arrow); (B) the second pair of forceps gently grabs the stalk. (C) By gently pulling the second pair of forceps away from the first, (D) the follicle layers begin to peel off as evidenced by the appearance of a constriction band on the oocyte (arrow). (E) Continual pulling away of the second pair of forceps results in the constriction band moving down the oocyte surface. (F) The oocyte is almost completely removed from the follicle. If the oocyte remains attached to the follicle, it can be gently teased out with the forceps. Note the sheet of follicle tissue (arrow) which remains after successful defolliculation.

to stage VI, requires at least 8 months, but progression through the various stages does not occur at a constant rate. Once oocytes have reached stage VI, they do not continue to increase in size but can remain at this point for some time before undergoing atresia. Animals with ovaries containing significant numbers of atretic oocytes usually have not ovulated for several months. Such ovaries also usually contain "white-banded" stage VI oocytes.

The rate of progression through oogenesis is influenced by a variety of environmental factors, such as food and crowding, as well as by hormonal stimulation of animals. Keem et al. (1979) have estimated that oocyte growth from late stage III to stage VI in unstimulated females (not previously induced to ovulate) is 16–24 weeks but only 9–12 weeks in animals recently (1–2 weeks) induced to ovulate with human chorionic gonadotropin (HCG). The practical implication of these observations is that in animals with few full grown (stage VI) oocytes, oocyte growth can be stimulated by the hormonal induction of ovulation or by treatment of animals with peptide hormones, such as follicle-stimulating hormone (FSH), which stimulate follicle growth but not ovulation (Wallace, 1985). Under ideal conditions, animals with healthy ovaries containing numerous stage VI oocytes can be obtained after only a few months of hormone injection. However, hormonal stimulation also alters endogenous metabolic activity (see below), a fact that should be considered when using animals treated in this manner.

It should be emphasized that the size difference between stage VI and stage V oocytes is not easy to discern by simple observation through a stereomicroscope; stage V oocytes have a diameter of 1000–1200 μm, whereas stage VI oocytes have diameters larger than 1200 μm. However, there are significant metabolic differences between late stage IV, stage V, and stage VI oocytes (Cicirelli and Smith, 1987; Taylor and Smith, 1987). These differences can have a major effect on the capacity of oocytes to translate injected messages and process proteins. Thus, depending on the experiment to be conducted, it may become important to size oocytes using an ocular micrometer. At the minimum, this approach helps reduce some of the normal variability one encounters in evaluating data.

III. Isolation of Individual Oocytes from the Ovary

A sexually mature animal contains tens of thousands of oocytes at all stages of oocyte growth, but most experiments involve the need for only a few hundred. Hence, we usually plan to obtain oocytes from the same frog on several occasions. To accomplish this, it is necessary to anesthetize animals and surgically remove one or more pieces of ovary. A convenient and effective method for anesthesia is to immerse the animal for an appropriate time (at least 30 minutes) in a solution of 0.1% tricane–methane sulfonate (MS222) in water. It should be emphasized that although MS222 is an effective anesthetic for cold-blooded vertebrates, the anesthetic itself can affect oocytes. Specifically, several tricaine anesthetics have been reported to induce full grown oocytes to undergo oocyte maturation (Smith, 1989). To avoid this potential complication, we also use hypothermia as an anesthetic. In practice, an animal is packed in an amphibian bowl in ice water for about 40–45 minutes and then placed on a bed of ice for subsequent surgery.

To remove ovarian tissue, a small incision (less than 1 cm) is made in the skin and body wall in the posterior ventral side of the animal (Colman, 1984b). Care should be exercised to make the incision away from the midline, which contains a major blood vessel. If the ovary is not readily visible after making the incision, gentle probing with relatively blunt forceps will expose it. The ovary consists of several lobes that are usually teased out and tied with sutures at the base. The lobes are then snipped off using scissors and placed in culture medium. The incision in the body wall is then sutured separately from the skin, which subsequently is sutured, and the animal is left to recover in shallow water. The recovery process is quite rapid (several minutes) at room temperature.

Isolated ovarian tissue should be washed well in culture medium to remove blood and then prepared to remove individual oocytes. We have routinely used two procedures for this purpose. For experiments that require several thousand oocytes, the ovarian tissue is digested with collagenase. The ovarian tissue is minced into small pieces with watchmaker's forceps and then incubated in 0.2% collagenase (type I, Sigma Chemical Co., St. Louis, MO) in 0.1 M sodium phosphate, pH 7.4. The exact time of treatment can be quite variable with different batches of collagenase and ovarian tissue from different females. Generally, we continually observe the material and, as individual oocytes (or groups of oocytes) are freed from their follicular tissue, they are removed from the enzyme and washed thoroughly in culture medium. Prolonged treatment with enzyme removes the vitelline envelope from oocytes, thus making the oocytes very fragile.

Treatment of ovarian tissue with collagenase releases oocytes at all stages of oogenesis. Small numbers of individual oocytes at different stages of oogenesis can be removed from one another with a pipette. If large numbers of oocytes at each of the different stages of oogenesis are desired, they can be separated in bulk by filtering oocytes through Nitex screens with different pore sizes. Alternatively, a population of oocytes at different stages can be swirled in a beaker or cylinder and allowed to settle. The largest oocytes settle first and can be separated from the smaller oocytes by several means (Dolecki and Smith, 1979).

We do not recommend the use of collagenase if only small numbers of oocytes (less than a few hundred) are needed for an experiment. Collagenase is detrimental to oocytes and, in our experience, can inhibit oocyte maturation induced by progesterone, an indicator of physiological health. In addition, collagenase can result in a marked depression in endogenous protein synthesis, as demonstrated in an experiment in which oocytes were injected with radioactive leucine at various times after collagenase treatment followed by incubation in OR2 medium (Wallace et al., 1973) (Fig. 2). The data show clearly that at least 8 hours is required for recovery when oocytes, treated for 1.5 hours at an enzyme concentration of 0.2%, are compared with oocytes manually removed from their follicles; enzyme treatment for longer periods or at a higher concentration should be followed by longer recovery times. The

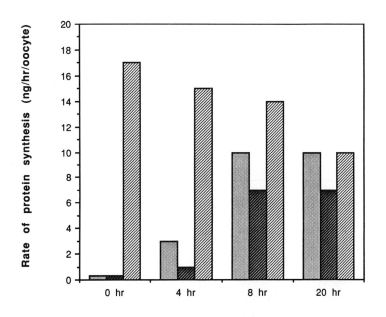

FIG. 2. Rate of protein synthesis after collagenase treatment. Stage VI oocytes were defolliculated in 0.2% type I collagenase (Sigma) in 0.1 M phosphate/buffer, pH 7.4. All oocytes were labeled by microinjection of 50 pmol of [^{3}H]leucine, with a final specific activity of 0.5 μCi/pmole, followed by incubation for 75 minutes. The incorporation of radiolabeled amino acid was linear over the time period examined. The actual amount of protein synthesized was calculated based on the leucine pool size of stage VI oocytes (Taylor and Smith, 1985) and the specific activity of radioactive label. Bars indicate the rate of protein synthesis from oocytes defolliculated by treatment with collagenase for 1.5 hours (▦) and 2.5 hours (▧) and for oocytes defolliculated manually (▨).

practical outcome of these observations is that use of collagenase-treated oocytes for mRNA injection experiments (or other experiments) too soon after enzyme treatment may account for many claims that animals frequently have "bad oocytes."

Although we use collagenase to obtain large numbers of oocytes for bulk extraction of nucleic acids and/or proteins, we generally remove individual oocytes from their ovarian follicles by manual dissection for experiments in which individual oocytes are injected with mRNAs (or other materials). Manual defolliculation is best accomplished with a piece of ovary containing several hundred or more follicles; it is very difficult to remove oocytes from single, isolated follicles. The major prerequisite for successful defolliculation is a sharp pair of watchmaker's forceps (Dumont No. 5). We do not prefer forceps that are stainless steel, since the tips are soft and easily bent. To defolliculate manually, one should observe first that individual follicles are attached to the ovary by a membranous "stalk." One pair of forceps is used to grab the stalk near the ovary (away from the follicle), and the second pair of forceps is then used to grasp ovarian tissue near the first one (Fig. 3). This second pair then is pulled upward toward the operator, much like ripping off the skin of a grape. Defolliculation is successful if no blood vessels are seen associated with the isolated oocyte. Defolliculated oocytes generally are less turgid than oocytes within their follicles. With practice, it is readily possible to remove 100–200 oocytes per hour using this approach.

Oocytes removed from their ovarian follicles are relatively simple to microinject and, with micropipettes of 15 μm diameter or less at the tip, rarely leak after microinjection. There are two potential disadvantages to this procedure. First, oocytes removed from their follicles and cultured *in vitro* appear to decrease in synthetic capacity over time (Fig. 2). This phenomenon has been observed previously (Dolecki and Smith, 1979). Second, manually defolliculated oocytes still contain follicle (somatic) cells. If removal of these cells is required, several procedures involving very brief enzymatic treatment are available (Horrell et al., 1987). Depending on the proposed experiment, we have found two approaches to be useful. First, immersion of individual oocytes in α-chymotrypsin (30 μg/ml in OR2) will remove follicle cells in 1–3 minutes. Alternatively, treatment of oocytes with β-glucosidase (200 units/ml) for about 30 minutes effectively removes follicle cells (R. L. Varnold and L. D. Smith, unpublished data).

Considering this discussion, many investigators routinely microinject oocytes still contained within their ovarian follicles. Injection in this case requires insertion of a micropipette through several somatic cell layers, including an ovarian epithelium and a thecal layer. Such injected follicles may be maintained for several days in culture with no obvious detrimental effects. However, it should be emphasized that, to our knowledge, no studies have been carried out to compare synthetic activities of such follicles as a function of time in culture. In addition, since somatic cells also are present when injected oocytes (follicles) are subsequently processed, we find it more convenient to work with defolliculated oocytes.

Procedures for the culture of *Xenopus* oocytes are described in detail in Chapter 6 of this volume. However, for purposes of the proceeding discussion, it seems desirable to outline some of the commonly used methods for defolliculated oocytes. Culture media for amphibian oocytes fall into four main groups, two of which were developed originally for the culture of *R. pipiens* eggs and embryos and/or *R. pipiens* oocytes. The first group includes a series of formulations originally reported by Barth and modified for use with *Xenopus* oocytes by Gurdon (1968). The second is based on the formula for amphibian Ringer's solution (Kimelman and Kirschner, 1987; Gelerstein *et al.*, 1988). Both types of media appear to be perfectly adequate for the culture of *Xenopus* oocytes. Based on attempts to maximize vitellogenin uptake by *Xenopus* oocytes *in vitro*, Wallace and colleagues (1973) tested a series of media and developed a formula (OR2), which we routinely use in experiments on defolliculated oocytes. This simple medium maintains oocytes in healthy condition for about 2–3 days as determined by capacity to undergo maturation in response to progesterone or by measurements of protein synthetic rate. If longer culture periods are required, we incubate oocytes in 50% Liebovitz L-15 medium and 15 mM HEPES (pH 7.8), which is the nutrient medium utilized by Wallace and Misulovin (1978) in development of *in vitro* conditions for the growth of *Xenopus* oocytes. We have used this medium to culture defolliculated oocytes for as long as a week. Since it is a nutrient medium, the solution and all glassware should be sterilized to prevent bacterial growth.

IV. Protein Synthesis during Oogenesis: Practical Hints

Full grown (stage VI) *Xenopus* oocytes contain a large store of maternal mRNA that equals a complex population of transcripts equivalent to many thousands of diverse messages (Davidson, 1986). This mass of putative mRNA is accumulated already by the end of oogenesis stage II. Nevertheless, during the entire period of oogenesis, oocytes contain active lampbrush chromosomes that exhibit transcription rates higher than those of typical somatic cells by several orders of magnitude (Anderson and Smith, 1977, 1978; Dolecki and Smith, 1979). As in somatic cells, most of the newly synthesized nuclear RNA does not enter the cytoplasm. However, because of the enormous rate of nuclear RNA synthesis, the actual amount of putative message that enters the cytoplasm in oocytes still is quite high. For example, Dolecki and Smith (1979) have estimated that the rate of entry of poly(A) RNA into the cytoplasm in stage III and stage VI oocytes is about 1.4 ng per day. However, this still is a small percentage of the stored maternal mRNA, and less than 10% of the newly synthesized cytoplasmic transcripts are found on polysomes (Dolecki and Smith, 1979). Perhaps this observation accounts in part for the variability often observed when injection of DNA into the oocyte nucleus is followed by translational assay for gene product (Colman, 1984a). More generally, such data show that protein synthesis during much of oogenesis is encoded for almost entirely by maternal transcripts.

The rate of protein synthesis in *Xenopus* oocytes increases over 100-fold between stage I and stage VI of oogenesis (Fig. 4). The content of ribosomal RNA increases similarly. Thus, the fraction of ribosomes engaged in protein

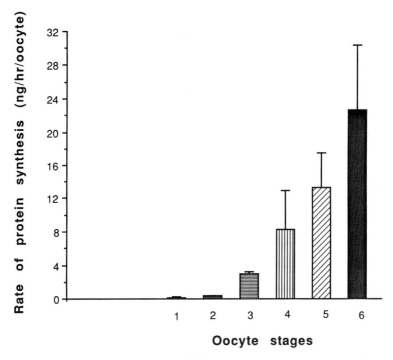

FIG. 4. Protein synthetic rates in stage I–VI *Xenopus* oocytes. Data are based on Taylor and Smith (1985). Oocytes were either incubated (stage I and II) or injected (stage III–VI) with [^{3}H]leucine. Bars represent standard deviation. For details of the method, see Taylor and Smith (1985).

synthesis remains approximately constant at about 2% throughout oogenesis (Taylor and Smith, 1985). The rate of protein synthesis increases an additional 2-fold in oocytes from unstimulated females when oocyte maturation is induced. However, in stage VI oocytes from stimulated females [injected either with HCG or pregnant mare serum gonadotropin (PMSG)], protein synthesis already is almost as high as that seen in maturing oocytes from unstimulated females, and progesterone injection results in much less of an increase (Wasserman et al., 1982). The main conclusion of these observations for those interested in using *Xenopus* oocytes for the translation of injected messages is that one can obtain higher rates of translation in recipient oocytes by using material from stimulated females. On the other hand, this approach involves certain trade-offs (see Section VI).

The original experiments of Gurdon et al. (1971) were based on the assumption that stage VI oocytes contain spare translational capacity, that is, injected mRNAs would be translated in addition to endogenous messages. Subsequent experiments showed clearly that mRNA injected into stage VI oocytes is translated only when it successfully competes with endogenous mRNAs for the translational machinery of the oocytes (Laskey et al., 1977; Asselbergs et al., 1979; Richter and Smith, 1981). Thus, when increasing amounts of a heterologous message such as globin mRNA are injected into stage VI oocytes, total protein synthesis remains relatively constant, globin synthesis increases, and synthesis of endogenous proteins decreases. However, Richter and Smith (1981) demonstrated that the translational capacity of the oocyte for different classes of messages is not uniform. Thus, for mRNAs translated on free cytosolic polysomes, increased amounts of the foreign protein were made as more message was injected. On the other hand, for messages translated on endoplasmic reticulum (ER), translational capacity was saturated at relatively low amounts of injected mRNA, and nontranslating message accumulated as more was injected. This difference appears to be due to a low content of ER in stage VI oocytes, since injection of heterologous ER into oocytes previously injected with excess message led to increased translation of the mRNA (Richter et al., 1984). The practical significance of these observations is that in studies designed to use the oocyte to identify mRNAs for proteins that are translated on ER, or to produce membrane or secreted proteins from such messages, the translational capacity of the oocyte is very limited; injecting increasing amounts of message in an effort to obtain more product is not likely to be fruitful.

The rate of protein synthesis in stage IV oocytes is about one-third that in stage VI oocytes (Fig. 4). Nevertheless, Taylor et al. (1985) demonstrated that protein synthesis can be increased in stage IV oocytes to the level seen in stage VI by injection of increasing amounts of globin mRNA. The simplest conclusion from this observation is that stage IV oocytes do have spare translational capacity. This conclusion would indicate that stage IV oocytes should be a better system than stage VI oocytes to overproduce the translational product of injected messages. In actuality, this assumption was shown to be correct in recent experiments reported by Johnson et al. (1990). However, there is a caveat to this approach. In the experiments reported by Taylor et al. (1985), stage IV oocytes were shown to be even less efficient than stage VI oocytes in translation of membrane-bound messages. Presumably, if the content of ER limits translation of injected mRNAs in stage VI oocytes, it is even more limiting in smaller oocytes.

It should be emphasized that all of the aforementioned studies involved the injection of heterologous mRNAs containing a 3' poly(A) tail. The potential importance of a poly(A) tail for translation has been the subject of study for some time. A variety of studies, usually involving injection of mRNA into *Xenopus* oocytes, has suggested that a 3' poly(A) tail serves two functions: it increases message stability after injection and it enhances translational efficiency (Soreq, 1985; Drummond et al., Galili et al., 1988; Jackson and Standart, 1990). Drummond et al. (1985) initially suggested that the poly(A) tail functions primarily to enhance translational efficiency. These studies were extended by Galili et al. (1988) in showing that polyadenylated messages were translated 15- to 30-fold more efficiently after injection into stage VI oocytes compared with deadenylated messages, a conclusion that applied both to mRNAs translated on free polysomes and those translated on ER. Apparently, reinitiation of terminating ribosomes is far more efficient on messages with a 3' poly(A) tail (Galili et al., 1988). Although this conclusion may not apply to all mRNAs (McGrew et al., 1989), the use of polyadenylated mRNAs is routine in our laboratory for all experiments involving injection into *Xenopus* oocytes. This is accomplished with native messages by isolating poly(A) RNAs during extraction. In the case of messages transcribed *in vitro*, we routinely use vectors designed to add a poly(A) tail (Galili et al., 1988; Johnson et al., 1990).

V. Labeling of Oocytes

Procedures for labeling *Xenopus* oocytes involve both incubation with radioactive precursors and microinjection of the precursor directly into oocytes. For nucleic acids, we prefer the use of guanosine (incubation) or GTP (injection) as the precursor. The oocyte pool of GTP is not as large as that of other triphosphates, and the kinetics of both uptake into the endogenous pool and incorporation into RNA have been studied thoroughly (LaMarca et al., 1975; Anderson and Smith, 1977, 1978; Dolecki and Smith, 1979). The choice of an amino acid for labeling proteins is dependent in part on the size of endogenous amino acid pools (Taylor and Smith, 1987; Shiokawa *et al.,* 1986) and the relative abundance of that amino acid in the protein(s) under investigation; most studies have been performed with radioactive leucine or methionine.

Stage VI (and smaller) oocytes are permeable to exogenous precursors. Thus, for convenience, oocytes can be continuously incubated with the appropriate precursor, providing long-term incorporation into proteins (or nucleic acids). In practice, uptake of radioactive precursors from the medium is extremely variable among oocytes from different females. Thus, low levels of incorporation into macromolecules may reflect limited uptake into the precursor pool(s) rather than limited translation from injected mRNA. However, we usually can obtain highly radioactive proteins by continuous incubation in [^{35}S]methionine (Trans-label, ICN, Irvine, CA, 10 μCi/ml) for only 10–12 hours. It should be emphasized that although these approaches are useful for obtaining qualitative data, it is readily feasible to quantitate incorporation in oocytes incubated with precursors by monitoring uptake as well as incorporation into macromolecules. In practice this approach requires extracting small numbers of oocytes with acid [10% trichloroacetic acid (TCA) or 0.5 N perchloric acid (PCA)] at various times after addition of radioactive precursor to the medium. Acid-soluble radioactivity represents uptake into the precursor pool, whereas acid- precipitable radioactivity equates to that incorporated into proteins (or nucleic acids). When uptake has reached steady state (plateaued), the precursor pool is at steady state and incorporation from that time is linear. Protein synthetic rate is calculated based on the slope of incorporation and the specific activity of the pool.

Most experiments in our laboratory involve microinjection of radioactive precursors directly into oocytes. This procedure has the advantage that incorporation begins almost immediately, constant amounts of precursor can be administered, and absolute rate of synthesis can be determined without independent measurements of acid-soluble radioactivity (Shih *et al.,* 1978). Provided the precursor is concentrated prior to injection, extremely high levels of incorporation also can be obtained. The one potential disadvantage of this approach (other than mastering the technique of microinjection) is that incorporation is complete within relatively short time spans, at least when amino acids are injected. Depending on the degree of endogenous pool expansion, radioactive amino acids are depleted from the endogenous pool within a few hours at most (Shih *et al.,* 1978). Thus, if long-term incorporation over several days is required, multiple injections would become necessary. This requirement is offset in large part by the fact that incorporation of injected radioactive amino acids results in proteins labeled to a much higher activity than when oocytes are incubated with the precursor. Thus, the product of less than 1 ng of injected mRNA has readily been detected in small numbers of stage VI, as well as stage IV, oocytes injected later with labeled precursor and incubated an additional 1–3 hours (Richter and Smith, 1981; Taylor et al., 1985; Johnson et al., 1990).

VI. Oocyte Maturation

Most studies that involve injection of materials use stage VI oocytes as recipients. Stage VI *Xenopus* oocytes are arrested in the first meiotic cell cycle at the G_2/M transition and must mature to the second meiotic metaphase before fertilization can occur. This process of oocyte maturation can readily be induced *in vitro* by exposing oocytes dissected from their ovarian follicles to the steroid hormone progesterone (Smith, 1989). Thus, it is possible to inject stage VI oocytes with a variety of substances, induce them to mature *in vitro*, and subsequently fertilize them (see Chapter II, this volume). This procedure allows a variety of experiments to be carried out that are concerned with the fate and disposition of materials, namely, normal and mutated mRNAs and/or proteins, injected into oocytes and assayed in developing embryos.

The induction of oocyte maturation involves a number of morphological and biochemical changes. These changes include, among many alterations, breakdown of the oocyte nucleus or germinal vesicle (GVBD); an approximate doubling of the rate of protein synthesis; activation of an intracellular protein kinase, which regulates the G_2/M transition; changes in the adenylation of mRNAs; and a major reduction in permeability (Smith, 1989). It is not our intention to discuss these events in detail at this time. However, some of the changes have obvious impact on the design of the previously discussed experiments, and the nature of that impact needs to be pointed out.

Breakdown of the oocyte nucleus (GVBD) is the most obvious external indication that oocytes are maturing, and GVBD represents the approximate time at which many other changes occur. For example, maturing oocytes have become essentially impermeable at GVBD, the rate increase in protein synthesis has occurred by this time, and changes in the adenylation of maternal mRNAs begins; up to 50% of existing sequences become deadenylated (Smith, 1989). From a practical viewpoint, the time at which GVBD occurs after progesterone exposure can vary widely, depending on a number of environmental parameters. For example, the injection of female frogs with gonadotropins, either to induce ovulation (HCG) or to stimulate follicle (oocyte) growth (PMSG), can dramatically alter the time at which GVBD occurs in stage VI oocytes exposed to progesterone. In oocytes from such "stimulated" females, GVBD is apparent within 2 hours after steroid exposure, whereas GVBD in similar stage VI oocytes from "unstimulated" females may occur at greater than 6–8 hours after exposure to progesterone (Smith, 1989). The timing of GVBD also varies as much as 20% among oocytes maintained in different culture media (R. L. Varnold and L. D. Smith, unpublished data) and tends to become longer as a function of time in culture media.

The fertilization of oocytes matured *in vitro* involves the transfer of maturing oocytes into the body cavity of a female previously induced to ovulate. The maturing oocytes enter the host female oviducts, complete maturation as they acquire jelly layers, and then become fertilizable by conventional techniques (see Chapter 11). However, oocytes that already have matured prior to transfer into the "foster" female, or which would complete maturation shortly after transfer, frequently activate during passage down the host female oviducts. Thus, the practical outcome of the previous discussion on maturation is that conditions which maximize the synthetic efficiency of injected oocytes (i.e., oocytes from stimulated females) are not the most efficient for subsequent fertilization procedures. We suggest that for most manipulations, it is best to design conditions such that GVBD occurs at about 6–8 hours after progesterone exposure. This suggests the use of "unstimulated" females and incubation of oocytes in a culture medium like OR2.

VII. Concluding Comments

The *Xenopus* oocyte has become an important and powerful research system to study the translation of almost any exogenous mRNA and the proteins encoded by injected messages. The development of successful protocols for synthesis of transcripts *in vitro* has accelerated use of the oocyte as a tool for such studies. However, use of the oocyte now extends well beyond its utility solely for the translation of injected messages, as indicated by this volume. In our laboratory, we frequently use *Xenopus* oocytes at several stages of growth as an assay system to study transcription, translation, and processing of the proteins encoded by injected molecules. However, our major emphasis has been, and remains, on understanding the endogenous biosynthetic processes associated with oogenesis and oocyte maturation. This approach has led us to develop methodologies that maximize oocyte physiology. In the present chapter, we have attempted to summarize these experimental approaches, which also appear to maximize the probability of success in measuring the product(s) of injected messages.

Acknowledgments

Original research was supported by a research grant from the National Institutes of Health (HD 04229) awarded to L.D.S.; W.X. is supported by a training grant from the NIH.

References

Anderson, D. M., and Smith, L. D. (1977). Synthesis of heterogeneous nuclear RNA in full grown oocytes of *Xenopus laevis*. *Cell (Cambridge, Mass.)* **11**, 663–671.

Anderson, D. M., and Smith, L. D. (1978). Patterns of synthesis and accumulation of heterogeneous RNA in lampbrush stage oocytes of *Xenopus laevis*. *Dev. Biol.* **67**, 274–285.

Asselbergs, F. A. M., Van Venrooij, W. J., and Bloemendal, H. (1979). Messenger RNA competition in living *Xenopus* oocytes. *Eur. J. Biochem.* **94**, 249–254.

Briggs, R., and King, T. J. (1952). Transplantation of lining nuclei from blastula cells into enucleated frogs' eggs. *Proc. Natl. Acad. Sci. U.S.A.* **38**, 455–463.

Ceriotti, A., and Colman, A. (1989). Protein transport from endoplasmic reticulum to the Golgi complex can occur during meiotic metaphase in *Xenopus* oocytes. *J. Cell Biol.* **109**, 1439–1444.

Ceriotti, A., and Colman, A. (1990). Trimer formation determines the rate of influenza virus haemagglutinin transport in the early stages of secretion in *Xenopus* oocytes. *J. Cell Biol.* **111**, 409–420.

Cicirelli, M. F., and Smith, L. D. (1987). Do calcium and calmodulin trigger maturation in amphibian oocytes? *Dev. Biol.* **121**, 48–57.

Colman, A. (1984a). Expression of exogenous DNA in *Xenopus* oocytes. *In* "Transcription and Translation: A Practical Approach" (B. D. Hames and S. J. Higgins, eds.), pp. 49–69. IRL Press, Oxford.

Colman, A. (1984b). Translation of eukaryotic messenger RNA in *Xenopus* oocytes. *In* "Transcription and Translation: A Practical Approach" (B. D. Hames and S. J. Higgins, eds.), pp. 271–302. IRL Press, Oxford.

Davidson, E. H. (1986). "Gene Activity in Early Development," 3 Ed. Academic Press, New York.

Dolecki, G. I., and Smith, L. D. (1979). Poly(A)$^+$ RNA metabolism during oogenesis in *Xenopus laevis*. *Dev. Biol.* **69**, 217–236.

Drummond, D. R., Armstrong, J., and Colman, A. (1985). The effect of capping and polyadenylation on the stability, movement and translation of synthetic messenger RNAs in *Xenopus* oocytes. *Nucleic Acids Res.* **13**, 7375–7392.

Dumont, J. N. (1972). Oogenesis in *Xenopus laevis* (Daudin). *J. Morphol.* **136**, 153–180.

Ecker, R. E., and Smith, L. D. (1968). Protein synthesis in amphibian oocytes and early embryos. *Dev. Biol.* **18**, 232–249.

Fischberg, M., Gurdon, J. B., and Elsdale, T. R. (1958). Nuclear transfer in Amphibia and the problem of the potentialities of the nuclei of differentiating tissues. *Exp. Cell Res., Suppl.* **6**, 161–178.

Froehner, S. C., Luetje, C. W., Scotland, P. B., and Patrick, J. (1990). The postsynaptic 43K protein clusters muscle nicotinic acetylcholine receptors in *Xenopus* oocytes. *Neuron* **5**, 403–410.

Galili, G., Kawata, E. E., Smith, L. D., and Larkins, B. A. (1988). Role of the 3′-poly(A) sequence in translational regulation of mRNAs in *Xenopus laevis* oocytes. *J. Biol. Chem.* **263**, 5764–5770.

Gelerstein, S., Shapora, H., Dascal, N., Yekuel, R., and Oron, Y. (1988). Is a decrease in cyclic AMP a necessary and sufficient signal for maturation of amphibian oocytes? *Dev. Biol.* **127**, 25–32.

Gurdon, J. B. (1960). The developmental capacity of nuclei taken from differentiating endoderm cells of *Xenopus laevis*. *J. Embryol. Exp. Morphol.* **8**, 505–526.

Gurdon, J. B. (1968). Changes in somatic cell nuclei inserted into growing and maturing amphibian oocytes. *J. Embryol. Exp. Morphol.* **20**, 401–414.

Gurdon, J. B., Lane, C. D., Woodland, H. R., and Marbaix, G. (1971). Use of frog eggs and oocytes for the study of messenger RNA and its translation in living cells. *Nature (London)* **233**, 177–182.

Horrell, A., Shuttleworth, J., and Colman, A. (1987). Transcript levels and translational control of hsp70 synthesis in *Xenopus* oocytes. *Genes Dev.* **1**, 433–444.

Jackson, R. J., and Standart, N. (1990). Do the poly(A) tail and 3′ untranslated region control mRNA translation? *Cell (Cambridge, Mass.)* **62**, 15–24.

Johnson, A. D., Cork, R. J., Williams, M. A., Robinson, K. R., and Smith, L. D. (1990). H-ras^{val12} induces cytoplasmic but not nuclear events of the cell cycle in small *Xenopus* oocytes. *Cell Regul.* **1**, 543–554.

Kawata, E. E, Galili, G., Smith, L. D., and Larkins, B. A. (1988). Translation in *Xenopus* oocyte of mRNAs transcribed *in vitro*. *Plant Mol. Biol.* **B7**, 1–22.

Keem, K., Smith, L. D., Wallace, R. A., and Wolf, D. (1979). Growth rate of oocytes in laboratory-maintained *Xenopus laevis Gamete Res.* **2**, 125–135.

Kimelman, D., and Kirschner, M. (1987). Synergistic induction of mesoderm by FGF and TGF-beta and the identification of an mRNA coding for FGF in early *Xenopus* embryo. *Cell (Cambridge, Mass.)* **51**, 869–877.

LaMarca, M. J., Strobel-Fidler, M. C., Smith, L. D., and Keem, K. (1975). Hormonal effects on RNA synthesis by stage 6 oocytes of *Xenopus laevis*. *Dev. Biol.* **47**, 384–393.

Lane, C. D. (1983). The fate of genes, messengers, and proteins introduced into *Xenopus* oocytes. *Curr. Top. Dev. Biol.* **18**, 89–116.

Laskey, R. A., Mills, A. D., Gurdon, J. B., and Partington, G. A. (1977). Protein synthesis in oocytes of *Xenopus laevis* is not regulated by the supply of messenger RNA. *Cell (Cambridge, Mass.)* **11**, 345–351.

Masui, Y., and Markert, C. L. (1971). Cytoplasmic control of nuclear behavior during meiotic maturation of frog oocytes. *J. Exp. Zool.* **177**, 129–146.

McGrew, L. L., Dworkin-Rastl, E., Dworkin, M. B., and Richter, J. D. (1989). Poly(A) elongation during *Xenopus* oocyte maturation is required for translational recruitment and is mediated by a short sequence element. *Genes Dev.* **3**, 803–815.

Richter, J. D., and Smith, L. D. (1981). Differential capacity for translation and lack of competition between mRNAs that segregate to free and membrane-bound polysomes. *Cell (Cambridge, Mass.)* **27**, 183–191.

Richter, J. D., Anderson, D. M., Davidson, E. H., and Smith, L. D. (1984). Interspersed poly(A) RNAs of amphibian oocytes are not translatable. *J. Mol. Biol.* **173**, 227–241.

Schuetz, A. W. (1967). Mechanism of progesterone- and pituitary-induced germinal vesicle breakdown in oocytes of *Rana pipiens*. *J. Cell Biol.* **35**, 123A.

Shih, R. J., O'Connor, C. M., Keem, K., and Smith, L. D. (1978). Kinetic analysis of amino acid pools and protein synthesis in amphibian oocytes and embryos. *Dev. Biol.* **66**, 172–182.

Shiokawa, K., Kawazoe, Y., Nomura, H., Miura, T., Nakakura, N., Horiuchi, T., and Yamana, K. (1986). Ammonium ion as a possible regulator of the commencement of rRNA synthesis in *Xenopus laevis* embryogenesis. *Dev. Biol.* **115**, 380–391.

Smith, L. D. (1989). The induction of oocyte maturation: Transmembrane signaling events and regulation of the cell cycle. *Development (Cambridge, U.K.)* **107**, 685–699.

Smith, L. D. (1991). Translational regulation of maternal messenger RNA. *Adv. Dev. Biochem.* **1** (in press).

Smith, L. D., Ecker, R. E., and Subtelny, S. (1968). *In vitro* induction of physiological maturation in *Rana pipiens* oocytes removed from their ovarian follicles. *Dev. Biol.* **17**, 627–643.

Soreq, H. (1985). The biosynthesis of biologically active proteins in mRNA-injected *Xenopus* oocytes. *CRC Crit. Rev. Biochem.* **18**, 199–238.

Taylor, M. A., and Smith, L. D. (1985). Quantitative changes in protein synthesis during oogenesis in *Xenopus laevis*. *Dev. Biol.* **110,** 230–237.

Taylor, M. A., and Smith, L. D. (1987). Induction of maturation in small *Xenopus laevis* oocytes. *Dev. Biol.* **121,** 111–118.

Taylor, M. A., Johnson, A. D., and Smith, L. D. (1985). Growing *Xenopus* oocytes have spare translational capacity. *Proc. Natl. Acad. Sci. U.S.A.* **82,** 6582–6589.

Wallace, R. A. (1985). Vitellogenesis and oocyte growth in nonmammalian vertebrates. *In* "Developmental Biology: A Comprehensive Synthesis, Volume 1: Oogenesis" (L. W. Browder, ed.), pp. 127–177. Plenum, New York.

Wallace, R. A., and Misulovin, Z. (1978). Long-term growth and differentiation of *Xenopus* oocytes in a defined growth medium. *Proc. Natl. Acad. Sci. U.S.A.* **75,** 5534–5538.

Wallace, R. A., Jared, D. W., Dumont, J. N., and Sega, M. W. (1973). Protein incorporation by isolated amphibian oocytes. III. Optimum incutation conditions. *J. Exp. Zool.* **184,** 321–334.

Wasserman, W. J., Richter, J. D. and Smith, L. D. (1982). Protein synthesis during maturation promoting factor- and progesterone-induced maturation in *Xenopus* oocytes. *Dev. Biol.* **89,** 152–158.

Yu, L., Blumer, K. J., Davidson, N., Lester, H. A., and Thorner, J. (1989). Functional expression of the yeast alpha-factor receptor in *Xenopus* oocytes. *J. Biol. Chem.* **264,** 20847–20850.

Chapter 5

Early Embryonic Development of Xenopus laevis

RAY KELLER

Department of Molecular and Cell Biology
University of California at Berkeley
Berkeley, California 94720

I. Introduction
II. Origins of the Gastrula: Cleavage and Blastula Stages
    A. Determination of Axes
    B. Cleavage and Structure of the Blastula
    C. Pregastrular Movements
    D. Fate Maps
III. Gastrulation and Neurulation
    A. Fate Map and Movements of Gastrulation and Neurulation
    B. Structure of Gastrula and Neurula As Seen in the Stereomicroscope
    C. Common Misconceptions
    D. Superficial Epithelial Cells and Deep Nonepithelial Cells
    E. Staging the Gastrula
IV. Investigating Cell Behaviors
    A. Bottle Cells
    B. Convergence and Extension of Involuting and Noninvoluting Marginal Zones: Cell Intercalations
    C. Migrating Mesodermal Cells: Analysis of Individual and Aggregate Mesodermal Cell Behavior
    D. Epiboly of the Animal Cap
V. Experimental Design and Context-Dependent Function of Cells and Tissues
VI. Microsurgical Methods, Tools, and Manipulations
    A. Tools
    B. Methods
    C. Explants
    D. Solutions and Healing
    E. Sterilization and Antibiotics

VII. Microscopy, Image Processing, and Recording: Epi-illumination and Low-Light Fluorescence Microscopy
    A. Epi-illumination of Explanted Cells and Tissues
    B. Use of Fluorescein Dextran Amine and DiI for Low-Light Fluorescence Microscopy
    C. Microscopy, Image Processing, and Recording
    References

I. Introduction

The objective of this chapter is to provide a practical guide to understanding and manipulating the early development of *Xenopus laevis*, the African clawed frog. I describe the structure of the embryo and the locations, movements, and changes in shape and position of the major prospective tissues through the blastula, gastrula, and neurula stages. I illustrate these phenomena with diagrams and videomicrographs to show how specific features of the embryo will appear in the investigator's own stereomicroscope. I summarize what is known about the cell behaviors driving these morphogenetic movements and illustrate them with higher resolution light and scanning electron micrographs, with the goal of deepening our understanding of the embryo and at the same time describing methods of obtaining this type of information. I describe methods of identifying, isolating, and culturing cells and tissues from specific regions. Where useful I give a brief history of the methods, terminology, and concepts. I hope this information will be a useful guide for understanding where important cells and tissues are located, what they are doing, and how to use them in cell and developmental biology.

This chapter builds on the detailed description of the normal development and the staging of *Xenopus* presented in the normal table by Nieuwkoop and Faber (1967). This work is complete, detailed, and accurate and should be consulted for additional information on development of *X. laevis*. Staging is according to the Nieuwkoop and Faber staging series, and the times referred to here in assume development at 23°C.

II. Origins of the Gastrula: Cleavage and Blastula Stages

A. Determination of Axes

1. Expression of Polarity

The major axes, prospective tissues, and morphogenetic movements of *Xenopus* early development are organized progressively by pattern-forming processes previously described in detail (Gerhart, 1980; Gerhart *et al.,* 1989).

Unfertilized eggs have an animal–vegetal polarity expressed on the surface by a pigmented animal half and unpigmented vegetal half (Figs. 1a and 2a) and in the cytoplasm by differences in yolk, position of the nucleus and other features along the same axis (see Chapters 4, 6, and 7, this volume). Fertilization frees the embryo within the vitelline envelope, and the less dense pigmented half rotates to the top Fig. 1b). Following fertilization the cortex rotates relative to the deep region and sets in motion a cascade of events (see Chapters 11 and 14, this volume) that result in formation of the future dorsal side of the embryo opposite the sperm entry point (SEP) (Fig. 1b–f; Fig. 2b, arrow). Because of the cortical rotation, the future dorsal side tends to be lighter in color than the ventral side, but this difference is an imprecise and unreliable indicator of where the future dorsal side will form. This lighter dorsal pigmentation may reflect similar underlying events but is not usually considered equivalent to the "gray crescent" of some other amphibians.

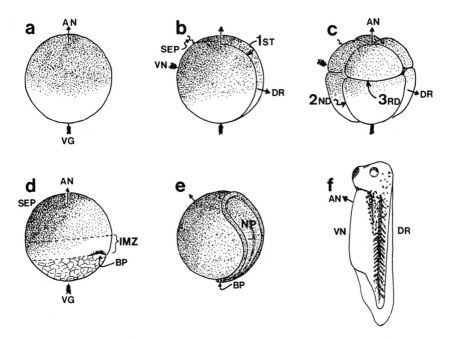

FIG. 1. Schematic diagrams illustrate early *Xenopus* development. (a) Unfertilized egg showing animal (AN) and vegetal (VG) poles. (b) The dorsal side (DR) forms opposite the sperm entry point (SEP), and the first cleavage is meridional but not necessarily through the SEP and future dorsal midline (through the future plane of bilateral symmetry); VN indicates the ventral side. (c) The second cleavage is meridional and perpendicular to the first, and the third cleavage is horizontal above the equator. (d) A dorsolateral view of the early gastrula shows the involuting marginal zone (IMZ) and the blastopore (BP), which forms opposite the SEP. (e, f) The neural plate (NP) forms on the dorsal side (e) and elongates to form the nervous system on the dorsal side of the larva (f). (Modified from Keller, 1986.)

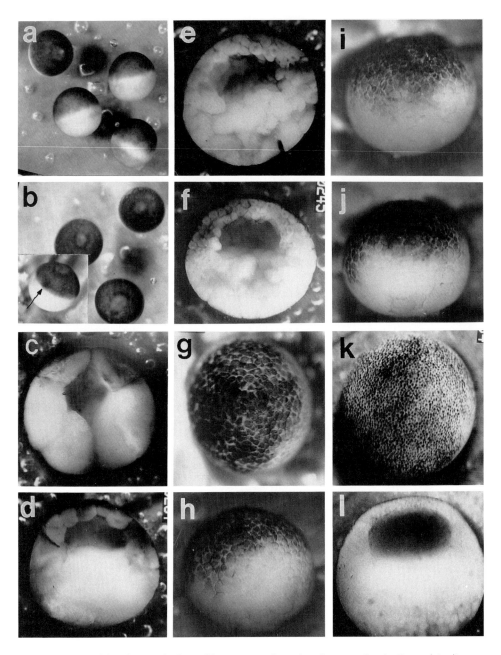

FIG. 2. Light micrographs from videotape recordings show features of early *Xenopus* development as they appear in the dissecting stereomicroscope. (a) Unfertilized eggs. (b) Fertilized eggs showing the sperm entry point (arrow) on the future ventral side. (c–e) Sagittal dissections of embryos at the 8-cell stage (c), the 16 to 32-cell stage (d), and stage 7 (e). (f–j) Stage 8 is shown in sagittal dissection (f), from the animal (g), the lateral (h), the dorsal (i), and the ventral sides (j). (k, l) Stage 9 blastula is shown from the animal pole (k) and in midsagittal dissection (l) Dorsal is to the right in (g), (h) and (k).

2. Marking the Dorsal Side

It is important for a number of experiments to predict where the dorsal side will form before it does so. To predict the position of the future dorsal side reliably, the embryos are tipped 90° after fertilization, and the uppermost side, which will become dorsal, is marked with a wand of vital dye (Kirschner and Hara, 1980). This method is described in Appendix A, this volume.

B. Cleavage and Structure of the Blastula

The first cleavage follows fertilization by about 100 minutes (23°C) and begins with a meridional cleavage (Fig. 1b), followed by second cleavage, which is meriodional and perpendicular to the first (Fig. 1b), and a third, which is horizontal and the same distance above the equator (Fig. 1c). Although the first cleavage passes through the SEP and future dorsal side, it bears a nearly random relationship to these features and *does not* define the future plane of bilateral symmetry (Danilchik and Black, 1989).

The beginning of each cleavage occurs by furrowing of the primary or oocyte membrane, and the furrow deepens by insertion of vesicles of nascent or secondary membrane into the walls of the furrow (Fig. 3a). A junctional complex, consisting of tight junctions, desmosomes, and gap junctions, forms early in cleavage, near the boundary of oocyte membrane and the new membrane, binding the apices of the blastomeres together into an epithelium (Fig. 3b–d) (reviewed in Gerhart, 1980). The blastocoel enlarges by osmotic uptake of water. Vesicles fusing with the cleavage furrow wall empty matrix material into the blastocoel, and the secondary membrane contains sodium/potassium pumps, which pump sodium into the blastocoel, raising its osmotic pressure (reviewed in Gerhart, 1980).

After the first cleavage, which occurs at about 100 minutes, subsequent cleavages occur at approximately 30-minute intervals. Cleavage planes are initially perpendicular to the surface of the embryo, forming a single-layered embryo (Figs. 2c,d and 3c) until stage 7, when cleavage planes parallel to the surface form two layers of cells: a layer of superficial, epithelial cells and a layer of deep, nonepithelial cells (Figs. 2f and 3c). Continued cleavage forms multiple layers of smaller cells in all regions, with the vegetal ones remaining larger. The cleavages are synchronous until cycle 13, at which point they become asynchronous (Newport and Kirschner, 1982). The embryo can often be oriented since the dorsal side is often lighter in color in animal view (Fig. 2g), with the transition of pigment more gradual and closer to the animal pole on the dorsal side (Fig. 2h,i) than on the ventral side (Fig. 2j). Compared with the early gastrula, the late blastula (Fig. 2k,l) has a larger blastocoel and a thinner blastocoel roof, consisting of several layers of small cells.

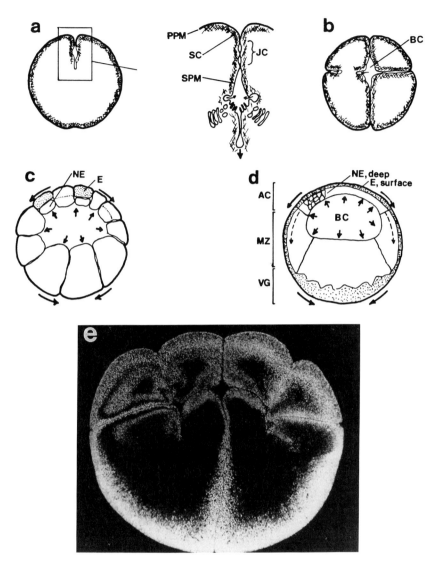

FIG. 3. Cleavage is illustrated diagrammatically. (a) The primary or oocyte membrane (thick line, PPM) furrows to initiate cleavage, and then the furrow deepens as secondary membrane (thin lines, SPM) is added by fusion of vesicles to the walls of the furrow. A junctional complex (JC) ultimately forms between the apical ends of the blastomeres. SC indicates subcortical cutoplasm. (b) The blastocoel (BC) forms in the animal half, and the SC ingresses with the furrows (cross-hatching). (c) At stage 7, two categories of cells are formed—the deep, nonepithelial cells (NE) and the superficial epithelial cells (E)—as a result of cleavages oriented parallel to the surface (compare with Fig. 2e). (c, d) As cleavage proceeds, the animal cap (AC), or blastocoel roof, thins and spreads in epiboly (arrows) owing to osmotic forces within the blastocoel (arrows, BC). The vegetal region (VG) contracts (arrows) and the marginal zone (MZ) moves vegetally. (e) Movement of the SC with cleavage furrows is illustrated by labeling yolk platelets with a fluorescent dye, trypan blue (Danilchik and Denegre, 1991). [(e) is from Danilchik and Denegre, 1991; all others modified from Keller et al., 1986.]

C. Pregastrular Movements

Movements of gastrulation are preceded by pregastrular movements that have been characterized quantitatively and in some detail by time-lapse recording (Keller, 1978) and morphometric analysis of scanning electron micrographs (Keller, 1980). The blastocoel increases in volume and the animal cap (blastocoel roof) expands uniformly in a movement called epiboly. The diameter of the embryo increases during blastulation, largely because of the expansion of the blastocoel roof. As it spreads, the blastocoel roof also thins (Keller, 1980) (compare Figs. 2c–f, with Fig. 2l; Fig. 3c with Fig. 3d). Meanwhile, the vegetal region contracts over the same period (Fig. 3d) (Ballard, 1955; Keller, 1978). As a result of these movements, the marginal zone, where the prospective mesodermal tissues lie, is displaced from its initial position above the equator to a subequatorial position at the onset of gastrulation (compare Figs. 4 and 5).

D. Fate Maps

Several fate maps of prospective tissues have been made of cleavage stages, including ones by Nakamura and Kishiyama (1971) and Cooke and Webber (1985) (Fig. 4), among others (Jacobson and Hirose, 1979a,b; Dale and Slack, 1987; Moody, 1987). Fate maps made of blastula stages are limited in resolution by the size of blastomere and provide only a rough idea of where tissues actually originate. Although they are more difficult to present concisely, the data in Dale and Slack (1987) are a better guide to what one can actually expect to develop from a given region. Note that the cleavage pattern

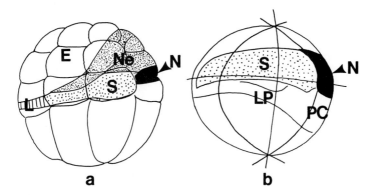

FIG. 4. Fate maps of the 32-cell stage (a) and an idealized 16-cell stage (b) of *Xenopus* show the prospective epidermis (E), neural tissue (Ne), notochord (N), somitic mesoderm (S), prechordal region (PC), lateral plate (LP), lateral mesoderm in the caudal region (L), and vegetal endoderm (bottom, unshaded). Dorsal is to the right in both diagrams. [(a) is redrawn from Nakamura and Kishiyama, 1971; and (b) from Cooke and Webber, 1985.]

and the blastomeres themselves do not predict fate, since the cleavage pattern is variable and does not determine cell fate (Danilchik and Black, 1989). The animal cap, composed of prospective ectodermal tissues, is relatively small, and the marginal zone, containing the prospective mesodermal tissues, is compact, thick, and located above the equator (Fig. 4), compared to its location below the equator at the onset of gastrulation (compare with Fig. 5a). These regions are moved vegetally by the pregastrular morphogenetic movements mentioned above.

III. Gastrulation and Neurulation

A. Fate Map and Movements of Gastrulation and Neurulation

The structure, fates, and morphogenetic movements of prospective tissues through gastrulation and neurulation are geometrically complex, but an understanding of these events is fundamental to the understanding of nearly all aspects of early development. The mass movements of cell populations comprise the highly integrated behavior of cells such that the specific functions of genes, molecules, and cell behaviors in morphogenesis are defined largely in terms of the cellular and tissue level biomechanics. Thus it is critical that endeavors at other levels be done with knowledge of morphogenetic process.

Movements of prospective tissues during gastrulation are illustrated in Fig. 5, which is based on vital dye mapping (Keller, 1975, 1976) and subsequent time-lapsed recording work (see Keller, 1986; Wilson et al., 1989; Keller

---

FIG. 5. Prospective fates and morphogenetic movements of the gastrula and neurula are shown in several views. Four horizontal rows show development in chronological order, from top to bottom, at the late blastula (a, e, i, and m), the early gastrula (b, f, j, and n), the late gastrula (c, g, k, and o), and the late neurula (d, h, l, and p). The vertical columns show different views which are, from left to right, the lateral surface, with dorsal to the left (a–d), a midsagittal view (e–h), the ring of deep prospective mesodermal cells (i–l), and the overlying ring of suprablastoporal endoderm (m–p). Special features illustrated include the animal pole (AP); archenteron (A); the blastocoel (BLC); the bottle cells (BC); the blastopore (pointers); the head mesoderm (Hd); the heart mesoderm (Ht); the involuting marginal zone (IMZ), composed of deep, nonepithelial mesoderm (i–l) and superficial, epithelial endoderm (m–p); the lateral and ventral mesoderm (LV); the noninvoluting marginal zone (NIMZ); and the vegetal pole (VP). Prospective tissues shown are epidermis (light blue), neural tissue (darker blue), dorsal NIMZ (blue green), notochord (red), somitic mesoderm (red orange), migrating mesoderm at the leading edge of the mesodermal mantle (orange), suprablastoporal endoderm (yellow), a special region of suprablastoporal endoderm known as the bottle cells (green), and vegetal, subblastoporal endoderm (yellow, divided into cells). Movements are indicated by arrows. Dorsal is to the left in all cases.

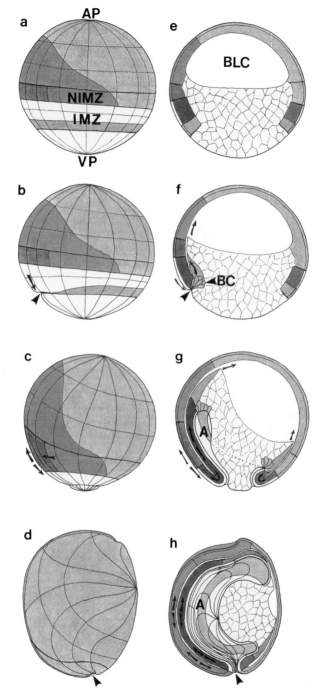

Fig. 5.

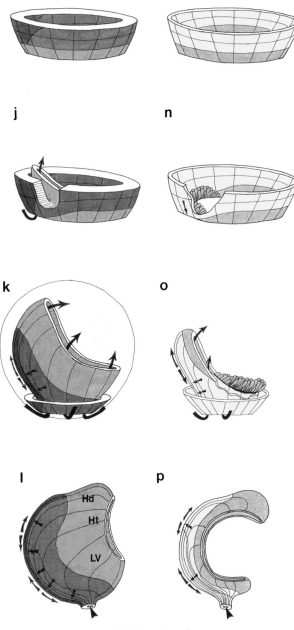

Fig. 5. *(continued)*

et al., 1991b). During gastrulation, the prospective ectodermal tissues expand in area and spread vegetally in a continuation of the epiboly begun in the blastula (Fig. 5a–c). As the prospective ectodermal tissues spread vegetally, they replace the endodermal tissues, which disappear inside the gastrula by mechanisms discussed below (Fig. 5a–c). As gastrulation proceeds, the prospective neural tissue (Fig. 5a–c, dark blue) narrows toward the dorsal midline, a movement called *convergence*, and lengthens in the anteroposterior direction, a movement called *extension* Fig. 5c, (arrows). The lower edge of the ectodermal tissue is called the noninvoluting marginal zone (NIMZ). It converges during gastrulation, constricting circumferentially and closing around the blastopore as the endodermal tissue moves inside. Convergence and extension movements are particularly strong in the dorsal, vegetal region of the NIMZ (Fig. 5a–c, arrows and blue-green area), which will form the posterior midline of the neural plate and ultimately the ventral aspect of the spinal cord. This region is a special part of the NIMZ that will autonomously converge and extend in culture, a property important in the mechanism of gastrulation and neurulation. During neurulation the posterior neural tissues continue extension, particularly at the midline, roll into a tube, and thus disappear from the surface (Fig. 5d). The neural tissues continue to extend anteroposteriorly during neurulation [Fig. 5g–h; note the exceptional extension of the dorsal NIMZ in sagittal sections (arrows, blue-green area)].

The movements of the endodermal and mesodermal tissues are much more complex and are best shown in several views. The tissues moving inside during gastrulation can be divided into the involuting marginl zone (IMZ) (Fig. 5a,e, yellow and green areas) and the large sub-blastoporal endoderm (SBE) (Fig. 5a,e, yellow area, divided into cells). The IMZ consists of an epithelial layer of prospective endoderm that will form the roof of the archenteron (Fig. 5a,e,m, yellow area). The vegetal edge of this superficial layer consists of prospective bottle cells (Fig. 5a,e,m, green area). Deep to the prospective bottle cells are four or five layers of prospective migrating mesodermal cells that will subsequently migrate into the interior and form the leading edge of the mesodermal mantle (Fig. 5e,i, orange area). Deep to the remaining superficial endodermal cells are four or five layers of prospective converging and extending mesoderm (Fig. 5e,i, red area). This mesoderm will converge and extend during gastrulation to form the axial mesoderm, including the prospective notochord (Fig. 5e,i, red area) on the dorsal side and prospective somitic mesoderm (Fig. 5e,i, reddish orange area), extending from the notochord to the midventral line. The vegetal region consists of the sub-blastoporal endoderm that will form the future floor of the archenteron (Fig. 5a,e, yellow area, divided into cells). This region is derived from the yolky vegetal region of the egg and thus consists of large cells, whereas the remainder of the embryo consists of smaller cells.

Gastrulation begins, on the surface, with the formation of the bottle cells. The vegetal-most five or six rows of cells of the endodermal epithelium of the IMZ contract their apices in the animal-vegetal direction and thus compress a large area of the superficial endodermal epithelium vegetally against the large sub-blastoporal endodermal cells; meanwhile the epithelium of the IMZ immediately animal to the contracting bottle cells stretches and increases in area (Fig. 5a-b, e-f, and m-n). As their apices contract, cytoplasm is forced deep, owing to mechanical interactions with surrounding tissues, and the "bottle" shape emerges (Hardin and Keller, 1988). As the bottle cells form, the mechanics of the local tissue environment result in formation of an initial invagination, the blastoporal groove, and the vegetal end of the IMZ is rotated above it, initiating involution (Fig. 5b,f). As the apices of the bottle cells constrict, the pigment in this region is concentrated and appears as a black line called the blastoporal pigment line (Fig. 5b,f). As the bottle cell epithelium is contracted into a very small area, the deep mesodermal cells associated with its inner surface, the migrating mesoderm, lose their association with the overlying epithelium and move inward and upward, where they begin their migration across the roof of the blastocoel (Fig. 5e-f and i-j). This behavior is also part of the process of initiating involution of the mesodermal mantle. A discontinuity appears between the migrating mesoderm moving animally and the overlying wall of the gastrula moving vegetally. This structure is a continuation of the inner surface of the blastocoel roof, sometimes called Brachet's cleft (Ballard and Ginsberg, 1980), and it represents the line of shear between involuting and uninvoluted tissue. Most likely it is actually formed by movement of the outer gastrular wall vegetally and movement of the migrating mesoderm animally on its inner surface (Hardin and Keller, 1988), but there is no definitive evidence eliminating the possibility that delamination might also play a role.

The formation of bottle cells, along with the associated involution of mesoderm, begins on the dorsal side, proceeds laterally, and occurs ventrally by the late gastrula stage (Fig. 5f-g, j-k, and n-o). At the midgastrula stage, the bottle cells respread, beginning mid-dorsally (Fig. 5g,o), progressing laterally, and reaching the midventral line by the late neurula-early tailbud stage to form a large area at the periphery of the archenteron (Fig. 5h,p). Their order of respreading is the same as the order of their formation, and both processes are autonomous in culture (Hardin and Keller, 1988).

Migrating mesoderm involutes first but is followed immediately by converging and extending mesoderm (prospective notochordal and somitic mesoderm) (Fig. 5e-h and i-l). The migrating mesodermal cells spread across the blastocoel roof by a process of migration (discussed in more detail below), and in so doing form the leading edge of the mesodermal mantle (Fig. 5j-l). As the converging and extending mesoderm involutes, it converges toward the dorsal midline and extends in the anteroposterior direction (Fig. 5g-h and k–l, arrows). The roof of the archenteron is directly superficial to this mesoderm and converges and extends with it, probably because it is firmly attached at its margins by the bottle cells, which are more adherent to the mesoderm than the remaining endodermal cells (Keller, 1981) (Fig. 5o–p, arrows). When the bottle cells respread, they do so near the lateral edge of the somitic mesoderm. The converging and extending portions of the endoderm (the archenteron roof) and the mesoderm (the notochord and somite) thus are more or less coincident (Fig. 5i–l compared with 5n–p). The convergence and extension of the notochordal and somitic mesoderm squeeze the blastopore shut, carrying these tissues and the associated endoderm to the dorsal side of the embryo and establishing their elongate shape (Fig. 5j–l and m–p, arrows).

FIG. 6. Major features of gastrulation are shown as they appear in the dissecting stereo microscope. (a–c) Onset of gastrulation is marked by progressive darkening of the constricting apices of the bottle cells to form the blastoporal pigment line at the lower edge of the IMZ (pointers). (d) A sagittal view of (b) shows the initial blastoporal groove (pointer). The numbers correspond to the same positions in later embryos (f, g, and q). (e) Shortly thereafter, the invagination deepens the blastoporal groove (white pointer), and Brachet's cleft, indicating the line separating the involuted, leading edge mesoderm from the roof of the blastocoel, appears on the dorsal side above the bottle cells (black pointers). (f, g) vegetal view of stage 10.5 shows that the blastoporal pigment line has reached the midventral line (pointers, f), and a sagittal view of a slightly later stage shows the archenteron (A) and bottle cells at its tip marked by pigmentation (black pointer, g). The advance of the migrating mesoderm is indicated (white pointer, g). (h) A dissection of the late midgastrula, stage 11, shows the endodermal archenteron roof (AR), the pigmented apices of the bottle cells at the blind end of the archenteron (white pointer) and the subblastoporal (vegetal) endoderm (SBE) forming the archenteron floor. (i) In the same view of a later stage, the bottle cells have largely respread (pointer). (j, k) A vegetal view of a late gastrula, stage 12.5, shows the small yolk plug (pointer, j), and a dorsovegetal view of the same stage (k) shows streaks of pigment (black pointers), drawn out by convergence and extension toward the blastopore (white pointer), that mark the dorsal side. (l) A midsagittal dissection of a late gastrula shows the archenteron (A), the blastocoel (BL), the ventral diverticulum of the archenteron (large pointer), the thickened prospective brain region (small pointers), and the yolk plug (YP). (m) A dorsal view of a midneurula shows accumulation of pigment at the neural plate midline (pointers). (n, o) Removal of the neural plate reveals the notochord (N) and somitic mesoderm (SM) beneath (n) and the thin prospective floor plate on the underside of the neural plate where the notochord was tightly adherent (pointers, o). (p) The thickened neural folds of the late neurula (black pointers) will fuse middorsally to form the neural tube; the wider brain region is located anteriorly (white pointer). (q) A sagittal dissection of a late neurula shows the thin archenteron roof (AR), the notochord (N), the spinal cord portion of the neural tube (SC), the inner surface of the forebrain where the optic vesicle will form (small white pointer), and the blastopore (large white pointer). (r) A view into the posterior archenteron shows the slit-shaped blastopore (pointer) surrounded by the thick ring of prospective posterior somitic mesoderm on the lateral and ventral sides of the blastopore. The transverse view of the notochord and neural tube is seen at the top. (s) A similar view directed anteriorly shows the pharyngeal region with the ridge formed dorsally (lower black pointer) by the floor of the brain pressing the head mesoderm against the dorsal surface of the endodermal layer and bulges of the somites into the roof of the archenteron (small black pointers). The notochord (single pointer, top) and closing neural tube (double pointer) are shown.

72 RAY KELLER

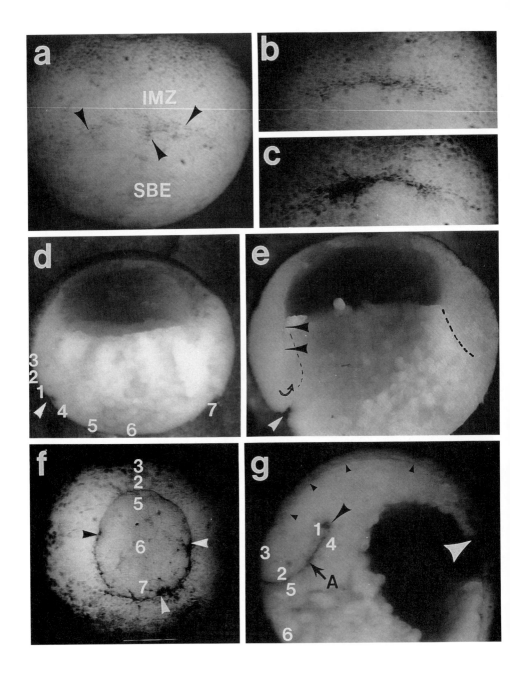

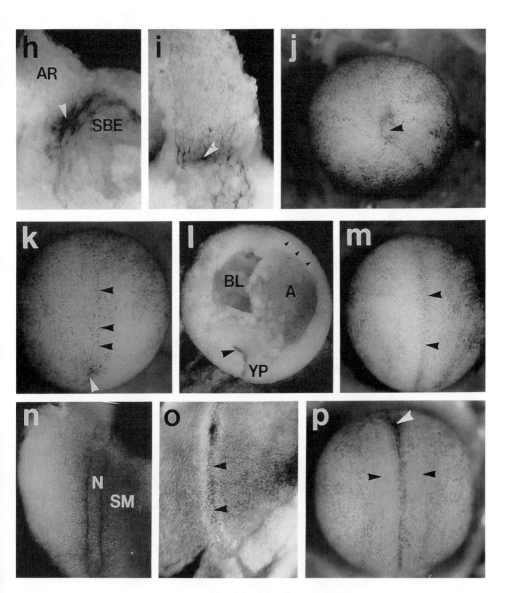

FIG. 6. (continued)

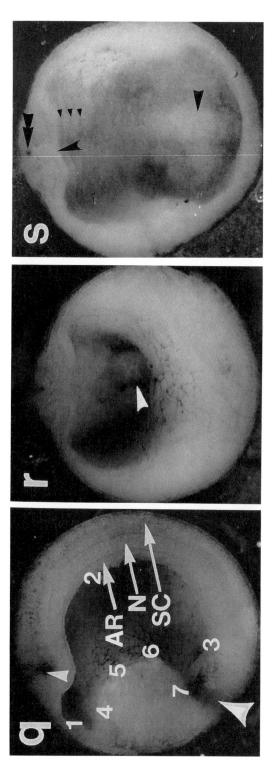

FIG. 6. (continued)

B. Structure of Gastrula and Neurula As Seen in the Stereomicroscope

It is one thing to visualize the movements of gastrulation in idealized diagrams and quite another to do the same when confronted with a relatively low-contrast, featureless expanse of living embryo in which few landmarks can be recognized. As a practical aid for finding one's way around in the gastrula, the major features of embryonic development will be illustrated by images of living or fixed embryos, made with a stereomicroscope and video camera, stored on an optical disk, and photographed from the video monitor. Nieuwkoop and Faber (1967) should be consulted for more detail. Here 1 focus on key issues and problems in orienting embryos and identifying specific cells and tissues, using illustrations comparable to what can be observed with a stereomicroscope.

The first superficial sign of gastrulation is the appearance of faint, gray pigment in the constricting apices of the forming bottle cells (Fig. 6a, pointers). Apical constriction progressively concentrates pigment, making the apices progressively darker as constriction proceeds and forming the blastoporal pigment line (Fig. 6b,c). It may or may not be symmetrical in size, and it often does not straddle the midline but sometimes lies 5–10° left or right. The blastoporal pigment line will form the smaller suprablastoporal endodermal cells that lie just above the larger sub-blastoporal endoderm cells. These two types of endodermal cells differ by a factor of 3 or 4 in apical area, and the boundary between the two types is best located with low-angle epi-illumination, which will shadow the cell boundaries and reveal their size. A sagittal dissection shows the site of initial invagination, the *blastoporal groove*, produced by apical constriction of the bottle cells (Fig. 6d). The depth of this groove, and thus the progress of gastrulation, can be determined by rolling the embryo on its side so that the depth of the groove can be seen. As apical constriction nears completion, a blastoporal groove deepens (Fig. 6e, white pointer), and involution of the migrating mesoderm (Fig. 6e, curved arrow) and formation of the cleft of Brachet occur (Fig. 6e, dorsal pointers). Identification of this cleft is important for a variety of procedures, since it separates what has involuted from the overlying blastocoel roof. If it is not seen immediately, *in vivo*, its presence can be detected by attempting to pull the wall of the blastocoel outward; if the cleft is present, it will open up. This procedure is described in older stages below. Note that on formation of the cleft of Brachet, the blastocoel roof–floor junction is no longer concave but squared off where the floor meets the cleft of Brachet. Note that the ventral side lacks a cleft of Brachet and retains the concave blastocoel roof–floor junction. The inner surface of the involuted, migrating mesoderm where it bounds the sub-blastoporal endoderm is a region of transition in cell size that is difficult to see, and its exact position is approximate (Fig. 6e, fine dashed line). On the ventral side, the inner surface of the marginal zone where it bounds the sub-blastoporal endoderm has the same characteristics (Fig. 6e, heavy dashed line).

The appearance of a well-developed blastoporal pigment line and bottle cells is usually taken as the onset of gastrulation, but, in fact, these features represent gastrulation movements that have been underway for some time. The faint appearance of gray apices of incipient bottle cells (Fig. 6a) is really the earliest indication of gastrulation and is detectable as early as 30 minutes before stage 10, on careful inspection.

Bottle cell formation proceeds laterally and ventrally, extending the blastoporal pigment line to the ventral midline by stage 10.5 (Fig. 6f). Involution of the dorsal IMZ moves the dorsal bottle cells inside and toward the animal pole. The suprablastoporal endodermal epithelium follows along behind the bottle cells, forming the roof of the archenteron, and the large sub-blastoporal endodermal cells fall inward to form the archenteron floor (Fig. 6g). The white numbers 1–3 indicate anterior-to-posterior positions on the future archenteron roof, and the numbers 4–7 show anterior-to-posterior positions on the floor. Note that by stage 10.5, both 1 and 4 have disappeared from the surface (Fig. 6f,g). In fact, the blastoporal groove has formed everywhere above the black pointer on the left and the white pointer on the right (Fig. 6f). The archenteron appears as a line rather than a space because the floor and roof are pressed against one another. Its anterior end can be located by the presence of bottle cell pigment until the late midgastrula stage when the bottle cells have begun to respread, perhaps to the point of being indistinguishable by shape or by pigment concentration. The leading edge of the mesodermal mantle has advanced across the animal cap (Fig. 6g, pointer). By stage 11–11.5, the archenteron has deepened, and by cutting parasagittally about 0.2 mm on either side of the midline, the lateral edges of the archenteron roof can be freed and the dorsal sector of the gastrula turned upward, revealing the endodermal epithelium lining the roof of the archenteron (Fig. 6h, AR), the blind anterior end of the archenteron where the pigmented bottle cell apices are respreading (Fig. 6h, pointers), and the sub-blastoporal endodermal cells forming the archenteron floor (Fig. 6h, AF). By late gastrula, the dorsal bottle cells usually have respread (Fig. 6i, pointers). The outer wall of the gastrula thins greatly through gastrulation (compare Fig. 6e with 6g), which reflects its expansion in area (epiboly) (Fig. 5a–d).

By late gastrula (stage 12–12.5), the blastopore is nearly closed and the yolk plug is small (Fig. 6j, pointer). The dorsal side of the embryo at this stage can be distinguished from the ventral side by its lighter pigmentation and gradual change in pigmentation in the animal–vegetal direction (Fig. 6j). Moroever, the powerful convergence and extension movements of the dorsal NIMZ will stretch any local pigment irregularity in the animal–vegetal direction, forming long streaks aligned toward the blastopore on the *dorsal* side (Fig. 6k, pointers). This may happen on the ventral side as well, but the streaks will be very short. A sagittal view shows that the archenteron has deepened and expanded tremendously Fig. 6l, (A), and the blastocoel is decreased in size and crowded over to the anteroventral side (Fig. 6l, BL). The yolk plug (Fig. 6l, YP) obscures the lateral surface of the posterior archenteron and the ventral diverticulum of the archenteron involuted to its maximum extent (Fig. 6l, single pointer). The anterior neural plate is thickened in the region that will form brain (Fig. 6l, triple, small pointers). At the midneurula stage, a neural groove forms, marked by pigment concentration, as in the case of the gastrula bottle cells, as the bottle-shaped cells of the neural midline contract their apices (Fig. 6m). If one peels back the neural plate at this stage, a central notochord is revealed (Fig. 6n, N), bounded on both sides by somitic mesoderm that is unsegmented (Fig. 6n, SM). The thinner floor plate region of the neural plate is seen from the underside of the neural plate (Fig. 6o, pointers). The floor plate is more tightly attached to the notochord than the somitic mesoderm is to the thickened lateral parts of the neural plate. Often many midline cells are torn off and remain adherent to the notochord as the neural plate is removed surgically.

By the late neurula stage, the neural groove has deepened and the neural folds (Fig. 6p, pointers) approach the midline, where they will fuse. Both the neural folds and the neural groove of the future brain region (Fig. 6p, white pointer) are broader than the same features in the spinal cord region. A midsagittal view at this stage shows the archenteron roof (Fig. 6q, AR), notochord (Fig. 6q, N), and neural tissue, including future brain, optic vesicle (Fig. 6q, small white pointer), and spinal cord (Fig. 6q, SC) regions. By comparing the numbers on the roof and floor of the archenteron at the late neurula stage (Fig. 6q) with their earlier positions (Fig. 6d,f,g), one can see the tremendous extension of the archenteron roof (1–3) and the movements of the archenteron floor (4–7). The ventral diverticulum has now straightened out and is no longer visible, and the yolk plug has been pushed inside and flattened on the floor of the archenteron (Fig. 6q, large pointer).

Looking posteriorly into a transverse fracture of the same stage reveals the blastopore (Fig. 6r, pointer), flanked on both sides by thickened regions of mesoderm that later swing around both sides of the posterior notochord to form the posterior somites (Keller *et al.*, 1991b; Wilson *et al.*, 1989; Keller, 1976). (This process is described in more detail in Fig.12). A similar view facing anteriorly shows the pharyngeal region of the archenteron, where the base of the brain depresses the anterior midline of the archenteron roof (Fig. 6s, lower pointer). The head mesoderm lies beneath the brain and just under the thin sheet of endodermal archenteron roof; it can be exposed by peeling off the endodermal epithelium. Note that the anterior archenteron curves ventrally (Fig. 6q), and if a transverse fracture passes through this anterior region, the archenteron appears very large (Fig. 6s). It is considerably smaller in a transverse section at the midbody (Fig. 6r). The incipient neural tube (Fig. 6s, double pointer), the endodermal roof of the archenteron (Fig. 6s, single pointer, top), and notochord between them are visible at the surface of the transverse cut (Fig. 6s). The intersomitic furrows show up on the roof of the archenteron as depressions (Fig. 6s, small pointers).

C. Common Misconceptions

There are several misconceptions, even among established developmental biologists, that are common and severe enough in their consequences to warrant special attention. It is often believed that bottle cells continue to form at the dorsal lip of the blastopore, and thus one can find them there at the early, mid, and late gastrula stage. This is not so; they form at the onset of gastrulation and are transported to the interior by movements of the mesoderm, as described in the preceding sections.

Invagination and involution are also confused. An invagination *is a local bending of a cell sheet*, and the initial formation of the blastoporal groove is an example of this process. An involution *is a rolling of a sheet of tissue around an inflection point*, and the involution of the IMZ is a good example of this process. Most of the depth of the archenteron is not formed by invagination but by involution and the coordinate extension of its roof posteriorly. Invagination and involution involve grossly different tissue movements and very different cellular mechanisms (Keller, 1986; Hardin and Keller, 1988; Keller *et al.*, 1991b).

A companion misconception is the idea that the length of the archenteron and the dorsal mesoderm is generated by movement of the mesoderm and attached archenteron toward the animal pole. In fact, however, more than half of the depth (length) of the archenteron is generated because the dorsal NIMZ and IMZ extend vegetally across the yolk plug; only the remaining depth is generated by movement of its anterior end toward the animal pole. To visualize the tremendous effect of convergence and extension on archenteron formation, consider an early gastrula lying on its animal pole, with its vegetal pole exactly uppermost and its animal–vegetal axis exactly vertical. Assume that a time-lapse recording of such an embryo gastrulating is made from the vegetal view and assume that the dorsal side will form at the top of the screen. The recording will reveal the dorsal IMZ involuting, and as it does so it will converge, constricting the blastopore and extending toward the animal pole on the inside, where it cannot be seen. The dorsal NIMZ will extend vegetally as the IMZ disappears inside. As the NIMZ extends vegetally and converges, the dorsal blastoporal lip will pass over the yolk plug, which does not move, and will close far on the ventral side. The neural plate will form and the neural folds rise and fuse, right above where the sub-blastoporal endoderm (the yolk plug) was located. The anterior end of the brain will form just out of sight on the curvature of the embryo above and the posterior end of the spinal cord will do the same at the bottom.

It is sometimes not realized that the lips of the blastopore are dynamic. The blastoporal lip is a locus of involution and is occupied by progressively more posterior cells as time passes.

Finally, it should be clearly understood that the IMZ is made up of two layers prior to involution (superficial and deep) and that these layers involute together such that, after involution, the deep layer of the IMZ faces the deep layer of the NIMZ. The distinction between deep and superficial is not changed by the process of involution. Moreover, one still speaks of the lining of the archenteron as a superficial layer, regardless of the fact that it is now out of sight from the exterior. The cavity of the archenteron is continuous with the outside space through the blastopore and is lined with an epithelium that is continuous over the lip of the blastopore with the epithelium covering the "outside" of the embryo. By contrast, the blastocoel is not lined with an epithelium, has no continuity with the exterior, and is truly an internal cavity. One can always know whether one has cut into the blastocoel or into the archenteron by looking for the tightly apposed apices of the epithelium lining the latter as opposed to the loosely connected boundaries of the nonepithelial cells lining the former.

D. Superficial Epithelial Cells and Deep Nonepithelial Cells

In many situations, beyond the one just mentioned, it is important in manipulating the gastrula, or any other stage, to be able to distinguish between an epithelial cell layer and deep cells with the stereomicroscope. The epithelial layer can be cut through by using the tip of an eyebrow hair, inserting it just under the epithelium and raising the tip, and peeling the layer back (Fig. 7a, white pointer). Beneath this layer is the deep layer, which is more delicate and more easily torn than the epithelium (Fig. 7a, D). The relation of the superficial epithelial (white pointer) and deep mesenchymal layers (dark pointer) is shown in sectional view (Fig. 7b). The epithelial layer of most nonvegetal regions is easily removed at the early gastrula stage but is progressively more difficult to remove cleanly both before and after this stage. At higher resolution, using the compound microscope and epi-illumination, the boundaries of the epithelial cells are closely apposed and bound circumapically by junctions to form a smooth pavement (Fig. 7c). By contrast, the deep cells are not so closely apposed and are outlined by shadows in low-angle epi-illumination, revealing more relief (Fig. 7d). Corresponding views of the epithelium (Fig. 7e) and deep regions (Fig. 7f) are shown with scanning electron microscopy. Epithelial cells are bound circumapically into a continuous sheet, whereas deep cells have spaces between them and are bound together by focal contacts at the ends of protrusions of various sorts.

E. Staging the Gastrula

Staging of the early gastrula is done, in part, by the progress of bottle cell formation. We distinguish a stage 10— when a very faint pigment appears dorsally (Fig. 6a) and internally only a very few bottle cells have begun to form. At this point, the embryo is still a half hour or so from the stage 10 of Nieuwkoop and Faber, when the pigment line is dark but still straight and no groove has formed. We distinguish a stage 10+, when a groove has formed and the blastoporal pigment line is wider and beginning to curve ventrally. At 10.25 this line has extended to 90° from the dorsal midline, and at 10.5 it has reached the ventral midline. At this stage, the dorsal side has a sizable groove (Fig. 6f, lying above the pointers) and the blastoporal pigment line has disappeared inside.

Stage 10.5 is long and can be divided into an early and late form (Nieuwkoop and Faber, 1967). The difference between early and late is determined by the onset of the powerful convergence and extension movements in the dorsal sector. In the early form, the circular profile formed by the blastoporal lip dorsally and the pigment line ventrally tends to be egg shaped with the pointed end dorsal; this occurs because migration of the mesoderm tends to pull the lip upward on the inside, and convergence and extension have not yet begun to relieve the tension. In the late stage the convergence and extension movements push the lip vegetally across the yolk plug, despite the pulling of the migrating mesoderm on the inside, and the profile of the blastopore rounds up; in some cases the dorsal sector will protrude across the yolk plug in advance of the dorsolateral regions. A blastoporal groove is not clearly visible ventrally until stage 11.5. As the blastoporal groove forms, the dark apices of the bottle cells are moved inside. At this point, the blastoporal groove is not darkly pigmented and will only appear dark if shadowed

FIG. 7. (a) Dissection shows the superficial epithelium of the animal cap (white pointer) pulled off the deep cells (D). (b) The superficial epithelium (white pointer) and the underlying deep cells (black pointer) are viewed from their edges. (c, d) Higher-resolution views, using a compound microscope and epi-illumination, show the close apposition of the apices of superficial cells, in this case the endodermal roof of the archenteron (c), and the less closely apposed, shadowed outlines of deep cells, in this case the inner surface of the mesoderm (arrow d). (e, f) Scanning electron micrographs show in more detail the structure of an epithelium, in this case endoderm (e), and deep cells, in this case the inner surface of the blastocoel roof (f).

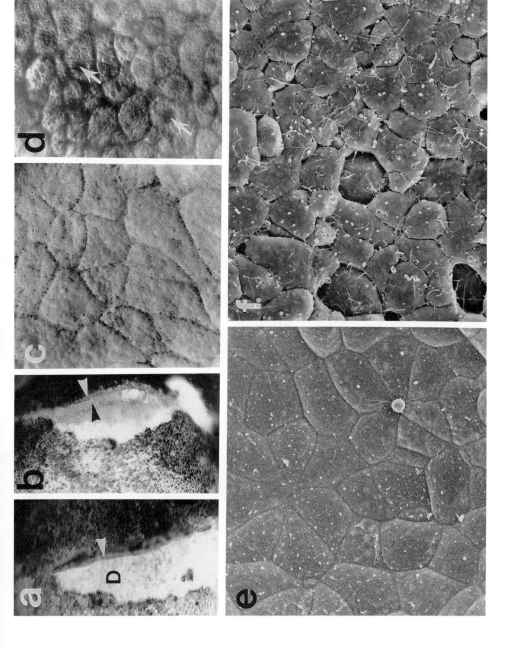

by low-angle illumination. The absence of a dark line dorsally and the presence of one ventrally at stage 11 has led some investigators to confuse the dorsal and ventral sides, based on the mistaken notion that the dorsal lip is always darker.

Staging based on the appearance of the blastoporal pigment line, the groove, and the size of the yolk plug (Nieuwkoop and Faber, 1967) is strongly correlated with internal events, but the correlation is not assolute. The correlation coefficient of yolk plug circumference and archenteron depth is 0.87 (Keller, 1974, unpublished data), and thus one must always expect some variation inside among embryos that appear the same outside.

IV. Investigating Cell Behaviors

It is important to understand in some detail the regional cell behaviors driving gastrulation movements, both to investigate them further and because other studies of early development will eventually require consideration of these behaviors. Although *Xenopus* embryos are opaque, their ease of culture and manipulation has allowed resolution of morphogenetic cell behavior second to none, and they offer the added advantage of microsurgical manipulation to test tissue interactions regulating morphogenetic behavior and tissue differentiation as well as to test ideas about mechanics. I will describe enough of what is known about these behaviors to deepen our understanding of the embryo and enable the reader to visualize in more detail what was described above with diagrams (Fig. 5) and low-resolution microscopy (Fig. 6). At the same time, this discussion will demonstrate several kinds of preparations used to isolate and analyze the function of specific populations of cells in this laboratory. A detailed discussion of how cells generate the forces producing morphogenesis has been discussed previously (reviewed in Keller, 1986; Keller *et al.*, 1991a,b; Keller and Winklbauer, 1990).

A. Bottle Cells

Bottle cells display a simple behavior whose function is modified by the mechanics and geometry of the surrounding tissues (Hardin and Keller, 1988). *In vivo*, bottle cells apices contract mostly in animal–vegetal direction, pulling the epithelium of the IMZ vegetally against the large sub-blastoporal endodermal cells as described previously (Fig. 8a, arrows). The elongate shape of the cell apices and their alignment around the large sub-blastoporal endoderm reflect this anisotropic contraction (Fig. 8b, inset). As their apices

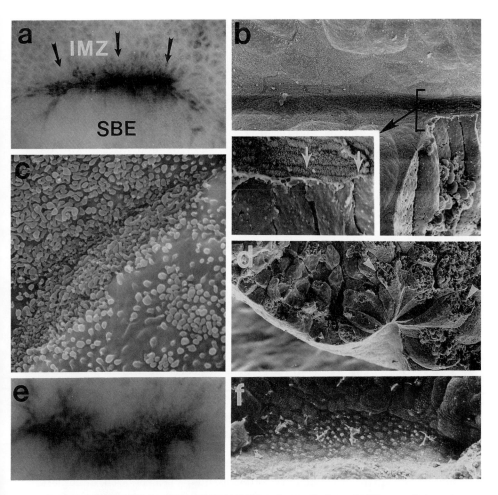

FIG. 8. Details of bottle cell formation. (a) A light micrograph shows the blastoporal pigment line representing the constricted apices of the bottle cells. The involuting marginal zone (IMZ) and sub-blastoporal endoderm (SBE) are indicated. (b) Scanning electron micrographs show the same region at low magnification and at high magnification (inset), revealing the elongate apices of the bottle cells and the microfolds of membrane (c) on their surfaces. (d) A sagittal view of a similar stage shows the characteristic bottle cell shape. (e, f) Bottle cells in isolation contract uniformly in all directions, forming circular apices, seen with light (e) and electron microscopy (f). [(a), (b), (e) and (f) are reprinted and modified from Hardin and Keller, 1988.]

contract, the apical membrane is thrown into microfolds as the apparent area of the apices decreases (Fig. 8b,c), and the bottle cells become elongated in the apical–basal direction (Fig. 8d). In culture, unattached to any other tissue, their contraction is intrinsically isotropic, producing rounded apices and minimal apical–basal elongation (Fig. 8e,f). *In vivo*, this isotropic contraction is directed vegetally because the vegetal endoderm is stiffer than the IMZ and will not deform as much. As a result, the outer part of the vegetal end of the IMZ is pulled vegetally, and the IMZ is bent inward, initiating involution (Fig. 5e,f). Although much is known about how the bottle cells distort the surrounding tissue during their formation, their role in initiating involution has not been tested directly because of technical problems (Hardin and Keller, 1988). Removal of bottle cells after their formation does not stop gastrulation (Keller, 1981). The process of respreading has been characterized in detail (Keller, 1981; Hardin and Keller, 1988).

Apical constriction concentrates pigment and anything associated with the apical membrane of bottle cells; this is why the pigment darkens in bottle cells in direct proportion to their contraction. The few pigment granules in apices of endodermal cells (Fig. 7c) can be concentrated into a very small area during bottle cell formation. The same process would also concentrate anything else associated with the membrane, and some researchers have wrongly concluded that markers appearing at the site of invagination have something to do with the invagination process; in fact, the marker is associated with the apical surface over the entire embryo and is concentrated, along with the pigment, by the constriction process. It should be noted that the sides of cuboidal or columnar embryonic cells, including epithelial cells, can appear to be heavily labeled by fluorescent markers or pigment markers when viewed from the edges, when in fact the marker on these surfaces are no denser than on any other.

Bottle cells in *Xenopus* are not equivalent to some of those in other amphibians. In the urodele embryos studied thus far (Keller, 1986; Lundmark, 1986), and in other anurans (Vogt, 1929; Purcell, 1990), much of the mesoderm is initially on the surface and ingresses during gastrulation and neurulation in the form of bottle-shaped cells. These bottle cells resemble those of *Xenopus* but have the very different morphogenetic function of *ingression*, and they belong to a different germ layer, namely, the prospective mesoderm.

B. Convergence and Extension of Involuting and Noninvoluting Marginal Zones: Cell Intercalations

The capacity of the marginal zone to narrow (converge) and extend (lengthen) is a fundamental process in gastrulation and neurulation of *Xenopus* and probably other vertebrates as well (Keller, 1987; Keller *et al.*,

1991a,b). These movements have been called by a variety of other names, including stretching and elongation, but the traditional names for these movements, discovered long ago and ignored until recently, are convergence and extension (Keller et al., 1991b). These movements are regionally autonomous and involve several types of *intercalations* of cells. Cell intercalation is the movement of cells between one another in a fashion that changes the shape of the tissue (Keller et al., 1985a,b, 1991b). I describe the procedures by which these phenomena were demonstrated, since this further illustrates the organization and character of various regions of the embryo and how they can be manipulated.

1. Sandwich Explants

The traditional way of demonstrating regional autonomy is to make explants of specific regions of the gastrula to learn if their normal movements will occur in isolation from the rest of the gastrula (Spemann, 1938; Schechtman, 1942). By sandwiching the deep surfaces of two dorsal marginal zones together to make "sandwich explants," we showed that the convergence and extension processes in the dorsal sector of the gastrula were autonomous (Keller et al., 1985a,b; Keller and Danilchik, 1988). The dorsal sector of the gastrula is excised at 45° or so to either side of the dorsal midline and across the animal cap at stage 10+. This flap of tissue comprising the entire gastrula wall is peeled outward and teased away from any mesodermal cells that have begun migration upward on its inner surface (Fig. 9). In doing this, a common problem is being able to distinguish the outer superficial layer and the deep layer from the involuted mesoderm. In Fig. 10a, the white pointer marks the outer epithelium and the black pointers mark the interface between the overlying deep layer of the blastocoel roof and the involuted mesoderm. As the gastrular wall is peeled outward (Fig. 10a, white arrows), involuted mesodermal cells may adhere to the inner surface (Fig. 10b). These cells are easily identified as larger, yellowish cells on the gray background of the preinvolution material, which is either preinvolution mesoderm or the inner surface of the NIMZ–animal cap (Fig. 10c, arrows). At the lower edge, the preinvolution material must be cut away from migrating mesodermal cells that have begun involution; at stage 10 this cut will emerge at the bottle cell region (Fig. 9). Two such explants are prepared and their inner surfaces sandwiched together with edges matching (Fig. 9).

These explants show two separate regions of convergence and extension: one in the dorsal IMZ, which differentiates into a normal array of somites on both sides of a central notochord; and another in the dorsal NIMZ, which differentiates into neural tissue (Keller et al., 1985a,b; Keller and Danilchik, 1988) (Fig. 9). These movements are specific to the dorsal marginal zone and

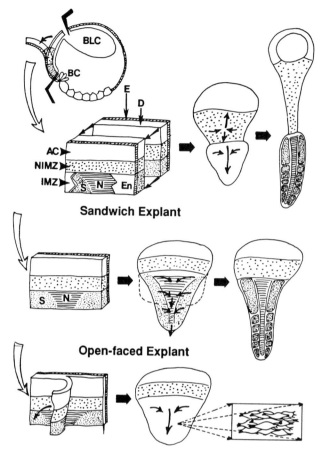

Fig. 9. A schematic diagram shows the development of sandwich, open-faced, and shaved open-faced explants. The dorsal sector of the early gastrula, depicted in sagittal section at upper left, is excised on both sides, 45–60° from the midline, at the bottle cells (BC) and near the animal pole. This explant consists of a superficial epithelial region (E) and a deep, nonepithelial (mesenchymal) region (D) and is divided into an animal cap region (AC), a noninvoluting marginal zone (NIMZ), and an involuting marginal zone (IMZ). In normal development the AC and NIMZ have a uniform tissue fate in both layers, which is neural ectoderm, with the AC contributing to the brain and the NIMZ contributing to the posterior hindbrain and spinal cord. The superficial and deep regions of the IMZ have different fates; the superficial layer forms endoderm (En) and the deep region forms mesoderm, specifically notochordal (N) and somitic (S) mesoderm. The sandwich explants show two regions of convergence and extension, one in the NIMZ and one in the IMZ (arrows). The NIMZ develops into a tangle of neurons, and the IMZ develops into notochord and somites. The open-faced explants are cultured in Danilchik's solution under coverslip conditions in which the IMZ will converge, extend, and differentiate into somites and notochord. Shaved explants are made and cultured like open-faced explants, but the innermost layers of deep cells are shaved or peeled off with an eyebrow hair knife, exposing the deep cells next to the endodermal epithelium to observation and video recording for analysis of cell behavior at high resolution (box, lower right). (Figure and legend modified from Keller et al., 1991a.)

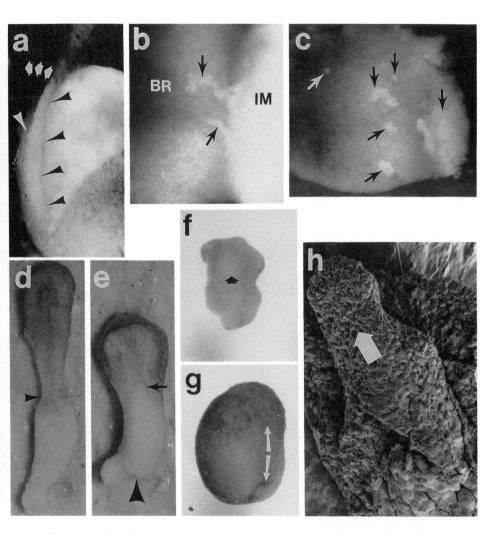

FIG. 10. (a) A sagittal dissection of the dorsal sector of the midgastrula shows the outer epithelial layer (white pointer) and the inner surface of the deep layer (dark pointers) where it bounds the involuted mesoderm. To separate the blastocoel roof from the involuting mesoderm, the animal cap is pulled outward (white arrows) and teased along the interface (black pointers) with an eyebrow hair. This process, viewed from the animal aspect (b), shows the blastocoel roof (BR) peeled off the involuted mesoderm (IM) and some of the mesodermal cells adhering to the roof (pointers, b). These cells are easily identified as lighter, larger cells on the darker background of the roof (c). (d) Normal explants show two regions, an ectodermal NIMZ (above pointer) and a mesodermal–endodermal IMZ (below pointer). (e) Large numbers of involuted mesoderm cells left on the inner surface of the explant migrate toward the animal cap and redirect the converging and extending mesoderm to extend toward the animal pole, inside the ectodermal NIMZ. The anterior extent of the invading mesoderm is marked by thickening (arrow), and the posterior end is marked by a site of involution (pointer). (f–h) Such explants are short (f) compared with the corresponding part of the neurula (g), and inside the redirected mesoderm (arrow, h) can be seen with scanning electron microscopy.

mimic the movements of the axial mesoderm and the posterior medial neural plate in normal development. The role of these movements in producing gastrulation has been discussed previously (Keller *et al.,* 1985a,b, 1991a; Keller, 1986; Wilson and Keller, 1991). Because there are two regions in the marginal zone that converge around the blastopore, we distinguished between the one that involutes (IMZ) and the one that does not (NIMZ). In these preparations, the entire explant is covered by an epithelial sheet, and thus the deep cells are protected from the external environment.

2. Use of Sandwich Explants in Analyzing Tissue Interactions

Convergence and extension of the dorsal NIMZ and its subsequent formation of neurons in sandwich explants was a surprise, since the ectoderm has only edgewise or planar interactions with the putative inducer, the mesoderm, in this configuration of explant. This observation implied that these neural-specific morphogenetic movements occured in response to a signal moving from the IMZ through the plane of the tissue to the adjacent NIMZ (Keller and Danilchik, 1988). It is clear that this is the case (Keller *et al.,* 1991b). Phillips and colleagues (Akers *et al.,* 1986; London *et al.,* 1988; Savage and Phillips, 1989) showed that early planar interaction was sufficient to repress expression of the epidermis-specific antigen Epi-1 in the prospective neural ectoderm, and there is more evidence that planar interactions have an important role in neural induction (Sater *et al.,* 1991; Chapter 18, this volume). The sandwich explant has become a popular preparation for investigation of these planar signals (Dixon and Kintner, 1989).

One of the assumptions in using a sandwich explant for analysis of planar signals is that the IMZ deep cells (mesodermal in fate) do not move up inside the explant, as they normally would have inside the gastrula, and thus provide vertical signals in addition to the planar ones. This migration is easily prevented, and the prevention can be documented in the following manner. Any migrating mesoderm that has begun migration across the inner surface of the explant is identifiable, even with the stereoscope (Fig. 10c), and must be removed (Keller and Danilchik, 1988). If so, all the converging and extending mesoderm goes in one direction and all the neural tissue in another, forming an explant with two clearly identifiable regions (Fig. 10d, pointer). However, if the explant is thick short, and not divided into two regions (Fig. 10e), it is likely that mesoderm has moved up inside the NIMZ. This can usually be verified by thickening of the NIMZ up to the point of mesodermal invasion (Fig. 10e, arrow), and often a site of involution is seen at the vegetal end (Fig. 10e, pointer). Such explants will be short (Fig. 10f), compared with control neurulae (Fig. 10g), and dissection will reveal that mesoderm left on the inner surface of the explant has succeeded in turning the vegetal end of the IMZ back on itself, inside the explant (Fig. 10h, arrow). Thus the mesoderm lies beneath the overlying NIMZ (ectoderm), producing vertical interactions that should have had only planar ones between mesoderm and ectoderm. These interactions can be prevented by cleaning off all invasive migrating mesoderm when making the explant, and the absence of migrating head mesoderm cells, or any other kind, can be shown by labeling the IMZ cells (Keller et al., 1991c).

3. Direct Analysis of Cell Behavior: Open-faced Explants

The aforementioned method is a good assay for autonomous regional behaviors and tissue interactions, but it does not allow direct analysis of deep cell motility during morphogenesis. Since it is the deep cell population that probably produces most of the forces for convergence and extension, the open-faced explant system, in which deep cells are exposed to direct view of the microscope while they carry out their morphogenetic motility functions, was developed. These open-faced explants are made by excising the dorsal marginal zone as in the case of sandwich explants (Fig. 9), placing it between a culture dish and a restraining coverslip, and culturing it in Danilchik's solution. The rationale of the culture method and solution will be described below. Under these conditions the IMZ converges, extends, and differentiates into notochord and somites (Fig. 9). The patterns of movement, paths, and gross aspects of cell behavior can be seen with epi-illuminated light microscopy, revealing that convergence and extension occur by two processes (Wilson and Keller, 1991). First, the deep cells show radial intercalation, moving between one another along the radius of the embryo to form a longer, thinner explant in the first half of gastrulation (Fig. 11, left). In the second half of gastrulation, the deep cells undergo mediolateral intercalation, moving between one another along the mediolateral axis to form a longer, narrower, and somewhat thicker array (Fig. 11, right).

The dorsal explant shows development of the entire notochord and the first six to eight somites, depending on how much lateral tissue was included in the explant (Fig. 12a–c). The prospective posterior somites are located further ventrally and are not included in this explant (Fig. 12c). Sandwich explants can be made of the entire dorsal-to-ventral marginal zone, and this preparation is useful for some purposes, particularly in the sandwich configuration (Keller and Danilchik, 1988); in the open-faced configuration, however, the distortion of tissues in this preparation is difficult to interpret.

Analysis of the development of all the notochordal and somitic mesoderm, together, is best done in explants of the dorsal region of the late gastrula or early neurula. The dorsal tissues of these embryos can be exercised and the endodermal archenteron roof removed to reveal the axial and paraxial mesoderm (Fig. 12d,e). Such explants will extend and converge (Fig. 12f,g) and

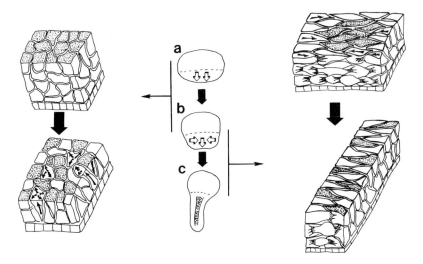

FIG. 11. Movements of an open-faced explant (center), made as described in Fig. 9, and the underlying cell behaviors in early gastrulation (top right) and late gastrulation (bottom right) are shown diagrammatically. The explant is shown with its deep surface uppermost. The IMZ (below dashed line) thins and extends without much convergence in the first half of gastrulation (arrows, a–b), and it converges and extends in the second half of gastrulation (arrows, b–c). Thinning and extension in the first half of gastrulation are accomplished by radial intercalation, in which cells move along the radius of the embryo to form fewer layers of greater area (left). Convergence and extension in the second half is accomplished by mediolateral intercalation, in which cells move among one another along the mediolateral axis to form a longer, narrower, and somewhat thicker array (right). (Modified from Keller et al., 1991a.)

normal development of the notochord and somitic mesoderm will follow, including segmentation of the somitic mesoderm and vacuolation of the notochord (Fig. 12h). The cell behaviors occurring in this preparation and their mediolateral and anteroposterior patterns of expression have been analyzed in detail (Wilson et al., 1989; Wilson, 1990). The notochord and somitic mesoderm continue to converge, extend and thicken, but they do so differently. The notochord extends throughout its length first by mediolateral intercalation to produce a longer, narrower array (Keller et al., 1989) and, after stage 21, by swelling of the individual notochord cells by osmotic inflation of vacuoles (Adams et al., 1989). It pushes posteriorly against the blastopore (Fig. 12f, straight open arrow). The somitic mesoderm elongates primarily by addition of prospective somitic mesoderm from the ventral and ventrolateral circumblastoporal regions and extension of this material by radial intercalation alongside the posterior notochord (Fig. 12f, curved open arrows). Further anteriorly, mediolateral intercalation results in

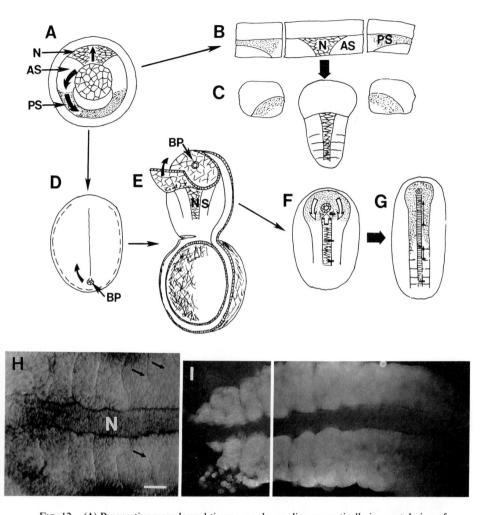

FIG. 12. (A) Prospective mesodermal tissues are shown diagrammatically in vegetal view of the early gastrula as they would appear without the overlying superficial endodermal IMZ (B, C) Open-faced explants of the dorsal sector (B) include notochord (N) and anterior somitic mesoderm (AS), which converges, extends, and differentiates (C). Isolate posterior somitic mesoderm (PS, shaded) from the ventral sector of the gastrula does none of these things. (D, E) Open-faced explants of the dorsal tissues of the late gastrula and early neurula are made by cutting along the edge of the neural plate into the periphery of the archenteron (D), folding the dorsal tissues anteriorly, and peeling off the endodermal archenteron roof (E) to reveal the underlying notochord (N) and somitic mesoderm (S). (F, G) With further development the notochords of such explants converge and extend posteriorly (straight arrow) by mediolateral intercalation of cells (Keller et al., 1989), pushing the blastopore (BP) posteriorly. At the same time, the posterior somitic mesoderm (shaded) swings around both sides of the posterior notochord (curved arrows) and begins a sequence of specific behaviors and anteroposterior zones, beginning with radial intercalation posteriorly, mediolateral intercalation further anteriorly, and segmentation anteriorly (Wilson et al., 1989). (H) A light micrograph made from a video recording shows the notochord (N) and segmentation in such an explant (arrows). (I) Somitic mesoderm stained with monoclonal antibody 12–101. The arrows and bars indicate the shear of somitic mesoderm posteriorly (see text). Bar equals 100 μm. [(A)–(G) are modified from Keller et al. 1991b; H is from Wilson et al., 1989.]

convergence and thickening rather than extension, which results in the buttresses of somitic mesoderm forming on the ventral sides of the neural folds (Schroeder, 1970). The somitic mesoderm then undergoes a sequence of cell behaviors that bring about segmentation (Fig. 12h) and rotation of the myotome cells (Wilson et al., 1989). Because the notochord extends along its full length and the somitic mesoderm only posteriorly, the notochord shears posteriorly with respect to the somitic mesoderm (Keller et al., 1989; Wilson et al., 1989) (Fig. 12g, arrows). The notochord and overlying neural plate become progressively more tightly attached to each other after stage 11.5 and soon do not move with respect to one another. Thus, the neural plate moves along with the notochord, shearing posteriorly with respect to the somitic mesoderm (Keller, 1976).

High-resolution videomicroscopy reveals a characteristic sequence of behaviors in the notochord (Keller et al., 1989), the somitic mesoderm (Wilson et al., 1989), and their precursors in the gastrula stage (Wilson and Keller, 1991; Shih and Keller, 1991b,c) which provided very fine markers of spatial and temporal patterning in both the anteroposterior and mediolateral direction. However, these high-resolution markers require videomicroscopy and image processing in time-lapse mode. In the absence of these methods, the somitic mesoderm can be identified and some anteroposterior progression of development revealed by staining with a somitic mesoderm–specific monoclonal antibody 12-101 (Kintner and Brockes, 1984) (Fig. 12i) (the method is described in Appendix A). The posterior end of the mesoderm will show a graded decrease in staining, provided the explant contains somitic mesoderm in this period of its differentiation. For the full complement of anteroposterior zones of cell behavior, see Wilson et al. (1989).

It is a common misconception that the ventral side of the gastrula contains no dorsal tissues. In fact, the prospective posterior somitic mesoderm lies in this region, and the anteroposterior axis of the somitic mesoderm curves around from dorsal to ventral in the IMZ (Fig. 12a, curved arrows) (Keller et al., 1991b).

4. DIRECT ANALYSIS OF CELL BEHAVIOR: "SHAVED" EXPLANTS

Fine protusive activity cannot be seen by epi-illumination of the open-faced explant, and the most important protusive activity is that of the outermost layer of deep cells, those next to the overlying epithelium, since at least some aspects of mediolateral intercalation are organized by this epithelial layer (Shih and Keller, 1991a). To resolve details of outer deep-cell activity, the open-faced explant was made, but the inner layers of deep cells were shaved off with an eyebrow hair knife (Fig. 9), making a thin, flat explant that allows better resolution of cell behavior under epi-illumination (Fig. 13a). To resolve fine protrusive activity, cells are obtained from embryos labeled with fluorescein dextran amine (FDX) (see Chapter 15, this volume), seeded into the corresponding regions of an unlabeled explant, and their behavior recorded with low-light, fluorescent microscopy by the method described in Section VII (Fig. 13b). Alternatively, cells in the unlabeled explant can be labeled with the fluorescent lipid dye, DiI (Molecular Probes, Eugene, OR), which intercalates into the plasma membrane and brightly labels individual cells (Keller *et al.*, 1989). These methods have been used to record and trace patterns of cell intercalation (Fig. 13c), lineage relationships, division patterns, and the protrusive activity thought to drive mediolateral intercalation of dorsal mesoderm of the gastrula (Shih and Keller, 1991b) and of notochord cells in the neurula (Keller *et al.*, 1989, 1991b).

Using these methods it is now possible to do precise, direct, and detailed analyses of motile activities that are specific to regional morphogenetic tissue movements in an explant system useful for experimental manipulation (Keller *et al.*, 1991b). Moreover, the spatial and temporal progressions of specific cell behaviors revealed by these methods can be seen very early, are highly organized, and are very detailed and fine-grained reflections of patterning. Thus, they should be useful as markers in tissue interaction and pattern formation studies (Wilson *et al.*, 1989; Keller *et al.*, 1989, 1991b; Shih and Keller, 1991b,c).

5. Time-Lapse Recordings of Surface Cells

Time-lapse recordings can be made with film or video recording methods, using a compound microscope and epi-illumination as described in Section VII, to reveal behavior of superficial epithelial cells in whole embryos (Keller, 1978). This is somewhat easier than looking at deep-cell behavior, since, in most cases, it is not necessary to make explants. Using these methods, it was shown that the superficial epithelial cells accommodate convergence and extension by mediolateral intercalation (Keller, 1978). Such time-lapse records allow direct tracing of cell lineages and quantitative analysis of changes in cell shape, size, number, division, rearrangements, and paths of movement (Keller, 1978, 1981; Hardin and Keller, 1988).

6. Tracing of Deep-Cell Intercalation by Cell Labeling

Movements of deep cells or superficial cells can be traced in whole embryos or in explants by grafting a patch of cells from an embryo labeled with FDX to an unlabeled embryo and fixing and sectioning the embryo or clearing whole mounts, using methods described below, to learn where the labeled cells have gone. Mediolateral intercalation of deep cells was first demonstrated with this method. A coherent block of labeled deep cells, grafted into an unlabeled embryo, intercalate with the unlabeled cells on both sides during convergence and extension of the dorsal marginal zone (Keller and Tibbetts, 1989) (Fig. 13d). However, this technique is indirect and requires some care in interpretation (see Fig. 1 of Keller and Tibbetts, 1989). It is not a sensitive method when an unlabeled population is abutted on a labeled population along the axis of extension; doubling the length of the explant involves mediolateral intercalation by only one cell diameter, and thus the labeled–unlabeled boundary will, at best, show each cell type crossing into the other's territory by an average of a cell diameter. This is at once a testament to the efficiency of the mediolateral intercalation process in producing extension and to the insensitivity of the method. Second, convergence and extension tend to straighten irregularities along the axis of extension. Thus if one uses large blocks of labeled cells and does not look carefully with sufficient high resolution to see individual cells clearly, the boundaries of labeled and unlabeled patches can appear smoother after intercalation and extension than in nonintercalating regions.

7. Movements of Neurulation

Gastrulation grades directly into neurulation. The posterior neural plate. converges dorsally and extends, complementing these movements in the IMZ and resulting in blastopore closure (Keller and Danilchik, 1988). The convergence and extension of the posterior neural plate is autonomous in sandwich explants (Figs. 9 and 10d) and depends only on planar contact with the mesoderm (Keller and Danilchik, 1988; Keller *et al.*, 1991c). These movements do not require the mesoderm to be underneath in vertical contact with the neural ectoderm as in normal development (Fig. 5). Columnarization of the lateral neural plate and thinning and wedging of the floor plate at the midline do not occur in sandwich explants and thus must involve vertical signals (Keller and Danilchik, 1988). Columnarization of the lateral regions

---

FIG. 13. Methods of documenting the cell behavior during mediolateral intercalation in open-faced shaved explants, made as described in Fig. 9, are shown. (a) *Xenopus* deep mesodermal cells are seen under epi-illumination and video-image processing. (b) Details of protrusive activity, including the finest filiform and lamelliform protrusions, are visualized by low-light fluorescence microscopy of fluorescein dextran-labeled cells grafted into an unlabeled explant and imaged with a silicon-intensified target camera and image processor. (c) Mediolateral cell intercalation is documented from tracings of cells in video recordings over 6.5 hours of extension of the explant to the left (arrow). (d) A coherent block of labeled prospective notochord cells intercalated with unlabeled ones on both sides during extension of the notochord in gastrulation and neurulation. [(a) and (b) are provided by John Shih; (c) is from Keller *et al.*, 1991a; and (d) is from Keller and Tibbets, 1989.]

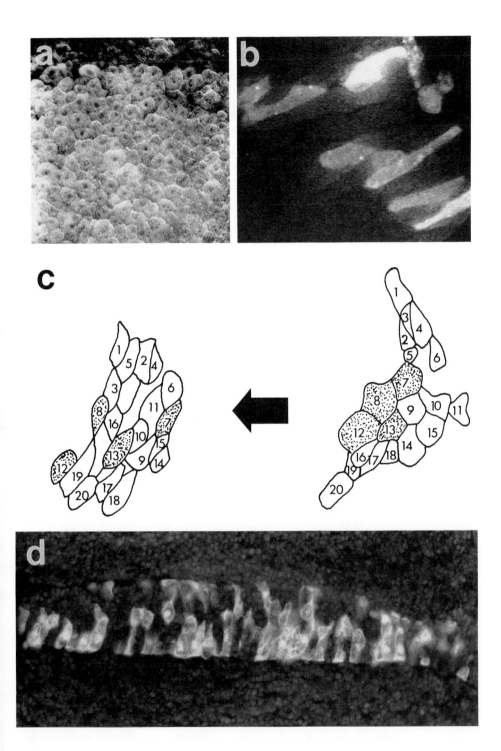

of the plate is probably due to interaction with somitic mesoderm and thinning of the floor plate owing to interactions with notochord (Holtfreter and Hamburger, 1955), a finding that holds for other vertebrates as well (Smith and Schoenwolf, 1989; Jessell et al., 1989; see Chapter 18, this volume). Thus, the convergence and extension movements and the thickening and rolling of the neural plate are independent, the first being regulated by planar interactions with the mesoderm and the last by vertical interactions with the mesoderm. Changes in cell shape involved in the process of thickening the neural plate in *Xenopus* and rolling it into a tube have been described in detail by Schroeder (1970).

Labeling deep-cell populations of the dorsal NIMZ and following the cells as described above show that they undergo sequential radial and mediolateral intercalations, as in the IMZ. Radial intercalation first thins the NIMZ without much convergence and then mediolateral intercalation produces convergence and extension, as in the IMZ (Keller et al., 1991c). Superficial cells of the NIMZ intercalate mediolaterally during convergence and extension (Keller, 1978). Thus converging and extending neural and mesodermal tissues show similar underlying cell behaviors, and these behaviors occur in the same sequence with qualitatively the same effects on the tissue. It is not known whether the protrusive activity of neural intercalating cells is the same as the mesodermal ones. It is possible that neural and mesodermal tissue share a fundamental cell behavior despite following different paths of cell differentiation.

C. Migrating Mesodermal Cells: Analysis of Individual and Aggregate Mesodermal Cell Behavior

1. Migrating Mesodermal Cells

The cells lying beneath the forming bottle cells constitute a special population of mesodermal cells, the migrating mesoderm. As the bottle cells form, they begin migration animally, along the inner surface of the blastocoel (Figs. 5 and Fig. 14). The inner surface of the blastocoel roof has a fibrillar matrix, not as highly developed as in urodele embryos, but substantial and containing fibronectin (Nakatsuji and Johnson, 1983a). The fibronectin in this matrix is synthesized by many cells (Lee et al., 1984) and appears to be localized to the blastocoel roof by an RGD-sensitive process, most likely binding by the integrin fibronectin receptor (Keller and Winklbauer, 1990; DeSimone et al., 1991). Adhesion and migration of mesodermal cells on the roof of the blastocoel do not require interaction with fibronectin, but if allowed to interact with fibronectin, the cells move with greater persistence

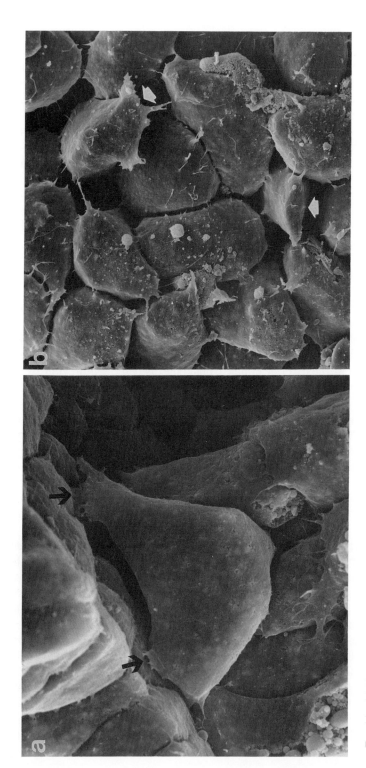

FIG. 14. (a) The migrating mesodermal cells are attached to the overlying roof of the blastocoel by protrusions that may be involved in their migration on this substrate (arrows). (b) Viewed from the outer surface, these mesodermal cells bear large flattened areas bounded by protrusions where they were formerly attached to the roof of the blastocoel (pointers). (From Keller and Schoenwolf, 1977.)

(Winklbauer, 1990). The role of fibronectin in this system seems to be to impart a more persistent motile activity to the cells. Individual migrating mesodermal cells move randomly on this matrix, but aggregates of mesodermal cells move directionally toward the animal pole (Winklbauer et al., 1991). The leading-edge mesoderm cells need the overlying blastocoel roof and its matrix to spread properly and differentiate, but migration on the roof of the blastocoel does not generate much of the total force responsible for gastrulation. Physical removal of the entire blastocoel roof will not stop most of gastrulation in *Xenopus*. Blastopore closure, convergence and extension, and development of the axial mesoderm occur very well; only the leading-edge mesoderm is affected in that it has no substratum to spread on (Keller et al., 1985a,b; Holtfreter, 1933).

2. Methods of Isolation, Identification, and Culture

Winklbauer developed methods to analyze the behavior of individual mesodermal cells and aggregates of mesodermal cells of specific type, culturing them as individuals and as aggregates on the roof of the blastocoel or on fibronectin in culture dishes (Winklbauer, 1990). The wall of the gastrula is peeled down at the early or midgastrula stage, as described previously for making explants (Figs. 9 and 10b,c), and laid flat (Fig. 15a,b). Along the vegetal–animal extent of this tissue, one can isolate regions on the basis of mesoderm cell type, including the migrating mesoderm, converging and extending mesoderm, cells from the posterior converging and extending region of the neural ectoderm, and cells from the anterior prospective brain region. One can also isolate tissue on the basis of whether it has involuted or not, at any stage, either by cutting the two regions apart at the blastoporal lip or by keeping track of the point of involution by making tick marks at the margins of the explant with a needle and then cutting the regions apart once they are laid out flat.

To make aggregates, the desired mesoderm is cut out with an eyebrow knife and cultured in one of several ways. To isolate individual cells, the appropriate region is excised and the cells dissociated in Winklbauer's dissociation buffer (Winklbauer, 1988, 1990; the recipe is given in Appendix A). The entire aggregate can be explanted onto a culture dish (Fig. 15c,d) or returned to the inner surface of the blastocoel roof and cultured as an open-faced explant. For analysis of individual cells, the desired region is excised, placed in dissociation buffer, and allowed to dissociate until loosened, up to 50 minutes. The loosened cells are pipetted into a culture dish (Fig. 15e) or on the inner surface of a blastocoel roof, and the entire preparation is treated as an open-faced explant (Fig. 15f). Using these methods, Winlkbauer and associates have shown the axial mesoderm, which converges and extends, and migrating mesoderm, which crawls on the blastocoel roof, have different motile

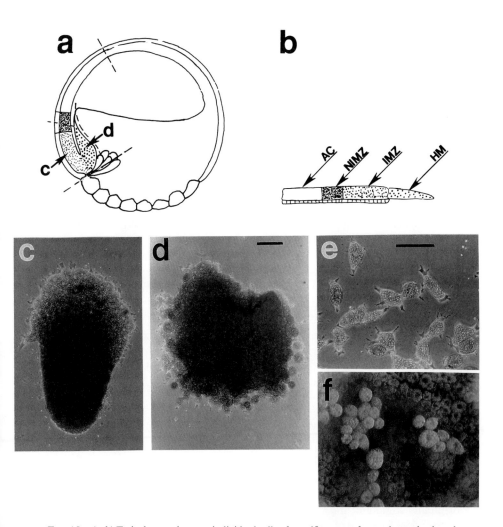

FIG. 15. (a, b) To isolate explants or individual cells of specific types of mesoderm, the dorsal tissues are cut away from surrounding tissues (dashed lines) and peeled outward (a), as described in Fig. 10, separating what has involuted from what has not, and the tissue is laid out flat, as in preparation for making a sandwich explant (b). In the early gastrula, the entire prospective anteroposterior axis can be laid out flat (b) and the cell type chosen by position, including the migrating mesoderm, located anterior to the bottle cells [in the dorsal sector this is prospective head mesoderm (HM)]; converging and extending mesoderm (IMZ), located within 12–15 cell diameters of the bottle cells; dorsal NIMZ that will converge and extend as posterior neural tissue (NIMZ), located another 5–7 cell diameters above the IMZ; and the prospective brain region in the animal cap (AC). (c, d) Large explants of preinvolution tissue (c) and postinvolution tissue (d) from the early gastrula (a) are shown after 3 hours of culture on fibronectin (100 μg/ml). The preinvolution explant consists of prospective posterior neural tissue at its trailing end (bottom) and prospective converging and extending mesoderm at its anterior end (top). (e, f) Dissociated prospective head mesodermal cells are shown after culture on fibronectin (100 μg/ml) for 1 hour (e) and reseeded onto the inner surface of an explanted blastocoel roof, in this case in the presence of 1 mg/ml of GRGDSP peptide (f). Dissection procedures are based on techniques developed by Rudolf Winklbauer in the author's laboratory (see Winklbauer, 1990; Winklbauer et al., 1991). Bars equal 100 μm. [(a) and (c–f) are modified from Winklbauer, 1990.]

behaviors with respect to fibronectin (Fig. 15c,d). This matrix may function in controlling the protrusive activity of mesoderm, in culture, on fibronectin and on the blastocoel roof (Winklbauer, 1990, Winklbauer *et al.*, 1991).

3. CONDITIONING SUBSTRATA

Behavior of isolated mesodermal cells or explants can be analyzed on the blastocoel roof directly, on purified matrix such as fibronectin or collagen (Nakatsuji and Johnson, 1983b; Winklbauer, 1988, 1990), or on substrata conditioned with extracellular matrix found on the inner surface of the blastocoel (Nakatsuji and Johnson, 1983c; Shi *et al.* 1989; Winklbauer *et al.*, 1991). An explant of the roof is made and pressed against an appropriate culture dish or coverslip for a specific time. It should be remembered that plastic culture dishes, petri dishes, and glass and plastic coverslips from different manufacturers, and sometimes between batches from one manufacturer, differ in binding extracellular matrix, and thus results using different dishes may not be the same. Coverslip differences can be alleviated to some degree by a uniform washing procedure. Evidence from substratum conditioning experiments argue that the matrix deposited by the animal cap directs mesodermal cell migration (Nakatsuji and Johnson, 1983b) and explant migration (Shi *et al.*, 1989) toward the animal cap in urodeles and in *Xenopus* (Winklbauer *et al.*, 1991).

D. Epiboly of the Animal Cap

The animal cap region continues the epiboly that began during cleavage into gastrulation and spreads uniformly in all directions (Keller, 1978, 1980). This movement occurs by radial intercalation of deep cells to form fewer layers of greater area, and the cells of the superficial epithelium divide and spread (Keller, 1980). (Fig. 16). It is not certain that the epiboly of the animal cap is

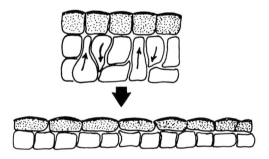

FIG. 16. Epiboly of the animal cap is illustrated schematically. The superficial epithelial cells (shaded) divide and flatten, and the deep cells (unshaded) undergo radial intercalation to form fewer layers of greater area.

autonomous, though some evidence suggests that it might be. The fact that isolated animal cap becomes corrugated has been taken as evidence that it continues epiboly independent of an intact, pressurized blastocoel and attachment to the rest of the gastrula (Spemann, 1938; Keller, 1986). However, the corrugation may be due to thickening, and no one, to my knowledge, has ever documented spreading in these circumstances by the appropriate time-lapse recordings.

V. Experimental Design and Context-Dependent Function of Cells and Tissues

Although most of this chapter is directed at practical aspects of experimentation on *Xenopus* cells, it should be noted that enough is now known about the nature of mass movements of tissues and their cellular basis to sharpen experimental designs and expectations when probing the behavior of cells and tissues isolated from the gastrula. Much of the behavior and function of cells and tissues in the gastrula is context dependent. A good example is bottle cell formation, where both the details of how the cells behave and their probable function are dependent on mechanical interactions in the embryo (Hardin and Keller, 1988). Likewise, the mediolateral intercalation behavior of deep cells involves a movement of cells transverse to the direction of ultimate cell displacement. In this case, mediolaterally directed protrusive activity is transduced by the geometric and mechanical properties of the cell population to produce force for extension in the perpendicular direction (Keller *et al.*, 1991b). Moreover, this behavior is controlled by complex tissue interactions (Keller *et al.*, 1991b). In the case of mesodermal cell migration, cells in large aggregates display properties that differ from those of individual cells in isolation (Winklbauer *et al.*, 1991). The consequences of vacuolation and swelling of notochord cells are dependent on the mechanical properties of the notochord sheath and the arrangement of cells in the notochord (Koehl *et al.*, 1990).

In all these examples (and there will be more to follow), disassembly of the system results in the loss of mechanical and chemical constraints on the cells, and their behavior consequently is altered, usually in the direction of being less specific and more general. Moreover, their behavior usually makes sense only in terms of a geometric and biomechanical context within the tissue. Fortunately in *Xenopus*, researchers now have methods of analyzing cell behavior and function in whole tissues, and by progressively breaking down the organization of the tissue into simpler systems—and then reassembling it progressively, noting at each level what new properties are lost or emerge—they will be able to resolve the mechanisms of morphogenesis.

VI. Microsurgical Methods, Tools, and Manipulations

Various guides have been published on microsurgery of amphibians, including Hamburger (1960), Rugh (1962), and Jacobson (1967). Here I describe a few tools and methods that may be particularly useful for *Xenopus* embryos. *Xenopus* embryos are different from those of most other amphibians in popular use. They do not cut well. They are stickier and offer more mechanical resistance than the common urodele embryos, such as *Ambystoma mexicanum* (Mexican axolotl) and *Taricha torosa* (California newt), and other anurans, such as *Ceratophrys ornata* and *Bombina orientalis*. However, they are mechanically hardy and survive control manipulations as well or, in most cases, better than any other embryo, enabling one to perform experiments not possible in other animals.

A. Tools

The best instrument for cutting blastula, gastrula, and neurula embryos is the eyebrow hair knife. This tool is made by pulling out eyebrow hairs and choosing ones uniform of curvature and sharp of point. These hairs are then inserted into the tapered end of a disposable pipette that has been pulled out to a small diameter and cut off evenly by scoring with a diamond pencil. The hair is held in place by immersing the end of the pipette in melted wax and allowing the wax to cool. The excess wax on the eyebrow hair is removed by heating a forceps handle, laying a Kimwipe over it, and briefly putting the hair on top, melting the wax and wicking it into the paper. Other cutting tools include tungsten knives and needles, made from wire sharpened by oxidation in a flame, in a nitrate bath, or by electrolysis (Dossel, 1958). Glass needles can also be made by pulling out fine filaments off a pipette or rod and then pulling a lateral extension from this fine rod (Hamburger, 1960; Rugh, 1962). In all cases, the cutting end should be about 120° off the axis of the handle for maximum versatility. The tip should be sharp and the shaft tapered and thus inflexible.

For older tissues, which offer more resistance, glass needles and tungsten needles offer the sharpness and stiffness needed to cut cleanly, the former having the disadvantage of low visibility and the advantage of not being sticky, and the latter having the reverse of these properties. For neurula, and earlier stages as well, small diamond knives are particularly effective in making clean cuts of small blocks of tissue. These instruments are very sharp and can be made in a variety of shapes (L.A.B. Instruments, Carson City, NV).

For moving tissue about, a hair loop should be made in the same fashion as eyebrow hair knives. Both ends of a piece of baby hair are inserted into the same kind of pipette, until the loop is of the desired size, and fastened in place using the wax method. A long hair is used since friction on the sides of the pipette is greater in this case and thus the loop can be made very small. Transfer pipette can be made by pulling a disposable pipette to the desired diameter and then bending it to 120°. An inside diameter three or four times that of the cells is useful for precise and quick handling of cells.

For transfer of small explants and cells, a Spemann pipette is the instrument of choice. It is made by sealing the end of a pipette in a flame and then heating a small spot on one side near the taper, until soft; the soft glass is blown out, making a hole. After fire-polishing of the hole, the end of the pipette is pulled out to the desired diameter, length, and angle. A pipette bulb is placed on the large end and a soft rubber tube over the hole. Crude control is done with the pipette bulb and fine control is done with the rubber membrane over the hole.

The best forceps to use in removing membranes and manipulating tissue is No. 5 Dumont watchmaker's forceps. They come in tool steel, which takes a fine edge but corrodes in saline; stainless steel, which does not take an edge as easily but is corrosion proof; and titanium, which is lighter and corrosion proof and takes a decent edge, although it is not as strong as steel. For removal of vitelline envelopes, as described in Section VI.B.1, one pair should be blunt but meeting precisely. The other pair should have sharply tapered points that meet exactly, sharpened on an "Arkansas" hard sharpening stone.

Tissues will stick to instruments or dishes, usually when not desired. Several methods can be used to reduce sticking. Coating with a silicon compound, such as Sigmacote, which is in a heptane solution (Sigma Chemical Co., St. Louis, MO), and others, will reduce sticking. One of the easiest and most effective methods is to have purified bovine serum albumin (BSA), which coats surfaces and reduces nonspecific sticking, in the medium. The plastic poly HEMA has also been used (Gimlich and Gerhart, 1984).

B. Methods

1. REMOVAL OF JELLY COATS AND VITELLINE ENVELOPE

A cysteine solution is used to dejelly the embryos, as described in Appendix A. Trypsin can be used to remove the vitelline membrane (Okamoto, 1972), but perhaps with some damage to the embryo. Mechanical removal is done by grabbing the vitelline envelope with forceps having broad, squared-off tips that meet precisely. Usually the envelope cannot be seen, but it can be grasped by bringing the forceps points together while pressing them perpendicularly against the embryo. Next to the point of grasp, a second pair of pointed forceps is used to grasp and quickly tear a hole in the membrane. The torn membrane is discarded some distance away since it can interfere with subsequent work. This is difficult, and work will go faster if the animal cap is punctured slightly when grabbing the envelope with the second forceps. This puncture has little effect, since the hole heals quickly if it is not too large.

2. GRAFTING

Grafting of embryonic tissues has been described previously (Hamburger, 1960; Rugh, 1962; Jacobson, 1967). Thus, I discuss what may aid in working with *Xenopus* in particular. Operations on *Xenopus* embryos are easiest to do on a substratum to constrain and hold the embryo or its parts in various positions. A 2-4% (w/v) agarose bed in a plastic petri dish, made with the solution used for the culture of embryos, serves this purpose. Holes of the size of the embryo can be melted in this substratum prior to filling with solution by heating a glass ball smaller than the hole desired in an alcohol flame, applying it to the agarose surface, and melting a hole. Beds made of various waxes can be used in the same way and may be darkened with lampblack to improve contrast (Rugh, 1962). I have found colored clay, made for use as childrens' toys and thus having low toxicity, to be the most useful substratum. (One of the best currently available products is from American Art Clay Co., Indianapolis, IN.) Dark colors are better than light ones as they cut down on glare and improve contrast. All should be tested for toxicity by culturing embryos overnight in a dish lined with clay. In any case, it is best not to leave embryos on clays longer than necessary for healing. It should be kept in mind that the reflectivity of the clay will change exposures on automatic cameras.

Using the clay substrata, embryos can be positioned and grafts made with great precision. A hemispherical depression is made in the clay with a ball-tip pen and the embryo placed in the depression. Small points of clay can then be moved against the embryo, preventing its rolling one way or the other without undue pressure or interference with movements. Any side can be placed uppermost with this method of retention. A hole is cut to receive the graft, or the graft is cut out of another embryo first and the hole to receive it made in the second, depending on which operation can be done the fastest. The slower operation is always done first. After the graft site and the graft are prepared, the latter is placed in the former, and the edges are checked for matching. The graft site will initially get larger, as the edges retract, and then smaller, as healing begins, whereas the graft will round up, usually getting progressively smaller in outline. After the graft is in place, forceps are used to pull up a peak of clay on either side of the embryo. A small coverslip fragment, about $6-8 \times 4$ mm, is cut with a diamond pencil and placed across the peaks of clay with forceps. The coverslip is tapped down on both ends until it presses the graft in place, and care is taken to adjust the angle of the coverslip to press directly normal to the surface of the graft to keep it from slipping sideways.

Large early blastomeres have been successfully grafted from one embryo to another (Gimlich and Gerhart, 1984), but they require special handling (R. Gimlich, 1984, personal communication). The blastomeres are loosened with low-calcium medium and then teased apart with an eyebrow hair knife and slickened with poly HEMA. Needless to say, one does not want to puncture the blastomere of interest. The blastomeres are recombined in the desired configuration and the embryo held in a round hold melted in an agarose bed. It is important to provide the spherical constraint normally provided by the vitelline envelope.

One problem in healing of grafts is often due to the wound being coincident in both the superficial and deep layers. If this is the case, and the graft does not fit the site very well, the free edges of the epithelia will migrate down into the wound and then finally meet at the bottom, much too late for the purposes of some experiments. The way to avoid this is to make slightly overlapping joints. In making the graft, a cut is made through the epithelium about three cells outside of the desired graft size, using the tip of the eyebrow hair knife, and the edge of the epithelium is lifted back across the graft about three cell diameters. After all sides of the epithelium are cut, the deep layer is cut by placing the hair back under the epithelium and pressing down, cutting the deep layer beneath the epithelium. The hole to receive the graft is made in the complementary way. After the graft is inserted, the result is nonoverlapping boundaries in deep and superficial layers and much better healing. This technique applies as well to combining tissues in explants and is used and illustrated in detail in Keller *et al.* (1991c).

In making very small grafts of deep and superficial layers near the early gastrula stage, the epithelium will tend to come off the deep layer entirely, if upward strokes of the eyebrow hair are used. Under these circumstances, it is best to use short sewing-machine strokes of the hair tip to saw downward through the epithelium, or else the epithelium will be lifted completely off the deep region and become unmanageable.

Orientation of grafts is aided somewhat by sloppiness. If the graft is irregular in outline, its orientation is easy to remember when the much-desired reinsertion occurs some minutes later. However, the graft can often be made to stick to the tip of an eyebrow hair by one edge and transported to its new home without falling, obviating the need to remember which end was which.

C. Explants

The making of explants has been described diagrammatically in this chapter, but several practical tips are in order. First, if sandwich explants of the marginal zone are to be used in tissue interactions or in analysis of convergence and extension, it is important to remove the involuted cells. Second, speed is important. If one hesitates, both halves of the sandwich will curl concavely at their deep surfaces, making them tend to turn each other crosswise. Third, they are best made on clay substrates and held together by moving a coverslip, borne on peaks of clay, down normal to their surface to hold them together, as described in the section on grafts.

Open-faced explants are made and cultured several ways. An explant is made by placing it deep surface down on the bottom of a plastic culture dish, which is flooded to above the brim with solution. The explant is held in place by putting a small amount of silicone high-vacuum grease on either end of a coverslip fragment and pressing the fragment down gently on top of the explant. The bottom or top of a larger culture dish or glass plate is then placed across the top of the dish, squeezing out the excess fluid. The whole preparation is righted and placed under a standard microscope. If use with an inverted microscope is desired, the explant is cultured in the same way but less solution is used and the lid of the culture dish is used in the normal way. In cases demanding use of high-resolution (large numerical aperture) objectives, a hole about 1 cm in diameter is made in the bottom of the plastic dish by heating a cork borer and pressing it through the plastic. The edges are trimmed smooth and a coverslip (No. 1.5, 0.17 mm thick) is glued across the hole with silicone grease.

D. Solutions and Healing

Grafting, isolating, and manipulating embryos, explants, and cells require use of solutions that will support health and normal development. Holtfreter (1943) worked out the salt concentration (about 120 mOsM; 0.385% salt concentration) in his *standard solution*, now called Holtfreter's solution, on the basis of cell health, motility, and lack of osmotically induced changes in cell volume. Many solutions used for early embryonic work hover around the composition of this solution, mainly differing in buffers, but others vary widely, based on what individual investigators thought to be experimentally useful. Recipes of the common solutions used for *Xenopus* are given in Appendix A.

Holtfreter learned that the epithelial layer (referred to in his work as the layer bearing a syncytial "surface coat") is destabilized by high pH, low calcium, and high salinity. Normally it has dilute saline (pond or stream water) on the outside and higher salinity inside. By contrast, deep cells thrive at high pH, higher calcium, and higher salinity. In low pH, low salinity, and high calcium, deep cells can form an epithelium. This is also true of *Xenopus* (R. Keller, 1979, unpublished data). High salinity causes the epithelium to weaken and even break down. Indeed, high salt concentrations break down the epithelial integrity, resulting in abnormal ingression of surface cells, thickening of the blastula or gastrula wall, and exogastrulation. Thus, all embryos should be held and reared in dilute solutions once healing is accomplished (1/10 to 1/3 strength of normal solutions). On the other hand, the healing following operations generally is best accomplished in full-strength saline. Because of the different requirements of deep and superficial cells, not all cells will be completely suited when the embryo is cut open, no matter what solution is used.

To support normal deep cell movement in an explant without a covering epithelium, we developed Danilchik's solution (Keller et al., 1985a,b), which is based on the measurements of ions in the blastocoel fluid of *Xenopus* (Gillespie, 1983). It has a high pH (8.3) and relatively low calcium and low chloride concentrations, of which the latter appears to be toxic at the higher levels normally found in amphibian media. We have since modified this medium several times (Wilson et al., 1989; Wilson and Keller, 1991). Two versions are currently used in this laboratory, one modified by Shih (Shih and Keller, 1991a,b) and one modified by Sater (see Appendix A for compositions), both of which are more stable and allow better development and morphogenesis. Shih-modified Danilchik's medium contains 1% BSA, which makes all surfaces less sticky.

Dissociation of cells at the gastrula stage can be done in calcium- and magnesium-free solutions. Several solutions are described in Appendix A. Some have cation chelators and others do not. Winklbauer's dissociation medium is one that is particularly good for gastrula cells, allowing fast recovery of normal behavior. At the neurula stage, proteases become necessary for separation of tissue boundaries and cells, depending on the cell type and tissue of interest. Collagenase is useful for removing the endodermal epithelium of the archenteron roof in the neurula stage (Wilson et al., 1989). Sometimes cocktails of several proteases are useful in separating tissue boundaries while permitting adhesions between individual cells within tissues (Adams et al., 1990). Most protocols for separating tissues or cells enzymatically are based on experience, and investigators settle on what works. There are several things to be kept in mind. The degree of purification and biological source of the enzyme and even the lot number may make a difference; sometimes it is not the primary enzyme in the preparation but a contaminating one that has the desired effect. Varying the concentration of specific ions may be necessary to balance the activity of one enzyme against another. It is often not useful to dissociate cells from one another, but only tissue masses from one another, and certain proteases can be useful for this (Keller and Spieth, 1984; Adams et al., 1990).

E. Sterilization and Antibiotics

Procedures for sterilizing instruments have been described in detail (Jacobson, 1967) and are not reviewed here. All solutions contain 5 ml of antibiotic solution (10,000 units penicillin, 10 mg streptomycin, 25 μg amphotericin B, 25 mg gentamicin/ml) per liter, or 20 mg oxytetracyline per liter, which degrades rapidly and should be added before use. Solutions are sterilized by filtration, and some containing BSA, such as Danilchik's, are frozen in 50-ml aliquots.

VII. Microscopy, Image Processing, and Recording: Epi-illumination and Low-Light Fluorescence Microscopy

A. Epi-illumination of Explanted Cells and Tissues

The easiest way to track large populations of cells during development of whole embryos or explants is by epi-illuminated light microscopy. In this method, the explant or embryo is illuminated from the side with a high-intensity fiber optic light, usually at low angles (Figs. 7d and 13a), which casts shadows at the boundaries of cells. Experimentation with individual culture chambers and lighting is required to maximize contrast. This method is further improved with frame-averaging and shading correction, which are common features on image processors (see below).

B. Use of Fluorescein Dextran Amine and DiI for Low-Light Fluorescence Microscopy

We have used two methods for visualizing the delicate protrusive activity of cells in explants. The first consists of injecting the embryo at the one-cell stage with fluorescein dextran amine, as described by Gimlich (Chapter 15, this volume), and allowing it to develop to the appropriate stage. Then, individual cells or groups of cells are grafted from a labeled embryo to an unlabeled one, or from a labeled embryo to an explant. The number of cells and the shape of the graft are varied depending on what process is being studied (Keller and Tibbetts, 1989; Shih and Keller, 1991b; Keller *et al.,* 1991a). The labeled cells are then visualized using a fluorescence microscope. Alternatively, the explant is made and individual cells labeled with DiI (Molecular Probes) (Honig and Hume, 1986), a lipid fluorescent compound that is applied by drying it onto a glass fiber and then applying that to the surfaces of the cells (Keller *et al.,* 1989). These methods are extremely powerful and have been developed by a number of investigators, including Harris *et al.* (1987), for use with materials at later stages of development.

C. Microscopy, Image Processing, and Recording

With DiI- or FDA-labeled cells, the region of the explant is found by epi-illumination and then low-level fluorescence illumination is provided, either by a variable-intensity quartz halogen illuminator or by a mercury vapor lamp with illumination reduced by neutral density filters. The image is formed by a silicon-intensified target (SIT) camera and processed via an image processor. We use Image I software run on an IBM AT computer (Universal Imaging Corporation, Media, PA). The processing consists of averaging about 8 to 16 frames and contrast control, and sometimes more extensive processing (Keller *et al.,* 1989; Shih and Keller, 1991b). Images are stored on an optical disk recorder (Panasonic, high-resolution black and white) and copied for publication and illustration by photographing the screen with a Nikon macro lens and 35-mm camera, using Pan X film, or by using a video printer. For an extended and excellent discussion of video microscopy, including use of video cameras, monitors, recording devices, and image processing, refer to the work of Inoue (1986).

Acknowledgments

I thank Amy Sater for valuable comments on the manuscript and John Shih, Paul Wilson, Amy Sater, Rudi Winklbauer, and Mike Danilchik for advice and materials.

References

Adams, D., Keller, R., and Koehl, M. A. R. (1990). The mechanics of notochord elongation, straightening, and stiffening in the embryo of *Xenopus laevis. Development (Cambridge, U.K.)* **110,** 115–130.

Aker, R. M., Phillips, C. R., and Wessels, N. K. (1986). Expression of an epidermal antigen used to study tissue induction in the early *Xenopus* embryo. *Science* **231,** 613–616.

Ballard, W. (1955). Cortical ingression during cleavage of amphibian eggs, studied by means of vital dyes. *J. Exp. Zool.* **129,** 77–98.

Ballard, W., and Ginsburg, A. (1980). Morphogenetic movements in acipenserid embryos. *J. Exp. Zool.* **213,** 69–103.

Cooke, J., and Webber, J. (1985). Dynamics of the control of body pattern in the development of *Xenopus laevis. J. Embryol. Exp. Morphol.* **88,** 85–112.

Dale, L., and Slack, J. M. W. (1987). Fate map of the 32 cell stage of *Xenopus laevis. Development (Cambridge, U.K.)* **99,** 527–551.

Danilchik, M., and Black, S. (1989). The first cleavage plane and the embryonic axis are determined by separate mechanisms in *Xenopus laevis. Dev. Biol.* **128,** 58–64.

Danilchik, M., and Denegre, J. M. (1991). Deep cytoplasmic rearrangements during early development in *Xenopus laevis. Development (Cambridge, U.K.)* (submitted for publication).

DeSimone, D., Smith, J., Howard, J. Ransom, D., and Symes, K. (1991). The expression of fibronectins and integrins during mesodermal induction and gastrulation in *Xenopus. In* "Gastrulation: Movements, Patterns, and Molecules." (R. Keller, W. Clark, and F. Griffin, eds.), in press. Plenum, New York.

Dixon, J., and Kintner, C. (1989). Cellular contacts required for neural induction in *Xenopus* embryos: Evidence for two signals. *Development (Cambridge, U.K.)* **106,** 749–757.

Dossel, W. E. (1958). Preparation of tungsten microneedles for use in embryologic research. *Lab. Invest.* **7,** 171–173.

Gerhart, J. (1980). Mechanisms regulating pattern formation in the amphibian egg and early embryo. *In* "Biological Regulation and Development, Volume 2: Molecular Organization and Cell Function" R. F. Goldberger, ed.), pp. 133–293. Plenum, New York.

Gerhart, J., Danilchik, M., Doniach, T., Roberts, S., Rowning, B., and Stewart, R. (1989). Cortical rotation of the *Xenopus* egg: consequences for the anteroposterior pattern of embryonic dorsal development. *Development (Cambridge, U.K.)* **107** (Suppl.), 37–51.

Gillespie, J. I. (1983). The distribution of small ions during the early development of *Xenopus laevis* and *Ambystoma mexicanum* embryos. *J. Physiol (London)* **344,** 359–377.

Gimlich, R., and Gerhart, J. (1984). Early cell interactions promote embryonic axis formation in *Xenopus laevis*. *Dev. Biol.* **104,** 117–130.

Hamburger, V. (1960) "A Manual of Experimental Embryology" (Revised Ed.) University of Chicago Press, Chicago, Illinois.

Hardin, J., and Keller, R. (1988). The behavior and function of bottle cells during gastrulation of *Xenopus laevis*. *Development (Cambridge, U.K.)* **103,** 211–230.

Harris, W., Holt, C., and Bonhoeffer, F. (1987). Retinal axons with and without their somata, growing to and arborizing in the tectum of *Xenopus* embryos: A time-lapse video study of single fibers *in vivo*. *Development (Cambridge, U.K.)* **101,** 123–133.

Holtfreter, J. (1933). Die totale Exogastrulation eine Selbstablosung Ektoderm von Entomesoderm. *Arch. Entwicklungsmech. Org.* **129,** 669–793.

Holtfreter, J. (1943). Properties and function of the surface coat in amphibian embryos. *J. Exp. Zool.* **93,** 251–323.

Holtfreter, J., and Hamburger, V. (1955). Embryogenesis: Progressive differentiation, amphibians. *In* "Analysis of Development" (B. H. Willier, P. A. Weiss, and V. Hamburger, eds.), pp. 230–296. Saunders, Philadelphia, Pennsylvania.

Honig, M., and Hume, R. (1986). Fluorescent carbocyanine dyes allow living neurons of identified origin to be studied in long-term culture. *J. Cell Biol.* **103,** 171.

Inoue, S. (1986). "Video Microscopy." Plenum, New York.

Jacobson, A. (1967). Amphibian cell culture, organ culture, and tissue dissociation. *In* "Methods in Developmental Biology" (F. Wilt and N. Wessels, eds.), pp. 531–542. Thomas Y. Crowell, New York.

Jacobson, M., and Hirose, G. (1979a). Clonal organization of the central nervous system of the frog. I. Clones stemming from individual blastomeres of the 16-cell and earlier stages. *Dev. Biol.* **71,** 191.

Jacobson, M., and Hirose, G. (1979b). Clonal organization of the central nervous system of the frog. II. Clones stemming from individual blastomeres of the 32- and 64-cell stages. *J. Neurosci.* **1,** 271.

Jessell, T., Bovolenta, M., Placzek, M., Tessier-Lavigne, M., and Dodd, J. (1989). Polarity and patterning in the neural tube: The origin and function of the floor plate. *In* "Cellular Basis of Morphogenesis" (Ciba Foundation Symposium), pp. 257–282. Wiley, New York.

Keller, R. E. (1975). Vital dye mapping of the gastrula and neurula of *Xenopus laevis*. I. Prospective areas and morphogenetic movements of the superficial layer. *Dev. Biol.* **42,** 222–241.

Keller, R. E. (1976). Vital dye mapping of the gastrula and neurula of *Xenopus laevis*. II. Prospective areas and morphogenetic movements of the deep layer. *Dev. Biol.* **51,** 118–137.

Keller, R. E. (1978). Time-lapse cinemicrographic analysis of superficial cell behavior during and prior to gastrulation in *Xenopus laevis*. *J. Morphol.* **157,** 223–248.

Keller, R. E. (1980). The cellular basis of epiboly: An SEM study of deep cell rearrangement during gastrulation in *Xenopus laevis*. *J. Embryol. Exp. Morphol.* **60**, 201–234.

Keller, R. E. (1981). An experimental analysis of the role of bottle cells and the deep marginal zone in gastrulation of *Xenopus laevis*. *J. Exp. Zool.* **216**, 81–101.

Keller, R. E. (1986). The cellular basis of amphibian gastrulation. *In* "Developmental Biology: A Comprehensive Synthesis, Volume 2: *The Cellular Basis of Morphogenesis*" (L. Browder, ed.), pp. 241–327. Plenum, New York.

Keller, R. E. (1987). Cell rearrangement in morphogenesis. *Zool. Sci.* **4**, 763–779.

Keller, R. E., and Jansa, S. (1991). Gastrulation in *Xenopus* embryos without the blastocoel roof. (submitted for publication).

Keller, R. E., and Danilchik, M. (1988). Regional expression, pattern and timing of convergence and extension during gastrulation of *Xenopus laevis*. *Development. (Cambridge, U.K.)* **103**, 193–210.

Keller, R. E., and Schoenwolf, G. C. (1977). An SEM study of cellular morphology, contact, and arrangement, as related to gastrulation in *Xenopus laevis*. *Wilhelm Roux's Arch. Dev. Biol.* **181**, 165–182.

Keller, R. E., and Tibbetts, P. (1989). Mediolateral cell intercalation is a property of the dorsal, axial mesoderm of *Xenopus laevis*. *Dev. Biol.* **131**, 539–549.

Keller, R., and Winklbauer, R. (1990). The role of the extracellular matrix in amphibian gastrulation. *Semin. Dev. Biol.* **1**, 25–33.

Keller, R. E., and Spieth, J. (1984). Neural crest cell behavior in white and dark larvae of *Ambystoma mexicanum*. *J. Exp. Zool.* **229**, 109–126.

Keller, R. E., Danilchik, M., Gimlich, R., and Shih, J. (1985a). Convergent extension by cell intercalation during gastrulation of *Xenopus laevis*. *In* "Molecular Determinants of Animal Form" (G. M. Edelman, ed.), pp. 111–141. Alan R. Liss, New York.

Keller, R. E., Danilchik, M., Gimlich, R., and Shih, J. (1985b). The function of convergent extension during gastrulation of *Xenopus laevis*. *J. Embryol. Exp. Morphol.* **89** (Suppl.), 185–209.

Keller, R., Cooper, M. S., Danilchik, M., Tibbetts, P., and Wilson, P. A. (1989). Cell intercalation during notochord development in *Xenopus laevis*. *J. Exp. Zool.* **251**, 134–154.

Keller, R., Shih, J., and Wilson, P. A. (1991a). Cell motility, control and function of convergence and extension during gastrulation of *Xenopus*. *In* "Gastrulation: Movements, Patterns, and Molecules" (R. Keller, W. Clark, and F. Griffen, eds.), in press. Plenum, New York.

Keller, R., Shih, J., Wilson, P., and Sater, A. (1991b). Pattern and function of cell motility and cell interactions during convergence and extension in *Xenopus*. *In* "Cell–Cell Interactions in Early Development" (49th Symposium of the Society for Developmental Biology, (J. Gerhart, ed.), Wiley, New York.

Keller, R., Shih, J., Sater, A., and Moreno, C. (1991c). Regulation of neural plate convergence and extension by the involuting marginal zone. in preparation.

Kintner, C. R., and Brockes, J. P. (1984). Monoclonal antibodies identify blastemal cells derived from differentiating muscle in newt limb regeneration. *Nature (London)* **308**, 67–69.

Kirschner, M., and Hara, K. (1980). A new method of local vital dye staining of amphibian embryos using Ficoll and crystals of nile red. *Microskopie* **36**, 12–15.

Koehl, M. A. R., Adams, D., and Keller, R. (1990). Mechanical development of the notochord in *Xenopus* early tail-bud embryos. "Biomechanics of Active Movement and Deformation of Cells" (N. Akkas, 3ed.), NATO ASI Series, Vol. H 42, pp. 471–485. Springer-Verlag, Berlin.

Lee, G., Hynes, R., and Kirschner, M. (1984). Temporal and spatial regulation of fibronectin in early *Xenopus* development. *Cell (Cambridge, Mass.)* **36**, 729–740.

London, C., Akers, R., and Phillips, C. (1988). Expression of Epi-1, an epidermis-specific marker in *Xenopus laevis* embryos, is specified prior to gastrulation. *Dev. Biol.* **129**, 380–389.

Lundmark, C. (1986). Role of bilateral zones of ingressing superficial cells during gastrulation of *Ambystoma mexicanum*. *J. Embryol. Exp. Morphol.* **97**, 47–62.

Moody, S. (1987). Fates of the blastomeres of the 16-cell stage *Xenopus* embryo. *Dev. Biol.* **119**, 560–578.

Nakamura, O., and Kishiyawa, J. (1971). Prospective fates of blatomeres at the 32 cell stage of *Xenopus laevis* embryos. *Proc. J. Acad.* **47**, 407–412.

Nakatsuji, N., and Johnson, K. (1983a). Comparative study of extracellular fibrils on the ectodermal layer in gastrulae of five amphibian species. *J. Cell Sci.* **59**, 61–70.

Nakatsuji, N., and Johnson, K. (1983b). Cell locomotion *in vitro* by *Xenopus laevis* gastrula mesodermal cells. *J. Cell Sci.* **59**, 43–60.

Nakatsuji, N., and Johnson, K. (1983c). Conditioning of a culture substratum by the ectodermal layer promotes attachment and oriented locomotion by amphibian gastrula mesodermal cells. *J. Cell Sci.* **59**, 43–60.

Newport, J., and Kirschner, M. (1982). A major developmental transition in early *Xenopus* embryos: I. Characterization and timing of cellular changes at the midblastula stage. *Cell (Cambridge, Mass.)* **30**, 675–686.

Nieuwkoop, P., and Faber, P. (1967). "Normal Table of *Xenopus laevis*." North-Holland Publ., Amsterdam.

Okamoto, M. (1972). A method for removal of the jelly and the vitelline membrane from the embryos of *Xenopus laevis*. *Dev. Growth Differ.* **14**, 37–41.

Purcell, S. (1990). A different type of anuran gastrulation and morphogenesis as seen in *Ceratophrys ornata*. *Am. Zool.* **29**, 85a.

Rugh, R. (1962). "Experimental Embryology. Techniques and Procedures." 3rd Ed. Burgess Publ., Minneapolis, Minnesota.

Sater, A., Uzman, J. A., Steinhardt, R., and Keller, R. (1991). Neural induction in *Xenopus*: Induction of neuronal differentiation by either planar or vertical signals. Submitted for publication.

Savage, R., and Phillips, C. (1989). Signals from the dorsal blastopore lip region during gastrulation bias the ectoderm toward a nonepidermal pathway of differentiation in *Xenopus laevis*. *Dev. Biol.* **133**, 157–168.

Schechtman, A. M. (1942). The mechanics of amphibian gastrulation. I. Gastrulation-producing interactions between various regions of an anuran egg (*Hyla regilla*). *Univ. Calif. Publ. Zool.* **51**, 1–39.

Schroeder, T. (1970). Neurulation in *Xenopus laevis*. An analysis and model based upon light and electron microscopy. *J. Embryol. Exp. Morphol.* **23**, 427–462.

Shi, D.-L., Delarue, M., Darribere, T., Riou, J.-F., and Boucaut, J.-C. (1987). Experimental analysis of the extension of the marginal zone in *Pleurodeles waltl* gastrulae. *Development (Cambridge, U.K.)* **100**, 147–161.

Shi, D.-L., Darribere, T., Johnson, K. E., and Boucout, J.-C. (1989). Initiation of mesodermal cell migration and spreading relative to gastrulation in the urodele amphibian *Pleurodeles waltl*. *Development (Cambridge, U.K.)* **103**, 639–655.

Shih, J., and Keller, R. (1991a). The organizer function of the epithelial layer of the dorsal marginal zone of *Xenopus*. in preparation.

Shih, J., and Keller, R. (1991b). Protrusive activity and cell behavior during mediolateral intercalation in *Xenopus* gastrulation. in preparation.

Shih, J., and Keller, R. (1991c). Patterning of cell behavior during convergence and extension in gastrulation of *Xenopus*. in preparation.

Smith, J., and Schoenwolf, G. (1989). Notochordal induction of cell wedging in the chick neural plate and its role in enural tube formation. *J. Exp. Zool.* **250**, 49–62.

Spemann, H. (1938). "Embryonic Development and Induction." Yale Univ. Press, New York.

Vogt, W. (1929). Gestaltanalyse am Amphibienkein mit ortlicher Vitalfarbung. II. Teil. Gastrulation und Mesodermbildung bei Urodelen und Anuren. *Wilhelm Roux' Arch. Entwicklungsmech. Org.* **120,** 384–706.

Wilson, P. A. (1990). The development of the axial mesoderm in *Xenopus laevis*. Ph.D. Dissertation, University of California, Berkeley, California.

Wilson, P. A., and Keller, R. E. (1991). Cell rearrangement during gastrulation of *Xenopus*: Direct observation of cultured explants. *Development (Cambridge, U.K.)* **112,** 289–300.

Wilson, P. A., Oster, G., and Keller, R. E. (1989). Cell rearrangement and segmentation in *Xenopus*: Direct observation of cultured explants. *Development (Cambridge, U.K.)* **105,** 155–166.

Winklbauer, R. (1988). Differential interaction of *Xenopus* embryonic cells with fibronection *in vitro Dev. Biol.* **130,** 175–183.

Winklbauer, R. (1990). Mesodermal cell migration during *Xenopus* gastrulation. *Dev. Biol.* **142,** 155–168.

Winklbauer, R., Selchow, A., Nagel, M., Stoltz, C., and Angres, B. (1991). Mesoderm cell migration in the *Xenopus* gastrula. *In* "Gastrulation: Movements, Patterns, and Molecules" (R. Keller, W. Clark, and F. Griffen, eds.), in press. Plenum, New York.

Part II. Oocytes

Chapter 6

Vitellogenin Uptake and in Vitro Culture of Oocytes

LEE K. OPRESKO

Department of Pathology
University of Utah
Salt Lake City, Utah 84112

I. Introduction and History
II. Methods
   A. Handling of Ovarian Tissue and Oocytes
   B. Oocyte Culture Media
   C. Vitellogenin Isolation
   D. Labeling of Vitellogenin and Other Proteins
   E. Performing Protein Uptake Experiments
III. Discussion
   References

I. Introduction and History

The oocytes of nonmammalian vertebrates are specialized for the endocytic incorporation of proteins (Wallace, 1978). Oocytes grow primarily via the specific internalization of a serum protein, vitellogenin (VTG), which is the precursor to the yolk (Wallace and Dumont, 1968; Bergink and Wallace, 1974). VTG is synthesized in the liver under the influence of estrogen, then secreted into the bloodstream (Wallace and Jared, 1968). The circulating levels of VTG in healthy, mature females is approximately 5 mg/ml of serum, and it is internalized exclusively by the ovary (Follet *et al.*, 1968; Wallace and Jared, 1969). At maturity over 80% of the protein of the oocyte is contained in the yolk platelets, which are made up of proteins derived from VTG (Callen *et al.*, 1980).

The study of protein uptake by *Xenopus* oocytes was initiated in the 1960s using *in vivo* systems and electron microscopic examination of cells (Wallace and Dumont, 1968). To facilitate more detailed experimentation, it was necessary to develop reliable *in vitro* maintenance and culture systems. Such systems initially consisted of a modified amphibian saline optimized for the internalization of VTG by oocytes (solution OR2) and were later extended to nutritive media designed for extended maintenance and growth of oocytes (Wallace et al., 1973; Wallace and Misulovin, 1978).

The active study of amphibian embryology over the past seven decades resulted in the development of a number of different media suitable for the maintenance of various stage embryos from a variety of species. With the enhanced interest in investigating oogenesis, many of these media were used or adapted for use with oocytes. Prior to the 1960s, *Xenopus* was not a widely used experimental animal, so the early solutions were designed for different amphibian species. During the 1950s several groups formulated modified salines for different species. These include Barth's medium (Barth and Barth, 1959) which was designed for *Rana* spp. and Niu and Twitty solution (Niu and Twitty, 1953) which would maintain embryonic tissue from *Triturus* and *Ambystoma*. The criteria used to judge the performance of these solutions was the maintenance of cell viability and morphology during the differentiation of embryonic explants. It was not until the late 1960s that a saline specific for *Xenopus* oocytes was developed (Jared and Wallace, 1969). Called solution O, this saline resembled Barth's solution except that it lacked phosphate and used Tris rather than sodium bicarbonate to buffer the medium. This solution was subsequently optimized by the substitution of HEPES buffer for Tris, by the addition of phosphate, and by minor alterations in ionic composition (Wallace et al., 1973); the formulation of this saline solution OR2 is included below.

Early oocyte internalization experiments used *in vivo* labeled *Xenopus* serum derived from hormonally stimulated (estrogen-treated) animals as a source of VTG. These sera were added to different media at relatively high concentrations (up to 50%), and protein incorporation was measured over an extended period of time (1–25 hours) because of the low specific activity of the *in vivo* labeled VTG (Jared and Wallace, 1969; Wallace et al., 1970, 1973). Because estrogen treatment increases serum VTG concentrations up to 160 mg/ml, the major labeled component of such serum was VTG regardless of the radioactive amino acid employed. However, if the serum was labeled with ortho [^{32}P]phosphate then VTG was the only labeled protein. More recently, labeling procedures have been developed that utilize purified VTG and yield relatively high specific activities; these are described below.

VTG purification methods were developed in the early 1970s. The first of these used chromatography of vitellogenic serum on TEAE cellulose to obtain a preparation of high purity (Wallace, 1970). Two other less arduous but less stringent procedures were developed the following year (Ansari et al., 1971; Redshaw and Follett, 1971). Currently, the most popular VTG purification method is selective precipitation from the serum using a Mg–EDTA complex (see below). The availability of purified VTG made it possible to determine the specificity of VTG internalization by oocytes. Such studies demonstrated that VTG was internalized by oocytes 25 to 50-fold faster, on a molar basis, than other labeled proteins (Wallace and Jared, 1976).

Once *in vitro* maintenance of oocytes was possible, the next step was to develop a system that allowed for oocyte growth in culture. During the 1960s several different mammalian cell media were used to grow amphibian cells of different types. *Rana* cells were grown in a 55% dilution of Eagle's minimal essential medium supplemented with 10% fetal calf serum and 10% chicken egg ultrafiltrate (Wolf and Quimby, 1964). In 1966 Balls and Ruben cultured cells from *Xenopus* spleen, liver, and kidney in a 50% dilution of Liebovitz-L-15 medium containing 10% fetal calf serum. A similar medium was used to support the *in vitro* synthesis of radiolabeled VTG by *Xenopus* liver slices (Wallace and Jared, 1969). The first attempt to formulate a nutrient-containing medium specifically to support *Xenopus* oocytes was in 1976 (Eppig and Dumont, 1976). This medium, DNOM (defined nutrient oocyte medium), could support normal oocyte morphology for over 2 weeks. However, during that time there was a steady decrease in the rates of VTG internalization, uridine incorporation, and protein synthesis, all indicative of suboptimal conditions for long-term growth. In addition, DNOM was complex and had a short shelf life. In 1978 Wallace and Misulovin developed a supplemented 50% L-15 medium capable of supporting long-term oocyte growth and differentiation. The formulation of this medium is included below; not surprisingly, the ingredient that is required for oocyte growth is VTG.

II. Methods

A. Handling of Ovarian Tissue and Oocytes

In order for oocytes to internalize protein *in vitro*, they must be divested of their surrounding thecal layers. Removal of this material can be accomplished in two ways. The first is manual dissection from the ovary using two pairs of watchmaker's forceps. The second is ovarian digestion with 0.2% collagenase (Eppig and Dumont, 1976). This method suggests the use of Ca^{2+}-free medium and elevated temperatures to facilitate cell dissociation. Both of these conditions can compromise oocyte viability and inhibit protein internalization even after the cells are returned to Ca^{2+}-containing medium. In addition only impure preparations of collagenase which contain other proteases are able to remove the theca effectively. To ensure that the oocyte plasma membrane remains undamaged, it is best to use only manually dissected cells unless the cells are allowed abundant time to recover (> 2 days).

Immediately upon removal from the animal, the ovary is placed into sterile saline solution OR2 in a 100-mm petri plate. If a laminar flow hood is unavailable the saline should be supplemented with antibiotics. We generally use gentamicin sulfate at 50 μg/ml; however, penicillin and streptomycin are also good choices. After removal, ovarian tissue is cut into pieces about 2 cm in length and transferred to fresh saline in 35-mm dishes. Ovarian pieces should be agitated in the original saline to remove any cellular debris that might be adhering to the tissue prior to transfer. Smaller amounts of ovary are more readily examined for the presence of oocytes of particular sizes or stages and usually remain visible for the entire time of dissection. Breakage of cells is very common and rapidly causes high levels of turbidity.

Once the oocytes (and the surrounding follicle cells) have been freed from the ovarian theca, care must be taken to prevent sticking. Oocytes isolated from different ovaries vary in the ease with which they adhere to either each other or to plastic surfaces. The presence of a small amount (< 1 mg/ml) of protein will generally be sufficient to alleviate this problem. If the cells have already become adherent, they can be rescued by the addition of a drop of serum or 1–2 mg of powdered bovine serum albumin (BSA) to the petri dish.

B. Oocyte Culture Media

In general, oocytes that have been dissected from the ovary are placed in saline overnight prior to their use in experiments. This initial incubation permits an evaluation of the viability of the cells from a particular ovary. If a large proportion (> 10%) of the cells die overnight or appear morphologically abnormal, it is usually best to start again using a different ovary. This is particularly true if the experiment requires the long-term survival of the cells. In my laboratory I use saline solution OR2 for all short-term experiments. OR2 was formulated for optimal VTG uptake by *Xenopus* oocytes and is a good medium to use for any protein uptake experiment (Wallace *et al.*, 1973).

1. PREPARATION OF OR2

Prepare stock solution I by dissolving the following in 1 liter of water:

NaCl	48.2 g
KCl	1.85 g
Na_2HPO_4	1.4 g
HEPES	11.9 g
NaOH to pH 7.8	

Prepare stock solution II by dissolving the following in 1 liter of water:

$CaCl_2 \cdot 2H_2O$	1.45 g
$MgCl_2 \cdot 6H_2O$	2.05 g

For 1 liter of OR2, add 100 ml of stock solution I to 700 ml of distilled water mix, and add 100 ml of stock solution II. Next add water to 1 liter. If solution I is not sufficiently diluted before adding solution II, precipitation may occur. Check that the pH is 7.8 before use.

If oocytes are required to survive beyond 2–3 days in culture, it is best to transfer them to a nutritive medium. The medium of choice was developed in 1978 (Wallace and Misulovin) and is made as follows.

2. Preparation of L-15 Medium

To prepare 100 ml of L-15 medium assemble and mix the following ingredients:

L-15 medium	50 ml
300 mM HEPES buffer, pH 7.9	5 ml
100 mM Glutamine solution	1 ml
50 mg/ml Gentamicin	0.1 ml
Water to 90 ml	

To prepare the final working solution, add 9 ml of the above medium to 1 ml of solution containing either 25 mg of VTG and 1 µg/ml insulin *if oocyte growth is required* or 0.5 ml dialyzed calf serum *if oocyte maintenance is required*. The source of the VTG can either be the purified protein or serum from a frog treated with estrogen to stimulate VTG synthesis. Because vitellogenic serum works as well if not better than purified VTG, it is the preferred additive. The only drawback is that serum can have greater toxic effects than purified VTG so serum samples are not pooled unless they have been tested for toxicity. Under all conditions, the inclusion of insulin is not a requirement but does partially prevent the decline in protein uptake that is normally seen in cultured cells following a period of days *in vitro*.

3. Culture Vessel

The choice of culture vessel is dictated by the number of cells to be maintained and the duration of the incubation period. For overnight incubations in OR2, either 35- or 60-mm petri dishes are used. If cells are placed in L-15 medium, use a 35-mm petri dish or a 24-well plate to separate out groups of cells. For long-term culture (>1 week) the cells are incubated individually in 96-well plates with 100 µl of medium per oocyte. This is essential because the cells will stick avidly to each other after several days and will not be separable without breaking. For long-term culture the medium is changed every 3 days.

C. Vitellogenin Isolation

VTG is obtained from the serum of estrogen-treated frogs. Animals (usually females because they are larger) are injected with 17β-estradiol dissolved in propylene glycol (10 mg/ml) at a dose of 1–1.5 mg/100-g frog on day 1, followed by a second injection on day 3. The animal is kept for a total of 10 days and then subjected to cardiac exsanguination following anesthesia in 0.2% Tricaine (3-aminobenzoic acid ethyl ester) in *tap* water for 30–60 minutes. This terminal procedure is done by carefully opening the thoracic cavity via a V-shaped cut following the distal outline of the sternum. The sternum is then partially removed to reveal the heart in the pericardium. Using fine tweezers and forceps, the pericardial membranes are cut, and the heart is gently expressed from the chest cavity by application of pressure on either side of the organ. A paper towel with a small diamond-shaped opening is placed over the exposed heart to prevent contamination of the blood. The animal is next placed, ventral side down, over a sterile, 15-ml conical centrifuge tube. The heart is cut, using fine scissors, at the tip of the ventricle and then placed over the opening of the tube. Five to ten milliliters of blood can be collected from a single female.

Use of these methods yields vitellogenic serum that is suitable for cell culture. If uncleaved VTG is required it is best to collect plasma, include protease inhibitors in the blood collection tube, and follow the isolation procedure of Wiley *et al.* (1979) using either selective precipitation or column chromatography. Selective precipitation is very fast, requires a minium of equipment, and yields a very concentrated VTG preparation that can be diluted to the desired level. The procedure for VTG precipitation follows.

1. Isolation of Vitellogenin from Vitellogenic Serum

1. Plasma is collected in sterile 50-ml tubes containing 1 ml Ca^{2+}-free 50% phosphate-buffered saline (PBS) and 70 mM sodium citrate.
2. Plasma is centrifuged for 15 min at 2500 g and the supernatant carefully removed.
3. Repeat Step 2.
4. Add 20 ml of 20 mM EDTA, pH 7.7, for every 5 ml of plasma. Gently mix the contents of the tube by inversion and add 1.6 ml of 0.5 M $MgCl_2$. Mix again.
5. Centrifuge at 2500 g for 15 min; discard supernatant.
6. Dissolve the pellet in 3 ml of 1 M NaCl, 50 mM Tris, pH 7.5 (buffer A).

Use a glass stirring rod and avoid forming bubbles—do not vortex! The pellet can be quite sticky, and patience is required to obtain total dissolution.
   7. Centrifuge the VTG solution for 30 min at 2500 g and decant.
   8. Add 25 ml distilled water and centrifuge as in Step 2.
   9. Dissolve the pellet as in Step 6.
   10. Dialyze the VTG against several changes of buffer A and then switch to the desired dialysis solution.

Although this single-step purification does not yield 100% purity, it does produce a preparation that is at least as pure as that derived by column chromatography. The only requirement for efficient precipitation is an initial VTG concentration of at least 50 mg/ml in the plasma. Such starting concentrations are not difficult to achieve. To obtain highly pure VTG, the preparation produced by selective precipitation is chromatographed on DEAE cellulose as described by Wiley et al. (1979). The dilute VTG sample that is obtained is concentrated by vacuum dialysis. Concentration can also be effected by coating the dialysis tubing containing the VTG sample with solid polyethylene glycol. The advantage of vacuum dialysis is that the sample is gently concentrated into a relatively small area, minimizing the loss of protein on the sides of the tubing. Concentration by ultrafiltration is not recommended because VTG is denatured by the stirring required.

For purposes of cell culture, there is no need to use pure VTG, nor is it totally necessary to use VTG that is cleavage free. VTG exists as a dimer (Wallace, 1970) which can remain in solution even if partially proteolyzed. As shown by Wallace et al. (1980b) vitellogenic serum can be used to support oocyte growth in culture. If VTG-containing serum is only to be used for this purpose, it is possible to increase the yield (in terms of both volume and concentration) by allowing the frog to synthesize VTG for 3–4 weeks after a large injection of estrogen (4 mg/animal). The collected blood is allowed to clot at room temperature for several hours, refrigerated overnight, and then spun at 3000 rpm for 15 minutes. The green, VTG-containing serum is removed, sterile filtered (using a 1.0-μm prefilter if necessary), and the VTG titer determined as follows (from Wallace et al., 1980b).

2. Measurement of Vitellogenin Titer in Serum

1. To 50–100 μl of vitellogenic serum (depending on the estimated VTG concentration) add 400 μl of 20 mM EDTA and 40 mM MgCl$_2$. Mix and allow the precipitate to form at 0°C for 10 minutes.

2. Spin down at 5000 rpm for 10 minutes, discard the supernatant, and dissolve the pellet in a few drops of 1 M NaCl, 50 mM Tris, pH 8.0.

3. Bring up the volume to 5 or 10 ml (again depending on the estimated concentration of VTG).

4. Read the absorbance at 280 nm. The extinction coefficient of pure VTG is 0.75 liter/g·cm. A correction factor of 2.3 mg/ml is then added to the measured value to arrive at the absolute VTG concentration (Wallace et al., 1980b). Thus, the calculations would be as follows: (a) For an absorbance of 0.75 at 280 nm using a 10-ml final liquid volume and a 100-μl starting sample, (b) multiply 0.75 by 100 (0.1 ml sample diluted to 10 ml = 75; (c) divide 75 by 0.75 (VTG extinction coefficient) = 100; (d) add 2.3 mg/ml to arrive at the actual VTG concentration, 102.3 mg/ml.

After the titer of VTG is determined, that volume of serum equivalent to 2.5 mg/ml is added to the L-15 culture medium. Additional serum (dialyzed fetal calf or calf serum) can be added to bring the total serum concentration to 5%. The VTG-containing serum should be aliquoted and frozen at $-20°C$ for storage (typically 12.5- and 25-mg aliquots for the formulation of 5–10 ml of media). VTG is an easily denatured protein (vortexing, freeze–thawing, and lyophilization) and should not be subjected to more than one round of freezing. Even under optimal conditions many batches of sera are cloudy after thawing owing to lipoprotein precipitation. A second round of sterile filtration after medium formulation will eliminate the denatured lipoprotein and render the serum preparation usable for culture (but not for labeling, see below).

D. Labeling of Vitellogenin and Other Proteins

Vitellogenin can be labeled in a variety of ways. Prior to the 1980s it was chiefly labeled *in vivo* by injection of radioactive amino acids into an estrogen-treated frog. This yielded labeled VTG that retained the greatest amount of native conformation but with a low specific activity. To alleviate this problem, labelling of VTG in organ culture using liver slices from estrogen-treated animals was also utilized (Brummett and Dumont, 1977). The specific activities obtained by this method were much higher, but it was tedious and involved sacrifice of an animal to obtain the starting material for each preparation.

In vitro labeling of VTG primarily involved three different procedures. The first *in vitro* labeling was done by Wallace and Jared (1976) using [^{14}C]formaldehyde and reductive methylation. This yielded very mild labeling conditions but relatively low specific activities. A modification of this procedure using NaB^{3}H$_3$ as the source of label enabled Opresko et al. (1980a) to achieve a much higher specific activity without a substantial reduction in the native conformation of the molecule. However, both of these procedures suffer from the drawback that oocytes are able to reutilize the dimethyllysine formed by the labeling reaction.

A second labeling method was developed in 1980 which involved the phosphorylation of VTG *in vitro* through the use of a protein kinase (Opresko *et al.*, 1980b). The enzyme was initially isolated from *Xenopus* liver; a similar enzyme was later purified from chicken liver (Opresko and Wiley, 1984). There were two major advantages to this technique. One advantage was the high specific activities that could be achieved through the use of [γ-^{32}P]ATP (>3000 Ci/mmol). The reaction had an efficiency of 50–60%, and the amount of protein labeled could be adjusted to yield the required specific activity. The second advantage was that VTG could be labeled in serum, thus minimizing the possibility of denaturing the protein. This was possible because the kinase was a phosphoprotein protein kinase that would only label phosphorylated substrates. Since VTG is the only phosphoprotein in the serum, it is the only protein labeled. After labeling, the VTG can be used without further purification following dialysis to remove the unreacted ATP. Use of VTG purified by selective precipitation is not recommended with protein kinase labeling because the VTG will precipitate under the ionic conditions required by the enzyme. The major drawbacks to this labeling procedure are the necessity of purifying the enzyme and the short half-life of the isotope. The limitation of the 14-day half-life can be alleviated through the use of ATP$_\gamma$S (L. K. Opresko, personal observation), which permits labeling with ^{35}S. However, there is a substantial loss of labeling efficiency when using this ATP derivative.

Both reductive methylation and selective phosphorylation suffer from the limitation that the label can be reutilized if the parent molecule is degraded. A further limitation is the necessity of using a phosphoprotein if the protein kinase technique is used. Ideally, an *in vitro* labeling procedure should yield a labeled protein that has high specific activity, a reasonable radioactive half-life, and the ability to be degraded into products that cannot be reincorporated into other molecules. Iodination using Na^{125}I is the method of choice for protein labeling since it meets all of these requirements and can be used for any protein that contains tyrosine. Historically, however, there have been two reasons why iodination of VTG as well as other proteins has been avoided for oocyte internalization studies. The first is that VTG is quite susceptible to denaturation, and the oxidative conditions normally found in iodination reactions can be damaging. The second is that oocytes possess a potent deiodination activity that rapidly removes up to 90% of the iodine from the labeled molecule (Opresko *et al.*, 1980a). In addition, this free iodine is not released from the oocyte and thus necessitates the use of precipitation protocols to ensure that the parent molecule is the labeled species being monitored.

Use of iodinated proteins is now feasible for oocytes since we discovered a potent inhibitor of their deiodination activity. This compound, sodium ipodate or 3-{[(dimethylamino)methylene]amino}-2,4,6-triiodobenzenepropanoic acid (Oragrafin, Squibb), consists of an aromatic ring structure that contains three atoms of iodine. Pretreatment of oocytes for 1 hour in 15 μg/ml of ipodate inhibits deiodination by over 85%, and the compound has no apparent toxic effect on the cells. The drug is dissolved in dimethyl sulfoxide (DMSO) at a concentration of 20 mg/ml and is included in the formulation of the radioactive medium as well as in the pretreatment solution. Long-term exposure of oocytes to ipodate does not inhibit VTG uptake. With the discovery of this compound, iodination has become the labeling method of choice for oocyte studies.

The destructive potential of protein oxidation during iodination reactions has been reduced by the advent of immobilized oxidizing agents. We routinely use Iodobeads (Pierce, Rockford, IL) to label a number of different proteins including VTG. In the case of VTG, we iodinate 3–5 mg with 1 mCi of iodine using the protocol of Pierce as follows:

1. Place the Iodobead in 1 ml of 100 mM Tris, pH 7.0, for a few seconds and then remove the bead.

2. Add 100 μl of Tris buffer and 1 mCi of Na^{125}I (10 μl; 100 mCi/ml) to the bead in an Eppendorf tube; wait 5 minutes.

3. Add VTG (purified or high titer, nonlipemic serum) and water to bring the volume to 200 μl; wait 15 minutes at room temperature.

4. Terminate the reaction by removing the protein solution from the bead; rinse once with 100 μl of OR2.

The labeled preparation is then either dialyzed against 1–2000 volumes of OR2 overnight or is passed over a G-25 column previously equilibrated with OR2 plus 1 mg/ml BSA. Chromatographic separation has the advantage of selectively eliminating denatured molecules (they adhere to the column) but the disadvantage that some protein is lost and the sample is diluted. Since iodine is able to partition into lipid it is important that any serum used for iodination be free of excess lipid and, since there is no selectivity in this labeling procedure, it should also contain a high titer of VTG. Alternately, purified VTG can be used for iodination.

E. Performing Protein Uptake Experiments

Once it has been determined that the overnight viability of a batch of oocytes is acceptable, the cells are ready for use. If the experiment to be done entails quantitation of internalized or surface-associated proteins an additional requirement is that the oocytes be of similar size. Sizing of cells actually begins with the removal from the ovary of only those oocytes that fall within a particular range of diameters. If there are enough cells of a single size, then the entire experiment can be done with those. This is the usual case if stage VI (Dumont, 1972) oocytes are being used since they all have similar diameters (1.2–1.3 mm) and are the largest and most obvious oocytes in the ovary. If stage IV or V are to be used, the cells are measured with an ocular micrometer mounted in a dissecting microscope and placed into groups whose diameters are within 0.05 mm of each other. The closer the size range, however, the more accurate the results. The size classification is done on the day before the experiment since it will dictate the size composition of the test groups. If a range of sizes must be used, it is important to make sure that the size composition of each experimental group is comparable. A convenient way to size large numbers of cells is to first remove about 40 cells from the total and then separate out all the cells that seem to be of similar size. Measure the cells with the ocular micrometer to ensure that they do fall within a similar size range. With practice, separation becomes more accurate. Cells of a similar diameter are usually placed into a well (or 2–3 wells, depending on numbers) of a 24-well multiwell plate.

On the morning of the experiment, all dead or abnormal oocytes are discarded. If the number of cells within each well is recorded on the plate, it will be easy to determine the number of cells remaining. The appropriate number of oocytes from each size class are then transferred to the protein-containing solutions. We generally use 24-well plates for experiments. The oocytes can be incubated in as little as 700 μl in these plates and are easy to remove from the wells, as long as the extra-deep well plates that are also available are not used. In addition, each well will comfortably hold 50 stage IV or 30 stage VI cells. If more cells are needed either more wells can be used or other containers utilized.

In preparing protein solutions for experiments either 5% dialyzed serum or 5 mg/ml BSA is added. This ensures that the oocytes do not stick to each other and can be easily removed from the culture dishes. In addition, if iodinated proteins are used the cells are pretreated for at least 1 hour with sodium ipodate. Depending on the experiment, protease inhibitors may also be added. The protease inhibitors pepstatin, chymostatin, aprotinin, and antipain are each added to a concentration of 100 μg/ml. To initiate protein uptake the oocytes are transferred to the labeled protein solution and maintained at 20°–22°C for the desired times. The labeled protein is usually present in milligram quantities. From 2.5 to 3.0 mg/ml of VTG is required to saturate surface receptors (Opresko and Wiley, 1987a), and uptake rates of other proteins are much lower. To terminate the incubation the cells are removed from the solution and washed in solution OR2 containing 0.1% BSA. The oocytes are washed 3 times sequentially in 100-mm petri dishes. This is done by swirling the cells (by aspiration of the liquid with a Pasteur pipette) 10 times before transfer to the next dish. Transfers are done with the minimum possible amount of liquid carryover. With a little practice, oocytes (5–20) can be washed and transferred to their appropriate destinations in about 1.5 minutes.

A smaller number of cells takes less time, whereas a larger number is more difficult to handle. Time points should therefore be planned to take this sample processing time into account.

To determine the amount of protein binding to the oocyte plasma membrane, incubations are done at 0°C. This temperature is required to totally inhibit protein internalization (Wallace et al., 1970). Incubations at this temperature are most conveniently done in test tubes which can be set into an ice bucket. Removal of cells from the tubes can present a problem since it is difficult to see both the pipette and the cells and still maintain the tube at the proper temperature. Oocyte removal can be facilitated by (1) marking the end of the Pasteur pipette used to retrieve the cells with a colored (Sharpie) marker (do not use black since oocytes are partly black). This will enable you to see where the cell and the pipette are in relation to each other. In addition, (2) set the tube in a beaker filled with some ice and a lot of ice water. The beaker and tube can be moved so the tube stays at 0°C. Removing the condensation from the side of the beaker permits viewing the contents of the tube, as long as it is translucent or transparent.

In experiments where it is necessary to differentiate surface-associated from internalized protein it is also necessary to wash the cells at 0°C (to inhibit dissociation during the wash steps). This is done by setting petri plates containing cold saline and some powdered BSA on top of a metal tray that is placed on top of a quantity of ice. Aluminum cafeteria trays are excellent for this purpose and can be obtained from a restaurant supply company. The ice can be in either a styrofoam shipping box or a large, polyethylene busboy tray (also available from restaurant supply companies). The cells are washed in the same manner as described above and are then transferred to microtiter plates, also kept at 0°C, for removal of surface-bound ligand. Microtiter plates are kept at the required temperature by placing them in a metal tray containing ice and ice water; the tray is set into a quantity of ice in a styrofoam chest. Free access of the water to the entire bottom of the plate is ensured by cutting a notch in the outer edge of each side of the plate using a small, electric grinding wheel.

VTG bound to the oocyte surface is removed by the following solution: 25 mM EDTA, 0.5 M NaCl, 0.1 M glycine–NaOH buffer, pH 9.5, and 5 mg/ml VTG in serum. The presence of vitellogenic serum in the surface stripping solution only slightly enhances its stripping efficiency (about 2%) and is optional (Opresko and Wiley, 1987a). Groups of 3–5 cells are placed in 50–100 μl of this solution in a single well of the microtiter plate for 30–40 minutes. For enhanced visibility the solution may be colored with dye like pyronin Y or methyl green. The solution is removed using a drawn-out Pasteur pipette whose tip has been colored so it is more visible, and the cells are rinsed with an equal volume of solution lacking VTG. Following stripping, the cells are dissolved in formic acid (100 μl per 3–5 cells) and then counted. Dissolution takes about 1 hour and should be done in a well-ventilated area with the plate placed in a container that is fairly airtight, like a Ziploc plastic bag.

Dissolution of oocytes prior to counting in a γ counter is only necessary if surface-associated ligand is also being quantitated. The surface-bound ligand will always be removed in an aqueous medium. To ensure that the counting efficiencies of the surface-bound and internalized material are similar, they should both be counted in a similar environment (liquid). For other experiments washed oocytes can be placed into the bottom of a counting tube (with as little liquid as possible) and counted without further processing.

The following is an outline of the procedures required to perform a protein uptake experiment:

Day −2 Plan your experiment and determine the number of cells required and which solutions are needed
Day −1 Operate on frog and remove ovary
Remove oocytes and culture
Label proteins
Day 0 Examine oocytes and remove abnormal cells
Pretreat cells in sodium ipodate if iodinated compounds are being used
Prepare the protein solutions
Set up wash solutions and strip solutions if needed
Begin experiment, leaving adequate time between time points for sample handling

Once the amount of radioactivity in each sample has been determined, the quantity of protein taken up by the oocytes can be calculated. Since rates of VTG uptake vary with oocytes from different ovaries, only cells from the same ovary can be directly compared. When such cells are used, the rate of protein uptake is a direct function of surface area within the range of 0.8–1.05 mm in diameter (Wallace et al., 1970). Protein uptake may be expressed as molecules/cell/hour if all cells are the same size or as molecules/mm^2/hour if different sizes are used. In the case of VTG, uptake rates have been reported as high as 346 ng/mm^2/hr (Wallace et al., 1978). However, uptake rates for oocytes derived from unstimulated frogs (no prior treatment with human chorionic gonadotropin or pregnant mare gonadotropin) are usually much lower (Wallace et al., 1978, 1980a). In addition, rates of VTG uptake will vary depending on the VTG concentration and the level of receptor saturation (Opresko and Wiley, 1987b), whereas the rate of internalization of nonspecifically incorporated proteins will be directly proportional to the external concentration in the medium (Wallace and Jared, 1976).

III. Discussion

The choice of oocyte culture conditions is dictated by the need for either cell growth or maintenance. For investigations in which oocyte growth and/or differentiation is required, 50% L-15 medium containing VTG is the only choice. Studies have shown that cultured oocytes grow at rates that are much faster than cells *in vivo* (Wallace and Misulovin, 1978; Keem et al., 1979). However, the cells still maintain the relationship between diameter, differentiation state, and protein content under these conditions of accelerated growth. In addition, processes other than protein synthesis depend on oocyte growth. For example, the proper localization of the mRNA for Vg1, normally found in the oocyte vegetal hemisphere, requires the presence of VTG in the culture medium of oocytes between late stage III and mid stage IV (Yisraeli et al., 1989). To study any process in which cells pass from one stage of development to another, the oocytes must be incubated in VTG.

In the case of oocyte maintenance there are two choices of media available. The first is saline solution OR2 which is capable of supporting morphologically normal cells for several days to over a week. Once again, oocytes obtained from different ovaries will display different levels of viability. This is true for both nonnutritive saline as well as for L-15 medium. The second medium that can be used is 50% L-15 containing 5 mg/ml BSA or 5% dialyzed calf serum. This solution contains amino acids as well as an energy source and is used when cells are required to maintain high levels of protein synthesis, for example, after cell microinjection with mRNA. This is not to say that mRNA-injected oocytes cannot be kept in OR2. Uninjected cells maintained in OR2 for several days do not show any reduction in synthetic capacity when compared to initial measurements (L. K. Opresko, personal observation). However, it has been shown that injection of mRNA can elevate the total rate of oocyte protein synthesis (Johnson et al., 1990). The inclusion of amino acids in the culture medium should ensure that such synthesis occurs at an optimal rate by maintaining oocyte amino acid pools at high levels.

For protein uptake or binding experiments the labeled protein should be placed into solution OR2 containing 5 mg/ml BSA or 5% serum. In general, dilution of OR2 up to 5% will not affect the cells; if further dilution is anticipated, the medium can be formulated using 10 × stocks. If the protein is iodinated, the cells should be pretreated for 1 hour with sodium ipodate to prevent nonspecific deiodination. This is most important when examining proteins other than VTG since deiodination is more severe at low levels of protein incorporation (Opresko et al., 1980a). Protein uptake experiments are usually performed at 20° to 22°C, although temperatures up to 25°C can be used without compromising cell viability (Wallace et al., 1970). In contrast, protein binding experiments require a temperature of 0°C to totally inhibit internalization.

References

Ansari, A. Q., Dolphin, P. J., Lazier, C. B., Munday, K. A., and Akhtar, M. (1971). Chemical composition of an estrogen-induced calcium binding glycolipophosphoprotein in *Xenopus laevis*. *Biochem. J.* **122**, 107–113.

Balls, M., and Ruben, L. N. (1966). Cultivation *in vitro* of normal and neoplastic cells of *Xenopus laevis*. *Exp. Cell Res.* **43**, 694–695.

Barth, L. G., and Barth, L. J. (1959). Differentiation of cells of the *Rana pipiens* gastrula in unconditioned medium. *J. Embyol. Exp. Morphol.* **7**, 210–222.

Bergink, E. W., and Wallace, R. A. (1974). Precursor–product relationship between amphibian vitellogenin and the yolk proteins lipovitellin and phosvitin. *J. Biol. Chem.* **249**, 2897–2903.

Brummett, A. R., and Dumont, J. N. (1977). Intracellular transport of vitellogenin in *Xenopus* oocytes: An autoradiographic study. *Dev. Biol.* **60**, 482–486.

Callen, J. C., Tourte, M., Dennebuoy, N., and Mounolou, J. C. (1980). Mitochondrial development in oocytes of *Xenopus laevis*. *Biol. Cell.* **38**, 13–18.

Dumont, J. N. (1972). Oogenesis in *Xenopus laevis* (Daudin). I. Stages of oocyte development in laboratory maintained animals. *J. Morphol.* **136**, 153–180.

Eppig, J. J., and Dumont J. N. (1976). Defined nutrient medium for the *in vitro* maintenance of *Xenopus laevis* oocytes. *In Vitro* **12**, 418–427.

Follett, B. K., Nicholls T. J., and Redshaw, M. R. (1968). The vitellogenic response in the South African clawed toad (*Xenopus laevis* Daudin). *J. Cell. Physiol.* **72** (Suppl.), 91–102.

Jared, D. W., and Wallace, R. A. (1969). Protein uptake *in vitro* by amphibian oocytes. *Exp. Cell Res.* **57**, 454–457.

Johnson, A. D., Cork, R. J., Williams, M. A., Robinson, K. R., and Smith, L. D. (1990). H-ras[val12] induces cytoplasmic but not nuclear events of the cell cycle in small *Xenopus* oocytes. *Cell Regul.* **1**, 543–554.

Keem, K., Smith, L. D., Wallace, R. A., and Wolf, D. (1979). Growth rate of oocytes in laboratory-maintained *Xenopus laevis*. *Gamete Res.* **2**, 125–135.

Niu, M. C., and Twitty, V. C. (1953). The differentiation of gastrula ectoderm in medium conditioned by axial mesoderm. *Proc. Natl. Acad. Sci. U.S.A.* **39**, 985–989.

Opresko, L. K., and Wiley, H. S. (1984). An enzymatic method for radiolabeling vertebrate vitellogenin. *Anal. Biochem.* **140**, 372–379.

Opresko, L. K., and Wiley, H. S. (1987a). Receptor-mediated endocytosis in *Xenopus* oocytes 1. Characterization of the vitellogenin receptor system. *J. Biol. Chem.* **262**, 4109–4115.

Opresko, L. K., and Wiley, H. S. (1987b). Receptor-mediated endocytosis in *Xenopus* oocytes 2. Evidence for two novel mechanisms of hormonal regulation. *J. Biol. Chem.* **262**, 4116–4123.

Opresko, L., Wiley, H. S., and Wallace, R. A. (1980a). Proteins iodinated by the chloramine-T method appear to be degraded at an abnormally rapid rate after endocytosis. *Proc. Natl. Acad. Sci. U.S.A.* **77**, 1556–1560.

Opresko, L., Wiley, H. S., and Wallace, R. A. (1980b). Differential postendocytotic compartmentation in *Xenopus* oocytes is mediated by a specifically bound ligand. *Cell. (Cambridge, Mass.)* **22**, 47–57.

Redshaw, M. R., and Follet, B. K. (1971). The crystalline yolk-platelet proteins and their soluble plasma precursors in an amphibian, *Xenopus laevis*. *Biochem. J.* **124**, 759–766.

Wallace, R. A. (1970). Studies on amphibian yolk IX. *Xenopus* vitellogenin. *Biochim. Biophys. Acta* **215**, 176–183.

Wallace, R. A. (1978). Oocyte growth in nonmammalian vertebrates. *In* "The Vertebrate Ovary" (R. E. Jones, ed.), pp. 469–502. New York.

Wallace, R. A., and Dumont, J. N. (1968). The induced synthesis and transport of yolk proteins and their accumulation by the oocyte in *Xenopus laevis*. *J. Cell. Physiol.* **72** (Suppl.), 73–102.

Wallace, R. A., and Jared, D. W. (1968). Studies on amphibian yolk VII. Serum phosphoprotein synthesis by vitellogenic females and estrogen-treated males of *Xenopus laevis*. *Can. J. Biochem.* **46**, 953–959.

Wallace, R. A., and Jared, D. W. (1969). Studies on amphibian yolk VIII. The estrogen-induced hepatic synthesis of a serum lipophosphoprotein and its selective uptake by the ovary and transformation into yolk platelet proteins in *Xenopus laevis*. *Dev. Biol.* **19,** 498–526.

Wallace, R. A., and Jared, D. W. (1976). Protein incorporation by isolated amphibian oocytes V. Specificity for vitellogenin incorporation. *J. Cell Biol.* **69,** 345–351.

Wallace, R. A., and Misulovin, Z. (1978). Long-term growth and differentiation of *Xenopus* oocytes in a defined medium. *Proc. Natl. Acad. Sci. U.S.A.* **75,** 5534–5538.

Wallace, R. A., Jared, D. W., and Nelson, B. L. (1970). Protein incorporation by isolated amphibian oocytes I. Preliminary studies. *J. Exp. Zool.* **175,** 259–270.

Wallace, R. A., Jared, D. W., Dumont, J. N., and Sega, M. W. (1973). Protein incorporation by isolated amphibian oocytes III. Optimum incubation conditions. *J. Exp. Zool.* **184,** 321–334.

Wallace, R. A., Misulovin, Z., Jared, D. W., and Wiley, H. S. (1978). Development of a culture medium for growing *Xenopus laevis* oocytes. *Gamete Res.* **1,** 269–280.

Wallace, R. A., Deufel, R. A., and Misulovin, Z. (1980a). Protein incorporation by isolated amphibian oocytes VI. Comparison of autologous and xenogeneic vitellogenins. *Comp. Biochem. Physiol.* **65B,** 151–155.

Wallace, R. A., Misulovin, Z., and Wiley, H. S. (1980b). Growth of anuran oocytes in serum-supplemented medium. *Reprod. Nutr. Dev.* **20,** 699–708.

Wiley, H. S., Opresko, L., and Wallace, R. A. (1979). New methods for the purification of vertebrate vitellogenin. *Anal. Biochem.* **97,** 145–152.

Wolf, K., and Quimby, M. C. (1964). Amphibian cell culture: Permanent cell line from the bullfrog (*Rana catesbeiana*). *Science* **144,** 1578–1560.

Yisraeli, J. K., Sokol, S., and Melton, D. A. (1989). The process of localizing a maternal messenger RNA in Xenopus oocytes. *Development (Cambridge, U.K.)* **1989** (Suppl.), 31–36.

Chapter 7

Biochemical Fractionation of Oocytes

JANICE P. EVANS AND BRIAN K. KAY

Department of Biology
University of North Carolina at Chapel Hill
Chapel Hill, North Carolina 27599

I. Introduction
II. Methods
    A. Manual Isolations
    B. Biochemical Fractionations
III. Discussion
    References

I. Introduction

To appreciate the biochemical composition of *Xenopus* oocytes, one should understand some basic facts about oogenesis. During oogenesis, the oocyte is programmed to accumulate enough macromolecules and organelles to last into early embryogenesis. This accumulation is important because the biosynthetic activities of the embryo do not begin immediately on fertilization. For example, transcription of the zygotic genome does not begin until the midblastula transition (Newport and Kirschner, 1982), approximately 5 hours into development. Synthesis of 18 S and 28 S rRNA does not become detectable until gastrulation (Brown and Littna, 1964). Conversely, translation gradually begins to increase through oocyte maturation (3-fold) and fertilization (2-fold), levels off to a steady rate through cleavage, and finally accelerates at the tailbud stage (stage 25). Despite this steady increase in protein synthesis, all the maternal ribosomes are not recruited onto polysomes for translation until stage 42, right before feeding begins, indicating that the oocyte has enough ribosomes to last until the third day of embryogenesis (Woodland, 1974).

Oogenesis in *Xenopus* takes approximately 3 months, progressing through a series of developmental stages termed I to VI. During this time, one of the most obvious changes in the oocyte is its increase in size, from a diameter of 50 μm to 1.2 mm (Dumont, 1972). The main cause for this increase in size is the accumulation of yolk proteins, which begins in stage II. The yolk protein precursor, vitellogenin, is synthesized by the estrogenic liver and then secreted into the bloodstream. This protein is taken up by the oocyte via receptor-mediated endocytosis, routed into endosomes, proteolytically processed, assembled into the mature yolk proteins lipovitellin and phosvitin, and finally packaged into yolk platelets. During embryogenesis, the yolk proteins are degraded and the amino acids used for *de novo* protein synthesis (Tata, 1976; Wahli *et al.*, 1981).

In addition to accumulating yolk proteins, the oocyte is synthesizing protein of its own as well. Protein synthesis proceeds at an essentially steady pace through oogenesis. A full-size oocyte (stage VI) is capable of synthesizing 200–400 ng of protein per day, averaging 17.5 ng/hour (Wassarman *et al.*, 1982), although slightly higher estimates (19–25 ng/hour) have been obtained by measuring the recruitment of ribosomes to polysomes (Woodland, 1974). Transcription is also an ongoing process during oogenesis. Rosbash and Ford (1974) calculated that the total RNA content of an oocyte increases from 0.04 μg in previtellogenic (stage I) oocytes to at least 4.3 μg in a mature (stage VI) oocyte. Of this, 0.7–1.0%, or 40 ng, is poly(A)$^+$ RNA, representing 20,000 different species of mRNA (Perlman and Rosbash, 1978). Ribosomal RNA makes up the major portion (80–90%) of the rest of the total RNA, with some direct measurements indicating that the oocyte contains up to 4 μg rRNA (Rosbash and Ford, 1974). The result of all these biosynthetic processes, showing the various macromolecules and organelles that accumulate during oogenesis (adapted from Laskey, 1974), is as follows:

	Approximate excess over amount in larval cells
Mitochondria	100,000
RNA polymerases	60,000–100,000
DNA polymerases	100,000
Ribosomes	200,000
tRNA	10,000
Histones	15,000
dNTPs	2500

In this chapter, we describe a few of the protocols for separating oocytes into various subcellular fractions. Oocytes can be fractionated in two ways. Because of their large size, it is possible to dissect manually some cellular components away from others. There are also a number of biochemical fractionation techniques that have been previously applied to cultured cells or whole tissues. Although we present protocols devised particularly for analysis of fractions by gel electrophoresis, it would be possible to modify them for other uses. It is also worth noting that many of the protocols presented also work on *Xenopus* ovary, eggs, and embryos.

II. Methods

A. Manual Isolations

Because of the large size of a *Xenopus* oocyte, it is quite easy to dissect the cell manually into various components. The secrets to manual isolations are patience and good forceps. The most popular are Dumont No. 5 forceps (Fine Science Tools, Inc., Belmont, CA; Cat. No. 11252-30), but any sharp, fine-tip watchmaker's forceps will do. In general, most of these isolations are done with the oocytes immersed in a saline buffer [OR2, Ringer's solution, modified Barth's saline (MBS), oocyte culture medium (OCM), etc.]; recipes for these preparations in Appendix A, this volume.

One of the easiest manual dissections is enucleation (Fig. 1). Place the oocytes in a plastic petri dish containing oocyte medium under a dissecting microscope and view with optic fiber illumination. Puncture the animal pole (the darkly pigmented hemisphere) of the oocyte with one tip of the forceps, then gently squeeze the oocyte with another pair of forceps. The nucleus, which is also called the germinal vesicle (GV), should emerge cleanly from the hole in the plasma membrane. The nucleus is large, approximately 0.4 mm in diameter, and opaque or whitish in appearance. To prevent any leakage of its contents, the GV should be isolated within 30 seconds of puncturing the oocyte and transferred via a yellow-tipped pipettor to a small tube containing 95% ethanol. If yolky cytoplasm is still attached to the GV, the GV can be cleaned up by repeated pipetting up and down in the OR2 solution; discard the nucleus if the nuclear membrane ruptures. (*Note*: The nucleus will lyse if it strikes an air–water interface.) It is possible to fix the oocyte before manual enucleation: Expose the oocyte to ice-cold 2% trichloracetic acid (TCA) for about 15 minutes, then wash with oocyte medium prior to dissection. Another tip to slow down loss of material from the nucleus during dissection is to include 1.5% polyvinylpyrolidine (PVP; Sigma Chemical Co., St. Louis, MO;

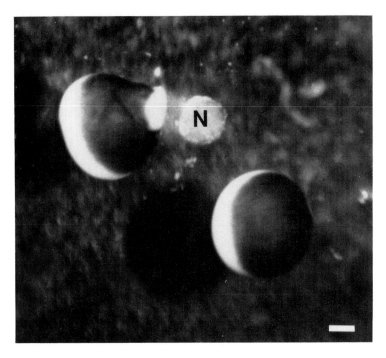

FIG. 1. Two oocytes, one intact (lower right), showing size and pigmentation. The dark hemisphere (which ranges in color from light to dark brown) is the animal pole; the lighter hemisphere (which is pale yellow in color) is the vegetal pole. The other oocyte (upper left) has been enucleated by piercing the animal pole. The nucleus (N) is to the right. Bar equals 0.3 mm.

Cat. No. PVP-40) in the oocyte medium. It should be noted that there may be difficulties with trying to enucleate an unfertilized *Xenopus* egg; these are described by Elsdale *et al.* (1960) in their discussion of the application of the technique of Briggs and King (1952) for enucleating the eggs of *Rana pipiens*.

With care, it is also possible to recover the remainder of the oocyte. For protein or RNA extractions, it would be best to remove the enucleated oocyte from the dish and freeze it in a tube on dry ice until ready for extraction. These fractions can be analyzed later after solubilization in sodium dodecyl sulfate (SDS)–dithiothreitol (DTT) loading buffer by polyacrylamide gel electrophoresis (PAGE) (Fig. 2).

It is also possible to remove the plasma membrane and cortical regions of cytoplasm from the rest of the oocyte. This procedure involves grasping the surface of the oocyte with two pairs of forceps and tearing to rip the plasma membrane away from the cytoplasm. The plasma membrane and cortical

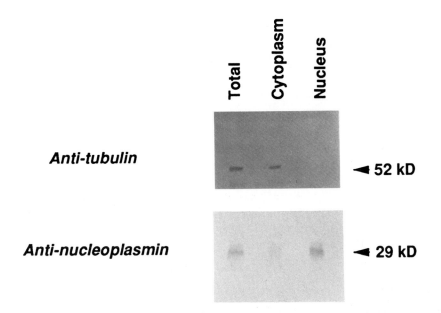

FIG. 2. Oocytes were left intact (for "total" samples) or manually dissected into nucleus and cytoplasmic fractions. Total and cytoplasmic proteins were prepared by Freon extraction (see Section II,B,1); nuclear proteins were extracted by homogenization in 1× SDS–DTT loading buffer. Three oocytes' equivalents of each sample were separated by SDS-PAGE, transferred to nitrocellulose, and reacted with either antitubulin (top blot) or antinucleoplasmin (lower blot) antibodies. Nucleoplasmin, being the most abundant nuclear protein in *Xenopus* oocytes (Mills, et al., 1980), separates primarily to the nuclear fraction; the region of the blot presented here shows the monomer (Laskey et al., 1978). Tubulin is detected in the cytoplasmic protein fraction.

cytoskeleton are fairly tough, so they can be pipetted up and down through a yellow pipette tip; the force of the passing buffer should wash any residual cytoplasm (primarily yolk granules) away from the cortical regions. (See Chapter 13 this volume, for variations on this type of fractionation.)

A similar technique can also be used to isolate manually the ovarian envelope (an acellular layer of glycoproteins than surrounds the oocyte, precursor of the vitelline envelope; see Chapter 12, this volume) from isolated oocytes. Follicular sheaths, including follicle cells and other ovarian accessory cells, may similarly be peeled from the surfaces of oocytes that are still contained in intact ovarian tissue. Finally, animal and vegetal hemispheres can be separated from each other simply by freezing the oocyte on a piece of aluminum foil on dry ice and cutting it with a sharp scalpel or razor blade (Rebagliati et al., 1985); the pieces can then be transferred quickly (before they thaw) to a tube on dry ice for protein or RNA isolation.

B. Biochemical Fractionations

1. Freon Extraction to Remove Yolk Proteins from Oocyte Lysates

Of all the proteins in the oocyte, the yolk proteins are the most abundant, accounting for over 80% of total protein. Vitellogenin (~250 kDa) is cleaved within the oocyte to give the yolk proteins lipovitellin I (~120 kDa), phosvitin (~35 kDa), lipovitellin II (~30 kDa), and the smaller phosphoproteins phosvettes (15–21 kDa) (Wiley and Wallace, 1981). The abundance of yolk polypeptides in the oocyte can be especially troublesome when preparing samples for SDS-PAGE and immunoblotting, as these proteins can often nonspecifically react with antibodies or mask a signal. Fortunately, there is a simple method (Gurdon and Wickens, 1983) for extracting yolk from protein lysates (Fig. 3: in lane 2 note the lack of the broad band around 110–130 kDa in the Freon-extracted sample).

Protocol

1. Homogenize oocytes in ice-cold 15 mM Tris (pH 6.8) plus 150 µg/ml phenylmethylsulfonyl fluoride (PMSF). (*Note*: The original protocol suggests extracting 10–30 oocytes in 1 ml of buffer, although we find that 10–20 µl per oocyte works well.

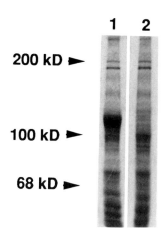

FIG. 3. Oocyte proteins were prepared for SDS-PAGE by either simple homogenization (lane 2) or by Freon extraction (lane 1). Two oocytes' equivalents of protein were loaded in each lane. Samples were run on a 6% SDS-polyacrylamide gel and stained with Coomassie blue. Note that the prominent band of yolk proteins at 110–130 kDa in lane 2 is absent from the protein sample in lane 1.

2. Add an equal (or greater) volume of Freon (1,1, 2-trichlorotrifluoroethane). Vortex well.
3. Spin for 10 min at 10–14 krpm in a Sorvall SS34 rotor or in a microcentrifuge.
4. Recover the upper phase and transfer it to a clean tube. (*Note*: The interface between the two phases will be coated with dark pigment granules and yolk protein; avoid taking up any of this material. The supernatant can also be spun a second time to clarify it.)
5. Add 2–10 volumes of cold acetone to the supernatant. Let sit on ice 25 min. Spin in a microcentrifuge for 25 minutes and discard supernatant. Air-dry the pellet. Resuspend in an appropriate volume (allow 1–3 µl per oocyte) of 1X SDS–DTT sample buffer. (*Note*: The proteins extracted from 1–2 oocytes can easily be detected by Coomassie blue staining in a lane of an SDS-polyacrylamide gel.)

Comments

1. The original reference states that you can analyze the supernatant directly by SDS-PAGE. However, in some instances it may be preferable to precipitate the proteins from the solution with acetone and resuspend them in a smaller volume of SDS–DTT loading buffer (see below). As a bonus, the acetone precipitation seems to clean up the sample a little. However, acetone precipitation is optional and may not be ideal for some applications (e.g., preparing samples for immunoprecipitation).
2. Phenylmethylsulfonyl fluoride (Sigma, Cat. No. P-7626) is a serine protease inhibitor. A 10 mg/ml stock solution is prepared in 95% ethanol and stored at $-20°C$. This compound is very toxic, so appropriate precautions should be taken.
3. 1,1,2-Trichlorotrifluoroethane is commercially known as Freon (Sigma, Cat. No. T-5271).
4. We use an SDS–DTT-containing buffer for our sample for SDS-PAGE. The protein sample buffer described by Laemmli (1970) is also acceptable.

5 × SDS–DTT Sample Loading Buffer (10 ml)

0.5 M Tris base	0.61 g
8.5% SDS	0.85 g
27.5% Sucrose	2.75 g
100 mM DTT	0.154 g
0.03% Bromophenol blue	3.0 mg

Add water to 10 ml. It will take time and a little bit of heating to solubilize all the components. Store 500-µl aliquots at $-20°C$. (*Note*: DTT may be omitted for nonreducing gel conditions.)

2. Heat Extraction for Cytoskeletal Proteins

The following protocol was adapted from one by Feramisco and Burridge (1980) for purifying α-actinin and filamin from chicken smooth muscle. It works on *Xenopus* oocytes as well, specifically for fractionation of microfilament-associated proteins (Fig. 4A shows a Coomassie-stained gel of the proteins extracted at various steps of the protocol).

Protocol

1. Homogenize oocytes in 10 volumes of ice-cold doubly distilled water with 1 mM PMSF and 1 mM leupeptin present. Use approximately 10–20 µl per oocyte.
2. Spin for 10 minutes at 10,000 rpm (or in a microcentrifuge) at 4°C.
3. Discard the supernatant or acetone-precipitate it and run it on a gel (lane 1) to compare the fraction of soluble proteins with that of cytoskeletal

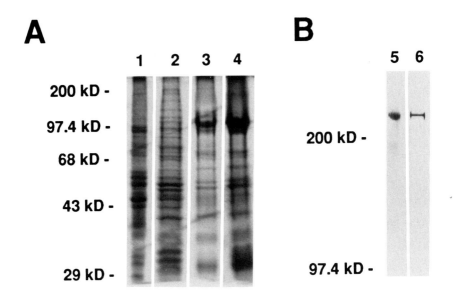

FIG. 4. (A) Ovarian protein samples from each step of the heat-extraction protocol, separated by 10% SDS-PAGE and stained with Coomassie blue. Lane 1 indicates first supernatant (Step 3); lane 2, second supernatant (Step 5); lane 3, heat-extracted cytoskeletal proteins (Step 7); and lane 4, proteins extracted from final pellet (Step 8). (B) Proteins from the first supernatant (5 oocytes' equivalents; lane 5) and heat-extracted proteins (10 oocytes' equivalents; lane 6) separated by 6% SDS-PAGE, transferred to nitrocellulose, and reacted with antifilamin antibodies. Filamin migrates at 240 kDa.

proteins. This fraction also serves as a good control to monitor the intactness of the initial crude homogenate.

4. Repeat the homogenization after solubilizing the pellet in 10 volumes of ice-cold water with 1 mM PMSF and 1 mM leupeptin present. Spin for 10 minutes at 10,000 rpm (or in microcentrifuge) at 4°C.
5. Again, discard this supernatant or acetone-precipitate it and check it by gel electrophoresis (lane 2).
6. Homogenize the pellet in 2 mM Tris-HCl, 5 mM EGTA, pH 9.0 (*Note*: pH is very important. Check the buffer before adding to the pellet and after homogenization and adjust if necessary.) Let the homogenate incubate at 37°C for 30–90 minutes, mixing periodically by shaking, stirring, or vortexing (every 5–15 minutes). (*Note*: The temperature is also very important; if possible, check with a thermometer or let the sample incubate for an extended time to make sure the temperature of the homogenate reaches 37°C.)
7. Spin for 10 minutes at 10,000 rpm (or in microcentrifuge) at 4°C.
8. The supernatant from Step 7 should contain cytoskeletal proteins; aliquot the supernatant and acetone-precipitate it (lane 3). Discard the pellet or homogenize it in 6 M urea, spin for 5–10 minutes and save the supernatant (acetone precipitation is recommended yet again to get a manageable volume and to clean the sample up a bit). Urea-extracted proteins from the pellet are run in lane 4 (Fig. 4A).

Comments

1. Leupeptin is a protease inhibitor (Sigma, Cat. No. L-2884) and can be made as a 10 mg/ml stock solution in water. Store at −20°C.

2. As noted in Step 6, the pH and temperature of the heat extraction are critical. [See the miniprint of Molony *et al.* (1987) for additional details.] Some attention must be directed toward optimizing the pH conditions and incubation times. A time-course analysis of samples from different incubation times may be done (e.g., compare what is extracted after 30; 60; and 90-minute time points).

3. When analyzing protein samples from each step of the fractionation procedure (Fig. 4A), it should be pointed out that sometimes the second supernatant (Step 5, shown in lane 2) may contain much less protein than is shown; this difference is dependent on the degree of completeness of protein solubilization from the first homogenization. In addition, since the heat-extracted fraction is such a minor component of the total protein mixture, more oocyte equivalents need to be loaded than those loaded of the first supernatant. [For example, the first supernatant (lane 1 of Fig. 4A) shows 0.05% of the protein extracted at this step. The second supernatant (lane 2) shows 0.75% of the protein extracted. In contrast, the lane loading of the heat

extraction (lane 3) is 2% of the protein extracted by that step. The pellet fraction (lane 4) shows 1% of the protein extracted by solubilization in urea.]

4. *Makers for a successful heat extraction.* The heat-extracted fraction from chicken gizzard should contain actin, desmin, filamin, α-actinin, vinculin, and talin (Feramisco and Burridge, 1980; Molony et al., 1987). Antibodies for some of these and other cytoskeletal proteins are available commercially (Sigma). (Figure 4B shows an example of an immunoblot using antifilamin antibodies.)

5. It is worth noting that this protocol may not be sufficient to solubilize intermediate filament proteins. Pondel and King (1988) have devised a protocol involving a high salt extraction that solubilizes cytokeratins and vimentin.

6. This protocol also has not been checked with regard to tubulin fractionation. Methods for isolating polymeric and free tubulin subunits from *Xenopus* oocytes have been published (Jessus et al., 1987; Gard and Kirschner, 1987).

3. FRACTIONATION OF OOCYTES INTO SOLUBLE CYTOSOLIC AND VESICLE PROTEINS

The following protocol was devised by Lane et al. (1979) as an alternative to a more complicated protocol, based on density gradient centrifugation, to separate soluble cytosolic proteins from proteins contained in membranous vesicles; this vesicle fraction is thought to contain endoplasmic reticulum (Zehavi-Willner and Lane, 1977). The protocol initially solubilizes cytosolic proteins while the vesicles are pelleted; the vesicles are later solubilized with detergent.

Protocol

1. Homogenize oocytes in ice-cold 50 mM NaCl, 10 mM MgCl$_2$, 20 mM Tris-HCl (pH 7.6), 1 mM PMSF, and 1 mM leupeptin (using 10–20 μl per oocyte).
2. Layer this crude homogenate on a 400-μl cushion of 20% sucrose, made up in the same buffer as in Step 1, in a 1.5-ml microcentrifuge tube.
3. Spin for 30 minutes in a microcentrifuge at 4°C.
4. Transfer the supernatant to a clean tube (aliquot to several tubes, if necessary); this fraction represents soluble cytoplasmic proteins. Acetone-precipitate to prepare for SDS-PAGE.
5. Homogenize for pellet(s) in 200 μl of phosphate-buffered saline (PBS) with 1% Nonidet P-40 (NP-40) and 1 mM PMSF. Spin for 10 minutes in a microcentrifuge at 4°C.
6. Transfer the supernatant from the extraction to a clean tube and acetone-precipitate it. This fraction represents the vesicle fraction.

Comments

1. There are two minor differences from the original protocols. First, the original protocol calls for magnesium acetate instead of magnesium chloride in the homogenization buffer. Second, the original protocol spun the homogenate on the sucrose cushion at 10,000–17,000 g (depending on the specific reference).
2. Nearly any cytosolic protein could be a marker for the soluble fraction, such as actin monomers. Markers for the vesicle fraction include various enzymes (Zehavi-Willner and Lane, 1977) or secretory or integral membrane proteins.
3. An alternative to this method is published in Bordier (1981). This protocol uses Triton X-114 instead of NP-40, but it is somewhat more difficult with respect to preparation of the detergent and fractionation of the protein samples.

4. RNA Extraction

It is relatively easy to isolate RNA from oocytes compared with other cell and tissue types. The RNA obtained is suitable for Northern blot analysis (Fig. 5), poly(A)$^+$ isolation, and cDNA synthesis.

Protocol

1. Homogenize oocytes in RNA extraction buffer (see *Comments*), allowing 50 µl per oocyte. [Small numbers may be homogenized with a Teflon dounce or by pipetting through a small-bore pipette tip. Larger numbers may be homogenized with a Tissuemizer (Tekmar Co., Cincinnati, OH) or Polytron (Brinkmann Instruments, Westbury, NY)]. All materials should be properly cleaned and RNase free.
2. Incubate at 37°C for 1 hour.
3. Spin at 10,000 rpm for 10 minutes.
4. Recover the supernatant. Add an equal volume of phenol–chloroform; vortex well (at least 15 seconds). Spin to separate phases (5 minutes at 10,000 rpm). There will be a white precipitate at the interface; be sure not to take any of it when withdrawing the upper aqueous phase to a clean tube. Repeat the phenol–chloroform extraction on the upper phase. After the centrifuge spin, extract the upper aqueous phase with an equal volume of chloroform.
5. Transfer the upper phase to a clean tube. Add 1/10 volume of 3 M sodium acetate and 2 volumes of ice-cold 100% ethanol. Precipitate nucleic acids overnight at $-20°C$ (or for a short time at $-80°C$, or on dry ice).

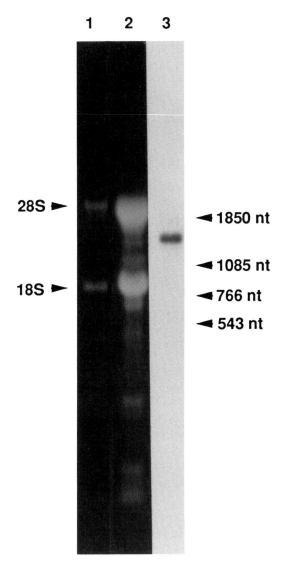

FIG. 5. Total RNA from tadpoles (stage 41) was run on a 1% formaldehyde–agarose gel and stained with ethidium bromide. Lane 1 shows 1 μg of RNA, and lane 2 shows 10 μg of RNA. The two prominent bands represent the 28 S and 18 S rRNAs. Lane 3 shows 1 μg of RNA hybridized with a probe corresponding to chicken α-actin cDNA.

6. Spin for 15 minutes at 5000 rpm or for 10 minutes at 10,000 rpm.
7. Let the pellet dry and then resuspend in diethyl pyrocarbonate (DEPC)-treated water equal to at least 25% of the volume used in Step 1. Add an equal volume of 8 M LiCl (Auffray and Rougeon, 1980). Leave at $-20°C$ overnight.
8. Spin for 10 minutes at 10,000 rpm. Wash the pellet with ice-cold 100% ethanol and repeat the spin.
9. Resuspend the pellet (after air-drying it) in DEPC-treated water. Store as an ethanol precipitate (i.e., add 1/10 volume of 3 M sodium acetate and 2 volumes ethanol) at $-20°C$ or $-80°C$.

Comments

RNA extraction buffer

	For 50 ml	Stock
50 mM NaCl	0.5 ml	5 M
50 mM Tris-HCl, ph 7.5	2.5 ml	1 M
5 mM EDTA	0.5 ml	0.5 M
0.5% SDS	2.5 ml	10%

Note: Add 200 µg/ml fresh proteinase K (Boehringer-Mannheim, Indianapolis, IN, Cat. No. 745-723) when doing the homogenization.

Preparation of stocks

Water, 5 M NaCl, and 0.5 M EDTA should be treated with 0.1% DEPC Sigma, Cat. No. S-5758) and then autoclaved (for at least 1 hour to remove the DEPC).

1 M Tris should be autoclaved alone. (DEPC cannot be used with Tris-containing buffers because it will react with primary amines.)

A solution of 10% SDS should be incubated at 65°C for 60 minutes.

Cleaning dounce or tissue homogenizer

Rinse first in 95% ethanol, then with 10 mM NaOH (to denature proteins), and finally with sterile water.

For alternatives to this protocol, refer to Gurdon and Wickens (1983), in which they describe their protocol and reference four others. Chirgwin *et al.* (1979) give a very different method using guanidinium to denature RNases. Yet another option is that of Chomcynski and Sacchi (1987), which uses extractions in water-saturated phenol to remove proteins and separate RNA from DNA.

5. Isolation of Organelles by Density Gradient Centrifugation

There are a number of published protocols on isolating various organelles from oocytes, eggs, and embryos; in general, these procedures involve gentle homogenizations and separation of subcellular components on sucrose gradients. Specific references include plasma membranes (Bretzel et al., 1986), nuclei and nuclear fractions (Bretzel and Tiedemann, 1986), mitochondria (Chase and Dawid, 1972), polysomes (Woodland, 1974).

III. Discussion

There are numerous examples in the literature describing how manual and biochemical fractionations of *Xenopus* oocytes and embryos have provided insight into a number of questions. One of the best examples was the separation by Melton and colleagues of animal and vegetal hemispheres of oocytes, extraction of RNA, and the identification of mRNA species unique to each hemisphere (Rebagliati et al., 1985). The best known of these is Vg1, which has sequence homology to transforming growth factor-β, one of the family of growth factors implicated in mesoderm induction in embryos (Weeks and Melton, 1987). Somewhat related to this discovery was the finding of Pondel and King (1988) that Vg1 RNA could be extracted from oocytes in a biochemical fraction that contains cytokeratins.

In our laboratory (Evans et al., 1990), the cytoskeletal protein vinculin was found to be absent from the cortical region of oocytes by immunofluorescence. This finding was verified biochemically in fractionation experiments: By immunoblotting, vinculin was found to be absent from cytoskeletal extracts of oocytes, while it was present in cytoskeletal extracts from other cell types.

These protocols can be applied to microinjection experiments as well, checking the fates of injected proteins or protein products of injected RNAs. An example of this (Colman et al., 1981) is the finding that when chicken oviduct mRNA was injected into oocytes, three ovalbumin polypeptides were synthesized and secreted into the incubation medium. When the injected oocytes were fractionated into soluble and vesicle fractions, newly synthesized ovalbumin was found in the vesicle fraction, with the exception of one polypeptide species that was miscompartmentalized to the cytosol. This ovalbumin species was neither glycosylated nor secreted, presumably because it did not get packaged in the endoplasmic reticulum.

A major advantage of the oocyte system is the relative ease with which it may be fractionated into various subcellular components. It is possible in some cases to mass-isolate certain components, such as nuclei (Scalenghe et al., 1978). We have presented a handful of useful protocols and references for separating components of interest from oocytes.

Acknowledgments

We thank Carl Feldherr for the antibody to nucleoplasmin, Ted Salmon for the antibody to tubulin, Keith Burridge for the antibody to filamin, Vickie Fowler for the preparation of material for Fig. 2, and Vickie Fowler and Bronwen Nishikawa for critical reading for the manuscript.

References

Auffray, C., and Rougeon, F. (1980). Purification of mouse immunoglobulin heavy chain messenger RNAs from total myeloma tumor RNA. *Eur. J. Biochem.* **107**, 303–314.

Bordier, C. (1981). Phase separation of integral membrane proteins in Triton X-114 solution. *J. Biol. Chem.* **256**, 1604–1607.

Bretzel, G., and Tiedemann, H. (1986). Neural-inducing activity of nuclei and nuclear fractions from *Xenopus* embryos. *Roux's Arch. Dev. Biol.* **195**, 123–127.

Bretzel, G., Janeczek, J., Born, J., Manorama, J., Tiedemann, H., and Tiedemann, H. (1986). Isolation of plasma membranes from *Xenopus* embryos. *Roux's Arch. Dev. Biol.* **195**, 117–122.

Briggs, R. and King, T. J. (1952). Transplantation of living nuclei from blastula cells into enucleated frogs' eggs. *Proc. Natl. Acad. Sci. U.S.A.* **38**, 455–463.

Brown, D. D., and Littna, E. (1984). RNA synthesis during the development of *Xenopus laevis*, the South African clawed toad. *J. Mol. Biol.* **8**, 669–687.

Chase, J. W., and Dawid, I. B. (1972). Biogenesis of mitochondria during *Xenopus laevis* development. *Dev. Biol.* **27**, 504–518.

Chirgwin, J. M., Przybyla, A. E., MacDonald, R. J., and Rutter, W. J. (1979). Isolation of biologically active ribonucleic acid from sources enriched in ribonuclease. *Biochemistry* **18**, 5294–5299.

Chomcynski, P., and Sacchi, N. (1987). Single-step method of RNA isolation by acid guanidinium thiocynanate–phenol–chloroform extraction. *Anal. Biochem.* **162**, 156–159.

Colman, A., Lane, C. D., Craig, R., Boulton, A., Mohun, T., and Morser, J. (1981). The influence of topology and glycosylation on the fate of heterologous secretory proteins made in *Xenopus* oocytes. *Eur. J. Biochem.* **113**, 339–348.

Dumont, J. N. (1972). Oogenesis in *Xenopus laevis* (Daudin): I. Stages of oocyte development in laboratory maintained aanimals. *J. Morphol.* **136**, 153–180.

Elsdale, T. R., Gurdon, J. B., and Fischberg, M. (1960). A description of the technique for nuclear transplantation in *Xenopus laevis*. *J. Embryol. Exp. Morphol.* **8**, 437–444.

Evans, J. P., Page, B. D., and Kay, B. K. (1990). Talin and vinculin in the oocytes, eggs, and early embryos of *Xenopus laevis*: A developmentally regulated change in distribution. *Dev. Biol.* **137**, 403–413.

Feramisco, J. R., and Burridge, K. (1980). A rapid purification of α-actinin, filamin, and a 130,000-dalton protein from smooth muscle. *J. Biol. Chem.* **255**, 1194–1199.

Gard, D., and Kirschner, M. (1987). Microtubule assembly in cytoplasmic extracts of *Xenopus* oocytes and eggs. *J. Cell Biol.* **105**, 2191–2201.

Gurdon, J. B., and Wickens, M. P. (1983). Use of *Xenopus* oocytes for the expression of cloned genes. *In* "Methods in Enzymology" (R. Wu, L. Grossman, and K. Moldave, eds.), Vol. **101**, pp. 370–386. Academic Press, New York.

Jessus, C., Thibier, C., and Ozon, R. (1987). Levels of microtubules during meiotic maturation of the *Xenopus* oocyte. *J. Cell Sci.* **87,** 705–712.

Laemmli, U. K. (1970). Cleavage of structural proteins during the assembly of the head of the bacteriophage T4. *Nature (London)* **227,** 680–685.

Lane, C., Shannon, S., and Craig, R. (1979). Sequestration and turnover of guinea-pig milk proteins and chicken ovalbumin in *Xenopus* oocytes. *Eur. J. Biochem.* **101,** 485–495.

Laskey, R. A. (1974). Biochemical processes in early development. *In* "Comparison to Biochemistry" (A. T. Bull, J. R. Lagnado, J. O. Thomas, and K. F. Tipton, eds.), Vol. 2, pp. 137–160. Longman, London.

Laskey, R. A., Honda, B. M., Mills, A. D., and Finch, J. T. (1978). Nucleosomes are assembled by an acidic protein which binds histones and transfers them to DNA. *Nature (London)* **275,** 416–420.

Mills, A. D., Laskey, R. A., Black,P., and DeRobertis, E. M. (1980). An acidic protein which assembles nucleosomes *in vitro* is the most abundant protein in *Xenopus* oocyte nuclei. *J. Mol. Biol.* **139,** 561–568.

Molony, L., McCaslin, D., Abernathy, J., Paschal, B., and Burridge, K. (1987). Properties of talin from chicken gizzard smooth muscle. *J. Biol. Chem.* **262,** 7790–7795.

Newport, J., and Kirschner, M. (1982). A major developmental transition in early *Xenopus* embryos: I. Characterization and timing of cellular changes at the mid-blastula transition. *Cell (Cambridge, Mass.)* **30,** 675–686.

Perlman, S., and Rosbash, M. (1978). Analysis of *Xenopus laevis* ovary and somatic cell polyadenylated RNA by molecular hybridization. *Dev. Biol.* **63,** 197–212.

Pondel, M., and King, L. (1988). Localized mRNA related to transforming growth factor β mRNA is concentrated in a cytokeratin-enriched fraction from *Xenopus* oocytes. *Proc. Natl. Acad. Sci. U.S.A.* **85,** 7612–7616.

Rebagliati, M. R., Weeks, D. L., Harvey, R. P., and Melton, D. A. (1985). Identification and cloning of localized maternal RNAs from *Xenopus* eggs. *Cell (Cambridge, Mass.)* **42,** 769–777.

Rosbash, M., and Ford, P. J. (1974). Polyadenylic acid-containing RNA in *Xenopus laevis* oocytes. *J. Mol. Biol.* **85,** 87–101.

Scalenghe, F., Buscaglia, M., Steinheil, C., and Crippa, M. (1978). Large-scale isolation of nuclei and nucleoli from vitellogenic oocytes of *Xenopus laevis*. *Chromosoma* **66,** 299–308.

Tata, J. R. (1976). The expression of the vitellogenin gene. *Cell (Cambridge, Mass.)* **9,** 1–14.

Wahli, W., Dawid, I. B., Ryffel, G. U., and Weber, R. (1981). Vitellogenesis and the vitellogenin gene family. *Science* **212,** 298–304.

Wassarman, W. J., Richter, J. D., and Smith, L. D. (1982). Protein synthesis during maturation-promoting factor- and progesterone-induced maturation in *Xenopus* oocytes. *Dev. Biol.* **89,** 152–158.

Weeks, D. L., and Melton, D. A. (1987). A maternal mRNA localized to the vegetal hemisphere in *Xenopus* eggs codes for a growth factor related to TGF-β. *Cell (Cambridge, Mass.)* **51,** 861–867.

Wiley, H. S., and Wallace, R. A. (1981). The structure of vitellogenin. Multiple vitellogenins in *Xenopus laevis* give rise to multiple forms of the yolk proteins. *J. Biol. Chem.* **256,** 8626–8634.

Woodland, H. R. (1974). Changes in the polysome content of developing *Xenopus laevis* embryos. *Dev. Biol.* **40,** 90–101.

Zehavi-Willner, T., and Lane, C. (1977). Subcellular compartmentation of albumin and globin made in oocytes under the direction of injection messenger RNA. *Cell (Cambridge, Mass.)* **11,** 683–693.

Chapter 8

Lampbrush Chromosomes

JOSEPH G. GALL AND CHRISTINE MURPHY

Department of Embryology
Carnegie Institution
Baltimore, Maryland 21210

HAROLD G. CALLAN

Gatty Marine Laboratory
University of Saint Andrews
Saint Andrews, KY16 8LB Scotland

ZHENG'AN WU

Institute of Developmental Biology
Academia Sinica
Beijing 100080, People's Republic of China

I. Introduction
II. Protocol for Lampbrush Chromosomes of *Xenopus*
   A. General Comments
   B. Detailed Protocol
III. Composition of Isolation and Dispersal Media
   A. Isolation Medium
   B. Dispersal Medium
IV. Alternatives If Germinal Vesicle Contents Fail to Disperse
V. Use of Lampbrush Chromosome Preparations
   A. Morphological Observations
   B. Immunofluorescence
   C. *In Situ* Hybridization
VI. Solutions
   A. Stock Solutions
   B. Working Solutions
   C. Mounting Medium for Immunofluorescence

VII. Materials
   A. Subbed Slides
   B. Dispersal Chamber
   C. Tungsten Needles
References

I. Introduction

Xenopus has many outstanding features as an experimental organism, but its lampbrush chromosomes are not among them. To begin with, the diploid number of *Xenopus laevis* is 36, so there are 18 lampbrush bivalents to contend with. Furthermore, the genomic DNA content (C value) is 3.1 pg, one of the lowest among the Amphibia. A small genome has two consequences. First, the chromosomes are short and have small chromomeres. But more importantly for lampbrush chromosome analysis, even at maximal extension the lateral loops are short relative to loops from species with higher genomic DNA values. Just why there is a correlation between loop length and C value is not intuitively obvious, although it probably has to do with the spacing of genes along the DNA (Gall *et al.*, 1983). In any event, if one wants to look at spectacular lampbrush chromosomes, one should start with salamanders having C values in the 30- to 50-pg range such as *Notophthalmus viridescens*, *Triturus cristatus*, or *Pleurodeles waltl*. Even in frogs such as *Rana*, with C values around 10 pg, the lampbrush loops are much more amenable to study than those of *Xenopus*. However, if *Xenopus* chromosomes are required for some special reason, good preparations can be made by attention to the factors discussed in the guidelines in this chapter. It should be mentioned that the relatively small lampbrush chromosomes do not detract from the usefulness of *Xenopus* for the study of extrachromosomal organelles in the oocyte nucleus or germinal vesicle (GV), including nucleoli (Miller, 1966), snurposomes (Gall, 1991; Wu *et al.*, 1991), and spheres (Gall and Callan, 1989).

Amphibian oocytes, like most animal oocytes, undergo their entire growth period during prophase of the first meiotic division. The earliest meiotic stages—leptotene, zygotene, pachytene, and early diplotene—occur before any significant enlargement. The chromosomes then "arrest" in diplotene while the nucleus and the cell undergo dramatic growth. During the initial period the nucleus and cytoplasm enlarge more or less proportionately, but after the deposition of yolk begins, the nucleus does not keep pace. Consequently the GV from a medium sized oocyte of a frog or newt will be almost as large as one from a fully mature oocyte.

Just when the lampbrush stage is reached is a matter of definition. Transcriptionally active chromosomes are present from the beginning of the growth phase (Hill and Macgregor, 1980), and in this sense lampbrush chromosomes are present throughout oogenesis. For technical reasons, however, only the most intrepid lampbrushologist would attempt a conventional GV spread from an oocyte with a diameter less than 0.5 mm. In most of the commonly studied amphibians, the lampbrush loops reach their maximal extension well before the oocyte reaches its final size. In the largest oocytes the loops have usually begun to regress, and the chromosomes as a whole have shortened in preparation for the first meiotic division.

In frogs and salamanders from temperate climates, the condition of the chromosomes depends strongly on the season, which, of course, governs the hormonal state of the animal. Thus, lampbrush chromosomes from the largest oocytes of the newt *N. viridescens* have extended loops in the fall and winter, but just before egg laying in spring, the loops regress as the chromosomes condense. In a frog like *Xenopus*, particularly when raised or held in the laboratory for a long time, the state of the oocytes will depend primarily on when ovulation last occurred (assuming the female has been well fed and otherwise cared for). The most dramatic effect of ovulation is the release of mature oocytes, but ovulation also stimulates growth of immature oocytes, with immediate changes in the lampbrush chromosomes. This fact is crucial for obtaining satisfactory lampbrush chromosomes from *Xenopus* (Callan *et al.*, 1987).

II. Protocol for Lampbrush Chromosomes of *Xenopus*

A. General Comments

Lampbrush chromosomes were first described in the latter part of the nineteenth century (Flemming, 1882; Rückert, 1892), but it was not until Duryee (1937, 1941, 1950) showed that they could be hand-isolated from living oocytes that their study took on special significance. Duryee realized the importance of Ca^{2+}-free solutions for maintaining the chromosomes in a lifelike state, and he made a number of significant observations under difficult working conditions. In particular, he did not have access to a phase-contrast microscope, and was thus limited to observations at relatively low numerical aperture. Over the years several important changes have been introduced into the technique of handling lampbrush chromosomes, of which the most valuable have been modifications of the isolation and spreading solutions, along with centrifugation as a means of attaching the chromosomes to the underlying substrate.

Although the following detailed instructions may appear complicated, the actual steps in making a lampbrush chromosome preparation are quite easy, and they can be learned in a few hours by anyone of moderate dexterity. The essential features are as follows:

1. Oocytes are removed from the frog and placed in a modified Ringer's solution (OR2), where they may remain for hours or days.

2. A GV is removed from an oocyte with forceps or needles in a Ca^{2+}-free medium designed to keep the nuclear contents as a gel indefinitely.

3. The nuclear envelope is removed with needles or forceps.

4. The gelled nuclear contents are transferred to a spreading solution in an appropriate chamber and allowed to disperse over a period of minutes or hours.

5. The preparation is centrifuged to attach the spread nuclear contents to the underlying substrate, usually a glass microscope slide.

6. The centrifuged material, now firmly attached to the substrate, is postfixed in whatever way desired.

Any further treatment of the GV material, such as for immunofluorescence or *in situ* hybridization, follows standard procedures that would be equally applicable to other types of squashes, spreads, or sections.

B. Detailed Protocol

In the following protocol the essential steps are given in italics, followed by explanatory notes. This protocol is based on the one given in Callan *et al.* (1987). Additional comments on technique can be found in *Working with Animal Chromosomes* (Macgregor and Varley, 1988) or in *Lampbrush Chromosomes* (Callan, 1986).

1. *Take a mature female Xenopus, chill it in ice water until it is immobile, and inject 200 IU of human chorionic gonadotropin into a lymph sac to the left or right of the dorsal midline.*

We find that the state of the lampbrush chromosomes depends critically on the hormone condition of the female, and that the best preparations (i.e., largest loops) come from stimulated animals. We normally keep mature animals in aquaria with the water temperature about 18°–20°C. It is probably advantageous to change to a higher temperature (25°C) for 2 or 3 days before injection, as suggested by Jamrich *et al.* (1983). During this period the female can be kept in a large bowl in the laboratory, unfed. Unless one has access to natural spring water, it is best to keep the animal in artificial spring water consisting of 1 mM CaCl$_2$, 1 mM NaHCO$_3$, and 1 μM Na$_2$EDTA.

After the gonadotropin injection, inspect the animal for ovulation over the next 2 days. If she ovulates, take an ovary sample within the following 2 days. If she does not ovulate, give a second injection of 100 IU, and again inspect daily for ovulation.

2. *After the female has ovulated, remove an ovary sample and place it in OR2 medium in a small petri dish (60 × 15 mm). Keep the sample refrigerated at about 4°C.*

Xenopus can be anesthetized by immersion for about 30 min in a 0.1% solution of MS222 (ethyl *m*-aminobenzoate methane sulfonate, Eastman 9671 Eastman Kodak, Rochester, NY). Before surgery the animal should be rinsed in artificial spring water and placed on her back on top of chopped ice. Alternatively, the anesthesia itself can consist of chilling in ice water until the animal is completely immobile. Because of the thick muscular wall of the abdomen, it is best to cut through the overlying skin first, and then make a second incision through the muscles. In a mature female the ovary occupies much of the abdominal cavity and is difficult to miss. An incision 2-3 cm to the right or left of the midline and about two-thirds the distance from front to hind legs will usually be satisfactory. A relatively small piece of ovary is adequate for lampbrush preparations. Close the incision with #3-0 silk suture; it may be easier to sew the musculature and skin separately. After the incision is closed, the animal can be kept for a day or two of observation in the laboratory before being returned to the main aquarium. *Xenopus* almost never becomes infected, probably because of natural antibiotics in the skin (Zasloff, 1987); sterile procedures and antibiotics are not required during or after surgery.

A curious feature of *Xenopus* oocytes, not evident in other species we have studied, is the gradual retraction of the lampbrush loops (presumably representing shutdown of transcription) after storage for more than a few hours in OR2. Thus, if one wants to make lampbrush preparations over a longer period, fresh oocytes must be removed periodically from the female.

3. *Remove a few oocytes to a small petri dish (35 × 10 mm) containing "5:1 + PO_4^{3-} + Mg^{2+}." Transfer through a second and then into a third dish of the same. The purpose of these washes is to remove all traces of Ca^{2+}.*

These and subsequent operations are performed under a binocular dissecting microscope, such as the Zeiss stereomicroscope with 0.8 × to 6.4 × zoom system, and a cold light source, such as the fiber optic Schott KL1500. Two pairs of stainless steel forceps (Dumont No. 4 or 5), or one pair of forceps and a pair of iridectomy scissors, are best for dissection of oocytes. The importance of good illumination cannot be overemphasized. If you cannot see what you are doing, particularly when dealing with the isolated GV, the situation is hopeless. You want dark-field illumination. If the dissecting microscope has feet below the stage and a reflecting substage mirror, remove them; the operating stage should rest directly on the table. Illuminate from the side and slightly above with one or preferably two bright sources, so that the specimen shines against a nearly black background.

The sample may be used for 30 minutes or so in "5:1 + PO_4^{3-} + Mg^{2+}," but oocytes do not keep well in a Ca^{2+}-free solution. They also distend rapidly in this solution, so that accurate sizing is not possible.

4. *Select an oocyte of 1.0 mm diameter. With two pairs of No. 5 forceps or one pair of forceps and a sharpened steel needle, make a large tear in the animal hemisphere of the oocyte. Find the GV within the protruding cytoplasm and roll it away.*

Although it is a relatively trivial matter to tear open an oocyte with forceps and find the GV, it is difficult to describe the operation in words. Do not try the method often used for enucleating mature oocytes for biochemical purposes, in which one simply pokes a small hole in the animal hemisphere and squeezes out the GV. This method removes the GV with minimal loss of cytoplasm but may damage the GV itself. The following details are offered by one of us (H.G.C.):

With Dumont No. 5 forceps, grasp the ovary wall where the oocyte is attached, holding the tips of the forceps steady on the floor of the petri dish. Take a mounted steel needle, sharpened on a carborundum stone, and slash the oocyte from a point just within the dark animal hemisphere toward the yellowish vegetal hemisphere. It is sometimes convenient to stab the needle through the oocyte and down on the floor of the petri dish, then rapidly pull the needle away. Cytoplasm will protrude from the tear and it should be stiff, that is, retain its shape. Now stroke the side of the needle over the protruding cytoplasm, and continue doing so until a persistent convexity, or a dark zone within the light reflecting yolk, indicates where the GV lies. Using a glass pipette with an inner diameter of about 0.8 mm, pull up some of the medium and gently pump away at the exposed surface of cytoplasm. Continue to do so after the envelope of the GV becomes visible, until the GV becomes detached from the cytoplasm. If the envelope of the GV has already become free from cytoplasm, and if the floor of the petri dish is reasonably clean, the GV will attach itself to the plastic surface; if the GV still has much attached cytoplasm, pick it up in the pipette and pump it in and out a few times until most of the GV surface is clean; then deposit the GV over a clean area of the floor of the petri dish, to which it should at once attach. Now gently pump medium over the GV, so forcing a greater area of envelope to attach itself to the plastic. During the entire procedure avoid air bubbles, and have enough isolation medium in the dish to prevent the GV from accidentally touching the air–water interface. Isolation and cleaning of a GV should not take more than a minute or two, but speed is not essential, because the nuclear contents do not disperse in the isolation medium.

If you find it difficult to work on a surface to which the GV sticks readily, you can "condition" the petri dish beforehand by filling it with 10% bovine serum albumin or 10% horse serum and rinsing well with water. Such conditioned dishes can be reused many times and will become even less sticky with time.

5. *Remove the nuclear envelope with two pairs of forceps or with one pair of forceps and a fine needle. The nuclear contents will remain indefinitely as a gelatinous ball in the isolation medium.*

If the GV is not attached to the petri dish, grasp it from above with one pair of forceps, forcing it down against the bottom of the dish. With a second pair of forceps or a tungsten needle, tear open the envelope, beginning the tear adjacent to the first pair of forceps. In most cases the GV contents will pop out of the envelope intact.

If the nucleus is attached to the bottom of the dish, the envelope can be removed as follows. Under a high magnification of the dissecting microscope, focus accurately on the plane where the GV is attached to the petri dish. In one hand pick up a sharpened pair of No. 5 Dumont forceps (preferably *not* stainless steel, since they can be ground sharper than stainless steel forceps) and in the other hand a very sharp tungsten needle. Approach the envelope carefully with the forceps, and nip it slightly above where it is attached to the plastic (a bite that includes about one-third of its diameter is appropriate). Lift the envelope, which will tear, over the top of the GV contents and tack it down on the floor of the petri dish on the opposite side. The tungsten needle may be required to assist the tacking down operation, or to tear more envelope along its line of attachment to the plastic if too little width of envelope was nipped initially by the forceps. The aim is to leave the contents of the GV undisturbed but freely exposed above. All of this can be done at leisure, since the GV contents of *Xenopus* remain as a stiff gel in the isolation medium. In organisms with more fluid GV contents, such as *Pleurodeles*, it is a hindrance to have the envelope stick to the plastic dish. In such cases it is best to work with conditioned plastic.

6. *With a nuclear pipette pick up the gelled GV contents and transfer to a small petri dish of dispersing medium. Drop the gel from just under the surface, so that it falls through the solution to the bottom of the dish. Remove the pipette, squirt out residual isolation medium, and fill the pipette with dispersing medium. Pick up and redrop the gel once. Proceed immediately to the next step.*

The purpose of this step is to transfer the nucleus from the isolation medium, in which it would remain indefinitely as a gel, to a medium in which the gel will slowly disperse. Since the final dispersal chamber on the microscope slide is quite small, one must first go through a dish of dispersal medium to ensure complete exchange. The nuclear gel without its envelope is more difficult to see than an intact GV, and it begins to change from opalescent to transparent as it sits in the dispersal medium. For this reason, carry out the washing step quickly, preferably under a low magnification of the dissecting microscope and with strong lateral illumination (i.e., maximal Tyndall effect).

7. *Transfer the GV gel to a dispersal chamber previously filled with 1/4 strength ($5:1 + PO_4^{3-}$) + 0.5 mM Mg^{2+} + 10 µM Ca^{2+} + 0.1% w/v paraformaldehyde + 0.5 mM dithiothreital (DTT), add an 18 mm² coverslip, and seal with petroleum jelly.*

The liquid in the well should have a convex surface. Drop the coverslip onto the liquid surface: do not slide it into place. Blot excess liquid with a piece of filter paper laid flat on top of the coverslip. Seal the edges of the coverslip with melted petroleum jelly. A convenient way to do this is to squeeze the petroleum jelly from a 10- to 20-ml syringe around the edge of the coverslip and then melt with a heated copper rod. Examine the preparation with a low magnification phase-contrast objective ($10 \times -20 \times$). Let the nuclear gel disperse for 1 hour or longer until the chromosomes are completely flat against the bottom of the well slide. Carry out the dispersal in a cold room or a refrigerator at about 4°C.

The biggest problem with *Xenopus* is obtaining adequate dispersal of nuclei that have well displayed lampbrush loops. Nuclei from the largest oocytes (>1.0 mm diameter) generally disperse rapidly, but the chromosomes are contracted and have short to nonexistent loops. Smaller oocytes (<1.0 mm diameter) have good chromosomes, but the GV contents often fail to disperse adequately in the medium suggested here. This is the reason for choosing oocytes of exactly 1.0 mm diameter. If your preparations still do not disperse, see Section IV for additional suggestions.

8. *Centrifuge the slides at 4000–5000 rpm (3100–4800 g) for 60 minutes at 5°–10°C, using special holders for the Sorvall HS-4 rotor; use the slow start.*

After centrifugation, the chromosome loops and all the extrachromosomal granules in the GV should be firmly attached to the glass slide, and you should not be able to detect Brownian motion. If you see movement at this point, material will partially or completely detach from the slide in subsequent steps.

The centrifugation step is the only one that requires special apparatus. Metal slide carriers similar to the ones we use have been described by Macgregor and Varley (1988). Alternatively, some large centrifuge buckets will accommodate microscope slides. For example, each bucket of the Sorvall RT-6000 bench-top refrigerated centrifuge will hold up to four slides. The bottom of the bucket is slightly curved; it should be fitted with a plastic plate that is flat on top to accommodate the slides but convex below to fit the bucket.

9. *To make permanent preparations, place the slide horizontally in a petri dish filled with 2% paraformaldehyde in phosphate-buffered saline (PBS). Push the coverslip away, and leave the slide undisturbed for 30–60 minutes. Transfer the slide carefully to a staining dish containing 70% v/v ethanol. After about 1 hour, split off the top chamber with a razor blade. The chromosomes will be on the glass slide surrounded by a more or less complete "dam" of petroleum jelly/paraffin.*

The slide can now be observed at higher magnification (in ethanol or buffer) after laying a coverslip on top of the preparation. Residual petroleum jelly/paraffin holds the coverslip above the plane of the chromosomes, preserving them from mechanical damage at this stage. The appearance of the chromosomes after fixation depends critically on how well the loops stuck to the slide. In the best slides, the loops will appear "relaxed" and well spread out from the main chromosome axis, the major difference after fixation being increased contrast of the specimen. If the chromosomes do not stick well to the slide, many of the loops will contract and become plastered over the axes of the chromosomes; those loops that remain extended will appear "jagged."

For some purposes you may want to avoid fixation with paraformaldehyde; for example, some antigens fail to give good immunofluorescence after aldehyde fixation. In this case you can go directly to 70% ethanol after centrifugation. The loops will be somewhat thinner without prior fixation in paraformaldehyde, but otherwise the preparation is satisfactory. Finally, depending on the use, chromosome preparations can be stored for days or weeks in 70% ethanol.

III. Composition of Isolation and Dispersal Media

Because a good deal of experimentation has gone into the development of the isolation and spreading media suggested here, we thought it would be valuable to discuss why the various ingredients are present and what effect changes in their concentrations have. We would be delighted if others can come up with improvements.

A. Isolation Medium

The isolation medium consists of 83 mM KCl, 17 mM NaCl, 10 mM PO_4^{3-}, and 1 mM Mg^{2+}, pH 7.2. Its purpose is (1) to maintain the morphology of the nuclear contents unchanged and (2) to maintain the GV contents as a gel during the initial manipulation, yet permit subsequent dispersal. The K/Na ratio makes no apparent difference; we use a 5:1 ratio because the actual ratio in the GV is in that range (Whitley and Muir, 1974). If the overall tonicity is lowered, the gel will tend to disperse. Phosphate has a specific gelling effect. For example, the GV will remain as a gel in 100 mM K/NaCl plus 10 mM PO_4^{3-}, but it will begin to disperse if the PO_4^{3-} is replaced by 10 mM Tris or MOPS. Similarly, 1 mM Mg^{2+} helps preserve the gel, and it has a dramatic stabilizing effect on the morphology of the nucleoli and other extrachromosomal structures. If the Mg^{2+} concentration in the isolation medium is raised too high (5 mM), the GV will not disperse in subsequent steps despite its otherwise normal appearance. The Mg^{2+} can be left out of the isolation medium, but it is required in the dispersal medium if good morphology of the extrachromosomal elements is desired.

B. Dispersal Medium

The dispersal medium contains K/NaCl and PO_4^{3-} at one-quarter the concentration in the isolation medium. It also contains 0.5 mM Mg^{2+}, 10 μM Ca^{2+}, and 0.1% paraformaldehyde. The lowered overall tonicity promotes gel dispersal. The 0.5 mM Mg^{2+} stabilizes the nucleoli and other extrachromosomal granules but counteracts the tendency to disperse. Contrary to expectation, based on other gels, Ca^{2+} dramatically destabilizes the nuclear contents at micromolar concentrations. However, at concentrations much above 10 μM, it also causes the loop matrix to disappear and the chromosomes to contract. These effects were first noticed by Duryee (1937, 1941, 1950), who therefore recommended Ca^{2+}-free media for the study of lampbrush chromosomes. Finally, a small amount of paraformaldehyde (0.1%) helps to stabilize the loops, which otherwise have a tendency to dissolve in low saline, particularly when traces of Ca^{2+} are present. Paraformaldehyde also destabilizes the nuclear gel—another unexpected result, since glutaraldehyde has exactly the opposite effect. However, the higher the paraformaldehyde concentration, the poorer the GV contents stick to the glass slide after centrifugation, possibly because the paraformaldehyde interacts with the gelatin in the subbing solution.

For organisms with more fluid GV contents, such as *Notophthalmus* and *Pleurodeles*, we recommend 1 mM Mg^{2+} and 0.01% paraformaldehyde in the isolation medium. The higher Mg^{2+} preserves the extrachromosomal structures better, and the lower paraformaldehyde ensures attachment to the glass slide. In the case of *Xenopus*, one is faced with a very stiff nuclear gel. The suggested Mg^{2+} and paraformaldehyde concentrations give the best compromise between dispersal, morphological preservation, and attachment to the slide.

In our earlier protocol for *Xenopus* lampbrush chromosomes (Callan et al., 1987), we did not include Mg^{2+} in either the isolation or dispersal solution. If one is interested in the chromosomes alone, Mg^{2+} is not necessary. In fact, the loop matrix is fuller and the overall length of the loops is greater without it. Miller spreads, which are carried out at very low ionic strength without divalent ions, represent the extreme in this direction. Without Mg^{2+}, however, most extrachromosomal structures (nucleoli, spheres, snurposomes) either dissolve or become extremely disrupted, and there is probably loss of protein from the loops as well.

IV. Alternatives If Germinal Vesicle Contents Fail to Disperse

As already mentioned, GVs from oocytes of less than 1.0 mm will usually not disperse well in the medium recommended above. Unfortunately, sometimes the same is true for GVs from larger oocytes. Under these circumstances, you will need to alter the dispersal medium. Success is often possible by simply diluting the dispersal medium in half, or by diluting just the K/NaCl and PO_4^{3-}, keeping the Mg^{2+}, Ca^{2+}, and paraformaldehyde at the original concentrations. Another possibility is to leave the Mg^{2+} out of the *isolation* medium but retain it in the spreading solution. The longer the GV remains in the isolation medium with Mg^{2+}, the more difficult it becomes to disperse. Those who are used to making lampbrush preparations can maneuver the nucleus into the dispersal medium quickly, but this may be difficult for beginners. As already mentioned, leaving Mg^{2+} out of both isolation and dispersal media is satisfactory if chromosomes are the only interest. Increasing the Ca^{2+} to much above 10 μM will cause the gel to disperse more rapidly but will usually have disastrous effects on the chromosomes; it might be a useful alternative for studying extrachromosomal structures.

If none of these methods work, and you have lost interest in seeing *Xenopus* lampbrush chromosomes, try a newt GV where dispersal is never a problem!

V. Use of Lampbrush Chromosome Preparations

Once the GV contents are attached permanently to a microscope slide, they may be treated like any other cytological preparation. The following notes will give a few suggestions for specific applications.

A. Morphological Observations

When the plastic square that forms the dispersal chamber is removed from the underlying microscope slide, some or all of the paraffin/petroleum jelly seal will remain on the slide. For many purposes this paraffin/petroleum jelly is useful, because it forms a shallow protective chamber around the GV contents. For instance, one can add a coverslip to the preparation with no fear of damaging the chromosomes. For phase-contrast or differential interference-contrast observations the GV contents can be mounted in PBS or 25% v/v glycerol in PBS. Glycerol preparations will last for weeks or months without drying out. Glycerol concentrations above 50% are usually not desirable, because the higher refractive index appreciably lowers the optical contrast. Preparations can be stained with a variety of fluorescent dyes for cytological examination. For instance, they can be stained for 1 or 2 minutes with DAPI (4′,6-diamidino-2-phenylindole) at 1 μg/ml in PBS, washed, and mounted, or the stain can be incorporated directly into the mounting medium. After DAPI staining, the chromomere axes of the chromosomes show up brilliantly, and the extrachromosomal DNA in the multiple nucleoli is usually obvious. Propidium iodide (1 μg/ml) stains intensely but somewhat less specifically. In addition to the chromosomes, the loops are moderately stained (some specific loops especially brightly), and the nucleoli are uniformly stained. For staining with Coomassie blue, see the comments in Section V,C.

B. Immunofluorescence

Antibody staining is particularly easy to carry out on centrifuged GV preparations, and the results are often spectacular because of the extremely low background. Once again the shallow chamber formed by the paraffin/petroleum jelly seal is advantageous, because it allows one to use as little as 5–10 μl of antibody. Our routine procedure is as follows:

1. If the slide is in 70% ethanol, transfer it through 50% and 35% ethanol to PBS. Wipe off excess PBS from the slide, leaving a small drop in the shallow well of paraffin/petroleum jelly. Add a drop of 10% horse serum/PBS to the well. Leave 10–15 minutes.

2. Replace the 10% horse serum with the first antibody. Leave 1 hour.

3. Rinse away the first antibody with several changes of 10% horse serum/PBS. Leave for a few minutes and rinse again.

4. Apply the second antibody, usually commercial rhodamine-or fluorescein-labeled goat antibody diluted 1:200 with 10% horse serum/PBS. Leave 1 hour.

5. Rinse away the second antibody with several changes of 10% horse serum/PBS, add about 8–10 μl of mounting medium, and apply a 22-mm

square coverslip (thickness #1). The mounting medium consists of 50% glycerol, 1 mg/ml p-phenylenediamine, and 0.02% w/v NaN_3.

The type of fixation (after centrifugation) will be dictated by the specific antigen under study. Most antibodies with which we have worked give best results when the GV contents are fixed for 30 minutes to 1 hr in 2% paraformaldehyde, followed by storage in 70% ethanol. A few antibodies fail to react after paraformaldehyde fixation, and in these cases the centrifuged preparations should be placed directly into 70% ethanol. Still others do better if the 70% ethanol step is omitted. However, some type of postfixation is necessary; otherwise, the material will come off the slide during the antibody steps, possibly because of proteases and/or nucleases in the horse serum.

Some antigens stain more intensely if the preparation is treated with urea before staining (Hausen and Dreyer, 1982). After the initial "blocking" in 10% horse serum/PBS, cover the preparation with 3 M urea for 5 min. Rinse in PBS and proceed to the primary antibody.

C. *In Situ* Hybridization

Spread GV preparations can be hybridized *in situ* with a wide variety of ^{3}H- or biotin-labeled probes (e.g., Callan *et al.*, 1987; Weber *et al.*, 1989). Since the hybridization steps are the same as for other types of cytological preparations, we mention only a few special features. Before hybridization it is necessary to remove the excess paraffin/petroleum jelly from the slides. This can be done by passing the slides from 70% ethanol through 95 and 100% ethanol (three quick changes) to xylene (three changes over a period of several hours). Return the slides to 100% ethanol (three quick changes) and then acetone. Remove the slides one at a time from the acetone, wave in the air to dry, and store for later use. For some reason the last traces of paraffin take a long time to dissolve, and if they are not removed completely, they will cause high background in the autoradiographic film. Carry out the *in situ* hybridization by standard procedures. Sodium dodecyl sulfate (SDS) is often used in hybridization solutions for nucleic acids on filters; it should not be used on cytological preparations, however, because it will damage the morphology.

Staining of autoradiographs has always presented problems. Many people have tried to stain chromosomes before applying the autoradiographic emulsion, only to learn that this does not work! Most histological stains are partially or completely removed by the development and fixation steps, but before their removal some cause an intense chemical "autoradiograph" and some desensitize the emulsion. For all practical purposes, staining of the specimen through the emulsion after development and fixation is the only alternative. Unfortunately, the emulsion itself consists of protein and may stain strongly. Giemsa stain is popular for autoradiographs because it stains the emulsion only weakly, but Giemsa staining of lampbrush chromosomes has always been erratic in our hands. More recently we have used Coomassie blue, which gives an intense and completely reliable stain of nuclear components; moreover, by adjusting the pH, one can completely eliminate stain in the emulson. For staining of autoradiographs proceed as follows:

1. After the autoradiograph has been fixed, washed, and dried, place it in 1% Coomassie blue R-250 in 50% methanol, 0.2 M NaCl, and 8% acetic acid. Add the acetic acid just before use. The solution is good for a few days, but the methanol gradually reacts with the acetic acid, the pH rises, and staining of the emulsion becomes a problem. Staining of the GV contents is very rapid: a few seconds to a minute is adequate. Alternatively one can overstain for 5–10 minutes and then destain in the next step.

2. Rinse for a few seconds in 50% methanol in 0.2 M NaCl (or longer, if the staining time was extended), then in 0.2 M NaCl, and finally in water. The purpose of the salt in the stain and washes is to prevent reticulation of the film, which is sometimes a problem in acid solutions of low molarity.

3. Coomassie blue stains so intensely that the silver grains in the autoradiograph may be difficult to see. There is a simple solution to this problem: use a blue filter in the microscope illumination system. The appropriate range is between 470 and 500 nm. The best is a variable-wavelength filter, which can be adjusted to give optimal contrast between the specimen and the silver grains (the grains will, of course, be completely opaque at all wavelengths). Such a filter is particularly advantageous for photography. The specimen can be photographed at about 470 nm to show only the silver grains, and then at a longer wavelength to show morphological details.

VI. Solutions

A. Stock Solutions

All of the working solutions can be made from the following stock solutions:

1 M KCl	0.5 M HEPES, pH 8.3
1 M NaCl	0.1 M Na$_2$HPO$_4$
1 M CaCl$_2$	0.1 M KH$_2$PO$_4$
1 M MgCl$_2$	5% paraformaldehyde

To make the paraformaldehyde solution, add 5 g paraformaldehyde powder to 100 ml of 1 mM Na$_2$CO$_3$, heat to 60°C until dissolved, cool, and filter. Store well stoppered.

B. Working Solutions

All the working solutions are sterilized by passage through a nitrocellulose filter (0.45-μm pore size) to prevent growth of bacteria. Some commercial filters contain traces of detergent to expedite wetting. Because the detergent causes loss of the lampbrush loop matrix, the filter should be washed repeatedly with distilled water before use.

1. OR2 Solution for Storage of Oocytes

For details, see Wallace et al. (1973).

Final concentration	Amount of stock solution
82.5 mM NaCl	82.5 ml
2.5 mM KCl	2.5 ml
1.0 mM CaCl$_2$	1.0 ml
1.0 mM MgCl$_2$	1.0 ml
1.0 mM Na$_2$HPO$_4$	10.0 ml
5.0 mM HEPES	10.0 ml
Distilled Water	893.0 ml

2. Isolation Medium without Mg^{2+} ("5 : 1 + PO_4^{3-}")

Final concentration	Amount of stock solution
83.0 mM KCl	83 ml
17.0 mM NaCl	17 ml
6.5 mM Na$_2$HPO$_4$	65 ml
3.5 mM KH$_2$PO$_4$	35 ml
Distilled water	800 ml

The pH should be 7.0–7.2. Dithiothreitol can be added as a reducing agent if desired (1 mM = 154 mg per liter). The solution can be stored for weeks in the refrigerator, but it should be filter sterilized just before use because it supports a detectable bacterial population.

3. Isolation Medium with Mg^{2+} ("5 : 1 + PO_4^{3-} + Mg^{2+}")

Same as above plus 1 mM MgCl$_2$.

4. Dispersal Medium

Take 250 ml of "5 : 1 + PO_4^{3-}," dilute with 720 ml of distilled water, and add 0.5 ml of 1 M MgCl$_2$ (0.5 mM final), 10 ml of 1 mM CaCl$_2$ (10 μM final), and 20 ml of 5% paraformaldehyde (0.1% final). Filter sterilize. Add 0.5 mM dithiothreitol, if desired.

C. Mounting Medium for Immunofluorescence

After immunofluorescent staining, preparations can be mounted in 50% glycerol with an antifade reagent such as *p*-phenylenediamine. Make a 10 mg/ml solution of *p*-phenylenediamine in water and adjust the pH to 8.5. Mix 5 ml glycerol, 4 ml water, 5 µl of 4% NaN_3, and 1 ml of the *p*-phenylenediamine solution, and aliquot into 1-ml portions in small tubes. Store in the freezer at $-20°C$ to retard oxidation (browning) of the *p*-phenylenediamine.

VII. Materials

A. Subbed Slides

Lampbrush preparations are generally made on standard 3 × 1 inch glass microscope slides. The GV contents will usually stick better to the slide if the surface has first been "subbed" or coated with a thin layer of gelatin. Take a box of slides and wash them in a warm detergent solution, wiping them individually with your fingers to remove surface contaminants. Place the washed slides in standard 10-slide carriers and rinse well with running water and finally distilled water. Dip the whole carrier in a 1× or 5× subbing solution and drain at an angle until dry. For most lampbrush work with other species we use a 1× subbing solution, but *Xenopus* chromosomes seem to attach better with the higher gelatin concentration. The stock 5× subbing solution is made as follows. Dissolve 5 g of gelatin, such as household Knox gelatin, in 1 liter of hot water, cool, and add 500 mg of chrome alum $[CrK(SO_4)_2 \cdot 12H_2O]$.

B. Dispersal Chamber

Spreading of GV contents takes place in a small well on top of the subbed slide. A simple well can be made from a 25-mm square of 1-mm-thick Plexiglas, in the middle of which is bored a 5 mm round hole. After the hole is bored, there will probably be a burr or lip around the edge, which must be removed by sanding down with a fine grade of emery paper. The plastic square is attached to the middle of a slide with a 1:1 mixture of petroleum jelly/paraffin wax. This can be done by placing a small amount of petroleum jelly/paraffin on a slide, pressing the plastic square into it, and melting on a warm plate. The amount of petroleum jelly/paraffin should be chosen so that it spreads evenly to the edges of the square without entering the hole. Subbed slides with plastic chambers can be made ahead of time and stored.

C. Tungsten Needles

For opening oocytes and removing the GV envelope, some workers prefer a tungsten needle and a pair of jeweler's forceps, rather than two pairs of forceps. Tungsten needles are made from 0.5-mm tungsten wire. Cut a 30- to 50-mm length of wire and attach it to a wooden handle or seal it in the end of a glass rod. Make a point on the end of the wire by repeatedly dipping into a dish of molten $NaNO_2$. This is a hazardous procedure because of the danger of sputtering or spilling the $NaNO_2$; it should be carried out in a hood with protective goggles.

Acknowledgments

Research was supported by National Institutes of Health Grant GM-33397 to J.G.G. and Science and Engineering Research Council Grant GR-E89926 to H.G.C. J.G.G. is an American Cancer Society Professor of Developmental Genetics.

References

Callan, H. G. (1986). "Lampbrush Chromosomes," pp. 1–254. Springer-Verlag, Berlin, New York.

Callan, H. G., Gall, J. G., and Berg, C. A. (1987). The lampbrush chromosomes of *Xenopus laevis*: Preparation, identification, and distribution of 5S DNA sequences. *Chromosoma* **95,** 236–250.

Duryee, W. R. (1937). Isolation of nuclei and non-mitotic chromosome pairs from frog eggs. *Arch. Exp. Zellforsch.* **19,** 171–176.

Duryee, W. R. (1941). The chromosomes of the amphibian nucleus. *In* "Cytology, Genetics, and Evolution," pp. 129–141. University of Pennsylvania Bicentennial Conference, Philadelphia, Pennsylvania.

Duryee, W. R. (1950). Chromosomal physiology in relation to nuclear structure. *Ann. N.Y. Acad. Sci.* **50,** 920–953.

Flemming, W. (1882). *in* "Zellsubstanz, Kern und Zelltheilung," pp. 1–424. F. C. W. Vogel, Leipzig, Germany.

Gall, J. G. (1991). Organelle assembly and function in the amphibian germinal vesicle. *In* "Advances in Developmental Biochemistry" (P. Wassarman, ed.), Vol. 1. JAI Press, Greenwich, Connecticut.

Gall, J. G., and Callan, H. G. (1989). The sphere organelle contains small nuclear ribonucleoproteins. *Proc. Natl. Acad. Sci. U.S.A.* **86,** 6635–6639.

Gall, J. G., Diaz, M. O., Stephenson, E. C., and Mahon, K. A. (1983). The transcription unit of lampbrush chromosomes. *Symp. Soc. Dev. Biol.* **41,** 137–146.

Hausen, P., and Dreyer, C. (1982). Urea reactivates antigens in paraffin sections for immunofluorescent staining. *Stain Technol.* **57,** 321–324.

Hill, R. S., and Macgregor, H. C. (1980). The development of lampbrush chromosome-type transcription in early diplotene oocytes of *Xenopus laevis*: An electron microscope analysis. *J. Cell Sci.* **44,** 87–101.

Jamrich, M., Warrior, R., Steele, R., and Gall, J. G. (1983). Transcription of repetitive sequences on *Xenopus* lampbrush chromosomes. *Proc. Natl. Acad. Sci. U.S.A.* **80,** 3364–3367.

Macgregor, H. C., and Varley, J. (1988). "Working with Animal Chromosomes," 2nd Ed., pp. 1–290. Wiley, Chichester, New York.

Miller, O. L. (1966). Structure and composition of peripheral nucleoli of salamander oocytes. *In* "The Nucleolus: Its Structure and Function" (W. S. Vincent and O. L. Miller, eds.), Vol. 23, pp. 53–66. National Cancer Institute Monographs, National Cancer Institute, Bethesda, Maryland.

Rückert, J. (1892). Zur Entwickelungsgeschichte des Ovarialeies bei Selachiern. *Anat. Anz.* **7,** 107–158.

Wallace, R. A., Jared, D. W., Dumont, J. N., and Sega, M. W. (1973). Protein incorporation by isolated amphibian oocytes: III. Optimum incubation conditions. *J. Exp. Zool.* **184,** 321–333.

Weber, T., Schmidt, E., and Scheer, U. (1989). Mapping of *situ* hybridization with biotin-labeled cDNA probes. *Eur. J. Cell Biol.* **50,** 144–153.

Whitley, J. E., and Muir, C. (1974). Determination of sodium and potassium in the nuclei of single newt oocytes. *J. Radioanal. Chem.* **19,** 257–262.

Wu, Z., Murphy, C., Callan, H. G., and Gall, J. G. (1991). Small nuclear ribonucleoproteins and heterogeneous nuclear ribonucleoproteins in the amphibian germinal vesicle: Loops, spheres, and snurposomes. *J. Cell Biol.* **113,** 465–483.

Zasloff, M. (1987). Magainins, a class of antimicrobial peptides from *Xenopus* skin: Isolation, characterization of two active forms, and partial cDNA sequence of a precursor. *Proc. Natl. Acad. Sci. U.S.A.* **84,** 5449–5453.

Chapter 9

Preparation of Synthetic mRNAs and Analyses of Translational Efficiency in Microinjected Xenopus Oocytes

MIKE WORMINGTON

Department of Biology
University of Virginia
Charlottesville, Virginia 22901

I. Introduction
   A. Choice of Transcription Vectors and Bacteriophage DNA-Dependent RNA Polymerases
   B. Capping and Polyadenylation of Synthetic mRNAs
II. Methods
   A. Solutions and Buffers
   B. Preparation of DNA Templates for *in Vitro* Transcription by Bacteriophage DNA-Dependent RNA Polymerases
   C. *In Vitro* Transcription and Isolation of Synthetic mRNAs
   D. Posttranscriptional Polyadenylation of Synthetic mRNAs with *Escherichia coli* Poly(A) Polymerase
   E. Analyses of Transcription Products and Microinjection of Synthetic mRNAs
   F. Analyses of Stability and Polysomal Association of Synthetic mRNAs in Microinjected Oocytes
References

I. Introduction

Functional mRNAs can be synthesized *in vitro* from cDNAs of interest cloned into a variety of plasmid or phagemid vectors which contain promoters recognized by the highly specific bacteriophage SP6, T3, and T7 DNA-dependent RNA polymerases. The ability to generate relatively large amounts of specific transcripts in combination with the efficient expression of these synthetic mRNAs in microinjected *Xenopus* oocytes provides a powerful and straightforward approach to address diverse problems ranging from the function and localization of a variety of proteins to the posttranscriptional and translational regulation of maternal mRNAs. A representative, although necessarily incomplete, summary of these applications includes the following:

1. The contribution of the 5′ terminal cap structure and 3′ poly(A) tail to mRNA stability and translational efficiency in oocytes (Krieg and Melton, 1984; Drummond *et al.*, 1985; Galili *et al.*, 1988).

2. The identification of cis-acting sequences required for altered polyadenylation states and translational activity of maternal mRNAs during oocyte maturation (Hyman and Wormington, 1988; McGrew *et al.*, 1989; Fox *et al.*, 1989; Varnum and Wormington, 1990; Fox and Wickens, 1990), and the endonucleolytic cleavage of a homeobox-containing maternal mRNA in stage VI oocytes (Brown and Harland, 1990).

3. Analyses of HIV-1 TAT-dependent activation of TAR-containing mRNAs (Braddock *et al.*, 1989), and trans-activation of DNA replication by a BPV-1-encoded gene product (Romanczuk and Wormington, 1989).

4. Analyses of the nuclear localization of a homeobox-containing gene product (Harvey *et al.*, 1986) and the targeting of a chromosomal protein to lampbrush chromosome loops (Roth and Gall, 1989).

5. Analyses of cell cycle regulatory proteins such as cyclin A (Swenson *et al.*, 1986) and the cytostatic factor, *c-mos* (Sagata *et al.*, 1989).

This chapter describes protocols which enable the synthesis and functional analysis of synthetic mRNAs in microinjected *Xenopus* oocytes. These procedures also address the requirements for and the ability to introduce a 5′ terminal cap structure and a 3′ poly(A) tail into synthetic mRNAs.

A. Choice of Transcription Vectors and Bacteriophage DNA-Dependent RNA Polymerases

A large selection of plasmid or phagemid vectors which contain promoters recognized by the highly specific bacteriophage SP6, T3, and T7 RNA polymerases are currently available (for a summary of representative transcription vectors, see Sambrook *et al.*, 1989). All three RNA polymerases are commercially available from many suppliers at comparable prices since these enzymes are now routinely purified from *Escherichia coli* strains overexpressing the relevant genes on high-copy plasmids. In general, SP6, T3, and T7 RNA polymerases have similar transcriptional efficiencies, although some variability may be encountered with individual enzyme lots purchased from different suppliers.

Although individual preference often governs the choice of a particular vector and RNA polymerase, the location of restriction endonuclease sites within the cDNA and available polylinker sequences in the vector must be considered. The 5' end of a transcript is dictated by the promoter, whereas the 3' terminus is determined by the restriction endonuclease cleavage site downstream of the insert. The cDNA must be subcloned in the sense orientation downstream of the promoter corresponding to the RNA polymerase to be utilized. It is advantageous to use a polylinker site which allows the 5' end of the cDNA to be inserted as close to the promoter as possible. This will minimize the presence of extraneous vector-encoded polylinker sequences which precede the translation initiation codon. These palindromic sequences have the potential to form stable secondary structures which can significantly reduce translational efficiency (Kozak, 1986). The sequences located between the promoter and the coding region within the cDNA should be scrutinized to ensure that no ATG codons are present. This sequence can cause artifactual translation initiation events to occur upstream of the legitimate start site within the cDNA (Kozak, 1989). When feasible, it is also advisable to remove as much of the 5' untranslated region as possible from the cDNA when it is subcloned into the transcription vector. This will remove potential cis-acting sequences which may negatively regulate translation such as those identified within the 5' untranslated regions of *c-myc*, ferritin, and ribosomal protein mRNAs (Lazarus *et al.,* 1988; Casey *et al.,* 1988; Mariottini and Amaldi, 1990).

The probability of efficiently translating a synthetic mRNA in microinjected oocytes can be increased by either of two ways during initial subcloning steps. The first and least labor-intensive approach is to subclone a cDNA which contains its own initiation codon into the SP6 vector, pSP64T (Krieg and Melton, 1984). Coding sequences inserted into pSP64T are flanked by 5' and 3'-untranslated regions derived from the *Xenopus* β-globin mRNA which is efficiently translated in microinjected oocytes. Alternatively, the sequences flanking the initiation codon within a cDNA can be modified by site-directed mutagenesis so that they conform to the optimal translation initiation sequence, GCCGCCA/GCCATGG (Kozak, 1989). For the vast majority of mRNAs, the presence of both a purine at position -3 and a G at position at $+4$ is sufficient to direct optimal translation provided that negative cis-acting untranslated sequences are absent. Naturally, if translational regulation of a synthetic mRNA is to be analyzed, its cognate 5' and 3' untranslated sequences should be retained in their entirety.

B. Capping and Polyadenylation of Synthetic mRNAs

The 5' terminal cap structure is a common feature of virtually all eukaryotic mRNAs. A number of studies have unambiguously established that the presence of a 5' cap structure is essential for the stability of synthetic mRNAs in microinjected *Xenopus* oocytes. For example, the presence of a 5' cap structure increases the stability of microinjected SP6 β-globin mRNA from less than 15 minutes to at least 48 hours (Krieg and Melton, 1984). Similarly, Drummond *et al.* (1985) demonstrated that capping improves the stability of several different synthetic transcripts over a 24-hour period by approximately 5-fold.

In addition to its role in mRNA stabilization, the 5' cap structure is recognized by eIF-4E (cap binding protein) which facilitates the efficient recruitment of an mRNA into a 43 S translation initiation complex (Rhoads, 1988). These interactions contribute significantly to the translational efficiency of synthetic mRNAs in microinjected oocytes (Krieg and Melton, 1984; Drummond *et al.*, 1985). In general, capping increases translational efficiency at least 5-fold in microinjected oocytes. All three of the bacteriophage promoters specify initiation at a G residue, and it is fortuitous that all three bacteriophage RNA polymerases will efficiently initiate transcription using either $m^7G(5')ppp(5')G$ or $G(5')ppp(5')G$ as primers. This permits a 5' terminal cap structure to be easily and efficiently added to synthetic mRNAs without the need for a subsequent posttranscriptional guanylyltransferase reaction (Krieg and Melton, 1984). Furuichi *et al.* (1977) have reported that oocytes will methylate the appropriate residues of injected capped (but unmethylated) RNAs. This means that synthetic mRNAs can be capped by inclusion of the significantly less expensive dinucleotide $G(5')ppp(5')G$ in transcription reactions.

In contrast to an essential requirement for a 5' terminal cap, the contribution of a 3' poly(A) tail to mRNA stability and translational efficiency in microinjected oocytes is somewhat more complicated. In general, polyadenylation improves the stability of at least some but not necessarily all synthetic mRNAs in injected oocytes (Drummond *et al.*, 1985; Galili *et al.*, 1988). The behavior of endogenous maternal mRNAs during *Xenopus* oocyte maturation clearly indicates that the poly(A) tail is a primary determinant of their translational efficiency. Moreover, these poly(A)-dependent changes in translational activity are not accompanied by alterations in mRNA stability. Thus, the translational activation of several mRNAs during maturation is coupled with their cytoplasmic polyadenylation (McGrew *et al.*, 1989; Fox *et al.*, 1989). Conversely, the deadenylation of a large number of maternal mRNAs results in their release from polysomes (Hyman and Wormington,

1988; Varnum and Wormington, 1990; Fox and Wickens, 1990). There is increasing evidence that the poly(A) tail enhances translation initiation via its interaction with poly(A) binding protein and the 60 S ribosomal subunit (Galili et al., 1988; Munroe and Jacobson, 1990; Sachs and Davis, 1989, 1990). Drummond et al. (1985) have reported that polyadenylation increases the translational efficiency of certain synthetic mRNAs in microinjected oocytes by as much as 20-fold. Thus, whereas it appears that polyadenylated RNAs are indeed preferentially translated relative to their deadenylated cognates, the exact increase observed for a given mRNA will depend on the contribution of the poly(A) tail to its intrinsic translation efficiency.

At least three SP6 vectors are available which contain oligo(dA-dT)$_{30-100}$ tracts inserted into their polylinker sequences (Baum et al., 1988; Munroe and Jacobson, 1990; Promega Biotec, Madison, WI). Transcripts synthesized from these templates contain 3' poly(A) tails which closely resemble those found on the corresponding native mRNAs although they are terminated by a few remaining nonadenylate residues of polylinker sequence. Similarly, RNAs transcribed from pSP64T contain a 3' poly(A)$_{23}$ tract terminated by a poly(C)$_{30}$ tail (Krieg and Melton, 1984). This "blocked" 3' terminus is refractory to deadenylation in mature oocytes, thereby enabling the translation of pSP64T-derived RNAs to be maintained following the meiotic maturation of microinjected stage VI oocytes (Varnum and Wormington, 1990; Fox and Wickens, 1990). The somewhat limited number of polylinker sites located upstream of the oligo(dA-dT) tracts within these vectors as well as the reported instability of the homopolymer sequences when propagated in bacterial hosts may preclude the use of one or more of these vectors. However, a 3' poly(A) tail containing approximately 100–325 adenylate residues can be easily and efficiently posttranscriptionally added to synthetic RNAs with *E. coli* poly(A) polymerase.

II. Methods

A. Solutions and Buffers

5 × Transcription Buffer
  200 m*M* Tris-HCl, pH 7.5, 30 m*M* MgCl$_2$, 50 m*M* NaCl, 10 m*M* spermidine. Autoclave for 15 minutes at 15 psi and store in aliquots at −20°C.
TE Buffer
  10 m*M* Tris-HCl, pH 7.5, 1 m*M* EDTA. Autoclave for 15 minutes at 15 psi and store in aliquots at room temperature.

10× "Cold" Ribonucleotide Mix
  10 mM each of ATP, CTP, UTP, and 0.5 mM GTP in sterile TE buffer. Neutralized 100 mM solutions of individual ribonucleotides can be purchased from Pharmacia LKB (Piscataway, NJ). These individual ribonucleotide stock solutions and the 10× "cold" ribonucleotide mix are stable for at least several months when stored in aliquots at $-20°C$.

10× "Hot" Ribonucleotide Mix
  10 mM each of ATP and CTP; 0.5 mM each of UTP and GTP in sterile TE buffer. Neutralized 100 mM solutions of individual ribonucleotides can be purchased from Pharmacia LKB. These individual ribonucleotide stock solutions and the 10× "hot" ribonucleotide mix are stable for at least several months when stored in aliquots at $-20°C$.

10× Cap Analog Solution
  10 mM G(5')ppp(5')G *or* 10 mM m^{7}G(5')ppp(5')G in sterile TE buffer. Store in aliquots at $-20°C$.

Dithiothreitol Solution
  100 mM dithiothreitol in sterile TE buffer. Store in aliquots at $-20°C$.

RNA Extraction Buffer (TNES)
  100 mM Tris-HCl, pH 7.5, 300 mM NaCl, 10 mM EDTA, 2% (w/v) sodium dodecyl sulfate (SDS). Autoclave for 15 minutes at 15 psi and store in aliquots at room temperature.

2× Poly(A) Polymerase Buffer
  100 mM Tris-HCl, pH 8.0, 500 mM NaCl, 20 mM MgCl$_2$, 2 mM MnCl$_2$. Autoclave for 15 minutes at 15 psi and store in aliquots at $-20°C$.

2× Injection Buffer
  176 mM NaCl, 10 mM Tris-HCl, pH 7.5. Autoclave for 15 minutes at 15 psi and store in aliquots at room temperature.

SDS–Proteinase K Buffer
  TNES plus 200 μg/ml of proteinase K. Add proteinase K from a 20 mg/ml stock made in sterile water stored at $-20°C$.

Polysome Buffer (PB)
  20 mM Tris-HCl, pH 7.4, 10 mM MgCl$_2$, 300 mM KCl, 4 μg/ml polyvinyl sulfate, 0.5% (v/v) Nonidet P-40. Autoclave for 15 minutes at 15 psi and store in aliquots at 4°C. Just before use, add 2 mM dithiothreitol, 25 units/ml placental ribonuclease inhibitor, and 10 μg/ml cycloheximide. *Note*: For control EDTA-release experiments, cycloheximide should be *omitted* from PB.

Sucrose Gradient Stock Solutions
  20% (w/v) *or* 15% (w/v) and 40% (w/v) sucrose in 20 mM Tris-HCl, pH 7.4, 10 mM MgCl$_2$, 300 mM KCl, 4 μg/ml polyvinyl sulfate, 0.5% (v/v) Nonidet P-40. Autoclave for 15 minutes at 15 psi and store in aliquots at 4°C. Just before use, add 2 mM dithiothreitol, 25 units/ml placental

ribonuclease inhibitor, and 10 μg/ml cycloheximide. *Note*: For control EDTA-release experiments, cycloheximide should be *omitted* from the sucrose gradient stock solutions, which should then be supplemented with 20 mM EDTA just before preparing the gradients.

B. Preparation of DNA Templates for *in Vitro* Transcription by Bacteriophage DNA-Dependent RNA Polymerases

To produce a transcription template, linearize the plasmid containing the cDNA of interest with an appropriate restriction endonuclease which cleaves at a single site within polylinker sequences located downstream of the insert. It is highly recommended that the integrity of the linearized DNA and completion of the restriction endonuclease digest be confirmed by electrophoresis of an aliquot of the reaction on nondenaturing gels before using the DNA as a template for *in vitro* transcription reactions. Following digestion, the DNA is extracted with phenol–chloroform, ethanol precipitated, and redissolved in TE buffer at a final concentration of 0.5–1.0 μg/μl. Linearized DNA templates should be stored at −20°C. Linearization of plasmids with restriction enzymes which leave 3′ protruding termini (e.g., *Kpn*I, *Pst*I, and *Sac*I) should be avoided if possible since bacteriophage RNA polymerases initiate aberrantly at these ends. This results in the synthesis of significant amounts of plasmid-length RNAs complementary to the desired mRNA (Schenborn and Mierendorf, 1985). If such an enzyme must be used, the linearized DNA will need to be treated with either T4 DNA polymerase or the Klenow fragment of *E. coli* DNA polymerase I to remove the 3′ protruding ends prior to *in vitro* transcription (Sambrook *et al.*, 1989).

C. *In Vitro* Transcription and Isolation of Synthetic mRNAs

Detailed procedures for *in vitro* transcription reactions using bacteriophage DNA-dependent RNA polymerases have been published elsewhere (Krieg and Melton, 1984; Melton, 1987; Sambrook *et al.*, 1989). The two protocols presented here are adapted from these published procedures and will permit the routine synthesis of 10–50 μg of capped mRNA per 50-μl reaction. These yields are sufficient for the microinjection of between 200 and 500 oocytes. The "cold" reaction described in Protocol 1 includes a trace amount of [^{3}H]UTP to facilitate determination of the exact yield of RNA synthesized (Sambrook *et al.*, 1989). The stability and polysomal association of these ^{3}H-labeled mRNAs can be ascertained by either Northern blot or nuclease protection analyses of RNA isolated from microinjected oocytes. The "hot" reaction described in Protocol 2 includes [α-^{32}P]UTP to uniformly label the synthetic mRNA to a relatively high specific activity. The stability and polysomal association of the ^{32}P-labeled mRNAs can be ascertained directly by electrophoretic fractionation of RNA isolated from microinjected oocytes followed by autoradiography. Either protocol can be easily scaled up or down to accommodate larger or smaller synthetic requirements, respectively. All steps are carried out using RNase-free plasticware, glassware, and solutions. A comprehensive description of methods used to eliminate nuclease contamination can be found in Sambrook *et al.* (1989). All reagents should be added to a sterile 1.5-ml microcentrifuge tube at room temperature in the order stated.

1. Synthesis of "Cold" mRNAs

Add the following to a sterile 1.5-ml microcentrifuge tube:

10.0 μl	5× transcription buffer
5.0 μl	100 mM dithiothreitol
50 units	placental ribonuclease inhibitor (e.g., Promega RNasin)
5.0 μl	10× "cold" ribonucleotide mix
5.0 μl	10× cap analog solution
1.0 μl	[5,6-^{3}H]UTP (1 μCi at 1 μCi/μl; 35–50 Ci/mmol)
5.0 μl	linearized DNA template at 0.5–1.0 μg/ul in TE

Add water to a final volume of 50 μl, then add 50 units of SP6, T3, or T7 RNA polymerase. Mix the components gently and incubate the reaction for 60 minutes at 37°–40°C. RNA yields can be improved by as much as 2- to 3-fold by the addition of a second aliquot of RNA polymerase followed by an additional 60-minute incubation at 37°–40°C.

2. Synthesis of "Hot" mRNAs

Add the following to a sterile 1.5-ml microcentrifuge tube:

10.0 μl	5× transcription buffer
5.0 μl	100 mM dithiothreitol
50 units	placental ribonuclease inhibitor (e.g., Promega RNasin)
5.0 μl	10× "hot" ribonucleotide mix
5.0 μl	10× cap analog solution
5.0 μl	[α-^{32}P]UTP (50 μCi at 10 μCi/μl; 800 Ci/mmol)
5.0 μl	linearized DNA template at 0.5–1.0 μg/μl

Add water to a final volume of 50 μl, then add 50 units of bacteriophage SP6, T3, or T7 RNA polymerase. Mix the components gently and incubate the reaction for 60 minutes at 37°–40°C. RNA yields can be improved by as much as 2- to 3-fold by the addition of a second aliquot of RNA polymerase followed by an additional 60-minute incubation at 37°–40°C.

3. REMOVAL OF DNA TEMPLATE AND RNA EXTRACTION

The DNA template should be removed from transcription reactions prepared as described in either Protocols 1 or 2 by digestion with RNase-free DNase I. The RNA is then extracted from the reactions and ethanol precipitated.

1. Add RNase-free DNase I (e.g., Promega RQ DNase I) to a final concentration of 1 unit/μg DNA template. Incubate at 37°C for 10 minutes.
2. Add 50 μl of TNES.
3. Extract with 100 μl of phenol–chloroform.
4. Extract with 100 μl of chloroform.
5. Remove two 1- to 2-μl aliquots from the final aqueous layer to measure the incorporation of radioactive UTP into the RNA product by either precipitation with trichloroacetic acid onto glass-fiber filters (e.g., Whatman GF/C, Clifton, NJ) or by adsorption onto DE-81 filters as described in detail by Sambrook *et al.* (1989). Both methods work efficiently and give comparable determinations of RNA yields.
6. Precipitate the RNA with 2.5 volumes of ethanol. The RNA can be stored under ethanol indefinitely at −20°C.
7. Recover the ethanol-precipitated RNA by centrifugation. Rinse the pellet twice with 70% ethanol. This will remove essentially all of the unincorporated nucleotides. Resuspend the RNA in sterile water and store at −70°C prior to microinjection or *in vitro* translation.

D. Posttranscriptional Polyadenylation of Synthetic mRNAs with *Escherichia coli* Poly(A) Polymerase

The role of the 3′ poly(A) tail in enhancing the translational efficiency of synthetic mRNAs in microinjected oocytes has been discussed previously. A 3′ poly(A) tract can be incorporated transcriptionally by the use of one of the SP6 vectors which contain oligo(dA-dT) tracts inserted into the polylinker sequences (Krieg and Melton, 1984; Baum *et al.*, 1988; Munroe and Jacobson, 1990; Promega Biotec). The somewhat limited number of polylinker sites located upstream of the oligo(dA-dT) tracts within these vectors as well as the reported instability of these homopolymer sequences when propagated in bacterial hosts may restrict the use of these vectors. Alternatively, a 3′ poly(A) tail containing approximately 100–325 adenylate residues can be easily and efficiently posttranscriptionally added to synthetic RNAs with

E. coli poly(A) polymerase. The protocol presented here is adapted from procedures described in detail elsewhere (McGrew et al., 1989; Sheets and Wickens, 1989). All reagents should be added to a sterile 1.5-ml microcentrifuge tube at 4°C in the order stated.

1. In Vitro POLYADENYLATION WITH Escherichia coli POLY(A) POLYMERASE

Add the following to a sterile 1.5-ml microcentrifuge tube at 4°C:

25 µl	2× poly(A) polymerase buffer
1 µl	100 mM dithiothreitol
50 units	placental ribonuclease inhibitor
1.0 µl	2.5 µg/µl bovine serum albumin in water
1.0 µl	2.5 mM ATP in TE
5.0 µl	synthetic RNA at 0.5–1.0 µg/µl in water

Add water to a final volume of 50 µl, then add 3 units of E. coli poly(A) polymerase. Mix the components gently and incubate the reaction for 5 minutes at 37°C.

2. EXTRACTION OF POLYADENYLATED RNA

1. Add 50 µl of TNES.
2. Extract with 100 µl of phenol–chloroform.
3. Extract with 100 µl of chloroform.
4. Precipitate the polyadenylated RNA with 2.5 volumes of ethanol. The RNA can be stored under ethanol indefinitely at −20°C.
5. Recover the ethanol-precipitated RNA by centrifugation. Rinse the pellet twice with 70% ethanol. This will remove essentially all of the unincorporated ATP. Resuspend the RNA in sterile water and store at −70°C prior to microinjection or in vitro translation. The poly(A) tail length can be measured by comparison of the polyadenylated and nonadenylated RNAs in denaturing polyacrylamide–urea or agarose–formaldehyde gels which contain the appropriate size standards. Transcripts can be visualized by staining with ethidium bromide or autoradiography.

E. Analyses of Transcription Products and Microinjection of Synthetic mRNAs

It is strongly recommended that the integrity and translational competence of a synthetic mRNA preparation be ascertained before undertaking analyses in microinjected oocytes. To determine if the RNA product is of the size predicted from the cDNA insert and choice of a 3' "runoff" site, a small aliquot (~1–2 μg) of the synthetic mRNA is electrophoresed with appropriate size standards on either a denaturing polyacrylamide–urea or an agarose–formaldehyde gel (Sambrook et al., 1989). The RNA is visualized by either staining with ethidium bromide (^{3}H-labeled mRNAs) or by autoradiography (^{32}P-labeled mRNAs). This will ensure that the synthetic mRNA is full length and that the preparation does not contain any of the three most common and potentially deleterious contaminants. First, residual DNA template, if even partially denatured, may hybrid-arrest translation in vitro and in microinjected oocytes. The second contaminant is plasmid-length RNAs which originate from either the promoter-specific transcription of uncut plasmid DNA template or aberrant initiation at protruding 3' termini. The plasmid-length 'sense' RNAs can compete with the bona fide mRNA of interest for the limited translational capacity of fully grown stage VI oocytes. Transcripts initiated at a protruding 3' terminus downstream of the cDNA insert will contain sequences complementary to the mRNA of interest. These antisense RNAs can hybrid-arrest translation of the synthetic mRNA in microinjected oocytes (Melton, 1985). Finally, partial-length RNAs which are due to either premature transcription termination or nucleolytic degradation will obviously not direct synthesis of the appropriate translation product.

The translational competence of a synthetic mRNA preparation can be easily determined by in vitro translation of an aliquot containing 0.5–1.0 μg of RNA in either the wheat germ extract or rabbit reticulocyte lysate system. It is crucial that the mRNA is redissolved in sterile water and *not* injection buffer since sodium ions are potent inhibitors of in vitro translation. Both wheat germ extract and rabbit reticulocyte lysate translate synthetic mRNAs efficiently and are commercially available from many suppliers. Detailed protocols for in vitro translation reactions and the analysis of protein products by SDS–polyacrylamide gel electrophoresis are routinely provided by most suppliers or can be found in Sambrook et al. (1989). Although the efficient translation of a synthetic mRNA in vitro does not necessarily guarantee its comparable utilization in microinjected oocytes, experience generally dictates that the *absence* of efficient translation in vitro typically correlates with poor expression in oocytes. In addition, the relevant in vitro translation product provides a valuable reagent with which to confirm synthesis of the cognate protein in microinjected oocytes using one or more of the criteria described in detail elsewhere in this volume.

Detailed procedures for the isolation, microinjection, in vitro maintenance, and metabolic labeling of oocytes at various stages of development are presented elsewhere in this volume. In general, the synthetic mRNA preparation is diluted with 2 × injection buffer to a final concentration of 1–2 μg/μl. Using a standard microinjection apparatus, a volume of 5–50 nl of solution containing 5–100 ng of synthetic mRNA is routinely delivered into the cytoplasm of individual oocytes. The translational capacity of a stage VI oocyte is saturated by approximately 100 ng of a typical non-membrane-associated mRNA (e.g., globin) and by about 20 ng of a typical membrane-associated mRNA (e.g., zein) (Richter and Smith, 1981; Galili et al., 1988). It is important to realize that the injected synthetic mRNA competes with endogenous oocyte transcripts for translation factors (Laskey et al., 1977; Richter and Smith, 1981). Thus, whereas as much as 50% of the newly synthesized protein in microinjected oocytes can be directed by the synthetic mRNA, endogenous proteins will continue to be expressed. This is in contrast to mRNA-dependent *in vitro* translation systems which exhibit little or no basal protein synthesis activity. However, given the high protein synthesis rate in stage VI oocytes, as much as 10 ng of protein per hour per oocyte may be synthesized from an efficiently translated injected mRNA (Taylor and Smith, 1985; Galili et al., 1988).

In general, most capped and polyadenylated synthetic mRNAs are relatively stable and efficiently translated over a period encompassing several days following microinjection (Krieg and Melton, 1984; Drummond et al., 1985; Galili et al., 1988). Protein synthesis directed by the injected mRNA can be detected within 30 minutes and usually reaches its maximal rate within 6 hours (Richter and Smith, 1981; Galili et al., 1988; Romanczuk and Wormington, 1989). Detailed procedures for the metabolic labeling and biochemical fractionation of microinjected oocytess to examine the subcellular localization and function of proteins expressed from synthetic mRNAs are presented elsewhere in this volume. The remainder of this chapter addresses analyses of transcript stability and translational efficiency of synthetic mRNAs in microinjected oocytes.

F. Analyses of Stability and Polysomal Association of Synthetic mRNAs in Microinjected Oocytes

The protocols presented in this section enable direct analyses of the stability and translational efficiency of synthetic mRNAs microinjected into *Xenopus* oocytes. In general, these analyses can be routinely performed using as few as five microinjected oocytes per sample. Protocol 1 describes the isolation of total oocyte RNA using a rapid SDS–proteinase K procedure similar to those described in detail elsewhere (Wormington, 1986; Sambrook et al., 1989). Similarly, the two protocols for the extraction of RNA from nonpolysomal and polysomal fractions are adapted from procedures described in detail elsewhere (Richter and Smith, 1981; Baum et al. 1988; Galili et al., 1988; Hyman and Wormington, 1988; McGrew et al., 1989; Varnum and Wormington, 1990). No more than 15 oocytes should be pooled per sample to avoid overloading the sucrose gradients in either protocol.

Polysomal mRNAs are separated from nonpolysomal transcripts by sedimentation through a sucrose cushion in Protocol 2. This protocol is recommended when several samples are to be analyzed simultaneously and when the relative distribution of a synthetic mRNA between polysomal and nonpolysomal fractions is to be determined. Translatable mRNAs as short as 165–200 nucleotides in length can be quantitatively recovered in the polysomal fraction using this protocol (Varnum and Wormington, 1990). In Protocol 3, the polysomes are fractionated by centrifugation through a linear sucrose density gradient, and RNA is extracted from individual fractions. This protocol provides a precise measurement of the ribosome loading on a given mRNA, thereby giving a more accurate assessment of its translational efficiency (Galili et al., 1988).

"Cold" ^{3}H-labeled synthetic mRNAs isolated from microinjected oocytes can be detected by Northern blot or nuclease protection assays using the appropriate hybridization probes (Sambrook et al., 1989). "Hot" ^{32}P-labeled synthetic mRNAs isolated from microinjected oocytes can be detected directly by autoradiography following electrophoresis on denaturing polyacrylamide–urea or agarose–formaldehyde gels. The exact amounts of synthetic mRNA recovered from oocytes can be determined by using serial dilutions of a stock solution of synthetic mRNA at a known concentration as standards. All steps are carried out using RNase-free plasticware, glassware, and solutions.

1. ISOLATION OF TOTAL RNA FROM MICROINJECTED OOCYTES

1. Pools of 5–15 oocytes are collected at the desired times following microinjection and placed into sterile 1.5-ml microcentrifuge tubes. Excess saline solution is removed with a drawn-out sterile pipette, and the oocytes are immediately frozen at $-70°C$, at which temperature they can be stored indefinitely before RNA isolation.
2. Remove tubes containing frozen oocytes and immediately add 50 μl of SDS–proteinase K buffer per oocyte. Mix thoroughly by vortexing. Incubate at 50°C for 60 minutes. Vortex intermittently.
3. Extract twice with an equal volume of phenol–chloroform.
4. Extract once with an equal volume of chloroform.
5. Precipitate the RNA with 2.5 volumes of ethanol. The RNA can be stored under ethanol indefinitely at $-20°C$.
6. Recover the ethanol-precipitated RNA by centrifugation. Rinse the pellet twice with 70% ethanol. Resuspend the RNA in sterile water or any other appropriate buffer and store at $-70°C$ prior to electrophoretic or other analyses using procedures described in detail elsewhere (Sambrook et al., 1989).

2. Isolation of Nonpolysomal and Polysomal mRNAs from Microinjected Oocytes by Centrifugation through Sucrose Step Gradients

1. Pools of 5–15 oocytes are collected at the desired times following microinjection and stored at $-70°C$ as described in Step 1 in Protocol 1. All subsequent steps are at $4°C$ unless indicated otherwise.
2. Remove tubes containing frozen oocytes and immediately add 50 μl of PB per oocyte. Homogenize the oocytes by pipetting several times with a sterile "yellow" tip (e.g., Gilson Pipetman P20 or P200). Mix thoroughly by vortexing.
3. Centrifuge the homogenate at 12,000 g for 15 minutes at $4°C$ to pellet pigment granules, yolk platelets, and mitochondria. Transfer the postmitochondrial supernatant to a sterile tube and discard the pellet.
4. Dilute the postmitochondrial supernatant with PB to a final volume of 2.5 ml and layer the solution over a 2.5-ml cushion of 20% (w/v) sucrose in PB. Pellet the polysomes by centrifugation at 149,000 g for 2 hours at $4°C$ in either a Beckman (Fullerton, CA) SW50.1 or SW55 rotor. Carefully transfer the pooled supernatant and sucrose cushion to a sterile tube and precipitate the nonpolysomal RNA at $4°C$ for 30 minutes with 2 volumes of ethanol. Proceed to Step 5 to extract polysomal RNA and Step 6 to extract RNA from the precipitated nonpolysomal fraction.
5. Resuspend the polysomal pellet in 500 μl of SDS–proteinase K buffer and proceed exactly as described in Steps 2 through 6 in Protocol 1.
6. Recover the ethanol-precipitated nonpolysomal fraction by centrifugation at 12,000 g for 15 minutes. Do not rinse or dry the precipitate. Resuspend the pellet in 500 μl of SDS–proteinase K buffer and proceed exactly as described in Steps 2 through 6 in Protocol 1.
7. Resuspend the RNA in sterile water or any other appropriate buffer and store at $-70°C$ prior to electrophoretic or other analyses using procedures described in detail elsewhere (Sambrook *et al.*, 1989). When analyzing the distribution of a synthetic mRNA between the polysomal and nonpolysomal fractions, it is important to compare aliquots of RNA corresponding to equivalent numbers of oocytes. Typically, 1–2 oocyte equivalents of RNA is sufficient for most analyses.

3. Isolation of Nonpolysomal and Polysomal mRNA from Microinjected Oocytes by Centrifugation through Linear Sucrose Density Gradients

1. Pools of 5–15 oocytes are collected at the desired times following microinjection and stored at $-70°C$ as described in Step 1 in Protocol 2.

9. SYNTHETIC mRNAs AND TRANSLATION IN OOCYTES

Homogenize the oocytes in PB and recover the postmitochondrial supernatant exactly as described in Steps 2 and 3 in Protocol 2.

2. Layer the postmitochondrial supernatant without further dilution over a 4.5-ml 15–40% (w/v) linear sucrose gradient in PB. Polysomal and nonpolysomal fractions are separated by centrifugation at 304,000 g for 1 hour at 4°C in either a Beckman SW50.1 or SW55 rotor.
3. Collect approximately 20 250-μl fractions from either the top or the bottom of the gradient into sterile 1.5-ml microcentrifuge tubes. Precipitate each fraction with 2 volumes of ethanol at 4°C for at least 30 minutes.
4. Recover the ethanol-precipitated fractions by centrifugation at 12,000 g for 15 minutes. Do not rinse or dry the precipitates. Resuspend each pellet in 500 μl of SDS–proteinase K buffer and proceed exactly as described in Steps 2 through 6 in Protocol 1.
5. Resuspend the RNA in sterile water or any other appropriate buffer and store at $-70°C$ prior to electrophoretic or other analyses using procedures described in detail elsewhere (Sambrook *et al.*, 1989). When analyzing the polysomal distribution of a synthetic mRNA it is important to compare aliquots of RNA corresponding to *equivalent amounts of each fraction*. Typically, 1–2 oocyte equivalents of RNA is sufficient for most analyses.

Note: An important control for these analyses is to demonstrate that mRNA which pellets through the sucrose cushion in Protocol 2 is indeed associated with polysomes. This can be easily ascertained by showing that the mRNA sedimenting in this pellet is "released" into the supernatant by treatment with EDTA as described below in Protocol 4. It is very important that *cycloheximide must be omitted from all solutions to be used for the control "release" studies.*

4. RELEASE OF mRNAs FROM POLYSOMES BY TREATMENT WITH EDTA

1. Pools of 5–15 oocytes are collected at the desired times following microinjection and stored at $-70°C$ as described in Step 1 in Protocols 2 or 3. Homogenize the oocytes in PB *which does not contain cycloheximide* and recover the postmitochondrial supernatant exactly as described in Steps 2 and 3 in Protocol 2.
2. Add EDTA to the postmitochondrial supernatant to a final concentration of 20 mM. Separate the nonpolysomal and polysomal fractions exactly as described in either Steps 4 through 7 in Protocol 2 or Steps 2 through 5 in Protocol 3. *Be certain that all sucrose solutions contain 20 mM EDTA and do not contain cycloheximide.*

In general, there are two likely explanations for why a given mRNA may still be associated with the polysomal pellet in EDTA-treated samples prepared using Protocol 2. First, it is possible that the gradients may have been overloaded with an excessive amount of postmitochondrial supernatant. This can usually be eliminated by reducing the number of oocytes pooled in an individual sample. Second, the mRNA of interest may be associated with large EDTA-resistant nonpolysomal ribonucleoprotein complexes which cosediment with polysomes through the 20% sucrose cushion in Protocol 2. This problem can be alleviated in some cases by either increasing the concentration of sucrose in the cushion to 35% or by using the linear sucrose density gradient procedure described in Protocol 3.

Acknowledgments

I am especially pleased to acknowledge the efforts of Ellen Baum, Linda Hyman, Brett Keiper, Rhonda O'Keefe, Helen Romanczuk, and Susan Varnum who optimized and successfully applied the procedures presented in this chapter throughout their graduate careers in my laboratory. I also thank Doug Melton and Joel Richter for many insightful discussions over the years and for generously communicating results and providing various plasmids prior to publication. Research in the author's laboratory is supported by National Institutes of Health Grant HD-17691.

References

Baum, E. Z., Hyman, L. E., and Wormington, W. M. (1988). *Dev. Biol.* **126,** 141–149.
Braddock, M., Chamber, A., Wilson, W., Esnouf, M. P., Adams, S. E., Kingsman, A. J., and Kingsman, S. M. (1989). *Cell (Cambridge, Mass.)* **58,** 269–279.
Brown, B. D., and Harland, R. M. (1990). *Genes Dev.* **4,** 1925–1935.
Casey, J. L., Hentze, M. W., Koeller, D. M., Caughman, S. W., Rouault, T. A., Klausner, R. D., and Harford, J. B. (1988). *Science* **240,** 924–928.
Drummond, D. R., Armstrong, J., and Colman, A. (1985). *Nucleic Acids Res.* **13,** 7375–7394.
Fox, C. A., and Wickens, M. P. (1990). *Genes Dev.* **4,** 2287–2298.
Fox, C. A., Sheets, M. D., and Wickens, M. P. (1989). *Genes Dev.* **3,** 2151–2162.
Furuichi, Y., LaFiandra, A., and Shatkin, A. J. (1977). *Nature (London)* **266,** 235–239.
Galili, G., Kawata, E. E., Smith, L. D., and Larkins, B. A. (1988). *J. Biol. Chem.* **263,** 5764–5770.
Harvey, R. P., Tabin, C. J., and Melton, D. A. (1986). *EMBO J.* **5,** 1237–1244.
Hyman, L. E., and Wormington, W. M. (1988). *Genes Dev.* **2,** 598–605.
Kozak, M. (1986). *Proc. Natl. Acad. Sci. U.S.A.* **83,** 2850–2854.
Kozak, M. (1989). *J. Cell Biol.* **108,** 229–241.
Krieg, P. A., and Melton, D. A. (1984). *Nucleic Acids Res.* **12,** 7057–7070.
Laskey, R. A., Mills, A. D., Gurdon, J. B., and Partington, G. A. (1977). *Cell (Cambridge, Mass.)* **11,** 345–351.
Lazarus, P., Parkin, N., and Sonenberg, N. (1988). *Oncogene* **3,** 517–522.
McGrew, L. L., Dworkin-Rastl, E., Dworkin, M. B., and Richter, J. D. (1989). *Genes Dev.* **3,** 803–815.
Mariottini, P., and Amaldi, F. (1990). *Mol. Cell. Biol.* **10,** 816–822.
Melton, D. A. (1985). *Proc. Natl. Acad. Sci. U.S.A.* **82,** 144–148.

Melton, D. A. (1987). *In* "Methods in Enzymology" (S. L. Berger and A. R. Kimmel, eds.), Vol. 152, pp. 288–296. Academic Press, New York.
Munroe, D., and Jacobson, A. J. (1990). *Mol. Cell. Biol.* **10**, 3441–3455.
Rhoads, R. E. (1988). *Trends Biochem. Sci.* **13**, 32–36.
Richter, J. D., and Smith, L. D. (1981). *Cell (Cambridge, Mass.)* **27**, 183–191.
Romanczuk, H., and Wormington, W. M. (1989). *Mol. Cell. Biol.* **9**, 406–414.
Roth, M. B., and Gall, J. G. (1989). *Proc. Natl. Acad. Sci. U.S.A.* **86**, 1269–1272.
Sachs, A. B., and Davis, R. W. (1989). *Cell (Cambridge, Mass.)* **58**, 857–867.
Sachs, A. B., and Davis, R. W. (1990). *Science* **247**, 1077–1079.
Sagata, N., Watanabe, N., Vande Woude, G. F., and Ikawa, Y. (1989). *Nature (London)* **342**, 512–518.
Sambrook, J., Fritsch, E. F., and Maniatis, T. (1989). *"Molecular Cloning: A Laboratory Manual,"* 2nd Ed. Cold Spring Harbor Laboratory, Cold Spring Harbor, New York.
Schenborn, E. T., and Mierendorf, R. C. (1985). *Nucleic Acids Res.* **13**, 6223–6229.
Sheets, M. D., and Wickens, M. P. (1989). *Genes Dev.* **3**, 1401–1412.
Swenson, K. I., Farrell, K. M., and Ruderman, J. V. (1986). *Cell (Cambridge, Mass.)* **47**, 861–870.
Taylor, M. A., and Smith, L. D. (1985). *Dev. Biol.* **110**, 230–237.
Varnum, S. M., and Wormington, W. M. (1990). *Genes Dev.* **4**, 2278–2286.
Wormington, W. M. (1986). *Proc. Natl. Acad. Sci. U.S.A.* **83**, 8639–8643.

Chapter 10

Use of Oligonucleotides for Antisense Experiments in Xenopus laevis Oocytes

CAROL PRIVES AND DIANA FOUKAL

Department of Biological Sciences
Columbia University
New York, New York 10027

I. Introduction
  A. Oligonucleotides Induce Specific Cleavage of mRNA in Oocytes
  B. Modified and Unmodified Oligodeoxynucleotides
  C. Antisense Experiments in Oocytes Reveal Insights into Functions of Membrane Proteins
  D. Protooncogenes and Oocyte Maturation
  E. Oligodeoxynucleotides Reveal Aspects of snRNA Functions in Oocytes
  F. Nonspecific Effects of Oligonucleotides in Oocytes: A Cautionary Note
II. Oligonucleotides Reveal Aspects of snRNA Structure and Function in Oocytes
  A. Methods
  B. Experimental Results and Discussion
  References

I. Introduction

The development of technologies for the synthesis of oligodeoxynucleotides (oligos) has opened yet another chapter in the extraordinary history of the *Xenopus* oocyte system. An extensive survey of the development of methods to synthesize normal and modified oligos and their use in a variety of biological systems other than oocytes is beyond the scope of this chapter, although several excellent reviews on this subject are available (e.g., Marcus-Sekura, 1988; Toulme and Helene, 1988; Cohen, 1990; Colman, 1990). Advantages of oocytes include their well-known large size and hardy nature that make microinjection experiments flexible in terms of site of introduction, the potential for multiple injections, as well as the diversity of small and large molecules that can be introduced into the cells. Moreover, oocytes possess two other features that are favorable for some antisense experiments. First, late stage oocytes express few of their endogenous genes that are normally transcribed by polymerase II. This is not due to limiting or inactive forms of either RNA polymerase II or relevant transcription factors because many genes, when introduced in appropriate form into oocytes, can indeed be expressed. Thus, it is possible to destroy specifically an existing RNA product irreversibly using the appropriate oligo or oligos. Second, normal, unmodified oligos are very unstable in oocytes and do not persist at effective concentrations for more than an hour at most (see Section I,B below). This makes it possible to introduce into oligo-injected oocytes various normal or mutant sources of RNA (either directly or as DNA that can be transcribed into the appropriate RNA) for subsequent study. These two attributes have led to several studies that have provided insight into basic cellular processes carried out by oocytes such as cell cycle-related events, mRNA splicing, and ribosome function. Furthermore, oocytes have been developed extensively as tools for studying membrane and neurotransmitter functions. Here, too, oligos have been used imaginatively in a manner that is more related to studying the function of the products of exogenously introduced RNAs than of RNAs normally present and functioning in oocytes.

This chapter consists of two parts: first, we present a general survey of relevant published papers in which oligos were used for antisense experiments in oocytes; second, we describe a set of experiments from our laboratory demonstrating some of the uses of oligos to study small nuclear RNAs (snRNAs) and their function in pre-mRNA splicing in oocytes. Many studies involving oligos have compared results obtained with *Xenopus* oocytes, eggs, and embryos. We focus herein primarily on the experiments concerned with oocytes, although, where relevant, the relative effects obtained with eggs and embryos are described. We would like to offer sincere apologies in advance to those whose work, through our ignorance, we have neglected to mention.

A. Oligonucleotides Induce Specific Cleavage of mRNA in Oocytes

Melton (1985), using complementary RNA, and Kawasaki (1985), employing oligos, were the first to explore the use of *Xenopus* oocytes for antisense experiments. Kawasaki injected mRNA from cultured cells that had been induced to synthesize interleukin-2(IL-2) either directly into oocytes or after having been annealed to several complementary oligos or DNA fragments. Production of IL-2, as monitored by measuring thymidine uptake into mouse cells after treatment with medium from injected oocytes, was sharply decreased from oligo-treated RNA samples. Dash et al. (1987) developed the potential of this approach further by showing that both exogenously introduced globin mRNA or calmodulin mRNA present within the oocyte were cleaved after injection of certain oligos. As cleavage of the targeted RNA occurred at the site of oligo–RNA hybrid formation, this indicated that oocytes contain an active RNase H-like activity. This study, importantly, showed that oligos complementary to different regions of their respective mRNAs varied significantly in their ability to induce efficient cleavage, a result that was confirmed and extended by several subsequent studies. Cazenave et al. (1987b) performed related experiments with oligo-targeted β-globin mRNA showing that the mRNA that was annealed to oligos was cleaved by RNase H both *in vivo* and *in vitro*. Shuttleworth and Colman (1988) compared several types of oligo-targeted RNA including endogenous and exogenous heat-shock mRNA, histone H4 mRNA, and Vg1 mRNA. The efficiency of RNA cleavage, when compared in oocytes, matured oocytes, and unfertilized eggs, was determined to vary in these different cells. Furthermore, the site of injection of the oligos as well as the abundance and location of the RNAs in oocytes were shown to affect the extent of cleavage.

In a second paper Shuttleworth et al. (1988) showed that oligos complementary to different regions of a given mRNA, in this case, histone H4 mRNA, elicited marked differences in the efficiency of RNA degradation ranging from as little as 5% to greater than 90%. As successive injections of each oligo did not induce further RNA cleavage, it was concluded that there are fractions of RNA ranging from very small to considerably larger that are inaccessible to oligos. It was also determined in that study that the minimum oligo length for effective RNA degradation is 12 nucleotides, although a 10-nucleotide oligo also induced the cleavage of a small amount of RNA. This group has provided evidence more recently that the relative resistance of RNA to different oligos in oocytes was mirrored in an *in vitro* assay, in which the same histone H4 oligos, when annealed to RNA and then cleaved with RNase H, produced similarly lesser or greater extents of RNA cleavage as they had after injection into oocytes (Baker et al., 1990). This suggests the possibility of a relatively simple assay to determine preliminarily the effectiveness of various oligos for efficient RNA cleavage in oocytes.

Oligo-targeted degradation requires knowledge of the sequence of the RNA to be cleaved. Jessus et al. (1988) have demonstrated that, for genes known to be highly conserved, the absence of direct sequence information about a *Xenopus* gene need not be prohibitive. Thus, it was possible to design an oligo complementary to a region of β-tubulin determined to be highly conserved within a wide variety of other species and to show that injection of this oligo resulted in near cessation of β-tubulin synthesis, while exerting no effect on the synthesis of α-tubulin. Their study showed that even as late as 24 hours after introduction of the oligo, very little β-tubulin synthesis was detected, consistent with prior experiments showing the paucity of ongoing transcription of most endogenous RNA polymerase II genes in oocytes. Somewhat surprisingly, such oligo-treated oocytes exhibited normal kinetics and appearance of maturation after treatment with progesterone. This indicates that oocytes are likely to contain highly stable stores of tubulin which are stockpiled to such extents that ongoing synthesis is not a requirement for oocyte maturation.

The vast majority of studies that have utilized oligos as antisense tools have demonstrated that loss of function of the targeted RNA is directly related to the extent to which it was cleaved. This has led to the assumption that a crucial requirement for effectiveness of oligos is their ability to induce efficient RNase H cleavage of the targeted RNA. One notable exception has been described in a paper by Saxena and Ackerman (1990) in which a series of oligos complementary to the α-sarcin loop of 28 S ribosomal RNA was injected into oocytes. Of such oligos only one that fully covered the α-sarcin loop region resulted in inhibition of oocyte protein synthesis. As evidence for cleavage of 28 S rRNA was not detected, and as the corresponding complementary oligoribonucleotide was also effective, this experiment represents an example of an effective oligonucleotide ablation of RNA function not mediated by RNase H.

Can oligos directed against mRNAs that are uniquely abundant during oogenesis and early embryogenesis be injected into oocytes and the subsequent effects of such injections on early amphibian development be followed? Shuttleworth et al. (1988) injected either an oligo complementary to the localized oocyte Vg1 mRNA or an unrelated oligo into oocytes and then followed the procedure of Holwill et al. (1987) in which oocytes are matured and then passed through laying frogs prior to subsequent fertilization of the injected oocytes in vitro. As embryonic development resulting from both specific and nonspecific oligo-injected oocytes was defective, they were unable to determine whether Vg1 RNA plays a role in early development. By contrast, Kloc et al. (1989) injected oligos complementary to another localized maternal mRNA, xlgv7, that is abundant in oocytes, eggs, and early embryos. Following an alternative in vitro fertilization protocol, they showed that although xlgv7 mRNA was reduced to undetectable levels, embryos resulting from oocytes that had been injected with either specific or nonspecific oligos developed normally. This result, although showing that destruction of the majority of the xlgv7 transcript did not interfere with early development, has provided a potentially important route for oligo-mediated antisense experiments examining the effects of either oocyte-specific endogenous mRNAs or exogenously introduced mRNAs. However, as discussed in Section I,F below and shown by

Shuttleworth et al. (1988), the potential for nonspecific deleterious effects of oligos will have to be monitored carefully in all such experiments.

B. Modified and Unmodified Oligodeoxynucleotides

Oocytes have proved to be an excellent testing system for the efficacy of several types of modified oligos. Several criteria must be met for a given type of oligo to be effective: these include chemical stability, water solubility, stability of the RNA–DNA hybrid formed, RNase H susceptibility, resistance to their other nucleases, effective concentration range, and toxicity. Most modifications consist of substitution of the nonbridging oxygen of the phosphodiester group (Cazenave et al., 1989; Mori et al., 1989; Baker et al., 1990; Dagle et al., 1990; Woolf et al., 1990) although dye conjugates (Cazenave et al., 1987b) and α-anomer forms (Cazenave et al., 1987a, 1989) of oligos have been explored as well.

Several of the aforementioned criteria have been compared in studies of modified and unmodified oligos. In general, unmodified oligos were shown to be significantly less stable than were the various modified oligos to which they were compared. However, estimates of the stability of unmodified oligos have varied among different groups. Estimates of half lives of unmodified oligos ranging from less than a few minutes (Dagle et al., 1990) to approximately 10–15 minutes (Cazenave et al., 1987a; Pan and Prives, 1988; Woolf et al., 1990) to greater than 60 minutes (Smith et al., 1988) suggest that there may be differences in experimental techniques (or in oocytes) that vary between laboratories. Whereas hybrids between RNA and modified oligo phosphothioates (Cazenave et al., 1989; Mori et al., 1989; Baker et al., 1990; Dagle et al., 1990; Woolf et al., 1990), phosphoselenates (Mori et al., 1989), and dye conjugates (Cazenave et al., 1987b) were substrates for RNase H, α-anomers were not (Cazenave et al., 1989). In general modified oligos that displayed both increased longevity in oocytes and the ability to induce RNase H cleavage were more effective at mRNA ablation than were the unmodified counterparts although such oligos were also frequently more toxic at high concentrations.

A potentially important approach to oligo modification has involved the synthesis of mixed-linkage oligos in which the internucleoside linkages at the 3' and 5' ends of the oligo are modified with phosphoramidates (Dagle et al., 1990) or methylphosphonates or phosphothioates (Baker et al., 1990) while the internal linkages are the natural phosphodiester form. The rationale for this design is to produce an oligo that exhibits reduced sensitivity to exonucleases and therefore increased stability while still providing an efficient substrate for RNase H. Such mixed-linkage oligos were found by Dagle et al. (1990) to be markedly more stable than were the normal form and also more effective at degrading targeted RNA. This approach may prove to be effective for mRNAs that are more resistant to cleavage with normal oligos. It is likely that further studies exploring the consequences of oligo modification for antisense experiments in oocytes will provide additional useful information.

C. Antisense Experiments in Oocytes Reveal Insights into Functions of Membrane Proteins

Xenopus oocytes have flourished as a source for studies on membrane proteins, in particular those involving neurotransmission (see Kushner *et al.*, 1989). The general approach has been to inject oocytes with purified or total mRNA derived from neural tissue and then characterize the protein products that have assembled within the oocyte membrane by methods such as voltage clamp experiments, in which a measurable electrical response is elicited by specific transmitters. Accordingly, the use of oligos has been exploited in some such cases to gain additional insights into the function of various membrane proteins. In one such case Akagi *et al.* (1989) used this approach to show that oligos specific for the mRNA encoding the strychnine-binding subunit of the glycine receptor isolated from adult rat brain were able to cause a block in the electrical response to glycine (or alanine) in oocytes injected with such RNA. By contrast, the same oligo had little or no effect on RNA isolated from neonatal rat spinal cord, indicating that different tissues or animals at different developmental stages express different classes of RNA encoding this receptor subunit.

Lotan *et al.* (1989) used oligos complementary to a cDNA encoding the dihydropyridine receptor to show that it is likely to be the channel-forming subunit of the voltage-dependent calcium channel and is distinct from the voltage- dependent sodium and potassium channels. Noguchi *et al.* (1990) used oligos to show that the β subunit of the *Torpedo californica* Na^+, K^+-ATPase must be present simultaneously with, or prior to, the α subunit in order to assemble into a functional ATPase in oocytes. Meyerhof and Richter (1990) reported that whereas microinjection of serotonin HT_2 receptor (SR) mRNA synthesized *in vitro* led to the acquisition of agonist-dependent membrane currents in oocytes, as measured by voltage clamp experiments, preincubation of such RNA with SR antisense oligos and RNase H prior to injection abolished this response to SR mRNA. The ramifications of this result for isolation of unidentified receptors are discussed by Meyerhof and Richter (1990).

D. Protooncogenes and Oocyte Maturation

Treatment of oocytes with hormones such as insulin and progesterone triggers a membrane-mediated response leading to a series of events that causes them to move from G_2 in interphase to meiosis II, a process accompanied by breakdown of the germinal vesicle (GVBD). The processes within the oocyte that are involved in this transition have been under scrutiny in recent years because of increasing interest in defining the molecules that control the cell cycle. A key player in cell cycle regulation in all eukaryotic organisms is the *cdc2* protein kinase, a highly conserved protein that has been detected in organisms ranging from yeast to humans. The identification of the *Xenopus* maturation-promoting factor (MPF) as the *cdc2* protein was a significant advance in tying together several disparate facts about cell cycle regulation (reviewed by Dunphy and Newport, 1988; Murray and Kirschner, 1989). The function of *cdc2*/MPF is regulated through its association with cyclins, a class of proteins whose abundance varies sharply in accordance with different stages of the cell cycle. Direct evidence for the importance of the cyclin–*cdc2*/MPF interaction was afforded by Minshull et al. (1989), who showed that addition of oligos complementary to *Xenopus* cyclin mRNAs to *Xenopus* cell-free egg extracts blocked entry into mitosis.

A protein whose function was shown to be important for oocyte maturation is the product of the protooncogene c-*mos*. Originally detected as the product of the Molony murine sarcoma virus, v-*mos*, its cellular homolog, c-*mos*, was subsequently identified in several different species. Sagata et al. (1988) cloned the *Xenopus* c-*mos* gene and showed that c-*mos* RNA is found in amphibian ovarian tissues, oocytes, and early embryos through the blastula stage. Whereas progesterone-matured oocytes contained the 39-kDa protein product of the c-*mos* RNA, immature oocytes did not express detectable levels of this polypeptide. Injection of c-*mos* antisense oligos, but not nonspecific oligos, caused a marked reduction in the number of oocytes that undergo maturation after progesterone treatment. Consistent with these observations, Freeman et al. (1990) reported that when c-*mos* or v-*mos* RNA was injected into oocytes GVBD was induced even in the absence of progesterone. Furthermore, confirming and extending the results of Sagata et al. (1988), these authors showed that although c-*mos* antisense oligos prevented progesterone-induced GVBD, oocyte maturation was rescued by subsequent injection with v-*mos* RNA. Taking a similar approach, Barrett et al. (1990) have shown that oocytes which normally undergo GVBD as a result of injection with Ha-*ras*, failed to do so when first injected with c-*mos* antisense oligos.

Roy et al. (1990) have provided evidence for a relationship between c-*mos* and MPF. They have shown that anti-c-*mos* antisera immunoprecipitates of extracts of either murine cells or *Xenopus* eggs were capable of phosphorylating added cyclin B2, a component of some forms of MPF. Furthermore, the ability of oocytes to phosphorylate cyclin was strongly reduced after injection with antisense c-*mos* oligos. Taken together these various studies have provided a body of evidence supporting the role of c-*mos* in regulation of oocyte maturation as well as in maintenance of matured oocytes in meiosis II. As documented above, many of the key experimental lines of evidence necessary to develop this model have relied heavily on oligo antisense experiments.

A recent report has provided evidence that another protooncogene, the product of the *ets-2* gene, is required for ooctye maturation (Chen et al., 1990). This study utilized oligos effectively to make that point. After cloning the *Xenopus ets-2* gene it was determining that *ets-2* mRNA, like c-*mos* RNA, is expressed in ovary, oocytes, eggs, and early embryos through but not past the blastula stage. Microinjection of several antisense oligos complementary to various regions of the *ets-2* RNA resulted in a significant reduction of the percentage of oocytes that underwent GVBD, whereas two *ets-2* sense oligos had little or no effect. The *ets-2* antisense oligo that elicited the greatest relative reduction in maturing ooctyes was that spanning the ATG region, whereas one complementary to the 3' noncoding region was virtually ineffective. The relationship of *ets-2* function to that of c-*mos* remains to be established.

E. Oligodeoxynucleotides Reveal Aspects of snRNA Functions in Oocytes

Xenopus oocytes have provided unique opportunities to gain insight into the process of pre-mRNA splicing in vertebrates. The use of oligos have provided definitive evidence for the involvement of U1 and U2 snRNAs (Kramer et al., 1984; Black et al., 1985; Krainer and Maniatis, 1985; Black and Steitz, 1986) as well as of U4 and U6 snRNAs (Black and Steitz, 1986; Berget and Robberson, 1986) in splicing of precursor RNAs in cell-free extracts. To further characterize the roles of snRNAs in splicing in oocytes, we began a program of injection of oligos complementary to regions of U1 and U2 snRNA into *Xenopus* oocytes both simultaneously with and prior to injection of SV40 DNA (Pan and Prives, 1988; Pan et al., 1989). We showed that simian virus 40 (SV40) late mRNA splicing is far more sensitive to oligo-directed U1 snRNA cleavage than is splicing of early SV40 mRNA (Pan and Prives, 1989). By contrast, it was observed that both early and late mRNA splicing required intact U2 snRNA. Our studies with U1 and U2 oligos also provided insight into the relative stability of U1 and U2 snRNAs in oocytes (Pan et al., 1989). Thus, athough several oligos complementary to different regions of U1 snRNA were injected into oocytes, only ones that were complementary to the 5' ends resulted in significant cleavage of U1 snRNA (C. Prives and Z. Q. Pan, unpublished data). This suggests that only the 5' terminus of U1 snRNA is exposed in the U1 snRNP. The cleaved U1 snRNA, moreover, was stable in oocytes. By contrast, we observed that injection of oligos complementary to either the 5' end or to the first loop region of U2 snRNA led first to cleavage and then to complete disappearance of U2 snRNA (Pan *et al.*, 1989). Interestingly, we found that oligos injected prior to the injection of SV40 DNA resulted in greater extents of U1 or U2 snRNA cleavage than when the same oligos were coinjected with the DNA. This somewhat surprising result suggests that some fraction of U1 or U2 small nuclear riboproteins (snRNPs) may associate with pre-mRNA in a manner that renders them inaccessible to oligos.

The efficacy of oligos as reagents to examine the roles of snRNAs in RNA processing in oocytes has also been demonstrated by other groups with other precursor RNAs. Mattaj and colleagues (Hamm *et al.*, 1989, Hamm *et al.*, 1990; Vanken *et al.*, 1990) showed that injection of oligos complementary to U1, U2, U4, or U6 snRNAs led to blockage of the splicing of RNA containing intron 1 of the adenovirus major late transcription unit (Ad1 RNA). Savino and Gerbi (1990) have reported that microinjection into oocytes of an oligonucleotide complementary to a region within U3 snRNA caused a marked reduction in 20 S and 32 S pre-rRNA. Their data also indicated that different *Xenopus* frogs display either one or two distinct rRNA processing pathways. In animals with two pathways, the U3 oligo led to increased quantities of the 36 S rRNA precursor that is characteristic of the second pathway.

One fruitful outcome of this line of research was our observation that oocytes deprived of their resident U1 or U2 snRNPs by oligo-targeted cleavage, if subsequently injected with either partially or highly purified preparations of HeLa cell U1 or U2 snRNA, were capable of forming functional snRNPs (Pan and Prives, 1988). This result was the first to provide evidence that the RNA and protein components of snRNPs are sufficiently highly conserved between amphibians and humans that hybrid snRNPs comprising human snRNA and amphibian snRNA binding proteins were functional in splicing. Since then Abelson and colleagues have extended this observation further showing that human U2 and U6 snRNAs can assemble into functional snRNP particles in oligo-targeted yeast splicing extracts (McPheeters *et al.*, 1989; Fabrizio *et al.*, 1989). Surprisingly, we observed that when U snRNAs transcribed *in vitro*, by SP6 or by T7 RNA polymerases, were injected in similar quantities as were HeLa snRNAs into oocytes, they were incapable of forming functional snRNPs even though they were incorporated into nuclear snRNP-like particles (Pan and Prives, 1989). The reasons for the discrepancy between the two sources of U1 or U2 snRNA are not yet clear. The *in vitro* transcribed RNAs contained 1–3 extra nucleotides at their 5′ end, perhaps this change interfered with the function of the snRNP particle. Alternatively, snRNA transcribed by a phage RNA polymerase may lack essential modifications that exist in natural *Xenopus* U1 and U2 RNA.

Of more general use was the observation that not only was purified HeLa U2 snRNA capable of rescuing U2 oligo-targeted oocytes, but injection of a plasmid expressing the *Xenopus* U2 gene subsequent to introduction of the U2 oligos into oocytes also restored splicing potential to oocytes. This opened the way for analysis of the snRNA sequences that are involved in splicing. Both Hamm *et al.* (1989) and Pan and Prives (1989) tested the ability of mutant U2 genes to rescue splicing in U2-depleted oocytes. Both groups determined that the region of U2 snRNA that was shown previously to be important for interactions with the U2-specific proteins, A' and B'', was, surprisingly, not necessary for U2 snRNA splicing function in oocytes. However, there were discrepancies in the results of the two laboratories concerning the requirement for sequences within the 5' proximal region of U2 snRNA. Thus, a mutant, ΔA, that deletes U2 nucleotides 14–26, was shown to mediate efficient splicing of Ad1 precursor RNA (Hamm *et al.*, 1989). By contrast, the same U2 mutant construct failed to rescue splicing of SV40 late pre-mRNA in U2 oligo-targeted oocytes (Pan and Prives, 1989). The reasons for this discrepancy are not yet established. Either differences in the requirements of different pre-mRNAs for sequences within snRNPs exist, or the different experimental protocols of the two groups employed to analyze splicing complementation are responsible for the different results obtained with the ΔA mutant U2.

Mattaj and colleagues have taken the oligo-mediated U RNA splicing complementation system further and have determined which regions of U1 and also of the U4 and U6 snRNAs are involved in splicing of Ad1 precursor RNA in oocytes. The experiments of Hamm *et al.* (1990) have identified domains within U1 snRNA whose interaction with U1-specific proteins or with the general snRNA binding proteins is clearly essential for the splicing function of the U1 snRNP. The effects of the mutant U1 snRNAs on assembly of splicing complexes in oocytes were consistent with the effects on Ad1 RNA splicing. Their study on complementation by U4 and U6 snRNAs has also been highly informative in defining several categories of domains within the U4–U6 complex (Vanken *et al.*, 1990). Mutations in U4 and U6 snRNAs which reduce the stability of interactions between U4 and U6, or those which affect interactions of these snRNAs with other participants in functional splicesomes, were shown to block splicing. It is likely that further experiments will provide more detailed information about the sequences within snRNAs and the roles they play in RNA processing. Future studies may also reveal whether and how different pre-mRNAs require different regions within snRNAs for efficient splicing.

F. Nonspecific Effects of Oligonucleotides in Oocytes: A Cautionary Note

That oligos can specifically induce cleavage of targeted RNAs in oocytes has provided the impetus for a wide variety of studies. Virtually all of the reports referred to in the preceding sections have included control experiments with nonspecific oligos. These studies have all tended to provide a coherent set of results supporting specificity of mRNA cleavage and subsequent response to such mRNA ablation. However, several authors have also noted cases where nonspecific oligos have induced deleterious effects. Cazenave et al. (1987b), in their study comparing normal or acridine dye-linked β-globin oligos, showed that although these reagents caused efficient cleavage of β-globin mRNA, an unrelated oligo led to partial degradation of that RNA as well. Jessus et al. (1988) observed that the synthesis of high molecular weight proteins in oocytes was inhibited by both specific and nonspecific oligos. Akagi et al. (1989) also described nonspecific inhibition of oocyte protein synthesis after injection of oligos into oocytes. The results of Shuttleworth et al. (1988) described in Section I,A provided evidence that injection of both nonspecific and specific oligos into oocytes produced abnormal embryos. It is worth mentioning that in our experiments on oligo-targeted U1 and U2 snRNAs we made the rather unexpected observation that although several control oligos did not decrease the efficiency of SV40 pre-mRNA splicing, the ratio of large T to small t spliced mRNA was reversed compared to that seen in oocytes that had not received any oligo (Pan et al., 1989).

Perhaps the most well-documented and carefully characterized study of the nonspecific effects of oligos was recently published by Smith et al. (1990). An earlier report had described experiments in which injection of oligos complementary to D7 mRNA, a maternal RNA that is detected almost exclusively in oocytes, eggs, and early embryos, caused a significant delay in progesterone-induced maturation (Smith et al., 1988). In that study evidence was provided that a nonspecific oligo did not affect the time course of oocyte maturation. By contrast, two oligos complementary to regions within the 5' half of D7 mRNA caused a large proportion ($>90\%$) of the oocytes to delay maturation. The more recent paper (Smith et al., 1990) has now provided extensive and convincing evidence that the delayed maturation of the oocytes that had been noted earlier is related closely to a variety of gross morphological changes in oocytes and a concomitant decrease in the synthesis of high molecular weight proteins that are caused by injection of several (but not all) nonspecific and specific oligos. The tendency of these "poisonous" oligos to inhibit oocyte protein synthesis suggests a useful parameter to include in future experiments. Perhaps most disturbingly, the authors have demonstrated that the tendency to induce such widespread deleterious effects in oocytes varied with different oligos in a manner that was not related to their method of preparation, sequence, or base composition. Their report should induce other workers to be extremely cautious in interpreting results with oligos, particularly with respect to processes involving oocyte function such as maturation.

II. Oligonucleotides Reveal Aspects of snRNA Structure and Function in Oocytes

Oligos have proved to be highly effective for studying pre-mRNA splicing in ocytes. They have provided a means both to assess the relative requirements of different pre-mRNAs for various snRNPs and to determine sequences within snRNAs that are required for snRNP function. As agents for analysis of snRNAs their potential is far from exhausted, and the parameters of their usefulness have not yet been fully defined. We have undertaken to explore further several aspects of oligo experiments related to snRNA structure and function in oocytes. A comparative analysis of the relative stability of oligos in the oocyte nucleus and cytoplasm have not been reported. More complete information on the concentration range and time courses of the effects of oligos on snRNA degradation than previously published are now presented. Finally the data we show in this study suggest that there are variations in the way different introns within the SV40 late precursor RNA transcript respond to oligo-targeted cleavage of U1 snRNA.

A. Methods

1. Preparation of *Xenopus* Oocytes

Excised ovaries of mature *Xenopus laevis* females (*Xenopus* I, Ann Arbor MI) are incubated at room temperature in OR2 medium (Eppig and Steckman, 1976) containing 0.15% collagenase (Worthington, Freehold, NJ) on an orbital shaker set at low speed. After 5–8 hours, ovarian tissue is dissociated and follicle cells digested, the OR2 medium is decanted, and oocytes are washed several times with Barth's solution (Gurdon, 1968) to remove any debris. Defolliculated oocytes are maintained for up to 4 days in modified Barth's solution at 19°–20°C.

2. Oligonucleotides

Oligonucleotides used in this study were chemically synthesized on an Applied Biosystems (Foster City, CA) DNA synthesizer. The oligos are dried to a powder in a desiccator, resuspended in 100 μl TEN buffer (100 mM NaCl, 10 mM Tris-HCl, pH 8.0, 1 mM EDTA), and passed through a Sephadex G-25 column (Pharmacia, Piscataway, NJ). Samples are ethanol precipitated and resuspended in 10 μl water; 40 μl of deionized formamide is then added, and the samples are incubated at 65°C for 15 minutes. Denatured oligos are fractionated on an 8 M urea–10% polyacrylamide gel. The most slowly migrating band is excised from the gel and soaked overnight at 37°C in 0.5 ml elution buffer containing 0.5 M ammonium acetate, 10 mM magnesium acetate, and 0.1% sodium dodecyl sulfate (SDS). The eluate is ethanol precipitated, resuspended in water, and the concentration determined by measuring A_{260}. Oligos are again ethanol precipitated and resuspended in water to a final concentration of 1 mg/ml. The sequence of the oligo used to target U1 for cleavage is complementary to the first 20 bases of the 5' end: 5'-CTCCCCTGCCAGGTAAGTAT-3'. Similarly, those used to target U2 are complementary to the first 15 nucleotides of *Xenopus* U2: 5'-AGGCCGAGAAGCGAT-3'. For oligonucleotide stability studies, the sequence of the single- and double-stranded oligos are 5'-TCGAGGAAGTGACTAACTGACCGCAT-3'. After end labeling oligos with T4 polynucleotide kinase and [$\gamma - {}^{32}$P]ATP they are purified from contaminating nucleotides by ethanol precipitation.

3. Microinjection

All injections are directed to the oocyte nucleus unless otherwise indicated. For nuclear injections oocytes judged to be stage VI according to Dumont (1972) are placed, animal pole up, into a 35-mm petri dish in 3 ml of modified Barth's buffer and centrifuged at 800 g for 5–8 minutes in a clinical centrifuge. The microinjection needle is then inserted into the center of the animal pole, and the appropriate solution is dispensed. Following injection ooctyes are maintained at 19–20°C. At the appropriate times, groups of four oocytes are rinsed briefly with water, quick frozen on crushed dry ice, and stored at -70°C until further use.

4. Nucleic Acid Preparation from *Xenopus* Oocytes

Groups of four frozen oocytes are lysed, without thawing, in 0.4 ml homogenization buffer [10 mM Tris, pH 7.5, 10 mM NaCl, 1 mM MgCl$_2$, 2% SDS, 1 mg/ml proteinase K (Sigma Chemical Co., St. Louis, MO)]. The homogenate is then incubated at 37°C for 30 minutes. Samples are extracted twice with an equal volume of chloroform–isoamylalcohol (24:1), then ethanol precipitated. The precipitate is resuspended in 12 μl water and stored at -80°C until further use.

5. Northern Blot Analysis

RNA equivalent to one oocyte, in a volume of 4μl water, is denatured by adding 16 μl deionized formamide and heating at 65°C for 15 minutes. Denatured RNA is immediately transferred to ice and loaded onto an 8 M

urea–10% polyacrylamide gel that has been prerun at 400 V/cm for 20 minutes. Samples are electrotransferred to Gene Screen Plus (Du Pont, Wilmington, DE) membranes (Zeitlin and Efstradiatis, 1984), and blots are then hybridized to plasmids containing *Xenopus* U1 or U2 DNA (Mattaj and DeRoberties, 1985) that are uniformly labeled with ^{32}P using the Random Primed DNA labeling kit (Boeringer Manneheim, Indianapolis, IN).

6. S1 Nuclease Mapping of 5' Ends of Spliced SV40 Late RNA

Oocytes that are to be analyzed for SV40 RNA expression are injected with 5 ng viral DNA and incubated for 18 hours prior to storage at $-70°C$ until further use. RNA is extracted from such oocytes as above, and S1 nuclease mapping is performed as described by Pan *et al.* (1989). For analysis of spliced viral RNA species, SV40 DNA probes are prepared by cleavage with *Eco*RI followed by treatment with calf intestine alkaline phosphatase; the DNA is then extracted, precipitated, and recut with BglI. The 1710-base pair (bp) SV40 fragment spanning nucleotides 1782 to 5235 is then purified on a 1% agarose gel, electroeluted, ethanol precipitated, and resuspended in water. This DNA is then 5' end-labeled with $[\gamma\text{-}^{32}P]ATP$ by T4 polynucleotide kinase (New England Biolabs, Beverly, MA) to a specific activity of at least 10^7 cpm/μg. The equivalent of one oocyte of RNA is annealed to 30 ng of DNA probe for 12 hours at 49°C in 80% formamide, 0.4 M NaCl, 40 mM PIPES (pH 6.5), and 1 mM EDTA, followed by digestion with S1 nuclease (Sigma) at a concentration of 500 units/ml for 30 minutes at 45°C. S1 nuclease-resistant fragments are fractionated on an 8 M urea–polyacrylamide gel, which is dried and exposed to film.

B. Experimental Results and Discussion

1. Oligonucleotides Are Unstable in *Xenopus laevis* Oocytes

Several studies have been published in which the stability of radioactively labeled oligos in oocytes was measured (see Section I,B). In one set of experiments performed in this laboratory we found that within 30 minutes the majority of end-labeled oligo had been converted to nucleotide form (Pan and Prives, 1988). As differences in the site of injection of the oligo might influence the rate of degradation, we compared the fate of labeled oligos injected into oocyte nuclei and cytoplasm. Most antisense experiments have involved injection of single-stranded oligos into the cytoplasm. However, double-stranded oligos have been injected into oocytes as well (Richter, 1989). Double-stranded oligos have potential uses in oocyte experiments such as competitive factor binding studies, and it was therefore of interest to compare their stability in nuclei and cytoplasm as well.

^{32}P-End-labeled double- and single-stranded oligos were injected into oocyte nuclei (Fig. 1) or cytoplasm (Fig. 2). Oocytes were then extracted at the time intervals indicated, and the labeled oligos were analyzed on urea–polyacrylamide gels followed by autoradiography. Although this assay allowed only the detection of oligos containing labeled 5′ ends, it is likely that the entire oligo was degraded because it was possible to detect a population of shortened, labeled molecules. We observed that the half-life of single-stranded oligos, after injection into the nucleus, was of the order of 10 minutes. At this time point, as well as at the 30- and even 60-minute time points, several labeled degradation products were observed with progressively greater proportions of labeled free nucleotide detected (see arrow). By 120 minutes no intermediate species were apparent. The stability of the labeled double-stranded oligo in the nucleus was significantly greater. In contrast to what was observed with the single-stranded oligo, it was possible to detect full-length double-stranded oligo even after 120 minutes, although we estimate that its half-life was between 30 and 60 minutes. Examining the relative stability of the two forms of oligos in the cytoplasm, it appeared that the double-stranded oligo exhibited a similar stability profile, whereas the

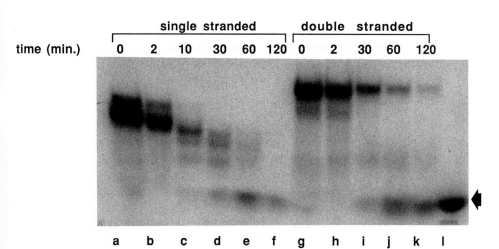

FIG. 1. Stability of oligos injected into oocyte nuclei. Ten nanograms of double-stranded (lanes b–f) and single-stranded (lanes h–k) oligos that had been 5′ end-labeled with ^{32}P were injected into the nuclei of groups of four oocytes. After incubation at 20°C for the indicated times, oocytes were extracted and 1 oocytes equivalent was fractionated on 8 M urea–10% polyacrylamide gels. Uninjected oligos (lanes a and g) were run in parallel. The position of free [γ-^{32}P]-ATP (lane l) is indicated by the arrow.

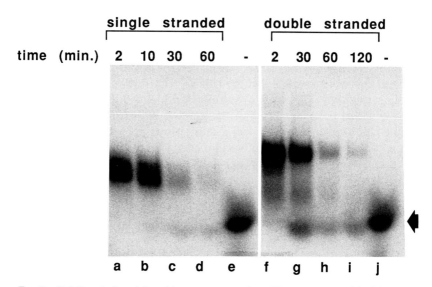

FIG. 2. Stability of oligos injected into oocyte cytoplasm. Ten nanograms of double-stranded (lanes a–d) and single-stranded (lanes f–i) oligos were 5′ end-labeled as in Fig. 1 and injected into the cytoplasm of groups of four oocytes. After incubation at 20°C for the indicated times, oocytes were extracted and 1 oocyte's equivalent was fractionated on an 8 M urea–10% polyacrylamide gel. The mobility of free [γ-^{32}P]ATP (lanes e and j) is indicated by the arrow.

single-stranded oligo was somewhat more stable than it was in the nucleus, such that its half-life was estimated between 10 and 30 minutes.

Although these studies do not reveal what enzymatic activities are involved in oligo degradation in oocytes, they do imply that such enzymes are more active in the oocyte nucleus than in the cytoplasm. Furthermore, as we did not determine the rate of equilibration of oligos between the nucleus and cytoplasm, it is not possible to state where in the cell degradation takes place. Consistent with this caveat we have not detected reproducible differences in the effects of oligos on snRNA cleavage after their injection into the nucleus or the cytoplasm (data not shown). We therefore infer from these experiments that the site of injection is not an important factor in the stability of natural oligos. However, our data suggest that double-stranded oligos are 3- to 5-fold more stable than are their single-stranded counterparts in oocytes.

2. U1 and U2 snRNAs Are Cleaved on Injection of Oligonucleotides Complementary to Portions of Their 5′ Ends

We and others have previously shown that injection of oligos complementary to regions within U1 or U2 snRNA into oocytes induces specific cleavage of these snRNAs. These reports, however, did not provide extensive information about either the effects of different concentrations of oligos on snRNA cleavage or the time course of RNA degradation following oligo injection. To determine these parameters in greater detail, we conducted both oligo concentration and time course experiments. Oligonucleotides complementary to the 20 5' nucleotides of *Xenopus* U1 snRNA were injected into the oocyte nucleus. As seen in Fig. 3 injection of 1 ng of the U1 oligo failed to induce significant amounts of extensively cleaved RNA, although a small proportion of a slightly shortened form of U1 snRNA was detected. Injection of either 5 or 10 ng of the U1 oligo caused more extensive cleavage of U1 at the 5' end of U1 snRNA. Interestingly, the extent to which the U1 snRNA was shortened depended on the amount of U1 oligo that was injected. Figure 3 (lanes a-c) shows a progressive, but slight, shortening of U1 snRNA as the amount of complementary oligonucleotide was increased. This degradation of U1 snRNA was selective, as no such cleavage was observed on injection of 20 ng of the U2 oligo (see below and Figure 4, lane f).

When the time course of U1 snRNA degradation was examined in oocytes that had been injected with the highest concentration of U1 oligo (10 ng) we observed that the extent to which U1 snRNA was shortened increased over the time period examined. Thus, although within 1 hour of injection of the U1 oligo virtually all of the U1 snRNA was cleaved, full shortening of U1 snRNA was not apparent until 10 hours after injection of the oligo. As the oligos are rapidly degraded in the oocyte nucleus, they must perform their function quickly. Thus, by the 1-hour time point most of the free oligonucleotide was no longer present (Fig. 1, lane e). The further shortening

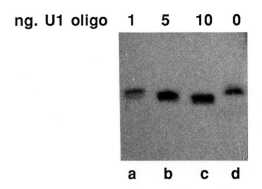

FIG. 3. U1 snRNA cleavage as a function of U1 oligo concentration. Groups of four oocytes were injected with the indicated amount of U1 oligo. Oocytes were then incubated at 20°C for 18 hours, RNA was extracted, and 1 oocyte's equivalent was run in each lane of an 8 M urea–10% polyacrylamide gel. After electrophoretic transfer of RNA to a nylon membrane, the Northern blot was probed with a ^{32}P-labeled plasmid containing the *Xenopus* U1 snRNA sequence.

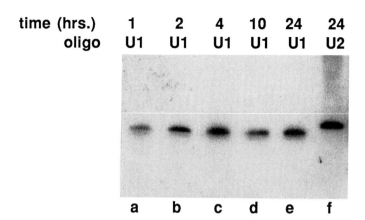

FIG. 4. Time course of oligo-targeted U1 snRNA cleavage in oocytes. Groups of four oocytes were injected with 10 ng of U1 oligo and incubated at 20°C for the indicated time periods. RNA was extracted, and 1 oocyte's equivalent was run in each lane of an 8 M urea–10% polyacrylamide gel. Similarly, four oocytes were injected with 20 ng U2 oligo, incubated at 20°C for 24 hours, extracted, and 1 oocyte's equivalent loaded in lane f. Northern blotting was as described in Fig. 3.

of U1 snRNA that took place between 1 and 10 hours (Fig. 3 compare lanes b and c to lanes d and e) is therefore likely to be the result of added removal of a few nucleotides in a manner that is not directly dependent on the oligo. However, consistent with our previous studies (Pan *et al.*, 1989), even after 24 hours there was no appreciable overall destabilization of U1 snRNA, and the amount of shortened U1 snRNA was equivalent to full-length U1 present in cells that either were not injected with oligo (not shown) or were injected with the U2 complementary oligo. These data confirm and extend earlier studies which indicate that the U1 particle is stable to the effects of removal of the exposed oligo-accessible 5′ end of U1 snRNA in HeLa cell extracts (Black *et al.*, 1985).

As described previously, the injection of the U2 oligo that is complementary to 15 nucleotides at the 5′ end of U2 snRNA resulted in markedly different effects on the integrity of U2 snRNA in oocytes than did the U1 oligo on U1 snRNA (Pan *et al.*, 1989). By 18 hours after injection with the U2 oligo, no detectable U2 snRNA forms were present. Figure 5 shows a gradual disappearance of detectable U2 snRNA as the amount of complementary oligo was increased (compare lanes a through d); U2 snRNA was completely degraded on injection of 20 ng U2 oligo (lane d). A time course revealed the progressive degradation of U2 snRNA (Fig. 6). After 1 hour (lane b), U2 was shortened by approximately 15 nucleotides (compare U2 snRNA from

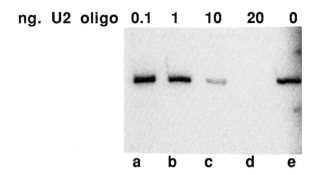

FIG. 5. U2 snRNA cleavage as a function of U2 oligo concentration. Groups of four oocytes were injected with the indicated amount of U2 oligo. Oocytes were then incubated at 20°C for 18 hours, RNA extracted, and 1 oocyte's equivalent was run in each lane of an 8 M urea–10% polyacrylamide gel. After transfer to a nylon membrane, the Northern blot was probed with a ^{32}P-labeled plasmid containing the *Xenopus* U2 snRNA sequence.

uninjected oocytes in lane g). This shortened form was relatively stable for 2 hours (lane c). After 4 hours (lane d), however, while most of this shortened form was still present, a smaller breakdown product became visible, suggesting that further cleavage of U2 snRNA had been initiated. By 10 hour U2 snRNA almost completely disappeared (lane e), and at the 24-hour time point no U2 snRNA was detectable (lane f). Pan *et al.* (1989) also injected an oligo complementary to nucleotides 28–42 in U2 loop I. This oligo also induced

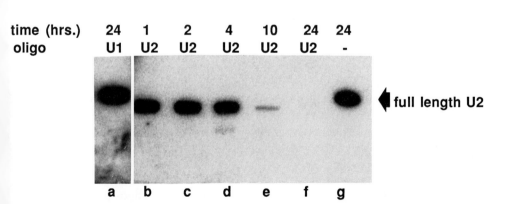

FIG. 6. Time course of oligo-targeted U2 snRNA cleavage. Groups of four oocytes were injected with 10 ng of U1 oligo. Oocytes were incubated at 20°C for the indicated time periods. As a control one group was injected with 20 ng U1 oligo and incubated for 24 hours (lane a). RNA was extracted, and 1 oocyte's equivalent from each sample was analyzed as in Fig. 5.

complete degradation of U2 and, in fact, did so more rapidly. Therefore our data show that oligos reveal markedly different stabilities of the U1 and U2 snRNPs in oocytes.

3. Injected U1 and U2 Oligonucleotides Inhibit Splicing of SV40 RNA in *Xenopus* Oocytes

Xenopus oocytes are capable of efficiently splicing SV40 late region-specific RNA transcribed from viral circular DNA (Wickens and Gurdon, 1983; Miller *et al.*, 1982; Fradin *et al.*, 1984). However, whereas oocytes generate correctly spliced mRNAs from SV40 primary transcripts, the efficiency with which oocytes from various viral spliced products differs considerably from that seen in most virus-infected or -transformed mammalian cell lines. For example, in infected monkey cells the major spliced late SV40 mRNA product is 16 S RNA. The spliced 19 S mRNA product comprises a much smaller proportion of the total late RNA such that the ratio of 16 S to 19 S RNA is approximately 10:1. By contrast, oocytes preferentially form the 19 S spliced product and produce relatively very small quantities of true 16 S mRNA, on the order of 5–10% of the 19 S species. These species and their detection by S1 nuclease analysis with the indicated probe are shown in Fig. 7.

We have noted that another form of 16 S RNA which we have called 16 S* is made in oocytes (Fradin *et al.*, 1984). This species most likely represents 16 S RNA from which the intraleader intron (nucleotides 294–435), but not the major 16 S intron (nucleotides 526–1463), has been removed. We previously showed that injection of U1 and U2 oligos that induce specific cleavage of these snRNAs resulted in reduced or blocked splicing of SV40 late pre-mRNA (Pan and Prives, 1988, 1989; Pan *et al.*, 1989). In those experiments we examined the products of S1 digestion on denaturing agarose gels. Such gels provided information about the relative ratios of unspliced RNA, as well as spliced 19 S and 16 SRNAs. We found that it was frequently difficult on such gels to resolve the 16 S* and 19 S species, which differ by only a few nucleotides (see Fig. 7). We have now repeated these experiments with a wider range of oligos using denaturing polyacrylamide urea gels. Our experiments confirm that oligos can provide insight into the potentially different requirements for snRNAs of different splice sites even within the same primary transcript.

When the effects of increasing concentrations of U1 oligos on late SV40 mRNA splicing were examined, several observations were made. First, as seen in Fig. 8 (lane 2), injection of 1 ng of the U1 oligo resulted in a pattern of spliced and unspliced RNA that was essentially unchanged from that seen with oocytes that had received no oligo. This is consistent with the very minor extent to which U1 snRNA was cleaved after injection of this amount of oligo.

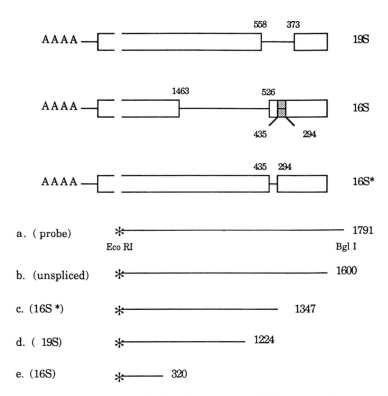

FIG. 7. Structure of SV40 late RNAs found in oocytes and their corresponding characteristic probe lengths after S1 analysis.

Second, when the amount of injected oligo was increased to 5 ng (lane 3), there was a marked increase in the amount of unspliced RNA detected. This was accompanied by significant but small decrease in the amount of spliced 19 S RNA. This slight decrease was more apparent in oocytes that had received 10 ng of the U1 oligo. Third and most striking, however, was the observation that both the 16 S and 16 S* RNA species were not detectable in oocytes that had received 5 or 10 ng of the U1 oligo. Thus the removal of both 16 S introns was far more sensitive to U1 targeting than was removal of the 19 S intron. As seen in Fig. 3, whereas both 5 and 10 ng of the U1 oligo resulted in virtual disappearance of detectable full-length U1 snRNA, the extent of U1 shortening was slightly greater with 10 ng than with 5 ng of the U1 oligo. Our splicing results suggest that this difference in length of the cleaved U1 snRNA is not responsible for the loss of the 16 S RNA species, but may contribute to the somewhat more reduced quantities of the 19 S RNA species detected at the higher concentration of oligo.

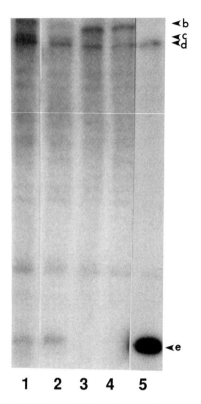

FIG. 8. U1 oligo inhibits SV40 late mRNA splicing in *Xenopus* oocytes. Aliquots of RNA samples shown in Fig. 3 from oocytes that had been injected with 0 (lane 1), 1 (lane 2), 5 (lane 3), and 10 (lane 4) ng of U1 oligo, followed by a second injection 4 hours later with 5 ng SV40 DNA, were analyzed. After incubation for 18 hours oocytes were frozen and RNA subsequently extracted and then subjected to S1 analysis using the probe shown in Fig. 7. For comparison, RNA from approximately 4×10^5 SV40-infected BSC-40 monkey cells was analyzed similarly (lane 5). Samples were fractionated on $8\,M$ urea–5% polyacrylamide gels and autoradiographed. Positions of SV40 RNAs (b–e) detected by the probe as described in Fig. 7 are indicated to the right of the autoradiogram.

When the S1 analysis of SV40 RNA in oocytes that had been injected with the U2-specific oligo was performed, we obtained results that were somewhat different than what we had observed with the U1 oligo (Fig. 9). With no oligo injected, in this S1 analysis of RNA synthesized in oocytes it was possible to resolve the probe (a) from the unspliced (b) and the differently spliced (c, d, and e) forms of SV40 late RNA as defined in the diagram in Fig. 7. As we and others have shown, and as seen in lane 1, it is difficult to detect either unspliced (band b) or 16 S* (band c) RNA species in RNA isolated from infected monkey cells.

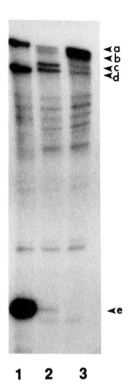

Fig. 9. U2 oligo inhibits SV40 RNA splicing in *Xenopus* oocytes. Groups of four oocytes were injected with no oligo (lane 2) or 20 ng U2 oligo (lane 3) followed by a second injection of 5 ng SV40 DNA 4 hours later. RNA prepared from the oocytes was subjected to S1 nuclease analysis along with RNA from SV40-infected BSC-40 cells (lane 1) as described in Fig. 8. Positions of SV40 RNA species (b–e) detected by the probe (a) described in Fig. 7 are indicated to the right of the autoradiogram.

After injection of 20 ng of the U2 oligo, there was a dramatic increase in the appearance of unspliced RNA. However, in contrast to what we had observed with the U1 oligo, all spliced RNA species (16 S*, 19 S, and 16 S RNAs) were strongly depressed. Although the 16 S species in oocytes was undetectable in U2 oligo-treated samples, the levels present in the absence of oligo in this experiment were themselves barely detectable.

These data extend our previous studies in which we showed that although intact U2 snRNA is required for splicing of both early and late mRNA, cleavage of the 5' end of U1 snRNA is far less deleterious to early mRNA splicing than it is to late mRNA splicing. In this study we have shown that we can identify conditions in which splicing of the 16 S SV40 late RNAs is far more dependent on full-length U1 snRNA than is the 19 S RNA splicing. It should be noted that some of our data in these experiments are not entirely consistent with our previous publications. For example, Pan et al. (1989) reported that low concentrations of the U1 oligo (e.g., 1 ng) blocked late SV40 19 S splicing. In that study, however, 1 ng of U1 oligo resulted in significantly greater quantities of cleaved U1 snRNA than in our current experiments. In our experience there are wide variations among frogs in many of the activities of oocytes, including their responses to oligos. Although our results have been for the most part both internally consistent and reproducible, differences in experimental protocols may have amplified more minor variations. With all of these caveats, it is clear that there is still much to be learned from experiments that employ oligos to study snRNA function and mRNA splicing in oocytes. This is also true in the more general sense when considering the wide variety of antisense and antifunction studies that can be performed using oligos in *Xenopus* oocytes.

ACKNOWLEDGMENTS

We are grateful to J. G. Gall for an illuminating conversation about nonspecific effects of oligos in oocytes during the course of writing this chapter. The experimental work that we have described herein was supported by U.S. Public Health Service Grant No. CA46121.

REFERENCES

Akagi, H., Patton, D. E., and Miledi, R. (1989). *Proc. Natl. Acad. Sci. U.S.A.* **86,** 8103–8107.
Baker, C., Holland, D., Edge, M., and Colman, A. (1990). *Nucleic Acids Res.* **18,** 3537–3543.
Barrett, C. B., Schroetke, R. M., Van der Hoorn, F. A., Nordeen, S. K., and Maller, J. L. (1990). *Mol. Cell. Biol.* **10,** 310–315.
Berget, S. M., and Robberson, B. L. (1986). *Cell (Cambridge, Mass.)* **46,** 691–696.
Black, D. L., and Steitz, J. A. (1986). *Cell (Cambridge, Mass.)* **46,** 697–704.
Black, D. L., Chabot, B., and Steitz, J. A. (1985). *Cell (Cambridge, Mass.)* **42,** 737–750.
Cazenave, C., Loreau, N., Thoung, N. T., Toulme, J.-J., and Helene, C. (1987a). *Nucleic Acids Res.* **15,** 4717–4736.
Cazenave, C., Chevrier, M., Thuong, N. T., and Helene, C. (1987b). *Nucleic Acids Res.* **15,** 10507–10521.
Cazenave, C., Stein, C. A., Loreau, N., Thuong, N. T., Neckers, L. M., Subasinghe, C., Helene, C., Cohen, J. S., and Toulme, J.-J. (1989). *Nucleic Acids Res.* **17,** 4255–4273.
Chen, Z.-Q., Burdett, L. A., Seth, A. K., Lautenberger, J. A., and Papas, T. S. (1990). *Science* **250,** 1416–1418.
Cohen, J., ed. (1990). "Oligonucleotide Antisense Inhibitors of Gene Expression." CRC Press, Boca Raton, Florida.
Colman, A. (1990). *J. Cell Sci.* **97,** 399–409.
Dagle, J. M., Walder, J. A., and Weeks, D. L. (1990). *Nucleic Acids Res.* **18,** 4751–4757.
Dale, L., Matthews, G., Tabe, L., and Colman, A. (1989). *EMBO J.* **8,** 1057–1065.
Dash, P., Lotan, I., Knapp, M., Kandel, E., and Goelet, P. (1987). *Proc. Natl. Acad. Sci. U.S.A.* **84,** 7896–7900.

Dumont, J. N. (1972). *J. Morphol.* **136**, 153–180.
Dunphy, W. G., and Newport, J. W. (1988). *Cell (Cambridge, Mass.)* **55**, 925–928.
Eppig, J. J., and Steckman, M. L. (1976). *In Vitro* **12**, 173–179.
Fabrizio, P., McPheeters, D. S., and Abelson, J. (1989). *Genes Dev.* **3**, 2137–2150.
Fradin, A., Jove, R., Hemenway, C., Keiser, H. D., Manley, J. L., and Prives, C. (1984). *Cell (Cambridge, Mass.)* **37**, 927–936.
Freeman, R. S., Kanki, J. P., Ballantyne, S. M., Pickham, K. M., and Donoghue, D. J. (1990). *J. Cell Biol.* **111**, 533–541.
Gurdon, J. B. (1968). *J. Embryol. Exp. Morphol.* **20**, 401–410.
Hamm, J., Dathan, N. A., and Mattaj, I. W. (1989). *Cell (Cambridge, Mass.)* **59**, 159–169.
Hamm, J., Dathan, N. A., Scherly, D., and Mattaj, I. W. (1990). *EMBO J.* **9**, 1234–1244.
Holwill, S., Heasman, J., Crawley C. R., and Wylie, C. C. (1987). *Development (Cambridge, U.K.)* **100**, 735–743.
Jessus, C., Cazenave, C., and Helene, C. (1988). *Nucleic Acids Res.* **16**, 2225–2233.
Kawasaki, E. S. (1985). *Nucleic Acids Res.* **13**, 4991–5004.
Kloc, M., Miller, M., Carrasco, A. E., Eastman, E., and Etkin, L. (1989). *Development (Cambridge, U.K.)* **107**, 899–907.
Krainer, A., and Maniatis, T. (1985). *Cell (Cambridge, Mass.)* **42**, 725–736.
Kramer, A., Keller, W., Appel, B., and Luhrmann, R. (1984). *Cell (Cambridge, Mass.)* **38**, 299–307.
Kressman, A., Clarkson, S. G., Teflod, J. L., and Brinstiel, M. L. (1977). *Cold Spring Harbor Symp. Quant. Biol.* **42**, 171–178.
Kushner, L., Lerma, J., Bennett, M. V. L., and Zukin, R. S. (1989). *In* "Methods in Neurosciences, Volume 1: Gene Probes" (P. M. Conn, ed.), pp. 3–29. Academic Press, New York.
Loke, S. L., Stein, C. A., Zhang, X. H., Mori, K., Nakanishi, M., Subasinghe, C., Cohen, J. S., and Neckers, L. M. (1989). *Proc. Natl. Acad. Sci. U.S.A.* **86**, 3474–3478.
Lotan, I., Goelet, P., Gigi, I., and Dascal, N. (1989). *Science* **243**, 666–669.
McPheeters, D. S., Fabrizio, P., and Abelson, J. (1989). *Genes Dev.* **3**, 2124–2136.
Marcus-Sekura, C. J. (1988). *Analy. Biochem.* **172**, 289–295.
Mattaj, I. W., and DeRobertis, E. M. (1985). *Cell (Cambridge, Mass.)* **40**, 111–118.
Melton, D. A. (1985). *Proc. Natl. Acad. Sci. U.S.A.* **82**, 144–148.
Meyerhof, W., and Richter, D. (1990). *FEBS Let.* **266**, 192–194.
Michaeli, T., and Prives, C. (1985). *Mol. Cell. Biol.* **5**, 2019–2028.
Miller, T. J., Stephens, D. L., and Mertz, J. E. (1982). *Mol. Cell. Biol.* **2**, 1581.
Minshull, J., Blow, J. J., and Hunt, T. (1989). *Cell (Cambridge, Mass.)* **56**, 947–956.
Mori, K., Boiziau, C., Cazenave, C., Matsukura, M., Subasinghe, C., Cohen, J. S., Broder, S., Toulme J.-J., and Stein, C. A. (1989). *Nucleic Acids Res.* **17**, 8207–8219.
Murray, A. W., and Kirschner, M. W. (1989). *Science* **246**, 614–621.
Noguchi, S., Higashi, K., and Kawamura, M. (1990). *Biochem. Biophys. Acta* **1023**, 247–253.
Pan, Z.-Q., and Prives, C. (1988). *Science* **241**, 1328–1331.
Pan, Z.-Q., and Prives, C. (1989). *Genes Dev.* **3**, 1887–1898.
Pan, Z.-Q., Ge, H., Fu, X.-Y., Manley, J. L., and Prives, C. (1989). *Nucleic Acids Res.* **16**, 6553–6568.
Richter, J. D. (1989). *Nucleic Acids Res.* **17**, 4503–4516.
Roy, L. M., Singh, B., Gautier, J., Arlinghaus, R. B., Nordeen, S. K., and Maller, J. L. (1990). *Cell (Cambridge, Mass.)* **61**, 825–831.
Sagata, N., Oskarsson, M., Copeland, T., Brumbaugh, J., and Vande Woude, G. F. (1988). *Nature (London)* **335**, 519–525.
Savino, R., and Gerbi, S. A. (1990). *EMBO J.* **9**, 2299–2308.
Saxena, S. K., and Ackerman, E. J. (1990). *J. Biol. Chem.* **265**, 3263–3269.

Scotto, K. W., Kaulen, H., and Roeder, R. G. (1989). *Genes Dev.* **3,** 651–662.
Shuttleworth, J., and Colman, A. (1988). *EMBO J.* **7,** 427–434.
Shuttleworth, J., Matthews, G., Dale, L. Baker, C., and Colman, A. (1988). *Gene* **72,** 267–275.
Smith, R. C., Dworkin, M. B., and Dworkin-Rastl, E. (1988). *Genes Dev.* **2,** 1296–1306.
Smith, R. C., Bement, W. M., Dersch, M. A., Dworkin-Rastl, E., Dworkin, M. B., and Capco, D. G. (1990). *Development (Cambridge, U.K.)* **110,** 769–779.
Toulme, J.-J., and Helene, C. (1988). *Gene* **72,** 51–58.
Vankan, P., McGuigan, C., and Mattaj, I. W. (1990). *EMBO J.* **9,** 3397–3404.
Wickens, M. P., and Gurdon, J. B. (1982). *J. Mol. Biol.* **163,** 1.
Wickens, M. P., and Gurdon, J. B. (1983). *J. Mol. Biol.* **2,** 1581–1594.
Wickstrom, E. (1986). *J. Biol. Chem. Biophys. Methods* **13,** 97–102.
Woolf, T. M., Jennings, C. G. B., Rebagliati, M., and Melton, D. A. (1990). *Nucleic Acids Res.* **18,** 1763–1769.
Zeitlin, S., and Efstradiatis, A. (1984). *Cell (Cambridge, Mass.)* **39,** 589–602.

Part III. Embryos

Chapter 11

Fertilization of Cultured Xenopus Oocytes and Use in Studies of Maternally Inherited Molecules

J. HEASMAN, S. HOLWILL, AND C. C. WYLIE

Department of Zoology
Cambridge University
Cambridge CB2 3EJ, England

I. Introduction
II. Protocol for Oocyte Fertilization Using the Host Transfer Technique
   A. Materials
   B. Isolation and Culture of Oocytes
   C. Manipulation of Oocytes
   D. Preparation of Host Female
   E. Maturation and Fertilization
   F. Rates of Success of Method
III. Protocol for Oocyte Fertilization Using *in Vitro* Fertilization Method
   A. Materials
   B. Isolation and Preparation of Oocytes
   C. Removal of Vitelline Membrane
   D. Fertilization
IV. Choice of Method
V. Applications of the Techniques
   A. Ultraviolet Irradiation Effects on Oocytes
   B. Introduction of Foreign DNA into Embryos by Microinjection of Oocytes
   C. mRNA Knock-out Experiments
VI. Summary
   References

I. Introduction

One question which has fascinated developmental biologists for over a century is the importance of maternal control in the development of the embryo. There is a great deal of evidence that in *Xenopus* embryos the maternal contribution to early embryogenesis is considerable. The fertilized egg develops to the late blastula stage (9 hours of development, 5000 cells) before significant transcription of the embryonic genome occurs (Bachvarova and Davidson, 1966, Newport and Kirschner, 1982). It follows, therefore, that cytoplasmic stores of proteins and messenger RNAs accumulated during oogenesis are necessary after fertilization for development of the embryo to the late blastula stage. This period in development is not simply the dividing of a single large cell into a hollow ball of smaller cells, but is a time when important developmental decisions are taken. For example, early steps in the establishment of the dorsal–ventral axis (Ruiz i Altaba and Melton, 1990) and in the determination of the endoderm and mesoderm germ layers (Heasman *et al.,* 1984; Smith, 1989) and primordial germ cell lineage (Smith *et al.,* 1983) occur during this period.

Although the concept of maternal inheritance is *Xenopus* has been generally accepted for many years, there is little concrete evidence for specifically which molecules are required for developmental decisions. It is obvious that large quantities of "housekeeping" proteins such as histones (Woodland, 1980) and cytoskeletal proteins (Woodland and Ballatine, 1980) are necessary. Even for these, we know little about the relative requirements for mRNA and protein. For many proteins, including cytoskeletal elements, transcription factors, and surface molecules, there have been few stringent tests of function. Also, some maternally inherited molecules are localized to specific areas of the cytoplasm and may be "determinants" in that they are inherited only by a small group of cells, and in some way direct the fates of these cells. The most likely candidates for determinants so far described in *Xenopus* are the "germ plasm," involved in establishment of the primordial germ cell lineage (Smith, 1966), and putative "dorsal" determinants (Elinson and Pasceri, 1989). However, the molecular identity of such determinants is still unknown.

One reason for the lack of information about maternal inheritance in *Xenopus* is the fact that there are no maternal effect mutants such as those available in invertebrates. The long life cycle and the tetraploid nature of the *Xenopus* genome make genetic studies difficult. However, *Xenopus* does have considerable advantages in the study of the control of maternal factors on development. For example, *Xenopus* oocytes and embryos are relatively large and easily manipulated, and they are resistant to damage by such techniques as microinjection and cytoplasmic transfer. They also afford a very useful time window during which, because embryonic transcription does not start until the midblastula transition, maternal mRNAs and proteins can be studied without confusion with embryonic transcripts (Bachvarova and Davidson, 1986; Newport and Kirschner, 1982). Finally, a considerable advantage of *Xenopus* is the fact that although its development is not well understood at the genetic level, there is a considerable body of information about embryogenesis at the molecular and cellular levels. *Xenopus* has been fate mapped more extensively than any other species (Moody, 1987a,b; Dale and Slack, 1987), and considerable progress has been made in understanding the origins of the three primary germ layers (Heasman *et al.,* 1984; Smith, 1989; Ruiz i Altaba and Melton, 1990).

Thus, a useful approach to follow may be to try to engineer "mutant" embryos by specifically removing or overexpressing particular mRNAs or proteins from the oocyte or fertilized egg, then studying the effects of these manipulations on the development of the embryo. The ovarian oocyte offers several advantages over the fertilized egg for such studies. In particular, the oocyte is far less sensitive to manipulation than the fertilized egg and can be maintained in culture for many hours without deterioration. In contrast, the fertilized egg cleaves after 80 minutes at 20°C (the first cell cycle can be extended to 3 hours if the eggs are kept at 15°C). Also, microinjections into eggs during the first half of the cell cycle lead to a high percentage of abnormal embryos, making the time available for manipulation of fertilized eggs severely restricted. Unfortunately, stage VI oocytes are not immediately fertilizable; maturing oocytes first need to be shed from the ovary into the body cavity and from there are wafted to the oviduct by ciliary action. Within the first part of the oviduct the vitelline membrane surrounding the oocyte is acted on by proteolytic enzymes which make this membrane penetrable to sperm at the time of fertilization.

Methods have been available for many years to overcome the problem of sperm penetration. Two general techniques have been tried. The first involves transferring manipulated oocytes into the body cavity of a host female, so that they pass down the oviducts of the host (Humphries, 1956, Applington, 1957; Arnold and Shaver, 1962; Rugh, 1962; Lavin, 1964; Smith *et al.,* 1968; Brun, 1975), and the second involves breaking down or removing completely the vitelline membrane (Subtelny and Bradt, 1961; Elinson, 1973; Katagiri, 1974). We have concentrated on refining the first technique which is described in detail below (Holwill *et al.,* 1987). Other laboratories have also had some success recently with the second method (Kloc *et al.,* 1989), in general using a procedure developed by Roberts and Gerhart (1991). This technique is also described below.

II. Protocol for Oocyte Fertilization Using the Host Transfer Technique

A. Materials

Oocyte Culture Medium, OCM (modified from Wallace and Misulovin, 1978)
200 ml Liebowitz medium (Flow Laboratories, McLean, VA)
200 ml Ultrapure water (double distilled)
0.16 g Bovine serum albumin (BSA) (Sigma Chemical Co., St. Louis, MO)
2 ml Glutamine from 200 mM stock (Sigma)
Make up fresh.

Modified Barth's Saline and HEPES, MBSH (Gurdon, 1976)

NaCl	88 mM
KCl	1.0 mM
NaHCO$_3$	2.4 mM
MgSO$_4 \cdot$7H$_2$	00.82 mM
Ca(NO$_3$)$_2 \cdot$4H$_2$O	0.33 mM
CaCl$_2 \cdot$6H$_2$O	0.41 mM
HEPES	10.00 mM

Adjust to pH 7.5 and filter. Store as 10× stock at 4°C.

Petri Dish Coated with 2% Agarose
Type 1, low EEO grade (Sigma) in MBSH. Store at 4°C.

Vital Dyes
0.25% Neutral red (RA Lamb) (Sigma N6634)
0.05% Nile blue A (Sigma N0766)
4% Bismark brown (Sigma B5263)

Keep as stock solutions at −20°C. These dyes are diluted in OCM just prior to labeling the oocytes.

B. Isolation and Culture of Oocytes

Female frogs (Xenopus I, Ann Arbor, MI) which have not been stimulated with human chorionic gonadotropin (HCG) during the previous 10 weeks are anesthetized in 2 g MS222 per liter water (aminobenzoic acid ethyl ester); Sigma 1501, and a piece of ovary is removed aseptically through a small incision in the ventral body wall. The incision is repaired by suturing the muscle and skin layers separately using sutures (Mersilk 4/0, Ethicon), and the frog is allowed to recover from the operation in tap water at 20°C.

The ovary is placed in oocyte culture medium (OCM) in a sterile petri dish and divided into small pieces with forceps. The full grown oocytes (stage VI) in the ovary fragments are examined under a dissecting microscope to assess their suitability for fertilization. In our experience, the status of the oocytes is critica for the success of the technique, although it is very difficult to predict with any certainty whether they will fertilize. Several criteria which we use include the following: (1) The oocytes must be full grown (i.e., 1.2 mm across). (2) Manually defolliculated oocytes should not be flaccid or wrinkled (e.g., if the surface is indented with forceps the indentation should disappear when the forceps are removed). (3) The animal hemisphere should be homogeneously pigmented; any speckling or inhomogeneity suggests that the oocytes are dying. When a suitable ovary is obtained, oocytes are manually defolliculated in OCM using watchmaker's forceps. Although defolliculation by collagenase treatment is quicker, we find that enzymatically treated oocytes will not fertilize. The oocytes are transferred to agarose-coated dishes [2% agarose in modified Barth's saline HEPES (MBSH)] in OCM and are maintained in an incubator at 18°C. Any damaged oocytes are discarded.

C. Manipulation of Oocytes

Oocytes are quite resistant to manipulations such as microinjection and cytoplasmic transfer, as long as the needle diameter is small enough to prevent cytoplasmic leakage. In our experiments, we use an automatic microinjector system (Picoinjector, Medical Systems, Inc.), a single micromanipulator (Leitz), and a dissecting microscope (Zeiss STEMI SV8). Injections are carried out in OCM, with the oocytes supported in small depressions melted in an agarose-coated petri dish with a Pasteur pipette. Injection volumes range between 2.5 and 40 nl. We find that injections of 40 nl of water are not deleterious to development. After injection, oocytes are again cultured at 18°C in OCM.

D. Preparation of Host Female

We have found that a useful yield of fertilized eggs can be obtained from ovarian oocytes by inserting them into the body cavity of a host female (Fig. 1). The host female must herself be stimulated to lay eggs by administration of 800–1000 units HCG (Profasi, Seralab), 8 hours before the experimental oocytes are introduced. She will then lay the transferred oocytes along with her own.

E. Maturation and Fertilization

After microinjection, the experimental oocytes are stimulated to advance to second meiotic metaphase by adding progesterone (1 μM final concentration) to the OCM. Oocytes are allowed to mature for 6 hours. In order to

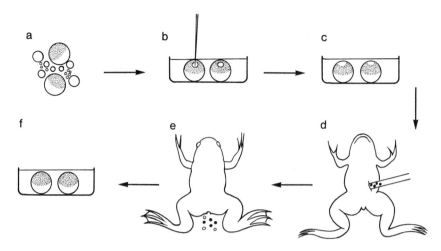

FIG. 1. Diagram of the host transfer technique. (a) Full grown stage VI oocytes are manually defolliculated and placed in OCM on agarose. (b) Oocytes are microinjected, as shown here in the nucleus. (c) Oocytes are matured with progesterone (1 μM) for 6 hours. (d) Oocytes are vital stained and then transferred into the body cavity of an anesthetized, HCG-stimulated host female. (e) Two hours later the host begins to lay the donor and host eggs. (f) Eggs are fertilized using a sperm suspension, then sorted.

distinguish experimental from host eggs, the injected oocytes are stained with vital dyes for 15 minutes followed by a brief wash in OCM. The dyes used, and their final concentrations are as follows: neutral red 0.0025%, nile blue sulfate 0.0005%, bismark brown 0.04%; and mauve which is a combination of 0.0025% neutral red and 0.0005% nile blue sulfate (Fig. 2).

A host frog that has already started to lay her own eggs is anesthetized, as described above. The donor oocytes are transferred aseptically into the abdominal cavity of the host through a small (0.5 cm) incision lateral to the midline of the lower ventral abdominal wall. The transfer is carried out with a fire-polished Pasteur pipette, and a minimum volume of OCM is introduced into the body cavity. After suturing, the host frog is allowed to recover for 2 hours at $22°-24°C$ in tap water. It is advisable to examine the tank for colored eggs during this time as, occasionally, the female may recover quickly and start to lay sooner.

FIG. 2. Vital stained oocytes and embryos: (a) matured oocytes stained with vital dyes as detailed in the text, (b) 8-cell embryos which were dyed as oocytes and fertilized by the host transfer technique (unstained host embryos are also present), and (c) stage 40 embryos derived from vital stained oocytes. Notice that some dye is retained in the yolky endoderm cells of the embryos.

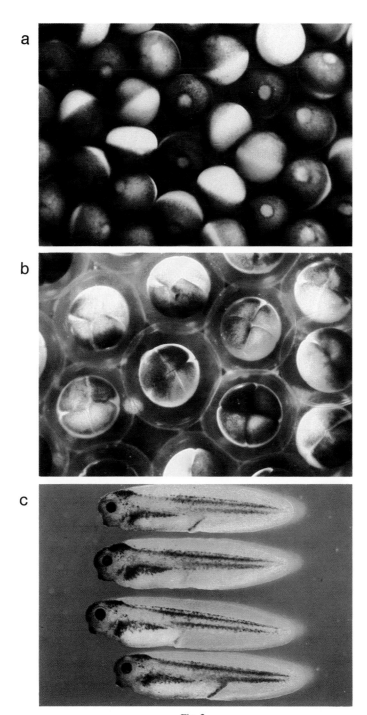

Fig. 2.

A sperm suspension is prepared by macerating a fragment of testis very finely with a pair of curved scissors in a petri dish without saline. MBSH (5–10 ml) is added to the dish. The host frog is then stimulated to lay eggs by gently squeezing at 30-minute to 11-hour intervals. The eggs are fertilized with the sperm suspension, and after 4 minutes they are transferred to 0.1 × MBSH. The colored eggs are sorted from the eggs of the host using blunt forceps. Eggs are then allowed to develop in 0.1 × MBSH.

F. The Rates of Success of Method

The percentage yield of fertilized eggs resulting from this method is exceedingly variable, ranging from 90 to 0% in control uninjected batches. We estimate that about 1 in 4 experiments yield sufficient embryos for analysis (where more than 50% of the control uninjected oocytes fertilize normally). This is in large part due to the inherent variability of the system. As all *Xenopus* users are aware, frogs cannot be relied upon to produce good batches of fertilized eggs on a regular basis, and, similarly, their yield of good oocytes is variable. This is presumably because the ovary at the time of HCG stimulation does not always contain oocytes at the peak of their growth. Unfortunately, we are unable to be certain that ovarian oocytes are fertilizable simply by observing them. In our experience, when a batch of oocytes are uniform in appearance and size and have homogeneously colored animal hemispheres, a high percentage will fertilize.

Many of the steps used in the method above have been chosen because they improve the likelihood of fertilization (Holwill, 1988). For example, OCM is better than MBSH as a culture medium, 6 hours is better than 12 hours as a maturation period, and culturing at 18°C is better than higher temperatures. Vital dying does not adversely affect the yield, nor in general does the status of the host's eggs. One exception to this is when the host lays eggs in very viscous jelly, in which strings of both host and donor eggs are crushed. Another problem we have experienced is when the host stops laying eggs after the first one or two manual squeezes, so that the entire batch of experimental eggs is lost.

III. Protocol for Oocyte Fertilization Using *in Vitro* Fertilization Method

The following procedure was provided by J. H. Roberts and J. C. Gerhart (University of California, Berkeley).

A. Materials

Modified Ringers (1 × MR)
  100 mM NaCl
  1.8 mM KCl
  1.0 mM $MgCl_2$
  2.0 mM $CaCl_2$
  5.0 mM Na-HEPES
  Stocks at pH 7.8 and at pH 6.5. Stocks are also required of 0.33 × MR, pH 7.8, and 0.1 × MR, pH 6.5.

1 × MR + Antibiotics
  1 × MR, pH 7.8, 1 mg/ml BSA, 50 µg/ml gentamycin, 50 µg/ml tetracycline

Agarose Plates
  2% agarose in 0.33 × MR, pH 7.8. Depressions are melted in the agarose with a fire-polished tip of a 25- to 50-µl micropipette (Drummond).

2% cysteine in 0.33 × MR, pH 7.8

Citrate–MR
  1 × MR, 2.5 mM sodium citrate, pH 4

Pepsin
  Stock of 0.15 mg/ml in citrate–MR, stored in 100-µl aliquots, frozen (−40°C).

Jelly Water
  Squeeze eggs from 3–6 females onto a preweighed culture dish. Remove excess water and arrange eggs in a monolayer. Weigh the dish and add 0.33 × MR (pH 7.8) in a ratio of 8 ml for every 3 g of eggs. Place the dish on a rocker plate at medium speed (15 cycles/min). Remove all the medium by tipping the plate and using a Pasteur pipette, after there has been at least a 30% loss in liquid volume as the egg swells (at −20°C, this takes 30–45 minutes). Aim to recover 5 ml of the 8 ml initially added to the 3 g of eggs. Measure the final extracted volume and add solid Ficoll (Sigma 400 DL) to a concentration of 10% (w/v). Store in 0.6-ml aliquots at −20°C in capped tubes. Aliquots retain activity for at least 1 month.

B. Isolation and Preparation of Oocytes

Frogs are injected with pregnant mare serum gonadotropin (50 units, Sigma) 12–15 hours before use. Pieces of ovary are removed and oocytes defolliculated as in the previous method. The oocytes are placed in 1 × MR + antibiotics or in OCM with antibiotics for longer storage.

Oocytes are matured with 1–3 µM progesterone in 1 × MR, pH 78. Germinal vesicle breakdown in 1 × MR, pH 7.8, begins 2–3 hours after progesterone exposure at 22°C. Oocytes are rinsed to remove progesterone when white spots appear. They are ready for fertilization 5–7 hours after stimulation and are processed in batches of 25–30 through the following steps.

C. Removal of Vitelline Membrane

Removal of the vitelline membrane is accomplished by a combination of pepsin and cysteine treatments, which loosen the membrane, followed by manual removal. Pepsin is sufficient to remove the membrane, but often has toxic side effects. These steps require the preparation of fresh pepsin (diluted immediately before use; 10 μl pepsin plus 40 μl citrate–MR, at room temperature in a microcentrifuge tube) and four petri dishes containing, in order, citrate–MR; 2% cysteine in 0.33 × MR; 1 × MR, pH 6.5; and 0.33 × MR, pH 7.8. A stock of 1 × MR, pH 7.8, is also needed for washes in between steps.

Oocytes are placed in the citrate–MR dish for 45 seconds and then transferred to the microcentrifuge tube containing pepsin, with a minimum of medium transfer. The oocytes are incubated for 45 seconds while carefully rotating the tube to prevent sticking of the oocytes to the tube wall. Oocytes are then washed with three 1-ml changes of 1 × MR, pH 7.8, by gently circulating the oocytes. Oocytes are transferred to the 2% cysteine dish for 30 seconds and then to a dish containing 1 × MR (pH 6.5) to rinse by gentle agitation. They are then transferred to 0.33 × MR (pH 7.8) and finally into agarose wells containing 0.33 × MR, pH 7.8.

Eggs arrive in the agarose wells with varying degrees of membrane removal, and care is required to pull off the remaining parts using fine forceps. Removal is easiest if started on the vegetal surface, and attempts should be made to finish with the animal surface upward. As sperm receptors are probably mainly in the animal hemisphere, it is most important to remove the membrane on this region.

D. Fertilization

First, a fragment of testis is macerated in an aliquot of jelly water. Next the 0.3 × MR solution is removed from the wells containing the demembranated eggs and is replaced by the sperm suspension. Care is needed not to make the fluid level too low. Activation occurs in 20–30 minutes, as indicated by contraction of the animal hemisphere, for development. The eggs are rinsed gently 3 or 4 times with 0.1 × MR, pH 6.5.

IV. Choice of Method

It is difficult to advise on which method is preferable when our laboratory has experience of only one of the techniques. However, from discussion with others it seems that both the oocyte transfer and the *in vitro* fertilization technique are moderately successful. The major problem of the first approach is in selecting a suitable ovary from which to take oocytes. A second problem is the lengthy nature of the procedure. We normally defolliculate and manipulate oocytes one day, then mature overnight and fertilize the next day. The advantage of the technique is that once suitable oocytes are placed in a host body cavity they can normally be fertilized and are then robust and easily handled as embryos. The advantage of the second technique is its *in vitro* nature; the oocytes can be monitored throughout the process. Its problems stem mainly from the delicate state of the demembranated oocytes: any careless manipulation causes lysis, and the oocytes are prone to infection.

V. Applications of the Techniques

One reason for persevering with these rather difficult methods is the large potential they offer for studying problems in the maternal control of development. Three possibilities which we have explored, although by no means exhaustively, are discussed here, namely, UV irradiation effects on oocytes, DNA injection, and RNA knock-out experiments. In these experiments we have used the host transfer technique. A number of other possibilities which are not considered further include the development of dominant-negative embryos, for example, by causing the egg and embryo to overexpress a mutated form of a protein, and protein depletion experiments using Fab fragments of specific antibodies.

A. Ultraviolet Irradiation Effects on Oocytes

It has long been documented that UV irradiation of the vegetal poles of fertilized eggs has a dose-dependent effect on two aspects of development, the formation of primordial germ cells (affected by low doses of UV) (Smith, 1966; Tanabe and Kotani, 1974) and the establishment of dorso–anterior structures (Grant and Wacaster, 1972; Malacinski *et al.,* 1975). The UV targets causing these effects are different (Thomas *et al.,* 1983). The vegetally localized "germ plasm" is considered a candidate for a primordial germ cell determinant. In contrast, considerable evidence exists that the UV target in disruption of the dorsal axis is the vegetal microtubular network. This network is responsible for a cortical cytoplasmic movement during the first cell cycle, which is absent in UV-irradiated eggs (Elinson and Rowning, 1988).

One question that can be resolved using the technique described above is whether this UV sensitivity is established before or after fertilization. We find that ovarian oocytes or fertilized eggs irradiated with $3-6 \times 10^{-4}$ J/mm^2 have similar reductions in the number of primordial germ cells, suggesting that the UV target is the same in both cases. However, oocytes are more sensitive than fertilized eggs to UV damage of the dorsal axis. Whereas exposure to $3-6 \times 10^{-4}$ J/mm^2 is sufficient to cause dorsal axis deficiency in oocytes, doses of 13×10^{-4} J/mm^2 using the same lamp are required for eggs (Holwill *et al.*, 1987). A further suggestion that the UV- sensitive factor present in oocytes is different from that in eggs is the work of Elinson and colleagues (Elinson and Pasceri, 1989), which shows that UV-irradiated oocytes, once fertilized, still undergo the normal cytoplasmic movement caused by the vegetal cortical microtubular network, even though they later go on to develop dorsal axis abnormalities. In preliminary cytoplasmic transfer experiments, we have shown that the effect on dorsal axis formation can be rescued by injecting UV-irradiated oocytes with vegetal cytoplasm from unirradiated oocytes (Holwill *et al.*, 1990). Although the putative "dorsal" determinant remains to be identified, possible candidates are growth factor type proteins such as the Vgl transcript (Weeks and Melton, 1987) and activins (Smith *et al.*, 1990).

B. Introduction of Foreign DNA into Embryos by Microinjection of Oocytes

The oocyte may be seen to have a number of important advantages over the fertilized egg in transgenic experiments. In particular, DNA injected into the germinal vesicle has contact with the genomic DNA and nuclear proteins for a relatively long period of time before maturation and fertilization, compared to the DNA injected into the fertilized egg.

Injecting DNA into the oocyte nucleus might increase the possibility of an integration event. We have explored the possibility of using ovarian oocytes to produce transgenic *Xenopus* by injecting certain DNA constructs into the germinal vesicles. We injected a recombinant with 3 kilobases (kb) of *Xenopus* cardiac actin gene upstream sequence coupled to the chloramphenicol acetyltransferase gene (CAT) and control plasmids that carry a short segment of the cardiac actin upstream sequence coupled to CAT (p3000 CAT and p56 CAT, gifts of Dr. T. Mohun; Mohun *et al.* 1986).

We find that, although CAT expression by circular DNA is barely detectable (when injections of 1–100 pg are used), CAT levels are high after fertilization when linear DNA is injected into oocytes in 100-pg amounts. The CAT enzyme is predominantly found in dorsal regions of embryos (Fig. 3), and by the feeding tadpole stage (stage 50) is mainly in the tail (Fig. 4), which essentially consists of somitic muscle. This suggests that the DNA is being expressed in a spatially correct fashion, consistent with the observations on egg-injected DNA.

However, the temporal pattern of expression of oocyte-injected DNA is different from that of eggs injected with actin−CAT enzyme constructs. We can detect CAT at the early gastrula stage (stage 10) when the DNA is injected into oocyte germinal vesicles (Fig. 5), whereas CAT expression in injected eggs coincides with that of the endogenous cardiac actin gene, which is switched on a few hours later at the early neurula stage (stage 13, (Mohun et al., 1984). It seems likely from these experiments that the exogenous DNA is switched on at mid-blastula transition (MBT), and that it is not under the same temporal control as the endogenous gene.

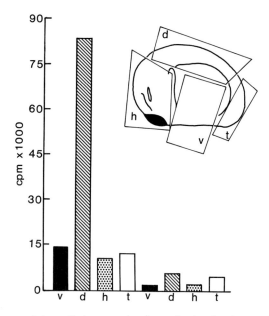

FIG. 3. Histogram of the preliminary results of a study of regional expression of the cardiac actin−CAT gene, after injection into oocytes. The inset shows the dissection of a late neurula embryo into four regions: d, dorsal; v, ventral; h, head; t, tail. Regions of six embryos were pooled for each assay. Controls (four right-hand bars) were injected with the bacterial CAT gene without the actin promoter sequence. Levels of acetyl [^{14}C]chloramphenicol synthesized were measured by scintillation counting after thin-layer chromatography (TLC).

11. FERTILATION OF CULTURED OOCYTES 225

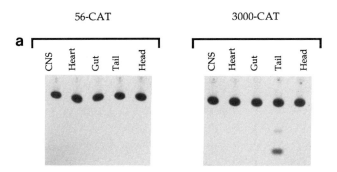

FIG. 4. Spatial expression of cardiac actin–CAT fusion genes. Oocytes were injected with 100 pg of either p56 CAT or p3000 CAT. The embryos were dissected at stage 50, and tissue fragments were pooled from 5 embryos for CAT assays. Results show that only the tail tissue of p3000 CAT-injected embryos has CAT activity (chloramphenicol is converted to chloramphenicol 3-acetate).

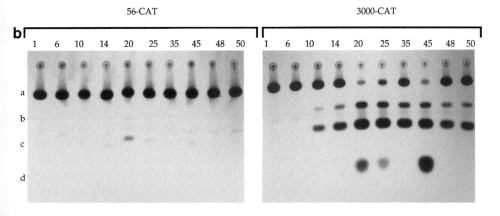

FIG. 5. CAT assays showing the temporal pattern of expression of cardiac actin–CAT fusion genes in embryos of different stages which were injected with 100 pg of p56 CAT or p3000 CAT as oocytes. Stages 1 to 45 represent material from a single embryo from a pool of 5. (a) Unchanged [^{14}C]chloramphenicol, (b) chloramphenicol 1-acetate, (c) chloramphenicol 3-acetate, (d) chloramphenicol 1,3-diacetate. b–d are the reaction products of CAT enzyme activity on a chloramphenicol substrate.

Although the experiments outlined above tell us about the spatiotemporal pattern of the expression of the injected DNA, they do not address the question of the fate of the DNA itself. A preliminary dot-blot analysis with a probe made from ^{32}P-labeled p3000 CAT DNA shows that this DNA is present at or below the 100-pg level the cleavage and blastula stages after

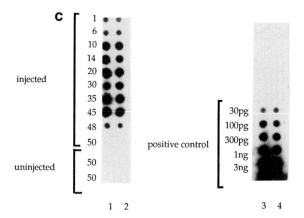

FIG. 6. DNA dot blots of DNA prepared from embryos of different stages derived from oocytes injected with 100 pg of p3000 CAT. Each dot contains 2 μg DNA prepared from embryos of each developmental stage: lane 1, stages 1–35 with 2 μg of DNA from a pool of 5 injected samples, stages 45–50 with 2 μg of DNA from a pool of 5 injected or uninjected samples; lane 2, duplicate of lane 1; lane 3, positive controls consisting of measured quantities of linear p3000 CAT made up to 2 μg with salmon sperm DNA; lane 4, duplicate of lane 3.

100 pg was injected into the oocytes, and increases to levels of 100–300 pg between gastrula and swimming tadpole stages (stage 10–45). By stage 50, the injected plasmid DNAs are barely detectable, suggesting that the DNA is lost by this time (Fig. 6). Also, Southern blot analysis shows that the DNA is arranged as random head–head and head–tail concatenates which are probably episomal.

Two major points which remain to be established are whether DNA injection into the oocyte can lead to integration in the embryonic genome, and whether the injected DNA is mosaically or evenly distributed during cleavage of the embryonic nuclei (see Chapter 19 this volume).

C. mRNA Knock-out Experiments

The approach of using antisense RNA to remove specific mRNAs from cells has not been very successful in *Xenopus* fertilized eggs (Melton and Rabagliati, 1986; Bass and Weintraub, 1987, 1988). A second approach to generate mRNA-minus "mutants" has been to inject short antisense oligodeoxynucleotides (oligos). These cause mRNA breakdown by annealing with the corresponding mRNA, and the RNA–DNA duplex is cleaved by an endogenous RNAse H activity (Dash *et al.,* 1987; Shuttleworth and Colman, 1988). It was shown by Colman and others that antisense experiments are more successful at removing messages when the oligo is injected into the cytoplasm of oocytes rather than eggs (Shuttleworth *et al.,* 1988). We have depleted RNA for intermediate filament proteins in oocytes using the oligo approach ourselves, and these results will be reported elsewhere.

VI. Summary

The methods described here of fertilizing stage VI oocytes are lengthy and quite difficult techniques. They would become more attractive if the success rate (i.e., the number of fertilizations compared to the numbers of matured oocytes) could be improved. An important step toward this for the host transfer technique would be to monitor carefully the status of mature *Xenopus* females ovaries in relation to cyclical HCG stimulation, so that we could predict more accurately whether stage VI oocytes are fertilizable. The *in vitro* technique would obviously be improved if oocytes could be fertilized without removing their membranes, perhaps by using oviduct extracts. So far, this approach has had only limited success.

It seems that the rewards of using these techniques could be great, in terms of understanding the maternal contribution to development. Although our experiments have not yet shown that oocyte injection of DNA has any advantage over egg injection, it is clear that it is possible to make "mRNA-minus mutants" by this approach. In the message depletion experiments mentioned here, we targeted the cleavage of an mRNA which is of low abundance in the full grown oocyte, but preliminary experiments have shown that we can deplete more abundant messages and produce specific phenotypes. Of course such experiments need to be controlled to show that the effect is specific, and the best proof that this is the case is to rescue the effect with injection of the appropriate mRNA. Finally, it seems likely that the method can be used to study the function of both localized molecules, such as the putative primordial germ cell (PGC) or dorsal determinants, and more ubiquitous molecules such as cytoskeletal elements.

Acknowledgments

Thanks are due to Dr. J. C. Gerhart for providing details of his methods. This work was supported by the Wellcome Trust and the Medical Research Council. We also thank Mrs. Niki Miller for typing the manuscript.

REFERENCES

Applington, H. W., Jr. (1957). The insemination of body cavity and oviductal eggs of Amphibia. *Ohio J. Sci.* **57**, 91–99.

Arnold, J. F., and Shaver, J. R. (1962). Interspecific transfer of eggs and ovaries in the frog. *Exp. Cell Res.* **27**, 150–153.

Bachvarova, R., and Davidson, E. H. (1966). Nuclear activation at the onset of amphibian gastrulation. *J. Exp. Zool.* **163**, 285–295.

Bass, B. L., and Weintraub, H. (1987). A developmentally regulated activity that unwinds RNA duplexes. *Cell (Cambridge, Mass.)* **48**, 607–613.

Bass, B. L., and Weintraub, H. (1988). An unwinding activity that covalently modifies its double-stranded RNA substrate. *Cell (Cambridge, Mass.)* **55**, 1089–1098.

Brun, R. (1975). Oocyte maturation *in vitro*: Contribution of the oviduct to total maturation in *Xenopus laevis. Experimentia* **31**, 1275–1276.

Dale, L., and Slack, J. M. W. (1987). Fate map for the 32-cell stage of *Xenopus laevis. Development (Cambridge, U.K.)* **99**, 527–551.

Dash, P., Lotan, I., Knapp, M., Kandel, E., and Geolet, P. (1987). Selective elimination of mRNAs *in vivo*; complementary oligodeoxynucleotides promote RNA degradation by an RNase H-like activity. *Proc. Natl. Acad. Sci. U.S.A.* **84**, 7896–7900.

Elinson, R. P. (1973). Fertilization of frog body cavity eggs enhanced by treatments affecting the vitelline coat. *J. Exp. Zool.* **183**, 291–302.

Elinson, R. P., and Pasceri, P. (1989). Two u-v sensitive targets in dorsoanterior specification of frog embryos. *Development (Cambridge, U.K.)* **106**, 511–518.

Elinson, R. P., and Rowning, B. (1988). A transient array of parallel microtubules: Potential tracks for a cortical/cytoplasmic rotation that forms the grey crescent of frog eggs. *Dev. Biol.* **128**, 185–197.

Grant, P., and Wacaster, J. F. (1972). The amphibian grey crescent region—A site of developmental information? *Dev. Biol.* **28**, 454–471.

Gurdon, (1976). The control of gene expression in animal development. Cambridge, Mass., Harvard University.

Heasman, J., Wylie, C. C., Hausen, P., and Smith, J. C. (1984). Fates and states of determination of single vegetal pole blastomeres of *X. laevis. Cell (Cambridge, Mass.)* **37**, 185–194.

Holwill, S. D. J. (1988). Manipulation of the oocyte of *Xenopus laevis* and its effects during early development. Ph.D. Thesis, University of London.

Holwill, S., Heasman, J., Crawley, C. R., and Wylie, C. C. (1987). Axis germ line deficiencies caused by u-v irradiation of *Xenopus* oocytes cultured *in vivo. Development (Cambridge, U.K.)* **100**, 735–743.

Holwill, S., Wylie, C. C., and Heasman, J. (1990). Dorsal axis specification in *Xenopus*: Another u-v sensitive target in the oocyte? *Dev. Biol. UCLA Symp. Mol. Cell. Biol. New Ser.* **125**, 79–84.

Humphries, A. A. (1956). A study of meiosis in coelomic and oviductal oocytes of *Triturus viridescens*, with particular emphasis on the origin of spontaneous polyploidy and the effects of heat shock on the first meiotic division. *J. Morphol.* **99**, 97–135.

Katagiri, C. (1974). A high frequency of fertilization in premature and mature coelomic toad eggs after enzymic removal of vitelline membrane. *J. Embryol. Exp. Morphol.* **31**, 573–587.

Kloc, M., Miller, M., Carrasco, A., Eastman, E., and Etkin, L. (1989). The maternal store of the xlg v7 mRNA in full grown oocytes is not required for normal development in *Xenopus. Development (Cambridge, U.K.)* **107**, 899–907.

Lavin, L. H. (1964). The transfer of coelomic eggs between frogs. *J. Embryol. Exp. Morphol.* **12**, 457–463.

Malacinski, G. H., Benford, H., and Ching, H. M. (1975). Association of an ultraviolet irradiation sensitive cytoplasmic localization with the future dorsal side of the amphibian egg. *J. Exp. Zool.* **191,** 97–110.

Melton, D. A., and Rebagliati, M. R. (1986). Anti-sense RNA injections in fertilized eggs as a test for the function of localized mRNAs. *J. Embryol. Exp. Morphol.* **97,** (Suppl.), 211–221.

Mohun, T. J., Brennan, S., Dathan, N., Fairman, S., and Gurdon, J. B. (1984). Cell type-specific activation of actin genes in the early amphibian embryo. *Nature (London)* **311,** 716–721.

Mohun, T. J., Garrett, N., and Gurdon, J. B. (1986). Upstream sequences required for tissue-specific activation of the cardiac actin gene in *Xenopus laevis* embryos. *EMBO J.* **5,** 3185–3193.

Moody, S. A. (1987a). Fates of the blastomeres of the 16 cell stage *Xenopus* embryo. *Dev. Biol.* **119,** 560–578.

Moody, S. A. (1987b). Fates of the blastomeres of the 32 cell stage *Xenopus* embryo. *Dev. Biol.* **122,** 300–319.

Newport, J., and Kirschner, M. (1982). A major developmental transition in early *Xenopus* embryos: I. Characterization and timing of cellular changes at the midblastula stage. *Cell (Cambridge, Mass.)* **30,** 675–686.

Roberts, R., and Gerhart, J. C. (1991). Manuscript in preparation.

Ruiz i Altaba, A., and Melton, D. A. (1990). Axial patterning and the establishment of polarity in the frog embryo. *Trends Genet.* **6,** 57–64.

Rugh, R. (1962). "Experimental Embryology: Techniques and Procedures." Burgess, Minneapolis, Minnesota.

Shuttleworth, J., and Coleman, A. (1988). Antisense oligonucleotide-directed cleavage of mRNA in *Xenopus* oocytes and eggs. *EMBO J.* **7,** 427–433.

Shuttleworth, J., Mathews, G., Dale, L., Baker, C., and Coleman, A. (1988). Antisense oligodeoxyribonucleotide-directed cleavage of maternal mRNA in *Xenopus* oocytes and embryos. *Gene* **7,** 267–275.

Smith, J. C. (1989). Mesoderm induction and mesoderm-inducing factors in early amphibian development. *Development (Cambridge, U.K.)* **105,** 665–677.

Smith, J. C., Price, B. M. J., Van Nimmen, K., and Huylebroeck, D. (1990). Identification of a potent *Xenopus* mesoderm-inducing factor as a homologue of activin A. *Nature (London)* **345,** 729–731.

Smith, L. D. (1966). The role of a 'germinal plasm' in the formation of primordial germ cells in *Rana pipiens*. *Devl. Biol.* **14,** 330–347.

Smith, L. D., Ecker, R. E., and Subtelny, S. (1968). In vitro induction of physiological maturation in *Rana pipiens* oocytes removed from their ovarian follicles. *Dev. Biol.* **17,** 627–643.

Smith, L. D., Michael, P., and Williams, M. A. (1983). Does a predetermined germ line exist in amphibians? *In* "Current Problems in Germ Cell Differentiation" (A. McLaren and C. C. Wylie, eds.), pp. 19–39. Cambridge Univ. Press, London and New York.

Subtelny, S., and Bradt, C. (1961). Transplantations of blastula nuclei into activated eggs from the body cavity and from the uterus of *Rana pipiens* II. Development of the recipient body cavity eggs. *Dev. Biol.* **3,** 96–114.

Tanabe, K., and Kotani, M. (1974). Relationship between the amount of the 'germinal plasm' and the number of primordial germ cells in *Xenopus laevis*. *J. Embryol. Exp. Morphol.* **31,** 89–98.

Tang, P., Sharpe, C. R., Mohun, T. J., and Wylie, C. C. (1988). Vimentin expression in oocytes, eggs and early embryos of *Xenopus laevis*. *Development (Cambridge, U.K.)* **103,** 279–287.

Thomas, V., Heasman, J., Ford, C., Nagajski, D., and Wylie, C. C. (1983). Further analysis of the effect of ultra-violet irradiation on the formation of the germ-line in *Xenopus laevis*. *J. Embryol. Exp. Morphol.* **76,** 67–81.

Wallace, R. A., and Misulovin, Z. (1978). Long-term growth and differentiation of *Xenopus* oocytes in a defined medium. *Proc. Natl. Acad. Sci. U.S.A.* **75,** 5534–5538.

Weeks, D. N., and Melton, D. A. (1987). A maternal mRNA localized to the vegetal hemisphere in *Xenopus* eggs codes for a growth factor related to TGF-β. *Cell (Cambridge, Mass.)* **51**, 861–867.

Woodland, H. R. (1980). Histone synthesis during the development of *Xenopus FEBS Lett* **121**, 1–7.

Woodland, H. R., and Ballantine, J. E. M. (1980). Paternal gene expression in developing hybrid embryos of *Xenopus laevis and Xenopus borealis J. Embryol. Exp. Morphol.* **60**, 359–372.

Chapter 12

Isolation of Extracellular Matrix Structures from Xenopus laevis Oocytes, Eggs, and Embryos

JERRY L. HEDRICK

Department of Biochemistry and Biophysics
University of California at Davis
Davis, California 95616

DANIEL M. HARDY

Howard Hughes Medical Institute
University of Texas Southwestern Medical Center
Dallas, Texas 75235

I. Introduction
II. Methods
    A. Induction of Ovulation and Obtaining Oviposited Eggs
    B. Preparation of Egg Jelly Coat Layers
    C. Sieving Methods for Isolation of Egg Envelopes
    D. Obtaining Oviposited Eggs for Vitelline Envelope Preparation
    E. Obtaining Activated or Fertilized Eggs for Fertilization Envelope Preparation
    F. Obtaining Coelomic Eggs for Coelomic Envelope Preparation
    G. Obtaining Oocytes for Ovarian Envelope Preparation
    H. Radioiodination of Isolated Envelopes
III. Discussion
    A. Jelly Coat Layers
    B. Envelopes
    References

I. Introduction

The extracellular matrix (ECM) surrounding amphibian eggs is composed of jelly coat layers, the egg envelope, and the fibrous elements of the perivitelline space (Fig. 2). The biological functions of the egg ECM in fertilization and development were first noted more than a century ago when Newport (1851) reported that unjellied eggs recovered from the coelom were not fertilizable. We now appreciate that the oviductally produced jelly coat layers of amphibian eggs are required for the process of sperm capacitation and/or induction of the sperm acrosome reaction (for review, see Hedrick and Nishihara, 1991; Katagiri, 1987). The oviposited egg envelope provides an initially penetrable structure to the sperm which is subsequently modified by products released in the cortical reaction to a sperm-impenetrable structure in the block to polyspermy reaction (for reviews, see Schmell *et al.,* 1983; Elinson, 1986). However, the ovulated egg recovered from the coelom possesses an envelope which is also impenetrable to sperm as originally shown in *Rana pipiens* and *Bufo japonicus* (Elinson, 1973; Katagiri, 1974). Thus, the egg envelope goes from a sperm-impenetrable state (the coelomic egg) to a penetrable state (the oviposited egg) and back to an impenetrable state (the fertilized egg or zygote).

At fertilization the egg envelope is chemically modified and its macromolecular permeability properties changed (discussed in Schmell *et al.,* 1983). This permeability change (hardening) results in an osmotically driven envelope elevation, owing to the influx of water into the perivitelline space. The jelly after fertilization functions as a "sticky substrate" for the adherence of the zygote to objects in its surroundings, protects the zygote against physical damage, and also provides a microbiological barrier (bacteria are rarely found within the jelly coat layers). Thus, after fertilization the ECM protects the developing embryo and functions as a barrier for the chemical and biological regulation of the embryo environment. This protective function of the ECM persists until the embryo develops into a free swimming tadpole and hatches from the ECM. The hatching process involves both physical and enzymatic mechanisms (Carroll and Hedrick, 1974). The ECM of the egg and the embryo, therefore, play significant roles in fertilization and development.

Contemporary approaches to the structure-function relations of the ECM and its component macromolecules using *Xenopus laevis* have utilized electron microscopic, biochemical, and immunological methods (for reviews, see Hedrick and Nishihara, 1991; Larabell and Chandler, 1991). The macromolecular compositions of the jelly coat layers (Table I) and the various forms of the egg envelopes have been determined (Fig. 1). Isolation of the individual glycoproteins composing the envelopes is currently in progress and

TABLE I

GLYCOPROTEIN COMPOSITION OF *Xenopus laevis* JELLY COAT LAYERS[a]

Jelly coat layer	Number of different glycoproteins				
			Electrophoresis		
	ID	SVC	IM	CA	SDS-A
J_1	2	2	3	2	3
J_2	2	2	2	2	2
J_3	4	4	4	5	4
Total	8[b]	8	9	9	9[c]

[a] Abbreviations: ID, immunodiffusion; SVC, sedimentation velocity centrifugation; IM, immuno; CA, cellulose acetate; SDS-A, sodium dodecyl sulfate–agarose. Adapted from Yurewicz et al., 1975.
[b] One additional component common to J_1 and J_3 was detected.
[c] Two additional components common to J_1 and J_3 were detected.

is a necessary prerequisite to the use of recombinant DNA methods for determination of the structure–function relations of the ECM glycoproteins (Gerton et al., 1982; Nishihara et al., 1986; Lindsay and Hedrick, 1988).

II. Methods

A. Induction of Ovulation and Obtaining Oviposited Eggs

The quality and quantity of eggs and the timing of ovulation are greatly improved by using a pregnant mare serum gonadotropin (PMSG) priming–human chorionic gonadotropin (HCG) ovulating protocol over that which uses HCG alone (Hedrick and Nishihara, 1991). Accordingly, females kept on a 12 hours light/12 hours dark schedule at 20°–23°C are injected into the dorsal lymph sac with 35 IU of PMSG dissolved in 1 ml of 0.15 M NaCl. After 96 hours, ovulation is induced by the injection into the dorsal lymph sac of 500–1000 IU of HCG. After 5–6 hours, eggs are stripped from the females into DeBoers solution. DeBoers solution is 110 mM NaCl, 1.3 mM $CaCl_2$, and 1.3 mM KCl, adjusted to pH 7.2–7.8 with $NaHCO_3$ (Katagiri, 1961). Normally, three or four strippings are done at 1.5- to 2-hour intervals and 1500–3000 eggs per female collected. Females are returned to their tanks and rested for at least 6 weeks before they are used again for egg production.

B. Preparation of Egg Jelly Coat Layers

1. Collection of Individual Jelly Coat Layers

Individual jelly coat layers can be microdissected from DeBoers-washed eggs using a pair of sharp watchmaker's forceps and a dissecting microscope (Yurewicz et al., 1975). A slight hydration of the jelly coat layers in DeBoers solution aids in their removal. This is usually accomplished by allowing the oviposited eggs to sit in DeBoers solution for 20–30 minutes at room temperature. The use of dilute salt solutions in this procedure is not possible as jelly coat layer J_3 becomes so sticky that manipulation of the eggs is difficult. However, for the recovery of eggs with only J_1 attached, this stickiness can be a benefit as the eggs adhere to the bottom of a culture dish and do not otherwise have to be held in place.

The soft outermost layer, J_3, can be cleanly removed from the tougher and more leatherlike J_2 layer by gently stripping away J_3 with one pair of forceps while using a second pair to hold the egg. Isolation of J_2 requires piercing layer J_2 with one pair of forceps and then teasing the torn J_2 layer away from the soft and highly hydrated J_1 layer with a second pair of forceps. J_2 is invariably dissected with small amounts of adhering J_1.

Because of its soft, sticky "chewing gum" nature, J_1 is most easily separated from the egg by solubilization in mercaptan solutions. Accordingly, J_3- and J_2-less eggs are immersed in 100 mM NaCl–50 mM Tris-HCl, pH 8.0, containing 2–5 mM dithiothreitol. The eggs are gently swirled, and dejellying is usually complete in 5–10 minutes at 22°C (as observed with the naked eye or with a dissecting microscope). The egg and vitelline envelope (VE) remain intact during this period. However, continued exposure to the dithiothreitol will dissolve the VE and lyse the egg. Decant the solution from the eggs and adjust the pH of the jelly solution to 6.5–7.0 for storage (oxidation of the sulfhydryl groups to disulfides occurs very slowly at neutral or acidic pH values; for long-term storage, 1 mM EDTA should also be added). A variety of mercaptans, such as mercaptoethanol, can be used instead of dithiothreitol, and DeBoers or other salt solutions can replace the NaCl–Tris solution. It should be noted that the rate of jelly dissolution is a function of the solution pH and mercaptan concentration (Gusseck and Hedrick, 1971). Lower pH values will slow the rate, and higher pH values will increase the rate of the dissolution reaction.

The isolated insoluble jelly coat layers J_3 and J_2 can be washed in appropriate salt solutions and then dissolved in dithiothreitol solutions as was done with J_1. The solubilized jelly coat layers are centrifuged at 12,000 g to remove particulate matter and dialyzed versus an appropriate buffer to adjust the pH and ion concentration. The jelly from a single egg contains approximately 41 μg of glycoprotein, with J_1, and J_2, and J_3 contributing 19, 5, and 17 μg, respectively.

2. Total Solubilization of Jelly

For the preparation of jelly coat glycoproteins from all the jelly coat layers or for the preparation of dejellied eggs, oviposited eggs can be treated with mercaptan solutions. Place a monolayer of washed eggs in a flat-bottomed dish such as a crystallization dish (e.g., 20–30 ml of eggs in a 125 × 65 mm dish). Add 80 ml of 10 mM Tris–DeBoers solution adjusted to pH 8.9 and containing 45 mM mercaptoethanol and gently rock the dish. Dissolution of the jelly coat layers can be followed by the naked eye and should be complete in 3–5 minutes. As stated previously, continued exposure of the eggs to the mercaptoethanol solution will dissolve the VE and lyse the eggs. Decant the solubilized jelly and adjust the pH to 6.5–7.0. If the eggs are also to be recovered, rinse them several times in 10 mM Tris–DeBoers solution at pH 7.0–7.8. From 1000 eggs, approximately 41 mg of glycoprotein is recovered with a composition of 39% protein and 61% carbohydrate.

3. Preparation of ^{35}S-Labeled Jelly

The egg jelly contains sulfate esters with the majority, if not all, of the sulfate located in jelly coat layer J_1 (Yurewicz et al., 1975). [^{35}S]Sulfate is readily incorporated into the J_1 jelly coat glycoproteins (Hedrick et al., 1974). Injection of 1–5 mCi into the dorsal lymph sac along with the HCG gives maximum incorporation into the egg ECM. Approximately 0.1% of the injected dose is incorporated with 3000–4500 cpm/egg of which 42% is associated with the jelly and 58% with the egg. The jelly contains both free and glycoprotein-bound [^{35}S]sulfate with the distribution ranging from 76 and 24%, respectively, from eggs collected in the first stripping to 33 and 67%, respectively, from eggs collected in the final stripping. The specific activity of the jelly is also a function of the stripping time, with the highest specific activity obtained in the last stripping. A second ovulation of the female 5–8 days later with an additional 2–5 mCi of [^{35}S]sulfate will give smaller numbers of eggs but with higher specific activity of the jelly glycoprotein. Some 93% of the [^{35}S]sulfate counts are incorporated into a single acidic glycoprotein in jelly coat layer J_1.

C. Sieving Methods for Isolation of Egg Envelopes

All forms of the egg envelope [VE, fertilization envelope (FE), coelomic envelope (CE), and ovarian envelope (OE)] are isolated using a common sieving method (Wolf et al., 1976). Jellyless eggs are poured into a large syringe (10–100 ml depending on the number of eggs used) and lysed by passage through an 18-gauge needle. The lysate is filtered through a 210-μm nylon screen suspended on the top of a wide-mouth bottle. Envelopes resting on the top of the screen are gently washed through the screen with a wash bottle containing ice-cold distilled water or 10 mM Tris–DeBoers solution, pH 7.0. This step removes the larger debris. The filtered envelope suspension is then poured through a 105-μm screen and the smaller particulate matter is washed through as with the above procedure. This sieving process is repeated (usually twice) until the envelopes are visually free from contaminating particulate material.

The purified envelopes are washed from the screen into a conical centrifuge tube and collected by a 1- to 2-minute centrifugation in a clinical centrifuge (room temperature, 1500 g). The major envelope contaminant is yolk granules, which are solubilized by overnight storage of the envelopes at 4°C in a salt solution (2 M NaCl, 0.2 M imidazole-HCl, 2mM $CaCl_2$, pH 7.0; Lindsay and Hedrick, 1989). The particulate envelopes are collected by centrifugation and washed with ice-cold water, DeBoers solution, or 0.15 M NaCl, depending on the subsequent use of the envelopes. The envelopes are stored at 4°C, pH 7.0, in high salt solutions (0.1 M NaCl) or $CaCl_2$ solutions to prevent swelling of the envelopes and solubilization of envelope glycoproteins, particularly the gp57 component of the VE and the FE (Bakos et al., 1990a). Envelopes can be solubilized by heating for 10 minutes at 80°C in water adjusted to pH 9 with Tris or NaOH. The envelopes remain soluble when the solution pH is subsequently lowered to pH 7 with dilute HCl or acetic acid.

D. Obtaining Oviposited Eggs for Vitelline Envelope Preparation

Oviposited eggs are collected using the hormonal stimulation methods stated above. The oviposited eggs are dejellied, lysed, and VEs collected by the sieving methods described above. Approximately 1 mg of glycoprotein (90% protein, 10% carbohydrate) is obtained per 1000 eggs. The VE is composed of seven glycoproteins as indicated in Fig. 1.

E. Obtaining Activated or Fertilized Eggs for Fertilization Envelope Preparation

Jellied oviposited eggs are activated by adding the calcium ionophore A23187 (2–5 μM final concentration) in 0.05 DeBoers or 0.05 Tris–DeBoers solution (Monk and Hedrick, 1986). The calcium ionophore A23187

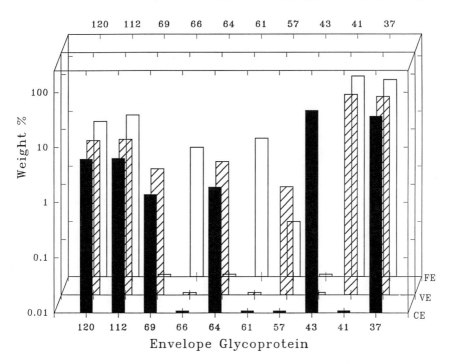

FIG. 1. The glycoprotein composition of egg envelopes. The glycoproteins are noted on the abcissa without the prefix gp. For graphing reasons (log of the weight %), 0% is equivalent to 0.01%. (The glycoprotein composition of CE is from Gerton and Hedrick, 1986b; that of VE and FE is from Gerton and Hedrick, 1986a; quantification by image analysis is from Hedrick and Nishihara, 1991.) The glycoprotein composition of the OE is qualitatively the same as the CE (data not shown).

stock solution (2–5 mM) is prepared by dissolving the ionophore in absolute ethanol or dimethyl sulfoxide. The activated eggs are gently swirled at room temperature, and after 20 minutes the medium is decanted. The activated eggs are washed, dejellied, lysed, and the activated FEs collected by sieving. Approximately 1.3 mg of glycoprotein (88% protein, 12% carbohydrate) is obtained per 1000 eggs (Wolf *et al.,* 1976). The FE is composed of at least nine glycoproteins as indicated in Fig. 1.

An alternative method for preparing the FE is to fertilize eggs (Wolf and Hedrick, 1971). A single testis is macerated in a conical centrifuge tube with a glass rod in 1 ml of DeBoers solution (the sperm concentration is approximately 1×10^7 cells/ml). DeBoers solution-washed eggs are spread in a monolayer in the bottom of a dish. The solution is decanted and replaced with 0.05 DeBoers or 0.05 Tris-DeBoers solution. Sperm are immediately added to a final concentration of at least 1×10^5/ml. The suspension is gently swirled, and, after 20 minutes at room temperature, the overlying solution is decanted, the eggs dejellied and lysed, and FEs collected by sieving.

The fertilization layer can be removed from the FE by solubilization to obtain the activated egg envelope moiety, VE* (the vitelline envelope component of the FE). Isolated FEs are suspended in 0.5 M galactose, 1.0 mM $CaCl_2$, and 10 mM Tris-HCl, pH 7.8, for 2 hours at room temperature (Nishihara et al., 1983). Alternatively, the fertilization layer can be solubilized with 5 mM EDTA, 10 mM Tris-HCl, 154 mM NaCl, pH 7.8 (Hedrick and Nishihara, 1991). The VE* is recovered by centrifugation (5000 g for 5 minutes), extracted an additional time with the galactose- or EDTA-containing solution, and finally washed by centrifugation in ice-cold water or a buffer solution. The extract contains the solubilized fertilization layer components, namely, the cortical granule lectin and its ligand (Nishihara et al., 1986).

F. Obtaining Coelomic Eggs for Coelomic Envelope Preparation

To obtain coelomic eggs, the oviducts are surgically ligated prior to the induction of ovulation (Bakos et al., 1990b). A female frog is anesthetized by immersion in 1 liter of MS222 (1.6 g of ethyl 3-aminobenzoate methane sulfonate/liter) at 0°C for 15–20 minutes. Arrange the frog on a cheesecloth-covered tray of ice (Fig. 3). By palpation, locate the posterior dorsal point of the scapula and make a 1-cm longitudinal incision through the skin about 1.5 cm posterior to the scapula and midway between the scapula and the lateral line (Fig. 3). The incision is best made by holding the skin taut with the fingertips and cutting with a scalpel.

After cutting through the skin, the underlying muscle layers are lifted with forceps and cut with a pair of scissors to avoid cutting the underlying lung. There are three muscle layers, with the inner layer being very thin; cutting through it will expose the lung (Fig. 4). While holding the inner muscle layer with forceps, with a second pair of forceps, grasp the innermost muscle layer inside the incision in a dorsal direction, gently pull the muscle externally, and transfer the grasp of the second forcep to a pronged hemostat. This procedure will prevent the inner muscle layer from retracting into the incision. Gently pull the inner muscle layer through the incision (evert) little by little until the attachment of the oviduct to the muscle layer and the ostium of the oviduct is located. The pars recta oviduct is easily differentiated from the pars convoluta oviduct as the pars recta is heavily vascularized relative to the white, opaque pars convoluta (Fig. 5).

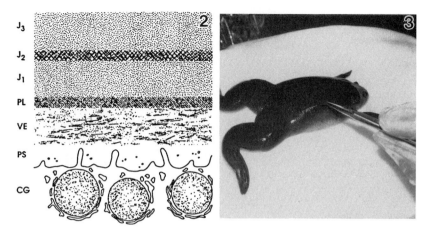

FIG. 2. Diagram of the extracellular matrix of the oviposited *Xenopus laevis* egg. The ECM structures in the diagram are not to scale. CG, Cortical granules; PS, perivitelline space; VE, vitelline envelope; PL, prefertilization layer; J_1, J_2, J_3, individual jelly coat layers.

FIG. 3. Location of the incision site for oviduct ligation.

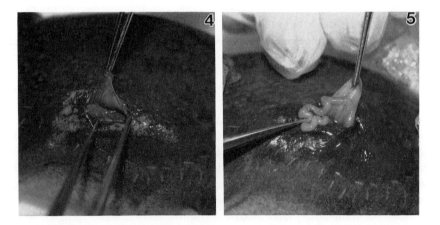

FIG. 4. Exposure of the pleuroperitoneal cavity. The three muscle layers have been cut and parted to expose the underlying lung.

FIG. 5. The pars recta and pars convoluta oviduct. The pars recta is heavily vascularized compared to the pars convoluta (forceps at left). The pars recta is attached to the body wall (forceps at top) and leads to the ostium.

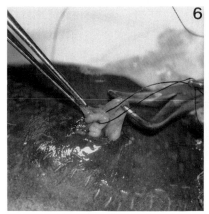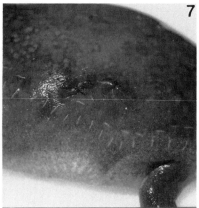

FIG. 6. Ligating the oviduct.
FIG. 7. The sutured incision at the completion of the ligation procedure.

Use 5–0 silk suture or single-use hemostatic silver clips to ligate the oviduct (Fig. 6). It is important to ligate the oviduct as close to the ostium as possible. Release the pronged hemostat and tuck the ligated oviduct back into the body cavity by lifting the body wall with forceps. Suture the three muscle layers of the body wall with 4–0 gut or close the incision with surgical staples. The three layers do not have to be separately sutured, but they must be well sutured to prevent leakage of ovulated eggs through the incision. Suture the skin with 3–0 silk suture (Fig. 7). If the frog begins to recover prematurely from the anesthesia, place some flakes of ice on its head. Repeat the process to ligate the other oviduct.

After the surgery is complete, place the frog in shallow water, keeping its skin moist and its nose exposed to prevent drowning. Allow the animal to recover slowly from the anesthesia for temperature acclimation before transferring to a larger volume of water at room temperature. If the lung is accidentally damaged during the surgery, the frog will suffer gas exchange problems and show substantial gaseous edema. The frog can be used for egg production immediately after recovery from the anesthesia or returned to its tank to be used at a later time.

The normal protocol for the induction of ovulation is used in obtaining coelomic eggs. Eggs can be recovered from the coelomic cavity of a surgically ligated animal 12 hours after HCG injection. The frog is anesthetized as before and laid ventral side up on a cheesecloth-covered tray of ice. Cut away the abdominal skin in the form of a three-sided flap (two ventral and one posterior cut) and drape the flap over the animal's head. Make lateral cuts in the muscle

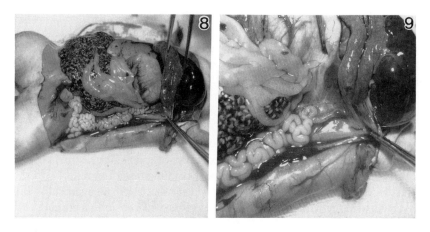

FIG. 8. The body cavity of a female showing the morphology of the internal organs and the oviduct.

FIG. 9. The pars recta–pars convoluta oviduct junction and the ostium of the oviduct where coelomic eggs tend to accumulate.

of the body wall on each side of the skinned area. Do not cut through the midline of the body wall muscle as the anterior abdominal vein will be severed and blood will contaminate the pleuroperitoneal cavity. With a spoon-shaped spatula, scoop the eggs from the body cavity and place them in DeBoers or Tris–DeBoers solution. Most of the eggs will be located in the anterior portion of the body cavity in the vicinity of the ostium (Figs. 8 and 9). The animal can be sacrificed by severing its spinal column or removing its heart. The recovered coelomic eggs are washed with DeBoers solution, lysed, and the CEs isolated by sieving. The CE is composed of six glycoproteins as indicated in Fig. 1.

G. Obtaining Oocytes for Ovarian Envelope Preparation

To obtain oocytes, ovaries are excised from adult females 96 hours after injecting 35 IU of PMSG. The females are sacrificed by severing the spine. The ovaries are washed with Tris–DeBoers solution, pH 7.4, to remove blood; fat and connective tissue are dissected away. The tissue is then processed through an industrial meat grinder (5 mm diameter holes in the template) to rupture the follicles. The ground tissue, containing numerous individual oocytes, is sieved through a 10-mesh stainless steel screen by gently shaking the screen while washing the oocytes with Tris–DeBoers solution, pH 7.4. Connective tissue clumps of oocytes are retained by the screen. The sieved fraction containing individual oocytes is washed free of egg lysate by suspending the cells in buffer, sedimenting them at unit gravity, and decanting the supernatant solution. The washed oocytes are then crudely fractionated according to size by repeatedly resuspending the cells in buffer, allowing them to settle briefly, and decanting the larger, more buoyant cells. In this way, a fraction enriched in late stage oocytes is obtained.

Although digestion of ovarian tissue with collagenase is a commonly used method of preparing *Xenopus laevis* oocytes, we found this enzymatic method unsuitable for preparing oocytes for OE isolation. Clostripain is a contaminant of collagenase preparations, and it proteolyzes gp43 of the OE (D. M. Hardy and J. L. Hedrick, unpublished observations). Thus, OEs obtained from collagenase-disrupted ovaries appear compositionally similar to the VE, an artifact which is avoided if the ovaries are disrupted mechanically. Inhibiting clostripain with thiol-reactive reagents such as iodoacetamide does not solve this problem as the clostripain activity appears to be necessary for effective digestion of the ovary into individual cells.

Using the tissue grinding method, we isolated OEs from up to 106 frogs (1.93 kg of ovaries) with a yield of 40 ml of packed envelopes representing 140 mg of envelope protein.

H. Radioiodination of Isolated Envelopes

Radiolabeling of the envelope glycoproteins with ^{125}I is useful for various experiments which require high sensitivity detection methods such as the topological location of envelope glycoproteins, Western blotting, and sperm–envelope binding experiments (Nishihara et al., 1983; Lindsay and Hedrick, 1988). The envelopes are readily labeled using IODO-GEN® (Pierce, Rockford, Il) and ^{125}I by the method of Markwell and Fox (1978). The envelopes can be iodinated in the form of intact particulate envelopes, heat-solubilized envelopes (large molecular weight supramolecular complexes of envelope glycoproteins), or as dissociated solubilized individual glycoproteins (SDS or guanidine-HCl dissociated).

Envelopes are solubilized by heating an envelope suspension to 70°–80°C in water or an appropriate buffer adjusted to pH 8–9 for 10 minutes. After cooling, the solution is neutralized with dilute HCl or acetic acid. Dissociation of the envelopes is effected by heating an envelope suspension at 100°C for 90 seconds in a buffer solution containing 1–2% SDS or 6 M guanidine-HCl at pH 7–8. The envelope suspension/solution (250 μg of protein) is transferred to a 10 × 75 mm culture tube plated with 50 μg of IODO-GEN® (Pierce, Rockford, Il). Approximately 500 μCi of carrier-free ^{125}I is added to initiate the reaction. After 15–20 minutes at room temperature, the free ^{125}I is separated from ^{125}I-derivatized glycoproteins by washing particulate envelopes with ice-cold water 5–6 times by centrifugation (1500 g for 5 minutes). For solubilized envelopes, the free ^{125}I is removed by gel-filtration chromatography using BioGel P-2 or Sephadex G-25 (room temperature and an appropriate buffer, e.g., DeBoers or Tris–DeBoers, pH 7.8).

III. Discussion

A. Jelly Coat Layers

The initial report on the number of *Xenopus laevis* egg jelly coat layers stated the presence of three layers (Freeman, 1968). More recent light microscopic studies indicated the possible presence of a subdivision of jelly coat layer J_3 into two separate layers in dilute DeBoers solution (Yoshizaki, 1985). This report awaits confirmation by other methods and has apparently not been pursued further. Fine structure studies on the jelly coat layers using contemporary methods remain to be done.

The gross chemistry and glycoprotein composition of the jelly coat layers has been determined (Yurewicz et al., 1975). Analysis using gel electrophoretic, immunological, and ultracentrifugal methods have established that at least nine different glycoproteins are present in all three layers with J_1, J_2, and J_3 having two or three, two, and four or five macromolecules, respectively (Table I). Isolation of individual glycoproteins has yet to be accomplished, which is a necessary preliminary for structure–function studies employing recombinant DNA methods and determining the chemical structures of functionally important oligosaccharides. In addition the location of individual glycoproteins within jelly coat layers (they are undoubtedly not uniformly distributed) using immunocytochemical or conjugated lectin methods has not been done. The specific roles of jelly coat glycoproteins in fertilization and early development is largely speculative although the importance of the jelly coat to fertilization was experimentally demonstrated more than 100 years ago (Newport, 1851). A specific hypothesis for the biological function of jelly coat glycoproteins was proposed for *Bufo japonicus*, namely, that the glycoproteins serve to bind Ca^{2+} and Mg^{2+}, ions which are necessary for the sperm acrosome reaction (for discussion, see Katagiri, 1987). The generality of this hypothesis in other amphibians, and in particular *Xenopus laevis*, lacks experimental support, however.

B. Envelopes

Ultrastructural differences in the various forms of the egg envelope correlate with functional changes in the envelope in regard to sperm penetration.

The organizational elements of the envelope are predominantly fibrous in nature (Hedrick and Nishihara, 1991; Larabell and Chandler, 1991). The OE is constructed of fasiculated fibers which are penetrated by the macrovillar processes of the cumulus cells and the microvillar processes of the oocyte. The villar processes are important for transfer of macromolecules into the oocyte, for example, vitellogenin. The ovulated egg envelope, the CE, lacks the villar processes but retains the fasiculated or bundled fibrous appearance and is impenetrable to sperm (Grey et al., 1977). The VE of the oviposited egg has a dispersed fibrous appearance, as though the fiber bundles were untied and scattered. The VE is penetrable by sperm. The most notable ultrastructural change in the FE of the fertilized egg is the appearance of the fertilization layer on the outer aspect of the envelope. The envelope is again rendered impenetrable to sperm by the VE to FE transformation.

Recent transmission electron microscopy studies have identified several new structures in or associated with the envelopes and one structural reorganization event of envelope fibers. A cloudlike prefertilization layer was identified by Yoshizaki and Katagiri (1984) situated between the outer aspect of the VE and the inner aspect of jelly coat layer J_1 (Fig. 2). The prefertilization layer was proposed as being a secretory product of the pars recta oviduct and to function as a precursor to the fertilization layer (Yoshizaki, 1984). Presumably, it is the ligand for the cortical granule lectin which, together with the lectin, forms the fertilization layer (Nishihara et al., 1986). However, this layer or the glycoprotein(s) of which it is composed (and found in mercaptan-solubilized jelly; Wyrick et al., 1974) have not yet been isolated and demonstrated to be the ligand for the cortical granule lectin. Interestingly, the jelly coat layers of *Bufo japonicus* eggs are immunologically related to the jelly coat layers of *Xenopus laevis* and also contain a ligand for the *Xenopus laevis* cortical granule lectin (Hedrick and Katagiri, 1988).

Using quick freeze, deep etch replica methods, Larabell and Chandler have identified a new structure in oviposited eggs (for review, see Larabell and Chandler, 1991). A horizontal filament layer located at the tips of the perivitelline space microvilli and in intimate contact with the inner aspect of the VE was observed in oviposited eggs. The function of this layer and its chemical/macromolecular nature has not yet been determined. It is not known if the layer is associated with isolated VEs. Upon fertilization, the horizontal filamentous layer is converted into a smooth layer, still located on the tips of the egg microvilli. Again, as with the horizontal filament layer, the function and macromolecular composition of the smooth layer are unknown. A rearrangement of the fibers composing the VE component of the FE, VE*, into concentric sheets interconnected by fine filaments occurs. The function of this fiber rearrangement is unknown, but since the VE* also is impenetrable to sperm (Grey et al., 1976), this structural rearrangement may relate to the block to polyspermy functions of the FE. Larabell and Chandler also described an intricate network of filaments within the perivitelline space which interconnects the microvilli with each other, the egg plasma membrane, and the VE. The presence of organized structures within the perivitelline space was hinted at, but could not be conclusively demonstrated by, thin section studies (Grey et al., 1974).

The molecular mechanisms involved in envelope conversions have been defined. Presumably conversion of the OE to the CE, when the villar processes retract from the envelope, is a purely physical phenomenon since no macromolecular differences between the envelopes have been detected and since the three-dimensional netlike structure of the fibrous envelope is flexible and deformable (Bakos et al., 1990a).

The CE to VE conversion takes place as the egg transits the oviduct. The molecular mechanisms involved are addition of a new glycoprotein, gp57 (of unknown cellular source and unknown function), limited proteolysis at the C-terminal end of gp43 to gp41, and subsequent conformational changes in the envelope glycoproteins as shown by dye binding, solubility, chemical modification, and deformability studies (Fig. 1; Bakos et al., 1990a,b). The pars recta-secreted protease was isolated and shown to be a 66K serine active site protease with homology to the chymotrypsin family of proteases (Hardy and Hedrick, unpublished observations). The isolated protease will convert the CE to the VE as shown by SDS-polyacrylamide gel electrophoresis of the envelope glycoproteins, ultrastructural changes as determined by electron microscopy, and alteration of envelope solubility.

The VE to FE conversion takes place at fertilization and involves factors derived from the cortical granules. The molecular mechanisms involved are limited proteolysis of the gp69 and gp64 envelope components at their C-terminal ends by serine active site proteases specific for Arg and/or Phe amino acid residues (Fig. 1; Lindsay and Hedrick, 1989). The chymotrypsin-like protease appears to be located in an inactive form associated with the fibers of the perivitelline space. It is activated by the presumably cortical granule-associated trypsinlike activity released in the cortical reaction (L. Lindsay, C. Larabell, and J. Hedrick, unpublished observations). This limited proteolysis of gp69 and gp64 is accompanied by or causes a subsequent conformational change in the envelope components (perhaps the formation of concentric sheets in the VE* observed by Larabell and Chandler, 1991), which is experimentally demonstrated by the same methods used to detect conformational changes in the CE to VE transformation (Bakos et al., 1990a). The fertilization layer is formed by a lectin–ligand binding reaction between the cortical granule lectin and its ligand (Nishihara et al., 1986). The lectin–ligand complex presumably binds to an envelope glycoprotein which prevents the fertilization layer from being lost during isolation (Hedrick and Nishihara,

1991). The VE* glycoprotein involved in adhering the fertilization layer to the envelope has not yet been identified.

It is apparent that the functional properties of the egg ECM are related to the ECM structure from the level of the macromolecule to the supramolecular complex. Isolation and determination of the structures of the individual glycoproteins of the egg jelly coat layers and the egg envelopes and relation of their glycoprotein structures to their gamete or zygote functions are necessary processes in understanding the role of the extracellular matrix in *Xenopus laevis* fertilization and development.

ACKNOWLEDGMENTS

Previously unpublished work reported here was supported by U.S. Public Health Service Research Grant HD04906 (J.L.H.) and a National Research Service Award, HD07088 (D.M.H.). We thank N. J. Wardrip and M. N. Oda for assistance with Fig. 1 and Fig. 2.

REFERENCES

Bakos, M., Kurosky, A., and Hedrick, J. L. (1990a). *Biochemistry* **29**, 609–615.
Bakos, M., Kurosky, A., and Hedrick, J. L. (1990b). *Dev. Biol.* **138**, 169–176.
Carroll, E. J., and Hedrick, J. L. (1974). *Dev. Biol.* **38**, 1–13.
Elinson, R. P. (1973). *J. Exp. Zool.* **183**, 291–302.
Elinson, R. P. (1986). *Int. Rev. Cytol.* **101**, 59–100.
Freeman, S. B. (1968). *Biol. Bull.* **135**, 501–513.
Gerton, G. L., and Hedrick, J. L. (1986a). *J. Cell. Biochem.* **30**, 341–350.
Gerton, G. L., and Hedrick, J. L. (1986b). *Dev. Biol.* **116**, 1–7.
Gerton, G. L., Wardrip, N. J., and Hedrick, J. L. (1982). *Anal. Biochem.* **126**, 116–121.
Grey, R. D., Wolf, D. P., and Hedrick, J. L. (1974). *Dev. Biol.* **36**, 44–61.
Grey, R. D., Working, P. K., and Hedrick, J. L. (1976). *Dev. Biol.* **54**, 52–60.
Grey, R. D., Working, P. K., and Hedrick, J. L. (1977). *J. Exp. Zool.* **201**, 73–84.
Gusseck, D. J., and Hedrick, J. L. (1971). *Dev. Biol.* **25**, 337–347.
Hedrick, J. L., and Katagiri, C. (1988). *J. Exp. Zool.* **245**, 78–85.
Hedrick, J. L., and Nishihara, T. (1991). *J. Electron Microsc. Technique* **17**, 319–335
Hedrick, J. L., Smith, A. J., Yurewicz, E. C., Oliphant, G., and Wolf, D. P. (1974). *Biol. Reprod.* **11**, 534–542.
Katagiri, C. (1961). *J. Fac. Sci. Hokkaido Univ. Ser. VI, Zool.* **14**, 607–613.
Katagiri, C. (1974). *J. Embryol. Exp. Morphol.* **31**, 573–581.
Katagiri, C. (1987). *Zool. Sci.* **4**, 1–14.
Larabell, C. A., and Chandler, D. E. (1991). *J. Electron Microsc. Technique* **17**, 294–318.
Lindsay, L. L., and Hedrick, J. L. (1988). *J. Exp. Zool.* **245**, 286–293.
Lindsay, L. L., and Hedrick, J. L. (1989). *Dev. Biol.* **135**, 202–211.
Markwell, M. A. K., and Fox, C. F. (1978). *Biochemistry* **17**, 4807–4817.
Monk, B., and Hedrick, J. L. (1986). *Zool. Sci.* **3**, 459–466.
Newport, G. (1851). *Philos. Trans. R. Soc. London* (Part 1, First Series) **141**, 169–242.
Nishihara, T., Gerton, G. L., and Hedrick, J. L. (1983). *J. Cell. Biochem.* **22**, 235–244.

Nishihara, T., Wyrick, R. E., Working, P. K., Chen, Y., and Hedrick, J. L. (1986). *Biochemistry* **25,** 6013–6020.
Schmell, E. D., Gulyas, B. J., and Hedrick, J. L. (1983). *In* "Mechanism and Control of Animal Fertilization" (J. F. Hartmann, ed.), pp. 365–413. Academic Press, New York.
Wolf, D. P., and Hedrick, J. L. (1971). *Dev. Biol.* **25,** 348–359.
Wolf, D. P., Nishihara, T., West, D. M., Wyrick, R. E., and Hedrick, J. L. (1976). *Biochemistry* **15,** 3671–3678.
Wyrick, R. E., Nishihara, T., and Hedrick, J. L. (1974). *Proc. Natl. Acad. Sci. U.S.A.* **71,** 2067–2071.
Yamaguchi, S., Hedrick, J. L., and Katagiri, C. (1989). *Dev. Growth Differ.* **31,** 85–94.
Yoshizaki, N. (1984). *Dev. Growth Differ.* **26,** 191–195.
Yoshizaki, N. (1985). *J. Morphol.* **184,** 155–169.
Yoshizaki, N., and Katagiri, C. (1984). *Zool. Sci.* **1,** 255–264.
Yurewicz, E. C., Oliphant, G., and Hedrick, J. L. (1975). *Biochemistry* **14,** 3101–3107.

Chapter 13

Analysis of Cellular Signaling Events, the Cytoskeleton, and Spatial Organization of Macromolecules during Early Xenopus Development

DAVID G. CAPCO AND WILLIAM M. BEMENT

Department of Zoology
Arizona State University
Tempe, Arizona 85287-1501

I. Introduction
   A. Meiotic Resumption
   B. Egg Activation
II. Methodology
   A. Application of Signal Mimetics and Antagonists
   B. Spatial Fractionation by Dissection
   C. Preparation of Detergent-Resistant Cytoskeleton of *Xenopus* Oocytes, Eggs, and Embryos
III. Conclusions
   References

I. Introduction

Development of the prophase I-arrested amphibian oocyte into the metaphase II-arrested egg is a complex process accompanied by changes in both cell function and responses to extracellular stimuli. For example, oocytes accumulate stores and maintain a state of readiness for response to progesterone. Progesterone triggers a cascade of intracellular signaling events which convert the oocyte to the egg (Smith, 1989; Bement and Capco, 1990a). In contrast, eggs are prepared for early embryogenesis and maintain a state of readiness for response to spermatozoa. Contact with a spermatozoan triggers another cascade of signaling events which "activate" the egg and initiate the program of early development (Charbonneau and Grandin, 1989; Bement and Capco, 1990b). Though oocyte and egg are separated by only a few hours within the same developmental continuum, the oocyte will not respond to sperm or other activating stimuli, such as a rise in intracellular free calcium ($[Ca^{2+}]_i$), and, conversely, the egg will not respond to progesterone. Thus, the round of intracellular signals experienced by the oocyte in response to progesterone converts the structural and biochemical makeup of the cell, such that the second round of signals which occur at fertilization elicit a new set of responses. Consequently, amphibian oocytes and eggs offer ideal model systems for study of interrelationships between cell signaling pathways, cell architecture, and cell function.

A. Meiotic Resumption

Progesterone elicits a relatively well-defined series of intracellular signaling and structural events which appear to be causally related (Fig. 1) (Bement and Capco, 1990a). Initially, progesterone causes a rapid drop in cAMP levels (Maller et al., 1979; Cicirelli and Smith, 1985) which causes movement of the cortical granules away from the plasma membrane (Bement and Capco, 1989a). Later, cAMP levels rise, and the cortical granules are repositioned immediately beneath the plasma membrane. Progesterone also causes a rapid decrease in the concentration of intracellular diacylglycerol (DAG), the endogenous activator of the calcium, phospholipid-dependent enzyme, protein kinase C (PKC) (Varnold and Smith, 1990). This decrease is followed by a later increase, which peaks late during the meiotic resumption (Varnold and Smith, 1990). The peak in PKC activity which presumably occurs at this time triggers formation of the cortical endoplasmic reticulum (Bement and Capco, 1989a), an organelle which stores calcium for release at the time of fertilization (Han and Nuccitelli, 1990). Meiotic resumption is also accompanied by cytoplasmic alkalinization (Cicirelli et al., 1983), which results in disruption of annulate lamellae (Bement and Capco, 1989a), organelles that resemble stacks of nuclear envelope. As meiotic resumption progresses, the various signaling events somehow activate maturation-promoting factor (MPF), a phosphoprotein complex which drives cells into M phase (Murray and Kirschner, 1989). MPF activation triggers nuclear envelope breakdown, chromatin condensation, and spindle formation (Smith, 1989). In addition, during meiotic resumption there is a redistribution of localized tubulin (Perry and

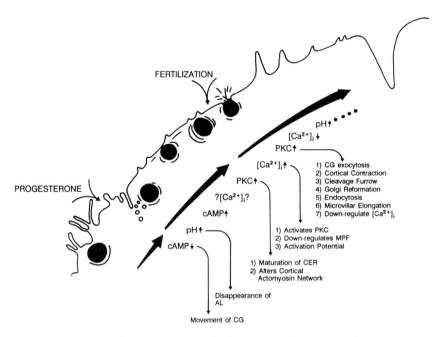

FIG. 1. Sequence of intracellular signals which accompany *Xenopus* meiotic resumption and egg activation. Arrows indicate potential interactions between specific signaling events and cellular reorganizations. For simplicity, this diagram omits extracellular layers such as the follicle cells, the theca, the vitelline envelope, and the jelly coat. CG, Cortical granules; CER, cortical endoplasmic reticulum; AL, annulate lamellae.

Capco, 1988) and Vg1 mRNA (Melton, 1987), which, at least in the case of tubulin mRNA, is triggered by cessation of a calcium-dependent chloride efflux (Larabell and Capco, 1988).

B. Egg Activation

Sperm penetration initiates a large, wavelike increase in $[Ca^{2+}]_i$ which propagates around the egg from the site of sperm penetration (Busa and Nuccitelli, 1985). How the calcium wave is generated is unclear, but it has been shown that fertilization is accompanied by rapid hydrolysis of phosphatidylinositol 4,5-bisphosphate (PIP_2) (Le Peuch et al., 1985), which results in formation of inositol 1,4,5-trisphosphate (IP_3) and DAG (Berridge, 1987). IP_3 causes release of calcium from intracellular stores (Busa et al., 1985), which may account for the $[Ca^{2+}]_i$ rise. It is also likely that the $[Ca^{2+}]_i$

rise, in combination with the DAG generated from PIP_2 hydrolysis (Le Peuch et al., 1985), results in PKC activation (Kikkawa and Nishizuka, 1986). Fertilization also inactivates MPF (Gerhart et al., 1984) and increases intracellular pH (Webb and Nuccitelli, 1981). These various signaling events regulate numerous physiological changes entailed by egg activation, with calcium acting at the top of the regulatory cascade (Fig. 1). For example, the rise in $[Ca^{2+}]_i$ apparently activates PKC, which subsequently triggers cortical granule exocytosis, cortical contraction, and downregulation of $[Ca^{2+}]_i$ (Fig. 1) (Bement and Capco, 1989b, 1990c). The rise in $[Ca^{2+}]_i$ also activates the calcium-dependent protease, calpain, which degrades the c-*mos* gene product, thereby inactivating MPF and contributing to transit into interphase (Sagata et al., 1989). Other events, such as pronuclear formation, migration, and fusion (Stewart-Savage and Grey, 1982), and further changes in mRNA and protein synthesis localization (Perry and Capco, 1988; Capco and Mecca, 1988), have yet to be linked to a particular signal.

Thus, a variety of intracellular signaling events herald important cellular transitions during early *Xenopus* development. The challenge, then, is to link individual signals to specific cellular events. Toward this end, this chapter details methods for manipulation of intracellular signaling pathways in *Xenopus* oocytes and eggs and describes methods for assaying cellular changes resulting from such manipulations.

II. Methodology

A. Application of Signal Mimetics and Antagonists

1. MICROINJECTION

Many signal agonists and antagonists require microinjection because they are insufficiently membrane permeable to reach the appropriate intracellular concentration after application. Descriptions of convenient microinjection techniques are described in appendix B of this volume; here we present only those approaches which allow injection of the metaphase II egg without causing injection-induced activation, a phenomenon which is undesirable when studying the process of activation itself. One way to avoid injection-induced activation is to inject oocytes while they are actually undergoing meiotic resumption. In this case, full grown oocytes are treated with progesterone, and, when the white spot appears in the center of the animal hemisphere, the oocytes can be microinjected. It is not until 1 to 2 hours after the white spot appears that such cells become activation competent (i.e., capable of being activated in response to local increases in calcium). Metaphase II eggs, in contrast, are fully activation competent; hence, precautions must be taken to prevent injection-induced increases in $[Ca^{2+}]_i$. This is accomplished by including in the injectate a calcium chelator such as BAPTA or EGTA at a concentration of 0.5–1 mM *in the injection buffer*, which prevents injection-induced activation but still allows subsequent activation in response to a calcium ionophore or other stimuli (Karsenti *et al.*, 1984; Bement and Capco, 1990c). Alternatively, the egg can be injected immediately after immersion in medium containing 10 mM chlorobutanol, an anesthetic which lowers $[Ca^{2+}]_i$ (Busa and Nuccitelli, 1985).

2. External Application

External application of signal agonists and antagonists is simpler than microinjection, but it is not without its own pitfalls. First, the agent being applied must be sufficiently membrane permeant to penetrate the cell and achieve the required intracellular concentration. Second, the oocyte or egg must be stripped of follicle cells (Bement and Capco, 1989b) and extracellular investments such as the egg jelly or vitelline envelope (Varnold and Smith, 1990). Third, caution must be exercised when interpreting membrane-related events, since the very fact that an agent can penetrate membranes means it may also perturb events involving membrane fusion. Finally, in contrast to microinjection, with external application one can never be certain as to the concentration of a given agent inside the oocyte or egg.

3. Preparation

To prepare oocytes, ovaries are removed from an incision in the ventral wall of a sexually mature female *Xenopus* (anesthetized by hypothermia) and placed in 1× OR2 medium (82.5 mM NaCl, 2.5 mM KCl, 1.0 mM Na$_2$HPO$_4$, 3.8 mM NaOH, 1.0 mM MgCl$_2$, 1.0 mM CaCl$_2$, 5 mM HEPES, pH 7.4; Wallace *et al.*, 1973) in a plastic petri dish. Ovaries are cut into clumps of 20–40 oocytes, which are subjected to a 1-hour collagenase treatment [1% type I collagenase (Sigma Chemical Co., St. Louis, MO) in 1× OR2] in a spinner flask. Collagenase in OR2 can be stored at −20°C and reused. After collagenase treatment, oocytes are washed 3 times in fresh OR2 and incubated overnight at 18°C. The following morning oocytes are transferred to fresh 1× OR2, and, using a dissecting microscope, Dumont No. 5 forceps, and a wide-bore pipette, full grown (1.1–1.3 mm) oocytes are transferred to a petri dish containing fresh 1× OR2. The follicle cells adhere to the plate and can be removed by rolling the oocyte across the plate with forceps. This technique removes all follicle cells from the oocyte, as assessed by electron microscopy. The oocytes can be used immediately or stored for 12 hours or more at 18°C. At times beyond 12 hours, however, oocytes begin to deteriorate and become less responsive to most agonists, including progesterone.

To prepare eggs, full grown oocytes are obtained as described above, then incubated in $1 \times$ OR2 containing 2 μg/ml progesterone (progesterone is stored in a stock solution of 1 mg/ml in 100% ethanol). Oocytes which successfully undergo meiotic resumption develop a white spot on the animal pole. This process takes from 3 to 12 hours, depending on the animal, the time of year, and whether or not the frog has been "primed" by prior injection with pregnant mare serum gonadotropin (PMSG, 35 IU). One to two hours after the white spot has appeared, the egg is fully competent to undergo activation in response to pricking. For *in vivo* matured eggs, mature frogs are injected with 800 IU of human chorionic gonadotropin (Sigma). Within 8–12 hours of injection, eggs can be stripped from females into plastic petri plates. To remove jelly, eggs are treated with 0.15% dithiothreitol (DTT) in $0.1 \times$ OR2 (pH 8.3) for 5 minutes. Eggs can be stored for 1–3 hours at 18°C in $1.5 \times$ OR2. To trigger activation, eggs are transferred to $0.1 \times$ OR2 and allowed to equilibrate for 15 minutes prior to application of the activating stimulus. For the treatments which follow, external application of signal mimetics and antagonists is accomplished by transferring oocytes, eggs, or embryos to the appropriate medium with a wide-bore pipette. Microinjection is accomplished as described in Appendix B, this volume, taking into account the caveats mentioned above.

4. Procedures

a. Protein Kinase A. Protein kinase A (PKA) is dependent on cAMP; thus, upregulation can be accomplished by treatments which raise intracellular cAMP levels. To elevate intracellular cAMP, oocytes or eggs are incubated in medium containing 5 mM dibutyryl-cAMP (Sigma), prepared immediately before use, for 5 minutes or longer. Alternatively, phosphodiesterase inhibitors such as theophylline block cAMP degradation, and thereby raise intracellular cAMP levels. Theophylline (Sigma) is prepared immediately before use and applied externally at a concentration of 1 mM for a minimum of 5 minutes. PKA downregulation is achieved by microinjecting oocytes or eggs with the regulatory subunit of PKA. Microinjection of 0.2 pmol of the regulatory subunit (Sigma) significantly decreases PKA activity and induces meiotic resumption in 100% of injected oocytes (Maller and Krebs, 1980).

b. Protein Kinase C. PKC activity requires calcium and DAG, but DAG activates PKC at resting levels of $[Ca^{2+}]_i$. A convenient form of DAG is 1,2-dioctanoylglycerol (DiC8) which can be purchased in 1-mg aliquots from Molecular Probes (Eugene, OR). To prepare DiC8, 20 µl of dimethyl sulfoxide DMSO is added directly to the DiC8 container, which is then vortexed for 1 minute. After vortexing, 100 µl of OR2 (1 × for oocytes, 0.1 × for eggs) is added to the container and vortexed for 1 minute. The container contents are transferred to a 15-ml test tube containing 5 ml OR2, which is then sonicated for 30 seconds using a probe sonicator. This gives a final DiC8 concentration of 200 µg/ml, which activates oocytes and eggs when applied for 10 minutes or longer. DiC8 is subject to oxidation and must be used immediately after dilution.

Phorbol esters activate PKC by mimicking DAG and are simple, inexpensive alternatives to DiC8. The most potent phorbol ester is phorbol 12-myristate 13-acetate (PMA; Sigma). PMA is stored as a 1 mM stock solution in DMSO at −20°C. The stock is stable for at least a year, and, once diluted into OR2, the PMA is stable for at least 2–3 days. When applied at 100 nM in OR2, PMA elicits both cortical granule exocytosis and cortical contraction in oocytes and eggs (Fig. 2). Caution must be exercised, however, when employing PMA. First, PMA is a potent tumor promoter; thus two pair of gloves should be worn when working with it, and PMA waste should be diluted 1:1 with bleach. Second, unlike DiC8, which is quickly metabolized,

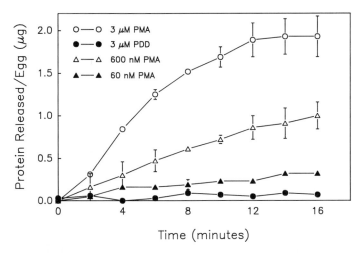

FIG. 2. Cortical granule exocytosis in *Xenopus* eggs treated with the PKC agonist PMA. Low PMA concentrations trigger linear release of cortical granule contents, whereas a high PMA concentration triggers a rapid and complete release of cortical granule components. PDD (4-α-phorbol didecanoate), a phorbol ester which does not activate PKC, has no effect. [Reproduced from W. M. Bement and D. G. Capco, *J. Cell Biol.* **108**, 885–892 (1989), by copyright permission from The Rockefeller University Press.]

PMA is stable within cells and, hence, results in long-term activation of PKC, instead of the transient activation which occurs *in vivo*. Consequently, protracted PMA treatment (at 100 nM, beyond 80 minutes) results in extreme cortical contraction, which ultimately leads to cell lysis. It is therefore often necessary to apply PMA for only short periods of time and then wash oocytes or eggs extensively in OR2. A useful control for PMA is the structurally similar but biologically inactive 4-α-phorbol 12,13-didecanoate. This agent is handled and applied in the same manner as PMA.

Several PKC antagonists are available, of which at least two are capable of inhibiting oocyte and egg PKC when applied externally. H7 (Hidaka *et al.*, 1984), when applied in OR2 at concentrations of 10–100 μM for 3 hours, inhibits PKC, although not completely. H7 is also known to inhibit the calcium/calmodulin-dependent protein kinase (CPK); consequently, a useful control agent is W7, which is structurally similar to H7 but inhibits CPK instead of PKC. Both H7 and W7 (Molecular Probes) are stored in 100 mM stock concentrations in DMSO at $-20°C$ and diluted into OR2 immediately before use. Sphingosine (Hannun *et al.*, 1986), applied at 100 μM for 1 hour, also partially inhibits PKC. Sphingosine (Sigma) is stored in a stock solution at a concentration of 100 mM in DMSO at $-20°C$. After thawing, the stock solution is vortexed to disperse the sphingosine within the DMSO, and 10 μl of the stock is then transferred to 10 ml OR2, which is sonicated for 1 minute. OR2 containing sphingosine is used immediately after preparation. In addition to H7 and sphingosine, a "pseudosubstrate" peptide which blocks the PKC active site (House and Kemp, 1987) is commercially available (Gibco BRL, Grand Island, NY). PKC pseudosubstrate is much more specific than either H7 or sphingosine; however, it is also much more expensive and must be applied by microinjection.

c. *Intracellular Free Calcium.* The most direct means to raise $[Ca^{2+}]_i$ is to simply microinject calcium into the oocyte or egg. However, since injected calcium is rapidly sequestered by the cell, it is preferable to inject calcium buffers, rather than calcium alone. In this approach, which allows reasonably accurate calculation of the final $[Ca^{2+}]_i$, buffers are prepared using known concentrations of calcium and the calcium chelator BAPTA (Molecular Probes). The calcium concentration is calculated according to the formula $[Ca^{2+}] = K_d[Ca^{2+} BAPTA]/[BAPTA]$, where $[Ca^{2+}]$ is the free calcium concentration of the injectate, K_d is the dissociation constant for BAPTA (0.1 μM), $[Ca^{2+} BAPTA]$ is the concentration of calcium-bound BAPTA, and [BAPTA] is the total concentration of BAPTA (calcium bound + unbound) (Kline, 1988). The aqueous egg volume is approximately 450 nl; however, since this is only an estimate, all calculations with respect to the final concentration of any injectate should be considered approximations. With this in mind, and given that resting oocyte $[Ca^{2+}]_i$ is approximately 100 nM and resting egg

$[Ca^{2+}]_i$ is approximately 400 nM, to raise $[Ca^{2+}]_i$, a buffer is formulated such that $[Ca^{2+}]$ surpasses these concentrations. As long as 10–50 nl is injected, the injectate swamps out endogenous calcium buffering systems, and the calculated $[Ca^{2+}]$ should reflect an approximation of $[Ca^{2+}]_i$ (Kline, 1988).

$[Ca^{2+}]_i$ can also be raised by treating oocytes or eggs with calcium ionophore A23187. A23187 is a mobile carrier which shuttles calcium across membranes, increasing $[Ca^{2+}]_i$ at the expense of both calcium in the external medium and intracellular calcium stores. A23187 (Sigma) is stored in a 10 mM DMSO stock solution at $-20°C$. To raise $[Ca^{2+}]_i$ with A23187, oocytes or eggs are exposed for 5 minutes to A23187 diluted to 100 nM in OR2. A23187 treatment is a simple, rapid means to raise $[Ca^{2+}]_i$; however, it has several shortcomings. First, short of measuring $[Ca^{2+}]_i$ after treatment with A23187, it is not possible, as it is with injection of calcium buffers, to estimate the rise in $[Ca^{2+}]_i$ triggered by A23187 treatment. Second, A23187 is light sensitive and unstable; hence, it must be stored in the dark and used soon after purchase. Because of this, it is important to perform control experiments in which eggs are treated with increasing concentrations of A23187 and monitored for cortical contraction, thereby allowing assessment of the potency of the A23187 stock solution. Finally, long incubations in A23187 (20 minutes or more) may result in egg lysis, necessitating transfer of treated eggs to fresh OR2.

Another approach to raising $[Ca^{2+}]_i$, more physiological than either buffer microinjection or ionophore treatment, is microinjection of IP_3. IP_3 is produced as a result of PIP_2 hydrolysis (see above) and is considered a natural stimulus for intracellular calcium release. IP_3 (Sigma) is made up in a stock solution of 1 mM in water and stored at $-20°C$. Injection of 1 fmol of IP_3, is sufficient to trigger an increase in $[Ca^{2+}]_i$ and activation in eggs (Busa et al., 1985), and 10–100 fmol will trigger $[Ca^{2+}]_i$ increases in oocytes (Lupu-Meiri et al., 1988).

To block rises in $[Ca^{2+}]_i$ or to lower resting $[Ca^{2+}]_i$, oocytes and eggs can simply be injected with the calcium chelators EGTA or BAPTA such that *intracellular* concentrations greater than 1 mM are achieved, and maintained in calcium-free medium, thereby limiting both intra- and extracellular calcium. For example, injection of BAPTA to a final intracellular concentration of 10 mM is sufficient to block egg activation in response to treatment with A23187 in calcium-free medium (OR2 without calcium and with 10 mM EGTA). For transfer of cells to calcium-free medium, cells are washed twice in calcium-free medium, to remove calcium bound to the cell surface, prior to the final incubation in calcium-free medium. It is also important that any wounds made by microinjection have healed prior to transfer into calcium-free medium, as calcium-free medium inhibits wound healing.

As with raising $[Ca^{2+}]_i$, it is often desirable to be able to estimate the final $[Ca^{2+}]_i$. In this case, calcium buffers are injected as described above, but instead of formulating the buffers such that $[Ca^{2+}]$ is above that of resting $[Ca^{2+}]_i$, buffers are prepared such that $[Ca^{2+}]$ is lower than $[Ca^{2+}]_i$. After injection, the calcium buffering provided by the injectate maintains $[Ca^{2+}]_i$ at the calculated $[Ca^{2+}]$.

d. Intracellular pH. To raise intracellular pH, oocytes or eggs are incubated in OR2 (pH 7.6 for oocytes, pH 7.8 for eggs) containing 10 m*M* procaine, for a minimum of 5 minutes (Houle and Wasserman, 1983). Procaine (Sigma) is a weak base which, when buffered to neutral pH, diffuses across the plasma membrane and on entering the cytosol acts as a proton acceptor. Procaine is dissolved into the medium immediately prior to use. Alternatively, oocytes and eggs can be incubated in OR2 (pH 8.0) containing 20 m*M* NH$_4$Cl for a minimum of 5 minutes (Charbonneau and Webb, 1987). OR2 containing NH$_4$Cl can be stored indefinitely at $-20°C$, but the pH must be checked after thawing. To lower intracellular pH, eggs or oocytes are incubated in OR2 (pH 6.8) containing 10 m*M* sodium acetate (Wasserman *et al.*, 1986). This can be stored at $-20°C$, but the pH should be checked after thawing.

5. IMPORTANT NOTES

a. Strategies for Determining Sequences of Related Signaling Events. What follows is a brief example, based on our results (Bement and Capco, 1989b, 1990c), of how the approaches described above can be used in combination to establish relationships between signaling events in the *Xenopus* system. Does PKC act upstream or downstream of the rise in egg $[Ca^{2+}]_i$ to induce egg activation events? To answer this question, it was first necessary to activate PKC while preventing rises in $[Ca^{2+}]_i$ and then to raise $[Ca^{2+}]_i$ while inhibiting PKC. To activate PKC in the absence of $[Ca^{2+}]_i$ rises, eggs were injected with BAPTA (10 m*M* final intracellular concentration) and then treated with 100 n*M* PMA in calcium-free 0.1 × OR2. This treatment lowers $[Ca^{2+}]_i$ while still resulting in PKC activation. Such eggs underwent cortical granule exocytosis and cortical contraction (Fig. 3), indicating that a rise in $[Ca^{2+}]_i$ was not required for PKC-induced egg activation. To raise $[Ca^{2+}]_i$ while inhibiting PKC activity, eggs were treated with H7 or sphingosine as described above and then transferred to 0.1 × OR2 containing 100 n*M* A23187 to raise $[Ca^{2+}]_i$. Exocytosis and cortical contraction were inhibited in such eggs, indicating that calcium-induced egg activation is dependent on PKC. Finally, full grown oocytes undergo exocytosis and cortical contraction in response to PMA but not A23187. This shows that PKC and activation events are coupled in oocytes, whereas calcium and activation events are not. It therefore follows that PKC is more proximal to activation events than calcium.

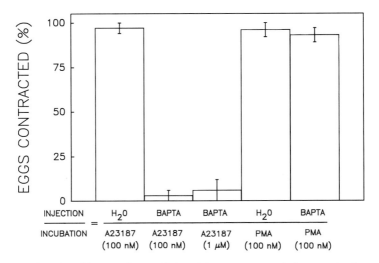

FIG. 3. Clamping of intra- and extracellular calcium prevents cortical contraction in response to treatment with calcium ionophore but not the PKC agonist PMA. Eggs were injected with BAPTA and then challenged in calcium-free medium with either calcium ionophore or PMA. [Reproduced from W. M. Bement and D. G. Capco, *Cell Regulation* 1, 315–326 (1990), by copyright permission from The American Society for Cell Biology.]

b. *Analysis of Signal Effects.* After determining the sequence of particular intracellular signaling events, it is important to determine the effect of those events on the physiological phenotype of the cell. Following are descriptions of two methods specialized for monitoring physiological changes in *Xenopus* oocytes, eggs, and embryos. One of these approaches takes advantage of the large cell size to gain information not readily obtainable from other cell types, that is, it provides a spatial map of the distribution of macromolecules within the cell; conversely, the other approach is designed to overcome difficulties entailed by the large cell size, that is, it allows preparation of reproducible, detergent-resistant cytoskeletons from oocytes, eggs, and embryos. Either of these methods can be applied after conducting any of the manipulations described above (e.g., PKC activation), thereby allowing assessment of the role specific intracellular signals play in controlling the cytoskeleton or the distribution of macromolecules in oocytes, eggs, and zygotes. Similarly, methods for analysis of other changes which may be induced by specific signals (e.g., ultrastructural reorganizations) are described in other chapters in this volume, and they can be applied after induction or inhibition of specific intracellular signals as described above.

The two approaches described below, namely, spatial fractionation by dissection and preparation of the detergent-resistant cytoskeleton, are useful on several levels. First, they provide quantitative, biochemical confirmation of data more typically obtained by microscopy. For example, identification of localized populations of mRNA is often determined by light microscopic analysis of the distribution of radiolabeled probes applied to sectioned material, a technique which can produce variable results and which does not easily lend itself to quantification. Spatial fractionation by dissection, on the other hand, employs conventional biochemical techniques which allow relatively precise quantification (Capco and Mecca, 1988; Perry and Capco, 1988). Second, and more importantly, these two approaches provide information which cannot be obtained by other techniques. For example, standard biochemical quantification of the amount of RNA begins with cell homogenization, a manipulation which prevents analysis of spatial differences in RNA distribution. Such differences, however, are retained by first spatially fractionating the cell. Moreover, detergent extraction has proved to be a powerful tool for analysis of the somatic cell cytoskeleton, but for reasons described below somatic cell protocols for detergent-extraction cannot be employed on amphibian eggs and embryos. The specialized detergent extraction protocol (see below), in contrast, circumvents problems associated with using standard somatic cell detergent extraction procedures, thereby allowing biochemical analysis of amphibian oocyte, egg, and embryo cytoskeletons and their associated components.

B. Spatial Fractionation by Dissection

Oocytes, eggs, zygotes, or embryos can be spatially fractionated into different regions by first hardening the cytoplasm and then cutting it into appropriate regions. When oocytes are used, the follicle cell sheath is removed as described above or during spatial fractionation along with the extracellular matrix. When naturally ovulated eggs are used, jelly layers must be removed prior to spatial fractionation using the approach described above.

1. Preparation

A microscalpel, forceps, and a microspoon are used to fractionate the oocytes, eggs, or zygotes. Scalpels are purchased from Ernest Fullum Inc. (Lanthum, NY; Wheeler Dissecting Knife, Cat. No. 13110). The microspoon is constructed by flattening a small stainless steel rod and making a depression in the flat area with a metal punch. Fractionations are conducted within petri plates coated with a thick layer of Siliclad, which prevents damage to the tips of forceps, microspoons, and microscalpels while acting as a pliable cutting surface. The Siliclad coating (Sylgard, 184 Silicone elastomer, Dow Corning, Phoenix, AZ) is prepared by mixing the catalyst and polymer, pouring the mixture into petri plates, and removing air bubbles under reduced pressure. Polymerization occurs within 24 hours.

To prevent cross-contamination of the various cytoplasmic regions, several Siliclad-coated plates are used to separate the different fractions. Oocytes, eggs, and zygotes are treated in small batches to determine the time required for ethanol treatment to harden specimens without becoming brittle. The following detailed description for fractionating oocytes into four regions, namely, animal hemisphere periphery, animal hemisphere center, vegetal hemisphere periphery, and vegetal hemisphere center, can also be used to divide eggs and zygotes into the same four regions, or into other regions, as desired by the investigator. In addition to oocytes, eggs, and zygotes, we have used this spatial fractionation by dissection technique successfully on multicellular embryos.

2. Procedures

If the germinal vesicle is to be removed from the oocyte it should be removed manually before placing the oocyte in ethanol, as the germinal vesicle cannot be distinguished from the rest of the cytoplasm following immersion in ethanol. All manipulations described below are conducted at room temperature, unless otherwise stated. This procedure requires about 2 hours for a single individual to process 30 eggs.

1. Two to four oocytes are placed in a Siliclad-coated plate containing 100% ethanol for approximately 4 minutes (Fig. 4A). Oocytes are then positioned with forceps and cut as desired with a microscalpel along the equator to separate the pigmented animal hemisphere from the nonpigmented vegetal hemisphere (Fig. 4B). This manipulation would be more

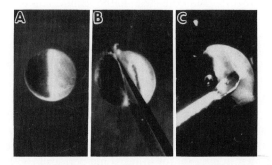

FIG. 4. Micrographs demonstrating the technique of spatial fractionation by dissection. (A) An intact *Xenopus* oocyte following submersion in ethanol. (B) A submerged oocyte cut with a microscalpel along the equator, dividing the cell into animal and vegetal hemispheres. (C) An isolated vegetal periphery, with a microspoon shown in the area formerly occupied by the vegetal central region. [Reproduced from D. G. Capco and M. Mecca, *Cell Differ.* **23,** 155–164 (1988), by copyright permission from Elsevier Scientific Publishers.]

difficult with albino oocytes; however, if the desire is to isolate the entire cell periphery from the entire central region, the initial bisection need not be positioned with reference to the animal–vegetal axis.

2. Animal and vegetal hemispheres are transferred to different plates containing ethanol using wide-bore pipettes to prevent cross-contamination of cytoplasms from the different hemispheres. Drying is prevented by aspirating the hemispheres in pipette partially filled with ethanol.
3. Hemispheres are held with forceps while the central region is scooped out with a microspoon (Fig. 4C) and transferred to an additional plate containing ethanol.
4. The peripheral region is isolated by holding its edge with one forceps and inverting the apex of the hemisphere with the other. The peripheral cytoplasm (which contains the outer 5 μm of the cell, referred to as the cortex, as well as an additional 100 μm) crumbles away from the extracellular matrix (i.e., the coelomic or vitelline envelope) and can then be removed from the petri plate.
5. Steps 1–4 are repeated for groups of three to four cells until the desired number of oocytes, eggs, or zygotes have been fractionated. The spatial fractions from the groups are individually transferred to microcentrifuge tubes as described in Steps 6 and 7.
6. Central regions are collected by gently swirling the petri plate with a circular motion to cause the larger fragments of cytoplasm to accumulate at the center of the plate. Larger fragments are removed with a Pasteur pipette, placed in a microcentrifuge tube, and concentrated by centrifugation. The remaining ethanol, containing smaller fragments of cytoplasm, is poured into centrifuge tubes and concentrated by centrifugation. Pellets from the centrifugations are then pooled.
7. Peripheral regions are collected in the same manner as described in Step 6.
8. Spatial fractions are stored in ethanol at $-20\,°C$ or processed further as described below.

3. Important Notes

Depending on the shape of the spatial fractions, the amount of time the specimen spends in ethanol can be varied. For example, oocytes, eggs, or embryos can be left in ethanol for longer time periods (4–30 minutes) if the goal is only to cut off the pole of the egg or to cut it in half. In cases where halves are further divided, however, timing must be more precise to prevent specimens from becoming too brittle. In these cases (Capco and Mecca, 1988; Perry and Capco, 1988) it may be necessary to employ two individuals: one to conduct artificial fertilizations and remove jelly or perform other preliminary manipulations at the appropriate time intervals, the other to fractionate specimens.

Another procedure for spatial fractionation by dissection which works as well as the one described above is to place the oocytes in 10% paraformaldehyde in OR2 at pH 7.4. This procedure has the advantage that the egg does not become so brittle. It has the disadvantage that the experimenter must be protected from paraformaldehyde vapors through use of a fume hood. Spatial fractionation is conducted as for ethanol fractionation; the only difference is that instead of the specimens becoming brittle with time, they become more rubbery. Analyses conducted with paraformaldehyde have provided the same results as those conducted with ethanol (D. G. Capco, 1988, unpublished results). However, owing to less toxic conditions, we use ethanol preferentially.

An ocular micrometer should be placed in the optical path of the dissecting microscope so estimates can be made concerning the size of various regions. Size estimates are then used to determine the volume of the spatial fractions. This is an important consideration for any type of spatial fractionation by dissection because the volume of cytoplasm can differ significantly between regions. Standardization of areas of different volume (based on volume) provides an estimation of the relative concentration of the component of interest in that region. Assuming that the peripheral region is 0.1 mm thick and the diameter of a full grown oocyte is 1.3 mm, the volume of each peripheral and central region is 1.43×10^8 and 5.75×10^8 μm^3, respectively (Perry and Capco, 1988).

4. Analysis of Fraction Components

Once spatial fractionation is completed, each region can be assayed for components of interest (Fig. 5). The type of spatial fractionation performed will influence the type of analysis that can be done. For example, if lipids are the components of interest, spatial fractionation conducted using the ethanol method for fractionation would be inappropriate as ethanol can extract lipids from membranes. In this case, spatial fractionation using paraformaldehyde is more appropriate. For assay of RNA and protein in the different spatial fractions, the ethanol method for preparing spatial fractions is adequate. For RNA extraction, the cytoplasm is pelleted by centrifugation, the ethanol is removed from the pellets, and the pellets are subjected to a procedure to purify RNA. We have used the phenol–chloroform method of extraction for RNA, although other methods can be used. Calculation of the amount of RNA lost during spatial fractionation by dissection relative to extraction beginning with

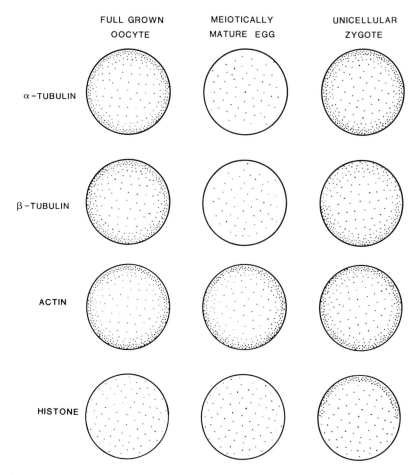

FIG. 5. Diagram demonstrating the distribution of tubulin, actin, and histone mRNA in *Xenopus* oocytes, eggs, and fertilized eggs. Stippling denotes mRNA localizations as identified by dot-blot analysis of spatial fractions obtained as described in the text.

living embryos has been found not to exceed 6% (Perry and Capco, 1988). For analysis of proteins from spatial fractions, the cytoplasm is pelleted by centrifugation, the ethanol is removed, and the pellet is dissolved into electrophoresis sample buffer (62.5 mM Tris-HCl, pH 6.8, 2% SDS, 10% glycerol, 5% 2-mercaptoethanol), sonicated, and heated briefly. For one-dimensional gels using SDS sample buffer, the buffer contains 4 M urea to solubilize intermediate filaments; in this case the molecular weight standards should also be dissolved in sample buffer with 4 M urea to assure equal conditions of migration during electrophoresis. Prior to electrophoresis samples dissolved in the sample buffer are microcentrifuged (10,000 rpm) for 5 minutes to pellet out remnants of the yolk which remain insoluble and would otherwise overload the gel.

Analysis of synthesis sites for cellular components such as proteins in oocytes and eggs is complicated by the fact that proteins may diffuse from their site of synthesis during the period of radiolabeling. Thus, short incubation times must be used such that newly synthesized proteins remain as nascent chains or do not have enough time to move from their site of synthesis. Taylor and Smith (1985) demonstrated that the transit time for ribosomes along mRNAs in *Xenopus* oocytes range from 4 to 17 minutes; hence, incubation times of 4–17 minutes are appropriate for study of the distribution of newly synthesized protein. If study is conducted on later stage embryos, such considerations are not so important because the plasma membrane of individual cells will block randomization of newly synthesized protein within the embryo.

C. Preparation of Detergent-Resistant Cytoskeleton of *Xenopus* Oocytes, Eggs, and Embryos

The detergent-resistant cytoskeleton of somatic cells is typically prepared by immersing those cells in medium containing a nonionic detergent and other components which stabilize the cytoskeleton (Capco et al., 1982). However, as traditionally applied, detergent extraction has some important limitations. For example, although detergent extraction is well suited to cultured cells (Capco and Penman, 1983; Capco et al., 1984) and small eggs and embryos (Capco and McGaughey, 1986; McGaughey and Capco, 1989), it does not work well on large cells, blocks of tissue composed of multiple cell layers, or cells surrounded by extensive extracellular matrices (Penman et al., 1981). These conditions impede the flow of detergent into the cell and prevent uniform extraction. Such drawbacks pose significant problems for preparation of the detergent-resistant cytoskeleton of *Xenopus* oocytes, eggs, and embryos, as they are extremely large and surrounded by an extracellular matrix.

To circumvent these problems, we have developed an approach that allows detergent extraction of blocks of tissue (Capco et al., 1987), which we have successfully applied to *Xenopus* oocytes, eggs, and embryos (Hauptman et al., 1989). This process entails rapid freezing, to minimize ice crystal-induced damage, followed by cryosectioning specimens to a thickness of 20 μm. This thickness approximates one or two layers of somatic cells and thus allows rapid detergent penetration when the sections are thawed into warmed detergent extraction medium.

1. Preparation

The following materials are needed: dry ice; cryotome; Freon 22 (obtainable from any local air conditioner or refrigerator supply company); Dumont No. 5 forceps; liquid nitrogen and 2 dewars; a copper cup (1.5 cm diameter, 1 cm, in depth); a metal rod; test tubes; test tube holders; camel hair brush; razor blade; detergent extraction medium; protease inhibitors. The extraction medium used for *Xenopus* oocytes (Hauptman *et al.*, 1989) was designed to resemble the intracellular ionic composition of *Xenopus* oocytes based on the report of Barish (1983). The extraction medium contains 92.5 mM KCl, 6.2 mM NaCl, 5 mM EGTA, 2 mM $MgCl_2$, and 12 mM HEPES, pH 7.4. Immediately before use the protease inhibitor phenylmethylsulfonyl fluoride (PMSF) is added to a final concentration of 200 μg/ml from a stock made in ethanol. The ethanol stock is stored at $-20°C$ for a maximum of 7 days. PMSF becomes rapidly inactivated in aqueous solutions and is therefore used within 30 minutes after addition to the extraction medium. In addition, the detergent Triton X-100 is added to the extraction medium at a final concentration of 1% (well above the critical micelle concentration).

2. Procedures

1. Suspend the copper cup in a dewar filled with liquid nitrogen. After the nitrogen settles fill the copper cup with Freon 22, which will solidify. Keep a second dewar of liquid nitrogen nearby to replenish the supply of nitrogen.
2. The Freon is melted with the metal rod to create a slurry, in which follicle-free oocytes or jelly-free eggs and embryos are immersed. To transfer specimens into Freon without transferring medium, specimens are lifted with two pair of forceps which are then tapped on the side of the copper cup to knock the specimen into the slurry. Several specimens can be transferred before the slurry begins to refreeze. A pair of forceps, wrapped with tape for insulation, is cooled in liquid nitrogen and used to transfer specimens from the slurry into a precooled cryogenic ampule for storage under liquid nitrogen.
3. Prior to sectioning, the cryotome chamber is packed with dry ice to cool the chamber to $-78.6°C$. The test tubes, insulated forceps, test tube holders, camel hair brush, and cryotome knife should be placed in the cryotome to be precooled to $-78.6°C$. The cryotome chuck should be precooled in a separate container on a block of dry ice. A layer of distilled water is applied to the chuck to build a platform away from the metal surface of the chuck. As the last portion of this platform solidifies, specimens should be removed from under liquid nitrogen and placed in

this droplet which will solidify instantly. Excess ice around specimens should be trimmed away with a razor blade to prevent dilution of the detergent extraction medium during thawing. The chuck should then be rapidly transferred into the cryotome. Sections are cut to a thickness of 20 μm; in thinner sections, more cytoskeletal filaments will be cut in the section plane, which may promote more rapid disassembly of what would otherwise be insoluble microtubules and actin filaments. During sectioning the knife angle should be set to about 10°. Sections should collect on top of the knife blade; these are swept with a precooled camel hair brush into a precooled test tube. Sections from no more than 20 oocytes, eggs, or embryos should be placed in 10 ml detergent extraction medium.

4. After sectioning, protease inhibitors are added to the detergent extraction medium (see above for the detergent extraction medium recipe), and the medium is placed in a water bath heated to 37°C. Rapidly, the prewarmed detergent extraction medium is poured into the tube containing sections as the tube is removed from the cryotome. The tube is placed briefly in the water bath and agitated until all ice has disappeared. Detergent extraction is continued for 5 minutes, and then the cytoskeleton is separated from solubilized components by centrifugation at 3660 g for 3 minutes. The supernatant containing the soluble components is removed and precipitated in 4 volumes of 100% ethanol at $-20°C$. The pellet, which remains insoluble, can be further fractionated (see below).

5. The detergent-resistant cytoskeleton can be further subdivided into high salt-soluble and high salt-insoluble fractions by placing it in a medium containing 250 mM ammonium sulfate or 1 M sodium chloride, agitating gently at room temperature, and allowing incubation for an additional 20 minutes. The high salt-soluble components of the cytoskeleton are separated from the high salt-insoluble components by an additional centrifugation step (4350 g, 3 minutes, room temperature). High salt-soluble components in the supernatant are precipitated using 4 volumes of ethanol at $-20°C$.

6. The detergent-soluble fraction and the detergent- and high salt-soluble fraction are left overnight at $-20°C$ to precipitate and then isolated by centrifugation.

3. Important Notes

Technically, this is a relatively simple procedure which can be performed in less than 2 hours. However, there are two useful stopping points. The first is after freezing the cells; the second is during precipitation. For analysis of proteins associated with each of the fractions, each pellet is solubilized in sample buffer for electrophoresis. (Urea [$4M$] should be present promote solubilization of intermediate filaments. The treatment described above does not solubilize yolk, and yolk platelets will be present in the detergent- and high salt-resistant fraction. Once solubilized in sample buffer it is possible to remove a large portion of the yolk by pelleting in a microcentrifuge for 5 minutes. Depending on the salt concentrations used in the high salt solubilization process it may be necessary to desalt the detergent and high salt-soluble fraction prior to applying it to one-dimensional gels, otherwise the salt may cause distortion of the lane. Desalting can be performed by microdialysis against standard detergent extraction medium for 10 minutes. If RNA is to be extracted from any fraction then at least one RNase inhibitor should be present in every solution; 10 mM vanadyl riboside complex works well for this purpose. When vanadyl is used, EGTA is omitted from the extraction medium to prevent inactivation by EGTA.

The further subdivision of the detergent-resistant cytoskeleton with high salt is useful in that the higher ionic strength medium solubilizes actin filaments and microtubules and their associated proteins, leaving behind largely intermediate filaments and nuclear matrix components (Capco *et al.,* 1982). This adds a further level of resolution to analysis of the cytoskeleton and associated components. This approach has revealed, in both amphibian and mammalian embryos, developmental changes in patterns of protein synthesis, as well as transition of specific proteins from detergent- and high salt-labile forms to detergent- and high salt-stable forms (McGaughey and Capco, 1989; Hauptman *et al.,* 1989). This chemical dissection method has the advantage of allowing the proteins of a cell to be separated into three reproducible fractions. The separation therefore exposes low-abundance proteins which would otherwise be masked by more prominent cellular proteins of the same molecular weight.

III. Conclusions

When attempting to unravel pathways of cell regulation, at least two general questions are involved. First, what is the sequence of signaling events which govern a given cellular transition? Second, what specific effects on the cell do those signaling events have? As model systems, *Xenopus* oocytes, eggs, and embryos are unsurpassed in terms of their utility for asking, and obtaining answers to, both of these kinds of questions.

The large size of *Xenopus* oocytes, eggs, and embryos confers a number of technical advantages to the researcher. First, single cells can be manipulated with forceps and a dissecting microscope. Second, markers of cellular transitions, such as cortical contraction which denotes egg activation, can be observed with the dissecting microscope, allowing rapid determination of whether that transition has occurred. Third, microinjection can be performed simply and rapidly. Fourth, biochemical analyses which require large amounts of material, such as measuring intracellular concentrations of signaling molecules, are feasible. Fifth, some approaches, such as the spatial fractionation by dissection technique described in this chapter, are absolutely dependent on large size. Sixth, some cellular phenomena, such as mRNA localizations, that can only be observed with difficulty in somatic cells (Lawrence and Singer, 1986) are clearly apparent in *Xenopus* oocytes and eggs (Capco and Jeffery, 1982; Melton, 1987; Perry and Capco, 1988).

Another significant advantage of the *Xenopus* system is that an entire developmental continuum can be triggered and observed *in vitro*. Comparison of cell responses to stimuli at different stages within the continuum provides clues as to how signaling pathways are coupled or uncoupled during development. Importantly, the continuum contains natural arrest points, such as prophase I and metaphase II. This makes it possible to obtain relatively synchronized cell populations without having to resort to treatments with drugs that prevent cell cycle transitions. Moreover, transit through these arrest points is associated with clear changes in cellular function; hence, the importance of such transitions can be readily assessed. In conclusion, *Xenopus* oocytes, eggs, and embryos provide fruitful material for analysis of how cellular signaling events are integrated and transduced into specific cellular phenomena.

ACKNOWLEDGMENT

This work was supported by National Institutes of Health Grants HD23686 and HD00598.

REFERENCES

Barish, M. E. (1983). *J. Physiol. (London)* **342**, 309–325.
Bement, W. M., and Capco, D. G. (1989a). *Cell Tissue Res.* **255**, 183–191.
Bement, W. M., and Capco, D. G. (1989b). *J. Cell Biol.* **108**, 885–892.
Bement, W. M., and Capco, D. G. (1990a). Transformation of the amphibian oocyte into the egg: Structural and biochemical events. *J. Electron Microsc. Tech.* **16**, 202–234.
Bement, W. M., and Capco, D. G. (1990b). Synthesis, assembly, and organization of the cytoskeleton during early amphibian development. *Semin. Cell Biol.* **1**, 383–389.
Bement, W. M., and Capco, D. G. (1990c). *Cell Regulation* **1**, 315–326.
Berridge, M. J. (1987). *Annu. Rev. Biochem.* **56**, 159–193.
Busa, W. B., and Nuccitelli, R. (1985). *J. Cell Biol.* **100**, 1325–1329.
Busa, W. B., Ferguson, J. E., Joseph, S. K., Williamson, J. R., and Nuccitelli, R. (1985). *J. Cell Biol.* **101**, 677–682.

Capco, D. G., and Jeffery, W. R. (1982). *Dev. Biol.* **89,** 1–12.
Capco, D.G., and McGaughey, R. W. (1986). *Dev. Biol.* **115,** 446–458.
Capco, D. G., and Mecca, M. (1988). *Cell Differ.* **23,** 155–164.
Capco, D. G., and Penman, S. (1983). *J. Cell Biol.* **96,** 896–906.
Capco, D. G., Wan, K. M., and Penman, S. (1982). *Cell (Cambridge, Mass.)* **29,** 847–858.
Capco, D. G., Krochmalnic, G., and Penman, S. (1984). *J. Cell Biol.* **98,** 1878–1885.
Capco, D. G., Munoz, D. M., and Gassman, C. J. (1987). *Tissue Cell* **19,** 606–616.
Charbonneau, M., and Grandin, N. (1989). *Cell Differ. Dev.* **28,** 71–94.
Charbonneau, M., and Webb, D. J. (1987). *J. Cell Sci.* **87,** 205–220.
Cicirelli, M. F., and Smith, L. D. (1985). *Dev. Biol.* **108,** 254–258.
Cicirelli, M. F., Robinson, K. R., and Smith, L. D. (1983). *Dev. Biol.* **100,** 133–146.
Gerhart, J., Wu, M., and Kirschner, M. (1984). *J. Cell Biol.* **98,** 1247–1255.
Han, J. K., and Nuccitelli, R. (1990). *J. Cell Biol.* **110,** 1103–1110.
Hannun, Y. A., Loomis, C. R., Merrill, A. H., and Bell, R. M. (1986). *J. Biol. Chem.* **261,** 12604–12609.
Hauptman, R. J., Perry, B. A., and Capco, D. G. (1989). *Dev. Growth Differ.* **31,** 157–164.
Hidaka, H., Ignaki, M., Kawamoto, S., and Sasaki, Y. (1984). *Biochemistry* **23,** 5036–5041.
Houle, J. G., and Wasserman, W. J. (1983). *Dev. Biol.* **97,** 302–312.
House, C., and Kemp, B. E. (1987). *Science* **238,** 1726–1728.
Karsenti, E., Newport, J., Hubble, R., and Kirschner, M. (1984). *J. Cell Biol.* **98,** 1730–1745.
Kikkawa, U., and Nishizuka, Y. (1986). *Annu. Rev. Cell Biol.* **2,** 149–178.
Kline, D. (1988). *Dev. Biol.* **126,** 346–361.
Larabell, C. A., and Capco, D. G. (1988). *Roux's Arch. Dev. Biol.* **197,** 175–183.
Lawrence, J. B., and Singer, R. H. (1986). *Cell (Cambridge, Mass.)* **45,** 407–415.
Le Peuch, C. J., Picard, A., and Doree, M. (1985). *FEBS Lett.* **187,** 61–64.
Lupu-Meiri, M., Shapira, H., and Oron, Y. (1988). *FEBS Lett.* **240,** 83–87.
McGaughey, R. W., and Capco, D. G. (1989). *Cell Motil. Cytoskeleton* **13,** 104–111.
Maller, J. L., and Krebs, E. G. (1980). *Curr. Top. Cell. Regul.* **16,** 271–311.
Maller, J. L., Butcher, F. R., and Krebs, E. G. (1979). *J. Biol. Chem.* **254,** 579–582.
Melton, D. A. (1987). *Nature (London)* **38,** 80–82.
Murray, A. W., and Kirschner, M. W. (1989). *Science* **246,** 614–621.
Penman, S., Fulton, A., Capco, D., Ben Ze'ev, A., Wittelsberger, S., and Tse, C. F. (1981). *Cold Spring Harbor Symp. Quant. Biol.* **41,** 1013–1028.
Perry, B. A., and Capco, D. G. (1988). *Cell Differ. Dev.* **25,** 99–108.
Sagata, N., Watanabe, N., Vande Woude, G. F., and Ikawa, Y. (1989). *Nature (London)* **342,** 512–518.
Smith, L. D. (1989). *Development (Cambridge, U.K.)* **107,** 685–699.
Stewart-Savage, J., and Grey, R. D. (1982). *Roux's Arch. Dev. Biol.* **191,** 241–245.
Taylor, M. A., and Smith, L. D. (1985). *Dev. Biol.* **110,** 230–237.
Varnold, R. L., and Smith, L. D. (1990). *Development (Cambridge, U.K.)* **109,** 597–604.
Wallace, R. A., Jared, D., and Sega, M. W. (1973). *J. Exp. Zool.* **184,** 321–334.
Wasserman, W. J., Penna, M. J., and Houle, J. G. (1986). *In* "Gametogenesis and the Early Embryo" (J. G. Gall, ed.), pp. 111–130. Alan R. Liss, New York.
Webb, D. J., and Nuccitelli, R. J. (1981). *J. Cell Biol.* **91,** 562–567.

Chapter 14

Generation of Body Plan Phenotypes in Early Embryogenesis

KEN KAO

S. Lunenfeld Research Institute
Mount Sinai Hospital
Toronto M5G IX5, Canada

MIKE DANILCHIK

Department of Biology
Wesleyan University
Middletown, Connecticut 06457

I. Introduction
II. Axis Perturbations in Early Development
   A. First Cell Cycle: Establishment of Dorsal–Ventral Polarity
   B. The Blastula: Mesoderm Formation
   C. Gastrulation: Morphogenetic Movements and Neural Induction
III. Scoring of Phenotypes
IV. Methods
   A. Ultraviolet Light
   B. Deuterium Oxide
   C. Lithium
   D. Trypan Blue, Suramin, Heparin, and Retinoic Acid
   E. Germinal Vesicle Sap
V. Summary
   References

I. Introduction

The generation of aberrant embryonic phenotypes is a useful way to study developmental processes. By knowing when, where, and at what organizational level a perturbation has its effect, we can gain insight into the normal mechanisms of developmental regulation. Because general features of the body plan, such as germ layers and body axes, are established before specific tissues and organs, interference with the earliest developmental steps can lead to profound disruptions of embryonic organization. Also, because developmental information exists in many forms, it is possible for apparently dissimilar kinds of interference to result in surprisingly similar morphologies or phenotypes. In the case of *Xenopus*, a number of seemingly unrelated agents and treatments can be used to interfere with the specification or some later aspect of the development of the dorsal–ventral axis. In this chapter, we outline a variety of methods to generate body axis perturbations in *Xenopus* embryos and give some general guidelines for scoring and analyzing the aberrant phenotypes.

II. Axis Perturbations in Early Development

Each axis perturbation is most effective at a specific stage or window of development. A diagram showing the sensitive period for a number of treatments is given in Fig. 1. The following brief description of *Xenopus*

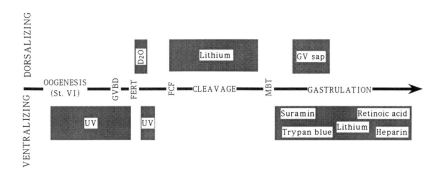

FIG. 1. Agents that affect body axis development. Shaded bars indicate periods in early development when *Xenopus* embryos are sensitive to various treatments. Treatments above the time line "dorsalize" or "hyperdorsoanteriorize," and those below "ventralize" or "posteriorize." UV, Ultraviolet irradiation of the vegetal pole; D_2O, deuterium oxide; GVBD, germinal vesicle breakdown; Fert, fertilization; FCF, first cleavage furrow; MBT, midblastula transition; GV sap, germinal vesicle sap.

early development indicates the different treatments and their timings as appropriate.

A. First Cell Cycle: Establishment of Dorsal–Ventral Polarity

The formation of the body plan in *Xenopus* depends on dorsal–ventral polarization of the fertilized egg, achieved by a microtubule-dependent rotation of the egg cortex with respect to the cytoplasmic core shortly after sperm entry (for review, see Elinson, 1989). Because the rotation is necessary to specify the embryonic axis, perturbations to the machinery of rotation, particularly microtubules, result in profound alterations in the body plan. The rotation can be inhibited by briefly exposing the vegetal poles of zygotes to short wavelength ultraviolet (UV) light (Grant and Wacaster, 1972). The embryos that develop consist of an endodermal mass surrounded by ciliated epithelia but completely lack the dorsal and anterior structures that make up the body axis (Fig. 2A). The only mesodermal tissue that develops is blood. As well as UV light, cold shock and high hydrostatic pressure can produce similar defects in dorsoanterior development (Scharf and Gerhart, 1983).

Conversely, fertilized eggs that are exposed to deuterium oxide (D_2O) develop exaggerated dorsoanterior structures, such as a radial eye and cement gland, along with large amounts of notochordal tissue (Scharf *et al.*, 1989), suggesting that D_2O has an axis-promoting effect on the mesoderm. D_2O also alleviates the effect of UV light (Scharf and Gerhart, 1983), reinforcing the idea of an axis-promoting effect. D_2O is most effective if applied before the vegetal cortical–cytoplasmic rotation; its microtubule-stabilizing property is suspected of being involved in dorsoanterior enhancement.

The effect of UV light can easily be reversed by tilting the eggs obliquely with respect to gravity, mimicking the cytoplasmic rearrangements that would have normally occurred. This rescue experiment indicates that the rotation is necessary for normal axis formation (Scharf and Gerhart, 1980; Chung and Malacinski, 1980). Recently, it has been shown that UV irradiation before the rotation can also cause axis deficiencies. Irradiated prophase I oocytes, when matured and fertilized, develop axis deficiencies similar to those caused by the later UV irradiation (Holwill *et al.*, 1987). However, the eggs cannot be rescued by tilting (Elinson and Pasceri, 1989), suggesting that the earlier irradiation disrupts the components required for dorsoanterior differentiation rather than the mechanism necessary for their asymmetric distribution.

B. The Blastula: Mesoderm Formation

Following fertilization, the egg undergoes a series of rapid cleavage divisions to form a blastula, a hollow ball of cells. Formation of the mesoderm

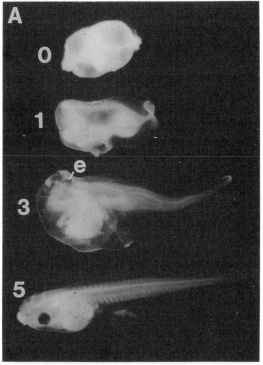
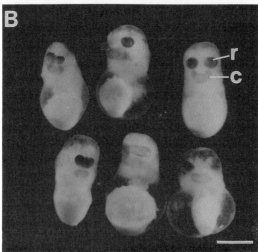

Fig. 2. Examples of axis-perturbed phenotypes caused (A) by UV irradiation before first cleavage and (B) by lithium treatment of early blastulas. The numbers in (A) refer to the dorsoanterior index (DAI) values assigned to the phenotypes of the embryos directly to the right of the numbers. The grade 5 embryo is normal, the grade 3 embryo is cyclopic (one eye, e), the grade 1 embryo has a tail fin and somites but no head (acephalic), and the grade 0 embryo has no dorsoanterior structures at all. In (B), all of the embryos were irradiated with UV light prior to first cleavage and would have developed as grade 1 or 0 embryos. They were treated with lithium at the 32-cell stage. Note the exaggerated anterior facial structures such as multiple eyes or bands of retinal pigment (r) and enlarged cement gland tissue (c). Bar equals 0.5 mm.

in the equatorial region is thought to require cell interactions in the blastula for normal development. Dorsoanterior mesodermal cell types require interactions between vegetal dorsal cells and the adjacent equatorial cells, whereas more ventral mesoderm requires interactions with more ventral vegetal cells (Gimlich and Gerhart, 1984; Gimlich, 1986; for reviews, see Gurdon, 1987; Smith, 1989).

It is possible to alter the body plan by exposing early blastulas to lithium for short periods of time (Backström, 1954; Kao et al., 1986; Breckenridge et al., 1987). Pulse treatment of early cleaving embryos up until the twelfth cleavage division (Yamaguchi and Shinagawa, 1989) for just a few minutes makes them develop exaggerated dorsoanterior structures (Fig. 2B). The spectrum of phenotypes caused by lithium is remarkably similar to that generated by D_2O prior to first cleavage (Scharf et al., 1989).

Lithium likely causes ventral cells of the blastula to behave like dorsoanterior, "organizer" cells (Regan and Steinhardt, 1988; Kao and Elinson, 1988), because UV-irradiated embryos, which usually develop only ventral structures, can be made to respond to lithium in the same way as normal embryos. It probably acts by altering the interactions between animal and vegetal cells that are needed for mesoderm formation so that only dorsoanterior mesoderm is produced (Slack et al., 1988; Kao and Elinson, 1989). One attractive hypothesis for the biochemical mode of its action is that it causes a rundown in polyphosphoinositide (PI) turnover by inhibiting the formation of free *myo*-inositol (Berridge et al., 1989), as the effects of lithium are alleviated by coinjection of *myo*-inositol (Busa and Gimlich, 1989). This would suggest a role for PI turnover in the maintenance of dorsoventral polarity in the pregastrula embryo.

Lithium also has effects on reducing anterior structures when applied after the midblastula transition. The transition occurs in *Xenopus* after the twelfth cleavage division and is associated with the initiation of zygotic transcription and asynchronous cell divisions (Newport and Kirschner, 1982a,b). When embryos are exposed to lithium after this point in development and until late gastrulation, they develop dorsoanterior defects similar to those seen after UV irradiation or exposure to retinoic acid (RA) (Lombard, 1952; Yamaguchi and Shinagawa, 1989; Sive et al., 1990).

C. Gastrulation: Morphogenetic Movements and Neural Induction

At about 8–10 hours following fertilization, the blastula undergoes the morphogenetic movements of gastrulation (for review, see Keller 1986, and Chapter 5, this volume). Externally, the first visible sign of gastrulation is the formation of the dorsal blastoporal lip in the subequatorial region of the embryo. As the yolk is internalized through the blastopore, the ectodermal covering stretches over the entire embryo by epiboly. Concomitant with epiboly, the marginal zone undergoes involution. In the dorsal region, the cells undergo the most extreme amount of involution and then migrate toward the animal pole, to underlie the future neural plate. The extent of this migration is correlated with differentiation of the anterior–posterior axis. Perturbations such as UV irradiation, or the addition of polysulfonated substances such as trypan blue (TB), suramin, or heparin, limit the extent of migration, and the dorsoanterior region of the embryo is truncated (Gerhart *et al.*, 1984, 1989; Mitani, 1989; Fig. 3). An interesting feature of TB and suramin is that their effectiveness diminishes progressively when applied at later stages of gastrulation. This suggests a possible important causal role for the extent of mesodermal migration in determining the degree of anterior development.

Embryos can also be treated with RA, which causes them to have dorsoanterior reduction (Durston *et al.*, 1989; Sive *et al.*, 1990). The exact

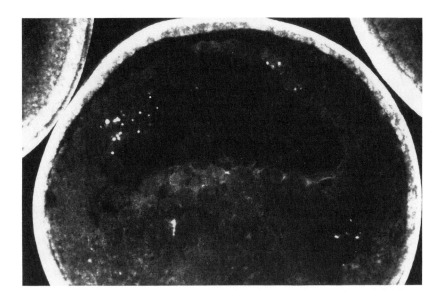

FIG. 3. Gastrulation movements affected by injection of TB. A stage 9.5 embryo was injected in the blastocoel with 20 nl of 40 μM TB, fixed at stage 13 in Bouin's fixative, washed with alkaline 50% ethanol, dehydrated and cleared with 2:1 benzyl benzoate–benzyl alcohol, and examined by confocal scanning microscopy (rhodamine excitation). Note the brightly fluorescent deposits of TB between cells of mesodermal mantle and floor of blastocoel (arrows). Mesoderm has clearly migrated up along blastocoel roof, but little involution has occurred. The archenteron is not evident in any plane of section. Animal cap and marginal zone cells are visible because they autofluoresce at the wavelength used. Embryos affected this severely develop with a DAI averaging about 1.

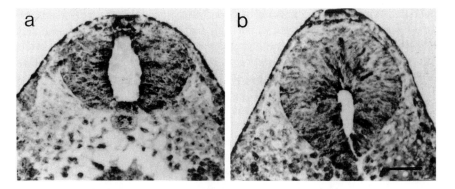

FIG. 4. Macrocephaly produced by injection of germinal vesicle sap. (a) Cross section at the hindbrain of a control embryo. (b) Cross section at the hindbrain of a macrocephalic embryo. Note the enlarged neural tube. Bar equals 150 μm.

period of sensitivity to RA is not clear; however, as with TB, embryos display the phenotype after treatment as early as the 64-cell stage, and they become resistant to RA by stage 15. RA appears to act at a different level than the polysulfonated compounds, since morphogenetic movements of gastrulation, particularly involution and migration of the mesoderm, seem to not be inhibited directly (Sive *et al.,* 1990). Thus, RA appears to transform the head into posterior structures, rather than cause anterior truncation.

An essentially opposite phenotype can be obtained also at the late blastula or early gastrula stage. By injecting the contents of an oocyte germinal vesicle into the blastocoel, embryos develop a characteristic macrocephalic condition, whereby the head becomes enlarged at the expense of more posterior structures (Kao and Elinson, 1985) (Fig. 4).

III. Scoring of Phenotypes

There has been much variation in the literature regarding the scoring and grading of axis-defect phenotypes. In most cases, a scale or index of grades of axis-deficient phenotypes is devised to suit the needs of the particular experiments that are described. The main point of such a scale is to quickly and conveniently assay the effect of a treatment in a semiquantitative way. This is most important for experiments in which one is attempting to determine a precise window of sensitivity to a particular treatment (Black and Gerhart, 1985, 1986; Scharf and Gerhart, 1983) or to map the location of axis-forming activity by blastomere transplantation (Gimlich and Gerhart, 1984;

Kao and Elinson, 1988). Because such assays involve only visual inspection of intact embryos, they are straightforwardly performed. On the other hand, they may not by themselves be useful for analyses, say, of specific gene expression. In such cases, region-specific markers, for example, homeobox gene products (Cho and DeRobertis, 1990; Sive et al., 1990), may be more appropriate, albeit less conveniently applied. In any case, some kind of visual assessment of the perturbation is crucial to put it into a context with developmental processes.

It is well known that dorsoanterior defects produced by various, seemingly unrelated treatments display an apparently continuous range of phenotypes. This observation suggests that one developmental process is interfered with, directly or indirectly, by the disparate treatments. The dorsoanterior index (DAI) was devised to describe a complete spectrum of such phenotypes ranging from severe axis deficiencies caused by UV irradiation (grades 0 to 4) to severe hyperdorsoanterior enhancements caused by lithium or D_2O (grades 6 to 10) (Kao and Elinson, 1988). The criteria for grades 0 to 5 derived from earlier work with UV irradiation (Grant and Wacaster, 1972; Scharf and Gerhart, 1983). The anatomical defects or enhancements which form the basis of the DAI result from modifications to the pattern of mesoderm specification in the pregastrula embryo: UV-irradiated embryos lack dorsal organizer mesoderm, whereas lithium and D_2O embryos are overcommitted to forming organizer mesoderm. Sive et al. (1990) modified the DAI to accommodate some detailed differences in head structures between embryos treated with different amounts of RA, whereas Cooke and Smith (1988) used a completely different index describing dorsoanterior enhancement caused by lithium. Scharf et al. (1989) also used, with some qualifications, Kao and Elinson's DAI for describing D_2O-treated embryos. Although the morphogenetic basis for defects induced by TB or suramin is probably different from that for, say, RA, nevertheless they can be scored by essentially the same criteria as used in the DAI. The anatomical criteria used for the DAI and related indices can be found in Kao and Elinson (1988), Scharf et al. (1989), and Sive et al. (1990).

IV. Methods

A. Ultraviolet Light

Axis-deficient embryos are easily obtained by irradiating the vegetal poles of fertilized eggs. Essential equipment includes (1) a short wavelength UV light source, either a hand-held illuminator (UV Light Products, San Gabriel, CA) or a transilluminator used for viewing ethidium bromide-stained gels; (2) a quartz cuvette or slide (Hellma, NY) on which to place the eggs; (3) a proper support on which to place the cuvette over the UV light source at variable distances; and (4) routine glassware such as embryo transfer pipettes and petri dishes.

To ensure successful results, a few key points should be kept in mind. First, the UV irradiation must be delivered onto the vegetal surface to be effective. Select only fertilized eggs that reorient with the animal poles upward. It is always a good idea to remove all unfertilized or abnormally orienting eggs. Fertilized eggs should not stick to the cuvette or slide, and they should immediately reorient "animal pole up" when tilted. Second, irradiation must occur before the cortical–cytoplasmic rotation. The sensitive period is approximately within the first third of the cell cycle. This means that the eggs must be dejellied thoroughly within about 30 minutes of insemination. Generally, we find that dejellying this early has no detrimental effects, but sometimes the vitelline envelope overinflates, causing the unsupported eggs to flatten. Unirradiated (control) eggs treated this way nevertheless develop normally. Third, it is absolutely essential to maintain the eggs in their normal orientation after irradiation. Simply allow the eggs to settle and leave undisturbed at least until first cleavage.

The UV lamp must be calibrated before a reliable yield of axis-deficient embryos can be produced. An empirical approach is useful here, because, in general, the sensitivity among different batches of eggs varies to a certain extent for a given UV dose. Finding the correct UV dosage to give a desired yield of axis-deficient embryos is a simple but nontrivial task. Too little UV light will result in embryos that show a partial phenotype (although this may be desirable for some purposes), and too much UV light can lead to embryos having difficulty cleaving their yolk-laden vegetal halves. These frequently have abortive cleavage furrows, and the embryos fail to incorporate their endoderm during gastrulation. As a first estimation, we usually try a range of dosages varied by altering the duration of irradiation and the distance from the lamp. A good range to start is from 1 to 3 cm from the light source for 30 seconds to 2 minutes. It is common to observe a wide range of individual responses to all but the most severe treatments. Thus, it is important to have sufficiently large sample sizes (e.g., $n \geq 20$), to obtain a reliable measure of the dose–response.

After irradiation, the embryos should be left undisturbed at least until after first cleavage. Survival of irradiated embryos is often improved by culturing at temperatures below 18°C. The first signs of axis reduction can be seen during gastrulation. The embryos develop a blastoporal lip that appears at the same time radially around the embryo and at a later time from when the dorsal lip appears in a normal embryo. The blastopore closes symmetrically, and it is not possible to identify a neural plate, as in normal embryos at the end of gastrulation.

B. Deuterium Oxide

A complete spectrum of dorsoanterior enhancement, ranging from mild macrocephaly (DAI of 6) and "janus twin" configurations (DAI of 7–8) to cylindrically symmetric forms with multiple or radial cement glands and eyes (DAI of 9–10) are easily obtained by briefly exposing eggs to D_2O in the culture medium (Scharf et al., 1989). Dejellied eggs are transferred to culture medium containing from 20 to 70% D_2O. Because D_2O is relatively dense, eggs float until they exchange sufficient D_2O for water; when transferring eggs to the medium, we expel them deep under the surface and then swirl gently to speed the exchange process. We generally use 66% D_2O and expose for 4 minutes (J. Denegre, 1991; personal communication). A broad spectrum of dorsal-enhanced phenotypes invariably results; increasing the length of exposure will increase the DAI.

C. Lithium

The most extreme phenotype of dorsoanterior enhancement (DAI of 10) can be obtained by treating dejellied embryos (either normal or UV irradiated) for 5 to 10 minutes with 0.3 M lithium chloride. The effect is stage dependent, with the maximal lithium sensitivity occurring between the 32- and 128-cell stages. After the midblastula transition, about 6 hours from fertilization, the same treatment causes dorsoanterior reduction (Yamaguchi and Shinagawa, 1989).

The lithium solution is made up in regular saline solution such as 20% Steinberg's solution (Rugh, 1956) or 1/10 modified Barth's saline (Gurdon, 1977). It is best to try a range of treatment durations, since there is considerable variability in the response to lithium depending on the batch of embryos used. Embryos must be washed thoroughly following treatment and raised at normal temperatures, but they should be placed in dishes at relatively low density. It also helps to change the medium at regular intervals following treatment, as this improves the survival rate.

The phenotype is first visible during gastrulation, when a darkly pigmented, radially symmetrical blastoporal lip appears. The blastopore is circular and closes rapidly. In extreme phenotypes, a circular "neural plate" and cement gland tissue surround the blastopore soon after gastrulation (Breckenridge et al., 1987; Kao and Elinson, 1988, Klein and Moody, 1989). In other cases, a large proboscis extends several hundred microns from the vegetal pole (as is true for D_2O-treated embryos; Scharf et al., 1989).

A supernumerary axis can be made to form in embryos by microinjecting lithium into single cells (Kao et al., 1986; Busa and Gimlich, 1989). A concentrated solution of lithium (typically 0.2–0.5 nl of 0.3 M LiCl) in distilled water can be microinjected into a vegetal ventral cell (tier 4; Kao et al.,

1986) or equatorial ventral cell (tier 2; Busa and Gimlich, 1989) at the 32-cell stage. An interesting experiment is to rescue axis-deficient UV-irradiated embryos by microinjecting lithium into a single vegetal or equatorial cell at the 32-cell stage. The embryos show a remarkable rescue of dorsoanterior structures and restoration of normal development (Kao et al., 1986). Microinjection of small volumes of relatively high concentrations of lithium into single cells is preferred; for this reason, air pressure injection systems rather than syringe-type systems are favorable.

D. Trypan Blue, Suramin, Heparin, and Retinoic Acid

A full range of dorsoanterior axis deficiencies, almost indistinguishable from those produced by UV irradiation, can be generated by bathing embryos in solutions of TB or suramin (millimolar concentrations) or RA (micromolar concentrations). For more uniform results, one may microinject small amounts of TB, heparin, or suramin into the blastocoel. The response to each of these agents is dose dependent. For example, TB in the micromolar range produces minor defects such as mild microcephaly (DAI of 4–4.5) and complete ventralization (DAI of 0–1) at about 10 μM (Danilchik, 1986). Slightly more than twice the molar concentration of suramin as TB is needed to elicit an equivalent dorsoanterior defect (T. Doniach, 1991; personal communication). In general, the response range of individual embryos to a given blastocoelar concentration is very narrow: equivalently treated embryos develop with about the same phenotype. In contrast, vegetal UV irradiation generally results in a much more heterogeneous response to a given dose. Maximal sensitivity to TB occurs well before the onset of gastrulation; injection any time between stages 6 and 10 results in about the same response. After stage 10, however, the response to a given dose diminishes, and embryos become completely resistant to interstitial TB from stage 14 on (Danilchik et al., 1991).

Variability in response arising from the injection procedure itself can be minimized simply by ensuring that small, indentical volumes are delivered to each embryo. Embryos tolerate injected volumes of as much as 20% of the blastocoel volume if the delivery rate is kept relatively slow, but smaller volumes are preferable, in order to minimize loss of the teratogen via leakage. A convenient, roughly 10-fold final dilution can be achieved by injecting embryos with a standard 20 nl (estimated as 10% of the blastocoel volume during stages 8–10; Slack et al., 1988). For situations such as dose–response curves, in which different final concentrations of teratogen must be introduced, several microinjection needles may be preloaded with different stock concentrations, and 20-nl injections of each can then easily be delivered to different groups of embryos.

Trypan blue also produces dorsal defects when applied externally. DAI values of 0–4 can be achieved by bathing embryos in millimolar concentrations of the dye (Waddington and Perry, 1956; Danilchik, 1986). Like lithium, and probably RA as well, the period of maximal sensitivity to external exposure of TB is during the early cleavage period, declining abruptly at about stage 7. The response of individual embryos is quite variable, unlike intrablastocoelar injections of the same material. Interestingly, embryos are completely resistant to even a high concentration (~ 10 mM) of externally applied TB if it is washed away thoroughly before first cleavage. These observations, plus the fact that TB does not traverse intact plasma membranes, suggest that the site of activity of the dye is between cells adjacent to the blastocoel and that it can get there along the early (before stage 7) cleavage furrows when applied externally. Figure 3 confirms that TB accumulates between mesodermal cells, probably complexed to extracellular matrix components.

Trypan blue should be dialyzed extensively against distilled water (molecular weight cutoff 15,000), following a recommendation by Beck and Lloyd (1966). The dye usually contains large amounts (sometimes 40% or more of the original dry weight) of salts and various contaminants (it does not diffuse through dialysis tubing despite its low molecular weight). TB can then be lyophilized and weighed out to make a 1% solution in distilled water, which should then be stored in the dark. Suramin (Naganol, Bayer AG) is made up as a 1% solution in distilled water and diluted as required for injection stocks (T. Doniach, 1991; personal communication). Heparin is made up as a 0.1 to 10 mg/ml solution in Steinberg's solution (Mitani, 1989). Sive *et al.* (1990) use a 0.1 M stock of *trans*-retinoic acid dissolved in dimethyl sulfoxide (DMSO), which is then diluted to the micromolar range in culture medium.

E. Germinal Vesicle Sap

The simplest way to generate macrocephalic embryos is to transfer the contents of a germinal vesicle (GV) with a microinjection needle into the blastocoel of a recipient late blastula stage embryo. If one needs large numbers of macrocephalics, it is more convenient to make an extract by isolating GVs manually in a suitable GV isolation medium (Merriam and Hill, 1976; we have used 10 mM Tris-HCl, pH 7.5, 100 mM KCl, 20 mM NaCl, 3 mM MgCl$_2$ with or without polyvinylpyrrolidone). Oocytes are manually defolliculated in calcium-free amphibian Ringer's solution (180 mM NaCl, 2 mM KCl, 1 mM EGTA, 5 mM NaHCO$_3$) and transferred to a dish of GV isolation medium where they can be cut in the animal pole with a needle and the GV squeezed out gently with forceps. Extracts are made by drawing up 50 to 100 GVs into a capillary tube, and the tube is heat sealed and centrifuged to rupture the GVs and separate the membranes from the nuclear sap. A piece of polyethylene tubing can be slipped over the centripetal end of the capillary tube to prevent crushing of the glass. The aliquots are stored at $-70°C$ until used.

V. Summary

We have presented a number of simple methods that can be used to interfere in the normal establishment and subsequent development of dorsal axial structures in *Xenopus*. It should be emphasized that, despite the striking similarity in phenotypes which result from these treatments, different developmental processes are being affected at the different windows of sensitivity. For example, UV light, known to damage RNA (for review, see Kalthoff, 1979), also disrupts microtubule polymerization. These activities may be important at different developmental times, relating to the distribution of maternal determinants (in the oocyte) or to the coordinated assembly of cortical microtubules (in the just-fertilized egg). The ventralizing and dorsalizing effects of the various late-acting agents (e.g., TB, suramin, RA, GV sap) undoubtedly stem from their interference with cellular behaviors during the critical morphogenetic period of gastrulation.

Acknowledgments

We thank J. Denegre and R. Savage for advice, and J. Wallingford and Dr. T. Doniach for sharing trypan blue and suramin results. M. D. is supported by National Science Foundation Grant DCB-8916614 and Basil O'Connor Starter Scholar Research Award No. 5-721 from the March of Dimes Birth Defects Foundation. K. K. is a fellow of the Medical Research Council (Canada).

References

Backström, S. (1954). *Arkiv. Zool.* **6**, 527–536.
Beck, F., and Lloyd, J. B. (1966). *Adv. Teratology* **1**, (Woollen, A. M., ed.), pp. 131–193. Acad. Press, New York.
Berridge, M. J., Downes, C. P., and Hanley, M. R. (1989). *Cell (Cambridge, Mass.)* **59**, 411–419.
Black, S. D., and Gerhart, J. C. (1985). *Dev. Biol.* **108**, 310–324.
Black, S. D., and Gerhart, J. C. (1986). *Dev. Biol.* **116**, 228–240.
Breckenridge, L. J., Warren, R. L., and Warner, A. E. (1987). *Development (Cambridge, U.K.)* **99**, 353–370.
Busa, W., and Gimlich, R. L. (1989). *Dev. Biol.* **132**, 315–324.
Cho, K. W. Y., and de Robertis, E. M. (1990). *Genes in Development* **4**, 1910–1916.
Chung, H.-M., and Malacinski, G. M. (1980). *Dev. Biol.* **80**, 120–133.
Cooke, J., and Smith, E. J. (1988). *Development (Cambridge, U.K.)* **102**, 85–99.

Danilchik, M. V. (1986). *J. Cell Biol.* **103**, 244a, Abstr. 912.
Danilchik, M., Doniach, T., and Gerhart, J. C. (1991). In preparation.
Durston, A. J., Timmermans, J. P. M., Hage, W. J., Hendricks, H. J. F., DeVries, N. J., Heideveld, M., and Nieuwkoop, P. D. (1989). *Nature (London)* **340**, 140–144.
Elinson, R. P. (1989). *BioEssays* **11**, 124–127.
Elinson, R. P., and Pasceri, P. (1989). *Development (Cambridge, U.K.)* **106**, 511–518.
Gerhart, J. C., Vincent, J.-P., Scharf, S. R., Black, S. D., Gimlich, R. L., and Danilchik, M. V. (1984). *Philos. Trans. R. Soc. London, Ser. B* **307**, 319–330.
Gerhart, J. C., Danilchik, M., Doniach, T., Roberts, S., Rowning, B., and Stewart, R. (1989). *Development (Cambridge, U.K.)* **107**, 37–52.
Gimlich, R. L. (1986). *Dev. Biol.* **115**, 340–352.
Gimlich, R. L., and Gerhart, J. C. (1984). *Dev. Biol.* **104**, 117–130.
Grant, P., and Wacaster, J. F. (1972). *Dev. Biol.* **28**, 454–471.
Gurdon, J. B. (1977). In "Methods in Developmental Biology" (F. H. Wilt and N. K. Wessells, eds.), pp. 75–84. Crowell, New York.
Gurdon, J. B. (1987). *Development (Cambridge, U.K.)* **99**, 285–306.
Holwill, S., Heasman, J., Crowley, C. R., and Wylie, C. C. (1987). *Development (Cambridge, U.K.)* **100**, 735–743.
Kalthoff, K. (1979). *Symp. Soc. Dev. Biol.*, 97–125.
Kao, K. R., and Elinson, R. P. (1985). *Dev. Biol.* **107**, 239–251.
Kao, K. R., and Elinson, R. P. (1988). *Dev. Biol.* **127**, 64–77.
Kao, K. R., and Elinson, R. P. (1989). *Dev. Biol.* **132**, 81–90.
Kao, K. R., Masui, Y., and Elinson, R. P. (1986). *Nature (London)* **322**, 371–373.
Keller, R. E. (1986). In "Developmental Biology: A Comprehensive Synthesis, Volume 2. The Cellular Basis of Morphogenesis" (L. Browder, ed.), pp. 241–327. Plenum, New York.
Klein, S. L., and Moody, S. A. (1989). *Development (Cambridge, U.K.)* **106**, 599–610.
Lombard, G. L. (1952). Ph.D. Thesis, Druk, Excelsiors, Gravenhage, The Netherlands.
Merriam, R. W., and Hill, R. J. (1976). *J. Cell Biol.* **69**, 659–668.
Mitani, S. (1989). *Development (Cambridge, U.K.)* **107**, 423–435.
Newport, J. W., and Kirschner, M. W. (1982a). *Cell (Cambridge, Mass.)* **30**, 675–686.
Newport, J. W., and Kirschner, M. W. (1982b). *Cell (Cambridge, Mass.)* **30**, 687–696.
Regen, C. M., and Steinhardt, R. A. (1988). *Development (Cambridge, U.K.)* **102**, 677–686.
Rugh, R. (1956). "Experimental Embryology," 3rd Ed. Burgess, Minneapolis, Minnesota.
Scharf, S. R., and Gerhart, J. C. (1980). *Dev. Biol.* **79**, 181–198.
Scharf, S. R., and Gerhart, J. C. (1983). *Dev. Biol.* **99**, 75–87.
Scharf, S. R., Rowning, B., Wu, M., and Gerhart, J. C. (1989). *Dev. Biol.* **134**, 175–188.
Sive, H. L., Draper, B. W., Harland, R. M., and Weintraub, H. (1990). *Genes Dev.* **4**, 932–942.
Slack, J. M. W., Isaacs, H. V., and Darlington, B. G. (1988). *Development (Cambridge, Mass.)* **103**, 581–590.
Smith, J. C. (1989). *Development (Cambridge, U.K.)* **105**, 665–677.
Waddington, C. H., and Perry, M. M. (1956). *J. Embryol. Exp. Morphol.* **4**, 110–119.
Yamaguchi, Y., and Shinagawa, A. (1989). *Dev. Growth Differ.* **31**, 531–541.

Chapter 15

Fluorescent Dextran Clonal Markers

ROBERT L. GIMLICH

Department of Anatomy and Cellular Biology
Harvard Medical School
Boston, Massachusetts

I. Introduction
   A. Clonal Analysis by Lineage Tracer Microinjection
   B. Fluorescent Clonal Markers
II. Methods
   A. Materials
   B. Procedures
III. Practical Considerations
   A. Controls and Precautions
   B. Dilution by Cell Proliferation
   C. Detection
   References

I. Introduction

A. Clonal Analysis by Lineage Tracer Microinjection

The ideal context for understanding the embryogenesis of an organism would include a complete knowledge of its cell lineages. Are they invariant between individuals? Where are the progeny of each cell from each embryonic stage, where are they at the end of embryogenesis, and to what cell types have they differentiated? At what developmental stage do cell lineages and tissue or organ anlagen begin to coincide? These questions have found a complete answer only in the case of the nematode *Caenorhabditis elegans*, which presents multiple advantages including a highly stereotyped cell division pattern, optical clarity, and relative simplicity and small cell number in the postembryonic worm (Sulston et al., 1983). In this case, and those of some other invertebrates, many of the cell lineages could be determined by direct microscopic observation (reviewed by Reverberi, 1971; Davidson, 1986).

Direct observation of cell lineage in other embryos is limited by small cell size, egg opacity, viviparous development, or all of the above. The presence of many cells in several layers complicates observation, particularly during periods of morphogenesis, when the geometry and timing of cell movements are complex. These difficulties have spawned methods for marking parts of an embryo so that the cells from each part can be followed. The methods fall into two classes which can be thought of as genetic and nongenetic. Genetic methods involve (a) the use of a mutagen to alter a clone genetically such that it differs visibly from neighboring clones (Ferrus and Garcia Bellido, 1976) and (b) transplantation of tissue from one species or genotype into an embryo of another. Examples of the latter are formation of chick–quail chimeric embryos (LeDourain, 1969) or hybrid *Xenopus laevis–Xenopus borealis* embryos (Thiebaud, 1983).

Nongenetic techniques include vital staining of embryonic parts with nontoxic dyes or precipitation of materials such as carbon particles on the embryonic surface. Marking with vital dyes or insoluble particles can give a general idea of the movement and developmental fate of a group of cells, particularly if the cells move and differentiate more or less as a unit (Keller, 1975, 1976). Since vital dyes seldom deeply and permanently mark individual small embryonic cells, however, they are usually inadequate for tracing cell lineages from early development to postembryonic stages. For this purpose, Weisblat, Stent, and colleagues pioneered the use of a microinjected enzyme tracer, horseradish peroxidase (HRP), in their studies of cell lineage in glosophoniid leeches (Weisblat et al., 1978). HRP was microinjected into the large teloblast cells early in development. The enzyme was passed on in catalytically active form to all of the descendants of the injected cell, and only to those descendants. HRP was detected using a chromogenic substrate on sections and whole mounts of larvae. The results verified that each teloblast is a stem cell which serially contributes progeny to found a segmental body plan. The progeny make stereotypical contributions to structures in each segment of the larval leech. These studies set the stage for experiments to determine whether this regularity is due to intrinsic properties of the stem cells or to geometrically regular patterns of cellular interaction among their descendants (reviewed by Shankland, 1987).

The HRP method was quickly adapted for use in vertebrate embryos, including *Xenopus laevis*. Jacobson and colleagues used HRP to map larval clones arising from identified blastomeres of cleavage stage *Xenopus* (Hirose and Jacobson, 1979). They extended their observations to clones originating as superficial layer cells as late as the 512-cell stage, producing detailed maps of clonal contributions to the larval central nervous system (Jacobson, 1983).

In contrast to the situation in the leech embryo, many clones marked very early in vertebrate development contribute progeny to most or all germ layers and to several tissue types (Dale and Slack, 1987; Moody, 1987; Kimmel and Law, 1985). The positions of early cleavage planes are not often stereotyped; a given clone may be widely dispersed in the later embryo, and its quantitative contribution to specific tissues may be highly variable (Kimmel *et al.*, 1990; Kimmel and Warga, 1987). Therefore, cell lineage is unlikely to play a crucial role in specifying cell fates until relatively late in vertebrate development. Nevertheless, an important use of clonal marking methods is, where possible, to identify blastomeres fated to populate known domains in the later embryo. One can then observe how fate specification changes in response to experimental perturbation. This depends on a marked cloned behaving as a cohort to occupy predictable parts of the larval body. The fact that many clones in *Xenopus* embryos from the early blastula on obey this rule (Wetts and Fraser, 1989) has facilitated studies of regional cellular interactions by rearrangement of marked pieces of the embryo (Gimlich and Cooke, 1983; Smith and Slack, 1983; Gimlich and Gerhart, 1984; Jacobson, 1984; Recanzone and Harris, 1985).

B. Fluorescent Clonal Markers

There are several disadvantages in the use of enzymatic markers such as HRP for cell lineage studies. Because the embryo must be fixed and stained to detect an enzyme, the method is limited to prospective clonal analysis, in which the distribution of the progeny of a founder cell is determined at only one stage in each individual embryo. The methods for enzyme detection are laborious and, in the case of HRP, require the use of toxic substrates. Finally, specimens must usually be sectioned to reveal the progeny not in the most superficial cell layers. For these reasons, several laboratories began to use microinjected fluorescent macromolecules for clonal analysis. Most notably, Zackson (1982) used fluorescently labeled peptides to study the patterns of cell division and cell interaction in the leech germinal plate.

Fluorescent markers have the advantage that their distribution can be determined immediately after microinjection, allowing the cells initially labeled to be identified and counted. This is important because embryonic sister cells may remain coupled by cytoplasmic bridges for much of a cell cycle; microinjection of a single cell often leads to labeling of two or more cells. For instance, Kimmel and Law (1985) were able to confirm sister cell labeling from single injections in zebrafish embryos using rhodamine-labeled HRP. If fluorescence exciting light is kept to a minimum, fluorescent tracers can be observed at intervals in living embryos; in a few cases this has allowed complete description of individual cell lineages (Kimmel and Warga, 1987) and detailed observation of cell movements *in situ* (O'Roarke and Fraser, 1986a,b; J. Shih and R. E. Keller, personal communication).

The use of fluorescent dextrans in studies of circulation, fluid phase pinocytosis, and cytoplasmic structure suggested a new class of macromolecular clonal marker for embryologists. Dextrans are relatively inert in cytoplasm, have an amorphous or globular structure, and are highly soluble under physiological conditions. Dextran is commercially available in many size classes. Dextran reacts to form covalent complexes with several fluorophore derivatives (deBelder and Granath, 1973) and, as described below, can be rendered histologically fixable by a simple technique.

Dextran-based tracers have now been used widely for prospective clonal analysis, fate mapping, and lineage tracing in invertebrate and vertebrate embryos. This chapter summarizes methods for synthesizing histologically fixable fluorescent dextrans and using them to mark frog embryonic clones by microinjection. The procedures presented here are probably not the most efficient, nor do they yield the most chemically stable tracers possible. Similar and potentially superior dextran markers are now sold by at least one vendor. However, the following protocols give the embryologist on a budget the means to produce bright, fixable clonal markers with a small investment of bench time.

II. Methods

A. Materials

Dextrans in several size ranges are available from various suppliers. Dextran of mean molecular weight 9,000–10,000 (Sigma Chemical Co., St. Louis, MO) is commonly used for preparation of clonal markers.

Fluorescein isothiocyanate (FITC) and tetramethylrhodamine-B isothiocyanate (isomer R; TRITC) were obtained from Sigma. Dichlorotriazinyl aminofluorescein dihydrochloride (DTAF) was from Research Organics (Cleveland, OH). Fluorophores were stored at $-20°C$ under dessication. Reaction of isothiocyanates with unmodified dextran was accelerated using the catalyst dibutyltin dilaurate, obtained from Fluka (Ronkonkoma, NY).

A standard for assays of lysine free amino group concentration was ε-amino-n-caproic acid (Sigma). Trinitrobenzene sulfonate was obtained as 5% (w/v) picrylsulfonic acid solution from Sigma. To provide a source of free amino groups for histological fixation, d-lysine (Sigma) was reacted with cyanogen bromide-activated fluorescent dextran. Cyanogen bromide was obtained from Sigma, stored at 4°C under dessication, and used soon after purchase.

B. Procedures

1. Conjugation of Dextran with Fluorophore Isothiocyanates

Fluorescein and rhodamine isothiocyanates react slowly with dextran in aqueous solutions. To accelerate these reactions, a procedure of deBelder and Granath (1973) is used. Dextran of 10,000 mean molecular weight is dissolved at 30 mg/ml in 3 ml of dimethyl sulfoxide (DMSO), along with FITC or TRITC at 7–10 mg/ml, to give at least a 5-fold molar excess of fluorophore to dextran molecules. Two drops of pyridine and 3 μl of dibutyltin dilaurate are added, and the solution is heated to 97°C with stirring in a water bath on a heating–stirring plate. After stirring for 2 hours, dextran is precipitated by pouring the solution gradually into 50 ml of stirred ice-cold ethanol. Precipitated material is collected by centrifugation or filtration through Whatman No. 1 paper, lightly dried, redissolved in water, and precipitated again from cold ethanol. The choice of further purification steps depends on the amount of conjugate produced, subsequent chemical modifications planned, and the nature of the application. For cell lineage tracing with unmodified fluorescent dextran in embryos, conjugate is dissolved in a convenient volume of water, and dialyzed exhaustively against water, using 3500 molecular weight cutoff tubing (Spectrum Medical, Los Angeles, CA). The product should be centrifuged after dialysis (at 10,000 g for 10 minutes) to remove insoluble aggregates. The conjugate is then lyophilized, weighed, and redissolved in injection solution at concentrations ranging from 10 to 100 mg/ml.

2. Conjugation of Dextran with DTAF

A simpler, though more expensive, synthesis of fluoresceinated dextran follows another procedure of deBelder and Granath (1973). This reaction can be scaled up to produce multigram quantities of fluorescein–dextran (P. McNeil, Harvard Medical School, personal communcation). For a small preparation, dextran is dissolved at 30 mg/ml in water along with DTAF at about 8 mg/ml, to give at least a 5-fold molar excess of fluorophore. The pH is adjusted to 10–11 by addition of 1 N sodium hydroxide, and the solution is incubated at room temperature for 2 hours. Conjugate is collected by repeated precipitation in an excess of ethanol and purified by dialysis, centrifugation, and lyophilization as described above.

3. Addition of Amino Groups to Dextrans by Cyanogen Bromide Activation

Unmodified fluorescent dextrans usually remain water soluble after aldehyde or alcohol fixation of injected embryos (R. L. Gimlich, unpublished observations). The versatility of dextran tracers can be increased by using the following technique to make them aldehyde fixable (Gimlich and Braun, 1985). Fluorescent dextrans are "activated" with cyanogen bromide to render them reactive with nonprotonated primary amino groups. Cyanogen bromide reacts with vicinal diols of polysaccharides to generate several reactive intermediates, and coupling with a primary amine usually generates an isourea linkage to the amine-carrying molecule (Porath, 1974). Activated dextran is mixed with lysine under conditions of pH favoring reactivity the α-amino groups but not the ϵ-amino groups. ϵ-Amino groups are then free in the product to facilitate aldehyde fixation in tissue. The isourea linkage to lysine is apparently stable under the conditions described here, but it is reported to be labile in amine-containing buffers (Porath, 1974). This procedure can also be used to add amino groups to unmodified dextran for later reaction with fluorochrome isothiocyanates, sulfonyl fluorides, DTAF, or other adducts in aqueous solution.

The reaction should be carried out at room temperature in a fume hood with gentle stirring and continuous pH monitoring. Fluorescent or unmodified dextran (100 mg) is dissolved in 20 ml water, and the pH is adjusted to 10.7–10.9 by addition of a small amount of 0.1 N sodium hydroxide. Cyanogen bromide is weighed in a fume hood, as it releases volatile cyanide. Solid cyanogen bromide (25 mg) is added to the dextran solution, and the pH is maintained by repeated addition of small volumes of 1 N sodium hydroxide or use of a pH stat. After 15–20 minutes the pH stops falling, indicating complete reaction and hydrolysis of the cyanogen bromide. Sodium bicarbonate (170 mg) and d-lysine (370 mg) are immediately added, and the pH is adjusted to 8.4. The solution is cooled to 4°–8°C in an ice bath, and coupling is carried out with gentle stirring at 4°C overnight. The product is dialyzed extensively against water and then lyophilized or ethanol precipitated and vacuum dried.

4. Reaction of Fluorophores with Lysine–Dextran

Isothiocyanate, sulfonyl chloride (e.g., Texas Red; Titus *et al.*, 1982), and dichlorotriazinyl fluorophores can all be reacted with lysinated dextran under conditions similar to those used to label proteins. A 5-fold molar excess fluorophore derivative is mixed with lysine–dextran in borate or carbonate buffer at pH 9. After 1–2 hours, the dextran is concentrated and purified by dialysis and/or ethanol precipitation as described above. To produce fixable tracers by this method, the lysine–dextran should carry an average of four free amino groups (see assays below), and the fluorophore coupling conditions should be adjusted to consume an average of two of these. Clearly, such conditions may be difficult to achieve, so an alternative is to repeat the cyanogen bromide activation procedure if necessary after reaction of fluorochrome and lysine–dextran.

5. Assays for Free Dye Contamination, Degree of Substitution with Fluorophores, and Number of Free Amino Groups

a. Column Chromatography to Detect Residual Free Dye. Extensive dialysis should remove unreacted fluorophore from the product solution. This can be checked by chromatography of a small amount of product on a column of Sephadex G-25. Labeled dextran is dissolved in buffered saline along with a small amount of bromophenol blue. Column fractions are monitored for absorbance at a maximum for the dye in parallel with absorbance at 590 nm for bromophenol blue. A significant fraction of dye included in the gel with bromophenol blue indicates the need to further purify the labeled dextran. This can be accomplished by preparative filtration on Sephadex or by repeating dialysis. Dialysis against a buffered saline followed by dialysis against water may help to dissociate free dye noncovalently adsorbed to dextran.

b. Measurement of Degree of Substitution with Fluorophores. In general, the degree of substitution with fluorophores can be measured by comparing the dry weight of the product with the optical density at an absorption maximum for the dye. For accuracy, the absorbance spectrum should be obtained. Generally, a range of labeled dextran concentrations around 100 μg/ml gives accurate dye absorbance determinations. For instance, DTAF–dextran at 100 μg/ml with an absorbance at 494 nm of 1.44 has approximately two fluorescent adducts per dextran chain. Reported absorbance maxima and extinction coefficients for the common fluorophores are listed by Simon and Taylor (1986). These determinations are approximate as they do not take into account changes in the extinction coefficients of bound versus unreacted fluorophore derivatives. It has been observed that dye adducts at degrees of substitution greater than about three per dextran chain are less useful than expected; presumably this is due both to fluorescence quenching at high local dye concentrations and to increased tendency to aggregate at high degrees of substitution.

6. Assay for Degree of Substitution with Lysine

The number of free amino groups covalently bound to each dextran chain can be estimated using several assays. A convenient one is that described by Habeeb (1966). Trinitrobenzenesulfonic acid (TNBS) reacts specifically under mild conditions with free amino groups to give trinitrophenyl (TNP) derivatives, which absorb strongly at 335 nm. Lysinated dextran is dissolved at 1 mg/ml in 300 µl of water. Then 300 µl of 4% (w/v) sodium bicarbonate (pH 8.5) is added and 300 µl of 0.1% (w/v) picrylsulfonic acid solution. The mixture is incubated for 3 hours at 40°C, then 150 µl of 1.0 N HCl is added and absorbance is read at 335 nm. The reaction may be scaled down to one-third if microcuvettes are available for the spectrophotometer, and to one-tenth if a microtiter plate reader with ultraviolet light source is available. For standardization, ε-amino caproic acid is used as a lysine analog. Typically, absorbance of a lysinated dextran sample will fall within the standard curve of readings substituting 0.1 to 1.0 mM ε-amino caproic acid for the lysine–dextran. The free amino group concentration can also be estimated using the extinction coefficient of TNP amino acids ($\sim 1.0 \times 10^4\ M^{-1}\ \text{cm}^{-1}$; Habeeb, 1966).

If an insufficient number of free amino groups per dextran chain are obtained, the cyanogen bromide activation and lysination reactions may be repeated. Two to three free amino groups per chain has been found to be optimal for fixability and nontoxicity. Toxicity to embryonic cells increases greatly as the degree of substitution approaches 5, as does the tendency to form intracellular aggregates (M. Shankland, personal communication).

7. Tracer Microinjection

For pressure injection into embryonic cells, fluorescent dextran derivatives are dissolved at 20–100 mg/ml in water and loaded into an injection needle. To prevent tip clogging, the solution should be cleared by centrifugation at top speed for 1 minute in a tabletop microcentrifuge or equivalent. Microinjection needles and the hardware for pressure injection can be constructed as described in Appendix B, this volume. The amount of dextran needed per cell varies with the tracer batch. As little as 150 ng of lysinated fluorescein–dextran per *Xenopus* egg has been used to label donors brightly for cell transplantation experiments (Gimlich, 1986).

For injection of volumes of a few nanoliters or less, particularly into relatively small embryonic cells, smaller-tipped microelectrodes and higher injection pressures are preferable. Pipettes are pulled from Omega-Dot microcapillaries (World Precision Instruments, New Haven, CT) of 1.5 mm outer diameter on a vertical electrode puller. Tips are then broken to an outer diameter of 5–10 µm under observation on the stage of a compound microscope with eyepiece reticule. Solution can be back-loaded into these capillaries or taken up through the tip by applying reduced pressure (for detailed methods, see Purves, 1981; Colman, 1984). A good source for regulated pressure pulses is the Picospritzer II (General Valve, Fairfield, NJ). The rate of delivery of solution is calibrated by immersing the tip in mineral oil under a stereomicroscope with an eyepiece reticule, then measuring the diameter of a droplet expelled by a given pressure pulse.

Fluorescent dextrans and the lysinated derivatives can be introduced into cells by iontophoresis, the application of a current through an injection pipet containing the charged tracer (for practical considerations, see Purves, 1981). The sign of the injection current and the amount of current required to label cells will depend on the sign and charge density of the individual tracer. Unmodified FITC–dextran is anionic and injected using negative current pulses; the net charge of its lysinated derivatives will depend on the relative number of lysine and fluorescein adducts per dextran chain. Rhodamine dextrans are net neutral, and lysinated derivatives prepared as described are cationic. Wetts and Fraser (1988) iontophoretically injected lysinated rhodamine–dextran to trace individual cell lineages in the optic cup of *Xenopus* embryos for several days.

8. Sample Fixation

Lysinated fluorescent dextrans have been cofixed with cell materials in embryos of several species using standard aldehyde fixation procedures. Labeled *Xenopus* embryos are typically fixed for 12 hours at 4°C in 4% paraformaldehyde, in 0.1 M phosphate buffer or cacodylate buffer at pH 7.5. After extensive washing in buffer, samples are dehydrated through an alcohol series to 100% ethanol or methanol and cleared or embedded for histological sectioning in paraffin or methacrylate resin (Gimlich and Braun, 1985). Whole-mount labeled embryos cleared in 2:1 benzyl benzoate–benzyl alcohol (Dent *et al.*, 1989) can be viewed directly in the standard epifluorescence microscope or "optically sectioned" using confocal laser scanning technology (see Fig. 1).

III. Practical Considerations

A. Controls and Precautions

A common problem with fluorescent dextran clonal markers, particularly some from commercial sources, is cytotoxicity. Different tracer preparations vary in their toxicity, and embryos from different species vary in sensitivity. The important variables seem to be the number of free amino groups, the degree of product purity, and exposure to exciting light *in vivo*. Each batch must be tested by microinjection at a range of doses to find a nontoxic intracellular concentration with good fluorescence yield. Tracers with more than four free amino groups per chain have often been found to be toxic at low doses in *Xenopus*, and an average of two free amino groups yields tracers with low toxicity and good fixability. Toxicity may not be evident immediately after microinjection; subsequent cleavage divisions may b˜ altered in geometry or timing, or the injected cells may divide but later dissociate from the embryo and die. To detect possible deleterious effects in amphibian early embryos, a clone labeled by injection of a cleavage stage blastomere should be observed briefly at intervals using visible light to note the rate of division. The final distribution, size, and number of cells in the labeled clone should be carefully observed and the cell morphology compared to that in equivalent unlabeled regions. Bright aggregates of extracellular dye are often seen following injection of toxic doses. In experiments involving microsurgical manipulation, homotopic transplants of marked cells or tissue should be included to detect any effect of labeling on the viability and behavior of transplants.

The cell-lineage specificity of labeling should be tested for a new batch of fluorescent dextran. Observation immediately after microinjection of embryonic blastomeres in *Xenopus* should reveal one or at most two labeled cells, since cytoplasmic bridges sometimes persist into the next cleavage cycle. A graded distribution of fluorescence in several cells may indicate the persistence of free fluorochrome capable of diffusion through gap junctions.

Finally, the fraction of fluorescence fixable by aldehydes should be estimated for a new preparation of lysinated fluorescent dextran. A variable-intensity exciting light source or neutral density filters in the exciting light path can be used to determine the threshold of visual detection for a labeled sample before fixation. After fixation and extensive washing, any drastic change in this detection threshold can be noted (Gimlich and Braun, 1985). Typically, at least 70% of tracer fluorescence survives fixation and processing of frog embryo samples.

When illuminated with exciting light, intracellular fluorochromes serve as efficient photoablation reagents (Ettensohn, 1990). Therefore it is important to minimize light exposure to avoid perturbing the development of labeled cells. Several investigators have, however, used low levels of exciting light and image-intensifying video microscopy (Bright and Taylor, 1986) for "time-lapse" tracing of labeled lineages *in vivo*. Notably, O'Rourke and Fraser (1986a,b) observed the development of normal and experimentally altered retinotectal fiber projections in living *Xenopus* tadpoles partially labeled with lysinated fluorescent dextrans. In such experiments, it is important to compare

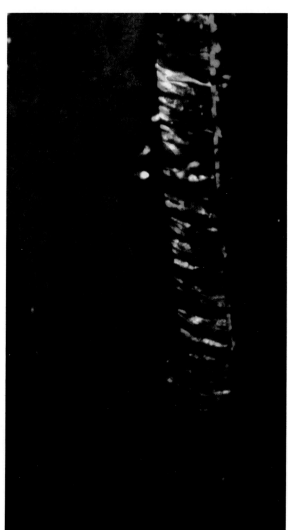

FIG. 1. Confocal laser scanning micrograph of labeled cells in the posterior trunk notochord of a stage 35 *Xenopus* embryo. Embryo was injected into adjacent blastomeres at the 32-cell stage with fluoresceinated lysine-dextran (cell B1; Dale and Slack, 1987) and rhodaminated lysine-dextran (cell C1). Section shows the characteristic "stack of coins" appearance of notochordal cells and the result of interdigitation of cells from the two clones during notochord elongation (Keller *et al.*, 1989).

the fates of observed cells with those in control embryonic samples not continuously observed.

B. Dilution by Cell Proliferation

An important limitation of any nongenetic method of marking clones is the dilution of label by cell growth and division. In *Xenopus laevis* and other animals with mesolethical eggs and holoblastic cleavage, early cell divisions are reductive; they do not dilute a clonal marker, they merely subdivide it. The first cell growth and proliferation takes place at the tailbud stage in the commonly studied amphibians. By that time, many of the progeny of embryonic blastomeres have begun differentiation (Nieuwkoop and Faber, 1975). Thus, a fate map of the embryonic progenitors of larval structures can be deduced. The spatial resolution of the fate map will depend on the stage at which progenitor cells are initially labeled, that is, on the size of the originally marked domains. Resolution also depends on the extent to which cleavage planes bear a relation to prospective areas. For this reason, embryos are often chosen for analysis which have a specific pattern of cleavage planes in relation to other markers (such as the point of sperm entry or the cortical pigmentation) where they exist. In *Xenopus laevis*, paratangenital divisions occur after the fifth or sixth cleavage, giving rise to deep and superficial blastomeres (Nieuwkoop and Faber, 1975); the 32- to 64-cell stage is the latest at which clonal marking of each blastomere can conveniently be used to construct a fate map for all the blastomeres. This approach has been taken by several laboratories using both HRP and fluorescent dextrans, and the results are in good agreement (Dale and Slack, 1987; Moody, 1987).

C. Detection

The one burdensome requirement for use of fluorescent dextran clonal markers is a fluorescence microscope with appropriate filter sets. Given the availability of this minimal equipment, much can be learned with these markers. They can be made bright enough to allow short-exposure photographic recording in living specimens without toxicity. They are retained through histological preparation and detectable in standard paraffin sections or in plastic sectioned as thin as a few microns. Dextran can be coupled to a variety of fluorophores, allowing double-labeling experiments or combined clonal marking and immunohistochemical staining. Enhancement techniques for low-light imaging and confocal scanning microscopy greatly extend the versatility of fluorescent dextran tracers.

References

Bright, G. R., and Taylor, D. L. (1986). *In* "Applications of Fluorescence in the Biomedical Sciences" (D. L. Taylor, A. S. Waggoner, R. F. Murphy, F. Lanni, and R. Birge, eds.), p. 257. Alan R. Liss, New York.

Colman, A. (1984). *In* "Transcription and Translation: A Practical Approach" (B. D. Hames and S. J. Higgins, eds.), p. 271. IRL Press, Oxford.

Dale, L., and Slack, J. M. W. (1987). *Development (Cambridge, U.K.)* **99,** 527–551.

Davidson, E. H. (1986). "Gene Activity in Early Development," 3rd Ed. Academic Press, New York.

deBelder, A. N., and Granath, K. (1973). *Carbohydr. Res.* **30,** 375–378.

Dent, J. A., Polson, A. G., and Klymkowsky, M. W. (1989). *Development (Cambridge, U.K.)* **105,** 61–74.

Ettensohn, C. A. (1990). *Science* **248,** 1115–1118.

Ferrus, A., and Garcia Bellido, A. (1976). *Nature (London)* **260,** 425–426.

Gimlich, R. L. (1986). *Dev. Biol.* **115,** 340–352.

Gimlich, R. L., and Braun, J. (1985). *Dev. Biol.* **109,** 509–514.

Gimlich, R. L., and Cooke, J. (1983). *Nature (London)* **306,** 471–473.

Gimlich, R. L., and Gerhart, J. C. (1984). *Dev. Biol.* **115,** 340–352.

Habeeb, A. F. S. A. (1966). *Anal. Biochem.* **14,** 328–336.

Hirose, G., and Jacobson, M. (1979). *Dev. Biol.* **71,** 191–202.

Jacobson, M. (1983). *J. Neurosci.* **3,** 1019–1038.

Jacobson, M. (1984). *Dev. Biol.* **102,** 122–129.

Keller, R. E. (1975). *Dev. Biol.* **42,** 222–241.

Keller, R. E. (1976). *Dev. Biol.* **51,** 118–137.

Keller, R. E., Cooper, M., Danilchik, M., Tibbetts, P., and Wilson, P. (1989). *J. Exp. Zool.* **251,** 134–154.

Kimmel, C. B., and Law, R. D. (1985). *Dev. Biol.* **108,** 94–101.

Kimmel, C. B., and Law, R. D. (1985). *Dev. Biol.* **108,** 75–85.

Kimmel, C. B., and Warga, R. M. (1987). *Dev. Biol.* **124,** 269–280.

Kimmel, C. B., Warga, R. M., and Schilling, T. F. (1990). *Development (Cambridge. U.K.)* **108,** 581–594.

LeDourain, N. (1969). *Bull. Biol. Fr. Belg.* **103,** 435–452.

Moody, S. A. (1987). *Dev. Biol.* **122,** 300–319.

Nieuwkoop, P. D., and Faber, J. (1975). "Normal Table of *Xenopus laevis* (Daudin)." North-Holland, Amsterdam.

O'Rouke, N. A., and Fraser, S. E. (1986a). *Dev. Biol.* **114,** 265–276.

O'Rourke, N. A., and Fraser, S. E. (1986b). *Dev. Biol.* **114,** 277–288.

Porath, J. (1974). *In* "Methods in Enzymology" (W. B. Jakoby and M. Wilchek, eds.), Vol. 34, pp. 13–30. Academic Press, New York.

Purves, R. D. (1981). "Microelectrode Methods for Intracellular Recording and Ionophoresis." Academic Press, New York.

Recanzone, G., and Harris, W. A. (1985). *Wilhelm Roux Arch. Dev. Biol.* **194,** 344–354.

Reverberi, G. (1971). "Experimental Embryology of Marine and Freshwater Invertebrates." North-Holland Publ., Amsterdam.

Shankland, M. (1987). *Curr. Top. Dev. Biol.* **21,** 31–63.

Simon, J. R., and Taylor, D. L. (1986). *In* "Methods in Enzymology" (R. B. Vallee, ed.), Vol. 134, pp. 487–507. Academic Press, New York.

Smith, J. C., and Slack, J. M. W. (1983). *J. Embryol. Exp. Morphol.* **78,** 299–317.

Sulston, J. E., Schierenberg, E., White, J. G., and Thomson, J. N. (1983). *Dev. Biol.* **100,** 64–119.
Thiebaud, C. H. (1983). *Dev. Biol.* **98,** 245–249.
Titus, J. A., Haugland, R. P., Sharrow, S. O., and Segal, D. M. (1982). *J. Immunol. Methods* **50,** 193–204.
Weisblat, D. A., Sawyer, R. T., and Stent, G. S. (1978). *Science* **202,** 1295–1298.
Wetts, R., and Fraser, S. E. (1988). *Science* **239,** 1142–1145.
Wetts, R., and Fraser, S. E. (1989). *Development (Cambridge, U.K.)* **105,** 9–15.
Zackson, S. L. (1982). *Cell (Cambridge, Mass.)* **31,** 761–770.

Chapter 16

Nuclear Transplantation in Xenopus

J. B. GURDON

Wellcome CRC Institute
University of Cambridge
Cambridge CB2 1QR, England

I. Introduction and Historical Background
II. Results Attainable with Nuclei Transplanted to Eggs
III. Methodology
    A. Media
    B. Equipment Needed
    C. Preparation of Donor Cells
    D. Genetic Markers
    E. Preparation of Recipient Eggs
    F. Nuclear Transfer Procedure
References

I. Introduction and Historical Background

Nuclear transplantation usually refers to the replacement of the haploid egg pronucleus by the diploid nucleus of a somatic cell. The aim of doing this is to determine to what extent a somatic cell nucleus can promote normal embryonic development with egg cytoplasm. There are three principal steps in this method of transplanting nuclei to eggs, namely, the preparation of donor cells, the preparation (including enucleation) of recipient eggs, and the transfer of a nucleus to the recipient egg. Genetic markers are essential in most experiments and are available in *Xenopus*. When transplanting nuclei from more specialized cells it can be very helpful to do serial nuclear transfers, which are discussed under the transplantation procedure.

Somatic cell nuclei can also be transplanted to oocytes. Since, unlike eggs, oocytes are not capable of development, nuclear transplantation experiments to oocytes have a different aim and are described elsewhere (Gurdon, 1977).

The first success with nuclear transplantation in eukaryotes was achieved by Briggs and King in 1952, who obtained some normal postneurula embryos from transplanted blastula nuclei in *Rana pipiens*. However, nuclei from late gastrula stages gave progressively much less normal development of nuclear transplant embryos (King and Briggs, 1955), and nuclei from tailbud stages did not support as normal a development as gastrula nuclei (Briggs and King, 1957). From these results and from serial nuclear transplantation (King and Briggs, 1956), they concluded that stable nuclear changes accompany normal development and the early differentiation of cells.

Nuclear transplantation in *Xenopus*, using a genetic nuclear marker, was first reported by Fischberg et al. (1958) and the production of genetically marked, sexually mature adults in the same year (Gurdon et al., 1958). Later experiments showed that it is still possible to obtain a few normal larvae and subsequently fertile frogs from nuclei of the larval intestine (Gurdon and Uehlinger, 1966). These experiments with *Xenopus* generated a different interpretation compared to work with *Rana*, and it was concluded that irreversible nuclear changes do not necessarily accompany early cell differentiation. Subsequent results with the nuclei of differentiated cells from *Xenopus* [e.g., larval intestine and adult skin (Gurdon, 1962; Gurdon et al., 1975) and antibody producing cells (Wabl et al., 1975)] as well as from *Rana pipiens* (e.g., Di Berardino, 1989) have supported the early conclusions from *Xenopus*. It is not clear why nuclear transplant embryos from adult cell nuclei have not so far survived significantly beyond metamorphosis, the most advanced *Xenopus* nuclear transplants from adult cells dying on completion of this phase (Gurdon, 1974). This might be because nuclear transplantation, which is known to generate chromosomal abnormalities (Di Berardino and Hoffner, 1971; Gurdon and Laskey, 1970), also generates small gene changes, many of which show their effects only during the growth (or feeding) stage onward. These and other aspects of nuclear transplantation in *Xenopus* have been reviewed in detail by Gurdon (1986).

For most kinds of nuclear transplantation experiments, *Xenopus* would appear to provide more convenient material than other amphibian species. For an equivalent complexity of procedure, *Xenopus* gives more normal results, and its eggs are available at all times of year, though sometimes of poor quality. Furthermore, several genetic markers are available in *Xenopus*, which also enjoys a rapidly growing list of early expressing genes.

II. Results Attainable with Nuclei Transplanted to Eggs

In *Xenopus*, nuclei from blastulas and from the endoderm of stages up to the neural folds yield swimming tadpoles in 30–50% of cases, and about one-half of these will proceed to sexually mature adults (reviewed in Gurdon, 1986). When slowly or nondividing differentiated cells are used as donors, the success rate declines greatly. From fully differentiated cells such as adult skin (Gurdon *et al.*, 1975), up to 5% of nuclear transfers reach the heartbeat stage (which contains differentiated cells such as muscle, nerve, lens, and blood) and nearly 20% reach the earlier muscular response stage (with functional muscle and nerve). Results with *Rana pipiens* are broadly similar (Hennen, 1970), though it has been found beneficial, with erythrocyte nuclei, to transplant nuclei to oocytes and to mature the oocyte to an egg about 1 day later (Di Berardino and Hoffner, 1983).

Very soon after a somatic nucleus is transplanted to egg cytoplasm it undergoes dramatic changes in morphology and activity. Within 30 minutes its volume increases many times, its chromatin becomes dispersed, and DNA synthesis is initiated (Graham *et al.*, 1966). As soon as general patterns of RNA synthesis can be measured at the late blastula stage, the daughter cells generated by a transplanted nucleus have become indistinguishable from cells of embryos derived from fertilized eggs (Gurdon and Woodland, 1969). The expression of individual genes seems also to be reversed as soon as it can be measured (Gurdon *et al.*, 1984).

III. Methodology

A. Media

Modified Barth's Saline (MBS)
   See Table I for composition (Gurdon, 1977). May be maintained as an autoclaved 10× stock solution, with antibiotics (e.g., penicillin and streptomycin each at 10 mg/liter) added just before dilution and use. This is a standard medium for embryo culture. It is helpful to add BSA (bovine serum albumin) at 0.1 mg/ml to MBS when using it for the maintenance of donor cells during transplantation. Some laboratories use other somewhat similar media for embryo culture (e.g., Normal Amphibian Ringer's see Table I). MBS appears to be at least as good as others so far described.

Ca, Mg-free MBS
   As above, but lacking Ca^{2+} and Mg^{2+} components. Used for dissociating donor cells. EDTA at 0.5 mM, and adjustment of pH to 8.0, will facilitate dissociation of more differentiated tissues into single cells.

0.1 × MBS

One-tenth concentration of MBS (above) is used for rearing embryos postgastrulation.

High salt MBS (Table I)

This is useful as a medium in which frogs will lay eggs, and in which eggs will remain unactivated and fertilizable for up to a few hours. Eggs laid in tap water or 0.1 × MBS deteriorate within 30 minutes and need to be used for nuclear transfer within this time.

B. Equipment Needed

Injection pipettes need very careful construction; those made for microinjection into oocytes are unsuitable. For nuclear transplantation, the end of the pipette needs to be parallel sided with no constriction near the tip. Its opening must be just too small for a donor cell to enter easily, thereby rupturing, or at least weakening, the cell membrane. On the other hand, the opening of the pipette must be smooth, so that cytoplasm is not separated from the nucleus which it surrounds and protects during transfer, and the pipette must have a

TABLE I

Composition of Salt Solutions Used in Nuclear Transplantation Experiments[a]

	Modified Barth's Saline (MBS) (mM)	High salt MBS (mM)	Normal amphibian ringer's (NAM) (mM)
NaCl	88	110	110
KCl	1	2	2
$CaCl_2$	0.41	—	1
$Ca(NO_3)_2$	0.33	—	—
$MgSO_4$	0.82	1	1
$NaHCO_3$	2.4	2	0.5
HEPES(NaOH), pH 7.4	10	—	—
Tris base, pH 7.6, acetic acid	—	15	—
EDTA	—	—	0.1
Sodium phosphate, pH 7.4	—	0.5	1
Penicillin, benzyl	10 mg/liter	—	100 units/ml
Streptomycin sulfate	10 mg/liter	—	60 units/ml
Nystatin	—	—	2 mg/liter

[a] For MBS, see Gurdon (1977). For NAM, see Slack and Forman (1980).

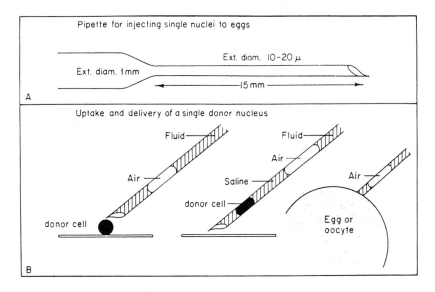

FIG. 1. (A) Diagram of a nuclear transplantation pipette suitable for *Xenopus* embryo nuclei. (B) Diagram of a procedure for drawing a donor cell into the pipette, so as to rupture the cell wall but not disturb cytoplasm from around the nucleus. (From Gurdon, 1977.)

sharp point to avoid damaging the egg (Fig. 1A). A number of commercially available pipette pullers and microforges can be used for the initial preparation of the pipette, but it is necessary to grind or microforge the tip to a hypodermic-like end (Fig. 1A). The same pipette can be used for several hundred transfers, and it can be reused on subsequent days after cleaning in detergent or by glass cleaning procedures.

The enucleation of *Xenopus* eggs is best achieved with a bacteriocidal ultraviolet light. A commonly available example is the Mineralite type UVS (Ultraviolet Products, Inc., San Gabriel, CA, Model UVSL-15). In addition it is very helpful to use a different wavelength UV lamp for weakening the very elastic membrane that immediately surrounds the egg. This can be done with a Hanovia quartz lamp, UVS-100, a medium-pressure mercury arc lamp (Hanovia, Slough, England).

For the transplantation procedure a standard stereomicroscope with a magnification range of about 25× is suitable. Figure 2 illustrates the equipment we have used with *Xenopus*. The transplantation pipette needs to be held by a low-power manipulator, of which the Singer is an especially suitable example (Singer Instruments, Roadwater, Watchet, Somerset, England). The pipette is connected by stiff plastic tubing to a syringe (e.g., Agla, Burroughs Wellcome, Beckenham, Kent, England), the whole system

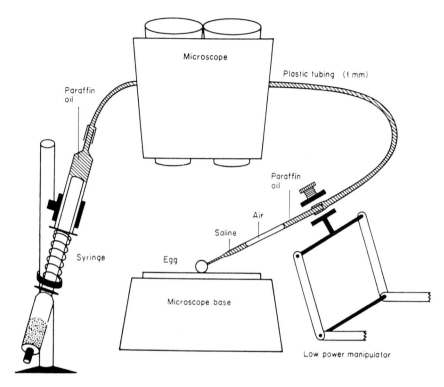

FIG. 2. Diagram of nuclear transplantation equipment. A syringe is connected by oil-filled tubing to a micropipette, which is held in a manipulator and viewed through a stereomicroscope. (From Gurdon, 1977.)

being filled with a light paraffin oil (0.85 g/ml). Air spaces in the syringe or tubing should be avoided.

C. Preparation of Donor Cells

Embryonic tissues dissociate into single cells very readily by incubation for about 10–20 minutes in Ca, Mg-free MBS in an agar covered dish. To help dissociate more differentiated cells, 0.5 mM EDTA should be included in the Ca, Mg-free MBS. For differentiated or adult tissue, any separation procedure that yields viable cells (e.g., trypsinization) will do. Once loosened by Ca, Mg-free solution, cells may be dispersed by pipette and can be kept in Ca, Mg-free MBS while being used for nuclear transplantation. It aids the smooth entry of donor cells into the transplantation pipette, and probably the survival of isolated cells, to add 0.1% BSA to the cell maintenance medium.

D. Genetic Markers

It is highly desirable to use a genetic marker in nuclear transfer experiments to prove noncontribution of the egg nucleus. In *Xenopus* a useful marker is the 1-nucleolate mutant (Elsdale *et al.*, 1960), in which all diploid nuclei have only one nucleolus instead of the usual two nucleoli seen in 30–50% of normal diploid nuclei (Fig. 3). This marker requires the recognition of a heterozygous 1-nucleolate donor embryo, and nucleoli can first be seen in nuclear transplant embryos at the gastrula stage. Another marker is an albino mutant, which is

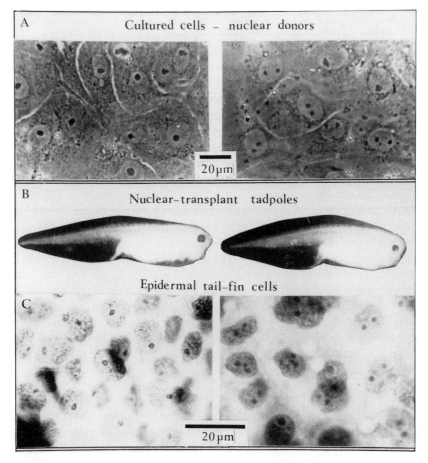

FIG. 3. Photographs of the 1-nucleolate nuclear marker of Elsdale *et al.* (1960) and of larvae obtained by serial transplantation of nuclei from a cultured line of *Xenopus* cells. (From Gurdon and Laskey, 1970.)

useful as a donor nuclear marker when nuclear transplant embryos survive to the feeding stage (stage 45); at this time maternal pigment is clearly reduced in albinos. In future experiments it may be possible to use donor nuclei from cultured cells which have been transfected with a marker such as β-galactosidase under the control of an appropriate promoter.

E. Preparation of Recipient Eggs

There is no need to use fertilized eggs as recipients, and enucleation is simplified with unfertilized eggs, from which only the egg pronucleus has to be eliminated. This is best done by ultraviolet light. The procedure we use (Gurdon, 1977) is to transfer unfertilized eggs, laid in high salt MBS, into $0.1 \times$ MBS. Each egg is then transferred, with forceps, in its jelly, to a dry glass slide, so that the egg spot is facing directly upward. A slide containing about four eggs (in the area of evenly lit UV light) is then exposed first to a bacteriocidal lamp for 30 seconds (UVS Mineralite, shield removed, and eggs placed ~ 5 cm from the bulb and delivering about 16,000 ergs/second/cm^2). Immediately after this, the slide is placed under the Hanovia lamp for about 10 seconds. The duration of exposure to the second UV light needs to be determined for each batch of eggs (typically 10 seconds), since sufficient UV light is required to make the elastic membranes penetrable, but not to so denude the egg as to impede healing of the egg cortex after removal of the injection needle. These treatments are shown diagrammatically in Fig. 4.

Both ultraviolet lights have the effect of killing the egg pronucleus, which is located on the surface of the egg as a second meiotic spindle. The UV dose is normally much more than is sufficient to kill the egg nucleus as judged from androgenetic haploidy tests. The second UV light has the function of providing penetrability while permitting good healing. As soon as irradiated, it is important to use recipient eggs for nuclear transfer at once, since the removal of jelly can cause the surface to dry quickly, and this is detrimental to egg survival. It would be possible to dejelly eggs by the standard cysteine procedure, but this would make UV enucleation difficult, since this must be done dry on account of UV absorption by water.

F. Nuclear Transfer Procedure

Nuclear transfer requires considerable skill, and practice seems to improve greatly the rate of success. The scope for skill seems to lie in obtaining the optimal distortion of a donor cell, so that the cell membrane is ruptured but cytoplasm around the nucleus disturbed as little as possible. It is generally best to give minimal distortion to the donor cell, even though this can result in a certain proportion of recipient eggs receiving unbroken donor cells which

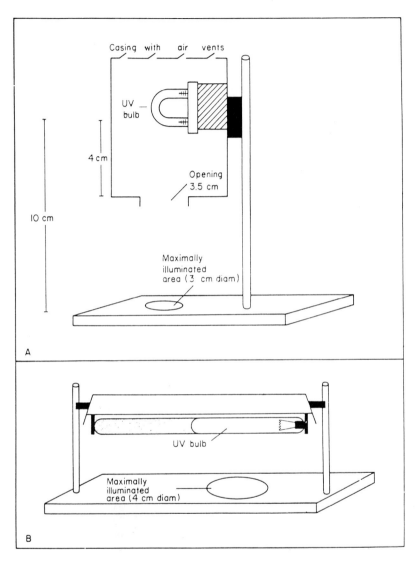

FIG. 4. Diagrams of ultraviolet light apparatus suitable for enucleating unfertilized *Xenopus* eggs. (A) Hanovia UVS-100. (B) Mineralite UVSL-15. (From Gurdon, 1977.)

do not therefore participate in development. It can be worthwhile, when transplanting nuclei from very small cells, to inject about three donor cells rather than one, since not all are easily ruptured to an appropriate extent. Figure 1B shows diagrammatically how an embryonic cell should be drawn into a pipette to an optimal extent.

The position in the egg where a transplanted nucleus is deposited is not critical, though nuclei will not promote development if placed in the vegetal part of the egg. To aim to deposit the nucleus at about one-third of the distance from the animal to vegetal poles will place it near the position which it should eventually occupy. The processes which bring the sperm and egg nuclei to the right position in the egg seem also to operate for a transplanted nucleus.

When transplanting nuclei from more specialized tissues, it is helpful to carry out serial nuclear transfers (Fig. 5), as first done by King and Briggs (1956). Experience with *Xenopus* (Gurdon, 1962; Gurdon *et al.,* 1975) has shown that the best results are obtained by selecting as donors for serial transplantation those first transfer blastulas which are only partially cleaved. It is known that nuclei from slowly dividing tissues often fail to replicate their chromosomes fully by the time the egg undergoes its first cytoplasmic cleavage, resulting in incompletely replicated chromosomes being pulled apart and damaged during the first mitosis (Gurdon and Laskey, 1970). Sometimes, a transplanted nucleus which has commenced replication fails to divide, and

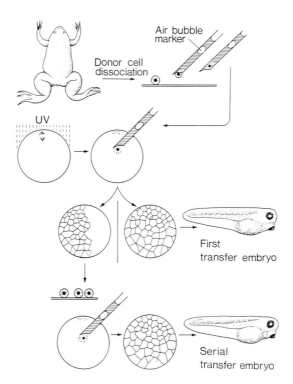

FIG. 5. Diagram of the serial nuclear transplantation procedure for *Xenopus*. (From Gurdon, 1986.)

moves as a whole into only one of the first two daughter cells. The blastomere which receives the nucleus continues to divide more or less normally, while the other (enucleate) blastomere remains undivided. Cells from the cleaved half of the resulting partial blastula tend to give good serial nuclear transfer results, possibly because the originally transplanted nucleus has had two division cycles in which to complete replication of its chromosomes.

References

Briggs, R., and King, T. J. (1952). *Proc. Natl. Acad. Sci. U.S.A.* **38,** 455–463.
Briggs, R., and King, T. J. (1957). *J. Morphol.* **100,** 269–312.
DiBerardino, M. A. (1989). *Dev. Biol.* **6,** 175–198.
DiBerardino, M. A., and Hoffner, N. (1971). *J. Exp. Zool.* **176,** 61–72.
DiBerardino, M. A., and Hoffner, N. J. (1983). *Science* **219,** 862–864.
Elsdale, T. R., Gurdon, J. B., and Fischberg, M. (1960). *J. Embryol. Exp. Morphol.* **8,** 437–444.
Fischberg, M., Gurdon, J. B., and Elsdale, T. R. (1958). *Nature (London)* **181,** 424.
Graham, C. F., Arms, K., and Gurdon, J. B. (1966). *Dev. Biol.* **14,** 349–381.
Gurdon, J. B. (1962). *J. Embryol. Exp. Morphol.* **10,** 622–640.
Gurdon, J. B. (1974). "The Control of Gene Expression in Animal Development." Oxford and Harvard Univ. Presses, London, New York, and Cambridge, Massachusetts.
Gurdon, J. B. (1977). *In* "Methods in Cell Biology" (G. Stein, J. Stein, and L. J. Kleinsmith, eds.), Vol. 16, pp. 125–139. Academic Press, New York.
Gurdon, J. B. (1986). *J. Cell Sci. (Suppl.)* **4,** 287–318.
Gurdon, J. B., and Laskey, R. A. (1970). *J. Embryol. Exp. Morphol.* **24,** 227–248.
Gurdon, J. B., and Uehlinger, V. (1966). *Nature (London)* **210,** 1240–1241.
Gurdon, J. B., and Woodland, H. R. (1969). *Proc. R. Soc. London, Ser. B* **173,** 99–111.
Gurdon, J. B., Elsdale, T. R., and Fischberg, M. (1958). *Nature (London)* **182,** 64–65.
Gurdon, J. B., Laskey, R. A., and Reeves, O. R. (1975). *J. Embryol. Exp. Morphol.* **34,** 93–112.
Gurdon, J. B., Brennan, S., Fairman, S., and Mohun, T. J. (1984). *Cell (Cambridge, Mass.)* **38,** 691–700.
Hennen, S. (1970). *Proc. Natl. Acad. Sci. U.S.A.* **66,** 630–637.
King, T. J., and Briggs, R. (1955). *Proc. Natl. Acad. Sci. U.S.A.* **41,** 321–325.
King, T. J., and Briggs, R. (1956). *Cold Spring Harbor Symp. Quant. Biol.* **21,** 271–290.
Slack, J. M. W., and Forman, D. (1980). *J. Embryol. Exp. Morphol.* **56,** 283–299.
Wabl, M. R., Brun, R. B., and DuPasquier, L. (1975). *Science* **190,** 1310–1312.

Chapter 17

Mesoderm Induction

IGOR B. DAWID

Laboratory of Molecular Genetics
National Institute of Child Health and Human Development
National Institutes of Health
Bethesda, Maryland 20892

I. Introduction
II. Mesoderm Induction Assay
    A. Age of Inducing and Responding Tissue and Size of Animal Cap Explants
    B. Equipment and Solutions, Culture Conditions, and Time of Exposure to Inducer
III. Assay End Points: Markers of Mesoderm Induction
    A. Morphological Criteria
    B. Antibody Detection
    C. Gene Regulation in Response to Mesoderm Induction
IV. Mesoderm-Inducing Factors
V. Specification of Cell Fate, Organization of Tissues, and Axis Determination
    A. Cell Differentiation versus Axis Determination
    B. Use of Dissociated Embryonic Cells in the Study of Induction
VI. Outlook
    References

I. Introduction

The elaboration of the body plan in all embryos, and particularly in vertebrate embryos, involves cell interactions. These interactions, together with information laid down in the egg during oogenesis, are required for the formation of body axes and the specification of diverse tissues. The amphibian embryo starts life with one axis, the animal–vegetal axis, already established, but the elaboration of dorsoventral and anteroposterior polarity and the specification of tissues that occurs during embryogenesis depend on cell communications. The earliest known interaction in the amphibian embryo is required for the differentiation of mesoderm, an event linked inextricably (if not fully understandably) to the establishment of the dorsoventral axis. The amphibian embryo has, historically and in recent times, been the prime object for the study of embryonic induction, and we know more about induction in these than in other animals (reviewed in Smith, 1989; Dawid et al., 1990). Although mesoderm is formed in similar ways in other vertebrate embryos, it is not clear to what extent the insights gained in the amphibian are more widely applicable; observations in chicken embryos point to both similarities and differences (Mitrani et al., 1990). Yet, although details may differ between various vertebrate classes, it is likely that similarities in factors and regulatory genes, which are known to occur, reflect significant similarities in mechanisms of embryogenesis in vertebrates.

The *Xenopus* embryo fate map shows that, in general, the animal region gives rise to ectoderm and the vegetal region to endoderm, whereas the mesoderm is elaborated by cells derived from the equatorial region or marginal zone (Dale and Slack, 1987a; Moody, 1987). On the basis of the work of Nieuwkoop (1973, and references therein) and subsequent supporting studies by others, we view the specification of mesoderm as requiring cell interactions in which a signal or set of signals emanating from the vegetal pole induces marginal zone cells toward a mesodermal fate (reviewed in Smith, 1989; Whitman and Melton, 1989a; Dawid et al., 1990). Animal pole cells are competent to respond to this induction but do not become exposed to it during normal embryogenesis, thereby preserving their cell-autonomous ectodermal specification. This competence is the basis of the widely used mesoderm induction assay described in Section II.

Specification of mesoderm is followed closely by the induction of the neural plate, which occurs during gastrulation. In this event, dorsal ectoderm is induced by influences derived from the dorsal lip and the advancing dorsal mesoderm to change its fate from an epidermal to a neural one, as discussed in Chapter 18 of this volume.

II. Mesoderm Induction Assay

A. Age of Inducing and Responding Tissue and Size of Animal Cap Explants

The assay system is based on the original work of Nieuwkoop and is summarized in Fig. 1. In its original form, induction is elicited by making recombinate explants from animal pole and vegetal pole cells ("recombinate

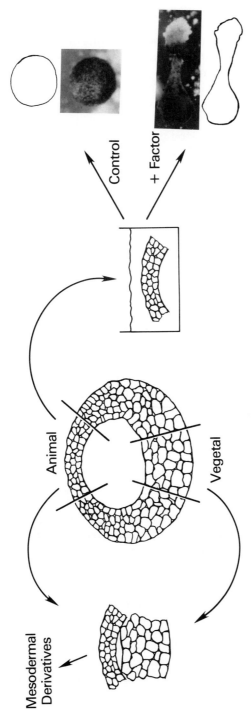

FIG. 1. Mesoderm induction assay. The left-hand side shows a recombinate prepared by culture of an animal cap explant together with vegetal tissue (Sudarwati and Nieuwkoop, 1971). The right-hand side illustrates the culture of an animal cap explant in test solution. The control, cultured in amphibian salt solution without inducing factors, forms a sphere of atypical epidermis. Exposure to a strong inducer leads to tissue elongation and to the differentiation of various cell types, as discussed in the text.

assay," left-hand side). The dorsal or ventral nature of the induced mesoderm depends on the dorsal or ventral origin of the inducing vegetal cells (Dale and Slack, 1987b). The right-hand side of Fig. 1 illustrates the form of the induction assay that tests the effect of soluble factors on animal cap explants ("factor assay"). This form of the assay has been used by Smith (1987) in the work that led to the discovery of the mesoderm inducing factor from XTC cells (XTC-MIF), in the initial work on the inducing activity of fibroblast growth factor (FGF), (Slack et al., 1987), and in many subsequent studies.

Studies using recombinate assays have established that animal explants ("animal caps") are competent to respond to the inducing stimulus between stages 6.5 and 10.5 (Gurdon et al., 1985; Jones and Woodland, 1987). It is more difficult to decide when vegetal cells become sources of inducing effects: Very early vegetal blastomeres do induce, but they divide and develop during the time of the assay. Vegetal cells from the 16-cell stage (stage 5) onward have been used as sources of the inducing stimulus in induction assays (Gimlich, 1986; Jones and Woodland, 1987). In these experiments, induction was assayed exclusively or primarily by the differentiation of muscle. It is not certain whether the times of competence for the induction of other mesodermal derivatives correspond exactly to those for muscle.

In the factor assay the age of the animal caps has been kept between stages 7.5 and 9 in most experiments, but no systematic studies of qualitative differences in outcome for explants of different age have been reported. The loss of competence of animal caps shows a rather rapid decline, with the exact time of loss depending on the inducer. Stage 10 is the last competent stage for a response to FGF and stage 11 for XTC-MIF (activin; see Section IV), using an assay of external morphology as end point (Green et al., 1990).

A second variable concerns the size of the animal cap that is dissected. In principle, any size is acceptable, as long as the control animal caps that are not exposed to inducer fail to display any mesoderm differentiation. In most experiments, the angle between the animal–vegetal axis and the line of dissection lies between 30° and 45°, and the resulting animal caps contain approximately one-tenth the mass of the embryo (e.g., Smith, 1987). In some cases larger explants, encompassing 100 cells of a 400-cell embryo, have been used (Sokol et al., 1990; see also Cooke et al., 1987).

B. Equipment and Solutions, Culture Conditions, and Time of Exposure to Inducer

During dissection experiments it is often convenient to slow development by keeping the embryos at low temperature; any temperature between 14° and 25°C can be used, allowing adjustment of developmental rate within a factor of about 3. Dissection of cysteine-dejellied embryos is carried out in plastic dishes in any one of several half- to full-strength amphibian salt solutions, e.g., NAM, MMR, GMB (see Appendix A, this volume). Removal of the vitelline membrane and dissection are done with sharpened forceps, although hairloops, fine glass needles, and electrolytically sharpened tungsten needles have also been used. Animal caps are transferred from the dissection dish with the help of glass capillaries (e.g., 50- to 100-μl pipettes), avoiding transfer of too much solution with the explant. Explants are cultured individually in agarose-coated wells of microtiter or Terasaki plates to prevent aggregation; 1% agarose in 0.1 × strength of one of the salt solutions is used to coat the wells. Use of Terasaki plates conserves precious factors, but problems of desiccation and dilution of culture fluid during transfer may arise. If factor availability is not severely limiting, it is convenient to use 100 to 150 μl culture medium for each explant in a 96-well microtiter dish. Culture of the explants, whether in the recombinate or factor versions of the assay, is carried out in half- to full-strength amphibian salt solution or in 67% strength L15 cell culture medium, supplemented with antibiotics (penicillin plus streptomycin, or gentamicin); conditioned media or factor preparations are diluted in one of these solutions (e.g., Gurdon et al., 1985; Smith, 1987; Rosa et al., 1988; Green et al., 1990; Sokol et al., 1990). When working with a purified peptide growth factor, bovine serum albumin (BSA) at a final concentration of 100 μg/ml must be present in all solutions. Diluted solutions of peptide factors should be used on the same day. Since all members of the transforming growth factor β (TGF-β) family are stored in 2 to 4 mM HCl (plus BSA, unless very concentrated), it is necessary to make sure that dilution into medium is sufficient to neutralize the acid.

The time of exposure of the animal explant to inducer that is required for induction has been studied in some detail. In the recombinate assay the minimal contact time required for the induction of accumulation of α-actin mRNA is 2.5 hours (Gurdon et al., 1985). Induction can be detected after exposure of animal caps to XTC-MIF (activin) for only 15 min (Cooke et al., 1987). When animal caps from an early blastula are treated with XTC-MIF for 1 hour followed by washing, they activate an early response gene 2 hours later at the midblastula transition (Rosa, 1989). These results suggest that animal cells rapidly bind the inducing factor so that it cannot be removed by washing, or record stably the inducing signal in some other way. Thus, exposure of competent cells to MIF acts as a rapid and stable signal that redirects the developmental fate of these cells. In practice, experiments with a molecular assay as end point are frequently incubated overnight, and one may expose the explant to the inducing medium for the entire culture period. When longer culture is planned (2–3 days), as is appropriate for observing extensive morphological differentiation, it is preferable to transfer the explants into fresh medium after exposure to inducer for several hours.

III. Assay End Points: Markers of Mesoderm Induction

A. Morphological Criteria

The simplest way of determining the effect of a mesoderm inducer is to observe the external morphology of the animal explant. Strong inducers (XTC-MIF, i.e., activin) have a rapid and profound effect on the animal explants in eliciting elongation movements that mimic the convergent extension (Keller et al., 1985) of dorsal mesoderm during gastrulation. As shown by Symes and Smith (1987), elongation is detectable in treated explants when control sibling embryos reach stage 11; this timing is only slightly dependent on the time at which exposure of the explant to MIF is initiated, suggesting that mesoderm induction follows an internal clock, a conclusion also supported by other experiments (Gurdon et al., 1985; Rosa, 1989).

Whereas the elongation of explants and the activation of several genes are fairly rapid responses to induction, observation of fully differentiated tissues requires culture of explants for 2 or more days. Cytological analysis employs staining protocols according to a variety of standard methods, for example, Feulgen light green–orange G (Cooke, 1979; Green et al., 1990), hematoxylin–eosin (Ruiz i Altaba and Melton, 1989c), or aniline blue–orange G (Grunz et al., 1988). Control uninduced animal explants form "atypical epidermis," a rather uniform group of cells that resembles epidermis but lacks the organization of the normal tissue. A good illustrated description of several mesodermal derivatives in induced explants is given by Green et al. (1990). Most easily recognized is muscle, often segmented in somitelike blocks, and notochord, characterized by vacuolated cells and, often, a rod-like structure; the identity of these tissues can be confirmed by antibody staining (see below). Unfortunately, no antibodies are available for mesodermal derivatives that are less easily identifiable from their morphology. The single-cell epithelium of the pronephros is quite distinct, but identification of mesenchyme and mesothelium has remained somewhat subjective. These two cell types have been illustrated by Dale and Slack (1987b) and by Green et al. (1990), who describe mesenchyme as a "loose network of fibroblast-like cells" and mesothelium as a "thin sheet of cells... just inside the... epidermis." These cell types are usually considered to represent ventral mesoderm.

Although the assignment of such cells to a particular region of the normal embryo is somewhat speculative, it is clear that the loose mesenchyme is quite distinct from the cells in uninduced animal explants (atypical epidermis). In addition to mesodermal cell types, induced animal explants often differentiate into neural tissue and cement gland (an anterior ectodermal derivative), possibly by a secondary induction mechanism. The identification of such products is discussed elsewhere in this volume (Chapter 18).

B. Antibody Detection

Tissue identification is greatly aided by the use of specific antibodies, but few antibodies for mesodermal derivatives are available (Table I). The tissue represented best is muscle; the availability of several antibodies allows the identification of small numbers of muscle cells within an explant, as illustrated, for example, by Gurdon (1988). The notochord-specific antibody listed in Table I recognizes keratan sulfate, a component of the extracellular matrix (Smith and Watt, 1985). Antibodies for epidermal markers may be useful in characterizing the cell types present in an explant, and two are listed in Table I. As discussed below (Section V,B) the expression of epidermal markers is turned off in animal pole cells induced under certain conditions. Antibody staining of explants is commonly done on sectioned material. It is likely that the whole-mount technique which is effective for embryos (Dent et al., 1989; Chapter 22, this volume) would also be useful in the study of induction.

C. Gene Regulation in Response to Mesoderm Induction

Induction elicits profound changes in gene expression in the responding tissue. These changes include the activation of many genes and the inactivation of others; genes that have been used in mesoderm induction studies are

TABLE I

ANTIBODIES USED IN IDENTIFICATION OF INDUCED TISSUES

Antibody	Tissue	Ref.
Activated markers		
Monoclonal 12/101[a]	Muscle	Kinter and Brockes (1984)
Monoclonal MZ15[b]	Notochord	Smith and Watt (1985)
Inactivated markers		
Monoclonal Epi 1	Epidermis	London et al. (1988)
Antipeptide Antibody 19/20	Epidermis (keratin)	Jamrich et al. (1987)

[a] This antibody recognizes intracellular components of muscle. Additional muscle-specific antibodies have been reported (e.g., myosin antibody by Slack et al., 1987).
[b] This antibody recognizes extracellular matrix.

TABLE II
Genes Affected by Mesoderm Induction

Gene	Tissue	Ref.
Activated genes		
α-Actin (cardiac actin)	Muscle	Mohunn et al. (1984, 1986, 1989); Sargent et al. (1986)
MyoD	Muscle (precursor)	Hopwood et al. (1989a)
Xtwi (homolog of twist)	Nonmuscle mesoderm[a]	Hopwood et al. (1989b)
Mix.1	Vegetal hemisphere[b]	Rosa (1989)
Xhox3	Posterior mesoderm	Ruiz i Altaba and Melton (1989a–c)
xsna (homolog of snail)	Entire marginal zone	Sargent and Bennett (1990)
Inactivated gene		
Keratins	Epidermis	Dawid et al. (1988)

[a] Includes the notochord and lateral plate mesoderm.
[b] The distribution of Mix.1 RNA is discussed in the text.

listed in Table II. A gene specific for a tissue that differentiates as a result of mesoderm induction is very useful, but it provides information about only the particular tissue rather than about mesoderm in general. In fact, a general marker for amphibian mesoderm is, so far, lacking. Of the various mesodermal tissues that arise as a result of gastrulation, only muscle is represented by a highly specific, conveniently used marker gene, which forms the basis of a quantitative assay: α-actin, a cardiac isoform that is expressed in axial muscle in embryogenesis, has been used widely in induction studies (Mohun et al., 1984; Gurdon et al., 1985; Sargent et al., 1986). A somewhat earlier marker for muscle is MyoD (Hopwood et al., 1989a), a helix–loop–helix DNA-binding factor involved in the regulation of muscle cell differentiation. Another helix–loop–helix factor, Xtwi, accumulates during embryogenesis in nonmuscle mesoderm including the notochord and lateral plate (Hopwood et al., 1989b). Thus a combination of Xtwi and MyoD may be representative of the entire mesoderm, but this point has not been established in detail.

The Mix.1 homeobox gene is activated as an immediate early response to induction, that is, it responds to XTC-MIF quickly and can do so in the presence of an inhibitor of protein synthesis and in dispersed cells (Rosa, 1989). This immediate response distinguishes Mix.1 from other marker genes. The distribution of Mix.1 mRNA during normal embryogenesis is something of a puzzle, however; it is found in the entire vegetal hemisphere, including much of the marginal zone and the entire endodermal precursor region. A possible role for Mix.1 in endoderm specification has been proposed but not established (Rosa, 1989). The homeobox gene Xhox3 responds somewhat later to induction, is expressed in posterior mesoderm in normal embryogenesis, and is involved in the determination of the anteroposterior axis during gastrula and neurula development (Ruiz i Altaba and Melton, 1989a–c). The expression of *Xhox3* is not specific for a tissue or even group of tissues but rather for a region of the embryo along one of its major axes.

It is interesting to note that, except for α-actin, all mesodermal marker genes available for studies in amphibian emrbryogenesis are regulatory molecules, and among these only *MyoD* is tissue specific. It appears that a further search for tissue-specific, early expressed markers may still be a promising exercise. The second point of interest emerges from a consideration of the expression patterns of the three non-tissue-specific, putative regulatory molecules Mix.1, Xtwi, and Xhox3. These molecules are expressed in patterns correlated with embryonic axes rather than individual tissues, and involvement in axis formation has been shown for Xhox3 (Ruiz i Altaba and Melton, 1989c). Yet the axis concerned is different for each: Mix.1, the earliest of these molecules to be expressed, aligns along the animal–vegetal axis; Xtwi, arising somewhat later, is distributed along the dorsoventral axis; and Xhox3, which accumulates at a similar or slightly later time than Xtwi, shows anteroposterior polarity. Thus, one might speculate that the major embryonic axis along which regulatory molecules align, and which they help determine, changes during the period from the blastula through the gastrula stages of embryogenesis.

IV. Mesoderm-Inducing Factors

A requirement for inductive interactions in the specification of mesoderm is indicated primarily by the work of Nieuwkoop and colleagues (see Nieuwkoop, 1973), who studied the development of animal–vegetal recombinates. Additional evidence for the requirement for induction in muscle differentiation derives from dispersion experiments (Sargent *et al.,* 1986), which are discussed further in Section V,B. The latter work, but especially transfilter induction experiments in recombinates (Grunz and Tacke, 1986; Gurdon, 1989), indicated that mesoderm-inducing factors are diffusible molecules. This implication was borne out by the identification of a powerful inducing substance secreted by the XTC cell line (Smith, 1987) and the discovery that previously characterized peptide growth factors are capable of mesoderm induction (Slack *et al.,* 1987; Kimelman and Kirschner, 1987; Rosa *et al.,* 1988).

Current knowledge about inducing factors, together with a brief reference to earlier work, is summarized in Table III. The identified inducers belong to

TABLE III

MESODERM-INDUCING FACTORS

Factor	Comments	Ref.
Identified growth factors		
Activin A, B	Highly potent; all types of mesoderm	Asashima et al. (1990); Smith et al. (1990a); van den Eijnden-Van Raaij et al. (1990); Thomsen et al. (1990); Chertov et al. (1990); Albano et al. (1990)
TGF-β3	Moderately potent; all types of mesoderm	Roberts et al. (1990)
TGF-β2	Less potent than TGF-β3	Rosa et al. (1988)
FGF[a]	Ventrolateral mesoderm	Slack et al. (1987); Kimelman and Kirschner (1987); Paterno et al. (1989)
Other inducing factors (a Selection)		
Vegetalizing factor, chick embryos (activin?)		Born et al. (1985); Tiedemann (1990)
Mesoderm inducing substance, carp swim bladder		Asashima et al. (1987)

[a] basic FGF (bFGF), acidic FGF (aFGF), and *kfgf* oncogene product (kFGF) are approximately equally active, whereas the *int-2* product is less active (Paterno et al., 1989).

two families of peptide growth factors. The transforming growth factor β (TGF-β) superfamily may be divided into several subfamilies according to sequence relationships; members of this superfamily have a remarkably broad range of effects on different cells and tissues (reviewed in Roberts and Sporn, 1990). The most effective mesoderm inducers are activin A and activin B, two related proteins that were identified initially as factors eliciting the secretion of follicle stimulatory hormone (FSH) by pituitary cells (reviewed in Vale et al., 1990). Activin A is also capable of stimulating erythroid cell differentiation and has been independently isolated as erythroid differentiation factor (EDF) (Murata et al., 1988). Mammalian activin A is a powerful inducer (Asashima et al., 1990; Chertov et al., 1990; Thomsen et al., 1990; van den Eijnden-Van Raaij et al., 1990), and so is recombinant activin B derived from a *Xenopus* activin B gene (Thomsen et al., 1990); further, the mesoderm-inducing factor from XTC cells (Smith, 1987) is a close relative of mammalian activin A and probably represents the *Xenopus* homolog of this molecule (Smith et al., 1990a). Activins are capable of eliciting mesoderm differentiation in animal explants at concentrations of 5 pM or higher (Thomsen et al., 1990). The type of response that is obtained depends on the activin concentration (Smith et al., 1988; Green and Smith, 1990; see also Section V,B). At high concentrations, activins induce dorsal mesodermal derivatives like notochord and segmented muscle, and neural tissue. Under certain conditions activins induce axial organization and anterior development in animal explants (Sokol et al., 1990; Thomsen et al., 1990).

Among the other members of the TGF-β superfamily two isoforms, TGF-β2 and TGF-β3, have been shown to act as mesoderm inducers, (Rosa et al., 1988; Roberts et al., 1990), albeit with lower effectiveness than activin A. It remains to be determined whether TGF-β2 and TGF-β3 act in a distinct, independent way from activin or whether they cross-react at lower efficiency with the activin receptor. Other members of the TGF-β superfamily, such as Vg1 (Weeks and Melton, 1987) have been considered as possible mesoderm inducers, but to date no confirmed positive results are available.

Several members of the fibroblast growth factor (FGF) family have mesoderm inducing activity, with bFGF, aFGF, and kFGF having similar potency, whereas the distantly related *int-2* product is less effective (Slack et al., 1987; Kimelman and Kirschner, 1987; Paterno et al., 1989). FGF induces "ventral" mesoderm, including mesenchyme, mesothelium, and some muscle. The inability of FGF to induce a full range of mesodermal tissues suggests that it cannot by itself account for *in vivo* mesoderm induction. However, bFGF and its mRNA are present in the embryo (Kimelman et al., 1988; Slack and Isaacs, 1989), and so is at least one form of FGF receptor and its mRNA (Gillespie et al., 1989; Musci et al., 1990; R. E. Friesel and I. B. Dawid, 1991). These observations indicate that FGF is likely to be active in mesoderm specification *in vivo* even if the additional involvement of at least one member of the TGF-β family, probably activin, is almost certainly required.

Additional factor preparations that are capable of inducing mesodermal tissues in competent ectoderm have been reported in the past (for review, see Gurdon, 1987). Among these, the most extensively studied has been the vegetalizing factor isolated from chick embryos (Born et al., 1985; Tiedemann, 1990). Since both its biological effects and its physical properties are similar to those of activins, and since activin B has been shown to be expressed in chick embryos (Mitrani et al., 1990), it appears possible that vegetalizing factor is a chicken activin. Vegetalizing factor is reported to induce endoderm in addition to mesoderm (see Tiedemann, 1990); whether this is also true for activins remains to be established.

V. Specification of Cell Fate, Organization of Tissues, and Axis Determination

One may ask what an inducer ultimately does at the biological level: Does it determine the differentiation of individual cells by turning a competent cell with an autonomous ectodermal fate into a muscle, kidney, or notchord cell? Or does it elicit the formation of organized tissues, for example, segmented muscle? Or does it establish dorsoventral and anteroposterior axes in an explant, with tissue differentiation a consequence thereof? There are indications that mesoderm inducers can do all of these things to some extent; which one is most apparent in a particular experiment depends on the experimental circumstances, especially the type and context of the target cells that are being induced.

A. Cell Differentiation versus Axis Determination

Most studies on mesoderm induction have dealt with the differentiation of cell types and have relied on morphological or biochemical identification of muscle, notochord, etc. Whereas the formation of segmented muscle has been often observed (e.g., Green et al., 1990), most induced explants are not obviously organized along a major axis (see references in Smith, 1989; Dawid et al., 1990). Sokol et al. (1990) and Thomsen et al. (1990) have emphasized their observation that animal cap explants treated with mammalian activin A or Xenopus recombinant activin B yield "embryoids" with a rudimentary anteroposterior axis and, often, anterior development resulting in brain cavities and eyes.

It is possible that the differences between these and earlier observations are due to the use of different factors. This interpretation is uncertain because most induction experiments, although using different factor preparations, are now known to have employed various forms of activin, and, when tested in any one laboratory, different activins appear to have very similar effects. Since the explants used by Sokol et al. (1990) are larger than those used by most other authors (e.g., Smith, 1987), it is possible that this fact contributes to the different results. If this were the case it follows that different animal hemisphere cells display different competence, and the differentiation achieved in response to a signal is modified by these differences. Clearly, explants from any region of the animal hemisphere can respond to activin by differentiating mesodermal cell types; it is possible, however, that only a certain region, perhaps tissue that is equator-proximal and dorsal, can respond to the signal by organizing an axis. If this speculation is correct it follows that animal hemisphere cells, which by themselves cannot differentiate any mesodermal tissue, nevertheless already exhibit polarity. Although it has been shown that activin can give a dorsalizing and anteriorizing signal (Green et al., 1990; Ruiz i Altaba and Melton, 1989b,c), it would seem that the polarity information in animal explants is not likely to be due to a localized store of activin, or the explants would self-differentiate. Thus, it is possible that a type of dorsal information, which is not an activin source, exists in blastula stage embryos. Such information could, for example, be a gradient of activin receptors, but other possibilities can be envisioned. It is clear that issues of cellular competence and polarity in the animal hemisphere are important questions for future study.

B. Use of Dissociated Embryonic Cells in the Study of Induction

Embryonic cells can be dissociated by removal of calcium ions from the medium (Gurdon et al., 1984). When the resulting nonadhesive cells were dispersed so that interactions were prevented, the eventual expression of the muscle-specific actin gene was totally suppressed, implying that cell interactions are a necessary event in muscle cell determination (Sargent et al., 1986). Epidermal keratin genes were expressed under dispersion conditions, suggesting that epidermal differentiation is a cell-autonomous function. It appeared likely that dispersion had prevented an inducing factor(s) secreted by vegetal cells from impinging on equatorial/animal cells, precluding their induction to muscle. To test this proposition, Symes et al. (1988) added XTC-MIF (activin) to dispersed animal cap cells. They found that these cells turned off an epidermal keratin gene but did not express their α-actin gene unless they were reaggregated. Reaggregation was carried out after the inducer was removed and at a time when the cells had lost their competence to respond to activin. Thus, one may envision a two-step process: Activin does induce individual cells away from ectoderm and toward muscle (and other mesoderm), but the cells cannot express this determination state unless they are reaggregated. This phenomenon may be related to the "community effect," the observation that very small groups of cells cannot be induced toward muscle even when surrounded by inducing vegetal cells, whereas larger groups respond (Gurdon, 1988).

The dispersed cell system has been utilized further by Green and Smith (1990) to show that animal cells respond with distinct differentiation events to different concentrations of activin, exhibiting sharp thresholds. These observations indicate that a gradient of inducer could generate an organized series of distinct tissues in the embryo, namely, an axis.

The application of inducing factors to dispersed cells may be more effective than application to whole explants because each responding cell is exposed to a uniform factor level. Problems of penetration into tissue are thus circumvented. Further, properties of single embryonic cells can be studied, for example, their behavior on an extracellular matrix (Smith et al., 1990b). In addition, this method allows one to study requirements for differentiation in addition to the inducing factors themselves.

VI. Outlook

It appears that one question in mesoderm induction, what are the primary inducers, has been or will shortly be answered. Yet many questions remain. Are there secondary signals for tissue differentiation in addition to inducing factors, as the dispersion experiments suggest? How are the signals given by the inducing factors received and transmitted? Cell surface receptors are just the first step in a putative hierarchy executing the response. Perturbation of known signal systems may provide information on the intermediate steps in mesoderm induction pathways, as suggested by the effects of injecting inositol into the embryo (Busa and Gimlich, 1989) and by expressing the polyoma middle T antigen (Whitman and Melton, 1989b). Regulation of gene expression is a major consequence of induction; studies of early response genes (Rosa, 1989) and of the control of expression of tissue-specific genes (Mohun et al., 1986, 1989; Hopwood et al., 1989a) indicate directions that will continue to yield useful results. An early response to induction is a change in cell shape and adhesivesness, followed by tissue deformation (Symes and Smith, 1987; Cooke et al., 1987). These observations suggest that studies on cell adhesion molecules and, possibly, cytoskeletal components during induction are likely to be illuminating.

Finally, it should be noted that mesoderm-inducing factor from chick embryos, named vegetalizing factor, has long been reported to induce endoderm in addition to mesoderm (reviewed in Tiedemann, 1990). The involvement of XTC-MIF (activin) in endoderm differentiation was suggested by Rosa (1989) on the basis of the expression of the early response homeobox gene *Mix.1* in the vegetal hemisphere, that is, in future endoderm. At the other extreme of the dorsoventral axis, activin is known to induce neural tissue in animal explants derived from blastula. This observation has usually been interpreted to point to an indirect effect in which mesoderm is induced first and in turn induces neural tissue in remaining ectoderm. In fact, Ruiz i Altaba and Melton (1989b) have shown that animal caps induced by activin or FGF are capable of acting as "organizers" to induce anterior or posterior differentiation, respectively, in host embryos. This suggests that inducing factors impart polarity information to mesoderm which in turn directs the formation of a nervous system with appropriate axial polarity. Other experiments, however, imply that activin might have direct neuralizing effects (see Sokol et al., 1990). The delineation of the entire range of biological effects of activins, and mesoderm inducing factors in general, is thus an important aim of future research.

Acknowledgments

I thank Peter Good and Mike Rebagliati for valuable comments on the text.

References

Albano, R. M., Godsave, S. F., Huylebroeck, D., Van Nimmen, K., Isaacs, H. V., Slack, J. M. W., and Smith, J. C. (1990). A mesoderm-inducing factor produced by WEHI-3 murine myelomonocytic leukemia cells is activin A. *Development (Cambridge, U.K.)* **110,** 435–443.

Asashima, M., Nakano, H., Matsunaga, T., Sugimoto, M., and Takano, H. (1987). Purification of mesodermal-inducing substances from carp swim bladders. I. Extraction and isoelectric focusing. *Dev Growth Differ.* **29,** 221–227.

Asashima, M., Nakano, H., Shimada, K., Kinoshita, K., Ishii, K., Shibai, H., and Ueno, N. (1990). Mesodermal induction in early amphibian embryos by activin A (erythroid differentiation factor). *Roux's Arch. Dev. Biol.* **198,** 330–335.

Born, J., Hoppe, P., Schwarz, W., Tiedemann, H., Tiedemann, H., and Wittmann-Liebold, B. (1985). An embryonic inducing factor: Isolation by high performance liquid chromatography and chemical properties. *Biol. Chem. Hoppe-Seyler* **366,** 729–735.

Busa, W. B., and Gimlich, R. L. (1989). Lithium-induced teratogenesis in frog embryos prevented by a polyphoshoinositide cycle intermediate or a diacylglycerol analog. *Dev. Biol.* **132,** 315–324.

Chertov, O. Y., Krasnoselskii, A. L., Bogdanov, M. E., and Hoperskaya, O. A. (1990). Mesoderm-inducing factor from bovine amniotic fluid: Purification and N-terminal sequence. *Biomed. Sci.* **1,** 499–506.

Cooke, J. (1979). Cell number in relation to primary pattern formation in the embryo of *Xenopus laevis.* I: The cell cycle during new pattern formation in response to implanted organisers. *J. Embryol. Exp. Morphol.* **51,** 165–182.

Cooke, J., Smith, J. C., Smith, E. J., and Yaqoob, M. (1987). The organization of mesodermal pattern in *Xenopus laevis*: Experiments using a *Xenopus* mesoderm-inducing factor. *Development (Cambridge, U.K.)* **101,** 893–908.

Dale, L., and Slack, J. M. W. (1987a). Fate map for the 32-cell stage of *Xenopus laevis. Development (Cambridge, U.K.)* **99,** 527–551.

Dale, L., and Slack, J. M. W. (1987b). Regional specification within the mesoderm of early embryos of *Xenopus laevis. Development (Cambridge, U.K.)* **100,** 279–295.

Dawid, I. B., Rebbert, M. L., Rosa, F., Jamrich, J., and Sargent, T. D. (1988). Gene expression in amphibian embryogenesis. *In* "Regulatory Mechanisms in Developmental Processes" (G. Eguchi, T. S. Okada, and L. Saxén, eds.), pp. 67–74. Elsevier, Ireland.

Dawid, I. B., Sargent, T. D., and Rosa, F. (1990). The role of growth factors in embryonic induction in amphibians. *Curr. Top. Dev. Biol.* **24,** 261–288.

Dent, J. A., Polson, A. G., and Klymkowsky, M. W. (1989). A whole-mount immunocytochemical analysis of the expression of the intermediate filament protein vimentin in *Xenopus. Development (Cambridge, U.K.)* **105,** 61–74.

Friesel, R. and Dawid, I. B. (1991). cDNA cloning and developmental expression of fibroblast growth factor receptors from *Xenopus laevis. Mol. Cell. Biol.* **11,** 2481–2488.

Gillespie, L. L., Paterno, G. D., and Slack, J. M. W. (1989). Analysis of competence: Receptors for fibroblast growth factor in early *Xenopus* embryos. *Development (Cambridge, U.K.)* **106**, 203–208.

Gimlich, R. L. (1986). Acquisition of developmental autonomy in the equatorial region of the *Xenopus* embryo. *Dev. Biol.* **115**, 340–352.

Green, J. B. A., and Smith, J. C. (1990). A *Xenopus* activin A homologue acts on embryonic cells like a morphogen in eliciting multiple fates separated by dose thresholds. *Nature (London)* **347**, 391–394.

Green, J. B. A., Howes, G., Symes, K., Cooke, J., and Smith, J. C. (1990). The biological effects of XTC-MIF: Quantitative comparison with *Xenopus* bFGF. *Development (Cambridge, U.K.)* **108**, 173–183.

Grunz, H., and Tacke, L. (1986). The inducing capacity of the presumptive endoderm of *Xenopus laevis* studied by transfilter experiments. *Wilhelm Roux's Arch. Dev. Biol.* **195**, 467–473.

Grunz, H., McKeehan, W. L., Knöchel, W., Born, J., Tiedemann, H., and Tiedemann, H. (1988). Induction of mesodermal tissues by acidic and basic heparin binding growth factors. *Cell Differ.* **22**, 183–190.

Gurdon, J. B. (1987). Embryonic induction—Molecular prospects. *Development (Cambridge, U.K.)* **99**, 285–306.

Gurdon, J. B. (1988). A community effect in animal development. *Nature (London)* **336**, 772–774.

Gurdon, J. B. (1989). The localization of an inductive response. *Development (Cambridge, U.K.)* **105**, 27–33.

Gurdon, J. B., Brennan, S., Fairman, S., and Mohun, T. J. (1984). Transcription of muscle-specific actin genes in early *Xenopus* development: Nuclear transplantation and cell dissociation. *Cell (Cambridge, Mass.)* **38**, 691–700.

Gurdon, J. B., Fairman, S., Mohun, T. J., and Brennan, S. (1985). Activation of muscle-specific actin genes in *Xenopus* development by an induction between animal and vegetal cells of a blastula. *Cell (Cambridge, Mass.)* **41**, 913–922.

Hopwood, N. D., Pluck, A., and Gurdon, J. B. (1989a). MyoD expression in the forming somites is an early response to mesoderm induction in *Xenopus* embryos. *EMBO J.* **8**, 3409–3417.

Hopwood, N. D., Pluck, A., and Gurdon, J. B. (1989b). A *Xenopus* mRNA related to *Drosophila twist* is expressed in response to induction in the mesoderm and the neural crest. *Cell (Cambridge, Mass.)* **59**, 893–903.

Jamrich, M., Sargent, T. D., and Dawid, I. B. (1987). Cell-type-specific expression of epidermal cytokeratin genes during gastrulation of *Xenopus laevis*. *Genes Dev.* **1**, 124–132.

Jones, E. A., and Woodland, H. R. (1987). The development of animal cap cells in *Xenopus*: A measure of the start of animal cap competence to form mesoderm. *Development (Cambridge, U.K.)* **101**, 557–563.

Keller, R. E., Danilchik, J., Gimlich, R., and Shih, J. (1985). The function and mechanism of convergent extension during gastrulation of *Xenopus laevis*. *J. Embryol. Exp. Morphol.* **89** (Suppl.), 185–209.

Kimelman, D., and Kirschner, M. W. (1987). Synergistic induction of mesoderm by FGF and TGF-beta and the identification of an mRNA coding for FGF in the early *Xenopus* embryo. *Cell (Cambridge, Mass.)* **51**, 869–877.

Kimelman, D., Abraham, J. A., Haaparanta, T., Palisi, T. M., and Kirschner, M. W. (1988). The presence of FGF in the frog egg: Its role as natural mesoderm inducer. *Science* **242**, 1053–1056.

Kintner, C. R., and Brookes, J. P. (1984). Monoclonal antibodies identify blastemal cells derived from dedifferentiating muscle in newt limb regeneration. *Nature (London)* **308**, 67–69.

London, C., Akers, R., and Phillips C. (1988). Expression of Epi 1, an epidermis-specific marker in *Xenopus laevis* embryos, is specified prior to gastrulation. *Dev. Biol.* **129**, 380–389.

Mitrani, E., Ziv, T., Thomsen, G., Shimoni, Y., Melton, D. A., and Bril, A. (1990). Activin can induce the formation of axial structures and is expressed in the hypoblast of the chick. *Cell (Cambridge, Mass.)* **63**, 495–501.

Mohun, T. J., Brennan, S., Dathan, N., Fairman, S., and Gurdon, J. B. (1984). Cell type-specific activation of actin genes in the early amphibian embryo. *Nature (London)* **311**, 716–721.

Mohun, T. J., Garrett, N., and Gurdon, J. B. (1986). Upstream sequences required for tissue-specific activation of the cardiac actin gene in *Xenopus laevis* embryos. *EMBO J.* **5**, 3185–3193.

Mohun, T. J., Taylor, M. V., Garrett, N., and Gurdon, J. B. (1989). The CArG promoter sequence is necessary for muscle-specific transcription of the cardiac actin gene in *Xenopus* embryos. *EMBO J.* **8**, 1153–1161.

Moody, S. A. (1987). Fates of the blastomeres of the 32-cell-stage *Xenopus* embryo. *Dev. Biol.* **122**, 300–319.

Murata, M., Eto, Y., Shibai, H., Sakai, M., and Muramatsu, M. (1988). Erythroid differentiation factor is encoded by the same mRNA as that of the inhibin βA chain. *Proc. Natl. Acad. Sci. U.S.A.* **85**, 2434–2438.

Musci, T. J., Amaya, E., and Kirschner, M. W. (1990). Regulation of the fibroblast growth factor receptor in early *Xenopus* embryos. *Proc. Natl. Acad. Sci. U.S.A.* **87**, 8365–8369.

Nieuwkoop, P. D. (1973). The "organizing center" of the amphibian embryo: Its origin, spatial organization, and morphogenetic action. *Adv. Morphog.* **10**, 1–39.

Paterno, G. D., Gillespie, L. L., Dixon, M. S., Slack, J. M. W., Heath, J. K. (1989). Mesoderm-inducing properties of Int-2 and kFGF: Two oncogene encoded growth factors related to FGF. *Development (Cambridge, U.K.)* **106**, 79–83.

Roberts, A. B., and Sporn, M. B. (1990). The transforming growth factors-β. *In* "Handbook of Experimental Pharmacology: Peptide Growth Factors and Their Receptors" (M. B. Sporn and A. B. Roberts, eds.), Vol. 95/I, pp. 419–472. Springer-Verlag, Heidelberg.

Roberts, A. B., Kondaiah, P., Rosa, F., Watanabe, S., Good, P., Danielpour, D., Roche, N. S., Rebbert, M. L., Dawid, I. B., and Sporn, M. B. (1990). Mesoderm induction in *Xenopus laevis* distinguishes between the various TGF-β isoforms. *Growth Factors* **3**, 277–286.

Rosa, F. (1989). Mix.1, an homeobox mRNA inducible by mesoderm inducers, is expressed mostly in the presumptive endodermal cells of *Xenopus* embryos. *Cell (Cambridge, Mass.)* **57**, 965–974.

Rosa, F., Roberts, A. B., Danielpour, D., Dart, L. L., Sporn, M. B., and Dawid, I. B. (1988). Mesoderm induction in amphibians: The role of TGF-beta2-like factors. *Science* **239**, 783–785.

Ruiz i Altaba, A., and Melton, D. A. (1989a). Bimodal and graded expression of the *Xenopus* homeobox gene *Xhox3* during embryonic development. *Development (Cambridge, U.K.)* **106**, 173–183.

Ruiz i Altaba, A., and Melton, D. A. (1989b). Interaction between peptide growth factors and homeobox genes in the establishment of antero-posterior polarity in frog embryos. *Nature (London)* **341**, 33–38.

Ruiz i Altaba, A., and Melton, D. A. (1989c). Involvement of the *Xenopus* homeobox gene *Xhox3* in pattern formation along the anterior–posterior axis. *Cell (Cambridge, Mass.)* **57**, 317–326.

Sargent, M. G., and Bennett, M. F. (1990). Identification in *Xenopus* of a structural homologue of the *Drosophila* gene *snail*. *Development (Cambridge, U.K.)* **109**, 967–973.

Sargent, T. D., Jamrich, M., and Dawid, I. B. (1986). Cell interactions and the control of gene activity during early development of *Xenopus laevis*. *Dev. Biol.* **114**, 238–246.

Slack, J. M. W., and Isaacs, H. (1989). Presence of basic fibroblast growth factor in the early *Xenopus* embryo. *Development (Cambridge, U.K.)* **105**, 147–153.

Slack, J. M. W., Darlington, B. G., Heath, J. K., and Godsave, S. F. (1987). Mesoderm induction in early *Xenopus* embryos by heparin-binding growth factors. *Nature (London)* **326**, 197–200.

Smith, J. C. (1987). A mesoderm-inducing factor is produced by a *Xenopus* cell line. *Development (Cambridge, U.K.)* **99**, 3–14.

Smith, J. C. (1989). Mesoderm induction and mesoderm-inducing factors in early amphibian development. *Development (Cambridge, U.K.)* **105,** 665–677.

Smith, J. C., and Watt, F. M. (1985). Biochemical specificity of *Xenopus* notochord. *Differentiation* **29,** 109–115.

Smith, J. C., Yaqoob, M., and Symes, K. (1988). Purification, partial characterization and biological effects of the XTC mesoderm-inducing factor. *Development (Cambridge, U.K.)* **103,** 591–600.

Smith, J. C., Price, B. M. J., Van Nimmen, K., and Huylebroeck, D. (1990a). Identification of a potent *Xenopus* mesoderm-inducing factor as a homologue of activin A. *Nature (London)* **345,** 729–731.

Smith, J. C., Symes, K., Hynes, R. O., and DeSimone, D. (1990b). Mesoderm induction and the control of gastrulation in *Xenopus laevis*: The roles of fibronectin and integrins. *Development (Cambridge, U.K.)* **108,** 229–238.

Sokol, S., Wong, G. G., and Melton, D. A. (1990). A mouse macrophage factor induces head structures and organizes a body axis in *Xenopus*. *Science* **249,** 561–564.

Sudarvati, S., and Nieuwkoop, P. D. (1971). Mesoderm formation in the anuran *Xenopus laevis* (Daudin). *Wilhelm Roux's Arch. Dev. Biol.* **166,** 189–204.

Symes, K., and Smith, J. C. (1987). Gastrulation movements provide an early marker of mesoderm induction in *Xenopus laevis*. *Development (Cambridge, U.K.)* **101,** 339–349.

Symes, K., Yaqoob, M., and Smith, J. C. (1988). Mesoderm induction in *Xenopus laevis*: Responding cells must be in contact for mesoderm formation but suppression of epidermal differentiation can occur in single cells. *Development (Cambridge, U.K.)* **104,** 609–618.

Thomsen, G., Woolf, T., Whitman, M., Sokol, S., Vaughan, J., Vale, W., and Melton, D. A. (1990). Activins are expressed early in *Xenopus* embryogenesis and can induce axial mesoderm and anterior structures. *Cell (Cambridge, Mass.)* **63,** 485–493.

Tiedemann, H. (1990). Cellular and molecular aspects of embryonic induction. *Zool. Sci.* **7,** 171–186.

Vale, W., Hsueh, A., Rivier, C., and Yu, J. (1990). The inhibin/activin family of hormones and growth factors. *In* "Handbook of Experimental Pharmacology: Peptide Growth Factors and Their Receptors" (M. B. Sporn and A. B. Roberts, eds.), Vol. 95/II, pp. 211–248. Springer-Verlag, Heidelberg.

van den Eijnden-Van Raaij, A. J. M., van Zoelent, E. J. J., van Nimmen, K., Koster, C. H., Snoek, G. T., Durston, A. J., and Huylebroeck, D. (1990). Activin-like factor from a *Xenopus laevis* cell line responsible for mesoderm induction. *Nature (London)* **345,** 732–734.

Weeks, D. L., and Melton, D. A. (1987). A maternal mRNA localized to the vegetal hemisphere in *Xenopus* eggs codes for a growth factor related to TGFβ. *Cell (Cambridge, Mass.)* **51,** 861–867.

Whitman, M., and Melton, D. A. (1989a). Growth factors in early embryogenesis. *Annu. Rev. Cell Biol.* **5,** 93–117.

Whitman, M., and Melton, D. A. (1989b). Induction of mesoderm by a viral oncogene in early *Xenopus* embryos. *Science* **244,** 803–836.

Chapter 18

Neural Induction

CAREY R. PHILLIPS

Bowdoin College
Brunswick, Maine 04011

I. Neural Induction: Historical Perspective
II. Recent Experimental Approaches to Neural Induction
    A. Assays for Competence of Ectoderm to Form Neural Tissue
    B. Developmental Stage When Ectoderm Is Normally Induced in *Xenopus*
    C. Developmental Time When Neurectoderm Is Regionally Specified
    D. Summary of Neural Competence, Timing of Induction, and Regional Specification
III. Signals Initiating Neural Induction
    A. Early Neural Bias
    B. Signals from Invaginated Dorsal Mesoderm
IV. Methods
    A. Einsteck Method
    B. Sandwich Method
    C. Exogastrula
    D. Edge-on Grafting or Aging of Tissues in Culture
    E. Keller Sandwiches
    F. Dissociation and Reaggregation
V. Summary
    References

I. Neural Induction: Historical Perspective

Studies on induction and pattern formation of the early neural system became a very popular pastime following the work of Hilda Mangold and Hans Spemann (Spemann and Mangold, 1924). Mangold found that a blastopore lip transplanted to the ventral side of a host embryo will induce a second dorsal axis. Subsequent studies showed that transplanted blastopore lip cells usually differentiate into a portion of the dorsal mesoderm structures appearing on the ventral side of the host embryo and that the remainder of the dorsal mesoderm structures derive from host ventral mesoderm. The neural tissue of the secondary axis originate almost entirely from the host ventral ectoderm, normally fated to become epidermal tissue (Gimlich and Cooke, 1983; Smith and Slack, 1983). Therefore, the transplanted blastopore lip tissue appears to have the ability to "organize" the surrounding host tissue to develop well-patterned dorsal structures, including a well-organized neural system (Spemann, 1938).

Spemann proposed two possible mechanisms to explain how the blastopore lip region induces ventral ectoderm to form neural tissue. One potential mechanism requires that the implanted dorsal blastopore lip region induces adjacent ventral marginal zone cells to become functional dorsal mesoderm. The dorsal mesoderm then undergoes the convergent extension movements of gastrulation and eventually comes to underlie the ventral ectoderm. The induced dorsal mesoderm, in turn, induces the overlying ventral ectoderm to become neural plate (Spemann, 1938). The second possibility discussed by Spemann was that the dorsal blastopore lip region sends a set of signals capable of inducing both the adjacent ectoderm and the adjacent mesoderm to form dorsal structures; the ventral mesoderm forms structures such as notochord and somites while the ventral ectoderm simultaneously forms neural plate. Spemann further suggested that signals directing the induction of neural plate might originate from the dorsal blastopore lip region and travel through the plane of the ectoderm (Spemann, 1938; see Hamburger, 1988).

Two of Spemann's students provided evidence suggesting that signals for neural induction were produced solely by invaginated dorsal mesoderm and that direct contact between dorsal mesoderm and overlying ectoderm was required in order for the signals to be transmitted. Holtfreter (1933) placed gastrulating embryos in a hypertonic solution (see Section IV), forcing them to exogastrulate. The ectoderm eventually forms a hollow sac, connected by only a thin isthmus of tissue to the mesoderm. The ectoderm produced no morphologically observable neural tissue. Holtfreter concluded that the induction of ectoderm to form neural tissue required direct contact with the underlying dorsal mesoderm and that an edgewise, or planar, contact was not sufficient to induce neural structures (Holtfreter, 1933).

Otto Mangold (1933) provided further experimental evidence bearing on the question of the origins of neural induction signals. Mangold (1933) found that by inserting portions of the invaginated dorsal mesoderm into the ventral blastocoel of a host *Urodele* embryo, a method often referred to as "einsteck," some secondary dorsal axes were induced. Mangold also demonstrated that the type of neural tissue induced was appropriate to the type of involuted dorsal mesoderm inserted into the blastocoel; posterior mesoderm induced spinallike structures, and anterior dorsal mesoderm induced headlike structures (Mangold, 1933). The "textbook" interpretation of these two experiments is that the invaginated dorsal mesoderm induces the overlying ectoderm to follow a neural pathway and that the mesoderm also controls the type and pattern of neural tissue which ultimately forms.

The organizing abilities attributed to the dorsal blastopore lip region attracted much scientific curiosity. Several laboratories became engaged in the quest for signals responsible for inducing and organizing the dorsal structures. The blastopore lip was found to be active as an organizer even when the cells were killed by alcohol, heat, or freezing. Many other animal tissues were also found to have organizing activity in urodeles. It was first suggested by Holtfreter that the induction was caused by a chemical substance. Various sterols were tested and found to be active inducers (Shen, 1939). Acids such as oliec, linolenic, nucleic, and adenylic acids were also found to have some inductive properties in urodeles (Fischer et al., 1935). Kaolin and p-quinone were reported to be active inducers (Okada, 1938; Fujii, 1945). Barth (1937, 1939) also reported that cephalin, a phospholipid from cow brain, had inducing capacity. There became so many foreign chemical substances which elicited a dorsal induction response that many in the field lost interest, feeling that this plethora of chemicals may simply act as nonspecific inductors.

The original experiments by Holtfreter and O. Mangold have led investigators to believe that invaginated dorsal mesoderm is the sole source of neural induction signals. However, recent advances in molecular technology have allowed for a reexamination of this question. An investigator is no longer required to rely on the differentiation of a morphologically recognizable structure in order to ascertain if an induction event has occurred, a process which frequently requires multiple steps, many interactions at the appropriate time, and often a lengthy period to complete. Investigators can now use early neural-specific molecular markers to study the initial or more intermediate events involved in the neural induction pathway. The synthesis of an early tissue-specific molecule is interpreted to mean that an early induction event has occurred. In this chapter, we examine methods used to address questions about the ability of ectoderm to respond to neural induction signals (competence) as well as the possible sources of neural induction signals and their modes of delivery.

II. Recent Experimental Approaches to Neural Induction

Neural induction can be divided into two very broad and interrelated sets of events. First, ectoderm, the tissue normally induced to become neural, must have the ability to respond to neural-inducing signals. This ability is defined as having "neural competence." The second event necessary for neural induction includes the interactions or signals, both cellular and molecular, causing competent ectoderm to follow a neural pathway. These interactions may include a series of events during which ectodermal cells become committed to a neural pathway and, finally, committed to becoming distinct tissue types within the neural pathway.

A. Assays for Competence of Ectoderm to Form Neural Tissue

Neural competence will be defined, for our purposes, as the ability of ectoderm to respond to an inductive influence by forming a neural plate-like structure or synthesizing neural-specific molecules. This discussion is therefore restricted to the events involved in the induction of the neural plate, at the expense of excluding discussions on the induction of other neurallike structures (sensory placodes). It should also be pointed out that the ability of a region of ectoderm to form neural structures does not preclude this same portion of ectoderm from possessing the competence to form other tissue types, a term referred to as potency by Slack (1983) and Jacobson and Sater (1988).

Neural competence of ectoderm was originally studied using two microsurgical transplantation techniques. In the earliest studies, a portion of the presumptive neural plate and adjacent presumptive epidermis was removed and rotated so that the presumptive epidermal region comes to lie within the region which is normally induced to form neural structures, while the presumptive neural region resides outside the area fated to form neural plate (Lehmann, 1926). Lehmann found that the ectodermal tissues of *Triton* embryos had lost the ability to form neural tissue by about midgastrula. Otto Mangold (1929) did a variation of this experiment by transplanting portions of the urodele neural plate into the ventral blastocoel of a host urodele embryo and scoring for the induction of neural tissues arising from the host ventral ectoderm. Mangold reported that the ability of the ventral ectoderm to respond to neural induction signals becomes abated by mid- to late gastrula.

A second method used to determine the state of neural competence utilized the dorsal blastopore lip as a source of neural induction signals. A dorsal blastopore lip was transplanted into the ventral region of a host embryo in order to test the ability of the adjacent ectoderm to make neural tissues. Again, it was found that ventral ectoderm loses neural competence by midgastrula. In addition, it was found that neural competence was first lost in the posteriormost regions of the ventral ectoderm and subsequently lost in the more anterior regions (Machemer, 1932; Schechtman, 1938). There is also a progressive loss of competence to respond to neural induction signals starting at the midventral portion of the ectoderm at midgastrula and proceeding to the dorsal lateral regions (Machemer, 1932). Holtfreter (1938) studied the loss of competence within the ventral ectoderm of *Triton alpestris*. Dissected ectoderm from early gastrula was incubated *in vitro* for varying lengths of time and then implanted into the presumptive neural plate region of either *Triton* or *Bombinator*. Holtfreter (1938) found that an *in vitro* incubation period in a neutral salt solution (Holtfreter's solution, see Appendix A, this volume) equivalent to midgastrula was sufficient for the ectoderm to have lost neural competence.

The question of neural competence in *Xenopus* has recently been reexamined. Tissue grafts were prelabeled with fluorescein–dextran (FLDx) cell lineage tracers, which allow the investigator to be more conclusive as to the source of the responding tissues. Antibodies for the expression of tissue-specific molecules were used instead of relying on morphology as an assay for the ability of tissues to respond to inductive signals. The developmental timing for the loss of neural competence in *Xenopus* ventral ectoderm was determined by grafting FLDx-labeled ventral ectoderm into the presumptive lens region, just lateral to the anterior neural plate, and assaying for the synthesis of a neural specific molecule (NCAM, neural cell adhesion molecule). Early gastrula ventral ectoderm expresses NCAM and exhibits a neural morphology when grafted into the presumptive lens region (Servetnick and Grainger, 1990). However, ventral ectoderm older than midgastrula fails to form a neural plate and forms lens tissue instead. Therefore, ectoderm has lost neural competence by midgastrula. Neural competence is also lost in ventral ectoderm isolated at stage 10 (early gastrula) and incubated in 100% Steinberg's solution (see Appendix A, this volume) for periods longer than the equivalent of midgastrula (Servetnick and Grainger, 1990). Therefore, it appears that the loss of neural competence in ventral ectoderm is an autonomous function of the ectoderm.

Thus, it appears that ectoderm autonomously loses the ability to respond to neural induction signals by midgastrula. It also appears that the loss is not uniform within the entire ectodermal cap and that the posterior ventral–midline regions lose the ability first. However, it has been reported that competence for the induction of neural-specific molecules which appear later in neural development (Xif 6, Xif 3, and Xlhbox6) can still be induced from ectoderm as late as midneurula (Sharpe and Gurdon, 1990).

B. Developmental Stage When Ectoderm Is Normally Induced in *Xenopus*

The developmental stage when neural plate becomes specified (can form the appropriate tissue when isolated and grown in culture; see Slack, 1983; Jacobson and Sater, 1988) was assayed by removing animal caps during successive stages of gastrulation. The explanted animal caps were cultured in a neutral salt solution until midneurula, fixed, and assayed for the expression of NCAM (a neural cell adhesion molecule). Animal cap ectoderm removed prior to stage 11 does not express NCAM when cultured, whereas those removed after midgastrula (Stage 11) will express NCAM in culture (Jacobson and Rutishauser, 1986; Phillips and Doniach, 1991). Caution should be exercised in experiments of this nature. The region of the animal cap which normally involutes over the dorsal blastopore lip [involuting marginal zone (IMZ) as described in Chapter 5, this volume] has the ability to induce the synthesis of NCAM in adjacent ectoderm even though it has not yet involuted (Phillips and Doniach, 1991). Therefore, care must be taken to ensure that the animal cap tissue being tested for specification of neural plate does not include any tissue fated to involute because this tissue can act as a neural inductor.

C. Developmental Time When Neurectoderm Is Regionally Specified

While a general neural induction has occurred by midgastrula, at least some areas of the neural plate have become regionally specified by early neural plate stage (stage 13) in *Xenopus*. Engrailed, a neural-specific molecule, is expressed at the early neural plate stage in a small group of cells where the junction between the midbrain and hindbrain will eventually form (Hemmati-Brivanlou and Harland, 1989). Therefore, some level of regional specification has already been established at the beginning of neural plate stage. In addition, Jacobson (1959) grafted regions of the presumptive forebrain ectoderm into the midbrain region of axolotl and demonstrated that many histological structures are already specified by early neural plate stages.

Therefore, it appears that the anterior neural plate is specified by late gastrula to early neural plate stage and that even changing the mesodermal environment does not effect the developmental pattern of the neurectoderm within the anterior neural plate (Kallen, 1965). However, it was also found that regions of the anterior neurectoderm moved to the spinal region will become spinallike instead of maintaining their anterior characteristics. These experiments suggest that the anterior neural plate is regionally specified by the early neural plate stage, independent of any information from underlying anterior dorsal mesoderm. It also appears, however, that general anterior–posterior characteristics of the neurectoderm are still subject to modification by underlying dorsal mesoderm as late as the neural plate stage. Therefore, it is possible that the underlying dorsal mesoderm provides only general anterior–posterior information during late gastrula and early neural plate stages and that regional specification of the neurectoderm requires other sources of information.

D. Summary of Neural Competence, Timing of Induction, and Regional Specification

The experiments discussed so far indicate that ectoderm loses its ability to form neural tissue (neural competence) by midgastrula. It also appears that the loss of neural competence in some amphibians does not occur simultaneously throughout the ectoderm, but begins in the posterior ventral region and progresses to the dorsal anterior regions. On the other hand, the neural plate, the involuted dorsal mesoderm, and the blastopore lip remain potent inducers of neural ectoderm long after neural competence has disappeared from the ectoderm tissue. The other issue of note is that the neural plate has already acquired a significant amount of regional specification by stage 13, and this regional specification is now independent of the anterior–posterior position of underlying dorsal mesoderm.

III. Signals Initiating Neural Induction

Signals initiating neural plate formation and subsequent regional specialization during early development are often referred to as inductive signals. The number and the physical nature of neural inductive signals are not yet known. However, experimental evidence from several laboratories suggest that many signaling events are involved in the neural induction process. These signaling events appear to begin by at least early cleavage stages, possibly between fertilization and first cleavage, and continue through gastrulation and early neurula. We are concerned with the origin of the tissues sending the signals and when and how the signals are being sent.

A. Early Neural Bias

The earliest neurallike distinction between dorsal ectoderm (presumptive neural) and ventral ectoderm (presumptive epidermal) was demonstrated to occur as early as the eight-cell stage. An antibody against an epidermal-specific cell surface marker, Epi 1, was used to determine when blastomeres have received information directing them to either a dorsallike or a ventrallike pathway. Normally, epidermal cells start synthesizing Epi 1 at stage 13, early neurula, and Epi 1 is not expressed in neurectoderm (Akers *et al.*, 1986). However, dorsal animal blastomeres isolated during early blastula stages and cultured in Ringer's solution (see Appendix A, this volume) were found to be biased against expression of Epi 1 (London *et al.*, 1988). These results suggest that there are events associated with neural induction by early cleavage stages. A second signaling event occurs at early gastrula that reinforces and extends the previous dorsal bias against Epi 1 expression. This event appears to be brought about by a signal which is sent through the plane of the ectoderm from the dorsal blastopore lip area (Savage and Phillips, 1989). Thus, there are at least two signals which control the expession pattern of a molecule differentially synthesized between neural and nonneural epithelium. Furthermore, both of these signals effect the pattern of expression prior to the movements of dorsal mesoderm during gastrulation.

The early neural ectoderm bias is also reflected in the observations that early dorsal ectoderm more readily responds to some neural induction signals than does ventral ectoderm. Dorsal ectoderm, cocultured in a sandwich configuration (see Section IV) with invaginated dorsal mesoderm, is induced to synthesize Xlhbox6 and NCAM more readily than is ventral ectoderm cultured under the same conditions (Sharpe et al., 1987; Phillips and Doniach, 1991). A dorsal–ventral difference in ectodermal response to activin B has also been observed. Dorsal ectoderm responds to incubation in activin B by forming embryoid structures with a significant amount of neural tissue, whereas incubation of ventral ectoderm with activin B induces few, if any, dorsal structures (Thomsen et al., 1990).

Dorsal ectoderm appears to be biased toward a neural pathway with respect to its ability to be induced. However, ectoderm removed from embryos prior to gastrulation and cultured do not make neural structures. Therefore, it would appear that an additional signal(s) is necessary to induce ectoderm to synthesize neural-specific molecules. The nature and the origin of these signals are an active area of research.

Spemann (1938) hypothesized that one possible source of neural induction signals might originate in the cells of the blastopore lip and that the signals might be propagated through the plane of the ectoderm. This possibility has been examined recently by several investigators. The expression of Epi 1 was found to be inhibited when blastopore lip cells were grafted onto the edge of ventral ectoderm (Savage and Phillips, 1989). Thus, it is possible for information to travel through the plane of ectoderm, which inhibits the expression of at least one molecule differentially expressed between neural and epidermal epithelium. The synthesis of NCAM can also be induced by edgewise or planar contact with blastopore lip material. The planar induction of NCAM was obtained by three different culture methods. First, the ectoderm of exogastrulated embryos was found to express both NCAM RNA and protein (Kintner and Melton, 1987; Dixon and Kintner, 1989). Construction and culture of Keller sandwiches (described in Chapter 5, this volume) also provide a configuration where the dorsal mesoderm, including cells of the blastopore lip, are in a planar configuration with ectoderm. Again, NCAM is expressed in these cultures even though the dorsal mesoderm has only had planar contact with the ectoderm (Dixon and Kintner, 1989; Phillips and Doniach, 1991; Sater et al., 1991).

The third method used to determine if planar signals are capable of eliciting a neural response is by grafting blastopore lip cells onto the edge of ectoderm and assaying for the expression of NCAM within the ectoderm. Contact between blastopore lip cells and ectoderm for as little as 2 hours is sufficient for induction of NCAM synthesis within the ectoderm tissue (Phillips and Doniach, 1991). In addition, experiments designed to study lens induction by grafting early gastrula stage ectoderm into the presumptive lens-forming region indicate that an NCAM-expressing neural plate-like structure will form, even though this tissue is not underlain by dorsal chordamesoderm (Servetnick and Grainger, 1990). The explanation for this observation is that a signal passes through the ectoderm to induce the grafted tissue to form neural tissue. The tissue grafted into the presumptive lens region is not required to be in direct contact with the neural plate in order to be induced, again suggesting that a signal is traveling through the plane of the ectoderm (Servetnick and Grainger, 1990).

Other neural-specific markers have also been used to show that induction can occur through the plane of the ectoderm. Signals traveling through the ectoderm which induce specific organs have also been reported for the formation of cement gland (Sive et al., 1989). The ability of isolated ectoderm to form cement gland in culture travels ahead of the involuting dorsal mesoderm (Sive et al., 1989). Xhox 3, an anterior neural marker, is also expressed in the apical ectoderm portion of an exogastrula embryo, a region which develops in the absence of any dorsal mesoderm (Ruiz i Altaba, 1990).

B. Signals from Invaginated Dorsal Mesoderm

Several investigators have observed that neural induction occurs when competent ectoderm is wrapped around a piece of involuted dorsal mesoderm. As mentioned previously, the dorsal mesoderm has often been implicated as the source of regional information (Mangold, 1933; Ragven and Kloos, 1945; Horst, 1948; Gimlich and Cooke, 1983; Smith and Slack, 1983; Recanzone and Harris, 1985; Sharpe et al., 1987; Sharpe and Gurdon, 1990; Jones and Woodland, 1989; Hemmati-Brivanlou and Harland, 1989; Sater et al., 1991; Phillips and Doniach, 1991). The methods most commonly used to assay for neural-inducing activity is by einsteck or sandwich cocultures, which are described in Section IV. The expression of both general neural and regional neural-specific molecules have been used as assays for neural induction signals associated with the invaginated dorsal mesoderm (Sharpe et al., 1987; Sharpe and Gurdon, 1990; Sater et al., 1991; Hemmati-Brivanlou and Harland, 1989;

T. Doniach, personal communications). Two very important concepts have emerged from these studies. First, the invaginated dorsal mesoderm has the ability to induce competent ectoderm to synthesize both neural-specific molecules and regionally specific neural molecules. Second, several of the authors reported that the regionally specific neural molecules were being synthesized at some distance from the implanted dorsal mesoderm within sandwich cocultures (T. Doniach, personal communications; Hemmati-Brivanlou and Harland, 1989; see also Sive et al., 1989). These results suggest that a direct imprinting by invaginated dorsal mesoderm to overlying ectoderm is not occurring. Instead, information may be passing from the mesoderm to the overlying ectoderm, and this information then travels through the plane of the ectoderm before a region of the ectoderm expresses the appropriate molecular response. As more region-specific neural markers become available, it will become possible to determine more accurately the time and the mechanisms by which regional specification occurs within the neural plate.

IV. Methods

A. Einsteck Method

The original einsteck method involved placing either involuted dorsal mesoderm or involuted dorsal mesoderm with overlying ectoderm into the ventral blastocoel cavity of a host embryo. The development of a second dorsal axis was used as an assay to show that the implanted dorsal tissues have the ability to induce the overlying ventral ectoderm to differentiate into neural tissue (Mangold, 1933). Similar einsteck experiments are also used to test for neural competence of host ectoderm. This assay can be adapted for several types of neural induction studies by implanting other tissues or any chemical substance impregnated into a small agar block into the blastocoel cavity.

The jelly coat is removed in 2.6% cysteine (pH adjusted to 8.0 with NaOH). The vitelline membrane is removed with sharp watchmaker's forceps. A small cut is made near the animal pole of the blastocoel roof. It is best if an advanced blastula or early gastrula is used as a host embryo, since the neural competence of the ectoderm diminishes quickly during gastrulation. Normally, the implanted tissue is placed on the ventral side of the blastocoel cavity, away from the influence of the endogenous blastopore lip region. Using a UV-irradiated embryo as a host can eliminate possible problems of competition with axial development of the host (Scharf and Gerhart, 1980). However, care must be taken to ensure that the endogenous level for development of a dorsal axis in the UV-irradiated embryo is known and low. The piece of test tissue is pushed to the ventral vegetal portion of the blastocoel cavity so that it comes to lie between the involuting mesoderm and the ectoderm during the process of gastrulation.

Three potential problems must be considered when using this approach. First, there is little control over where the implanted test tissue will ultimately reside within the blastocoel of the host. Second, there is no control over when the test piece actually makes sufficient contact with the ectoderm to permit a potential response like neural induction. This point is of critical importance because the ectoderm is competent for neural induction for only a relatively short time during gastrulation, and different portions of the ectoderm lose competence at different developmental times. The third potential problem involves the size of the inserted tissue. The inserted tissue may transport information from the vegetal cells to the ectoderm simply by spanning the space between them. This would result in a false positive. Therefore, it is important that the inserted tissue does not come in contact with the dorsal vegetal cells of the blastocoel floor and bridge the distance to the ectoderm.

B. Sandwich Method

Sandwich cultures have been used as assays to test for both the neural competence of ectoderm and the presence of neural inductors in either the involuted dorsal mesoderm or the blastopore lip. In either case, ectoderm is wrapped around another tissue, and the combination is cultured in a buffered salt solution. The cultures can be assayed for an observable morphology or used in whole-mount immunocytochemical assays for the expression of specific molecules (see Chapter 22, this volume).

For cell culture, the bottoms of small petri dishes are coated with 2% agar made in the culture medium of choice. The agar should be washed several times in ice-cold water containing 10 mM EDTA (pH 7.5), followed by several washes of ice-cold medium. Adjust the final pH of the agar solution as necessary. Antibiotics, 50 μg/ml each of penicillin and streptomycin, may be added when the agar cools, if desired. Small wells, slightly larger than the tissues to be cultured, are melted into the agar with a flame-heated Pasteur pipette. Mark the position of the wells on the underside of the petri dish with a magic marker, as the wells "disappear" when medium is added to the petri dish. In general, tissue explants will grow together and heal more quickly in a medium with a slightly higher salt concentration, like full-strength Ringer's. Tissues may be allowed to heal in higher salt medium and then transferred to 0.33 × Ringer's, or other lower salt culture medium, for long-term culture. Long-term culture of tissues whose superficial epithelial cells have been removed should be conducted in Danilchik's medium or a modified Danilchik's medium (see Appendix A, this volume). Long-term culture is usually most successful at temperatures between 14° and 21°C. In the event that contamination becomes a chronic problem, the embryos should be washed as follows: Remove the vitelline envelope with watchmaker's forceps and transfer to a dissociation medium (60% Ca^{2+}, Mg^{2+}-free Hank's saline, 0.4 mM EDTA, 10 nM HEPES, 20 mM Na_2CO_3, pH 7.2) for 1 to 2 minutes. The embryos are then placed in 33% modified Barth's saline (88 mM NaCl, 0.1 mM KCl, 0.24 mM $NaHCO_3$, 0.82 mM $MgSO_4$, 33 μM $Ca(NO_3)_2$, 0.41 mM $CaCl_2$, 1 mM HEPES, pH 7.4) with up to 100 IU/ml penicillin, 100 μg/ml streptomycin, 0.25 mg/ml amphotericin B, and 100 μg/ml ampicillin.

Animal cap ectoderm is removed from devitellinated embryos by cutting with sharpened watchmaker's forceps or by using a stainless steel knife. Good forceps can be obtained from Fine Science Tools (Belmont, CA; Bio No. 5). The forceps can be sharpened using fine granite stones. Knives can be prepared by pulling the end of a glass pipette to a small point over a flame. About 3 cm of 0.001 stainless steel wire (A & M Systems, Inc.) is inserted into the hole of the pulled pipette, leaving a few millimeters sticking out of the glass. The end of the pipette is melted shut over a small flame to hold the wire in place. The wire can be trimmed to any desired length; however, a shorter wire provides more strength. The dissected animal cap is placed superficial cell side down in one of the wells. A small piece of blastopore lip or involuted dorsal mesoderm placed in the ectoderm will become enclosed within the ectoderm in about 30 minutes, longer if incubated in lower salt medium. If necessary, a second piece of ectoderm can be placed over the top, which will then fuse into place.

C. Exogastrula

Exogastrulated embryos provide a situation in which the mesoderm does not involute into the blastocoel cavity. Instead, it undergoes the convergent extension movements of normal embryos, and the mesoderm is forced to the outside of the embryo. The result is that a sac of ectoderm is connected to the mesoderm by a thin isthmus of tissue. Exogastrula embryos have traditionally been used to study the effects of potential planar signals on inducing specific neural molecules.

The vitelline membrane is removed with a pair of sharp watchmaker's forceps during early blastula. The embryo is placed in a high salt medium from midblastula through the process of gastrulation. Full-strength Ringer's or 1–1.4× MMR (Newport and Kirschner, 1982) are often used. The blastocoel collapses in high salt solutions, forcing the mesoderm to evaginate instead of invaginating into the blastocoel cavity and under the ectoderm. As a result, the ectoderm is never in contact with the evaginating mesoderm. The mesoderm differentiates into notochord, somites, and pronephros. The ectoderm of exogastrulated embryos undergoes expansion by the movements of epiboly and becomes an empty sac. The ectoderm maintains a small connection to the mesoderm, and it is this contact which has been used to address the question of planar signals traveling from mesoderm to ectoderm. If there is not a thin isthmus of tissue between the ectodermal sac and the elongated mesoderm, or if the ectodermal sac shows elongation, than the embryo was not a complete exogastrula and a higher salt concentration should be used.

D. Edge-on Grafting or Aging of Tissues in Culture

It is often useful to control topologically the areas of contact between two or more tissues and to have one tissue labeled to show which tissue is responding to some induction event. In other circumstances, a tissue needs to be aged in culture for some subsequent experiment. Amphibian tissues tend to form a ball in culture, which is inconvenient if the tissue is to be grafted back into the embryo. Therefore, tissues to be aged in culture are often held in a flat configuration during the culture process.

Labeling of embryos with fluorescent dextran is described in Chapter 15 of this volume. Tissues are dissected in full-strength Ringer's or other high salt buffer. Ficoll (5%) can be added to the medium if early blastomeres are being removed for culture. Dissected tissues (see above for tool descriptions) are placed on a base of 1 to 2% agarose, washed as described above. A small piece of coverslip glass (previously washed in 100% ethanol and, if needed, siliconized) is placed over the explant to prevent the explants from rolling up. A thin 2% agar slab can be used instead of glass. Two or more pieces of tissue can be held flat and in contact with each other by putting them into a well and covering them with the cover glass or the agar slab. It takes approximately 30 minutes for two tissues to heal together, depending on the buffer, temperature, and the tissues being used in the experiment. Full-strength Ringer's or 100% Steinberg's solution is often used to promote healing.

The outer layer of ectoderm, or superficial layer, will often attempt to grow around the cells of the deeper layer, even in tissues held flat. If the cultured piece is to be transplanted back into the embryo or used to combine with another tissue after aging, the edge(s) should be trimmed with a glass or stainless steel knife prior to grafting the cultured piece of tissue either into a embryo or onto another piece of tissue. After healing, the explants can be cultured in a number of different media, depending on the experiment. However, it is best to remove them from the high salt buffer.

E. Keller Sandwiches

The procedures for making Keller sandwiches is described in detail in Chapter 5, this volume. In brief, these explants include the bottle cells, the involuting marginal zone which is fated to give rise to notochord and somites, and the noninvoluting marginal zone which participates in the formation of the posterior neural plate. Each explant can be as large as approximately 60° of the embryo circumference centered on the dorsal midline and extending to near the animal pole. Two such explants are sandwiched together so that the deep cells face each other. This configuration produces a situation similar to that of the exogastrula in that the dorsal mesoderm undergoes convergent extension movements without coming to underlie the ectoderm. However, a planar connection between the ectoderm and the dorsal mesoderm is maintained with the original blastopore lip cells at the junction between the two tissue types.

F. Dissociation and Reaggregation

Various investigators have indicated that induction often requires contact between the inducing and the responding tissues. In some cases, it has also been shown that a certain mass of cells is necessary for an induction response to occur. Therefore, dissociation and reaggragation of cells can be a useful tool for analyses of timing and cell interactions during an induction process.

To dissociate embryos, the vitelline membrane is removed with watchmaker's forceps, and the embryos are placed in Ca^{2+}, Mg^{2+}-free medium. Gentle shaking will dislodge the cells and prevent them from forming any further contacts. If early blastula stage embryos are disaggregated, then 5% Ficoll should be added to the Ca^{2+}, Mg^{2+}-free medium in order to support the large blastomeres. The Ficoll medium should be gradually replaced as the cells get smaller, about the equivalent of late blastula or early gastrula, because the cells do better in long-term culture without Ficoll present. Reaggregation can be initiated by adding calcium and magnesium to the media to a final concentration of 1 mM. Gentle swirling will bring the cells together in the middle of the dish. They will form an aggregate if they are held motionless.

V. Summary

The study of neural induction in *Xenopus* can be approached from two broad perspectives. One can study the competence of the ectoderm to respond to neural induction signals and any potential prepattern within the ectoderm.

The second area of study involves the neural induction signals, in terms of the chemical nature of the signals, their sources, and their method of delivery.

The neural competence of ectoderm has been studied by either grafting the ectoderm into areas which normally form neural tissue or by grafting tissues which normally induce neural structures onto the ectoderm to be tested. In general, it appears that ectoderm loses neural competence by midgastrula. However, there is some experimental evidence from various amphibian studies that the loss of neural competence in ectoderm does not occur simultaneously throughout all regions. It is not yet known if this phenomenon is also true in *Xenopus*.

There appear to be several signaling events involved in the process of neural induction. Molecular probes have made it possible to study early steps in the neural induction and patterning processes which were not possible to study using only the development of neural morphology as a marker for neural induction. Antibodies directed against early epidermal versus neural epithelium indicate that the dorsal animal blastomeres are biased toward a neural pathway during early cleavage. Another signaling event occurs at early gastrula and the resulting dorsal ectoderm now responds more readily to some neural induction events than does the ventral ectoderm. The source of the early gastrula signals has been studied by a variety of methods, including exogastrula embryos, Keller sandwiches, and grafting a blastopore lip to the edge of competent ectoderm. The blastopore lip can send signals through the plane of the ectoderm capable of inducing competent ectoderm to become neural tissue.

There are several issues relative to the process of neural induction which are not yet resolved. The major issue involves the mechanism of establishing pattern within the neural plate. Ectoderm appears to lose neural competence prior to the time when involuted dorsal mesoderm comes to underlie the anterior neural plate region. Several investigators have shown that information for expression of spatially restricted neural-specific molecules can travel through the ectoderm, independent of underlying dorsal mesoderm. A planar mechanism of induction may begin to answer some of the questions concerning the establishment of regionally specific areas within the neural plate. However, several questions concerning the quickness by which the pattern within the neural plate becomes determined (by stage 13) and the mechanism by which small areas of cells are restricted to a specific regional fate have yet to be addressed. Chaung (1938, 1939, 1940) and Holtfreter (1938) demonstrated that complex neural patterns can be elicited by foreign mesodermal sources in ectoderm of some amphibians. The explanations offered for this observation are that regional areas of the induced ectoderm are either responding to some prepattern within the ectoderm or that the ectoderm is regionally sorted by intracellular communication. In any event, the molecular nature of the neural induction response is a long way from being understood, along with the nature and means of delivering the neural induction signals.

REFERENCES

Akers, R. M., Phillips, C. R., and Wessells, N. K. (1986). Expression of an epidermal antigen used to study tissue induction in the early *Xenopus laevis* embryo. *Science* **231,** 613–616.

Barth, L. G. (1937). Chemical stimulation of the amphibian ectoderm. *Biol. Bull. (Woods Hole, Mass.)* **73,** 346–347.

Barth, L. G. (1939). The chemical nature of the amphibian organizer. III. Stimulation of the presumptive epidermis of *Ambystoma* by means of cell extract and chemical substances. *Physiol. Zool.* **12,** 22–29.

Chaung, H. H. (1938). Spezifische induktionsleistungen von leber und niere im explantatversuch. *Biol. Zentralbl.* **58,** 472–480.

Chaung, H. H. (1939). Induktionsleistungen vonfrischen und gekochten organteilen (Niere, leber) nach ihrer verpflanzung is explantate und verschiedene wirtsregionen von trtonkeimen. *Wilhelm Roux' Arch. Entwicklungsmech. Org.* **139,** 556–638.

Chaung, H. H. (1940). Weitere vershche uber die veranderung der induktionsleistrungen von gekochten organteilen. *Wilhelm Roux' Arch. Entwicklungsmech. Org.* **140,** 25–38.

Dixon, J. C., and Kintner, C. R. (1989). Cellular contacts required for neural induction in *Xenopus* embryos: Evidence for two signals. *Development (Cambridge, U.K.)* **106,** 749–757.

Fischer, F. C., Wehmeier, E., Lehmann, H., Juhling, L., and Hultzsch, K. (1935). Zur Kenntnis der Induktionsmittel in der Embryonal-Entwicklung. *Ber. Dtsch. Chem. Ges.* **68,** 1196–1199.

Fujii, T. (1945). Effects of acetylcholine and *p*-quinone on the presumptive ectoderm of the newt. *Zool. Mag.* **56,** 13–15.

Gimlich, R. L., and Cooke, J. (1983). Cell lineage and the induction of second nervous systems in amphibian development. *Nature (London)* **306,** 471–473.

Hamburger, V. (1988). "The Heritage of Experimental Embryology. Hans Spemann and the Organizer." Oxford Univ. Press, Oxford.

Hemmati-Brivanlou, A., and Harland, R. M. (1989). Expression of an engrailed-related protein is induced in the anterior ectoderm of early *Xenopus* embryos. *Development (Cambridge, U.K.)* **106,** 611–617.

Holtfreter, J. (1933). Die totale Exogastrulation, eine Selbstrablosung des Ektoderms von Entomesoderm. Entwicklung and funktionelles Verhalten nervenloser Organe. *Wilhelm Roux' Arch. Entwicklungsmech. Org.* **129,** 669–793.

Holtfreter, J. (1938). Veranderungen der Reaktionsweise im alternden isolierten Gastrulaektoderm. *Wilhelm Roux' Arch. Entwicklungsmech. Org.* **138,** 163–196.

Jacobson, A., and Sater, A. K. (1988). Features of embryonic induction. *Development (Cambridge, U.K.)* **104,** 341–357.

Jacobson, C. O. (1959). The localization of the presumptive cerebral regions in the neural plate of the axolotl larva. *J. Embryol. Exp. Morphol.* **7,** 1–21.

Jacobson, M., and Rutishauser, U. (1986). Induction of neural cell adhesion molecule (N-CAM) in *Xenopus. Dev. Biol.* **116,** 524–531.

Jones, E. A., and Woodland, H. R. (1989). Spatial aspects of neural induction in *Xenopus laevis*. *Development (Cambridge, U.K.)* **107,** 785–791.

Kallen, B. (1965). Early morphogenesis and pattern formation in the central nervous system. *In* "Organogenesis" (DeHaan and Ursprung, eds.). Holt, Rinehart Winston, New York.

Kintner, C. R., and Melton, D. A. (1987). Expression of the *Xenopus* N-CAM RNA in ectoderm is an early response to neural induction. *Development (Cambridge, U.K.)* **99**, 311–325.

Lehmann, F. E. (1926). Entwicklungsstorungen an der Medullaranlage von Triton, erzeugt durch Unterlagerungsdefekte. *Wilhelm Roux' Arch. Entwicklungsmech. Org.* **108**, 243–283.

London, C., Akers, R., and Phillips, C. R. (1988). Expression of Epi 1, an epidermal specific marker, in *Xenopus laevis* embryos is specified before gastrulation. *Dev. Biol.* **129**, 380–389.

Machemer, H. (1932). Experimentelle Untersuchung uber die Induktionsleistunger der oberen Unmundlippe in alteren Urodelenkeim. *Wilhelm Roux' Arch. Entwicklungsmech. Org.* **126**, 391–456.

Mangold, O. (1929). Experimente zur Analyse der Determination und Induktion der Medullarplatte. *Wilhelm Roux' Arch. Entwicklungsmech Org.* **117**, 586–696.

Mangold, O. (1933). Uber die Induktionsfahigkeit der verschiedenen Bezirke der Neurula von Urodelen. *Naturwissenschaften* **21**, 761–766.

Newport, J., and Kirschner, M. (1982). A major developmental transition in early *Xenopus* embryos. II. Control of the onset of transcription. *Cell (Cambridge, Mass.)* **30**, 687–696.

Okada, Y. K. (1938). Neural induction by means of inorganic implantation. *Growth* **2**, 49–53.

Phillips, C. R., and Doniach, T. (1991). Effects of the dorsal blastopore lip and the involuted dorsal mesoderm on neural induction in *Xenopus laevis*. In preparation.

Raven, C. P., and Kloos, J. (1945). Induction by medial and lateral pieces of the archenteron roof with special reference to the determination of the neural crest. *Acta Neerl. Morphol.* **4**, 348–362.

Ruiz i Altaba, A. (1990). Neural expression of the *Xenopus* homeobox gene Xhox3: Evidence for a patterning neural signal that spreads through the ectoderm. *Development (Cambridge, U.K.)* **108**, 595–604.

Sater, A. K., Uzman, J. A., Steinhardt, R. A., and Keller, R. (1991). Neural induction in *Xenopus* embryos: Induction of neuronal differentiation by either planar or vertical signals. In preparation.

Sato, S. M., and Sargent, T. D. (1989). Development of neural inducing capacity in dissociated *Xenous* embryos. *Dev. Biol.* **134**, 263–266.

Savage, R., and Phillips, C. R. (1989). Signals form the dorsal blastopore lip region during gastrulation bias the ectoderm toward a nonepidermal pathway of differentiation in *Xenopus laevis*. *Dev. Biol.* **133**, 157–168.

Schechtman, A. M. (1938). Competence for neural plate formation in *Hyla* and the so-called nervous layer of the ectoderm. *Proc. Soc. Exp. Biol. Med.* **38**, 430–433.

Servetnick, M., and Grainger, R. (1991). Changes in neural competence occur autonomously in *Xenopus* ectoderm. *Development (Cambridge, U.K.)*

Sharpe, C. R., and Gurdon, J. B. (1990). The induction of anterior and posterior neural genes in *Xenopus laevis*. *Development (Cambridge, U.K.)* **109**, 765–774.

Sharpe, C. R., Fritz, A., DeRobertis, E. M., and Gurdon, J. B. (1987). A homeobox-containing marker of posterior neural differentiation shows the importance of predetermination in neural induction. *Cell (Cambridge, Mass.)* **50**, 749–758.

Shen, S. C. (1939). A quantitative study of amphibian neural tube induction with a water-soluble hydrocarbon. *J. Exp. Biol.* **16**, 143–149.

Sive, H. L., Hattori, K., and Weintraub, H. (1989). Progressive determination during formation of the anteroposterior axis in *Xenopus laevis*. *Cell (Cambridge, Mass.)* **58**, 171–180.

Slack, J. M. W. (1983). "From Egg to Embryo: Determinative Events in Early Development." Cambridge Univ. Press, Cambridge.

Smith, J. C., and Slack, J. M. W. (1983). Dorsalization and neural induction: Properties of the organizer in *Xenopus laevis*. *J. Embryol. Exp. Morphol.* **78**, 299–317.

Spemann, H. (1938). "Embryonic Development and Induction." Yale Univ. Press, New Haven, Connecticut.

Spemann, H., and Mangold, H. (1924). Uber induktion von embryonanlagen durch implantation artfremder organisatoren. *Wilhelm Roux' Arch. Entwicklungsmech. Org.* **100,** 599–638.

ter Horst, J. (1948). Differenzierungs und Induktionsleistungen verschiedener Abschnitte der Medullarplatte und der Urdarmdaches von Triton im Kombinat. *Wilhelm Roux' Arch. Entwicklungsmech. Org.* **143,** 275–303.

Thomsen, G., Woolf, T., Whitman, M., Sokol, S., Vaughan, J., Vale, W., and Melton, D. (1990). Activins are expressed early in *Xenopus* embryogenesis and can induce axial mesoderm and anterior structures. *Cell (Cambridge, Mass.)* **63,** 485–493.

Chapter 19

Analysis of Class II Gene Regulation

THOMAS D. SARGENT AND PETER H. MATHERS

Laboratory of Molecular Genetics
National Institute of Child Health and Human Development
National Institutes of Health
Bethesda, Maryland 20892

I. Introduction
II. Brief Review of Class II Gene Transfer into Frog Embryos
III. Gene Selection
IV. Initial DNA Injection Experiments
V. Preliminary Mapping by Polymerase Chain Reaction Footprinting Analysis
    A. Recipes
    B. *In Vivo* Copper–Phenanthroline Digestion Procedure for *Xenopus* Embryonic Tissues
    C. *In Vitro* Copper–Phenanthroline Digestion Procedure
    D. *In Vivo* Dimethyl Sulfate Treatments for *Xenopus* Embryonic Tissues
    E. *In Vitro* Dimethyl Sulfate Treatment
    F. Primer 1 Extension and Linker Ligation Reactions
    G. Kinasing of Primers
    H. Taq Reactions for Genomic Footprinting
VI. Regulatory Element Functional Studies
    References

I. Introduction

One of the most frequently cited problems facing investigators interested in vertebrate development is the difficulty attending conventional genetic analysis with these species. This is an especially serious limitation with *Xenopus*, owing primarily to its long breeding cycle. Fortunately, there are alternatives to classical genetics that can be used to identify developmental control mechanisms. One approach, which has been followed by several laboratories, including our own, is to select a suitable gene whose expression responds to interesting embryonic signals and study its regulation, using the tools of molecular biology and protein chemistry to elucidate the control hierarchy. The initial step in this scheme is to map cis-regulatory elements, followed by the identification and cloning of the transcription factors that interact with these sequences. Given an astute choice of subject gene and a certain amount of favorable luck, these transcription factors will comprise the molecular components of an important developmental switch. Probes for these factors should make it possible to design experiments to reveal the relationship between differential gene expression and the control of embryogenesis. This general idea has motivated several groups to clone and study a variety of genes expressed in frog development. In this chapter we discuss the progress that has been made with some of these experimental systems and outline what we consider to be a productive approach to such research. We focus specifically on class II genes that can be studied by DNA transfer into embryos. Readers interested in injecting DNA into oocytes are referred to Gurdon and Wakefield (1986), which includes additional background information and details on buffers, animals, and equipment.

II. Brief Review of Class II Gene Transfer into Frog Embryos

The first clearly successful experiments exhibiting correct regulation of a gene introduced into frog embryos were carried out by Krieg and Melton (1985, 1987). These investigators identified a gene, *GS17*, that is activated at the midblastula transition (MBT) and expressed for a brief period thereafter. *GS17* DNA was injected in a closed supercoiled form, which replicated little, if at all, during cleavage. The injected DNA began and ended transcription at the correct times (i.e., midblastula transition and late gastrula, respectively). Deletion and DNA transfer experiments were used to narrow down the relevant control region to a small interval, a few hundred base pairs (bp), upstream from the transcriptional start site.

Temporal gene regulation is certainly important in development, but there is at least as much reason to be interested in the regulation of genes expressed in specific tissues. There are two examples of tissue-specific gene expression that have been successfully studied in *Xenopus* by the egg injection approach: α-actin, which is expressed only in muscle, and XK81A1, a keratin expressed only in early epidermis. In both cases it has been shown that the cloned genes are expressed predominantly, if not exclusively, in the appropriate tissues following injection of suitable constructions. Mohun and colleagues (Mohun et al., 1989; Taylor et al., 1989) found that activation of the α-actin gene in muscle requires the presence of a short element, the CArG box, which is part of the "serum response element," a short distance upstream from the initiation site. Less well-defined sequences further upstream are also required. Epidermal keratin expression appears also to be regulated by multiple factors; a control element located at -157 bp is required for high level expression, but its deletion does not result in ectopic expression or in complete inactivation of the gene. Other elements, not yet precisely mapped, are evidently involved in preventing nonepidermal expression (Jones et al., 1989; Snape et al., 1990). The work with epidermal keratin was carried out in our laboratory at the National Institutes of Health and most of the technical details following in this chapter are based on our experience with this gene.

The experiments outlined above clearly show that gene transfer into frog embryos is a practical approach. However, not all such experiments work equally well. There are two examples in the literature of partially aberrant expression of injected genes. Tadpole globin genes injected by Bendig and Williams (1984) were activated at about the correct time. However, no spatial distribution data were presented, so it is not possible to conclude that the introduced genes were active in the correct tissue type. Furthermore, in the same series of experiments injected copies of adult globin genes were expressed much earlier than the endogenous loci.

Another example of imperfect control of an injected gene comes from studies, carried out by Krone and Heikkila (1989), of the gene for the 30-kDa heat-shock protein (hsp30). These investigators demonstrated that injected hsp30 DNA responded correctly to heat shock. During frog development, however, the hsp30 gene is unresponsive to heat shock until tailbud stages. The injected copies could be activated as early as the midblastula transition, indicating that an important aspect of hsp30 gene control was not imposed on the introduced DNA.

One feature common to the globin and heat-shock genes is that neither is activated during early frog development (i.e., before the end of gastrulation). It may be true that genes activated later than gastrulation will in general prove more refractory to transient expression assays based on egg injection.

III. Gene Selection

The choice of a gene for study is the most important decision to be made. Obviously there can be no fixed rules for this, but we suggest the following guidelines. Ideally, the expression of the gene should respond to a known specification process, such as mesoderm or neural induction. The gene should also be regulated unambiguously, going promptly from a clearly "off" to a clearly "on" condition (or vice versa). If the gene is merely increasing or decreasing in activity, then, however important this may be to development, the injection experiments may be difficult to interpret and the regulatory element map corresponding elusive. In general, it is easier to study tissue-specific rather than time-specific gene expression. One problem with the latter derives from the transient and variable nature of DNA injected into frog eggs. If, as is quite common, a template is increasing or decreasing in concentration as much as a 100-fold during the period of the experiment, then it may be difficult to separate true temporal control from a change in the RNA accumulation rate that results from changing template concentration. Finally, as noted above, there is a suggestion that genes normally expressed during early development may have a greater chance of correct regulation in injection experiments than genes that are not transcribed until after gastrulation.

IV. Initial DNA Injection Experiments

Before attempting to identify regulatory elements, three fundamental questions need to be answered. (1) Will the injected gene be correctly regulated? (2) Is the regulation at the transcriptional level? (3) Approximately where are the critical regulatory elements (i.e., upstream, downstream, or internal)? It is also important to establish some basic properties the cloned DNA will exhibit in the embryo, such as the possible effects of linearizing DNA before injection and the tendency of the injected DNA to replicate. These can be addressed by a few fairly simple injection experiments. A gene fusion is especially helpful. Transcriptional regulatory elements tend to reside in 5'-flanking regions, so a few kilobases (kb) of upstream DNA "driving" a heterologous reporter gene will often confer the correct transcriptional activation time and tissue distribution. Of course, if the reporter sequence is substantially different from the original gene, then it is possible that mRNA and protein stabilities will have a major effect on the accumulation pattern.

An example of the gene fusion approach is shown in Fig. 1. The gene used in this experiment was constructed with about 6 kb of 5'-flanking DNA, including the transcriptional start signal, from the epidermal keratin gene, XK81A1, fused to the protein-coding portion of the human β-globin gene. Fertilized eggs were injected with the fusion gene, cultured to tailbud stage, and processed for immunocytochemistry with globin-specific antiserum. The data revealed that expression occurred exclusively in the correct tissue, epidermis. This result indicates that restricted expression of the keratin gene

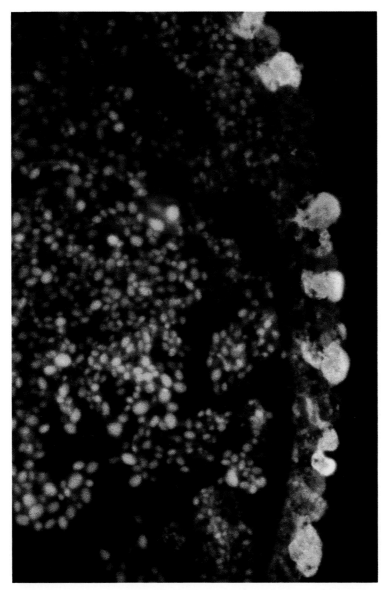

FIG. 1. Expression in epidermal cells of human β-globin driven by regulatory sequences from a *Xenopus* epidermal keratin gene. Embryos were injected with supercoiled KG5900 DNA and processed for immunocytochemistry as described by Jonas *et al.* (1989). Expression is mosaic but limited to epidermal tissues (A. Snape, unpublished).

must be essentially transcriptional, since tissue specificity is imposed on a heterologous reporter sequence, and that sufficient control elements resided in the 6 kb of 5'-flanking DNA. This experiment also reveals a property often (but probably not universally) associated with DNA injected into frog embryos: the expression tends to be highly mosaic.

Once the important regulatory region has been limited to a few kilobases, as in the above case, further characterization can be carried out relatively easily by preparing and injecting a terminal deletion series. An example of this for the XK81A1 keratin gene is shown in Fig. 2. In this experiment an intact keratin gene was modified by inserting 36 bp of heterologous DNA

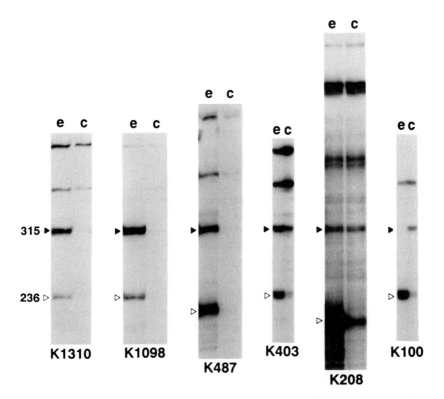

FIG. 2. Delimiting a regulatory region by terminal deletions. Different constructions of the XK81A1 keratin gene with varying lengths of 5'-flanking DNA were injected into fertilized eggs and the embryos dissected into epidermis (e) and carcass (c) fractions; RNA was isolated and assayed by ribonuclease protection. The 315-nucleotide (nt) band represents transcripts from the injected constructions, and the 236-nt band represents endogenous keratin mRNA. Tissue-specific regulation begins to break down when 403 bp or less of the 5'-flanking DNA remains. (Reproduced from Jonas et al., 1989.)

into the first exon, making it possible to distinguish injected gene transcripts from the endogenous keratin mRNA by hybridization with radiolabeled antisense RNA, followed by RNase mapping (Jonas *et al.,* 1989). Deletions carrying 1310, 1098, or 487 bp of flanking DNA showed correct epidermal specificity, whereas constructions with shorter 5'-flanking segments exhibited aberrant expression in nonepidermal tissue. This, in conjunction with the gene fusion approach, limited the important regulatory interval to about 500 bp, from -487 to the fusion point, $+26$ bp.

Most regulatory elements identified to date have been positive in nature: their removal results in reduced transcriptional efficiency. This is likely to be the case as well with sequence elements that control developmentally regulated genes. One potential problem that must be kept in mind when evaluating promoter strength data from injected frog embryos is that for some

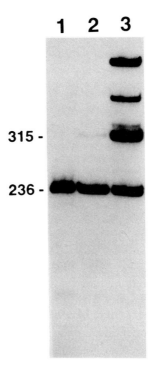

FIG. 3. Effect of DNA form on expression levels. A construction of the XK81A1 epidermal keratin gene having 1310 bp of 5'-flanking DNA was injected in superhelical (lane 2) or linear/concatenated (lane 3) form. Expression from the injected DNA is represented by the RNase protection band at 315 nt, and the 236-nt band represents endogenous keratin mRNA. Lane 1 shows material from uninjected embryos. (Reproduced from Jonas *et al.,* 1989.)

genes the form in which the DNA is injected can have a major effect on how actively transcribed it will be in the embryo. An example of this phenomenon is shown in Fig. 3. Lanes 2 and 3 show the result obtained by injecting 200 pg of DNA in superhelical or linearized form, respectively. In the experiment shown in Fig. 3 the linearized DNA was concatenated prior to injection by treatment with T4 DNA ligase, a step which we have found to be unnecessary (A. Snape, unpublished), presumably because linear DNA is concatenated in the fertilized egg (Marini *et al.*, 1988). For this keratin promoter, the relative transcript accumulation for supercoiled and linear forms differed by about 20-fold. The potential for confusion is exacerbated by the variable extent to which various constructions exhibit this differential; in our experience there is a tendency for longer constructions to be expressed strongly without linearization. The reason for this behavior is completely obscure, but it does not appear to be due to the presence of upstream positive control elements.

V. Preliminary Mapping by Polymerase Chain Reaction Footprinting Analysis

If the regulatory region of the chosen gene can be roughly localized to a few hundred base pairs of DNA, then a high-resolution map of protein contacts with potential regulatory elements can be generated using *in vivo* footprinting. Traditional *in vivo* footprinting protocols are impractical with *Xenopus* embryos (as well as most other species), owing to the relatively large genome size and the limited ability to get large quantities of purified cells. The recently developed ligation-mediated polymerase chain reaction (PCR) *in vivo* footprinting procedure (Mueller and Wold, 1989) takes major steps to alleviate both of these difficulties and has been used successfully to monitor protein–DNA contacts in both embryonic and adult *Xenopus* tissues (see Fig. 5).

We describe minor modifications to the basic procedure of Mueller and Wold, as well as protocols for application to *Xenopus* tissues. In addition, we present a newly developed procedure for *in vivo* digestion of DNA using copper–phenanthroline, along with the standard dimethyl sulfate modification. The procedure is diagrammed in Fig. 4.

There are several points to consider before beginning a footprinting analysis. First and foremost, it will be necessary to collect at least 10^5 cells, highly homogeneous for expression of the gene of interest, for each footprinting reaction lane. This number of cells is not based on signal considerations, which can be adjusted by varying the number of PCR cycles, but rather on the statistical need to represent accurately the abundance of each cleavage fragment

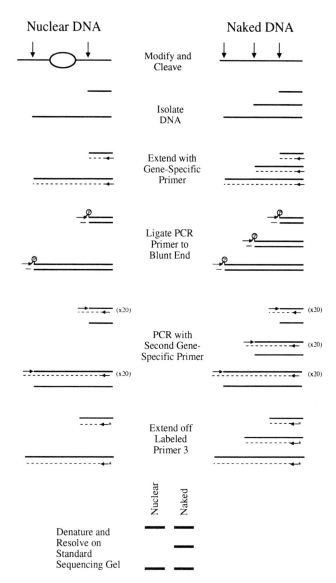

FIG. 4. PCR footprinting strategy. Nuclear (*in vivo*) and naked (*in vitro*) DNA are modified using one of several reagents (e.g., dimethyl sulfate, copper–phenanthroline). After the DNA is isolated, the first primer extension reaction is performed using the gene-specific primer 1 and serves to create a blunt-ended duplex. A linker primer duplex (Mueller and Wold, 1989) is ligated to this blunt end, providing a defined sequence for PCR amplification and preserving the random cleavage sites. The cleavage fragment carries a 5'-phosphate (Ⓟ) which allows the ligation reaction to occur. Using the top strand of the linker primer and a second gene-specific primer nested inside the 5' end of primer 1, exponential amplification can be obtained using the PCR. After 20 cycles of amplification, ^{32}P-labeled primer 3 is used in a primer extension reaction, again using Taq DNA polymerase and PCR conditions. After recovering the resulting DNA, the labeled fragments are resolved on an acrylamide gel and exposed to X-ray film. Reproducible regions of protection and enhancement of bands compared between nuclear and naked DNA show probable sites of protein binding.

prior to PCR amplification. Failure to start with a sufficient number of target sites could result in false protections and enhancements. Therefore, a typical reaction calls for 1–2 μg of purified DNA, which represents about $1.5-3 \times 10^5$ cells or $3-6 \times 10^5$ genome targets in *Xenopus*.

There is no definitive minimal value for how homogeneous the experimental cell population must be, but the cleaner the tissue, the stronger the footprint. Any amount of contaminating tissue will serve to obscure the footprints by contributing irrelevant fragments. Comparison of the footprint ladder from the tissue of interest to one from naked DNA and one or two control tissues should be made. The control tissues may be chosen for their ease of preparation, but it is necessary, of course, that they do not express the gene under study.

Finally, the position and composition of the oligonucleotide primers for each reaction are very important. The positioning of primers along the promoter is a balance between the amount of sequence expected to be read from a primer set, the base composition in the region under analysis, and the number of primer sets that can be practically analyzed. The extent of a sequence reaction depends on several factors, including the degree of digestion caused by the random modification, the signal intensity, and the gel running conditions. Although individual success rates may vary, the conditions presented should yield footprinting data over a 200- to 300-bp interval, starting about 10 bp from the end of the labeled primer. If the regulatory region is very large, multiple primer sets will be necessary, and this will increase the workload substantially. When designing multiple primer sets, it is advisable to space them by 150–250 bp to allow for some overlap as in standard sequencing strategies.

A primer set consists of three nested oligonucleotides with each subsequent primer having a higher melting temperature (T_m) and extending further 3' (see Fig. 4). Primer 1 is the Sequenase primer and should have a T_m over 47°C, the Sequenase extension temperature. Since T_m is based on G–C content as well as length, primers need to be carefully placed within the region to be analyzed. The equation for calculating T_m is $81.5°C + 16.6(\log M) + 0.41(\%GC) - (500 \div n)$, where M is the cation molarity and n is primer length. The cation concentration is 0.077 M for the Sequenase reaction and 0.047 M for the Taq reactions (Na$^+$ concentration plus 0.67 times Tris concentration). Therefore, a 20-mer at 40% G–C has a T_m of 54.4°C in Sequenase buffer and works well as primer 1. Primer 2 is the amplification primer and can accommodate a higher T_m thanks to the use of Taq DNA polymerase. The T_m for primer 2 should be in the 60°–65°C range; a 25-mer at 48–60% G–C works well. This will match the annealing temperatures for the linker primer and primer 2, which is necessary for driving the amplification. Primer 3, the labeling primer, should overlap primer 2 considerably in order to compete in the labeling step;

it helps if the T_m of primer 3 is 3°–4°C above that of primer 2. The general rule with these primers is to make the reaction as specific as possible by increasing the stringency, even if it means going above the calculated T_m. This will have to be determined empirically for each primer set, but it can be done on naked DNA instead of dissected tissue. All primers are purified on either thin-layer chromatography plates or polyacrylamide gels to ensure the correct length.

We describe below two different modification protocols for both *in vivo* and *in vitro* digestion. The standard dimethyl sulfate (DMS) reactions are basic Maxam and Gilbert (1980) chemistry, giving a G-specific cleavage pattern. The major advantage of using DMS is that it rapidly penetrates cells, eliminating the need to permeabilize. The drawback is its limited reactivity, especially in A–T-rich domains. Copper–phenanthroline shows cleavage patterns similar to those seen with DNase I or micrococcal nuclease, with some regional preference, but no specific base limitations. Because of the nature of the copper–phenanthroline complex, permeabilization of the cell is necessary, and it is achieved with a lysolecithin treatment. The DMS and copper–phenanthroline systems seem to complement each other in footprinting reactions, as can be seen in Fig. 5. Other modifying and cleavage reagents should also work well with this procedure. Phenanthroline, neocuproine, 3-mercaptopropionic acid, and lysolecithin are all commercially available from Sigma Chemical Co. (St. Louis, MO). Dimethyl sulfate and piperidine are available from Aldrich (Milwaukee, WI) and should be handled with extreme care as they are quite toxic. We present our most recent protocols describing digestion and footprinting conditions, worked out for analysis of the XK81A1 keratin promoter. We welcome any differing results, comments, or suggestions.

FIG. 5. Footprinting reactions comparing nuclear and naked DNA samples. A region upstream of the epidermal keratin gene, XK81A1, was analyzed using ligation-mediated PCR *in vivo* footprinting. Dimethyl sulfate and copper–phenanthroline were used as cleavage reagents. For the dimethyl sulfate lanes, embryonic epidermal DNA and adult skin DNA are compared after *in vivo* modification and amplification. The adult skin sample resembles the DMS sequence obtained from analyzing naked DNA (data not shown). For the copper–phenanthroline lanes, epidermal DNA cleaved *in vivo* after permeabilization with lysolecithin is compared to naked DNA cleaved *in vitro*. Brackets denote regions of protection in the epidermal samples, and asterisks denote regions of enhanced *in vivo* cleavage. Strong and continuous regions of protection are highlighted using the solid bars and are named for the approximate center point in the promoter sequence. The pattern seen *in vivo* at the −157 site closely resembles that observed for the *in vitro* binding of the keratin gene transcription factor, KTF-1 (Snape *et al.*, 1990). The −140 site is especially interesting in that it shows no protection in the dimethyl sulfate reaction (probably resulting from the limited number of G residues on this strand) but strong protection in the copper–phenanthroline reactions.

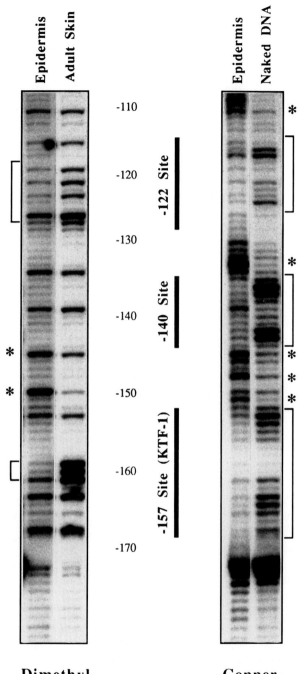

A. Recipes

Injection Buffer
88 mM NaCl, 10 mM HEPES, pH 6.8 (usual volume 5 nl; usual DNA dose 50–200 pg)

1× MMR
100 mM NaCl, 2 mM KCl, 1 mM $MgSO_4$, 2 mM $CaCl_2$, 5 mM HEPES, pH 7.4, 0.1 mM EDTA

1× PB Buffer
67 mM Na_2HPO_4, pH 7.6

1× CMFM
88 mM NaCl, 1 mM KCl, 2.4 mM $NaHCO_3$, 7.5 mM Tris, pH 7.6

1× Solution A
150 mM sucrose, 80 mM KCl, 35 mM HEPES, pH 7.4, 5 mM K_2HPO_4 pH 7.4, 5 mM $MgCl_2$, 0.5 mM $CaCl_2$

1× PEB
100 mM Tris, pH 8.0, 1% sodium dodecyl sulfate (SDS), 10 mM EDTA

1× GEB
4.2 mM guanidinium thiocyanate, 0.5% sarkosyl, 25 mM Tris, pH 8.0, and 0.7% 2-mercaptoethanol (added fresh)

DMS Stop
1.5 M sodium acetate, pH 7.0, 1 M 2-mercaptoethanol, 100 mg/ml tRNA

5× Mg-Free Sequenase Buffer
200 mM Tris, pH 7.7 (at room temperature), 250 mM NaCl

Sequenase Mix
20 mM $MgCl_2$, 20 mM dithiothreitol (DTT), 0.2 mM dNTPs, 4 units Sequenase version II (United States Biochemicals, Cleveland, OH) per reaction, freshly made

Ligation Mix
13.33 mM $MgCl_2$, 29.9 mM DTT, 1.67 mM rATP, 83.33 mg/ml bovine serum albumin (BSA), 2 units T4 DNA ligase per reaction, 100 pmol PCR linkers per reaction, and 27.8 mM Tris, pH 7.7, which is provided with the PCR linkers; freshly made

5× Taq Buffer
200 mM NaCl, 50 mM Tris, pH 8.9 (at room temperature), 25 mM $MgCl_2$, 0.05% gelatin

PCR Cocktail
3.3 × Taq buffer, 20 nmol each dNTP per reaction, 10 pmol primer 2 per reaction, 10 pmol linker primer (the 25-mer only) per reaction, and 1 unit Taq DNA polymerase (Perkin Elmer Cetus, Norwalk, CT) per reaction, freshly made; final amplification reaction should be 1 × Taq buffer

Labeling PCR Mix
  1× Taq buffer, 2 mM each dNTP, 1–2 pmol ^{32}P-end-labeled primer 3 per reaction, 2.5 units Taq DNA polymerase per reaction, freshly made

Taq Stop
  260 mM sodium acetate, 10 mM Tris, pH 7.5, 4 mM EDTA

Loading dye
  99% formamide, 0.1% xylene cyanol, 0.1% bromophenol blue, 5 mM EDTA

1× TBE
  100 mM Tris base, 40 mM boric acid, 20 mM EDTA

10× Kinase Buffer
  500 mM Tris, pH 7.6, 100 mM MgCl$_2$, 50 mM DTT, 1 mM spermidine, 1 mM EDTA

Gradient Buffer System
  0.5× TBE in top tank; 1× TBE in the gel; 0.67× TBE, 1 M sodium acetate, pH 7.8, in the bottom tank

B. *In Vivo* Copper–Phenanthroline Digestion Procedure for *Xenopus* Embryonic Tissues

1. Dissect embryos in 1× MMR. Transfer tissues to 200 μl of 1× PB buffer or 1× CMFM and wash by pelleting at 2000 rpm (2K) for 1 minute.
2. Resuspend cells in 200 μl of 1× CMFM and vortex lightly to fully dissociate.
3. Wash dissociated cells once or twice in 0.67× solution A (Miller *et al.*, 1978).
4. Resuspend cells in 75 μl cold 0.67× solution A. Add 25 μl of 1 mg/ml lysolecithin (type I) in 0.67× solution A (to 250 μg/ml) and incubate for 1 minute at 4°C (4°C is apparently 10-fold better than 37°C for permeabilization; Miller *et al.*, 1978).
5. Pellet cells to remove lysolecithin; drain the supernatant, and resuspend the pellet in 50 μl of 0.67× solution A at room temperature. Volumes can be scaled up to match system needs.
6. Add 5 μl of 1 mM phenanthroline (1,10-phenanthroline)/0.23 mM CuSO$_4$ and 5 μl of 58 mM 3-mercaptopropionic acid (MPA). [Copper digestion conditions are based on those described by Kuwabara and Sigman (1987)]. Incubate at room temperature for 1 to 5 minutes depending on the tissue. (The phenanthroline–copper mix is made fresh from 40 mM phenanthroline in ethanol and 9 mM CuSO$_4$ in water. This is probably important since the mixture will be unstable. MPA is

an 11.6 M stock, so this is a 1:200 dilution. For *Xenopus* neurula epidermis, 3 minutes seems to work well. For the resulting carcass, 1 minute is sufficient. Adult blood samples are prepared using 10 × volumes and a 5-minute digestion. *Note*: Everything is kept at the same ratio.)

7. Add 5 μl of 28 mM neocuproine (2,9-dimethyl-1,10-phenanthroline made up in ethanol) to quench the cleavage reaction. Dilute with 0.67 × solution A, pellet the cells, and remove the supernatant.
8. Add 200 μl of 1 × PEB, vortex to resuspend the cell pellet, and add 200 μl of 1 × GEB. Quick freeze on liquid nitrogen or dry ice. Store at −80°C until ready for phenol extractions.
9. Add 200 μl of saturated phenol, and vortex while thawing. Heat to 65°C for 2 minutes, vortex and add 200 μl chloroform, vortex, heat 65°C for 2 minutes, and spin at 15K for 2 minutes.
10. Remove the organic phase from the bottom and reextract in the same way, until the interface is minimal.
11. Remove the aqueous phase to a new tube and extract with 200 μl of chloroform; transfer the aqueous phase to a new tube.
12. Add 50 μl of 3 M sodium acetate, pH 5.5, and 550 μl 2-propanol.
13. Precipitate for 10 minutes on ice or overnight at −20°C.
14. Spin at 15K for 10 minutes, rinse with 70% ethanol, and lyophilize until dry.
15. Resuspend in 50 μl of 0.1 M NaOH at 65°C for 30 minutes, neutralize with 10 μl of 3 M sodium acetate, pH 5.5, 50 μl of 0.1 M HCl, and ethanol precipitate. Store at −20°C as a dried pellet.

C. *In Vitro* Copper–Phenanthroline Digestion Procedure

1. Dissolve naked DNA in 50 μl of 0.67 × solution A. (The amount of DNA is variable. The conditions here have been worked out for quantities of DNA from 2 to 100 μg. It may be necessary to change conditions slightly, varying time or concentration. Again, the solution A concentration should be adjusted to match *in vivo* conditions.)
2. Add 5 μl of 0.4 mM phenanthroline–0.09 mM $CuSO_4$ and 5 μl of 23.2 mM MPA. Incubate for 1 minute at room temperature. Quench the reaction with 5 μl of 28 mM neocuproine. [The conditions of Kuwabara and Sigman (1987) result in overdigestion for these purposes.]
3. Dilute with 35 μl water, add 10 μl of 3 M sodium acetate, pH 5.5, and ethanol precipitate.
4. If RNA has not already been removed, resuspend the precipitate in 0.1 M NaOH. Heat at 65°C for 30 minutes, neutralize with 10 μl of 3 M

sodium acetate, pH 5.5, 50 μl of 0.1 M HCl, and ethanol precipitate. Store at −20°C as a dried pellet.

D. *In Vivo* Dimethyl Sulfate Treatments for *Xenopus* Embryonic Tissues

1. Dissect embryos in 1× MMR. Transfer tissues to 200 μl of 1× PB buffer or 1× CMFM and wash by pelleting at 2K for 1 minute.
2. Resuspend the cells in 200 μl of 1× CMFM and vortex lightly to fully dissociate. Spin again and resuspend in 200 μl of 1× CMFM.
3. Add 5 μl of 10% DMS in 100% ethanol, vortex, and incubate for 0.5–3 minutes depending on tissue and digestion needs. [The time will have to be determined empirically to match *in vitro* digestion patterns. We treat neurula epidermis for 1.5 minutes, carcass for 30 seconds, and adult skin for 1 minute (in 500 μl of 1× CMFM). Because of higher cell numbers, we use 2 μl DMS for 3 minutes in 500 μl of 1× CMFM with adult blood.]
4. Add 1 ml ice-cold 1× CMFM, vortex, and spin at 2K for 1 minute to loosely pellet cells. Remove the supernatant. Repeat with another 1 ml of cold 1× CMFM.
5. Pellet the cells and resuspend in 200 μl of 1× PEB. Isolate the DNA as described above in the *in vivo* copper–phenanthroline procedure (Section V,B), Steps 8–14.
6. Resuspend in 50 μl of 1 M piperidine (stock is 10 M) and heat at 90°C for 30 minutes. [*Note*: The results of Brewer *et al.* (1990) suggest that the piperidine step can be eliminated since strand scission will occur in a subsequent denaturation step. We have not tested this as yet. Keep in mind that the RNA will still need to be removed.]
7. Transfer to ice, spin briefly, and then lyophilize in a Speed-vac.
8. Add 50 μl of water and lyophilize again. Store dry at −20°C until ready.

E. *In Vitro* Dimethyl Sulfate Treatment

The following conditions are basically those found in Maniatis *et al.* (1982).

1. Add the appropriate amount of naked genomic DNA to 1× CMFM for a final volume of 200 μl.
2. Add 1 μl of 10% DMS (in 100% ethanol), vortex, and incubate at room temperature for 1 minute.
3. Add 50 μl DMS stop and 750 μl ethanol, both ice cold. Store at −70°C for 3 minutes. Spin at 15K for 10 minutes and remove the supernatant.

4. Add 250 µl of 0.3 M sodium acetate, pH 7.0, and 750 µl ethanol, place at $-70°C$ for 3 minutes and spin at 15K for 10 minutes. Wash with 70% ethanol and dry under reduced pressure.
5. Resuspend in 50 µl of 1 M piperidine (stock is 10 M) and heat at 90°C for 30 minutes.
6. Transfer to ice, spin briefly, and then lyophilize in a Speed-vac.
7. Add 50 µl of water and lyophilize again. Store dry at $-20°C$ until ready.

F. Primer 1 Extension and Linker Ligation Reactions

1. Resuspend the DNA samples in water for measuring optical density. (The concentration should exceed 0.25 mg/ml, so at least 1–2 µg DNA can be put into the Sequenase reaction.)
2. Add 3 µl of primer 1 (0.3 pmol) and 3 µl of $5 \times$ Mg^{2+}-free Sequenase buffer to 9 µl of DNA (1–2 µg) in water. (The amount of DNA should represent *at least* 10^5 genomes.)
3. Denature for 5 minutes at 95°C.
4. Hybridize at 50°C for 30 minutes. (The temperature is dependent on the T_m of primer 1.) Cool and spin briefly to collect condensed water.
5. Add 9 µl Sequenase mix and incubate for 5 minutes at 47°C, then 5 minutes at 60°C to inactivate the polymerase.
6. Add 6 µl of 310 mM Tris, pH 7.7, and heat to $65°–70°C$ for 10 minutes. Cool on ice.
7. Add 45 µl ligation mix and transfer to 18°C overnight.
8. Heat inactivate the ligase for 10 minutes at 70°C, add 10 µg tRNA, 8.4 µl of 3 M sodium acetate, pH 5.5, and 220 µl ethanol. Precipitate as usual, wash with 70% ethanol, and lyophilize.
9. Resuspend the pellet in 70 µl water.

G. Kinasing of Primers

1. Add the following ingredients: 6 µl water, 1 µl primer 3 (10 pmol, but this can vary), 2 µl of $10 \times$ PNK buffer, 1 µl of T4 polynucleotide kinase, and 10 µl of $[\gamma\text{-}^{32}P]ATP$ (5000–6000 Ci/mmol, 10 mCi/ml).
2. Incubate at 37°C for 1–2 hours.
3. Add 180 µl NEN-SORB reagent A and purify the labeled primer over a NEN-SORB column (Dupont–NEN, Boston, MA) as per the manufacturer's instructions, using a 50% ethanol elution.
4. Collect drops 3 through 20 in a single fraction and lyophilize in a Speed-vac.
5. Store dry at $-20°C$. Resuspend in an appropriate volume of water for the labeling reaction.

H. Taq Reactions for Genomic Footprinting

1. Add 30 μl PCR cocktail to the 70 μl cleaved and ligated DNA.
2. Vortex and cover with 70 μl of light mineral oil. Spin briefly to separate.
3. Run 20 cycles of the following profile (manually or by thermal cycler):
   94°C for 1 minute—denaturation (The first cycle has a 2-minute denaturation step.)
   64°C for 2 minutes—annealing (Only a guideline! Use ∼2°C above calculated T_m.)
   76°C for 3 minutes—extension
   (*Note*: More than 20 cycles is usually not necessary and may lead to variations between samples owing to unequal amplification.)
4. Stop reaction on ice and spin briefly.
5. Add 5 μl of labeling mix and run 1 cycle of the following:
   94°C for 2 minutes
   68°C for 2 minutes (Again, only an example; use ∼2°C above calculated T_m.)
   76°C for 5 minutes
6. Add 100 μl of saturated phenol and 200 μl of chloroform to remove protein and mineral oil.
7. Vortex and remove the aqueous phase (top layer) to a new tube.
8. Add 295 μl Taq stop solution, 10 μg tRNA, and 1 ml 100% ethanol and precipitate overnight. Spin at 15K for 10 minutes, wash with 70% ethanol, and dry under reduced pressure.
9. Resuspend in 8 μl of 50% loading dye.
10. Denature for 5 minutes at 95°C. Load hot onto gel.
11. Run on a 6% acrylamide, 50% urea, 1 × TBE gel, with 0.5 mm spacers, 80 cm long by 40 cm wide, for about 8 hours at 90 W constant power using the gradient buffer system, usually until the xylene cyanol front has gone 60 cm.
12. Transfer the gel to Whatmann 3MM paper and expose to X-ray film with or without an intensifying screen. (Exposing without a screen gives slightly sharper bands. Gels can also be fixed and dried before exposing.)

VI. Regulatory Element Functional Studies

As shown in Fig. 5, *in vivo* footprinting can reveal sites of protein–DNA interaction within a regulatory region. These sites are likely to be important in regulating the associated gene. The functional nature of these domains can be analyzed by mutating the binding site *in vitro*, injecting the resulting constructions, and assaying the corresponding activity. Destruction of important regulatory domains should result in reduced or aberrant activity with respect to endogenous expression. It is a relatively simple matter to eliminate individual putative control elements, either by introducing base changes or by making small deletions. The point mutation method has the advantage of not disturbing the spatial relationships of other nearby regulatory sites. Several mutagenesis kits are commercially available and are provided with customized protocols.

Evaluating the effects of the point mutations needs to be done with care. As noted above, there are poorly understood phenomena, for example, the effects that linearization has on some constructions, that can have drastic effects on expression level. It is probably a good idea to inject reference constructions along with the wild-type and mutated versions of the promoter of interest. However, if the constructions need to be linearized before injection, a possible artifact is created: the control and experimental DNA will become concatenated in the fertilized egg. If there are control elements on one or both constructions that can act at a distance from a transcriptional start site, then the regulation of one construction could conceivably be imposed on the other. Probably the best advice in designing injection experiments is to use two or more different strategies. For example, an alternative to mixing control and experimental plasmids is to inject them separately into the two blastomeres after first cleavage, or even into separate embryos which can later be pooled. If similar results are obtained with different experimental designs, then it is reasonable to draw firm conclusions.

We have presented outlines for four important steps in the analysis of class II gene regulation: selection of a suitable gene, rough definition of regulatory regions by microinjection of deletion constructions, *in vivo* footprinting to identify protein binding domains, and, finally, mutagenesis of specific domains to test their functional importance. The process obviously does not end here, with further investigations possibly including transcription factor characterization and cloning, as well as functional disruptions of gene products using antisense and/or antibody injections. We hope that this chapter will help to guide those beginning a molecular approach to gene expression in *Xenopus* and will offer helpful aid to those currently involved in these studies.

References

Bendig, M. M., and Williams, J. G. (1984). Differential expression of the *Xenopus laevis* tadpole and adult β-globin genes when injected into fertilized *Xenopus laevis* eggs. *Mol. Cell. Biol.* **4,** 567–570.

Brewer, A. C., Marsh, P. J., and Patient, R. K. (1990). A simplified method for *in vivo* footprinting using DMS. *Nucleic Acids Res.* **18,** 5574.

Gurdon, J. B., and Wakefield, L. (1986). Microinjection of amphibian oocytes and eggs for the analysis of transcription. *In* "Microinjection and Organelle Transplantation Techniques: Methods and Applications" (J. E. Celis, A. Graessman, and A. Loyter, eds.), pp. 269–299. Academic Press, London.

Jonas, E. A., Snape, A. M., and Sargent, T. D. (1989). Transcriptional regulation of a *Xenopus* embryonic epidermal keratin gene. *Development (Cambridge, U.K.)* **106**, 399–405.

Krieg, P. A., and Melton, D. A. (1985). Developmental regulation of a gastrula-specific gene injected into fertilized *Xenopus* eggs. *EMBO J.* **4**, 3463–3471.

Krieg, P. A., and Melton, D. A. (1987). An enhancer responsible for activating transcription at the midblastula transition in *Xenopus* development. *Proc. Natl. Acad. Sci. U.S.A.* **84**, 2331–2335.

Krone, P. H., and Heikkila, J. J. (1989). Expression of microinjected hsp70/CAT and hsp30/CAT chimeric genes in developing *Xenopus laevis* embryos. *Development (Cambridge, U.K.)* **106**, 271–281.

Kuwabara, M. D., and Sigman, D. S. (1987). Footprinting DNA–protein complexes *in situ* following gel retardation assays using 1,10-phenanthroline–copper ion: *Escherichia coli* RNA polymerase–*lac* promoter complexes. *Biochemistry* **26**, 7234–7238.

Maniatis, T., Fritsch, E. F., and Sambrook, J. (1982). "Molecular Cloning: A Laboratory Manual." Cold Spring Harbor Laboratory Cold Spring Harbor, New York.

Marini, N. J., Etkin, L. D., and Benbow, R. M. (1988). Persistence and replication of plasmid DNA microinjected into early embryos of *Xenopus laevis*. *Dev. Biol.* **127**, 421–434.

Maxam, A. M., and Gilbert, W. (1980). Sequencing end-labeled DNA with base-specific chemical cleavages. *In* "Methods in Enzymology" (L. Grossman and K. Moldave, eds.), Vol. 65, pp. 499–560. Academic Press, New York.

Miller, M. R., Castellot, J. J., Jr., and Pardee, A. B. (1978). A permeable animal cell preparation for studying macromolecular synthesis. DNA synthesis and the role of deoxyribonucleotides in S phase initiation. *Biochemistry* **17**, 1073–1080.

Mohun, T. J., Taylor, M. V., Garrett, N., and Gurdon, J. B. (1989). The CArG promoter sequence is necessary for muscle-specific transcription of the cardiac actin gene in *Xenopus* embryos. *EMBO J.* **8**, 1153–1161.

Mueller, P. R., and Wold, B. (1989). *In vivo* footprinting of a muscle specific enhancer by ligation mediated PCR. *Science* **246**, 780–786.

Snape, A. M., Jonas, E. A., and Sargent, T. D. (1990). KTF-1, a transcriptional activator of *Xenopus* embryonic keratin expression. *Development (Cambridge, U.K.)* **109**, 157–165.

Taylor, M., Treisman, R., Garrett, N., and Mohun, T. (1989). Muscle-specific (CArG) and serum-responsive (SRE) promoter elements are functionally interchangeable in *Xenopus* embryos and mouse fibroblasts. *Development (Cambridge, U.K.)* **106**, 67–78.

Chapter 20

Assays for Gene Function in Developing Xenopus Embryos

PETER D. VIZE AND DOUGLAS A. MELTON

Department of Biochemistry and Molecular Biology
Harvard University
Cambridge, Massachusetts 02138

ALI HEMMATI-BRIVANLOU AND RICHARD M. HARLAND

University of California at Berkeley
Department of Molecular and Cell Biology
Division of Biochemistry and Molecular Biology
Berkeley, California 94720

I. Introduction
II. Methods
   A. Preparation of mRNA
   B. Preparation of DNA
   C. Preparation of Antibodies
   D. Preparation of Oligonucleotides
   E. Controls
   F. Microinjection
   G. Detection of β-Galactosidase
III. Results
   A. Overexpression from Injected mRNA
   B. Overexpression from Injected DNA
   C. Antisense RNA
   D. Injection of Antisense Oligonucleotides
   E. Ribozymes
   F. Injection of Antibodies
   G. Expression of "Dominant Negative" Proteins
   H. Commonly Encountered Injection Artifacts
IV. Conclusions
   References

I. Introduction

The investigation of vertebrate development is made difficult by a number of factors. The generation time and large genome size of vertebrates make genetic analysis difficult; furthermore, embryos are often inaccessible at early developmental stages. Among vertebrates, *Xenopus* has the obvious advantage of easy access to embryos at all stages of development. Although *Xenopus* development cannot be studied using conventional genetics, alternate methods have been devised which may circumvent this limitation. This chapter reviews some of these approaches and points out some potential benefits and pitfalls.

The strategies we discuss here are to over- or underexpress genes by injection of cloned DNA, synthetic mRNA, or antibodies. Although these approaches are still under development, we believe they show potential in obviating many of the problems encountered in the study of gene function in higher animals. The accessibility and rapid development of *Xenopus* embryos make it possible to assay the effect of the overexpression of cloned genes within days, and with further refinement, similar results with antisense-generated underexpression should be possible. The speed with which such experiments can be performed makes screening mRNAs for function in this way feasible. Transgenic methods are not yet available for *Xenopus*, and until such methods are developed one is limited to ectopic or overexpression of cloned genes in a transient manner, as well as transient blocking methods using antisense or antibody injection.

The overexpression of cloned genes can be achieved by the injection of *in vitro* synthesized mRNA during the first day of development, and expression from the mid blastula transition (MBT) (Newport and Kirschner, 1982) onward can be achieved by injecting plasmid DNA. Ablation of expression of endogenous genes can be achieved using antisense oligonucleotides and antisense RNA, by injecting antibodies, or by expressing proteins which block the activity of their active counterparts in a dominant fashion.

II. Methods

A. Preparation of mRNA

The *in vitro* production of capped mRNA from cloned cDNA sequences using bacteriophage RNA polymerase has been described in detail elsewhere (Melton *et al.*, 1984). It is important to include the 5' cap to ensure efficient translation (Krieg and Melton, 1984, 1987b). RNA for injection into embryos is synthesized by these protocols, usually incorporating a small amount of radioisotope (10 μCi) to trace the efficiency of the reaction. For microinjection experiments, it is important to purify the mRNA extensively. Following transcription and after DNase digestion, transcription reactions are extracted with phenol–chloroform and purified by chromatography over Sephadex G-50, followed by ethanol precipitation. The purified mRNA is resuspended in water at a concentration of 200 μg/ml, and stored at $-100°$C until used. Shortly before injection the mRNA is centrifuged in a microcentrifuge for 1 minute to pellet insoluble material. As with all work utilizing RNA, care should be taken at all steps to avoid contamination with RNases. Microinjection of mRNA at 200 μg/ml and at a lower concentration (e.g., 50 μg/ml) is suggested (see below).

B. Preparation of DNA

Plasmids are purified from *Escherichia coli* by alkaline lysis followed by banding in CsCl. Supercoiled DNA is recovered from CsCl bands by extraction with butan-1-ol (twice); 2 volumes of water and then added, and the solution is extracted with phenol–chloroform. DNA is then collected by the addition of 2 volumes of ethanol and centrifugation. DNA is rinsed with 70% ethanol, dried, and resuspended in TE. Purified supercoiled plasmid is linearized with a restriction enzyme, extracted with phenol–chloroform (1:1), and ethanol precipitated. The linearization site can be on either side of the promoter/test gene, and it can be cut with enzymes that generate blunt or sticky ends. The digested DNA is resuspended in water at a concentration of 10 μg/ml and stored at $-20°$C. It is useful to check the prepared DNA on an agarose gel to ensure that the concentration is correct and to verify the absence of contaminating RNA. The injection regimen is identical to that described above for mRNA, using injection volumes of 5–10 nl. Injecting higher concentrations of DNA results in higher levels of expression but causes a higher mortality. Lower concentrations should be used if a high mortality rate is observed. If supercoiled DNA is to be injected, DNA should be phenol–chloroform extracted and ethanol precipitated once again to ensure that no CsCl is present.

C. Preparation of Antibodies

Antibodies must be purified from serum before injection, as serum is toxic to the embryo (Oliver *et al.*, 1988). This can be achieved by any of the standard immunological protocols, such as ligand affinity chromatography, or by purifying immunoglobulins with protein A–Sepharose. Methods detailing each of these protocols are clearly explained in the book *Antibodies*, by Harlow and Lane (1988). The buffer containing the antibody should be exchanged for 88 mM NaCl, 1 mM KCl, 15 mM Tris-HCl, pH 7.5. Stocks can be stored as small aliquots at $-20°C$.

D. Preparation of Oligonucleotides

The materials used to synthesize oligonucleotides are highly toxic, and so care should be taken to carefully purify all oligonucleotides away from synthesis reagents. One method is polyacrylamide gel electrophoresis through native gels [15% acrylamide (30:1) in TBE] followed by elution, phenol–chloroform extraction, and ethanol precipitation. This method will not only remove traces of toxic chemicals, but will also separate incomplete products of the synthesis. Other methods employed include reversed-phase chromatography followed by purification over a cation-exchange column and gel filtration (see Woolf et al., 1990). It should be remembered that high-performance liquid chromatography (HPLC) solvents are also very toxic, so if this method is used to purify the oligonucleotide from synthesis products, care must then be taken to remove any traces of the purification reagents. The length of oligonucleotides used varies considerably, from 10 to 30 nucleotides. An intermediate length of between 18 and 20 nucleotides is a good starting point but may need to be optimized by trial and error. A discussion of the effects of varying oligonucleotide length can be found in Baker et al. (1990).

E. Controls

RNA, DNA, or sera for control injections should be prepared in parallel if possible. This may be important as variation in contaminants, in addition to differences in mRNA, DNA, or antibodies, could result in samples generating different results. For instance an mRNA preparation may contain different amounts of contaminating reagents from the *in vitro* synthesis, and antibody preparations different amounts of contaminating sera or azide. To interpret differences between control and test samples all samples are prepared in a similar fashion. Commonly used controls for mRNA injections are *Xenopus* β-globin or β-galactosidase (as presented in this chapter, see also Detrick et al., 1990). With toxicity controls it may be important to take into account the protein translated from injected mRNA, in addition to the toxicity of the RNA itself. More subtle controls in which specific amino acid are changed, so as to obliterate protein function, would in such instances be an advantage. Examples of such controls can be found in Harvey and Melton (1988) with *Xhox1*, McMahon and Moon (1989) with *int-1*, and Wright et al. (1989) with XlHboxl. One approach which may work especially well for secreted proteins is to alter cysteine residues to disrupt disulfide bond formation, yet conserve the remaining characteristics of the protein. (Because the cell is generally thought of as a reducing environment, disulfide bonds form when proteins are secreted.) This approach should only be necessary if nonspecific effects make interpretation of results difficult (see Section H). The injection of different concentrations of mRNA and showing a dose-dependent response are important controls.

One final method to control for embryo variability is to inject two-cell stage embryos in only one of the two blastomeres. This results in the injected mRNA only being present in one half of the embryo, usually only the left or right side, with the other half serving as a "control" (Danilchik and Black, 1988; Harvey and Melton, 1988; Klein, 1987; Masho, 1990). This could be a particularly useful control for proteins which dictate cell fate in some manner.

Controls for antisense oligonucleotides include annealing the antisense oligonucleotide to a sense oligonucleotide and testing the double-stranded form for activity. Control oligonucleotides with the base sequence shuffled also serve as controls to demonstrate the target specificity of the test oligonucleotide. Probably the best control possible is to inject *in vitro* synthesized mRNA corresponding to the target mRNA after the disappearance of the injected antisense oligonucleotide. If this experiment can "rescue" the injected phenotype, this is strong evidence that the phenotypic response was specific for the intended target mRNA (e.g., see Shuttleworth *et al.,* 1990).

F. Microinjection

The microinjection regimen is similar for the injection of mRNA, DNA, oligonucleotides, and antibodies. Fertilized eggs are dejellied with 3% cysteine-HCl (pH 7.9) immediately following cortical rotation. The cysteine is washed away thoroughly and embryos cultured at room temperature in $1 \times$ MMR. Approximately 60–70 minutes after fertilization embryos are transferred to a dish containing $1 \times$ MMR/3% (w/v) Ficoll 400. Droplets of this solution, each containing five embryos, are placed on siliconized glass slides for microinjection. An alternative method is hold embryos in place on a slide with a Nitex grid (1000 μm) for injections (see Harland and Misher, 1988). Approximately 20 nl of mRNA is injected into the vegetal pole of each embryo. Lower injection volumes, between 5 and 10 nl are used for injections into the marginal zone or animal pole of the egg. After injection, embryos are transferred to $1 \times$ MMR/3% Ficoll and incubated at $18°-24°C$ for 4 to 5 hours. Once embryos have reached stage 5 (Nieuwkoop and Faber, 1967), the medium is exchanged for $0.1 \times$ MMR, and the embryos are cultured until the required stage.

G. Detection of β-Galactosidase

To detect β-galactosidase expression (see also Detrick *et al.,* 1990) embryos are fixed in 2% paraformaldehyde (formaldehyde and glutaraldehyde also work well) 100 mM PIPES, 2 mM MgCl$_2$, 1.25 mM EGTA for 1 to 2 hours. Rinse with phosphate-buffered saline (PBS) plus 2 mM MgCl$_2$ (do not use alcohol fixatives). After rinsing twice with PBS, embryos are transferred to a staining solution [PBS plus 20 mM K$_3$Fe(CN)$_6$, 20 mM K$_4$Fe(CN)$_6$ · 3H$_2$O, 2 mM MgCl$_2$, 0.01% deoxycholate, 0.02% Nonidet P-40 (NP-40), with 5-bromo-4-chloro-3-indolyl-β-D-galactopyranoside (X-gal) added freshly to a concentration of 1 mg/ml] and incubate at 37°C for 2 to 24 hours. Following one rinse in PBS, then two in methanol, embryos are cleared in benzyl alcohol–benzyl benzoate (1:2) (A. Murray, personal communication). If pigmented embryos are used and staining is difficult to visualize, the endogenous pigment can be bleached out with hydrogen peroxide (Dent et al., 1989), after development in X-gal but before clearing. (The blue stain will disappear over the course of a few days if embryos are stored in clearing solution.) A permanent preparation can be obtained by whole-mount immunohistochemistry using an anti-β-galactosidase (anti-β-gal) antibody. The anti-β-gal monoclonal available from Promega (Madison, WI); Cat. No. Z3781) works well in these protocols, using standard fixation protocols (Dent et al., 1989; Hemmati-Brivanlou and Harland, 1989). If immunological detection is to be used, it may be necessary to perform the injections with albino embryos.

III. Results

A. Overexpression from Injected mRNA

If the cDNA that is being tested for developmental function is already cloned into an *in vitro* transcription vector, a simple experiment would be to transcribe the cDNA and inject the corresponding mRNA to test for an effect on development (Harvey and Melton, 1988). If no obvious phenotype is detected the following observations may be of interest.

A number of factors will influence the level of protein generated from injected mRNA. The major two factors are mRNA stability and translational efficiency. The half-lives of injected mRNAs have been examined by a number of researchers, and one comparative study has been published (Harland and Misher, 1988). Most injected mRNAs are present in the embryo for about 10 hours, up until the gastrula stage. mRNAs with longer half-lives, such as those for chloramphenicol acetyltransferase (CAT) and β-globin, can last well into the neurula stage, while some unstable mRNAs will only survive a few hours. The addition of a poly(A) tail at the 3' end of the synthesized mRNA will enhance the time of survival in the embryo, and this can be done in two ways. The first method is to clone the cDNA open reading frame (ORF) into a transcription plasmid that contains a stretch of dA residues 3' to the cloning site. The second method for adding poly(A) tails is enzymatic, using *E. coli* poly(A) polymerase to add a synthetic poly(A) tail to the 3' end of previously synthesized mRNAs (Harland and Misher, 1988). Longer poly(A) tails than are stable in plasmids are required to prolong mRNA half-lives, so the latter method may be necessary if rapid degradation still occurs.

The second factor influencing the amount of protein generated from an injected mRNA is the translational efficiency of the mRNA itself. The poly(A) tail which aids in mRNA stability is also important in this respect (Galili *et al.*, 1988). We routinely use pSP64T as a transcription vector, which contains a *Xenopus* β-globin cDNA cloned into pSP64 (Krieg and Melton, 1984). This cDNA has a "pseudo" poly(A) tail of 30 base pairs (bp) followed by a 30-bp poly(C) tail added during cloning. A number of restriction sites are present downstream of the poly(A-C) tail to enable linearization before *in vitro* transcription. Additional sequences, including those surrounding the AUG initiation codon and other sequences at both 5' and 3' ends of mRNA also greatly influence their effectiveness as templates for protein synthesis. The addition of 5' and 3' untranslated sequences from efficiently translated mRNAs, such as β-globin, onto reporter genes has been shown to increase the amount of protein generated from transfected mRNAs in cultured cells by a factor of 1000 (Malone *et al.*, 1989). Therefore, cloning the cDNA ORF you wish to express into pSP64T will enhance the translational efficiency of the RNA in many instances. We have found that transcripts generated from Bluescript vectors (Stratagene, La Jolla, CA) are particularly inefficient templates for generating proteins. In an experiment to test the relative efficiencies of translation of the CAT gene (which is usually translated quite efficiently) this gene was inserted into pSP64T, the *Sma*I site of pSP65, and the *Bam*HI site of a pBluescript KS$^+$ (in both orientations). The relative translational efficiencies were pBluescript (T7), 1 ×; pBluescript (T3), 42 ×; pSP64T, 270 ×; pSP65, 375 × (B. Brown and R. M. Harland, unpublished). The low translation of synthetic mRNAs generated off the pBluescript vector may be due to the polylinker sequences.

Two variables which can influence the response produced by an injected mRNA are the concentration of mRNA injected and the site of injection. Concentrations of mRNA much lower than 200 μg/ml are required for some messengers whose proteins have profound effects on development, for example, activin β_B, which at high doses causes gastrulation defects but at lower doses generates a secondary embryonic axis (Thomsen *et al.*, 1990). The site of injection is also important in some instances. For example, the injection of mouse *int-1* mRNA into the vegetal end of embryos results in a bifurcated anterior axis, whereas injection into the animal end of the embryo often results in exogastrulation (M. Whitman, personal communication). The overexpression of *Xhox3* exemplifies both of these variables. This mRNA shows both dose-dependent and site-dependent effects: either injections of high concentrations (200 μg/ml) or injections into the animal or equatorial region cause a higher frequency of anterior defects (Ruiz i Altaba and Melton, 1989).

Two examples of the utility of mRNA injection were investigations of *Xhox3* (Ruiz i Altaba and Melton, 1989) and *int-1* (McMahon and Moon, 1989). In both of these instances the phenotype generated from the injected mRNAs was extremely consistent: *Xhox3* caused anterior defects in up to 78% of injected embryos, and *int-1* induced a bifurcated axis in over 90% of injected embryos (Table I). This, in addition to the observation of Woodland and Jones (1987) that total poly(A)$^+$ selected RNA from cultured *Xenopus* cells or ovary can cause changes in cell fate in a large proportion of injected embryos, raises the possibility that this approach may be useful for screening for novel genes involved in morphogenesis. The injection of pools of RNA transcripts may thus allow the identification of transcripts with specific regulatory roles in early development.

Distribution of Injected mRNA

The distribution of injected mRNA can be examined by two procedures. The first is to stain the injected embryos immunologically for the protein translated from the injected RNA, under conditions which will not detect the endogenous protein or at times when the endogenous protein is not expressed. These experiments have shown that, in general, proteins from injected mRNAs are often distributed through about one-quarter of the embryo. Both the distribution and intensity of staining vary from embryo to embryo and experiment to experiment, probably because of differences in injection technique and the site of injection. It is therefore very important in each experiment to demonstrate that the mRNA is both being translated and widely distributed within the embryo. Simple protocols which enable the detection of proteins in whole-mount embryos are available (Dent *et al.*, 1989; Hemmati-Brivanlou and Harland, 1989; Chapter 22, this volume). An example of the type of result seen in such experiments is shown in Fig. 1 and 2 (see also Wright *et al.*, 1989). The observation that effects on development can

---

FIG. 1. Distribution of protein translated from injected mRNA visualized by immunostaining of nuclear *Xhox3* protein. (a) Animal view of blastula embryo injected with *Xhox3* mRNA, developed with an anti-Xhox3 antibody. (b) Control, uninjected embryo, animal view. (Data provided by A. Ruiz i Altaba.)

FIG. 2. Expression of nuclear β-galactosidase from injected mRNA. (Left) Group of embryos injected with a given concentration of RNA. (Right) Details of a representative inividual of the group. (a, b) Five nanograms of RNA was injected; (c, d) 500 pg of RNA injected; (e, f) 50 pg of RNA injected.

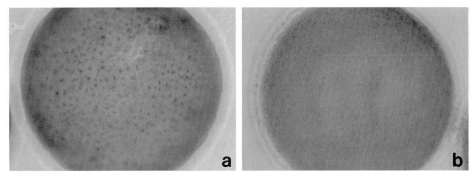

Fig. 1.

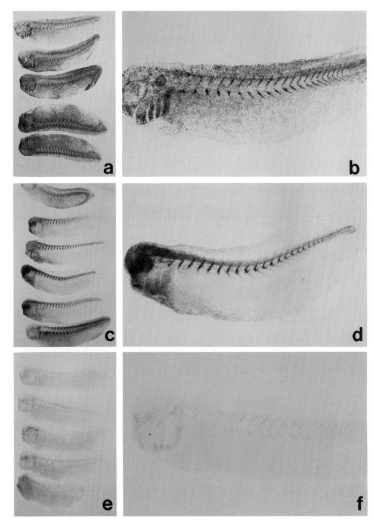

Fig. 2.

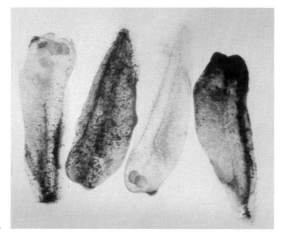

Fig. 3.

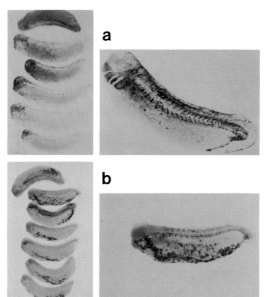

a

b

Fig. 4.

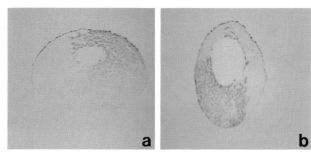

Fig. 5.   a   b

depend on the site of injection shows that even though protein distribution may appear to be widespread, a large proportion of the product concentrates in the cells derived from the injected region (see Kintner, 1988).

The second and often more simple approach to identify which cells have received the injected mRNA is to coinject the test mRNA along with a small amount of mRNA encoding a reporter such as β-galactosidase (Detrick et al., 1990). Histochemical detection, or staining with anti-β-gal antibodies, will then reveal which cells have received the mRNA. This is convenient, as an antibody against the experimental protein is not required, and the detection is fast and straightforward. This will allow the direct visualization of cells that have received the injected mRNA, but will not directly reflect the expression of the test RNA. We have quantitated the amount of β-galactosidase mRNA sufficient for use as a lineage tracer in such experiments and find that 500 pg of mRNA results in detectable immunological staining in injected embryos (Fig. 2). Controls to demonstrate that the injected mRNA is translatable include in vitro translation or in vivo translation in [^{35}S]menthionine-labeled oocytes (Harvey et al., 1986).

B. Overexpression from Injected DNA

The fate of DNA injected into embryos depends on whether linear or supercoiled DNA is used. Linear DNA is rapidly concatenated into high molecular weight complexes containing head to tail, head to head, and tail to tail fusions, and these concatamers are usually replicated up until at least the tailbud stages (Bendig, 1981; Etkin and Pearman, 1987; Etkin et al., 1984; Marini et al., 1988; Rusconi and Schaffner, 1981). However, it has also been observed that injected linear DNA can be concatenated but not replicated (Harland and Misher, 1988), and the replication of injected DNA may be concentration dependent (Rusconi and Schaffner, 1981). If supercoiled DNA

FIG. 3. Distribution of β-galactosidase activity expressed from microinjected plasmid DNA. An EF1 α promoter/β-galactosidase reporter plasmid was injected into one-cell embryos, which were stained for β-galactosidase enzyme activity at the tailbud stages. (Data provided by C. Jennings.)

FIG. 4. Region-specific expression of nuclear form of β-galactosidase from the Xenopus borealis cytoskeletal actin promoter (CSK). (a) Antibody detection of β-galactosidase expression from CSK/β-gal plasmid injections into the animal hemisphere. (b) Antibody detection of β-galactosidase expression from CSK/β-gal plasmid injections into the vegetal hemisphere. (Left) Group photograph of injected embryos. (Right) Higher resolution of a single embryo from each category.

FIG. 5. Distribution of injected antibody. An affinity-purified rabbit anti-Xhox3 polyclonal antibody was injected into one-cell embryos, and two of the resulting blastula were stained for the antibody with goat anti-rabbit immunoreagents. (Data provided by A. Ruiz i Altaba.)

TABLE I
Effects from Injected mRNAs and Antibodies

Microinjected RNAs mRNA	Class[a]	%	Developmental effect	References
Xhox1A	Hbox	51	kinking	Harvey and Melton (1988)
Xhox3	Hbox	>70	anterior deletion	Ruiz i Altaba and Melton (1989)
X1Hbox1(s)	Hbox	61	spinal cord defect	Wright et al. (1989)
X1Hbox1(l)	Hbox	67	kinking	Wright et al. (1989)
X1Hbox2	Hbox		none	A. Hemmati-Brivanlou and R. M. Harland, unpublished.
Xhox36	Hbox		none	R. M. Harland, unpublished.
En-2	Hbox		none	A. Hemmati-Brivanlou and R. M. Harland, unpublished.
int-1	GF	>90	bifurcated axis	McMahon and Moon (1989)
N-CAM	CS	75	kinking, epidermal defect	Kintner (1988)
N-cadherin	CS	47	ectodermal and neural tube defects	Detrick et al. (1990)
MyoD	DBP		none	Hopwood and Gurdon (1990)
Vimentin	CY		none	Christian et al. (1990)
Vg-1	GF		none	M. Whitman and D. Melton, unpublished.
activin	GF	>80	axial defect (duplications)	Thomsen et al. (1990)
XTC A$^+$ RNA		56	mesoderm induced in caps	Woodland and Jones (1987)
ovary A$^+$ RNA		73	mesoderm induced in caps	Woodland and Jones (1987)
middle T	KA	40	mesoderm induced in caps	Whitman and Melton (1989)
globin		5–13	none	Harvey and Melton (1989) and Ruiz i Altaba and Melton (1989)
raf	KA		none	M. Whitman, personel communication.
N-myc	DBP		none	P. Krieg, personel communication.

Antibody injection Antibody		%	Defect	References
X1Hbox1		29	spinal cord defects	Wright et al. (1989)
X1Hbox6			none	Wright et al. (1989)
X1Hbox8			none	Wright et al. (1989)
integrin		80	blocks gastrulation	Darriberre et al. (1990)
Xhox3		0–50	posterior defects	Ruiz i Altaba et al. (1991)
gap junction		63	defects in left/right asymmetry	Warner et al. (1984)

[a] CY, cytoskeleton; KA, kinase; DBP, DNA binding protein; CS, cell surface protein receptor; GF, growth factor; Hbox, homeobox.

is injected, it does not concatamerize and usually does not replicate (but see Rusconi and Schaffner, 1981), and it remains as stable supercoils up until the tailbud stages (Fu *et al.,* 1989; Harland and Misher, 1988; Krieg and Melton, 1985; Krone and Heikkila, 1989).

Expression of cloned genes occurs from both linear and supercoiled DNA, and the level of expression is dependent on the strength of the promoter/enhancer present in the plasmid. Transcription at a basal level occurs whether or not promoter sequences are present with linearized templates (Fu *et al.,* 1989). Promoter-driven transcription of circular or linear templates generally occurs at similar levels (Harland and Misher, 1988), although the replication of linear DNA can lead to more transcript being generated from linearized templates (e.g., Jonas *et al.,* 1989; Wilson *et al.,* 1986). Promoters from *Xenopus* genes which are normally expressed during early development are efficiently expressed following injection, and they are regulated in the correct temporal patterns. Examples of such promoters are GS17 (which is gastrula specific), EF1 α (ubiquitous), cardiac actin (muscle specific), cytoskeletal actin (ubiquitous), ribosomal protein L14 (ubiquitous), and keratin (epidermal). The regulation of these genes is discussed in more detail in Chapter 19 of this volume and by Harland and Misher (1988).

Promoters from tissue-specific genes seem to be restricted to the appropriate tissues only at low frequency, so, in general, tissue-specific expression may not possible in this form of transient assay. An exception is the expression of globin driven by the epidermal keratin promoter, which was found to be cell-type specific (Jonas *et al.,* 1989). Other promoters that may be useful are the *hsp70* promoter, which is induced by heat stress above 25°C and when activated generates high levels of transcription (Harland and Misher, 1988; Krone and Heikkila, 1989), and the murine sarcoma virus long terminal repeat (LTR), which is also efficiently expressed (Giebelhaus *et al.,* 1988) but only in a transient manner (Harland and Misher, 1988).

Injected DNA is not efficiently expressed until the midblastula transition, which combined with the stability and replication of injected DNA makes this approach useful for expressing cloned sequences in a later time window than is possible with *in vitro* synthesized mRNA experiments.

DISTRIBUTION AND EXPRESSION OF INJECTED DNA

Unlike mRNA, which is generally evenly translated in cells containing the injected RNA, microinjected DNA is expressed in a mosaic fashion. The level of protein generated from injected DNA constructs varies widely from cell to cell, resulting in a speckled pattern in embryos stained immunologically or cytochemically for protein expression. The number of cells per embryo which express the injected DNA varies widely, with some expressing the product in many cells, and some in very few. Figure 3 shows histochemical detection of cytoplasmic β-galactosidase (β-gal) in embryos directed from an EF1α promoter/β-gal plasmid in embryos, and Fig. 4 shows expression of nuclear β-galactosidase from an injected plasmid under the control of the cytoskeletal actin promoter detected using immunohistochemistry. Similar mosaic expression patterns have been observed for all promoters assayed; they are, therefore, not due to any promoter-specific effect. In addition to mosaicism, expression from injected DNA is localized to subregions of the embryo, depending on the site of injection, in a manner similar to injected mRNA. For example, injections into the animal hemisphere often result in mosaic expression in cell types arising from this region of the egg, for example, epidermis, nervous system, and somites, whereas injection into the vegetal hemisphere results in mosaic expression in the gut and somites (Fig. 4). Once again this result indicates that injections into different regions of the embryo should be tested for their effects on development.

Also, as with mRNA injection experiments, it is useful to examine the expression of the protein encoded by the injected gene directly, as the expression of genes with profound effects on development may differ considerably from that of nontoxic reporter genes. For example, expression of the *Xenopus En-2* homoeobox gene driven by the cytoskeletal actin (CSK) promoter consistently resulted in detectable *En-2* protein in far fewer embryos, and in far fewer cells per expressing embryo, than β-gal driven from the same promoter (A. Hemmati-Brivalou and R. M. Harland unpublished). In addition, whereas expression of CSK/β-gal is observed in all cell types, expression of CSK/*En-2* appears to occur only in a subset of epidermal cells. The staining of embryos for the specific protein being studied would, therefore, be very useful for other proteins with similar effects.

The mosaic expression pattern from injected DNA makes the likelihood of successful antisense RNA experiments driven from injected plasmids doubtful, although one case has been reported (Giebelhaus *et al.*, 1988). No other clear positive results have since been obtained, by other researchers or by these authors, despite numerous attempts. Some observations of limited interference by antisense plasmids against homeobox proteins have been made (Ruiz i Altaba *et al.*, in preparation; Melton *et al.*, in preparation), and a variable reduction of α-fodrin mRNA levels by antisense has been observed (D. H. Giebelhaus and R. T. Moon, personal communication), but the low success rate in these experiments makes it difficult to distinguish specific from nonspecific effects, as a whole range of abnormalities arise in control embryos at a low frequency (usually 5–10%). It is possible that the mosaicism of protein expression is due to postranscriptional effects, and the injected DNA and the RNA transcripts generated from it are more widely distributed. Whole-mount *in situ* hybridization analysis should resolve this question (Hemmati-Brivanlou *et al.*, 1990) and show whether antisense RNA experiments are feasible with transient expression systems.

One final point on mosaic expression is that it could be, under certain circumstances, quite useful. Consider, for example, the wealth of results obtained from *Drosophila* using X-ray-induced mosaic analysis. Unfortunately, the use of coinjected plasmids with markers, such as β-galactosidase, may not be useful in tracing cells containing and expressing injected DNA. If this approach is to be attempted, care should be taken to demonstrate that expression of the marker correlates with the expression of the test gene, as we have found that two coinjected reporters, β-galactosidase and CAT, under the control of the same promoter, were expressed in different patterns. Double-labeling experiments examining cell fate with marker antibodies against an overexpressed gene product and against markers of cell type-specific differentiation could be extremely fruitful if the exogenously expressed genes affect cell fate.

Efforts to develop a system that evenly expresses protein from injected DNA utilizing transposon, recombination, or retrovirus systems have not yet been successful but are still under development. It is hoped that such a system will be available soon.

C. Antisense RNA

The injection of antisense RNA molecules into oocytes appears to block the translation of target mRNAs quite effectively (Harland and Weintraub, 1985; Melton, 1985). However, as mentioned above, antisense RNA experiments have been largely unsuccessful in developing embryos. This could be due to a number of reasons. For example, the instability of injected antisense RNA may result in an insufficient amount of the antisense molecule to interfere effectively with the translation of the target mRNA. If antisense RNA is being expressed from an injected plasmid, the mosaicism of expression observed in these experiments would result in the antisense molecule being present in only some of the target cells, and thus limit its effectiveness. As a third possibility, the RNA–RNA "unwindase" activity present in early embryos may interfere with the blocking function of the antisense molecules (Bass and Weintraub, 1987; Rebagliati and Melton, 1987). However, the last point may not be relevant, as the base-modifying activity of the unwindase chemically alters the target RNA and must block its translation (Kimelman and Kirschner, 1989). The full exploitation of this approach may not be feasible until consistent transgenic technologies are devised to generate high levels of antisense RNA expression evenly, throughout the embryo.

D. Injection of Antisense Oligonucleotides

The injection of oligonucleotides complementary to mRNA sequences can result in the destruction of the target mRNA owing to the presence of RNase

H activity in the embryo. RNase H specifically degrades the RNA portion of a DNA-RNA hybrid. The variable effectiveness of different oligonucleotides appears to be due to target RNA secondary structure, and oligonucleotides against different regions of the same mRNA can work to very different extents (Baker et al., 1990). Oligonucleotides can also interfere with the translation or stability of their target mRNAs via other poorly understood mechanisms. As a practical tool for ablating mRNAs of choice, this method shows much promise. However, the short half-life of unprotected oligonucleotides in the embryo, usually of the order of 1 to 5 minutes, has so far made it possible to interfere with maternal mRNAs only. In addition, oligonucleotides are often toxic to the embryo at the high doses required to destroy mRNAs.

Modified oligonucleotides, which contain bases with "protective" chemical modifications have been tested. The first are phosphorothioate oligonucleoradation of target mRNA and have half-lives of up to 15 minutes. Two modifications have been tested. The first are phosphothiolate oligonucleotides. These have considerably longer half-lives than unmodified oligonucleotides, and direct the cleavage of targets with reasonably efficiency; however, as they are considerably more toxic this largely negates the advantage of greater stability. The second form of modifications are phosphoamidate oligonucleotides. These N-blocking groups seem to be less toxic and much more stable in embryos. However, in order to provide an efficient substrate for RNAse H a number of residues in the center of the oligonucleotide must be left unmodified. The 5' and 3' end-protected oligonucleotides are more resistant to exonucleases but can still target the cleavage of specific mRNAs. One anti-Vg1 oligonucleotide which worked particularly well was N-blocked on the four outermost residues on each side of a 17-mer (T. M. Woolf, D. L. Weeks, and D. A. Melton, unpublished), and, similarly, anticyclin and An2 oligonucleotides N-blocked on the outer residues were also effective cleavage agents (Dagle et al., 1990). These chemically modified oligonucleotides are available commercially from Integrated DNA Technologies Inc. (Iowa), or can be prepared as described by (Dagle et al., 1990).

The nonspecific effects of oligonucleotides in injected oocytes have been investigated by Smith et al. (1990), which should be consulted for cautionary comments on the use of oligonucleotides.

E. Ribozymes

Ribozymes, catalytic RNA nucleases whose structure is based on the self-cleaving RNA molecules of certain plant RNA viroids, are capable of catalyzing their own cleavage in the absence of any protein and are capable of cleaving the appropriate targets *in trans* (Haseloff and Gerlach, 1988).

Ribozymes are generated by synthesizing an RNA molecule designed to incorporate the ribozyme "hammerhead" sequence, along with flanking sequences homologous to the target mRNA. Ribozymes are usually synthesized *in vitro* with SP6 or T7 RNA polymerase, although they can also be generated *in vivo* by cloning the ribozyme sequence into an *in vivo* expression vector generating short processed RNAs, such as tRNA. This approach however suffers from the drawbacks associated with DNA-mediated overexpression, as discussed above.

Unfortunately, the cleavage of target mRNAs *in vivo* by ribozymes is much lower than that observed *in vitro* (usually performed at 50°C at pH 8) and rarely exceeds 10% destruction. In fact, ribozymes have not been found to function very effectively against targets *in vitro* under salt and pH conditions similar to those of the embryo (T. M. Woolf, C. G. B. Jennings, and D. A. Melton, unpublished). One example of *in vivo* destruction of an RNA in oocytes has been reported (Cotten and Birnstiel, 1989), but this was targeted against U7 snRNP RNA rather than a protein-coding mRNA. The usefulness of this approach in embryos remains to be confirmed.

F. Injection of Antibodies

The injection of antibodies has resulted in blocking protein function in a number of instances. Examples include anti-XlHbox1, where affinity-purified rabbit polyclonal antisera were utilized. The injection of this antibody caused spinal cord defects in approximately 29% of the injected embryos (Wright *et al.*, 1989). Work by the same authors with antibodies against XlHbox6 and XlHbox8, produced in a similar fashion as those against XlHbox1, failed to elicit any phenotypic alteration, indicating that this, as with the other methods discussed in this chapter, is by no means a guaranteed approach. Anti-Xhox3 antibodies also produce a variable phenotypic response, observed in 0–50% (average 27%) of injected embryos (Ruiz i Altaba *et al.*, in preparation).

Another example of success by antibody injection, although in this case performed in *Pleurodeles* rather than *Xenopus*, was obtained by Darribere *et al.* (1990). In these experiments Fab' fragments against the cytoplasmic domain of the integrin β_1 subunit were injected into one side of a two-cell embryo, and the embryos which developed formed a neural tube only on one side of the embryo. The inhibition of fibronectin fibril formation by the antibody in a dose-dependent manner was also demonstrated.

The distribution of antibodies in injected embryos is similar to that observed for injected mRNA, and an example of this is shown in Fig. 5. Antibodies appear to be very stable in developing embryos, lasting in the case of anti-XlHbox1 at least until stage 30/32 (Cho *et al.*, 1988), and they have excellent potential as a powerful method of blocking protein function.

G. Expression of "Dominant Negative" Proteins

It may be possible to block the function of an endogenous proteins by expressing a dominant mutant which inhibits the activity of the target. For example, if an excess of a mutant component of a multisubunit protein complex is overexpressed, it could quite effectively inhibit the formation of the functional form of that protein. Such targeted schemes for inactivating proteins have been reviewed by Herskowitz (1987). This method may be particularly applicable to the study of cytoskeletal proteins, as these large arrays of protein may be more easily disrupted by this approach (Albers and Fuchs, 1987; Wong and Cleveland, 1990). The only report of success with this approach in *Xenopus* embryos to date is by Christian *et al.* (1990), who demonstrated that expression of a mutant form of vimentin from injected mRNA inhibited the polymerization of wild-type vimentin subunits. In cultured cells a number of examples have demonstrated the usefulness of this approach, and it shows great promise as a method for disrupting functional gene products.

H. Commonly Encountered Injection Artifacts

In the products of any fertilization, defective embryos will be present, usually on the order of a couple of percent. The increased stress from microinjection and early dejellying will raise the number of "background" abnormalities to 5–10%. This is the figure observed for the injection of a control mRNA (Harvey and Melton, 1988; Ruiz i Altaba and Melton, 1989).

The most commonly encountered artifacts from injection are blocked gastrulation movements. This often manifests itself as embryos with protruding yolk plugs which look similar to embryos that have partially exogastrulated owing to development under high salt conditions. Sometimes the yolk plug is pushed upward and splits the neural tube, causing the "spinabifida" phenotype. It is not clear exactly why this happens and why some proteins, DNAs, or mRNAs cause this to happen more than others.

Kinked embryos are sometimes generated in embryos injected at the two-cell stage owing to abnormal somitogenesis on one side of the embryo (Table I). This phenotype has been generated by a diverse range of mRNAs, including those encoding homeobox and cell adhesion molecules, and although it is not necessarily an "artifact," it, like exogastrulation, does not help in assigning a developmental function. Similarly, blocking cell division is difficult to interpret, and this is a common observation in antisense oligonucleotide injection experiments. A final "artifact" which is quite common is death. If these artifacts occur, masking more subtle and interpretable phenotypes, it may be useful to attempt to alter the injection regimen as described above, such as the dose of injected molecule and the site of injection. Since abnormal embryos arise in injected control embryos, large numbers of embryos must be injected to draw any conclusions. In instances where the phenotype is not strong, hundreds of surviving embryos must be examined in order to obtain an accurate representation of the injected embryo phenotype.

IV. Conclusions

By far the most effective technique developed to date for manipulating gene expression in *Xenopus* is the injection of mRNA. This method results in fairly high levels of expression of cloned cDNAs in much of the embryo. Such overexpression has resulted in specific phenotypic effects with approximately one-third of the mRNAs so far tested, and the speed and ease of performing this analysis make it an excellent choice as a rapid functional screen. The major disadvantage of this method is that injected mRNAs are stable only up to the gastrula or neurula stage of development, and so this approach is limited to the analysis of genes acting very early in development.

The injection of DNA, although useful for the analysis of promoter function (e.g., Krieg and Melton, 1987a), is less well suited to the study of the role gene products play in early developmental decisions. The mosaic expression pattern observed in DNA-injected embryos may be useful for overexpressing extracellular proteins, such as growth factors, or for mosaic analysis, but until methods which result in a more even distribution of expression are developed this approach will probably be of only limited applicability. The perturbation of expression, using antisense oligonucleotides or antibodies, appears to show promise as a protocol for interfering with protein production and function. Antisense oligonucleotides may be limited in their usefulness to early stages of development owing to their short half-lives in embryos, but antibodies appear to be very stable and may allow the blocking of protein function for long periods of time. Both of these interference experiments will produce a reduction in the amount of functional protein present, but the generation of a truly null phenotype is yet to be demonstrated.

It should be remembered when performing experiments such as those described in this chapter that the expression of very high levels of proteins in cells could quite understandably generate consistent but meaningless phenotypes. Care must be taken in interpreting results, and, as with any *in vivo* experiment, it is important to be cautious in inferring a direct relationship between the injected molecule and the observed phenotype. It is quite possible that observed embryonic defects may be generated by unknown secondary interactions.

The most useful advance in these techniques would be a transgenic system, in which injected DNA is integrated in the one- or two-cell stage of development and results in even levels of expression, either in all cells under the control of a post-midblastula transition housekeeping promoter or in particular cell types under the control of a tissue-specific promoter. Approaches such as retrovirus- and transposon-mediated DNA transfer are being studied, and will, it is hoped, provide such systems in the near future.

Acknowledgments

We would like to thank Randy Moon and Dan Weeks for providing unpublished data, and Charles Jennings and Ariel Ruiz i Altaba for providing figures.

References

Albers, K., and Fuchs, E. (1987). The expression of mutant keratin cDNAs transfected in simple epithelial and squamous cell carcinoma lines. *J. Cell Biol.* **105**, 791–806.

Baker, C., Holland, D., Edge, M., and Colman, A. (1990). Effects of oligonucleotide sequence and chemistry on the efficiency of oligodeoxynucleotide-mediated mRNA cleavage. *Nucleic Acids Res.* **18**, 3537–3543.

Bass, B. L., and Weintraub, H. (1987). A developmentally regulated activity that unwinds RNA duplexes. *Cell (Cambridge, Mass.)* **48**, 607–613.

Bendig, M. M. (1981). Persistence and expression of histone genes injected into *Xenopus* eggs in early development. *Nature (London)* **292**, 65–67.

Cho, K. W. Y., Goetz, J., Wright, C. V. E., Fritz, A., Hardwicke, J., and DeRobertis, E. M. (1988). Differential utilization of the same reading frame in a *Xenopus* homeobox gene encodes two related proteins sharing the same DNA-binding specificity. *EMBO J.* **7**, 2139–2149.

Christian, J. L., Edelstein, N. G., and Moon, R. T. (1990). Overexpression of wild-type and dominant negative mutant vimentin subunits in developing *Xenopus* embryos. *New Biol.* **2**, 700–711.

Cotten, M., and Birnstiel, M. L. (1989). Ribozyme mediated destruction of RNA *in vivo*. *EMBO J.* **8**, 3861–3866.

Dagle, J. M., Walder, J. A., and Weeks, D. L. (1990). Targeted degradation of mRNA in *Xenopus* oocytes and embryos directed by modified oligonucleotides: Studies of An2 and cyclin in embryogenesis. *Nucleic Acids Res.* **18**, 4751–4757.

Danilchik, M. V., and Black, S. D. (1988). The first cleavage plane and the embryonic axis are determined by separate mechanisms in *Xenopus laevis*. I. Independence in undistubed embryos. *Dev. Biol.* **128**, 58–64.

Darribere, T., Guida, K., Larjava, H., Johnson, K. E., Yamada, K. M., Thiery, J.-P., and Boucaut, J.-C. (1990). *In vivo* analysis of integrin B_1 subunit function in fibronectin mediated matrix assembly. *J. Cell Biol.* **110**, 1813–1823.

Dent, J. A., Polson, A. G., and Klymkowsky, M. W. (1989). A whole-mount immunohistochemical

analysis of the expression of the intermediate filament protein vimentin in *Xenopus*. *Development (Cambridge, U.K.)* **105,** 61–74.
Detrick, R. J., Dickey, D., and Kintner, C. R. (1990). The effects of N-cadherin misexpression on morphogenesis in *Xenopus* embryos. *Neuron* **4,** 493–506.
Etkin, L. D., and Pearman, B. (1987). Distribution, expression and germ-line transmission of exogenous DNA sequences following microinjection into *Xenopus laevis* eggs. *Development (Cambridge, U.K.)* **99,** 15–23.
Etkin, L. D., Pearman, B., Roberts, M., and Baktesh, S. (1984). Replication, integration and expression of exogenous DNA injected into the fertilized eggs of *Xenopus laevis*. *Differentiation* **26,** 194–202.
Fu, Y., Hosokawa, K., and Shiokawa, K. (1989). Expression of circular and linearized bacterial chloramphenicol acetyltransferase genes with or without viral promoters after injection into *fertilized eggs, unfertilized eggs and oocytes of Xenopus laevis. Roux's Arch Dev. Biol.* **198,** 148–156.
Galili, G., Kawata, E. E., Smith, L. D., and Larkins, B. A. (1988). Role of the 3'-poly(A) sequence in translational regulation of mRNAs in *Xenopus laevis* oocytes. *J. Biol. Chem.* **263,** 5764–5770.
Giebelhaus, D. H., Eib, D. W., and Moon, R. T. (1988). Antisense RNA inhibits expression of membrane skeletal protein 4.1 during embryonic development of *Xenopus*. *Cell (Cambridge, Mass.)* **53,** 601–615.
Harland, R. M., and Misher, L. (1988). Stability of RNA in developing *Xenopus* embryos and identification of a destabilizing sequence in TFIIIA RNA. *Development (Cambridge, U.K.)* **102,** 837–852.
Harland, R. M., and Weintraub, H. (1985). Translation of mRNA injected into *Xenopus* oocytes is specifically inhibited by antisense RNA. *J. Cell. Biol.* **101,** 1094–1099.
Harlow, E., and Lane, D. (1988). "Antibodies: A Laboratory Manual." Cold Spring Harbor Laboratory, Cold Spring Harbor, New York.
Harvey, R. P., and Melton, D. A. (1988). Microinjection of synthetic Xhox-1A homeobox mRNA disrupts somite formation in developing *Xenopus* embryos. *Cell (Cambridge, Mass.)* **53,** 687–697.
Harvey, R. P., Tabin, C. J., and Melton, D. A. (1986). Embryonic expression and nuclear localization of *Xenopus* homeobox (Xhox) gene products. *EMBO J.* **5,** 1237–1244.
Haseloff, J., and Gerlach, W. L. (1988). Simple RNA enzymes with new and highly specific endoribnuclease activities. *Nature (London)* **334,** 585–591.
Hemmati-Brivanlou, A., and Harland, R. M. (1989). Expression of an engrailed-related protein is induced in the anterior neural ectoderm of early *Xenopus* embryos. *Development (Cambridge,* **106,** 611–617.
Hemmati-Brivanlou, A., Frank, D., Bolce, M. E., Brown, R. D., Sive, H. L., and Harland, R. M. (1990). Localization of specific mRNAs in *Xenopus* embryos by whole-mount *in situ* hybridization. *Development (Cambridge, U.K.)* **110,** 325–330.
Herskowitz, I. (1987). Functional inactivation of genes by dominant negative mutations. *Nature (London)* **329,** 219–222.
Hopwood, N. D., and Gurdon, J. B. (1990). Activation of muscle genes without myogenesis by ectopic expression of MyoD in frog embryo cells. *Nature (London)* **347,** 197–200.
Jonas, E. A., Snape, A. M., and Sargent, T. D. (1989). Transcriptional regulation of a *Xenopus* embryonic epidermal keratin gene. *Development (Cambridge, U.K.)* **106,** 399–405.
Kimelman, D., and Kirschner, M. W. (1989). An antisense mRNA directs the covalent modification of the transcript encoding fibroblast growth factor in *Xenopus* oocytes. *Cell (Cambridge, Mass.)* **59,** 687–696.
Kintner, C. (1988). Effects of altered expression of the neural cell adhesion molecule, N-CAM, on early neural development in *Xenopus* embryos. *Neuron* **1,** 545–555.

Klein, S. L. (1987). The first cleavage furrow demarcates the dorsal–ventral axis in *Xenopus* embryos. *Dev. Biol.* **120,** 299–304.

Krieg, P. A., and Melton, D. A. (1984). Functional messenger RNAs are produced by SP6 *in vitro* transcription of cloned cDNAs. *Nucleic Acids Res.* **12,** 7057–7070.

Krieg, P. A. and Melton, D. A. (1985). Developmental regulation of a gastrula-specific gene injected into fertilized *Xenopus* eggs. *EMBO J.* **4,** 3463–3471.

Krieg, P. A., and Melton, D. A. (1987a). An enhancer responsible for activating transcription at the MBT in *Xenopus* development. *Proc. Natl. Acad. Sci, U.S.A.* **84,** 2331–2335.

Krieg, P. A., and Melton, D. A. (1987b). *In vitro* RNA synthesis with SP6 RNA polymerase. *In* "Methods in Enzymology" Academic Press, New York (R. Wu, ed.), Vol. **155,** 397–415.

Krone, P. H., and Heikkila, J. J. (1989). Expression of microinjected hsp70/CAT chimeric genes in developing *Xenopus laevis* embryos. *Development (Cambridge, U.K.)* **106,** 271–281.

Malone, R. W., Felgner, P. L., and Verma, I. M. (1989). Cationic liposome-mediated RNA transfection. *Proc. Natl. Acad. Sci. U.S.A.* **86,** 6077–6081.

Marini, N. J., Etkin, L. D., and Benbow, R. M. (1988). Persistence and replication of plasmid DNA microinjected into early embryos of *Xenopus laevis*. *Dev. Biol.* **127,** 421–434.

Masho, R. (1990). Close correlation between the first cleavage plane and the body axis in early *Xenopus* embryos. *Dev. Growth Differ.* **32,** 57–64.

McMahon, A. P., and Moon, R. T. (1989). Ectopic expression of the protooncogene int-1 in *Xenopus* embryos leads to duplication of the embryonic axis. *Cell (Cambridge, Mass.)* **58,** 1075–1084.

Melton, D. A. (1985). Injected antisense RNAs specifically block mRNA translation *in vivo*. *Proc. Natl. Acad. Sci. U.S.A.* **82,** 144–148.

Melton, D. A., Krieg, P. A., Rebagliati, M. R., Maniatis, T., Zinn, K., and Green, M. R. (1984). Efficient *in vitro* synthesis of biologically active RNA and RNA hybridization probes from plasmids containing a bacteriophage SP6 promoter: *Nucleic Acids Res.* **12,** 7035–7056.

Melton, D. A., Ruiz i Altaba, A., Yisraeli, J., and Sokol, S. (1991). Localization and axis formation during *Xenopus* embryogenesis. *Ciba Found. Symp.* **114.**

Newport, J., and Kirschner, M. (1982). A major developmental transition in early *Xenopus* embryos 1. Characterization and timing of cellular changes at the midblastula transition. *Cell (Cambridge, Mass.)* **30,** 675–686.

Nieuwkoop, P. D., and Faber, J. (1967). "Normal Table of *Xenopus laevis* (Daudin)." North-Holland Publ., Amsterdam.

Oliver, G., Wright, C. V. E., Hardwicke, J., and DeRobertis, E. M. (1988). Differential antero-posterior expression of two proteins encoded by a homeobox gene in *Xenopus* and mouse embryos. *EMBO J.* **7,** 3199–3209.

Rebagliati, M. R., and Melton, D. A. (1987). Antisense RNA injections in fertilized frog eggs reveal an RNA duplex unwinding activity. *Cell (Cambridge, Mass.)* **48,** 599–605.

Ruiz i Altaba, A, and Melton, D. A. (1989). Involvement of the *Xenopus* homeobox gene Xhox3 in pattern formation along the anterior–posterior axis. *Cell (Cambridge, Mass.)* **57,** 317–326.

Ruiz i Altaba, A., Choi, and Melton, D. A. (1991). In preparation.

Rusconi, S., and Schaffner, W. (1981). Transformation of frog embryos with a rabbit β-globin gene *Proc. Natl. Acad. Sci. U.S.A.* **78,** 5051–5055.

Shuttleworth, J., Godfrey, R., and Colman, A. (1990). p40MO15, a *cdc*2-related kinase involved in negative regulation of meiotic maturation of *Xenopus* oocytes. *EMBO J.* **9,** 3233–3240.

Smith, R. C., Bement, W. M., Dersch, M. A., Dworkin-Rastl, E., Dworkin, M. B., and Capco, D. G. (1990). Nonspecific effects of oligodeocynucleotide injection in *Xenopus* oocytes: A reevaluation of previous D7 mRNA ablation experiments. *Development (Cambridge, U.K.)* **110,** 769–779.

Thomsen, G., Woolf, T., Whitman, M., Sokol, S., Vaughan, J., Vale, W., and Melton, D. A. (1990). Activins are expressed early in *Xenopus* embryogenesis and can induce axial mesoderm and anterior structures. *Cell (Cambridge, Mass.)* **63,** 485–493.

Warner, A. E., Guthrie, S. C., and Gilula, N. B. (1984). Antibodies to gap-junctional protein selectively disrupt junctional communication in the early amphibian embryo. *Nature (London)* **311,** 127–131.

Whitman, M., and Melton D. A. (1989). Induction of mesoderm by a viral oncogene in early embryos. *Science* **244,** 803–806.

Wilson, C., Cross, G. S., and Woodland, H. R. (1986). Tissue-specific expression of actin genes injected into *Xenopus* embryos. *Cell (Cambridge, Mass.)* **47,** 589–599.

Wong, P. C., and Cleveland, D. W. (1990). Characterization of dominant and recessive assembly-defective mutations in mouse neurofilament NF-M. *J. Cell Biol.* **111,** 1987–2004.

Woodland, H. R., and Jones, E. A. (1987). The development of an assay to detect mRNAs that affect early development. *Development (Cambridge, U.K.)* **101,** 925–930.

Woolf, T. M., Jennings, C. G. B., Rebagliati, M., and Melton, D. A. (1990). The stability, toxicity and effectiveness of unmodified and phosphothioate antisense oligodeoxynucleotides in *Xenopus* oocytes and embryos. *Nucleic Acids Res.* **18,** 1763–1769.

Wright, C. V. E., Cho, K. W. Y., Hardwicke, J., Collins, R. H., and DeRobertis, E. M. (1989). Interference with the function of a homeobox gene in *Xenopus* embryos produces malformations of the anterior spinal cord. *Cell (Cambridge, Mass.)* **59,** 81–93.

Chapter 21

Histological Preparation of Xenopus laevis Oocytes and Embryos

GREGORY M. KELLY, DOUGLAS W. EIB, AND
RANDALL T. MOON

Department of Pharmacology
University of Washington Medical Center
Seattle, Washington 98195

I. Introduction
II. Tissue Processing for Light Microscopy
   A. Fixation
   B. Embedding and Sectioning
   C. Staining
III. Light Microscope Level Immunohistochemistry
   A. General Considerations
   B. Staining
IV. Tissue Preparation for Electron Microscopy
   A. Fixation and Embedding
   B. Sectioning and Staining
   C. Scanning Electron Microscopy Preparation
V. Immunoelectron Microscopy
VI. Photomicrography
   A. Bright-Field Microscopy
   B. Fluorescence Microscopy
   C. Electron Microscopy
   References

I. Introduction

Light and electron microscopy have long served as powerful techniques for elucidating developmental events and mechanisms. Moreover, immunostaining techniques (Coons *et al.*, 1941) have provided developmental biologists with an additional, powerful analytical tool for identifying proteins which dictate spatiotemporally restricted events during embryogenesis. From the plethora of studies involving *Xenopus laevis* it is evident that these methods not only complement, but are a necessity for interpreting results from microinjection, cell/tissue transplantation, and other molecularly oriented studies. However, the relative size, amount of yolk, and degree of metabolic activity present in *Xenopus* oocytes and embryos are major impediments encountered during histological preparation.

Classic fixation protocols for routine light microscopy (e.g., Bouin's, Carnoy's) involve rapid cell and tissue penetration at the expense of uniform stabilization and preservation of cellular constituents. Historically, fixation problems have been overcome through the combination of different aldehydes within a fixative. Although glutaraldehyde-based fixatives have been paramount in the preservation of tissue for electron microscopy (Sabatini et al., 1963), their penetration rates are very slow. Karnovsky (1965) combined glutaraldehyde (GTA) with paraformaldehyde (PFA), creating a relatively rapidly penetrating fixative ideal for preserving ultrastructural elements. Later, Kalt and Tandler (1971) found that a PFA–GTA–acrolein mixture containing dimethyl sulfoxide (DMSO) was the optimal fixative for preparing early amphibian embryos for electron microscopy.

It is the intent of this chapter to outline the techniques which have been used successfully in our laboratory to examine morphologically oogenesis and particularly embryogenesis under normal and experimental conditions. Our modified Karnovsky fixative containing DMSO is ideal for preserving *Xenopus* tissue for transmission (TEM) and scanning electron microscopy (SEM), but it is not suitable for immunostaining. Fixatives specific for electron microscopy are generally not ideal for immunohistochemical analyses because (1) paraformaldehyde alone may maintain antigenicity but is not the most optimal fixative for preserving ultrastructural elements, and (2) glutaraldehyde, an excellent cross-linker, may severely reduce the antigenicity of many proteins. For immunostaining, we describe a fixative which meets the important criteria of balancing the maintenance of antigenicity within the tissue with the best possible retention of the normal cellular morphology.

A comparative analysis of the many methods for fixing, staining, and processing amphibian tissue for light and electron microscopy and their pros and cons is beyond the scope of this chapter. In addition to the numerous methods available for processing tissue for immunohistochemistry, there now exist a multitude of immunostaining variations, each having its advantages and disadvantages (see Sternberger, 1986; Larsson, 1988). We have outlined two such variations which not only illustrate differences but actually show results which complement each other. Cookbook type manuals, such as

Humason (1979) or Vacca (1985), should be consulted whenever difficulties and/or uncertainties are encountered. Excellent guides for electron microscopy are provided by Glauert (1977) and Hayat (1974, 1981b).

II. Tissue Processing for Light Microscopy

Based on the relatively hazardous nature of the methods outlined, we suggest working in a fume hood with gloves and masks whenever possible. Neophytes to histology and more particularly to electron microscopy are encouraged to follow general information and safety precautions presented by Smithwick (1985) and Vacca (1985).

Fixation should be considered the most important step in tissue processing. Whether a fixative solution is coagulative, noncoagulative, or a combination of both, it must act to stabilize proteins, lipids, carbohydrates, and other cellular components, in the shortest time possible. Furthermore, components must remain fixed so as to survive subsequent dehydration, clearing, and infiltration. Failing to achieve the most optimal fixation will invariably lead to changes in morphology. Fixation depends on several factors including tissue size, density, composition, and metabolic activity. Details regarding the chemistry of fixatives and factors influencing them are presented by Hayat (1981a) and Glauert (1977).

Temperature and duration of fixation may also be varied to attain uniform stabilization and preservation. Higher temperatures accelerate penetration rates but may also enhance enzymatic activities responsible for cell autolysis. Overfixation may not only influence staining, but may also reduce antigenicity. Long fixation periods, if necessary, should be performed at lower temperatures. One way in which to reduce the fixation time is to dissect tissue away from the bulk of unwanted material. As a general rule, and to prevent uneven fixation, always place tissue in a vial already containing fixative. Occasional mixing within the first 15–30 minutes will also aid in uniform fixation. Maintain the chemical conditions in the fixative by using a volume which is at least 10 times the tissue volume, and, unless otherwise specified, use a fixative within 1 week after being made.

Embryos, isolated using methods adopted from Dr. J. B. Gurdon and outlined in Moon and Christian (1989), must be dejellied before fixation. Fixing the jelly coat acts as a barrier and retards penetration. Transfer oocytes or embryos from the culture dish with a plastic pipette (Elkay Liquipette, Elkay Products, Shrewsbury, MA) directly into fixative, keeping fixatives away from culture work areas. Transfer samples in as little volume as possible and replace the fixative if there was too much dilution with culture medium. We outline below three fixatives used for routine light microscopy.

A. Fixation

1. Bouin's Fixative

Bouin's fixative has been used for years as an all-purpose fixative for paraffin embedding. It is the preferred fixative for histological investigations of oocytes and embryos if a laboratory is set up exclusively for paraffin embedding and sectioning. Bouin's is made by mixing 75 ml saturated picric acid solution (1.2% picric acid in glass-distilled water) with 5 ml glacial acetic acid and 25 ml concentrated formalin (37% formaldehyde). Fix samples for 30 minutes to 2 hours and wash for 3–6 hours in several changes of 70% ethanol. Optimal staining requires that picric acid crystals be removed from the tissue. Avoid lower alcohol concentrations as picric acid binding is reversible under more aqueous conditions.

Bouin's fixative will keep for an indefinite period, but since dried picric acid may explode it is imperative to keep containers well sealed to maintain an aqueous solution over the crystals. This fixative combines the qualities of a protein coagulant (picric acid) with those of a noncoagulant (formalin). Glacial acetic acid, another noncoagulant, counteracts the considerable shrinkage induced by picric acid. Yolk platelets become very brittle after even short periods in fixative. To improve sectioning quality, the more yolk-abundant stages fixed in Bouin's may require plastic embedment. Several staining protocols including hematoxylin and eosin and the trichromes work well with Bouin's-fixed tissue.

2. Carnoy's Fixative

Carnoy's, another example of a classic fixative, is easy to make (6 parts absolute ethanol, 3 parts chloroform, 1 part glacial acetic acid), keeps for an indefinite period, penetrates rapidly, and preserves nucleic acids. Fix oocytes and embryos for 30 minutes to 3 hours, transfer directly to absolute ethanol, and process immediately for paraffin embedment. Carnoy's fixative can be used for any developmental stage. It is especially useful for selectively identifying DNA by either Feulgen (bright-field) or Hoechst 33258 (fluorescence) stains (see below).

3. PARAFORMALDEHYDE-GLUTARALDEHYDE IN PHOSPHATE BUFFER

A fixative containing 3% PFA and 0.25% GTA in phosphate buffer (PB), modified after Karnovsky (1965), is our preference for fixing tissue for general light microscopy. It is a contemporary fixative which combines the penetrating power of PFA with the excellent cross-linking capabilities of GTA. The fix, which keeps for 1 week at 4°C, is made by diluting a 10% stock of PFA to 3% with 0.1 M PB, pH 7.4. Add biological grade glutaraldehyde (Ted Pella, Inc., Redding, CA) to 0.25% final concentration, and adjust the solution to pH 7.4. Since different fixatives used in our laboratory require different PFA concentrations, we find it convenient to make a 10% stock [10 g PFA, Baker (Phillipsburg, NJ) practical grade, in 100 ml glass-distilled water, heat to ~60°C, and add 1-2 pellets of NaOH to clear] which keeps for 2-3 weeks at 4°C. Fix the tissue for 30 minutes to 1 hour at room temperature, then wash overnight in 0.1 M PB. For immunostaining, the addition of sucrose to the wash buffer can leach out free aldehydes (Martinez-Hernandez, 1987) and may act to protect and prevent the loss of inadequately fixed antigens (Takamiya et al., 1980). We do not strongly recommend embedding aldehyde-fixed tissue in paraffin since many problems arise during sectioning. However, a method circumventing these problems, whereby tissue originally fixed with an aldehyde fixative suitable for electron microscopy is transferred to any light microscopic fixative prior to paraffin embedding, has been described by Tandler (1990).

With the excellent cross-linking capabilities of PFA-GTA, more structural detail and greater resolution can be attained when tissue is embedded in one of the many plastic resins. A favorable attribute of this fixative is that following tissue examination by routine bright-field microscopy, sections can be deplasticized, immunostained, and then reexamined by bright-field or electron microscopy (see below). For strictly nonimmunohistochemical, ultrastructural studies, however, we recommend the fixative described in Section IV,A.

B. Embedding and Sectioning

Paraffin embedding and sectioning is generally less expensive and much easier to learn than methods for plastic resins. In addition, many staining protocols to demonstrate specific cellular components are compatible with paraffin embedment (Humason, 1979; Vacca, 1985), making it more attractive to anyone who has considered using histological techniques. The most obvious disadvantages of paraffin include the inability to examine tissue at the ultrastructural level and the dramatically reduced resolution when compared with semithin (1–2 μm) plastic sections. Regardless, the advantages and disadvantages of each embedding medium must be weighed in accordance with the specific needs and equipment availability.

1. PARAFFIN

Dehydrate samples in an ascending ethanol series [70, 80, 95(2×), and 100% (3×), 15–30 minutes each]. Molecular sieve (Type 3A; Bio-Rad Microsciences Div., Cambridge, MA) is added to absolute ethanol stocks to absorb contaminating water. Remove 50–100 ml absolute ethanol from a newly opened 473-ml bottle, replace with an equivalent amount of molecular sieve, and allow particles to settle overnight. Use this as dry absolute ethanol until the level of alcohol is within 2–3 cm of the sieve. Following dehydration, clear samples by replacing 100% ethanol with xylene or toluene (2 times, 15–30 minutes each.) The purpose of the clearing agent is to introduce a solvent which is (a) miscible with the dehydrant and the embedding medium, thus allowing uniform infiltration, and (b) responsible for physically clearing tissue since its refractive index nears that of tissue. Xylene compared to toluene has a lower rate of evaporation but will harden tissue more than toluene (Vacca, 1985). To improve sectioning quality we use toluene as the clearing agent.

Infiltration is the gradual replacement of clearing agent with embedding medium. Transfer the tissue to a glass petri dish containing molten (58°–60°C) Paraplast X-tra (VWR Scientific, Seattle, WA) and infiltrate 15 minutes. Vacuum infiltration is unnecessary and may lead to artifacts owing to the presence of hollow compartments at particular embryonic stages, namely, the blastocoel or archenteron. To ensure complete removal of the clearing agent, repeat this step with fresh molten wax. Ideally Paraplast should not be heated to more than 60°C since this will alter its sectioning properties. After the second infitration fill a metal base mold approximately 5 mm deep with fresh molten Paraplast. Place the molds on a warming plate to keep the Paraplast molten, and use warmed, blunt forceps (or a pipette) to transfer the tissue and position it appropriately in the mold. It may be necessary to free tissue from the forceps by swirling it through the wax. To maintain samples in the correct orientation, allow the wax to solidify at room temperature, then attach a plastic embedding ring (VWR Scientific) and slowly fill the entire ring with wax. Blow on the wax to initiate solidification. Continue until the entire surface has solidified then leave to harden at room temperature. Remove the metal mold after 30 minutes and store the blocks at room temperature. If molds stick, place them in a $-20°C$ freezer for 15–20 minutes.

With numerous techniques available to trim and section paraffin blocks, we suggest consulting Humason (1979). We generally cut ribbons of 5–10-μm sections and then float them in a 45°C water bath to remove wrinkles. If wrinkled sections are a problem, either transfer the ribbons to a 70% ethanol bath for 1–5 minutes before the heated water bath or wave a cotton swab, laden with chloroform, over the floating sections. Collect the ribbons on subbed slides. Slides are subbed by first dissolving 1 g of gelatin in 1 liter of glass-distilled water using low heat. After cooling, mix with 0.1 g of chromium potassium sulfate and filter. Wash slides in 70% ethanol, air dry, dip in subbing solution, cover, and dry at 60°C. Freshly subbed slides provide the best adhesion of sections for immunostaining.

To collect ribbons, immerse a slide in the water bath, position it directly under sections, then slowly lift straight up. Alternatively, flood a subbed slide with water and heat to 45°C over a slide warmer. Place two 2.5 cm ribbons on the water and allow sections to spread. For both methods, dry the slides with a Kimwipe, taking precautions to avoid touching the ribbons. Leave slides for 1 hour at room temperature then heat in a 60°C oven until the Paraplast melts (~ 20 minutes). Slides can be stored for an indefinite period but are generally used within 1 month.

2. Plastic

There are numerous plastic resins available, each having its own advantages and disadvantages (see Glauert, 1977; Causton, 1984). Our personal preference is Eponate 12, a resin with sectioning qualities comparable to the more familiar, but now discontinued, Epon 812 resin. Tissue to be embedded in plastic is dehydrated as described for paraffin. After the last dry absolute ethanol step, clear samples in 2:1, 1:1, and 1:2 mixtures of absolute ethanol–propylene oxide, 15 minutes each. Two 15-minute changes of pure propylene oxide precede infiltration. Propylene oxide, a suspected carcinogen, may be replaced by acetone. After the last propylene oxide wash, fill half of the vial with fresh propylene oxide and add an equal volume of working mixture of Eponate 12 (Ted Pella, Inc.) containing DMP-30 [2,4,6-tri(dimethylaminomethyl)phenol]. Eponate 12 epoxy resin is made and stored as two solutions: (1) nadic methyl anhydride and Eponate 12 and (2) dodecynl succinic anhydride and Eponate 12. Mix the components as per the manufacturer's instructions and store individually at 4°C in well-sealed brown bottles. To make the working solution, allow the bottles to warm to room temperature, mix equal parts of (1) and (2), and then add 1.5% DMP-30 (w/v). Stir the mixture until streaks disappear and use immediately.

After adding the working solution to the propylene oxide, swirl the Wheaton snap-cap vial to ensure thorough mixing and cap. Propylene oxide is readily miscible in Eponate 12. The immediate settling of embryos is indication that tissue is dehydrated and well cleared. Later stage embryos, that is, tailbud and older, may sink slowly owing to their shapes and sizes rather than poor dehydration/clearing treatments. Remove the caps from the vials after 1 hour and leave the vials open overnight in the fume hood. As propylene oxide evaporates from the tissue it is replaced by resin. Make a new Eponate working solution the next day, add DMP-30, and mix very slowly with 3–4 applicator sticks. After degassing the mixture, fill flat embedding molds (Electron Microscopy Sciences, Fort Washington, PA) with plastic. To remove residual propylene oxide, pour the remaining resin onto Parafilm then transfer tissue with a sharpened applicator stick into the resin and gently mix. Place the tissue into the molds, position the samples below the surface, and cure the resin for 24–48 hours at 60°C. Positioning samples, immediately after placing them in the mold, is made easier by viewing them through a dissecting microscope.

Methods used to trim and section plastic blocks are as numerous as those for Paraplast. Consult electron microscopy technique books for tips on sectioning. Semithin Eponate 12 sections (1–2 μm), cut on glass knives, will float on water within the trough. Collect sections using a tungsten loop, eyelash brush, or fire-polished Pasteur pipette tip and transfer them to a drop of water on a subbed slide. For toluidine blue staining, sections do not have to be mounted on subbed slides. In either case, mount sections with low heat from a slide warmer. Wrinkles which develop within sections are usually the result of excessive heating.

C. Staining

There are many stains available to the histotechnologist, and although a comprehensive review is beyond the scope of this chapter, their existence and importance should not be deemphasized (Humason, 1979; Vacca, 1985). The basic protocol involves removing the embedding medium, rehydrating the tissue to water, staining, dehydrating the tissue to 100% ethanol, clearing, and coverslipping. The ultimate goal is to obtain a well-stained section which will serve as a permanent record. Below are the procedures we use most commonly: (1) a routine stain for morphological investigation; (2) a fluorescent stain to localize nuclei via DNA; and (3) a metachromatic stain that is ideal for rapid visualization of plastic-embedded tissue.

1. HEMATOXYLIN AND EOSIN

To remove paraffin, dip slides in three 5-minute changes of xylene. Rehydrate the tissue through a series of Coplin jars containing 100(3×), 95(2×), 80, 70, 50, and 35% ethanol, 3–5 minutes each, to glass-distilled water and then stain for 2–5 minutes in Gill's #3 hematoxylin (Polysciences,

Inc., Warrington, PA). An advantage of Gill's hematoxylin variant is that it keeps well and will not form the metallic scum often associated with other variants (Vacca, 1985).

Rinse slides in running tap water and differentiate for 2–3 minutes in Scott's tap water (dissolve 10 g of anhydrous magnesium sulfate in 1 liter glass-distilled water and add 2 g sodium bicarbonate). Rinse slides for 5–10 minutes in running tap water before counterstaining. Differentiating or "bluing" of nuclei may not be necessary in geographical areas having more alkaline tap water. Hematoxylin should be filtered periodically and checked by first adding a few drops to 100 ml glass-distilled water and then adding approximately 5 ml Scott's tap water. If the stain is good the solution will turn blue.

To counterstain the cytoplasm, dip slides in 70% ethanol for 1 minute, then place in working eosin Y solution for 1–5 minutes. A stock of eosin is made by dissolving 1 g eosin Y (C.I. 45380; Polysciences, Inc.) in 400 ml of 95% ethanol. Add 100 ml glass-distilled water and store in a brown bottle at room temperature. For staining, mix 25 ml of stock, 75 ml of 95% ethanol, and 0.35 ml glacial acetic acid. This solution can be reused until the color fades. Rinse slides in 100% ethanol (three changes, 20–30 seconds each), clear in xylene (three 5-minute changes), and coverslip. To coverslip, place a drop of Permount (VWR Scientific) on one edge of a coverslip that is sitting on the work table. Next, remove a slide from the xylene bath, dry the back with a Kimwipe, and lower the slide at a 45° angle until it contacts the mounting medium. By slowly decreasing the angle, the Permount will move across the surface of the coverslip. Invert the slide, allowing the Permount to harden overnight, before examining with oil-immersion objectives.

For staining Eponate sections, remove the resin by dipping slides in sodium ethoxide, a procedure modified after Lane and Europa (1965). To make this solution, mix 15 g NaOH in 100 ml of 100% ethanol for 1 hour, age 2–3 days until a light brown color appears, then decant solution into a Coplin jar or store in a well-sealed container. At room temperature the solution ages and becomes more caustic; check older solutions with test slides. Plastic from 1- to 2 μm sections of tissue embedded in resins such as Eponate 12 or Araldite 502 is removed in approximately 20 minutes, as monitored under the microscope. Polybed 812, Medcast, Spurr's medium, or Epon–Araldite combinations require 2–3 times longer for removal of plastic.

Remove deplasticized sections and immediately wash the slides in three changes of dry 100% ethanol (1–5 minutes each). Rehydrate sections following the 100% ethanol to glass-distilled water series described above. Subsequent staining procedures are comparable except that we double the staining times for both hematoxylin and eosin (Fig. 1A). Basophilic structures (nuclei) will appear blue, whereas acidophilic structures (cytoplasm) will appear pink.

2. TOLUIDINE BLUE O

As described above and as shown in Fig. 1A, hematoxylin and eosin staining provides excellent differentiation of cellular constituents. However, because of the number of steps involved in hematoxylin and eosin staining we routinely stain plastic sections first with toluidine blue (after Mercer, 1963) to evaluate fixation and section quality. Toluidine blue is a dye which can be used directly on plastic sections, yielding varying intensities of blue or blue-purple staining. Although toluidine blue has poorer qualities as a stain owing to its metachromatic staining properties, it provides faster staining than that of hematoxylin and eosin, (5 versus 60 minutes). By comparing Fig. 1A and 1B, one sees considerably less detail in the toluidine blue stained section than in the hematoxylin and eosin stained section.

To stain, cover sections with 1% toluidine blue [add 1 g of toluidine blue O (C.I. 52040; Polysciences, Inc.) to 100 ml of 1% sodium borate, filter before using, and store at room temperature in a well-sealed bottle] and heat slowly over a hot plate until first sign of evaporation. Wash slides with a gentle stream of glass-distilled water, air dry, and then coverslip with Aqua-Mount (Lerner Labs, Pittsburgh, PA).

3. BISBENZIMIDE

Bisbenzimide (Hoechst 33258) can be used with most fixatives. For paraffin studies we suggest using Carnoy's fixative, especially when examination of DNA-containing components supersedes the need to visualize other cellular constituents. Rehydrate paraffin sections to glass-distilled water (see Section II,C,1) and wash in 0.1 M PB containing 0.9% NaCl (phosphate-buffered saline, PBS). Stain for 15 minutes in 0.1 M PBS containing 1 mg/ml Hoechst 33258 (Sigma Chemical Co., St. Louis, MO), then rinse extensively in PBS. A Hoechst stock of 10 mg/ml can be made in glass-distilled water; this will last for months if stored at 20°C. Dehydrate slides, clear, and coverslip with Aqua-Mount (see Section II,C,1).

The Hoechst 33258 stain, depending on the pH of the solution, is DNA or RNA specific (Hilwig and Gropp, 1975) and relatively intolerant to photobleaching. Slides must be examined on a fluorescence microscope equipped with Hoechst filters (360 nm excitation, 470 nm emission). Investigators having limited access to such equipment should resort to using Feulgen stain (Vacca, 1985) to visualize DNA. We have had little success with Hoechst staining of deplasticized Eponate 12-embedded tissue and instead advise embedding the tissue in paraffin, JB-4 methacrylate (Fig. 2A), or any other water-soluble plastic resin.

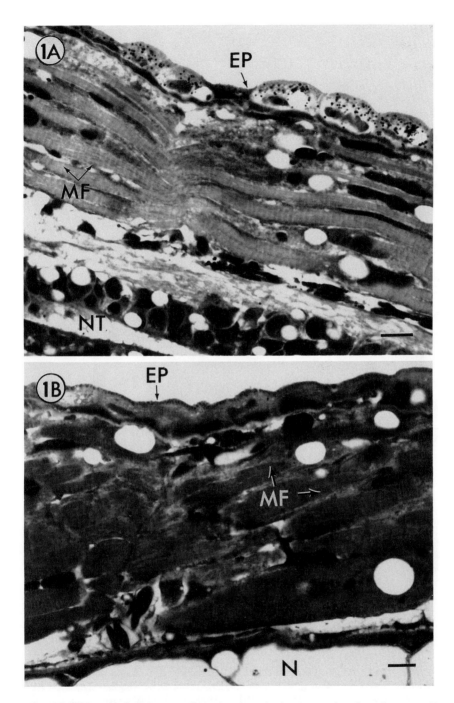

FIG. 1. (A) Longitudinal hematoxylin and eosin stained epoxy section through a stage 41 tadpole (Nieuwkoop and Faber, 1975) illustrating well-developed muscle fibers (see Muntz, 1975, for details). (B) Comparable section stained with toluidine blue. EP, Epidermis; MF, muscle fibers; N, notochord; NT, neural tube. The fixative was 3% PFA, 0.25% GTA. Magnification: (A) ×850, (B) ×800. Bars equal 10 μm.

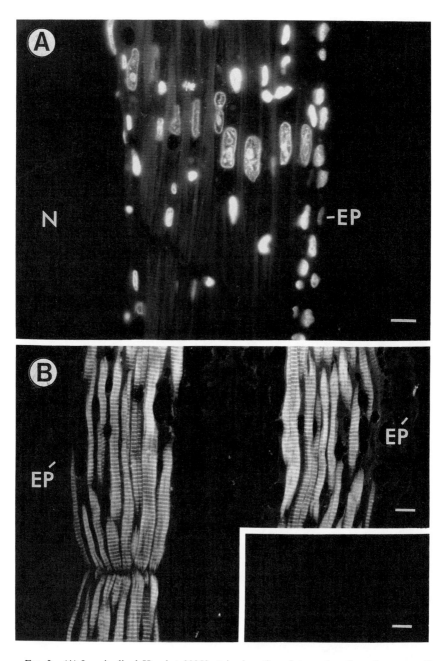

FIG. 2. (A) Longitudinal Hoechst 33258 stained methacrylate section through a stage 41 tadpole (Nieuwkoop and Faber, 1975) illustrating large central nuclei lying between muscle fibers. (B) Immunofluorescence micrograph of HHF35 stained longitudinal epoxy section through a stage 41 tadpole. HHF35 is a human muscle, actin-specific monoclonal antibody (see Tsukada et al., 1987). Note the well-defined muscle striations. (Inset) Control section (no primary antibody incubation). EP, Epidermis; N, Notochord. The fixative was 3% PFA, 0.25% GTA. Magnification: (A) ×670, (B) ×540, (inset) ×540. Bars equal 10 μm.

III. Light Microscope Level Immunohistochemistry

A. General Considerations

The many variations which exist for immunohistochemistry are staggering, each involving differences in detection methods and choice of markers to visualize antigen–antibody complexes. We have employed two detection systems for immunostaining tissue in deplasticized epoxy sections (compared in Fig. 2B and 3). Similar procedures are easily adapted to paraffin sections.

The first method (indirect labeling) involves using a fluorochrome-labeled secondary antibody (2° Ab) directed against immunoglobulins from the species which produced the primary antibody (1° Ab) (Fig. 2B). This method provides for some amplification of signal compared with direct labeling of the primary Ab, since each primary Ab can bind at least two labeled secondary antibodies. The indirect labeling method is also more time and cost efficient since theoretically only one labeled secondary Ab, directed against immunoglobulins of given species, is needed to detect all primary antibodies generated by this species. The major disadvantage of the indirect immunofluorescence method is loss of signal owing to photobleaching of label. Staining is not permanent, and weak signals may be difficult to detect by conventional fluorescence microscopy.

The second detection method utilized involves exploiting the high affinity of avidin for, and irreversible binding to, biotin (Guesdon et al., 1979) (Fig. 3). The first demonstration of this complex for immunohistochemistry employed fluorochromes (Heggeness and Ash, 1977), but since then many variants have been designed utilizing enzymatic markers whose reaction product can be visualized by bright-field microscopy (Warnke and Levy, 1980; Hsu et al., 1981; Van Noorden, 1986; Coggi et al., 1986). Our protocol involves a biotinylated secondary Ab which is used as a substrate to bind peroxidase-conjugated streptavidin. Peroxidase acts on the chromogenic hydrogen donor diaminobenzidine (DAB) in the presence of H_2O_2 to precipitate an insoluble, relatively stable, brown-colored reaction product (Graham and Karnovsky, 1966), corresponding spatially to the antigen–antibody complex. The products derived from this method are not susceptible to photobleaching, in most instances can be intensified (Newman et al., 1983; Van Noorden, 1986, Merchenthaler et al., 1989), and are electron dense, which allows for visualization at the electron microscope level. Background or nonspecific staining, especially prevalent in the early, more yolk-abundant developmental stages, is reduced when avidin is replaced by streptavidin. Furthermore, sections can be counterstained to enhance morphological detail. The colored precipitates and the counterstaining are further enhanced by viewing with a yellow filter (Fig. 3B has a filter, Fig. 3A has no filter).

Benefits far outweigh the disadvantages of this system. However, disadvantages to be considered include the facts that DAB is a suspected carcinogen, endogenous peroxidase activity must be blocked to prevent false positives, and it is possible for reaction products to diffuse away from the actual antigenic site (Novikoff et al., 1972).

B. Staining

Place 1 to 2-μm sections of tissue fixed with 3% PFA and 0.25% GTA in 0.1 M PB on a drop of filtered glass-distilled water on a subbed slide and dry overnight at 40°C. For orientation, circle sections on the back of a slide with a diamond pencil. Remove the resin in sodium ethoxide, rinse in two 5-minute changes of dry 100% ethanol, and rehydrate to glass-distilled water as described in Section II,C,1. Wash sections for 3 minutes, in 50 mM Tris buffer (TB), pH 7.5. For TB saline (TBS), dilute stock 0.5 M Tris (pH 7.5) to 50 mM and add NaCl to 0.9%. For immunoperoxidase staining, block endogenous peroxidase activity with 0.2% H_2O_2 in TB for 20 minutes at room temperature then rinse for 20 minutes in running distilled water. For all subsequent incubations, slides are kept in a moist chamber. Block nonspecific protein–protein interactions by incubating sections for 20 minutes in TB containing 2–3% bovine serum albumin (BSA, fraction V, Sigma) and 10% normal serum from the same species in which the secondary antibody was derived. Insufficient blocking is a major cause of high, nonspecific background staining. Blocking for longer periods is acceptable, providing precautions are taken to prevent the tissue from dehydrating. Furthermore, higher percentages of BSA may be necessary, particularly when using paraffin sections and/or polyclonal antisera.

Remove the blocking solution by capillary action with absorbent lens tissue, but do not rinse the sections. Apply 40–50 μl of primary antibody diluted in TBS and incubate at 4°C overnight in a moist chamber. To remove particulate matter which can cause high background, always spin diluted antibodies for 1–2 minutes in a microcentrifuge. Determination of an ideal primary Ab dilution is derived by testing a series of dilutions and different incubation periods on a positive control. Rinse the slides by immersing in four 5-minute changes of TBS, dry around the sections with lens tissue, and incubate for 1 hour at room temperature in either fluorescein-labeled goat, anti-mouse (Zymed Laboratories Inc., San Francisco, CA) or biotin-conjugated goat, anti-mouse (Zymed) secondary antibody diluted in TBS. Again, the proper dilution must be determined empirically. To prevent photobleaching, incubations with fluorochrome-conjugated antibodies should be done in the dark. Wash in TBS (1 hour for immunofluorescence or four 5-minute changes for peroxidase labeling) and either coverslip slides with the glycerol-based

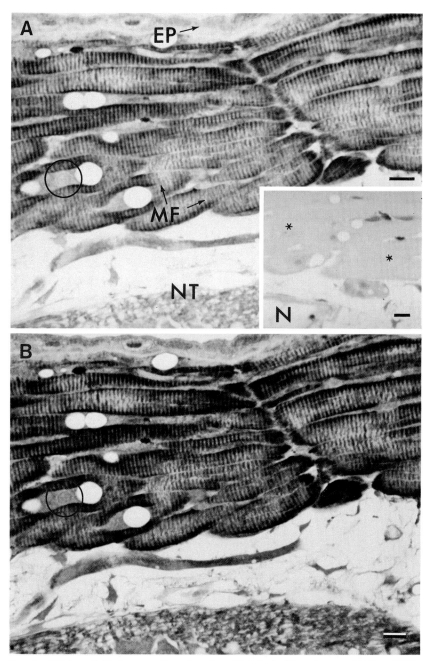

FIG. 3. (A) Longitudinal section through a stage 41 tadpole (Nieuwkoop and Faber, 1975) stained with HHF35, visualized with peroxidase, and hematoxylin counterstained. (B) Cellular details, especially nuclei (circled region), are enhanced when the same section is photographed with a yellow filter (Inset). Control section (no primary antibody incubation), demonstrating minimal nonspecific staining in muscle (asterisks). EP, Epidermis; MF, muscle fibers; N, notochord; NT, neural tube. The fixative was 3% PFA, 0.25% GTA. Magnification: (A) X600, (B) X600, (inset) X400. Bars equal 10 μm.

Aqua-Mount and examine with an epifluorescent microscope equipped with fluorescein filters (450–495 nm excitation, 520 nm emission) or proceed with the peroxidase tertiary incubation. It is noteworthy that other fluorochromes (e.g., Texas Red or rhodamine) may be used, providing the appropriate excitation and barrier filters are employed.

For immunoperoxidase staining, incubate sections for 1 hour at room temperature with streptavidin–horseradish peroxidase conjugate (Zymed), diluted in TBS. Wash extensively with TBS. Dissolve 60 mg DAB in a staining dish containing 200 ml TBS and mix on a magnetic stirrer. Just before use add 166 μl of 30% H_2O_2. Elevate the slide rack above the stir bar and incubate the slides for 5–7 minutes. Rinse the slides in TBS (5 minutes) at the first sign of background on control sections. Counterstain some slides with hematoxylin (1–3 minutes) to enhance cellular details and coverslip with Permount. Unused DAB, a suspected carcinogen, should be treated with bleach before disposal.

We have had the best success with immunostaining tissue fixed in 3% PFA and 0.25% GTA in phosphate buffer. Martinez-Hernandez (1987) states that most proteins greater than 40–50 kDa in size, when denatured by glutaraldehyde concentrations above 0.5%, will not react with their specific antibodies. However, we have found that with some antigens GTA concentrations may be increased in excess of 1%, allowing improved morphological preservation. We suggest varying the GTA concentration and antibody dilutions to obtain the best morphological preservation in concert with an optimal signal-to-noise ratio. Additional information concerning tissue preparation for immunohistochemistry is presented by Brandtzaeg (1982). Several techniques have been devised to enhance immunostaining of GTA-fixed tissue. The one we favor employs sodium borohydride. Treating sections for 10 minutes with $NaBH_4$ (0.5 mg/ml in PBS) following rehydration will not only reduce aldehyde groups responsible for inducing nonspecific fluorescence (Weber *et al.,* 1978), but will also reduce the possibility of antibody and/or antigen denaturation (Martinez-Hernandez, 1987).

The last issue to be considered for any immunohistochemical protocol is inclusion of positive and, more importantly, negative controls. Negative controls are necessary as they allow one to determine specific versus non-specific signals. One way to achieve this is by eliminating the primary antibody incubation. A significantly better control is to show that the signal is not obtained if the primary Ab is blocked with purified antigen. Replacement of primary Ab with either normal serum or, more appropriately, with preimmune serum is suggestive evidence that the observed signal is due to antibodies directed against the immunogen. Advantages of positive controls are obvious. For photomicrography, photographic exposures should be of the same duration for control and experimental sections.

IV. Tissue Preparation for Electron Microscopy

Much of what was discussed for processing tissue for light microscopy directly applies to electron microscopy. Eggs and embryos have always posed a problem for electron microscopists since there are many osmotically distinct cellular and subcellular compartments which makes complete and uniform fixation a difficult task (Morrill, 1986). Occasionally we find that a fixative ideal for preserving one cell type will not adequately fix juxtaposed cells. Shrinkage of the blastocoel and archenteron are often seen using our fixation protocol. We have not been able to eliminate shrinkage, but changing osmotic conditions and the length of time in fixative may minimize the problem at these early stages.

Another major hindrance has been the large amount of yolk, especially prevalent in early developmental stages. We routinely postfix with osmium tetroxide (OsO_4). Osmium is a heavy metal which after fixing lipids and proteins renders them electron dense (Nielsen and Griffith, 1979). We find that the regime of Kalt and Tandler (1971) of overnight osmication is adequate for preserving cell ultrastructure throughout development (Fig. 4); however, this length of time may be detrimental for preservation of actin filaments (Maupin-Szamier and Pollard, 1978). Whole, undissected embryos usually require 2 or 3 changes to ensure the osmium remains in its unreduced state. Osmication time can be reduced if embryos are dissected, thereby allowing greater penetration along more surface area.

A. Fixation and Embedding

A fixative we find ideal for TEM and SEM contains 3% GTA, 1% PFA, and 1% DMSO in 0.1 M cacodylate buffer (after Karnovsky, 1965; Kalt and Tandler, 1971). A 0.26 M cacodylate buffer, pH 7.2, is made by combining 26.68 g sodium cacodylate in 600 ml glass-distilled water with 50.4 ml HCl (1.66 ml conc. HCl in 100 ml glass-distilled water). This stock will keep for 3–4 months at 4°C. Since cacodylate is an arsenate we recommend wearing gloves whenever it is used. The fixative is made in two parts: (a) Heat 49 ml of glass-distilled water to 60°C and add 1 g PFA. While stirring add 1 N NaOH until the solution clears. (b) Mix 12 ml of 25% GTA with 1 ml DMSO and 38 ml of 0.26 M sodium cacodylate buffer, pH 7.2. When (a) cools mix with (b), then add 15 mg $CaCl_2$ and store at 4°C.

Fix oocytes and embryos overnight at 4°C then wash for 1 hour in 0.13 M sodium cacodylate. Since there is some hardening of tissue owing to fixation, if one wishes to embed only a specific structure or region, it is convenient to dissect and remove that region while the oocyte or embryo is in the wash buffer. Furthermore, more uniform osmication is achieved on smaller pieces of tissue. Immediately before osmicating, dilute 1 volume of 4% OsO_4 (Ted Pella, Inc.) with 1 volume of 0.26 M sodium cacodylate buffer, pH 7.2. Osmicate samples at 4°C and replace with fresh 2% solution at the first indication of color change (20–30 minutes). OsO_4 vapors are extremely volatile and unused and unreduced solutions containing osmium should be mixed with 95% ethanol and left at room temperature for 12 hours before disposal. Minimize exposure by handling open vials in a fume hood. Store osmium solutions at 4°C and in very well-sealed, light-tight containers.

After overnight osmication, rapidly dehydrate samples in an ascending series of −20°C ethanol (5 minutes each in 70, 80, 95, and dry 100%; repeat the 100% 3 times). Samples to be examined by SEM can be stored in −20°C acetone or processed as described below. For TEM, allow vials to warm to room temperature then replace the solution with dry absolute ethanol at 20°C; repeat 3 times, 15 minutes each. Clear, infiltrate, and embed as described in section II,B,2.

B. Sectioning and Staining

With limited space it is difficult to explain how to thin section for electron microscopy. Competency with thick sectioning paraffin-embedded and semithin sectioning plastic-embedded tissue is a definite asset. Manuals we refer to are good guides, but first-hand experience is a necessity. Never start out trying to learn thin sectioning with a diamond knife; mistakes are unforgiving and very costly. It is also advantageous to cut and stain a few semithin sections before investigating much time only to find out you have been thin sectioning in the wrong area. In addition, following first trimming (facing the block) it may be necessary to return blocks to a 60°C oven (1–2 hours) before uniform silver-colored sections (70–90 nm) can be obtained. The interference color of light reflected off the section floating in the water trough is used to estimate the thickness of the section. We use the low angle (35°) diamond knife of Diatome (Fort Washington, PA) for thin sectioning. This choice is a personal preference, but a low knife angle helps reduce tissue compression (Jésior, 1986), a frustrating problem which can occur in larger yolk platelets.

1. SECTIONING

Position silver sections, floating on water within the trough, with an eyelash glued to an applicator stick. Collect ribbons on either 50/75 or 75/300 slotted nickel or 200 hexagonal mesh copper grids (Electron Microscopy Sciences). The type of grid depends on the block face size. Grids may require cleaning, a mandatory prerequisite if used for electron microscopic immunohistochemistry. Cleaning involves sonicating grids, 1–2 minutes, in a beaker of 100% ethanol. Remove the ethanol and invert the beaker over a petri dish containing filter paper. Place this assembly into a 60°C oven and leave until all grids have fallen onto the paper.

Sections can be placed on either side of grid, but be consistent in order to keep track of them during handling. Anticapillary, self-closing forceps are very convenient for manipulating grids. Align ribbons in the water trough, then lower the grid until it touches them. It is not necessary to go through the water interface, and it is advisable to keep the grid away from the knife edge. Dry grids with the section side up on Ross optical lens tissue (Ted Pella, Inc.) and store them in special holders available from any electron microscopy supply house.

2. Staining

The heavy metal stains used for electron microscopy can be made in advance. To make uranyl acetate, heat a bottle containing a 7% aqueous solution of uranyl acetate under tap water, and then mix vigorously for 5 minutes. Allow the solution to stand for 24 hours and store it at room temperature. The stain has an indefinite shelf-life. Avoid shaking the bottle before use, and never take solution immediately next to the undissolved uranyl acetate crystals. Uranyl acetate is very toxic and slightly radioactive. Staining all the grids at once will usually increase the probability that there is something wrong with either the stain or technique. To stain a subset of grids, first place approximately 400 μl of uranyl acetate solution in a well of a Coor's porcelain plate. Float up to five grids per well, section side down, on uranyl acetate for 5 minutes. Pick up grids one at a time and remove excess stain by squirting a gentle stream of glass-distilled water along the forcep shaft. Dry the grids before lead staining.

To adsorb carbon dioxide, which will combine with lead to form an insoluble precipitate, fill a small container with fresh NaOH pellets and place it in a petri dish whose bottom is coated with dental wax; leave 5 minutes. Keep the dish covered as much as possible to minimize deposits from forming on sections. Grids are stained using the lead citrate recipe of Reynolds (1963); this stain is also very toxic. (a) Boil 50 ml of glass-distilled water to remove carbon dioxide. Once cool, take 30 ml and add 1.33 g lead nitrate and 1.76 g sodium citrate. (b) Shake the solution 5 minutes, let it stand 30 minutes, then add 8 ml of 1 N NaOH and dilute to 50 ml with the remaining boiled glass-distilled water. (c) Store at 4°C in a well-sealed vessel; the solution keeps for approximately 6 months.

Place 1–6 drops of lead citrate on dental wax, float grids section side down, and replace the petri dish lid. After 2–3 minutes, remove the stain by immersing the grid, held by fine forceps, up and down in a 100-ml beaker of glass-distilled water. Do not use these same drops to stain the next grids. Once dry, grids can be examined in the electron microscope. Figure 4 illustrates examples of a gastrula embryo (Fig. 4A,B) and skeletal muscle from a stage 41 tadpole (Fig. 4C) processed according to the above protocol and examined by TEM.

C. Scanning Electron Microscopy Preparation

SEM preparation is outlined in detail by Morrill (1986) and much of the methodology described for echinoderm embryos is applicable to *Xenopus* embryos. Vials containing specimens fixed as in Section IV,A and stored at $-20°C$ are brought to room temperature, and the acetone is replaced with fresh 20°C acetone. Immerse critical point drying (CPD) containers (Bio-Rad Microsciences Div.; or homemade, see Morrill, 1986, or Strout and Russell, 1990) in acetone. Transfer samples to containers and dry using CO_2 as the solvent (see Cohen, 1974, for explanation of CPD). Once dried, pour samples onto double sticky tape overlying a SEM stub. Details of internal structures are visualized by fracturing mounted specimens with a minutien pin (Fine Science Tools, Inc., Belmont, CA) held with fine forceps. In Fig. 5A,B, we show ectoderm and lateral mesoderm in a gastrula fractured after CPD. Somites are visualized after removing the overlying ectoderm from a fixed, stage 27 tadpole (Fig. 5C). Conditions for coating specimens with gold and palladium differ depending on available equipment. Larger specimens, such as intact oocytes and embryos, are susceptible to charging. To alleviate this, specimens may require two rounds of coating. Alternatively, a more laborious thiocarbohydrazide–osmium method has been used to reduce charging and enhance surface details on *Xenopus* embryos (Kelley *et al.*, 1975). Consult Morrill (1986) on methods to examine and photograph SEM preparations. Store unmounted specimens and stubs containing specimens in a desiccator.

V. Immunoelectron Microscopy

Immunoelectron microscopy is a powerful tool for localizing antigens in specific cellular compartments and/or organelles. The procedure we use for immunogold labeling on thin sections is modified after Mar and Wight (1989).

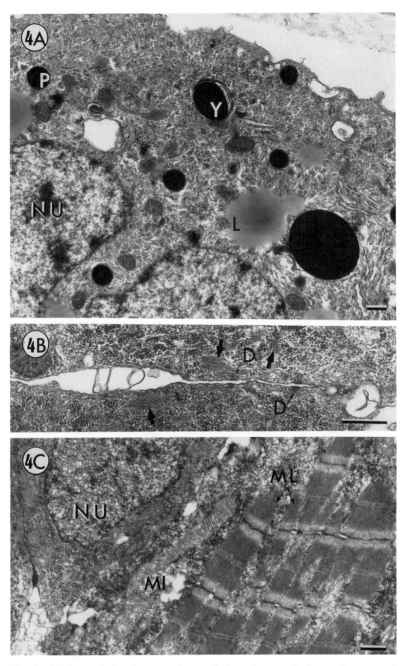

FIG. 4. (A) Transmission electron micrograph through one cell of gastrula stage embryo. Yolk, lipid, and pigment granules are major constituents within the cytoplasm. (B) Higher magnification, illustrating intermediate filament bundles (arrows) in two adjacent cells and desmosomes along plasmalemma. (C) Photomicrograph of developing muscle fibers in the trunk of a stage 41 tadpole. D, Desmosome; L, lipid; MI, mitochondria; ML, myofibrils; NU, nucleus; P, pigment granules; Y, yolk. The fixative was 3% GTA, 1% PFA, and 1% DMSO. Magnification: (A) ×11,210, (B) ×22,800, (C) ×13,775. Bars equal 0.5 μm.

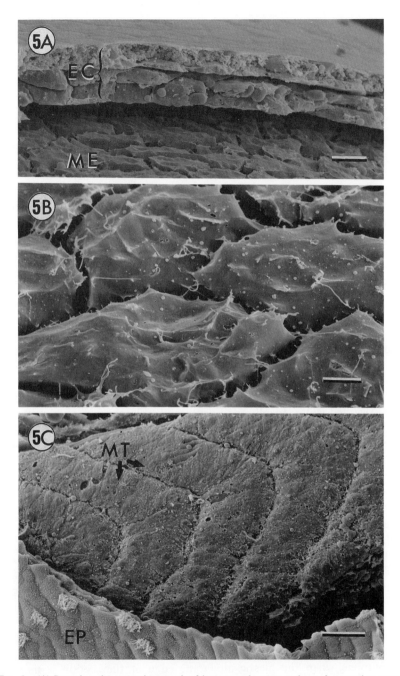

FIG. 5. (A) Scanning electron micrograph of late gastrula stage embryo, fractured to reveal ectoderm and lateral mesoderm. (B) Higher magnification of lateral mesodermal cells (for details, see Woo Youn et al., 1980; Keller et al., 1989). (C) Somites in the trunk of a stage 27 embryo (Nieuwkoop and Faber, 1975). Epidermis was removed to reveal underlying myotomes. EC, Ectoderm; EP, epidermis; ME, mesoderm; MT, myotomes. The fixative was 3% GTA, 1% PFA, and 1% DMSO. Magnification: (A) ×418, bar equals 25 μm; (B) ×1235, bar, 10 μm; (C) ×247, bar, 50 μm.

We suggest consulting the original reference before attempting this technique. Details on numerous permutations of the general methodology are presented by Roth (1983), Polak and Varndell (1984), Martinez-Hernandez (1987), and Bendayan *et al.* (1987).

For immunoelectron microscopy, we fix material according to the light level immunohistochemistry protocol presented in Section III,B. Although this fixation regime is not the most ideal for preserving fine structure (note lack of preserved detail by asterisk in Fig. 6B), it does allow one to screen tissue for immunoreactivity at the light level before investing time and expense incurred by immunoelectron microscopy techniques. In Fig. 6B we use this light level immunohistochemistry fixation protocol followed by immunogold localization of actin in skeletal muscle. Figure 6A illustrates a comparable embryo which had been fixed according to the TEM protocol (Section IV,A), demonstrating its superior preservation of structure.

For immunoelectron microscopy, sections must be mounted on Formvar-coated nickel grids. To coat grids, an ethanol-washed microscope slide is dipped on 0.25% Formvar (Ted Pella, Inc.) and dried under ethylene dichloride vapors, the solvent present in the Formvar solution. After 30 minutes remove the slide, air dry for 2 minutes, then etch the perimeter of the Formvar film on both sides of the slide with a razor blade. Fog the film by breathing on it, and then carefully lower the slide at a right angle into a staining dish filled with glass-distilled water. Ideally the Formvar film should release from the glass slide and float on the surface. Place precleaned 50/75 slotted nickel grids on the film then place a strip of Parafilm on top. After 1 minute, grasp the Parafilm from both sides, remove it, invert, and place it in a covered dish. The following day remove the grids from the Formvar sheet, invert and place one edge against a piece of double sticky tape firmly attached to Parafilm. The sheet containing 10–20 grids can be carbon coated to stabilize the Formvar. The carbon coating procedure, which will depend on the machine available, is detailed by Hayat (1981b).

After preparation of grids, immunoelectron microscopic localization as for skeletal muscle actin in Fig. 6B, can be accomplished using the following protocol. Gold sections (100–110 nm) cut from the Eponate 12 blocks used for light level immunocytochemistry (Section III,B) are collected, as described in Section IV,B,1, on the grid side opposite the carbon coating. Sections are deplasticized in sodium ethoxide diluted 1:3 with dry 100% ethanol. The time required for deplasticization is determined by examining test grids by TEM; it is complete when the plastic along the edge of the section is indistinct. Following deplasticization, rinse the grids by immersing them in three changes of dry 100% ethanol (1 minute each), then hydrate through a graded ethanol series to 35% ethanol (1 minute each). After blot drying the carbon-coated side, float the grids section side down on drops of 0.22-μm sieve-filtered TBS. All

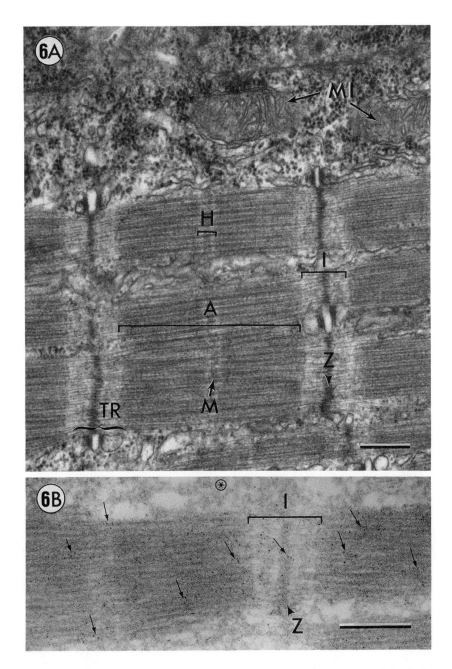

FIG. 6. (A) Transmission electron micrograph illustrating the fine structure of developing myofibrils. The tissue was fixed as in Fig. 4. (B) Electron micrograph of an immunogold, antiactin labeled (arrows) section from a tadpole originally fixed for light level immunohistochemistry (see legend to Fig. 3). This fixative, used to maintain antigenicity, results in reduced cellular preservation (asterisk). A, Anisotropic band; H, H band; I, isotropic band; M, M band; MI, mitochondria; TR, triad; Z, Z line. Magnification: (A) ×51,000, (B) ×77,500. Bars equal 0.25 μm.

incubations are done on Parafilm and in a humid, covered chamber. Grids can be kept at this step until all have been deplasticized and hydrated.

To block nonspecific protein–protein interactions, incubate sections on a drop of filtered 10% normal goat serum in TBS for 10 minutes, then rinse on 4 drops of filtered TBS for 2–3 minutes each. Dilute the primary antibody with TBS, filter, then incubate grids on drops for 12–16 hours at 4°C. Try several dilutions of antibody in order to obtain the optimal signal. Rinse grids on 6 drops of filtered TBS (2–3 minutes each), then incubate on drops of goat anti-mouse IgG conjugated to 5, 10, 15 or 40 nm colloidal gold particles (AuroProbe EM, Janssen, Ted Pella, Inc.). Secondary antibody is diluted with filtered TBS. After 1 hour of incubation at room temperature, rinse the grids on 10 drops of filtered TBS (2 minutes each), followed by 4 drops of 0.22-μm sieve-filtered 50 mM PBS (2–3 minutes each).

Postfix grids for 10 minutes on drops of 2% GTA in filtered 50 mM PBS, rinse on 4 drops of filtered PBS (2–3 minutes each), then osmicate for 10 minutes on drops of 1% OsO_4 in filtered PBS. To reembed, rinse the grids on 3 drops of filtered glass-distilled water and dehydrate through a graded ethanol series. Immerse grids for 2 minutes in 2% catalyzed Eponate 12 in 100% ethanol, blot dry on both sides with Whatman paper (#50), and polymerize overnight at 60°C. Once polymerized, stain the sections with uranyl acetate and lead citrate, but double the times that were described in Section IV,B,2. Grids are examined by routine transmission electron microscopy.

The reembedment protocol has it advantages and disadvantages. The major advantage, as described above, is that tissue can first be screened for antibody staining at the light level. Details regarding specific localizations are then possible using electron microscopy. One major problem encountered is trying to eliminate particulate matter which adheres to sections. All solutions are filtered and, whenever possible, made immediately before use. Each experiment should include enough grids such that 1–2 can be removed after key steps and examined by TEM. Increasing gold particle size and/or silver enhancement can also be employed for better signal visualization (Merchenthaler et al., 1989). It is noteworthy that conditions must be optimized for each antigen as there is no single ideal approach in immunocytochemistry (Bendayan et al., 1987).

VI. Photomicrography

With any morphological study the permanent record of the data is usually the photograph. The cost of photography, relative to time and money invested in sample preparation, is negligible. There is no excuse for not taking several photographs, a statement strongly endorsed by anyone familiar with pitfalls associated with photomicrography. The most obvious mistakes are those arising from a misaligned microscope. Kohler illumination (Szabo, 1967; Bradbury, 1986), a process designed to evenly illuminate the specimen, is critical for optimal photomicrography. It is not only a very easy method to adopt, but should become routine practice when using a compound microscope. Another common problem, which is strikingly obvious on photographs, occurs when the condenser diaphragm is closed to increase contrast and/or depth of field. It is aesthetically pleasing to the eye, but details on photomicrographs will be lost owing to introduced diffraction fringes. Incorrect choice of film and, more seriously, the use of color film to document all results are other problems associated with photomicrography. Certain applications demand color but in most instances black and white photomicrographs are less expensive, easier to produce, and of higher quality than their color counterparts. If color is a requirement, then tungsten balanced, reversal film should be used and the filament of the bulb should be saturated to its highest possible voltage rating. Additional details concerning color photomicrography are provided by Szabo (1967). Moreover, image quality can be improved using optical filters in conjunction with either black and white or color film (Bradbury, 1985; Fig. 3).

A. Bright-Field Microscopy

Bright-field micrographs of hematoxylin and eosin and toluidine blue stained sections are best recorded with Kodak Technical Pan 2415 film, ISO 125. Degree of contrast on the negative can be selected using different developers. For convenience we use Kodak D-19. A green filter, placed below the condenser, is used to enhance contrast. Immunostained (peroxidase) sections are also photographed with Technical Pan film; however, a yellow filter replaces the green filter. A direct comparison of the effects of adding this yellow filter when photographing peroxidase stained sections with color film (Kodak Ektachrome, 160T) is shown in Fig. 3B relative to Fig. 3A, where no filter was used. Gordon (1988) suggests the use of a BG38 and FITC-495 filter set to enhance the DAB reaction product. We have tried this combination but found no appreciable difference in our material (data not shown).

B. Fluorescence Microscopy

Immunofluorescent and fluorescent photomicrographs are taken with Kodak T-MAX film (ISO 400) and developed in Kodak HC-110, dilution B. Use of agents to retard photobleaching (Giloh and Sedat, 1982; Valnes and Brandtzaeg, 1985) is advisable for reducing exposure times. Additionally,

Schulze and Kirschner (1986) and Gard et al. (1990) describe using hypersensitized Technical Pan 2415 film to improve the resolution of highly magnified fluorescent images. If photobleaching is a problem we suggest using T-MAX 400 ISO film at 1600 ISO and push-processing at ISO 1600 with T-MAX developer. Photomicrographs will appear more grainy but will provide sufficient indication of the staining pattern.

C. Electron Microscopy

Transmission electron micrographs are taken on Kodak EM 4489 film and developed in Kodak D-19. The scanning electron microscope (Joel JSM-35C) we use has a camera which will accept either Polaroid Type 52 Polaplan ISO 400 or Kodak T-MAX ISO 100, 120 mm film. Polaroid film, though providing an immediate record, is more costly than the 120 mm film. TMAX film, developed with Kodak HC-110, dilution B, provides a high quality, fine-grained negative suitable for contact printing or enlargements.

Acknowledgments

The authors would like to thank Dr. E. Huebner, M. Collman, S. Lara, and H. Mar for providing technical assistance and J. Christian and B. Zelus for comments on the manuscript. We are grateful to the Department of Pathology for allowing us access to their electron microscopy facility. In addition, we thank Dr. A. Gown (Department of Pathology, University of Washington), for kindly providing the human muscle, actin-specific monoclonal antibody (HHF-35) (Tsukada et al., 1987). This work was supported by Public Health Service grants (R01-AR40089 and K04-AR01837) to R.T.M. G.M.K. was supported by a Postdoctoral Fellowship from the Medical Research Council of Canada.

References

Bendayan M., Nanci, A., and Kan, F. W. K. (1987). Effect of tissue processing on colloidal gold cytochemistry. *J. Histochem. Cytochem.* **35**, 983–996.
Bradbury, S. (1985). Filters in microscopy. *Proc R. Microsc. Soc.* **20**, 83–91.
Bradbury, S. (1986). Photomicrography. *In* "Immunocytochemistry: Modern Methods and Applications" (J. M. Polak and S. Van Noorden, eds.), pp. 225–242. John Wright and Sons, Bristol.
Brandtzaeg, P. (1982). Tissue preparation methods for immunohistochemistry. *In* "Techniques in Immunocytochemistry" (G. R. Bullock and P. Petrusz, eds.), Vol. 1, pp. 1–75. Academic Press, London.
Causton, B. E. (1984). The choice of resins for electron immunocytochemistry. *In* "Immunolabelling for Electron Microscopy" (J. M. Polak and I. M. Varndell, eds.) pp. 29–36. Elsevier, Amsterdam.
Coggi, G., Dell'Orto, P., and Viale, G. (1986). Avidin–biotin methods. *In* "Immunocytochemistry: Modern Methods and Applications" (J. M. Polak and S. Van Noorden, pp. 54–70. John Wright and Sons, Bristol.
Cohen, A. L. (1974). Critical point drying. *In* "Principles and Techniques of Scanning Electron

Microscopy: Biological Applications" (M. A. Hayat, ed.), Vol. 1, pp. 44–112. Van Nostrand-Reinhold, New York.
Coons, A. H., Creech, H. J., and Jones, R. N. (1941). Immunological properties of an antibody containing a fluorescent group. *Proc. Soc. Exp. Biol. Med.* **47,** 200–202.
Gard, D. L., Hafenzi, S., Zhang, T., and Doxsey, S. J. (1990). Centrosome duplication continues in cyclohexamide-treated *Xenopus* blastulae in the absence of a detectable cell cycle. *J. Cell Biol.* **110,** 2033–2042.
Giloh, H., and Sedat, J. W. (1982). Fluorescence microscopy: Reduced photobleaching of rhodamine and fluorescein protein conjugates by *n*-propyl gallate. *Science* **217,** 1252–1255.
Glauert, A. M. (1977). "Practical Methods in Electron Microscopy." North-Holland Publ., Amsterdam.
Gordon, S. R. (1988). Use of selected excitation filters for enhancement of diaminobenzidine photomicroscopy. *J. Histochem. Cytochem.* **36,** 701–704.
Graham, R. C., Jr., and Karnovsky, M. J. (1966). The early stages of absorption of injected horseradish peroxidase in the proximal tubules of mouse kidney: Ultrastructural cytochemistry by a new technique. *J. Histochem. Cytochem.* **14,** 291–302.
Guesdon, J.-L., Ternynck, T., and Avrameas, S. (1979). The use of avidin–biotin interaction in immunoenzymatic techniques. *J. Histochem. Cytochem.* **27,** 1131–1139.
Hayat, M. A., ed. (1974). "Principles and Techniques of Scanning Electron Microscopy: Biological Applications," Vol. 1. Van Nostrand-Reinhold, New York.
Hayat, M. A. (1981a). "Fixation for Electron Microscopy." Academic Press, New York.
Hayat, M. A. (1981b). "Principles and Techniques of Electron Microscopy: Biological Applications," 2nd Ed. University Park Press, Baltimore, Maryland.
Heggeness, M. H., and Ash, J. F. (1977). Use of the avidin–biotin complex for the localization of actin and myosin with fluorescent microscopy. *J. Cell Biol.* **73,** 783–788.
Hilwig, I., and Gropp, A. (1975). pH-dependent fluorescence of DNA and RNA in cytologic staining with "33258 Hoechst." *Exp. Cell Res.* **91,** 457–460.
Hsu, S.-M., Raine, L., and Fanger, H. (1981). Use of avidin–biotin–peroxidase complex (ABC) in immunoperoxidase techniques: A comparison between ABC and unlabelled antibody (PAP) procedures. *J. Histochem. Cytochem.* **29,** 577–580.
Humason, G. L. (1979). "Animal Tissue Techniques, "4th Ed." Freeman, San Francisco, California.
Jésior, J.-C. (1986). How to avoid compression. II. The influence of sectioning conditions. *J. Ultrastruct. Mol. Struct. Res.* **95,** 210–217.
Kalt, M. R., and Tandler, B. (1971). A study of fixation of early amphibian embryos for electron microscopy. *J. Ultrastruct. Res.* **36,** 633–645.
Karnovsky, M. J. (1965). A formaldehyde–glutaraldehyde fixative of high osmolality for use in electron microscopy. *J. Cell Biol.* **27,** 137A.
Keller, R., Cooper, M. S., Danilchik, M., Tibbetts, P., and Wilson, P. A. (1989). Cell intercalation during notochordal development in *Xenopus laevis*. *J. Exp. Zool.* **251,** 134–154.
Kelley, R. O., Dekker, R. A. F., and Bluemink, J. G. (1975). Thiocarbohydrazide-mediated osmium binding. A technique for protecting soft biological specimens in the scanning electron microscope. *In* "Principles and Techniques of Scanning Electron Microscopy: Biological Applications" (M. A. Hayat, ed.), Vol. 4, pp. 34–44. Van Nostrand-Reinhold, New York.
Lane, B. P., and Europa, D. L. (1965). Differential staining of ultrathin sections of epon-embedded tissues for light microscopy. *J. Histochem. Cytochem.* **13,** 579–582.
Larsson, L.-I. (1988). "Immunocytochemistry: Theory and Practise." CRC Press, Boca Raton, Florida.
Mar, H., and Wight, T. (1989). Correlative light and electron microscopic immunocytochemistry on reembedded resin sections with colloidal gold. *In* "Colloidal Gold Principles: Methods and Applications" (M. A. Hayat, ed.), pp. 357–378. Academic Press, New York.

Martinez-Hernandez, A. (1987). Methods for electron immunohistochemistry. *In* "Methods in Enzymology" (L. W. Cunningham, ed.), Vol. 145, pp. 103–148. Academic Press, New York.

Maupin-Szamier, P., and Pollard, T. D. (1978). Actin filament destruction by osmium tetroxide. *J. Cell. Biol.* **77**, 837–852.

Mercer, E. H. (1963). A scheme for section staining in electron microscopy. *J. R. Microsc. Soc.* **81**, 179–183.

Merchenthaler, I., Gallyas, F., and Liposits, Z. (1989). Silver intensification in immunocytochemistry. *In* "Techniques in Immunocytochemistry" (G. R. Bullock and P. Petrusz, eds.), Vol. 4, pp. 217–252. Academic Press, London.

Moon, R. T., and Christian, J. L. (1989). Microinjection and expression of synthetic mRNAs in *Xenopus* embryos. *Technique* **1**, 76–89.

Morrill, J. B. (1986). Scanning electron microscopy of embryos. *In* "Methods in Cell Biology" (T. E. Schroeder, ed.), Vol. 27, pp. 263–293. Academic Press, New York.

Muntz, L. (1975). Myogenesis in the trunk and leg during development of the tadpole of *Xenopus laevis* (Daudin 1802). *J. Embryol. Exp. Morphol.* **33**, 757–774.

Newman, G. R., Jasani, B., and Williams, E. D. (1983). Metal compound intensification of the electron-density of diaminobenzidine. *J. Histochem. Cytochem.* **31**, 1430–1434.

Nielsen, A. J., and Griffith, W. P. (1979). Tissue fixation by osmium tetroxide. A role for proteins. *J. Histochem. Cytochem.* **27**, 997–999.

Nieuwkoop, P. D., and Faber, J. (1975). "Normal Table of *Xenopus laevis* (Daudin), 2nd Ed. North-Holland publ., Amsterdam.

Novikoff, A. B., Novikoff, P. M., Quintana, N., and Davis, C. (1972). Diffusion artifacts in 3,3′-diaminobenzidine cytochemistry. *J. Histochem. Cytochem.* **20**, 745–749.

Polak, J. M., and Varndell, I. M., eds. (1984). "Immunolabelling for Electron Microscopy." Elsevier, Amsterdam.

Reynolds, E. S. (1963). The use of lead citrate at high pH as an electron opaque stain in electron microscopy. *J. Cell Biol.* **17**, 208–212.

Roth, J. (1983). The colloidal gold marker system for light and electron microscopic cytochemistry. *In* "Techniques in Immunocytochemistry" (G. R. Bullock and P. Petrusz, eds.), Vol. 2, pp. 217–284. Academic Press, London.

Sabatini, D. D., Bensch, K., and Barnett, R. J. (1963). Cytochemistry and electron microscopy. The preservation of cellular ultrastructure and enzymatic activity by aldehyde fixation. *J. Cell Biol.* **17**, 19–58.

Schulze, E., and Kirschner, M. (1986). Microtubule dynamics in interphase cells. *J. Cell Biol.* **102**, 1020–1031.

Smithwick, E. B. (1985). Cautions, common sense, and rationale for the electron microscopy laboratory. *J. Electron Microsc. Tech.* **2**, 193–200.

Sternberger, L. A. (1986). "Immunocytochemistry," 3rd Ed. Wiley, New York.

Strout, G. W., and Russell, S. D. (1990). A micro-sample critical point drying device for small SEM and TEM specimens. *J. Electron Microsc. Tech.* **14**, 175–176.

Szabo, D. (1967). "Medical Colour Photomicrography." Akadémiai Kiadó, Budapest.

Takamiya, H., Batsford, S., and Vogt, A. (1980). An approach to postembedding staining of proteins (immunoglobulin) antigen embedded in plastic: Prerequisites and limitations. *J. Histochem. Cytochem.* **28**, 1041–1049.

Tandler, B. (1990). Improved sectionability of paraffin-embedded specimens that initially are fixed in EM fixatives. *J. Electron Microsc. Tech.* **14**, 287–288.

Tsukada, T., Tippens, D., Gordon, D., Ross, R., and Gown, A. M. (1987). HHF35, a muscle-actin-specific monoclonal antibody. I. Immunocytochemical and biochemical characterization. *Am. J. Pathol.* **126**, 51–60.

Vacca, L. L. (1985). "Laboratory Manual of Histochemistry." Raven, New York.

Valnes, K., and Brandtzaeg, P. (1985). Retardation of immunofluorescence fading during microscopy. *J. Histochem. Cytochem.* **33,** 755–761.

Van Noorden, S. (1986). Tissue preparation and immunostaining techniques for light microscopy. *In* "Immunocytochemistry: Modern Methods and Applications" (J. M. Polak and S. Van Noorden, eds.), pp. 26–53. John Wright and Sons, Bristol.

Warnke, R., and Levy, R. (1980). Detection of T and B cell antigens with hybridoma monoclonal antibodies: A biotin–avidin-horseradish peroxidase method. *J. Histochem. Cytochem.* **28,** 771–776.

Weber, K., Rathke, P. C., and Osborn, M. (1978). Cytoplasmic microtubular images in glutaraldehyde-fixed tissue culture cells by electron microscopy and immunofluorescence microscopy. *Proc. Natl. Acad. Sci. U.S.A.* **75,** 1820–1824.

Woo Youn, B., Keller, R. E., Malacinski, G. M. (1980). An atlas of notochord and somite morphogenesis in several anuran and urodelean amphibians. *J. Embryol. Exp. Morphol.* **59,** 223–247.

Chapter 22

Whole-Mount Staining of Xenopus and Other Vertebrates

MICHAEL W. KLYMKOWSKY

Molecular, Cellular and Developmental Biology
University of Colorado at Boulder
Boulder, Colorado 80309-0347

JAMES HANKEN

Environmental, Population and Organismic Biology
University of Colorado at Boulder
Boulder, Colorado 80309-0334

I. Introduction
   A. Clearing Agents
   B. Bleaching
   C. Fixation
   D. Visualizing Binding of Antibodies
II. Methods
   A. Cortical Whole-Mount Immunocytochemistry
   B. Total Whole-Mount Immunocytochemistry
   C. Staining for Bone and Cartilage
   D. Combining Immunocytochemistry with Alcian Blue Staining
   E. Using Whole-Mount Methods Effectively
III. Formulations
IV. Conclusion
   References

I. Introduction

In toto or "whole-mount" imaging of organisms has a long history (see Campbell, 1986). Its advantages over section-based analysis are obvious. Whole-mount imaging provides an immediate and three-dimensional view of the stained components. Three-dimensional information can also be gained by serial section analysis, but only following an indirect, time-consuming, and rather tedious process. In addition, sectioning is subject to many preparation artifacts that can distort the morphology of the tissues under study. Early whole-mount methodology was largely restricted to staining for cartilage or bone. Recently, technical developments have made it possible to also use antibodies and nucleic acid probes in whole-mount preparations. At the same time, advances in light microscopy make it possible to section thick specimens optically. We are therefore now able to describe, at higher resolution and in the three-dimensional context of the intact organism, the temporal and spatial distribution of a wide range of cellular and extracellular components. Our aim here is to present a working guide to the use of these methods in *Xenopus*, other amphibians, and in vertebrate embryos in general. Recipes for stains, fixatives, etc., are provided in Section III.

A. Clearing Agents

The prerequisite for any whole-mount analysis is that we be able to see through the specimen. This means either that the specimen must be naturally transparent, a rare feature among higher metazoans, or that it must be possible to "clear" it. Clearing generally involves two steps: extracting material from the specimen, then matching the refractive index of the bulk of the specimen remaining, thereby rendering it transparent. Primary considerations in selecting a clearing agent are how closely it matches the refractive index (n_D) of the specimen, its inherent toxicity, and its compatibility with the staining reagents used. A wide range of clearing agents have been used over the last century. These range from potassium hydroxide/glycerol (Schultze, 1987; Mall, 1904) to methyl salicylate (artificial oil of wintergreen) (Spalteholz, 1914), methyl salicylate and benzyl benzoate (Ojeda et al., 1970), and carbon disulfide (Lundvall, 1904). With respect to the anuran amphibians *Xenopus*, *Bombina*, and *Eleutherodactylus*, the clearing agent devised by Andrew Murray and Marc Kirschner, a 1:2 mixture of benzyl alcohol ($n_D 1.54035$) and benzyl benzoate ($n_D 1.5681$) (BABB), is a close match to the refractive index of the oocyte and embryo. It is also compatible with fluorescence- and peroxidase-based immunocytochemistry (Dent et al., 1989; Dent and Klymkowsky, 1989) and polarization optics (Chu and Klymkowsky, 1989). BABB is an effective clearing agent for insects (R. B. Cary, personal communication), chick (J. B. Miller, personal communication), mouse (Wright et al., 1989a), shrimp (Fig. 1A) (P. Hertzler, personal communication), and zebrafish (Metcalfe et al., 1990) embryos. In addition, BABB works effectively on specimens stained for cartilage with Alcian blue (see Fig. 3E).

Benzyl alcohol and benzyl benzoate are both naturally occurring products (see *Merck Index*). Both are irritants, and care should be taken to avoid exposure to skin; we routinely use gloves when handling this reagent. Used BABB must be disposed of by incineration by the appropriate toxic waste disposal service. BABB suffers from two minor technical drawbacks. First, it renders the embryos brittle, and care must be taken when specimens are manipulated. Second, many of the commonly used chromogenic substrates are soluble in BABB (Dent et al., 1989).

CLEARING AGENTS FOR CARTILAGE AND BONE

A number of procedures for whole-mount staining of bone and/or cartilage call for clearing the specimen by prolonged immersion in potassium hydroxide followed by transfer to glycerol (e.g., Park and Kim, 1984; Wassersug, 1976). In our experience, using trypsin to macerate soft tissues, especially muscle, is more effective and faster, especially with large or dense specimens (Dingerkus and Uhler, 1977; Kelly and Bryden, 1983). Maceration must be done with care; the specimen eventually will decompose if left in the solution too long or if the solution is too concentrated. As with clearing with potassium hydroxide, maceration using trypsin is temperature dependent; the processing time for large or dense specimens can be shortened by placing the trypsin solution in a warm incubator. To determine whether maceration is complete, it often is helpful to advance the specimen to the next step in the staining procedure, that is, into the alizarin red solution; the degree of digestion of muscle is readily gauged against the background provided by the red bones and blue cartilage. If more maceration is needed, the specimen can simply be backed down into a fresh trypsin solution. In cases where specimens are to be stained for cartilage but not bone, BABB can be used to clear the specimens (see below).

B. Bleaching

In addition to their inherent opacity, many embryos are also highly pigmented. The presence of pigment generally interferes with the clear visualization of antibody and histochemical staining. Where naturally occurring albino variants are available, as in *Xenopus* (see Chapters 1 and 3, this volume), they can be used. However, maintaining a colony of albinos is an added burden (particularly as albinos are rather unattractive). In any case, it is possible to bleach most pigments using hydrogen peroxide (see Bechtol, 1948; Evans, 1948; Dent et al., 1989). For immunocytochemistry, we have found that 1 part 30% hydrogen peroxide in 2 parts Dent fixative bleaches even the darkest embryos within 2 to 7 days. Although some pigment may persist in the retina, it generally does not interfere with viewing the specimen. Bleaching also has the added benefit of destroying endogenous peroxidase activities that interfere with horseradish peroxidase-based secondary reagents. In our experience, bleaching has little affect on the antigenicity of a wide variety of monoclonal antibodies that we have tested. In fact, some antibodies that fail to react with their target molecule in simple methanol-fixed embryos react strongly following bleaching, suggesting that the bleaching step may sometimes uncover hidden epitopes.

Bleaching Bone–Cartilage-Stained Whole Mounts

Because bone–cartilage staining typically involves later embryonic or posthatching stages, naturally occurring pigment can be an even greater problem than in immunocytochemical whole mounts of early embryonic stages. Given that much of the pigment lies in the integument, it can be readily removed by skinning the specimen. Additional pigment, however, frequently lies beneath the dermis. It is not removed by either skinning or evisceration and must be bleached if it interferes with viewing the stained tissues. We use a 3% hydrogen peroxide solution, added during the early stages of glycerol infiltration, after maceration and staining. In our experience, the peroxide does not affect the intensity of Alcian blue- or alizarin red-stained tissues.

C. Fixation

Perfect fixation would immobilize every molecule of the specimen instantaneously. However, a perfect fixation would also pose a serious barrier to the diffusion of molecules into and out of the specimen. In the case of whole-mount immunocytochemistry, the sample must be permeable to molecules of the order of 150,000 (immunoglobulin G, IgG) to 900,000 (IgM) daltons. At the same time, to be useful the fixative must preserve cellular structure. Two basic types of fixatives have been used successfully for whole-mount immunocytochemistry: alcohol-based (Dent et al., 1989) and aldehyde-based fixatives (Patel et al., 1989). Alcohol-based fixatives reduce the dielectric constant of the solvent phase and result in the denaturation of proteins. In particular, the reduced dielectric constant of the solvent tends to disrupt hydrophobic interactions while stabilizing hydrogen bonds. Therefore, alcohol-based fixatives are expected to preserve secondary structure, while destabilizing tertiary interactions (Pearse, 1980). Antibodies directed against linear epitopes should be largely unaffected by alcohol-based fixatives. The denaturation of proteins by alcohol-based fixation is not necessarily permanent, and some local renaturation presumably occurs when the specimen is returned to aqueous conditions. Therefore, antibodies directed against some three-dimensional (as opposed to linear) epitopes should also be relatively unaffected by alcohol-based fixatives. Antibodies directed against three-dimensional antigenic determinants that cannot reform following alcohol denaturation will probably be lost. In addition, proteins and other components that are soluble in the alcohol fixative will be lost from the specimen and so not detectable.

Aldehyde-based fixatives act by chemically cross-linking proteins. Under the conditions usually employed (near neutral pH, 1–4% w/v formaldehyde) the cross-linking reaction is far from complete (Pearse, 1980). However, molecules that may be soluble in alcohols can be readily rendered insoluble by aldehyde-based fixatives. Aldehyde-based fixatives leave the plasma membrane largely intact, so they must be followed by extraction of membrane lipids with detergent or organic solvents. Antigenic sites that contain reactive groups may be chemically modified during aldehyde fixation and their antigenicity lost.

Typically, the choice of fixative is determined empirically. During the development of our whole-mount method, Joseph Dent examined a number of different alcohol-based fixatives, beginning with 100% methanol, a fixative used routinely for visualizing cytoskeletal systems in cultured cells. He found that antibody staining was largely restricted to exposed surfaces. By adding dimethyl sulfoxide (DMSO) to the methanol, he was able to arrive at a fixative (Dent fixative: 1 part DMSO to 4 parts methanol) that preserved cellular and tissue structure, while at the same time allowing antibody molecules to penetrate throughout the embryo. Dent fixative appears quite good for a range of antibodies against cytoskeletal, nuclear, and extracellular matrix components (Figs. 1A,B and 2) (see also Chu and Klymkowsky, 1989; Wright *et al.*, 1989a,b; McMahon and Moon, 1989; Jones and Woodland, 1989; Hanken *et al.*, 1990). However, Dent fixative is not always adequate.

The experiences of those who have tried to visualize the organization of microtubules in the *Xenopus* oocyte are particularly instructive. Palecek *et al.*, (1985) used Bouin–Hollande fixative and section-based immunocytochemistry and found a substantial accumulation of tubulin in the region of the mitochondrial mass (Balbiani body) of the early stage oocyte. Similar results were reported by Wylie *et al.* (1985). In contrast, we found no such accumulation of tubulin immunoreactivity in Dent-fixed oocytes (Dent and Klymkowsky, 1989). Electron microscopic analyses of glutaraldehyde-fixed oocytes reveal relatively few microtubules, particularly considering the known abundance of tubulin within the oocyte (see Dent and Klymkowsky, 1989, for a review). In contrast, Gard (1991) has found that fixation of *Xenopus* oocytes with either a formaldehyde/glutaraldehyde/taxol or formaldehyde/

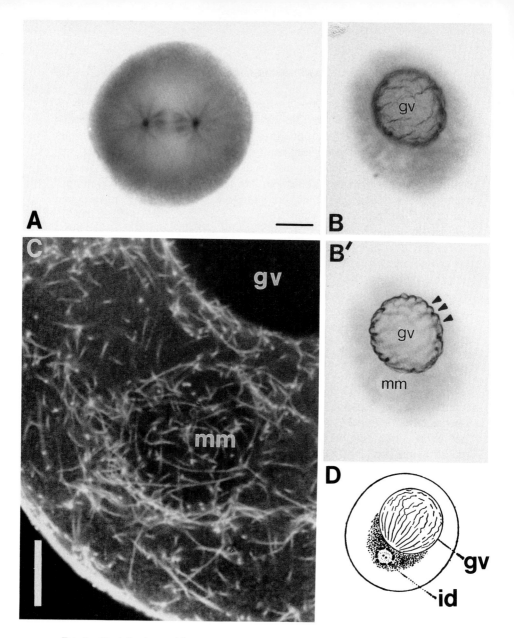

FIG. 1. Dent fixative and immunoperoxidase reagents provide clear low-resolution images of cytoskeletal components. In (A) the mitotic spindle of a brine shrimp egg is visualized using an antitubulin antibody. In small *Xenopus* oocytes, it is often possible to section the specimen optically using a standard microscope. Such "low-tech" optical sectioning of a stage I *Xenopus* oocyte stained with an antilamin antibody is shown in (B) and (B′); arrowheads point to indentations in nuclear envelope visible in one, but not the other "section." The mitochondrial mass (mm) can be made out as an unstained region adjacent to the nucleus. Higher resolution imaging requires confocal microscopy; (C) is such a confocal image of the microtubules of a stage I *Xenopus* oocyte. The interaction between microtubules and the mitochondrial mass is particularly evident (gv, germinal vesicle). The interaction between germinal vesicle and mitochondrial mass, known as the "idiozome" (id) in earlier literature, is illustrated diagrammatically in (D). [(A) was supplied by Phil Hertzler; (C) by Dave Gard; and (D) is from Wilson, 1925.]

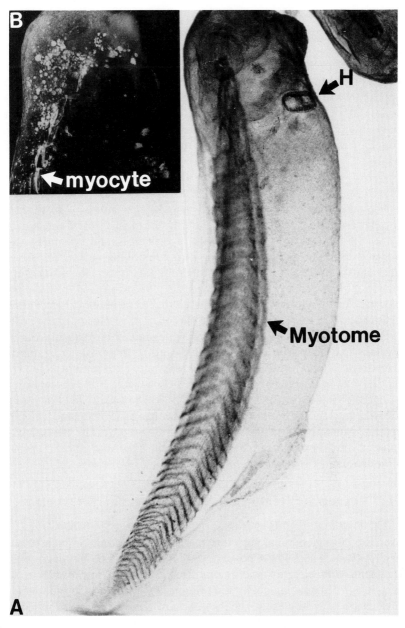

FIG. 2. Whole-mount immunocytochemistry can be used to visualize the specificity of an antibody. In this case, a commercial antidesmin antibody was tested. The pattern of staining (A) indicates that at the dilution used, the antibody was rather specific for myotomal (Myotome) and heart (H) muscle. Whole-mount methods can also easily be used to visualize the pattern of expression of various exogenous DNA or RNA molecules. For example, in (B) a fertilized egg was injected with DNA in which a *Xenopus* cytoplasmic actin promoter drove the expression of the bacterial protein β-galactosidase. Using an antibody against β-galactosidase, the pattern of cellular expression of the exogenous DNA can be easily visualized (head is oriented up). (B) was printed directly from a "positive" color slide and so is a negative image: stained cells are white.

glutaraldehyde fixative preserves a substantial microtubule system in all stages of oogenesis (Fig. 1C). In contrast to previous reports, microtubules appear to course around the mitochondrial mass but do not appear to be particularly concentrated there (Fig. 1C). So care must be taken, as dramatically different conclusions can arise from the use of different fixation protocols. Determining which reflects the "true" organization is often difficult.

Fixation for Bone Cartilage Staining

The stability of bone and cartilage make the choice of fixatives more straightforward. For whole-mount staining, the most effective and convenient fixative is buffered formalin. It is also possible to stain specimens fixed in buffered formalin and preserved in 70% ethanol, as is standard for many museum specimens, although with generally less favorable results. Cartilage staining, in particular, is often less intense following alcohol preservation. Nevertheless, we have found that the cartilages of specimens fixed in Dent fixative stain well (Fig. 3E).

Embryos, larvae, and adults whose integument is relatively porous to formalin may simply be immersed in the solution. In posthatching stages with a relatively impervious integument, for example, most amniotes, the fixative may have to be injected or perfused into the specimen, or the specimen may be skinned to ensure proper fixation. It is possible to bypass fixation and begin the staining procedure by immersing the freshly killed (and eviscerated) specimen directly into Alcian blue (Table II, Step 3). The efficacy of both staining and clearing are about the same in fixed and unfixed specimens, although unfixed specimens tend to curl.

D. Visualizing Binding of Antibodies

The binding of antibodies to specific components in fixed tissues can be visualized using colloidal gold, enzymatic, or fluorescent-conjugated secondary reagents. The highest possible resolution would be obtained with Fab fragments and electron microscopy, but it is often impossible to unambiguously recognize unlabeled Fab fragments. Therefore, immunogold conjugates are generally used. The resolution of these reagents is limited by the physical size of the primary and secondary reagents. At the level of light microscopy resolution is limited not by the reagents, but by the microscope itself. Fluorescent conjugates, chromogenic enzymes, and immunogold particles are all theoretically capable of generating images from structures below the resolution limit of the light microscope. They differ primarily in sensitivity. Because they create an image by emitting rather than absorbing light, fluorescent reagents are theoretically the most sensitive. Fluorescein-, rhodamine-, and Texas red-based fluorophores are the most widely used, and a wide range of secondary antibody reagents are commercially available. Fluorescein-based reagents are more sensitive to photobleaching under conditions of microscopic examination and photography than are rhodamine/Texas red-based reagents. However, the bleaching of fluorescein can be markedly reduced by the addition of "antifade" agents to the mounting media (Giloh and Sedat, 1982). We concoct our own mounting media for fluorescently stained specimens (see Section III).

Although extremely sensitive, fluorescent secondary reagents suffer from a number of practical limitations when used in whole-mount immunocytochemistry. First, their effective sensitivity is dependent on the numerical aperture (N.A.) of the lens used to examine the specimen. Unfortunately, most high N.A. lenses have very short working distances, making it impossible to focus completely through a thick specimen. Lower N.A., higher working distance lenses can be used, but these are effective only on rather robust signals. A second problem with fluorescent reagents arises from the fact that they form an image by emitting light. In *Xenopus*, there is a high level of autofluorescence, primarily from yolk platelets. This autofluorescence is not affected by clearing and obscures the signal from specific fluorescent reagents in most parts of the embryo, with the exception of thin, nonyolky regions, such as the tail (see Kay *et al.*, 1988). In addition, the excitation light illuminates more than just the focal plane currently being examined, and fluorophores both above and below the focal plane are excited and emit light. This generates out-of-focus fluorescence that decreases the signal-to-noise ratio and degrades the image. Autofluorescence from above and below the plane of focus also degrades the signal from the true focal plane.

These problems may be overcome either through the use of computational correction, which effectively removes the contaminating, out-of-focus noise (Agard *et al.*, 1989), or through the use of a confocal microscope. Gard (1991) used confocal microscopy to image the microtubule organization of *Xenopus* oocytes (Fig. 1C). The high signal-to-noise ratio of the confocal microscope provides a level of detail previously obtained only in relatively thin (5 to 10 μm thick) cultured cells. In addition, because confocal microscopes store data digitally, full serial reconstructions of the specimen can be generated, manipulated, and displayed. However, there still are serious limitations. Again, the working distance of high magnification/N.A. lenses generally is less than the thickness of the specimen (plus the cover glass). Under these conditions, the rather drastic measure of cutting the specimen in half (or into even smaller sections) must be taken. In addition, the high intensity of the excitation light that is used in many confocal microscopes tends to bleach the specimen (even in the presence of antifade reagents) in advance of its examination.

Enzyme-based secondary reagents work by generating a light-absorbing product from soluble substrate. Because the signal is formed by absorbing rather than emitting light, they are less sensitive than fluorescent reagents. Nevertheless, in many cases the lower sensitivity is more than compensated for by other characteristics. First, they are not affected by autofluorescence. Second, where the signal is relatively strong or tightly localized, such as within the nucleus, they are easy to visualize even at very low magnification (Fig. 1A,B, 2A, and 4). At low magnification it is possible to exploit the relatively narrow depth of field of light lenses to section the specimen optically (Fig. 1B,B'). At higher magnification/resolution, however, many of the same problems that plague fluorescent-based imaging affect enzyme-based imag-

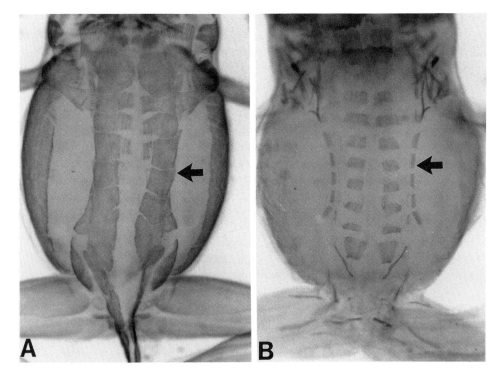

FIG. 4. Whole-mount images can dramatically illustrate both the specificity of particular antibodies and the differential distribution of specific molecules and anatomical components. Here, two posthatching *E. coqui* are stained with antibodies against either (A) fast-twitch or (B) slow-twitch fiber striated muscle myosins using monoclonal antibodies supplied by J. B. Miller (Massachusetts General Hospital) and Frank Stockdale (Stanford University). Note, for example, the different distribution of slow-versus fast-twitch fibers in the axial musculature (arrows).

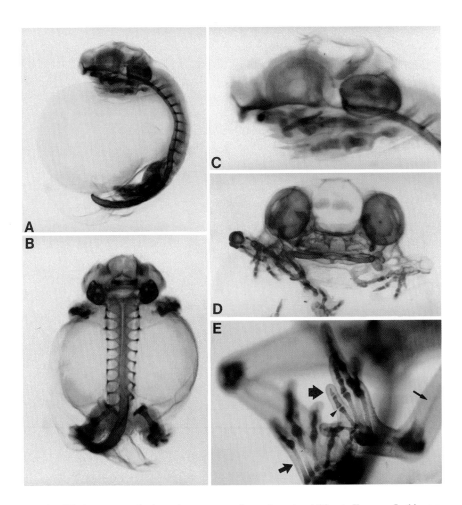

FIG. 3. Whole-mount methods work on a range of organisms, in addition to *Xenopus*. In this case, we have used a monoclonal antibody against type II collagen (provided by T. Linsenmayer, Tufts University) to describe cartilage development in the direct-developing frog *Eleutherodacytlus coqui*. Illustrated here are collagen-immunostained embryos, Townsend-Stewart (1985) stages 9 (A–C, different views) and 12 (D), as well as a posthatching specimen stained both immunocytochemically and with Alcian blue (E). The latter preparation effectively distinguishes cartilage in early stages of development (e.g., collagen-stained terminal phalange, large arrow) from mature cartilage (Alcian blue-stained metatarsal, medium arrow). As skeletogenesis proceeds, most of the cartilage is replaced by bone (small arrow), although some collagen is retained at joint surfaces (arrowhead). Color photographs were taken using Kodak Ektachrome (tungsten) ISO 50 or 160 film on a Wild M5A or M8 dissecting microscope equipped with fiber optic illumination. Black and white photographs were taken using Kodak TMAX 400 film.

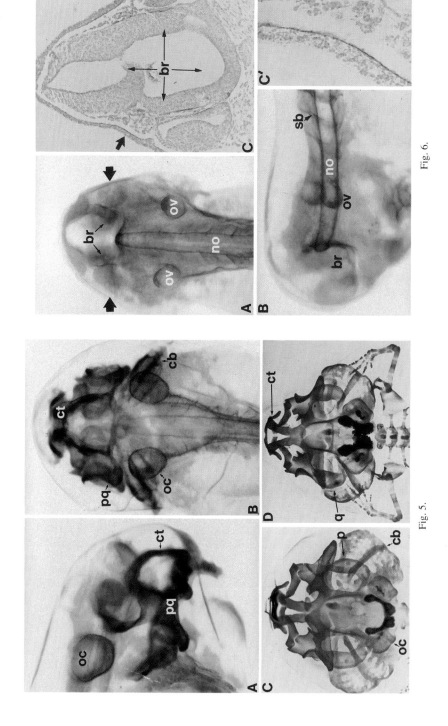

Fig. 5.

Fig. 6.

ing, namely, the need for high N.A., long working distance objectives and the degradation of the signal by out-of-focus information (Fig. 3A–C and 5A,B). Nevertheless, at present, enzyme-based visualization appears to be the best choice for most studies of protein/RNA localization during development.

II. Methods

A. Cortical Whole-Mount Immunocytochemistry

Fluorescence-based imaging of the cortical region of *Xenopus* oocytes and embryos is straightforward (Klymkowsky *et al.*, 1987). In the last few years, we have made only minor changes, simplifying the process. After being dejellied in 2% cysteine, specimens are fixed in Dent fixative overnight. They are rehydrated in Tris-buffered saline (TBS). Typically, we incubate 10 to 20 oocytes/embryos in 0.25 to 0.3 ml of primary antibody for 1 hour at room temperature or overnight at 4°C. During incubation, the samples are gently rocked; excessive movement can destroy the specimen completely, or remove its outer cortical layers. The specimens are then washed five times with TBS. Washing consists of aspirating the previous solution and adding TBS. When specimens have been stained for only an hour or two, the washes can be as short as 10 to 30 seconds each. For samples incubated overnight, each wash is for 2 to 5 minutes. Secondary antibody is then added and incubated (with rocking) for 2 hours at room temperature (or overnight at 4°C). Late stage *Xenopus* ooctyes and early embryos are approximately 1 to 1.2 mm in diameter. To examine cortical staining, we use 0.8-mm-deep depression slides

---

FIG. 5. Skeletogenesis may be followed in whole mounts from early stages of chondrogenesis through later stages of ossification, as in the development of the skull in the Oriental fire-bellied toad *Bombina orientalis*. (A, B) Hatchling tadpole, Gosner (1960) stage 24 (lateral and dorsal views, respectively), stained for type II collagen, which is particularly effective in early stages of cartilage development. (C) Mature tadpole (stage 36, ventral view) and (D) midmetamorphic tadpole (stage 39, dorsal view) differentially stained with alizarin red and Alcian blue; the latter stain is preferable for visualizing larger, mature cartilages. Note the changing morphology of several cartilages, including the otic capsule (oc), palatoquadrate (pq), ceratobranchials (cb), and cornu trabeculae (ct). Bone (red) is visible in the skull, vertebrae, and forelimbs in (D).

FIG. 6. Analysis of serial sections may be required to determine the complete distribution of specific molecules assayed initially in whole mounts. (A, B) In immunostained whole mounts of embryonic *B. orientalis* (Gosner stage 19; dorsal and lateral views, respectively), type II collagen is apparent in the notochordal sheath (no), otic vesicles (ov), segmental boundaries (sb), and the brain (br). (C, C′) In cross sections at the level of the developing eye (arrowheads in A), an additional subepidermal layer of collagen is also visible (C′ is higher a magnification view of the region denoted by the arrow in C).

(obtainable from most suppliers). Late stage oocytes, eggs, and embryos are placed in the center of these slides; earlier stage oocytes can be placed at the periphery. Mounting media (see Section III) is added, and a coverslip placed over the depression slightly flattens the specimens, bringing large expanses of their surface into a single focal plane (see Klymkowsky et al., 1987; Klymkowsky and Maynell, 1989) that can be viewed using high-resolution lenses. This approach has also worked well for microtubules (Elinson and Rowing, 1988; Dent and Klymkowsky, 1989).

B. Total Whole-Mount Immunocytochemistry

For complete whole-mount imaging we use peroxidease-conjugated secondary reagents and diaminobenzidine (DAB) (Table 1). Our current protocol includes some minor changes that reduce background and improve the penetration of antibodies into the specimen. In particular, the addition of DMSO to both the primary and secondary antibody incubations was suggested by Ben Szaro (National Institutes of Health). An important point to remember in all immunohistochemical staining is to control for the specificity of primary and secondary antibodies. Different antisera and secondary reagents can vary dramatically in titer and specificity. Each antibody *must* be titred to determine the most effective concentration with the lowest nonspecific, background staining (see below).

We have examined the usefulness of alkaline phosphatase-conjugated secondary reagents and various alternatives to DAB, such as cobalt/nickel intensification of DAB and the Histo-mark orange and black reagents from Kirkegaard & Perry, Inc., in the hopes of developing a workable double immunocytochemical staining method. Unfortunately, we have little success in this direction. The only alkaline phosphatase substrate we have found that is stable in BABB is black and difficult to distinguish from the brown DAB reaction product. In addition, we consistently find that alkaline phosphatase-conjugated secondary reagents have much higher nonspecific background staining than horseradish peroxidase-conjugated reagents, even in the presence of the phosphatase inhibitor levimisole (Dent et al., 1989). We also find very high backgrounds when the cobalt/nickel-intensified DAB reaction is used. The Histo-mark orange reagent is stable to BABB but produces relatively weak labeling in our hands.

C. Staining for Bone and Cartilage

A large number of whole-mount procedures for staining bone or cartilage have been published, beginning almost 100 years ago (Campbell, 1986). Only in the last 20 years, however reliable methods been available for dif-

TABLE I

WHOLE-MOUNT IMMUNOPEROXIDASE STAINING[a]

---

1. If a jelly coat is present, dejelly the specimen either chemically (2% cysteine, pH 8.0, for *Xenopus*) or maturally. Wash with Ringer's solution.

2. Fix overnight at room temperature in Dent fixative.

Alternatively, specimens can be fixed using aldehyde-based fixatives for 2 hours at temperature (Section III). Where aldehyde-based fixatives are used, extract the specimen overnight with 100% methanol or with Dent fixative following fixation.

3. Bleach pigment in 10% hydrogen peroxide (diluted from a 30% stock into Dent fixative).

Bleaching may take from 1 to 4 days; be patient!

4. Wash specimen in TBS for 15 minutes.

5. Incubate specimen overnight in primary antibody diluted into 95% calf serum/5% DMSO. (We add 0.1% thimoserol to serum as a preservative.) Rock gently.

Rocking should be carried out with rocker tilted to minimize turbulence.

6. Wash 5 times in TBS, 1 hour each.

7. Inclubate overnight in secondary antibody, diluted as for primary antibody.

8. Wash 5 times in TBS, 1 hour each.

9. React for 1 to 2 hours in 0.5 mg/ml DAB diluted into TBS plus 0.02% hydrogen peroxide. *Wear gloves*: DAB is carcinogenic!

We routinely make up stocks of DAB (10 mg/ml in water) and store them at $-20°C$, at which temperature they are stable indefinitely. Once diluted, unused DAB should be destroyed with bleach and discarded.

10. Stop reaction by dehydration with methanol (2 changes, 15 to 30 minutes each).

11. Clear in BABB (benzyl alcohol/benzyl benzoate).

---

[a] All steps can be carried out in either glass vials or microcentrifuge tubes. If clearing is not complete, specimens were not adequately dehydrated. They can be returned to 100% methanol and then cleared again. BABB causes specimens to become brittle; take care when manipulating them.

ferential and stimultaneous staining of bone cartilage in the same specimen (e.g., Dingerkus and Uhler, 1977; Park and Kim, 1984; Wassersug, 1976) (Table II). Several dyes have been used to stain cartilage, including Alcian blue, Alcian green, and toluidine blue. All have an affinity for acid mucopolysaccharides, one of the principal components of cartilage extracellular matrix (Bloom and Fawcett 1975; Ham and Cormack, 1979; Humason, 1979). The most reliable and effective of these stains for whole-mount double-staining techniques appears to be Alcian blue 8GX (e.g., Polysciences, Inc., Warrington, PA, Cat. No. 234). Because of the affinity of the stain for acid mucopolysaccharides, cartilage that has a poorly developed extracellular

TABLE II

Whole-Mount Staining for Bone and Cartilage[a]

1. Skin and eviscerate the specimen. During evisceration, be careful not to damage or even remove ventral skeletal structures, such as the limb girdles or the hyobranchial skeleton.

Skinning and evisceration may be omitted for embryonic, larval, and early posthatching stages in which the integument is poorly developed.

2. Rinse in distilled water (several changes) overnight. Blot on paper towels.

3. Immerse in Alcian blue for 6–24 hours, according to size (shorter interval for smaller specimens). Blot on paper towels.

4. Run the specimen through ethanol series (100, 100, 95, 70, 40, 15%) and into distilled water, 1–2 hours in each step (longer for large specimens). Specimens may be held overnight or longer in 70% ethanol.

5. Immerse in a solution of trypsin dissolved in 2% sodium tetraborate. Remove the specimen when it is limp and the muscles are translucent; it will eventually decompose if left too long. Begin with a 1% trypsin solution. and adjust the amount as necessary; the maceration rate is proportional to trypsin concentration. Embryos and early posthatching stages may be done after 1–2 hours; larger specimens may take up to several days. Change the trypsin solution daily.

6. Immerse in alizarin red working solution (10–15 drops alizarin red stock solution per 100 ml of 0.5% potassium hydroxide) overnight.

7. Run through a 0.5% potassium hydroxide/glycerol series (3:1, 1:1, 1:3) and then into pure glycerol. Allow at least 2 hours for each step; the specimen is ready to be moved when it sinks to the bottom of the container. Bleaching of pigment (e.g., integument, mesenteries) should be done in the first step (3:1), by adding a small amount of 3% hydrogen peroxide. Begin with a few drops of peroxide per 100 ml of solution; if the peroxide concentration is too high, oxygen bubbles will form within the specimen. Bleaching may require several days.

8. Store the specimens in pure glycerol. Thymol or phenol can be added to retard bacterial growth.

Cartilage stains deep blue. Bone, teeth, and other calcified tissues (e.g., calcified cartilage, calcified endolymph) stain red.

[a] Steps 5–8 are reversible, that is, specimens in glycerol may be backed down into alizarin for more intense bone staining or into trypsin for additional maceration, and then back up to pure glycerol. Small specimens should be scored for bone within a few days of completing the procedure, as extremely small or thin bones may destain and become invisible with time; restaining is now always effective. The cartilage stain (Alcian blue) is relatively stable.

matrix (e.g., fibrocartilage) or that is otherwise deficient in mucopolysaccharides will not stain well. This, however, is rarely a serious problem; typically, Alcian blue 8GX beautifully visualizes virtually the entire cartilaginous skeleton (Fig. 3E and 5C,D). Alizarin red S (alizarin sodium sulfonate; Sigma Chemical Co., St. Louis, MO, Cat. No. A 5533) is the preferred stain for bone; it stains calcium in the extracellular matrix, although it will react with other metals (Humason, 1979) (Fig. 5C,D). For obvious reasons, decalcified specimens will not give a positive reaction to the stain, nor will precalcified bone matrix (osteoid). Also extremely thin or otherwise small bones may be difficult to see in whole mounts because of the faintness of the red color.

D. Combining Immunocytochemistry with Alcian Blue Staining

BABB does not solubilize Alcian blue. It is therefore possible to combine immunocytochemistry with Alcian blue staining. This has the advantage, particularly in later stage embryos, where staining of cartilage can provide a reference with which to compare the pattern of antibody staining (Fig. 3E). Our procedure for this is simple (Table III) and essentially consists of the first part of the standard Alcian blue staining procedure (Table II) followed by standard whole-mount immunocytochemistry (Table I). The combination of these methods allows a direct bridge between immunocytochemistry and more classic cartilage and bone staining methods (Fig. 3). In our experience, however, specimens must first be fixed in Dent fixative.

E. Using Whole-Mount Methods Effectively

1. IMPORTANCE OF SECTIONING

Whole-mount methods must be used in conjunction with section-based methods to produce a complete and accurate view of the specimen. Judicious sectioning provides two types of information (Fig. 6). First, it reveals

TABLE III

ALCIAN BLUE/IMMUNOCYTOCHEMISTRY DOUBLE STAINING

1. Fix specimens in Dent fixative (Table I).
2. Rinse in distilled water (several changes) overnight. Blot on paper towels.
3. Immerse in Alcian blue (Table II). Blot on paper towels.
4. Bleach overnight in Dent bleach (Table I).
5. Rehydrate in TBS for 15 minutes.
6. Incubate in primary antibody, wash, and incubate in secondary antibody (Table I).
7. Wash and react for 1 to 2 hours in 0.5 mg/ml DAB diluted into TBS plus 0.02% hydrogen peroxide.
8. Stop the reaction with methanol and clear.

the exact position of the stained material with respect to surrounding cells and tissues (Dent et al., 1989; Chu and Klymkowsky, 1989). This type of information can be quite difficult to extract from whole-mount images. Second, weakly or diffusely stained components are more easily recognized in sections versus whole mounts. A particularly dramatic example of this comes from our work on the distribution of type II collagen in the fire-bellied toad, *Bombina orientalis* (Seufert et al., 1990). Whole-mount images of anti-type II collagen-stained embryo reveal the major collagen type II-containing regions associated with the notochord, axial segments, brain, and otic vesicles (Fig. 6A,B). Sectioning reveals additional type II collagen immunoreactivity in specific regions of the dermis (Fig. 6C,D). This dermal collagen was completely overlooked during the initial examination of whole-mounts specimens (Fig. 6). A similar caveat applies to the interpretation of bone- and cartilage-stained whole-mounts. Thin layers of calcified bone matrix that characterize the initial stages of ossification, and which are readily visualized in sections, often are invisible in whole mounts. For this reason, whole mounts cannot be used reliably to describe certain aspects of the early stages of bone development, such as the absolute timing of osteogenesis (Hanken and Hall, 1988). However, the combination of whole-mount and section-based analysis easily generates a complete description of the embryo.

Preparation of sections sometimes can be enhanced, and always simplified, by staining the specimen in whole mount. The specimen can also be examined first in whole mount and then sectioned if necessary, to provide a complete description (Table IV). It is difficult to counterstain sections once they have been cleared. If counterstaining is desired, the specimen can be stained in whole mount and then sectioned without prior clearing (Table IV). Originally we used a version of Steedman's polyethylene glycol:cetyl alcohol-based embedding medium for sectioning analysis (Dent et al., 1989). Such sections were fragile and difficult to handle. We have therefore switched to using Paraplast as an embedding medium (Table IV).

2. Whole-Mount Labeling in Practice

Once an antibody against a protein has been generated it must be characterized with respect to specificity. This is generally done by Western blot analysis, preferably of impure samples. If an antiserum is not completely specific, it can be made specific by affinity purification, either on nitrocellulose blots (Olmstead, 1981) or using antigen columns (Hudson and Hay, 1980). The specificity of monoclonal antibodies that are not originally monospecific cannot easily be improved. Moreover, the specificity of an antibody in Western blots does not necessarily guarantee its specificity in cytochemistry. Therefore, cytochemical specificity must be assayed directly. In addition,

TABLE IV
Sectioning Whole-Mount Immunostained Embryos[a]

1. Follow the protocol for whole-mount immunohistochemistry (Table I) through Step 9. Stop the reaction by washing embryos in TBS.

   Embryos that have been cleared in BABB can be transferred directly into Histoclear (Step 3).

2. Wash specimens in 70% ethanol for 1 hour and then in 95% ethanol for 1 hour, then wash twice in absolute ethanol for 30 minutes each time.

3. Wash twice in Histoclear, 1 hour each.

4. Immerse in molten Paraplast for 1 hour, replace paraplast and let sit for another hour, and then embed in Paraplast.

5. Cut as 6 to 10-μm serial sections using a microtome and mount on glass slides. We typically mount sections on albumin-coated slides.

6. Wash slides twice with Histoclear, each wash 3 minutes or longer.

   Note: BABB-cleared specimens do not stain with eosin, so Steps 7–10 can be omitted and the specimens mounted with cover glass following Step 6. Even in uncleared specimens, one can omit eosin staining if desired. In that case go directly to Step 11.

7. Wash slides twice in absolute ethanol, each wash 3 minutes or longer.

8. Wash slides in 95% ethanol and then 70% ethanol, each wash 3 minutes or longer.

9. Incubate in eosin for 20 seconds and then rinse in 70% ethanol for 2–3 seconds, followed by a 2- to 3-second rinse in 95% ethanol.

10. Wash twice with 100% ethanol (3 minutes each), then rinse twice in Histoclear (3 minutes each).

11. Mount cover glasss using Permount. Immunostained areas will appear dark brown. Nonstained regions will appear faint orange or pink (in eosin-stained specimens). Slides appear permanent; we have stored them for over 1 month without apparent deterioration of the specimens.

[a] Eosin is used as a counterstain to visualize background tissues and more effectively localize immunostained regions. Histoclear (National Diagnostics, Manville, NJ) is a nontoxic, low-odor histological clearing agent that may be substituted for traditional agents such as toluene and xylene. Permount is sold by Fisher Scientific, Paraplast by Monoject Scientific (St. Louis, MO). Additional details concerning procedures for infiltration, embedding, sectioning, and mounting can be found in standard histology manuals (e.g., Humason, 1979).

fixation conditions that preserve the immunoreactivity of the target protein and its normal distribution within the specimen must also be determined. Both of these goals are largely a matter of trial and error.

In the best case, the distribution of the target molecule in at least one stage of organismic development is already known. That stage can then be used to define appropriate fixation conditions and antibody specificity. In addition, this stage can be used to determine the appropriate working dilutions for both primary and secondary antibodies. It is particularly helpful if one has available an antibody whose staining properties are already fairly well defined. This antibody can be used to determine the appropriate working dilution for the secondary antibody. Determining the working dilution for the secondary antibody is particularly important; in our experience most difficulties in the successful use of whole-mount staining arise from the use of secondary antibodies at inappropriate dilutions. The optimal working dilutions of commercially available secondary reagents (from the same source) can differ by over 10-fold. Using a secondary reagent at too low a dilution will generate excessive background staining; too high a dilution will result in a weak or nonexistent signal. Once the working dilution of the secondary antibody has been determined, it is then critical to determine that the secondary antibody does not react with the specimen in the absence of primary antibody. Although fortuitous reactions of primary (Gordon et al., 1978) and secondary (Strome and Wood, 1982) antibodies can occasionally be quite helpful, unexpected reactions could be embarrassing if not recognized.

The working dilution of a particular secondary antibody will be the same for each primary antibody/antiserum used. This makes it possible to titer the primary antibody. Typically, we use monoclonal antibody supernatants at dilutions from 1:5 to 1:100; ascites fluids from 1:100 to 1:10,000; and rabbit antisera from 1:50 to 1:10,000. Again, as in the case of the secondary antibody, too low a dilution will generate excessive background, whereas too high a dilution will fail to stain the target tissue adequately.

When the distribution of a particular target molecule is known for a particular specimen, the staining pattern observed actually provides a further measure of antibody specificity. It is important to remember when comparing Western blots and immunocytochemical assays for antibody specificity that target molecule/antibody reactions can be quite different under these different assay conditions, and different classes of molecules may well react with the antibody. For example, a number of antibodies have been characterized that appear to be specific in immunocytochemistry but fail to react with their expected target molecule on Western blots. The reverse situation can also occur; in other words, a molecule can react on blots but fail to react in immunocytochemistry. Failure to react in immunocytochemistry can be due to destruction of the target epitope by fixation, loss of the target molecule from the specimen owing to poor fixation, or inaccessibility of the target epitope in the fixed specimen. Where unexpected reactivities occur, it is best to test other antibodies known to react with the same target molecule. If these are not available, care in interpretation of results must be taken, as it is possible that the target molecule is expressed in the unexpected position or that the reactivity is due to an unexpected but immunoreactive molecule. If sufficient material is available, the presence or absence of the target molecule can be determined directly by Western blots or immunoprecipitation. Unfortunately, when using monoclonal antibodies or affinity-purified antisera, absorption with the target molecule provides little additional information, since we know *a priori* that such absorption will remove all immunoreactivity.

For most studies, we find immunoperoxidase staining quite effective. It has very low background and easily reveals proteins of moderate abundance (Fig. 2–6). Low-power lenses can be used to provide a global overview of the distribution of the target molecule (Figs. 2A, 3A,B, and 4), whereas higher power lenses (Figs. 4C,D, 5A,B, and 6A,B) and sectioning (Fig. 6C,D) can be used to further define the exact distribution of the target molecule. When we were originally working on whole-mount immunocytochemistry, we were concerned about whether the size of the specimen would pose an insurmountable obstacle to the diffusion of antibodies into and out of the specimen. Over the years, however, we have applied the method to larger and larger specimens. In our latest work, we have examined the distribution of type II collagen and fast- and slow-skeletal muscle myosins in the direct-developing frog *Eleutherodactylus coqui* (Hanken et al., 1990). These embryos are 4–5 mm across, yet antibodies appear to penetrate throughout quite effectively (Figs. 3 and 4). We have not yet reached the upper limit for specimen size in whole-mount immunostaining.

3. Double Staining in Whole Mount

Reliable methods for double staining cartilage and bone are available (Fig. 5), and it is possible to double stain embryos with antibodies and Alcian blue (Fig. 3E). However, double-antibody labeling remains problematic. We have used DAB staining together with polarization optics to follow the interaction between outgrowing neurites and the muscle of the lateral myotome (Chu and Klymkowsky, 1989). However, only select structures can be visualized using polarization optics. Fluorescent-based secondary reagents may be quite useful, but in most cases they will need to be analyzed using a confocal microscope. Even with a confocal microscope, however, true *in toto* imaging of the embryo at the cellular and subcellular level will await the development of high-resolution, long-working distance lenses. When available, double fluorescence imaging of whole-mount stained specimens will open the oocyte and embryo as a readily accessible system in which to study the wide range of cell functions that underlie embryonic development.

III. Formulations

Mounting Medium (for fluorescently labeled speciemens)
Dissolve 10 g airvol 205 (polyvinyl alcohol; Air Products, Inc, Allentown, PA) in 40 ml 50 mM Tris, pH 8.0. This takes 24 to 48 hours at 37°C. Add

20 ml glycerol and 1.2 g n-propyl gallate. Aliquot and store at 4°C. This mounting medium will dry within 2 to 4 hours. The propyl gallate will reduce the rate of bleaching for fluorescein-conjugated antibodies.

Dent Fixative

1 part DMSO, 4 parts 100% methanol

Aldehyde Fixative

0.1 M MOPS, 2mM EGTA, 1 mM MgSO$_4$, 3.7% formaldehyde, pH 7.4 (Hemmati-Brivanlou and Harland, 1989).

Dent Bleach

1 part 30% hydrogen peroxide, 2 parts Dent fixative

BABB Clearing Agent

1 part benzyl alcohol, 2 part benzyl benzoate

Alcian Blue Working Solution

20 mg Alcian blue 8GX (C.I. 74240), 70 ml absolute ethanol, and 30 ml glacial acetic acid. Use at room temperature; store refrigerated. Discard after 6 months.

Alizarin Red Stock Solution (Humason, 1979)

5.0 ml alizarin red S (C.I. 58005; alizarin sodium sulfonate) saturated in 50% acetic acid, 10.0 ml glycerol, and 60.0 ml chloral hydrate, 1% aqueous. Keeps indefinitely at room temperature.

Buffered Formalin (Humason, 1979)

100 ml concentrated formalin (40% formaldehyde-saturated water), 900 ml distilled water, 4.0 g sodium phosphate, dibasic (monohydrate), and 6.5 g anhydrous sodium phosphate, monobasic.

Eosin Stock Solution (Humason, 1979)

1.0 g eosin Y (C.I. 45380), 1 liter 70% ethanol, and 5.0 ml glacial acetic acid. Dilute with an equal volume of 70% ethanol before use and add 2–3 drops of acetic acid.

IV. Conclusion

Whole-mount staining makes the analysis of normal and experimentally manipulated embryos much simpler. It can be used in the assay of cellular differentiation in induction and tissue recombination experiments (see Chapters 17 and 18 this volume). It should be possible not only to assay for the indication of specific tissues, but to characterize the three-dimensional relationships between the tissue types. Whole-mount staining greatly simplifies the characterization of the expressions patterns of exogenous DNAs. Similarly, the effects of injected antibodies, antisense reagents, or the ectopic expression of specific molecules (Chapter 23, this volume) on development can be analyzed rapidly.

Acknowledgments

We thank Joe Dent for his work in developing the original whole mount staining methods. Dan Seufert devised the procedure for Paraplast sectioning of immunostained whole mounts, and Nicole Ingebrigtsen provided technical support. We thank Dave Gard and Philip Hertzler for supplying photographs. We have been supported by the National Science Foundation and the Pew Biomedical Scholars Program.

References

Agard, D. A., Hiraoka, Y., Shaw, P., and Sedat, J. W. (1989) Fluorescene microscopy in three dimensions. *In* "Methods in Cell Biology" (D. L. Taylor and Y.-L. Wang, eds.), Vol. 30, pp. 353–377. Academic Press, New York.

Bechtol, C. O. (1948). Differential *in toto* staining of bone, cartilage and soft tissues. *Stain Technol.* **23**, 3–8.

Bloom, W., and Fawcett, D. W. (1975) "A Textbook of Histology." Saunders, Philadelphia, Pennsylvania.

Campbell, S. C. (1986). A bibliography on clearing and staining small vertebrates. "Proceedings of the 1985 Workshop on Care and Maintenance of Natural History Collections" (J. Waddington and D. M. Rudkin, eds.), pp. 115–116. Life Sciences Miscellaneous Publications, Royal Ontario Museum, Ontario, Canada.

Chu, D. T. W., and Klymkowsky, M. W. (1989). The appearance of acetylated α-tubulin during early development and cellular differentiation in *Xenopus*. *Dev. Biol.* **136**, 104–117.

Dent, J. A., and Klymkowsky, M. W. (1989). Whole-mount immunocytochemical analysis of cytoskeletal function during oogenesis and early embroyogenesis in *Xenopus*. In "The Cell Biology of Fertilization" (H. Shatten and G. Shatten, eds.), pp. 63–103. Academic Press, New York.

Dent, J. A., Polson, A. G., and Klymkowsky, M. W. (1989). A whole-mount immunocytochemical analysis of the expression of the intermediate filament protein vimentin in *Xenopus*. *Development (Cambridge, U.K.)* **105**, 61–74.

Dingerkus, G., and Uhler, L. D. (1977). Enzyme clearing of Alcian blue stained whole small vertebrates for demonstration of cartilage. *Stain Technol.* **52**, 229–232.

Elinson, R., and Rowning, B. (1988). A transient array of parallel microtubules in frog eggs: Potential tracks for a cytoplasmic rotation that specifies the dorso–ventral axis. *Dev. Biol.* **128**, 185–197.

Evans, H. E. (1948). Clearing and staining small vertebrates, *in toto*, for demonstrating ossification. *Turtox News* **26**, 42–47.

Gard, D. L. (1991). Organization, nucleation and acetylation of microtubules in *Xenopus laevis* oocytes: A study by confocal immunofluorescence microsopy. *Dev. Biol.* **143**, 346–363.

Giloh, H., and Sedat, J. W. (1982). Flourescene microscopy: Reduced photobleaching of rhodamine and fluorescein protein conjugates by *n*-propyl gallate. *Science* **217**, 1252–1255.

Gordon, W. E., Bushnell, A., and Burridge, K. (1978). Characterization of the intermediate (10 nm) filaments of cultured cells using an autoimmune rabbit antiserum. *Cell (Cambridge, Mass.)* **13**, 249–261.

Gosner, K. L. (1960). A simplified table for staging anuran embryos and larvae with notes in identification. *Herpetological* **16**, 183–190.

Ham, A. W., and Cormack, D. H. (1979). "Histophysiology of Cartilage, Bone, and Joints." Lippincott, Philadelphia, Pennsylvania.

Hanken, J., and Hall, B. K. (1988). Skull development anuran metamorphosis: l. Early development of the first three bones to form—the exoccipital, the parasphenoid, and the frontoparietal. *J. Morphol.* **195**, 247–256.

Hanken, J., Klymkowsky, M. W., Seufert, D., and Ingebrigtsen, N. (1990). Evolution of cranial patterning in anuran amphilibians analyzed using whole-mount immunohistochemistry. *Am. Zool.* **30**, 138a.

Hemmati-Brivanlou, A., and Harland, R. M. (1989). Expression of an engralied-related protein is induced in the anterior neural ectoderm of early *Xenopus* embryos. *Development (Cambridge, U.K.)* **106**, 611–617.

Hudson, L., and Hay F. C. (1980). "Practial Immunology," 2nd Ed. Blackwell, Oxford.

Humason, G. L. (1979). "*Animal Tissue Techniques*," 4th Ed. Freeman, San Franciso, California.

Jones, E. A., and Woodland, H. R. (1989). Spatial aspects of neural induction in *Xenopus laevis*. *Development (Cambridge, U.K.)* **107**, 785–791.

Kay, B. K., Schwartz, L. M., Rutishauser, U., Qiu, T. H., and Peng H. B. (1988). Patterns of N-CAM expression during myogenesis in *Xenopus laevis*. *Development (Cambridge, U.K.)* **103**, 463–471.

Kelly, W., and Bryden, M. M. (1983). A modified differential stain for cartilage and bone in whole formalin-fixed vertebrates. *Stain Technol.* **58**, 131–134.

Klymkowsky, M. W., and Maynell, L. M. (1989). MPF-induced breakdown of cytokertain filament organization during oocyte maturation in *Xenopus* depends upon the translation of maternal mRNA. *Dev. Biol.* **134**, 479–485.

Klymkowsky, M. W., Maynell, L. M., and Polson, A. G. (1987). Polar asymmetry in the organization of the cortical cytokeratin system of *Xenopus laevis* oocytes and embryos. *Development (Cambridge, U.K.)* **100**, 543–557.

Lundvall, H. (1904). Farbung des Skeltettes in durchsichtigen Weichteilin. *Anat. Anz.* **62**, 353–373.

McMahon, A. P., and Moon, R. T. (1989). Ectopic expression of the proto-oncogene *int-1* in *Xenopus* embryos leads to duplication of the embryonic axis. *Cell (Cambridge, Mass.)* **58**, 1075–1084.

Mall, F. P. (1904). On ossification centers in human embryos less than one hundred days old. *Am. J. Anat.* **5**, 433–458.

Metcalfe, W. K., Bass, M. B., Trevarrow, B., Myers, P. Z., Curry, M., and Kimmel, C. B. (1990). The pattern of expression of the L2/HNK-1 carbohydrate in embryonic zebrafish. *Development (Cambridge, U.K.)* **110**, 491–504.

Ojeda, J. L., Barbosa, E., and Gomez Bosque, P. (1970). Selective skeletal staining in whole chicken embryos: A rapid Alcian blue technique. *Stain Technol.* **45**, 137–138.

Olmstead, J. B. (1981). Affinity purification of antibodies from diazotized paper blots of heterogeneous protein samples. *J. Biol. Chem.* **256**, 1955–11957.

Palacek, J., Habrova, V., Nedivedek, J., and Romanovsky, A. (1985). Dynamics of tubulin structures in *Xenopus laevis* oogenesis. *J. Embryol. Exp. Morphol.* **87**, 75–86.

Park, E. H., and Kim, D. S. (1984). A procedure for staining cartilage and bone of whole vertebrate larvae while rendering all other tissues transparent. *Stain Technol.* **59**, 269–272.

Patel, N. H., Martin-Blanco, E., Coleman, K. G., Poole, S. J., Ellis, M. C., Kornbery, T. B., and Goodman, S. G. (1989). Expression of engrailed proteins in arthropods, annelids and chordates. *Cell (Cambridge, Mass.)* **58**, 955–968.

Pearse, A. G. E. (1980). "Histochemistry, Theoretical and Applied," 4th Ed. Churchill Livingstone, Edinburgh, Scotland.

Schultze, O. (1897). Uber Herstellung und conservirung durchsichtigen embryonen zum stadium der skeletbildung. *Anat. Anz.* **13**, 3–5.

Seufert, D. W., Hanken, J., Klymkowsky, M. W., and Ingebrigtsen, N. E. (1990). Early distribution of type ll collagen during development of the vertebrate head. *Am. Zool.* **30,** 83a.
Spalteholz, W. (1914). Uber das durchsichtigmachen von mensichlichen und tierischen praparaten. 2 Aufl., Hirzel, Leipzig, Germany.
Strome, S., and Wood, W. B. (1982). Immunofluorescence visualization of germ-line-specific cytoplasmic granules in embryos, larvae, and adults of *Caenorhabdities elegans. Proc. Natl. Acad. Sci. U.S.A.* **79,** 1558–1562.
Townsend, D. S., and Steward, M. M. (1985). Direct development in *Eleutherodactylus coqui* (Anura: Leptodactylidae): A staging table. *Copeia* **1985,** 423–436.
Wassersug, R. J. (1976). A procedure for differential staining of cartilage and bone whole formalin-fixed vertebrates. *Stain Technol.* **51,** 131–134.
Wilson, E. B. (1925). *"The Cell in Development and Heredity. "* Macmillan, New York.
Wright, C. V. E., Schnegelsberg, P., and De Roberties, E. M. (1989a). Vertebrate homeodomain proteins: Families of region-specific transcription factors. *Trends Biochem. Sci.* **14,** 52–56.
Wright, C. V. E., Cho, K. W. Y., Hardwick, J., Collins, R. H., and De Robertis, E. M. (1989b). Interference with function of a homeobox gene in *Xenopus* embryos produces malformations of the anterior spinal cord. *Cell (Cambridge, Mass.)* **59,** 81–93.
Wylie, C. C., Brown, D., Godsave, S. F., Quarmby, J., and Heasman, J. (1985). The cytoskeleton of *Xenopus* oocytes and its role development. *J. Embryol. Exp. Morphol.* **89** (Suppl.), 1–15.

Chapter 23

In Situ Hyridization

HEATHER PERRY O'KEEFE AND DOUGLAS A. MELTON

*Department of Biochemistry and
Molecular Biology
Harvard University
Cambridge, Massachusetts 02138*

BEATRIZ FERREIRO

*Department of Biology
University of California at San Diego
La Jolla, California 93128*

CHRIS KINTNER

*Molecular Neurobiology Laboratory
Salk Institute
La Jolla, California 93128*

I. Introduction.
II. *In Situ* Hybridization to Sectioned Tissue
   A. Reagents
   B. Preparation of Tissue Sections
   C. Prehybridization
   D. Hybridization
   E. Washing
   F. Autoradiography
   G. Staining and Mounting
   H. Microscopy
III. *In Situ* Hybridization to Whole-Mount Tissue
   A. Tissue Preparation
   B. Prehybridization
   C. Hybridization
   D. Immunohistochemistry
IV. Evaluation of Method
   References

I. Introduction

In situ hybridization is a widely used method to assay gene expression in both embryonic and adult tissues (Akam, 1987; Brahic and Haase, 1978; Gall and Pardue, 1971; Lynn *et al.*, 1983). The method has been applied in a variety of situations where the tissue distribution of gene transcripts needs to be determined with a reasonable degree of spatial resolution. In *Xenopus*, the method has been successfully used to localize the expression of transcripts during oogenesis, as well as during early and late stages of embryonic development (Dworkin-Rastl *et al.*, 1986; Jamrich *et al.*, 1984, 1987; Kintner and Melton, 1987; Melton, 1987). This chapter describes the various protocols that have been developed for localizing gene transcripts in both sectioned and whole-mount *Xenopus* tissue.

Because of its wide application, the methods used for *in situ* hybridization continue to evolve as protocols for hybridizing nucleic acid probes to tissue are refined and improved. Many of these improvements can be incorporated directly into the protocols used for *in situ* hybridization to frog tissue. Nonetheless, there are several special modifications that have been made to the *Xenopus* protocols which are essential for success, particularly when the technique is applied to oocytes and early embryos. These modifications were introduced to solve the problems that stem from the high yolk content and unusually large size of oocytes, eggs, and cells in early embryos.

The production of tissue sections for *in situ* hybridization with good morphology is made more difficult by the high yolk content of *Xenopus* tissue. Moreover, the methods that can produce good morphology with *Xenopus* tissue are not necessarily the best for retaining and preserving gene transcripts for *in situ* hybridization. As a result, modifications have been made to the *Xenopus* protocols that change either the conditions of fixation or the method for sectioning frog tissue. At tadpole stages, when cells lose their yolk and when the various differentiated tissues become more resilient, the problems associated with producing sections from *Xenopus* tissue are less critical.

Another special consideration in applying *in situ* hybridization to *Xenopus* is that the density of gene transcripts in oocytes, eggs, or cells in the early embryos is decreased by the displacement of the cytoplasm with yolk granules. This decrease in transcript density effectively reduces the hybridization signal in a situation where the detection of rare transcripts is already at the limit of the technique. What this means in practical terms is that extreme care must be taken to optimize the *Xenopus* protocol by observing the precautions, for example, in probe synthesis, which can maximize hybridization signal or minimize background noise. It should be borne in mind, however, that even when optimized, *in situ* hybridization to early *Xenopus* oocytes and embryos has been most successful in localizing transcripts that are highly concentrated in a subcellular location, or transcripts that are relatively abundant. Fortunately, rare transcripts become easier to detect when the embryo reaches tadpole stages and cells lose their yolk and decrease in size (Ruiz i Altaba and Melton, 1989).

The *in situ* hybridization methods that have been used for *Xenopus* tissue are described here in two parts. Section II describes protocols used to detect gene transcripts in tissue sections of oocytes or early embryos using radioactive probes and autoradiography. These protocols are derived extensively from the work of the Angerers and colleagues (Angerer and Angerer, 1981; Angerer *et al.*, 1984, Cox *et al.*, 1984; Lynn *et al.*, 1983; Venezky *et al.*, 1981). Section III describes a protocol more recently developed to detect gene transcripts in whole-mount tissues from later stage embryos using digoxygenin-labeled probes and immunohistochemistry (Hemmati-Brivanlou *et al.*, 1990; Tautz and Pfeifle, 1989).

II. *In Situ* Hybridization to Sectioned Tissue

The following flow chart outlines the procedures involved in *in situ* hybridization to sectioned tissue:

Preparation of tissue sections
    1. Oocytes and eggs: Ethanol–acetic acid–chromic acid fixation and Paraplast embedding
    2. Early embryos: Paraformaldehyde fixation and acrylamide embedding
Probe synthesis
    Generation of ^{32}P- and ^{35}S-labeled RNA probes
Prehybridization treatment
    Postfixation
    Protease digestion
    Acid extraction
    Acetic anhydride treatment
    Prehybridization (acrylamide sections)
Hybridization
Washing
Autoradiography

A. Reagents

The following stock solutions described in Maniatis (Sambrook et al., 1989) are required: 20× SSPE, 20× SSC, 0.5 M EDTA (pH 8.0), 1× Denhardt's, 1.0 M Tris-HCl (pH 7.5, 8.0, 9.5), and deionized formamide. Other stock solutions are as follows:

5.0% Paraformaldehyde
Heat 50 g of paraformaldehyde in 1 liter water to 60° and add 1 N NaOH dropwise until the paraformaldehyde dissolves. The final pH should be 7–8. Paraformaldehyde should be made fresh for fixing new tissue, but it will keep for 2 weeks for other uses.
2.5 M Triethanolamine, pH 8.0
10× APBS (Amphibian Phosphate Buffer)
1.03 M NaCl, 27 mM KCl, 1.5 mM KH_2PO_4, 7 mM Na_2HPO_4 (pH 7.2)
Proteinase K
25 μg/ml in water
50% Chromium Trioxide
Heparin
50 μg/ml in 4× SSC
PBT
1× APBS containing 0.1% Tween 20

B. Preparation of Tissue Sections

1. Oocytes and Eggs

The oocytes and eggs of *Xenopus* present special problems for tissue sectioning owing to their large size and high yolk content. To overcome these problems, attempts were made to identify a fixative that would provide good tissue morphology while retaining the RNA in a form which can hybridize probes. Fixatives that are known to give good morphology were tried, including alcoholic Bouin's, Bouin's, Smith's, Perenyi's, San Felice's (Humason, 1972), and 3.0% trichloroacetic acid. Fixatives were also tried which were known to work well for *in situ* hybridization in other organisms, including paraformaldehyde, glutaraldehyde, paraformaldehyde/glutaraldehyde in various ratios, or formaldehyde with or without acetic acid (Cox et al., 1984; Hafen et al., 1984; Jamrich et al., 1984). In general, the acid fixatives are more harsh and give better morphology, whereas the aldehyde fixatives are more likely to retain hybridizable RNA. After several modifications of existing protocols, the best fixative for oocytes and eggs was found to be a new concoction consisting of ethanol, acetic acid, and chromium trioxide (95:5:0.25%

by volume). This fixative, called BOSCO, penetrates yolky tissue rapidly and fixes RNA effectively. For optimal retention of hybridizable RNA, sections prepared with this fixation should be postfixed on the slides with a aldehdye fixative such as 4.0% paraformaldehedye (see below).

a. Fixation and Embedding. BOSCO is made up just before use by chilling acid alcohol (95% ethanol–5% acetic acid) on wet ice and adding chromium trioxide (prepared as a 50%, w/v, solution) to a final concentration of 0.25%. Oocytes, eggs, or early embryos are added to the fixative (20-ml glass scintillation vials work well) on wet ice with occasional mixing. The fixative is then diluted out by washing with increasing concentrations of ice-cold ethanol. After several washes, the tissue is taken through xylene, and embedded in Paraplast. The standard fixation and embedding protocol is as follows:

Fixation, 4°C	1 hour
95% ethanol, 2–3 changes, 4°C	20–30 minutes
100% ethanol, 2 changes, 4°C	20–30 minutes
50% ethanol–50% xylene, room temperature	10–15 minutes
100% xylene, room temperature	10 minutes
50% xylene–50% Paraplast, 50°C	1 hour
Paraplast, 50°C	Overnight (at least 6 hours)
Transfer to embedding molds	

Embryos, particularly at later stages of development, should be oriented in the embedding molds (Polysciences, Warrington, PA.) so that the sections can then be cut parallel to one of the embryonic axes. This is easily accomplished by manipulating the embryo with a warm dissecting needle under a stereomicroscope before the Paraplast hardens.

b. Sectioning. Mounting tissue sections of *Xenopus* oocytes and eggs can be troublesome because of their high yolk content. In our experience, *Xenopus* sections do not adhere tightly to slides that have been prepared by subbing with gelatin (Gall and Pardue, 1971) or poly(lysine). It should be kept in mind that during the prehybridization the tissue is subjected to a number of harsh treatments (such as protease digestion) and so the yolky tissue needs a particularly good surface to adhere to in order to remain attached. At the suggestion of the Angerers, we have used a modified protocol for treating microscope slides with 3-aminopropyltriethoxysilane (Gottlieb and Glaser, 1975). Slides are first washed in warm soapy water, rinsed multiple times in tap water, and then soaked overnight in chromic acid. The following day they are rinsed 8–10 times in distilled water and then rinsed with 95% ethanol and 0.1% acetic acid. After drying, the slides are dipped twice, 10 seconds each, in 2.0% silane (3-aminopropyltriethoxysilane) and made up in acetone, washed sequentially with acetone and distilled water, and then air dried. The silanized slides are soaked in 4.0% paraformaldehyde for 1 hour at room temperature, dried, and stored in slide boxes at room temperature. A white precipitate which usually forms can be wiped off the slides immediately before use.

Tissue sections are cut as ribbons on any standard rotary microtome set to a thickness of 5–10 μm. Ribbons must be allowed to spread out on diethyl pyrocarbonate (DEPC)-treated water that has been brought to 40°C on the surface of the slide using a slide warmer. Degassing the water beforehand helps to prevent air bubbles from forming between the glass slide and the sections. After the sections have fully spread, the excess water is carefully wicked away, and the sections are allowed to adhere tightly to the slide by incubation on a slide warmer overnight at 40°C.

The major obstacle encountered when sectioning oocytes and eggs is that the yolky tissue prevents the sections from flattening and attaching completely to the glass slide. A useful treatment to prevent this problem is to presoak the tissue within the block before sectioning. This is achieved by sectioning the block until 30–50 μm of tissue has been removed and then soaking the block, tissue side down, in 5% glycerol overnight. The glycerol solution seeps into the tissue so that it flattens nicely after sectioning.

2. Tissue Sections of Early Embryos Using Polyacrylamide Embedding

The fixation and paraffin sectioning described above works well with oocytes and eggs by producing both good morphology and hybridization signals for moderately abundant transcripts. For early embryos, however, better hybridization signals can be obtained when embryos are fixed with 3–4% paraformaldehyde, although the morphology of tissue fixed in this way is extremely poor when sectioned by standard techniques. Good morphology, however, can be obtained with gentle aldehyde fixatives by embedding embryos in polyacrylamide blocks which are then frozen and sectioned as described by Hausen and Dreyer (1981).

Fixation and Embedding. Embryos are fixed in 3.0% paraformaldehyde for 1–2 hours at room temperature. Saponin (Merck, Rahway, NJ; Cat. No. 7695) can be added to the fixative at a concentration of 0.025% in order to prevent the collapse of the blastocoel or gastrocoel. Fixed embryos are washed at room temperature in $1 \times$ APBS, 3–5 times, and then incubated overnight at 4–6°C in 10% acrylamide containing 0.7% TEMED (vv), pH 7.2. For embedding, the molds are first partially filled with 10% acrylamide containing 0.7% TEMED and 0.5% ammonium persulfate to allow a layer of polyacrylamide to form. With the embedding molds placed on ice, the embryos are transferred to a 4°C solution containing 10% acrylamide and 0.5% ammonium persulfate and then into the embedding molds. Polymerization takes place on ice. The polyacrylamide blocks are then frozen by immersion in 2-methylbutane which has been chilled in liquid nitrogen to a point where crystals of 2-methylbutane begin to form. In order to prevent cracking of the block, the freezing process should progress slowly from the outside of the block to the inside. Frozen blocks can be stored at $-20°C$ for several months.

Frozen sections are cut at a thickness of 6–10 μm at a temperature of $-24°$ to $-27°C$. Frozen sections are collected onto microscope slides prepared by silane treatment as described in Section II, B, 1. Frozen sections can be stored for 1–2 weeks at $-20°C$.

C. Prehybridization

The prehybridization protocol is as follows:

4.0% Paraformaldehyde (made up in 1 × APBS), room temperature	20 minutes
Wash with 2 × SSPE, room temperature	5 minutes
Proteinase K (1–3 μg/ml in 0.1 M Tris-HCl, pH 7.5, 1 mM EDTA), 37°C	5–30 minutes
0.2% glycine in 2 × SSPE, room temperature	15 seconds
Wash with 2 × SSPE, room temperature	5 minutes
0.2 N HCl, room temperature	15 minutes
Wash with 2 × SSPE, room temperature	5 minutes
Acetic anhydride, 0.25% in 0.1 M triethanolamine	10 minutes
Wash with 2 × SSPE, room temperature	5 minutes
Hold slides for hybridization	

1. RATIONALE

Tissue sections are prepared for hybridization by a series of treatments that are designed to expose target RNAs. It is generally a good idea not to allow the sections to dry out once they are hydrated. This is particularly true for oocytes and eggs where the yolky tissue is very hydrophobic and difficult to rewet once dry. *We repeat, the tissue sections should not be dried from the time the Paraplast is removed until the slides are dipped into emulsion.*

To rehydrate Paraplast sections, the slides are first dewaxed by immersion in 2 charges of xylene, 10 minutes each, and then dipped through an ethanol series consisting of 100, 95, 80, 70, and 40% ethanol for 30–60 seconds each. The slides are then immediately immersed in 4.0% paraformaldehyde made up in 1 × APBS for 20 minutes. Polyacrylamide sections do not need to be rehydrated, but are first washed in 1 × SSPE for 10 minutes and then incubated in paraformaldehyde as described above. From this point on, the paraplast and polyacrylamide sections are treated in the same manner.

2. Proteinase K Treatment

Slides are first treated with proteinase K after post fixation with paraformaldehyde. Among the prehybridization steps, this step has the greatest effect on the hybridization signals (Cox et al., 1984). Protease treatment is designed to remove enough protein so that the target RNA is accessible to hybridization, while not reaching a point where the sections are digested off the slide or the RNA digested free of the section. The extent of protease digestion, therefore, must be calibrated, and the optimal amount of digestion has been found to vary with different kinds of tissues, fixation, and sectioning methods. For acid fixation and Paraplast sectioning, we have found that small oocytes (stages I–III) (Dumont, 1972) need only 15 minutes of digestion, whereas large oocytes and eggs should be treated for 30–40 minutes. For aldehyde fixation and polyacrylamide embedding, a shorter protease treatment may be optimal. When using *in situ* hybridization for the first time, or with a new tissue, the extent of protease digestion should be varied to find an optimum. In addition, one should be aware that different tissues at later stages of development may react to the proteinase K at different rates, thus skewing the hybridization signals depending on the amount of digestion used.

Following protease digestion, the sections are treated with a glycine solution in order to terminate proteolytic activity rapidly. In some protocols, a second incubation in paraformaldehyde is added at this point in order to "refix" any tissue that has been dislodged by proteolysis. There is no indication that this helps or hurts.

3. Acid and Acetic Anhydride Treatment

Following protease digestion, the slides are briefly incubated in $0.2\ M$ HCl. This step is designed to remove basic proteins that are bound to RNA and thereby prevent hybridization. Finally, the slides are treated with acetic anhydride by placing them into $0.1\ M$ triethanolamine, adding acetic anhydride to 0.25%, and agitating the slides gently. The treatment is designed to neutralize positive charges on the glass which would otherwise bind nucleic probes by ion exchange (Hayashi et al., 1978).

D. Hybridization

1. Probe Synthesis

Consistent with the results of Cox et al. (1984), we find that single-stranded RNA probes are preferable to DNA probes in terms of sensitivity and signal-to-noise ratio. The methods for *in vitro* synthesis of RNA probes using bacteriophage RNA polymerases (Melton et al., 1984) are described extensively by Krieg and Melton (1987). Typical conditions used in probe synthesis consist of a 10-μl reaction volume containing 50 μCi of a labeled ribonucleotide ([^{32}P]UTP or -CTP, or [^{35}S]UTP, 400–1000 Cis/mmol, 10–20 μCis/μl). About 80–90% of the labeled nucleotide should be incorporated during the reaction, resulting in 10–100 ng of RNA depending on the specific activity of the label. There are several points worth noting when generating RNA probes for *in situ* hybridization.

First, the ^{35}S-labeled probes are theoretically more efficient in autoradiography than ^{32}P-labeled probes, although the better efficiency is somewhat offset by the fact that ^{35}S-labeled RNA produces higher backgrounds. The background problems with ^{35}S probes can be minimized by adding reducing agents during the hybridization and washes. If background problems are minimized, the ^{35}S probes are preferred, particularly when attempting to detect rate transcripts using very long exposure times. For many purposes, however, the efficiency of the ^{32}P-labeled probes is more than adequate to detect moderately abundant transcripts after a reasonably short exposure time.

Second, the generation of probes with high specific activity is carried out under conditions where the labeled nucleotide is included at concentrations near the K_m of the enzyme. In practical terms, this means that the conditions of the reaction do not favor the production of very long transcripts. Probe templates greater than 2 kilobases (kb), therefore, should be divided into smaller portions which are then transcribed individually. In all cases, the actual size distribution of probe transcripts should be checked by gel electrophoresis. Third, template sequences used for probe synthesis should be chosen to avoid poly(A) tracts or 3' untranslated regions which tend to contain homopolymeric sequences.

Fourth, a major source of background using antisense RNA probes is the generation of small amounts of contaminating sense strand RNA by inappropiate transcription off the wrong DNA strand. Even if a small fraction of probe is double stranded, there will be significant background problems because double-stranded RNAs are not digested by RNase A, a treatment which is used after hybridization to remove any probe that nonspecifically sticks to tissue. A common source of sense strand synthesis is the initiation of the bacteriophage RNA polymerase on the ends of the template DNA. This aberrant initiation at the ends of the DNA template is known to be more frequent when the ends are generated with restriction enzymes that leave a 3' overhang. In many cases, a probe that produces unworkable background problems can be improved by isolating the antisense transcripts away from contaminating sense transcripts by gel electrophoresis as described by Krieg and Melton (1987). Probes can also give background if they contain extensive amounts of secondary structure that permits intramolecular annealing. In these cases, a better probe might be obtained using different subregions of the probe sequences separately.

Finally, the optimal size of RNA probes is between 100 and 200 base pairs (bp). Larger probes can be reduce in size by hydrolysis in 40 mM NaHCO$_3$, 60 mM NCO$_3$, at 60°C (Lynn et al., 1983). RNAs of about 0.5 kb in length are treated for 42 minutes, 1 kb for 50 minutes, and 2 kb for 56 minutes. Following hydrolysis, the RNA is neutralized by adding acetic acid to a final concentration of 5.0% and then ethanol precipitated by adding sodium acetate to 0.3 M and 2 volumes of ethanol.

2. Hybridization Conditions

Hybridization to abundant target transcripts in sections reaches saturation when RNA probes are used at a concentration of 0.2–0.3 µg/ml/kb probe complexity (Cox et al., 1984). Decreasing the concentrating of probe will decrease the hybridization signal, whereas increasing the probe concentration will increase background hybridization without affecting the signal. Since the probe concentration necessary to saturate target RNAs will depend on the relative abundance of target transcripts, a range of probe concentrations should be tried in order to optimize the signal-to-noise ratio for a given transcript. The starting point for abundant transcripts is 3×10^7 cpm of probe in 50–200 µl of hybridization solution. Depending on the area of slide containing tissue, approximately 10–100 µl of hybridization solution is required for each slide. Typically, 40 µl of hybridization solution is added to the slide, which is then covered with a 25 × 25 mm coverslip.

Hybridization solution is prepared fresh and contains 50% formamide, 0.3 M NaCl, 10 mM Tris-HCl, pH 7–8, 10 mM NaH$_2$PO$_4$, 5 mM EDTA, 1× Denhardt's [0.02% bovine serum albumin (BSA), 0.02% Ficoll, 0.02% polyvinylpyrrolidone (PVP)], 10% dextran sulfate. In addition the hybridization solution should contain carrier nucleic acid consisting of 0.1–1.0 mg/ml of yeast tRNA. Some protocols also include 0.1–1.0 mg/ml of carrier DNA such as denatured salmon sperm DNA. Commerical sources of carrier nucleic acid are very impure and should be partially purified at least by phenol extraction. For ^{35}S probes, it is essential to add either 10 mM dithiothreitol (DTT) or 100 mM 2-mercaptoethanol to the hybridization solution to reduce sticking of probe to tissue. We have found that some batches of DTT can inhibit hybridization, and thus new batches need to be tested. The probe and carrier nucleic acids are first boiled at 80°C and then added to a stock solution containing the remaining components.

3. Prehybridization

Sections generated by polyacrylamide embedding *must* be prehybridized into order to prevent sticking of the probe to the polyacrylamide. This is accomplished after the prehybridization treatments by keeping the slides in a closed container kept moist with 1 × SSPE in 50% formamide. The container should contain a raised platform on which paper towels are placed with their ends immersed in 1 × SSPE in 50% formamide at the bottom. The slides are placed on the paper towel, and prehybridization solution (hybridization solution containing everything but the probe) is used to cover the tissue on each slide (200 µl/slide). The box is then placed at 50°C for 3 hours. Following this incubation, the prehybridization solution is gently removed before adding the probe.

4. Probe Addition

The hybridization solution containing the probe should be added to wet tissue. If the tissue has been prehybridized as described above, the probe solution can be added after pipetting most of the prehybridization solution away. A siliconized coverslip is then gently placed on top of the tissue, taking care not to trap air bubbles. For tissue that has not been treated with prehybridization solutions, the slides are removed from the 2 × SSPE (final) wash in the prehybridization treatment), dipped into DEPC-treated water, and then shaken 10 times to remove any excess water. The probe solution is then added to the tissue while it is still moist but not wet. This is critical since if the tissue is too wet the hybridization conditions will be altered, whereas if the tissue becomes dry the probe solution will not penetrate the tissue. After addition of the probe solution, siliconized coverslips are added as described above. The slides are then incubated at 50°C under mineral oil to prevent dehydration. When the slides are placed into mineral oil, the coverslips will not come off the slides unless too much hybridization solution has been added. One configuration is to place the slides in a slide rack in a staining dish that is maintained at 50°C in a oven or circulating water bath. In this configuration, the slides must rest up against a solid surface in order to keep the coverslips from sliding edgewise off the slide. Hybridization is reported to reach completion by 6 hours (Cox et al., 1984) but can be allowed to proceed overnight.

E. Washing

Reducing agents should be included in all washes for ^{35}S-labeled probes. The standard washing protocol is as follows:

Rinse slides 3 times in chloroform	5 minutes each
4 × SSPE, room temperature	1–2 hours

2× SSPE, room temperature	1 hour–overnight
RNase A, 20 μg/ml in 4× SSPE, 37°C	30 minutes
50% formamide, 2× SSPE, 50°C (^{35}S), or 0.1× SSPE, 60°C (^{32}P)	1–2.5 hours
Store in 2× SSPE	

Rationale

The slides are prepared for washing by first removing any trace of mineral oil by dipping the slides through three successive chloroform washes. The slides are then placed in 2× SSPE in a holder which allows the coverslips to slip edgewise off the slides. For ^{35}S-labeled probes, the wash solutions should contain either 10 mM DTT (tested batch) or 100 mM 2-mercaptoethanol to prevent sticking of the probe. Slides are then washed for 1 hour at room temperature in 2× SSPE in a volume of about 1 liter with several changes of the wash solution. A critical step for reducing background hybridization is treatment with RNase A which is carried out by placing the slides in RNase A at 20 μg/ml in 4× SSPE at 37°C. The slides are then again washed in 2× SSPE. The final, high stringency wash is done at 0.1× SSPE at 55°–60°C for 1 hour or, in the case of ^{35}S-labeled probes, in 50% formamide, 2× SSPE at 50°C for 1 hour.

F. Autoradiography

Paraplast sections of oocytes, eggs, and early embryos should *not* be dehydrated before coating with photographic emulsion. The dried yolky tissue will trap air bubbles that prevent proper clearing of the tissue at later stages. Instead of drying, the slides are transferred to 0.3 M ammonium acetate, 1.0% glycerol and then shaken just before dipping to remove excess moisture. The volatile salt is present to prevent dissociation of hybridized probe, while the glycerol present in the wash and in the emulsion acts to prevent drying artifacts that generate autoradiography grains. Polyacrylamide tissue sections can be dehydrated before dipping in emulsion by passing them sequentially through a series of increasing concentrations of ethanol containing 0.3 M ammonium acetate, 100% ethanol, and then xylene. The slides are then allowed to air dry.

Dipping

Stock solutions of Kodak NTB2 or Ilford K5 nuclear emulsion are melted at 42°C and aliquoted (5.0 ml) into plastic sealed cap tubes, which are wrapped in aluminum foil to exclude light and stored at 4°C. To prepare emulsion for dipping, an aliquot of emulsion is remelted in a water bath at 42°C for 15–20 minutes, then mixed (1:1) with 0.6 M ammonium acetate, 2.0% glycerol, also at 42°C. The diluted emulsion is placed in a dipping chamber within a 42°C water bath. Dipping chambers can be built from plexiglass using dimensions slightly larger than a microscope slide or purchased from Polysciences. The dipping procedure should be carried out in complete darkness or with a Wratten #2 red light filter with a low wattage bulb pointed away from the work area. Blank slides should be dipped and developed to determine if the light conditions of the darkroom are adequate and the background fogging of the emulsion is minimal.

Slides are dipped twice in emulsion and then stood on end in a large lightproof box where they dry for 1.0 hour. Slides are exposed at 4°C in slide boxes containing a small amount of desiccant. The boxes should be carefully sealed and wrapped with aluminum foil to be completely light proof. The slide boxes are moved to room temperature for 1 hour prior to removing the slides for developing. The developer is Kodak D19 diluted 1:1 in water at 15°–18°C. Higher temperatures increase background and produce very large silver grains. The slides are developed for 2.5 minutes, stopped in 0.2% acetic acid for 10 seconds, and fixed in Kodak fixer (not Rapidfix) for 5 minutes. The slides should be rinsed for 10 minutes in *cold* running tap water before staining and mounting.

G. Staining and Mounting

A variety of stains can be used, although it is important to note that stains which require an acid treatment can fog the emulsion. For most situations, a simple and effective stain is Giemsa (Polysciences). Slides are stained for 10 minutes in Giemsa diluted 1:30 (v/v) in 50 mM phosphate buffer, pH 6.5. Stained slides are rinsed in cold running tap water until the rinsing water is clear. The amount of stain required may vary depending on the tissue being stained. Overstaining will obscure autoradiography grains, whereas understanding will make the tissue difficult to view under bright-field illumination. Washed slides are dehydrated through a series of increasing ethanol concentrations, 100% ethanol, and then xylene. Coverslips are placed on slides using a mounting medium such as Permount (Fisher) or Canadian balsam in methyl salicylate.

H. Microscopy

Autoradiography grains are viewed under dark-field illumination. Pigment granules will appear the same as silver grains under dark-field conditions but can be removed by using a polarizing filter (Ellis et al., 1988) available for both Leitz and Zeiss microscopes. Alternatively, the pigment granules can be avoided altogether by using tissue from albino frogs.

III. *In Situ* Hybridization to Whole-Mount Tissue

The protocol for *in situ* hybridization to whole-mount tissue is adapted from a method originally described for *Drosophila* (Tautz and Pfeifle, 1989). These protocols use DNA probes that have been synthesized with a digoxigenin-labeled deoxynucleotide and that are detected after hybridization using a antidigoxigenin antibody coupled to akaline phosphatase. The synthesis of probes and their detection after hybridization are based on reagents provided in the "Genius" kit produced by Boehringer Mannheim Biochemicals (Indianapolis, IN; "Genius" kit, Cat. No. 1093 657). The methods described here use DNA probes, although the protocols can be easily adapted for the use of RNA probes labeled with digoxigenin as described by Hemmati-Brivanlou et al. (1990).

A. Tissue Preparation

Embryos are collected and dejellied using standard protocols. For early embryos, the vitelline membrane should be removed at this stage by manual dissection. The embryos are then fixed using 3.7% formaldehyde in 1 × APBS for 2 hours at room temperature. After fixation, the embryos are placed in methanol for 30 minutes and then in 1 × APBS, 0.5% Triton X-100 for 1 hour with gentle agitation. The embryos can be dehydrated through an ethanol series of 50, 70, and 95% and stored in 95% ethanol at $-20°C$ indefinitely. If desired, a specific tissue such as the nervous system can be dissected from embryos in order to decrease the mass of tissue handled.

B. Prehybridization

Embryos are treated in multiwell glass plates (sold as soil test plates) which hold about 1.0 ml in volume. In order to transfer the embryos at each step, they can be placed in a small wire mesh basket which is formed to the shape of the wells from a small piece of stainless steel mesh.

1. Postfix

The embryos are prepared for hybridization by incubation at room temperature through the following solutions:

100% ethanol	2–3 minutes
Ethanol–Xylene (1:1, v/v)	2–3 minutes
Xylene	2 hours
Ethanol–xylene (1:1, v/v)	2–3 minutes
100% ethanol, 2 times	2–3 minutes
Methanol–5% formaldehyde	2–3 minutes
Postfix in PBT–5% formaldehyde	20 minutes
Rinse 3 times in PBT	30 seconds each

2. Protease Digestion

The embryos are then digested with proteinase K in order to increase the accessibility of target RNA. Digestion is carried out in PBT containing 50 μg/ml proteinase K for 5–15 minutes. The length of digestion will depend on the batch of enzyme, and perhaps the developmental stage of the tissue being processed. The protease digestion is then stopped by rinsing the embryos twice with 2 mg/ml glycine in PBT. The tissue is then rinsed again with PBT and postfixed for a second time with PBT and 5% formaldehyde for 20 minutes. The embryos are rinsed extensively (5 times) with PBT. Before hybridization the embryos are prehybridized in a solution consisting of 50% formamide, 5× SSC, 100 μg/ml salmon sperm DNA, 100 μg/ml tRNA, 50 μg/ml heparin, and 0.1% Tween 20. After one rinse, the embryos are incubated in this solution for at least 1 hour in an air chamber with slight rotation at 48°C.

C. Hybridization

1. Probe Synthesis

DNA probes are generated by random primer synthesis in the presence of digoxigenin-labeled deoxynucleotides according to the specifications provided with the "Genius" kit. Briefly, gel-purified DNA inserts (0.25–1 μg) are digested with restriction enzymes that will produce small fragments of 100–500 bp in length. After digestion, DNA fragments are extracted with phenol–chloroform and ethanol precipitated. The digested DNA is resuspended in 5 μl of distilled water and added to a labeling reaction according to the manufacturer's protocol except that that the reaction is carried out overnight at 37°C. The digoxigenin-labeled DNA is purified by ethanol precipitation with tRNA as carrier (20 μg). The concentration of the probe is determined on dot blots by comparison with a labeled control DNA included in the kit as a standard. The probe is placed in a boiling water bath for 10 minutes immediately before use or is stored at −20°C. In addition, probes can be reused after hybridization if they are first redenaturated at 95°C for 10 minutes.

2. Hybridization

Hybridization solution contains 50% formamide, 5× SSC, 100 μg/ml salmon sperm DNA, 100 μg/ml tRNA, 50 μg/ml heparin, and 0.1% Tween 20, with the DNA probe at a concentration of 1 μg/ml. Hybridization is carried out overnight at 48°C with gentle rocking. Following hybridization, the embryos are washed at 48°C for 1 hour in hybridization solution, followed by one wash with PBT–hybridization solution (1:1, v/v) and then with 5 washes with PBT, all at 48°C.

D. Immunohistochemistry

1. Preabsorption of Antibody

The alkaline phosphatase-conjugated antidigoxigenin antibody provided by the manufacturer (Boehringer, Cat. No. 1093–274) should be preabsorbed on control embryos in order to remove nonspecific antibodies that can contribute to the background. This is achieved by incubating a 1:1000 dilution of antidigoxigenin–alkaline phosphatase antibody in PBT with a large volume of embryos such that the embryos make up about one-fifth of the total solution. The preabsorption is carried out with strong agitation for 2 hours at room temperature. Antibody is routinely preabsorbed just before use on embryos that have been treated (minus the probe) in parallel with the experimental embryos as described above.

2. Immunohistochemistry

Embryos are incubated in preabsorbed antibody for 2.0 hours at room temperature followed by extensive washing at room temperature with PBT (4 changes for 30–45 minutes each). The embryos are then rinsed twice with a solution used for the chromogenic reaction containing 100 mM NaCl, 50 mM $MgCl_2$, 100 mM Tris, pH 9.5, and 0.1% Tween 20. Finally, the embryos are developed in 1.0 ml of the solution above plus 4.5 µl nitro blue tetrazolium (NBT, 75 mg/ml in dimethylformamide) and 3.5 µl of 5-bromo-4-chloro-3-indolyl phosphate (BCIP, 50 mg/ml in dimethylformamide). The chromogenic reaction is allowed to proceed in the dark and should be checked at 5-minute intervals. Background staining usually begins to become a significant problem after 1.0 hour. The reaction is stopped by rinsing the embryos several times in PBT.

3. Mounting

The embryos should be cleared for viewing by dehydration through an ethanol series and then incubation in benzyl benzoate–benzyl alcohol (2:1). The tissue can also be processed further for tissue sections by incubation in xylene and Paraplast as described in Section II.

IV. Evaluation of Method

The relative sensitivity of the various protocols is likely to be equivalent, although this point is difficult to assess because there has not been a systematic comparison of the different methods. With perseverance, each method is capable of generating localization results for moderately abundant transcripts.

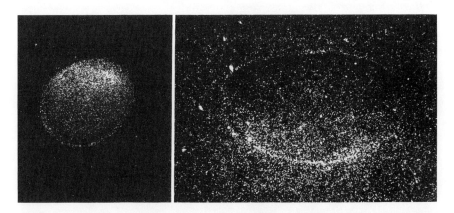

FIG. 1. *In situ* hybridization of the radiolabeled probes to Paraplast sections from oocytes fixed with BOSCO. (Left) Signal obtained with a probe for a transcript localized to the animal pole. (Right) Signal with a probe for a transcript localized to the vegetal pole. In both cases albino oocytes were used, and the animal pole is oriented upward.

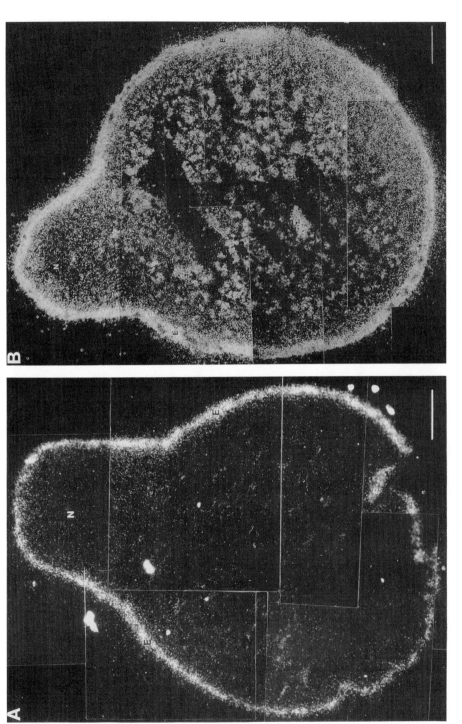

FIG. 2. Stage 22 embryo, probed with DG81, encoding epidermal cytokeratin. (A) With prehybridization; (B) without prehybridization.

The detection of rare transcripts, particularly in early embryos when the cytoplasm is diluted with yolk, is likely to be equally difficult for all of these methods. Judging when the method is appropriate for a given transcript is difficult because transcripts that are rare on a whole-embryo basis could be highly localized and therefore abundant in one region of the embryo. Conversely, moderately abundant transcripts may be difficult to detect if they are homogeneously expressed. As a rule of thumb, transcripts expressed in embryos at less than one part in 10^5 cannot be detected by *in situ* hybridization unless they are highly localized. In other words, it is not impossible that a given transcript, which can be detected in embryos by a sensitive method such as RNase protection assays, will be below the level of detection by *in situ* hybridization. The other cautionary point is that all of these methods are susceptible to hybridization artifacts which produce false positives. In the case of the whole-mount method, the localization of positive signals should be verified by sectioning in order to confirm that the signal comes from cells and not extracellular matrix. Any localization data obtained with *in situ* hybridization should be corroborated, if possible, by other means.

Examples of results obtained with the various protocols described in this chapter are shown in Figs. 1–3. Figure 1 shows an example of oocytes fixed

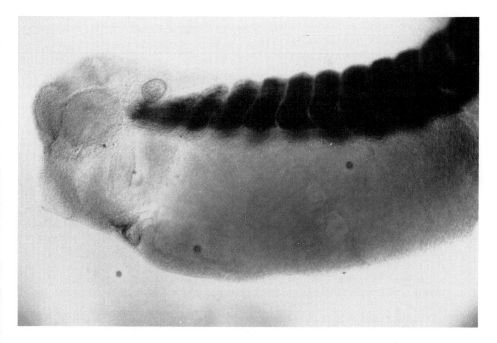

FIG. 3. *In situ* hybridization of a muscle-specific actin probe to a stage 32 albino embryo. The somites are darkly stained, and light staining can also be observed in the heart tissue.

in BOSCO, sectioned in Paraplast, and hybridized with a ^{32}P-labeled RNA probe against a vegetally localized message (Vg 1) or a probe to a message localized to the animal pole (An 2). Figure 2 shows an example of a tadpole embryo fixed in 3.0% paraformaldehyde, sectioned in polyacrylamide, and hybridized with a probe for DG81, an epidermal keratin (Jamrich et al., 1987; Jonas et al., 1985; Sargent and Dawid, 1983). Figure 3 shows an example of a whole-mount stage 30 embryo hybridized with a digoxigenin-labeled probe to muscle-specific actin. The whole-mount method has the obvious advantage that the three-dimensional distribution of a transcript can be easily assessed whereas a similar distribution would have to be reconstructed by a series of hybridized sections. Therefore, the whole-mount method will be particularly useful in localizing genes that are expressed in very limited regions of a late stage embryo, for example, homeobox genes which can have highly localized expression in the central nervous system. The whole-mount method, however, may not be appropriate for oocytes, eggs, and early embryos, where the large yolky cytoplasm may hinder penetration of the probe.

Acknowledgments

We are grateful to Kerstin Danker and Doris Wedlich for contributing the paraformaldehyde fixation and acrylamide embedding procedure, and for Figure 2. B. Ferreiro is grateful for funding from the Foundation Ramon Areces, and to Drs. William Harris and Volker Hartenstein for contributions to the whole-mount method.

References

Akam, M. (1987). The molecular basis for metameric pattern in the *Drosophila* embryo. *Development (Cambridge, U.K.)* **101,** 1–22.

Angerer, L. M., and Angerer, R. C. (1981). Detection of poly A$^+$ RNA in sea urchin eggs and embryos by quantitative *in situ* hybridization. *Nucleic Acids Res.* **9,** 2819–2840.

Angerer, L. M., DeLeon, D. V., Angerer, R. C., Showman, R. M., Wells, D. E., and Raff, R. A. (1984). Delayed accumulation of maternal histone mRNA during sea urchin oogenesis. *Dev. Biol.* **101,** 477–484.

Brahic, M., and Haase, A. T. (1978). Detection of viral sequences of low reiteration frequency by *in situ* hybridization. *Proc. Natl. Acad. Sci. U.S.A.* **75,** 6125–6129.

Cox, K. H., DeLeon, D. V., Angerer, L. M., and Angerer, R. C. (1984). Detection of mRNAs in sea urchin embryos by *in situ* hybridization using asymmetric RNA probes. *Dev. Biol.* **101,** 485–502.

Dumont, J. N. (1972) "Oogenesis in *Xenopus laevis*. *J. Morphol.* **136,** 153–180.

Dworkin-Rastl, E., Kelley, D. B., and Dworkin, M. B. (1986). Localization of specific mRNA sequences in *Xenopus laevis* embryos by *in situ* hybridization. *J. Embryol. Exp. Morphol.* **91,** 153–168.

Ellis, O., Bell, J., and Bancroft, J. D. (1988). An investigation of optimal gold particle size for immunohistological immunogold and immunogold–silver staining to be viewed by polarized incident light (EPI polarization) microscopy. *J. Histochem. Cytochem.* **36,** 121–124.

Gall, J. G., and Pardue, M. L. (1971). Nucleic acid hybridization in cytological preparations. *In* "Methods in Enzymology" (L. Grossman and K. Moldave, eds.), Vol. 21, pp. 470–480.

Gottlieb, D. I., and Glaser, L. (1975). A novel assay of neuronal cell adhesion. *Biochem. Biophys. Res. Commun.* **63**, 815–821.

Hafen, E., Kuroiwa, A., and Gehring, W. J. (1984). Spatial distribution of transcripts from the segmentation gene *fushi tarazu* during *Drosophila* embryonic development. *Cell (Cambridge, Mass.)* **37**, 833–841.

Hausen, P., and Dreyer, C. (1981). The use of polyacrylamide as an embedding medium for immunohistochemical studies of embryonic tissues. *Stain Technol.* **56**, 287.

Hayashi, S., Gillam, I. C., Delaney, A. D., and Tener, G. M. (1978). Acetylation of chromosome squashes of *Drosophila melanogaster* decreases the background in autoradiographs from hybridization with ^{125}I-labelled RNA. *J. Histochem. Cytochem.* **26**, 677–679.

Hemmati-Brivanlou, A., Frank, D., Bolce, M. E., Brown, B. D., Sive, H. L., and Harland, R. M. (1990). Localization of specific mRNAs in *Xenopus* embryos by whole-mount *in situ* hybridization. *Development (Cambridge, U.K.)* **110**, 325–330.

Humason, G. L. (1972) "*Animal Tissue Techniques.*" Freeman, San Francison, California.

Jamrich, M., Mahon, K. A., Gavis, E. R., and Gall, J. G. (1984). Histone RNA in amphibian oocytes visualized by *in situ* hybridization to methacrylate-embedded tissue sections. *EMBO J.* **3**, 1939–1943.

Jamrich, M., Sargent, T. D., and Dawid, I. B. (1987). Cell-type-specific expression of epidermal cytokeratin genes during gastrulation of *Xenopus laevis*. *Genes Dev.* **1**, 124–132.

Jonas, E., Sargent, T. D., and Dawid, I. (1985). Epidermal keratin gene expressed in embryos of *Xenopus laevis*. *Proc. Natl. Acad. Sci. U.S.A.* **82**, 5413–5417.

Kintner, C. R., and Melton, D. M. (1987). Expression of *Xenopus* N-CAM RNA is an early response of ectoderm to induction. *Development (Cambridge, U.K.)* **99**, 311–325.

Krieg, P. A., and Melton, D. A. (1987). *In vitro* RNA synthesis with SP6 RNA polymerase. *In* "Methods in Enzymology" (R. Wu, ed.), vol. 155, pp. 397–415. Academic Press, New York.

Lynn, D. A., Angerer, L. M., Bruskin, A. M., Klein, W. H., and Angerer, R. C. (1983). Localization of a family of mRNAs in a single cell type and its precursors in sea urchin embryos. *Proc. Natl. Acad. Sci. U.S.A.* **80**, 2656–2660.

Melton, D. A. Translocation of a localized maternal mRNA to the vegetal pole of *Xenopus* oocytes. *Nature (London)* **328**, 80–82.

Melton, D. A. Krieg, P. A., Rebagliati, M. R., Maniatis, T., Zinn, K., and Green, M. R. (1984). Efficient *in vitro* synthesis of biologically active RNA and RNA hybridization probes from plasmids containing a bacteriophage SP6 promoter. *Nucelic Acids Res.* **12**, 7035–7056.

Ruiz i Altaba, A., and Melton, D. A. (1989). Bimodal and graded expression of the *Xenopus* homeobox gene *Xhox3* during embryonic development. *Development (Cambridge, U.K.)* **106**, 173–183.

Sambrook, J., Fritsch, E. F., and Maniatis, T. (1989). "Molecular Cloning: A Laboratory manual," 2nd Ed. Cold Spring Harbor Laboratory, Cold Spring Harbor, New York.

Sargent, T. D., and Dawid, I. B. (1983). Differential gene expression in the gastrula of *Xenopus laevis*. *Science* **222**, 135–139.

Tautz, D., and Pfeifle, C. (1989). A non-radioactive *in situ* hybridization method for the localization of specific RNAs in *Drosophila* embryos reveals translational control of the segmentation gene hunchback. *Chromosoma* **98**. 81–85.

Venezky, D. L., Angerer, L. M., and Angerer, R. C. (1981). Accumulation of histone repeat transcripts in the sea urchin egg pronucleus. *Cell (Cambridge, Mass.)* **24**, 385–391.

Part IV. Model Systems Using Oocytes, Eggs, and Embryos

Chapter 24

DNA Recombination and Repair in Oocytes, Eggs, and Extracts

DANA CARROLL AND CHRIS W. LEHMAN

Department of Biochemistry
University of Utah Medical School
Salt Lake City, Utah 84132

I. Introduction
II. Methods
   A. Injection
   B. Preparation of Extracts
   C. Design of Substrates
   D. Incubation, Recovery, and Assay of DNA
III. Results and Dissussion
   A. Recombination
   B. Repair
   C. Prospects
   References

I. Introduction

Xenopus oocytes store large quantities of materials for use in the early stages of embryogenesis. In the first several hours after fertilization, a major metabolic activity of cleavage stage embryos is DNA synthesis. It has been, therefore, no great surprise to find that enzymes involved in DNA metabolism are abundant in oocytes and eggs. In the cases of some of the enzymes, activities are masked or suppressed in oocytes, since oocytes are not capable of replicative DNA synthesis (Benbow *et al.*, 1975; Zierler *et al.*, 1985).

In addition to replication (which is discussed in Chapter 29 of this volume), two other processes are usually included in the three R's of DNA metabolism: recombination and repair. For purposes of this chapter, it is useful to distinguish two styles of recombination. Homologous recombination depends on extensive sequence similarity between interacting chromosomes, whereas homology-independent (or illegitimate) recombination leads to the joining of unrelated sequences, frequently by ligating molecular ends. DNA repair is a collective term for processes that deal with DNA damage of many different kinds.

Recombination and repair have been recognized and investigated for many years. As is frequently the case, most progress in understanding mechanisms of these processes has been made with microorganisms, particularly bacteria and fungi (see Kucherlapati and Smith, 1988; Low, 1988; Friedberg, 1985). Genes have been identified whose produces are crucial for recombination and/or repair, the gene products have been purified in some cases, and their activities have been examined in detail. This has led to reasonably complete descriptions of some types of repair in *Escherichia coli* (Modrich, 1987) and to engaging models for homologous recombination in bacteria and yeasts (Orr-Weaver and Szostak, 1985; Thaler and Stahl, 1988).

The interest in recombination in higher organisms was stimulated in part by the observation that exogenous DNA could be delivered to cells and integrated into their chromosomes (see Roth and Wilson, 1988). Illegitimate events (i.e., integration at nonhomologous sites) predominate, but homologous events can also be observed. Clearly the purposes of genetic engineering would be best served by the ability to target integrations to homologous sites, or at least to choose the balance between homologous and nonhomologous events. Thus, an understanding of both types of recombination is desirable, for purposes of manipulation, as well as from the purist perspective of wanting to comprehend natural processes. DNA repair is certainly critical to our survival in the face of increasing exposures to environmental mutagens. Interest in repair has also been sharpened by the identification of several human diseases that are attributable to specific repair defects (see Friedberg, 1985). Some of these lead to increased incidences of cancer and emphasize the connection between mutation and carcinogenesis.

Because *Xenopus* oocytes and eggs had been shown to utilize injected DNAs effectively in other processes of DNA metabolism (Gurdon and Melton, 1981; Gurdon and Wickens, 1983), it occurred to a number of different investigators that they might also be able to accomplish recombination and/or repair. The manipulability of the system makes studies of mechanism quite feasible, and the large capacity for processing injected materials suggests the possibility of ultimately purifying and characterizing biochemically the activities responsible. In this chapter, we describe successful approaches to investigation of homologous (Carroll, 1983; Carroll *et al.,* 1986; Abastado *et al.,* 1987; Grzesiuk and Carroll, 1987; Maryon and Carroll, 1989, 1991a,b) and nonhomologous (Pfeiffer and Vielmetter, 1988; Thode *et al.,* 1990) recombination in oocytes and/or eggs, as well as the repair of mismatches (Brooks *et al.,* 1989; Varlet *et al.,* 1990), abasic sites (Matsumoto and Bogenhagen, 1989), and lesions induced by ultraviolet (UV) (Legerski *et al.,* 1987; Hays *et al.,* 1990) and X irradiation (Sweigert and Carroll, 1990). A key feature of these reports has been the design and construction of substrates that, after incubation in cells or extracts, will report reliably on the processes of interest.

II. Methods

A. Injection

Methods and apparatuses for injecting oocytes and eggs are described elsewhere in this volume (see Appendix B) and in previous reports (Stephens *et al.,* 1981). Because recombination and repair are processes of DNA metabolism, substrates have been injected directly into the nuclei (germinal vesicles, GVs) of oocytes (Kressmann *et al.,* 1978). The nuclear membrane breaks down during oocyte maturation, so nuclear injection is not meaningful in eggs; however, injection is generally targeted to the area beneath the animal pole where the nuclear contents have recently dispersed. Oocytes have been incubated in OR2 or modified Barth's solution (see Appendix A) with equivalent results.

B. Preparation of extracts

1. WHOLE-CELL EXTRACTS

The procedures for making egg extracts are based on methods developed for DNA replication (Lohka and Masui, 1983; Almouzni and Mechali, 1988) or chromatin assembly (Glikin *et al.,* 1984) and have not been designed specifically for recombination or repair activities. Fertilized or unfertilized eggs are dejellied and rinsed with extraction medium. Two similar buffers have been used in the studies described here: (1) from Glikin *et al.,* (1984), 30 mM Tris, pH 7.9, 90 mM KCl, 10 mM β-glycerophosphate, 2 mM EGTA, 1 mM dithiothreitol (DTT); (2) from Almouzni and Mechali (1988), 20 mM HEPES, pH 7.5, 70 mM potassium acetate, 1 mM DTT, 5% sucrose. All investigators stress the importance of removing dead or damaged eggs, which can be recognized by abnormal pigment distribution (often described as "marbling").

The eggs are placed in a centrifuge tube, with or without additional extraction buffer, and they are broken by centrifugation. Sometimes a low speed spin (12,000 g) is used to disrupt the eggs and the extract clarified at high speed (145,000 g) (Brooks et al., 1989), and sometimes only a single high speed spin (100,000–150,000 g) is used (Pfeiffer and Vielmetter, 1988; Matsumoto and Bogenhagen, 1989). The pellet containing pigment granules, yolk platelets, and other debris is removed; a lipid pellicle forms at the top of the tube and is also discarded. The extracts can be frozen directly and stored at $-70°$ or $-80°C$ for at least several months. Oocyte whole-cell extracts are prepared in essentially the same way.

Efforts are usually made to dilute the extracts as little as possible during preparation and use. This keeps conditions close to those inside the cell, and it may be very important to maintain high concentrations of multiple components that need to interact in carrying out complex processes. To date most studies have been done in crude homogenates. It is expected that fractionation of successful extracts to purify individual components will follow shortly. Some enzymes that may be involved in these processes have already been at least partially purified by investigators interested in replication or in the enzymes themselves (e.g., Kaiserman et al., 1988; Poll and Benbow, 1988).

2. Nuclear Extracts

Homologous recombination has not been observed in whole oocyte extracts, so it is necessary to isolate GVs as the starting material. This is usually done by manual dissection. Our procedure is similar to ones used for making transcriptionally active GV extracts (Birkenmeier et al., 1978; Lund and Dahlberg, 1989). Oocytes are placed in I buffer (see Appendix A, this volume) at room temperature shortly before dissection. A hole is made near the animal pole of each oocyte by scoring with a sharp syringe needle or by penetration with a pair of jeweler's forceps. The GV is then squeezed or teased out of the hole and pushed free of obvious cytoplasmic (yolk) material with forceps. Each GV is picked up with a micropipette in 0.5–1 μl of medium and kept on ice. After a few hundred GVs have been collected (an experienced dissector can isolate over 100 GVs per hour), they are disrupted by repeated passage through a micropipette tip, spun gently (500 g, 5 minutes) to remove particulates, then frozen in liquid nitrogen and stored at $-70°C$. GVs isolated in media other than I buffer were not effective in supporting recombination.

If purification of individual components is the ultimate goal, it would be extremely useful to have a bulk procedure for isolation of GVs. Some methods have been published, but they tend to be difficult to reproduce, and they yield nuclei of uncertain quality (Scalenghe et al., 1978; Ruberti et al., 1989). The challenge is to find a means to disrupt the investing layers of the oocyte (i.e., theca, vitelline envelope, follicle layers) without damaging the large, but delicate, GV. Homogenization of whole oocytes is clearly out of the question. The published procedures rely on proteases to digest the investing layers, but this risks hydrolysis of the proteins whose activities are of interest.

C. Design of Substrates

One of the great advantages of oocytes is that relatively large amounts of any desired DNA substrate can be delivered, by nuclear injection, directly to the site where it will be acted on. With available DNA technology and enzymology, molecules of almost any desired structure can be prepared for injection. The goals in designing substrates are as follows: first, there should be easy assays for products and intermediates derived from the substrates, and, second, they should offer some flexibility in structure, so various features of the process can be tested.

1. Recombination Substrates

In the earliest experiments with homologous recombination in injected oocytes, genetically marked pairs of bacteriophage λ DNA were used (Carroll, 1983; Carroll et al., 1986). These substrates offered large targets for crossing-over, since the λ chromosome is 48,500 base pairs (bp) long. In addition, this approach capitalized on the ability to package recovered DNAs in vitro, and thus made available the full range of λ genetics. Once it was clear that recombination in oocytes was quite efficient, it became preferable to work with smaller plasmid substrates, since the structure of the DNA could be manipulated and confirmed more easily (Carroll et al., 1986).

We now generally work with a family of plasmid substrates related to pDC10 (Fig. 1). This plasmid has all of the sequences of pBR322, plus a 1246-bp duplication of tet gene sequences, and a 1691-bp insertion of adenovirus 2 (Ad2) sequences between the tet repeats. Cleavage at the borders of the Ad2 insert with KpnI plus XhoI yields a linear molecule capable of intra- or intermolecular recombination in oocytes through the tet duplication. After the DNA has been recovered from oocytes, digestion with a restriction enzyme that cuts once within the pBR322 sequence yields one 4363-bp fragment for each recombination event that has taken place. One copy of the tet gene in pDC10 has eight single-base changes that create or destroy restriction sites. Thus, it can be determined which parental copy has contributed to a given recombination product at any site, and the distribution of effective crossovers can be mapped. Finally, if the DNA for injection is generated by cleavage at only one of the borders between the Ad2 and tet sequences and/

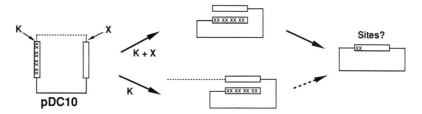

FIG. 1. Diagram of a substrate used to assay homologous recombination in oocytes. The plasmid pDC10 has two copies of the *tet* gene of pBR322 (boxes), one of which has been mutated at eight positions to create or destroy restriction enzyme sites (x's in left box). The solid line represents the remainder of pBR322; the dashed line is a segment of Ad2 DNA that has no homology to pBR322. Sites for *Kpn*I (K) and *Xho*I (X) are at the borders between the Ad2 sequences and the *tet* repeats. As indicated in the upper track, digestion with *Kpn*I plus *Xho*I generates a substrate with terminal direct repeats. Intramolecular recombination between these repeats in oocytes yields circular pBR322 (right) as the product (intermolecular recombination also occurs). The distribution of restriction sites within the single product *tet* gene reports on the locations of apparent crossovers. If digestion of pDC10 is with *Kpn*I alone prior to injection (lower track), the substrate has a long terminal nonhomology that interferes with recombination (the dashed arrow between substrate and product implies this lower efficiency).

or within the Ad2 sequences, the substrate will have terminal nonhomologies on one or both ends. In addition to direct physical analysis of products, bacterial transformation is sometimes used to assay recombination and to isolate individual recombination products for further characterization.

The substrates for illegitimate recombination do not contain extensive repeated sequences, so homologous recombination is not a possibility. The plasmid (pSP65) used by Pfeiffer and co-workers (Pfeiffer and Vielmetter, 1988; Thode *et al.*, 1990) has multiple sites for restriction enzymes that generate ends of different types. Digestions with different pairs of enzymes yield substrates with various combinations of ends to join (Fig. 2). Recombination is measured either by physical detection of joints across the original break or as the recovery of transformation efficiency. Transformation selects for circular recombination products, but these investigators have also been careful to isolate monomer circles from agarose gels for analysis. This is important since monomer circles can only be formed by joining the unlike ends on a single substrate molecule, whereas intermolecular products can include joints between like ends.

2. REPAIR SUBSTRATES

In the injection experiments described here, a plasmid DNA has simply been irradiated with UV light (Legerski *et al.*, 1987; Hays *et al.*, 1990) or with

Substrate type	Terminus configuration	Sequence
1) SmaI/SalI	blunt/5'PSS	C C C | T C G A C G G G | 5' G
2) SmaI/PstI	blunt/3'PSS	C C C | 3' G G G G | A C G T C
3) BamH1/PstI	5'PSS/3'PSS	G | 5' 3' G C C T A G | A C G T C
4) BamH1/SalI	5'PSS/5'PSS	G | 5'| T C G A C C C T A G | 5' G
5) SacI/KpnI	3'PSS/3'PSS	G A G C T | 3' C C | 3' C A T G G
6) KpnI/PstI	3'PSS/3'PSS	G G T A C | 3' G C | 3' A C G T C
7) SacI/PstI	3'PSS/3'PSS	G A G C T | 3' G C | 3' A C G T C

FIG. 2. Molecular ends used in the study of nonhomologous recombination in egg extracts. (From Pfeiffer and Vielmetter, 1988; by permission of Oxford University Press.) PSS is an abbreviation for protruding single strand, which may be either 3' or 5'.

X-rays (Sweigert and Carroll, 1990) prior to injection (Fig. 3a). This has the advantage that the types and levels of damage can be controlled and monitored quite precisely. Since the substrates are treated before injection, the equivalent of potentially lethal doses can be applied without compromising the capabilities of the cell in any way. In some cases, the plasmids carry antibiotic resistance genes. If the dose is high enough to inactivate such a gene in a substantial fraction of the molecules, repair can be assayed by simple reactivation of the ability to confer antibiotic resistance in a bacterial transformation assay. Physical analysis can monitor the repair of lesions (like strand breaks) that alter the electrophoretic behavior of circular DNAs. Since repair often requires the resynthesis of excised damaged segments, another approach has been to measure the incorporation of radiolabeled DNA precursors into injected DNA.

In their work with oocytes extracts, Matsumoto and Bogenhagen (1989) synthesized a specific oligonucleotide with an abasic site and ligated it into a plasmid vector (Fig. 3b). This approach, which is applicable to many types of DNA damage, has the advantage that a single lesion is placed in a specific sequence that is under the control of the experimenter.

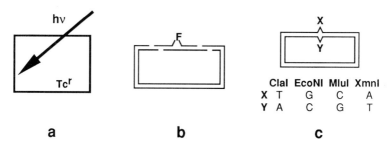

FIG. 3. Substrates used to study repair in oocytes and extracts. (a) Double-stranded plasmid DNA carrying a resistance marker (Tc^r) subjected to UV or X irradiation ($h\nu$) prior to injection. (b) Double-stranded oligonucleotide containing an abasic site (F, for furan) ligated into a plasmid vector (Matsumoto and Bogenhagen, 1989). The junctions between the oligonucleotide and the vector are shown as gaps in the drawing but were ligated prior to incubation in the extract. (c) Substrate used by Varlet *et al.* (1990) to study mismatch repair. All possible mismatched combinations of X and Y were constructed in double-stranded M13 DNA. The sequence around the variable position was

5'-TCGAXGCGT
3'-AGCTYCGCA

Correction to any Watson–Crick base pair was monitored by sensitivity to the four restriction enzymes given below the diagram with the combinations of X and Y that they recognize.

For studies of mismatch repair, Brooks *et al.* (1989) have used cloned variants of M13 phage that differ at a single position. Linear duplex DNA of one variant was denatured and annealed with circular single-stranded DNA of another. Following ligation, circular duplex DNA was isolated and used as the repairable substrate (Fig. 3c). With oligonucleotide-directed mutagenesis, they placed each of the four possible bases at the same test site and thus were able to create each of the 12 possible mismatches (Varlet *et al.*, 1990). Because the mismatches are within the sequences of four overlapping restriction sites, repair to any of the possible homoduplexes was evaluated by assaying sensitivity to the four enzymes, under conditions in which the mismatched heteroduplexes were resistant to digestion.

D. Incubation, Recovery, and Assay of DNA

In injection experiments, the conditions of incubation are determined to a large extent by the normal oocyte or egg milieu. These can be altered to some extent by injection of potential cofactors or inhibitors, in addition to substrates, but these additives are also subject to metabolism in the cells. With cell-free extracts, one has more control over components in the incubation mixture. This feature has been put to excellent use by Thode *et al.* (1990), who used dideoxynucleotides to define requirements for strand elongation in the end-joining reaction. We have also used this approach in determining the requirements for homologous recombination in GV extracts (Lehman and Carroll, 1991). Enzyme inhibitors can be employed both with injections and in cell-free systems (Legerski et al., 1987), but the usual caveats regarding inhibitor specificity apply.

In general, incubations in injected cells or extracts are carried out for tens of minutes to several hours. This is an indication of the relatively convenient time courses of recombination and repair reactions, which are slow enough to allow the isolation of intermediates, yet fast enough to be complete within a working day. If the ultimate products are circular, incubation in oocytes can be continued for a day or more, since circular DNAs are packaged into chromatin and are quite stable in oocyte nuclei, as long as the cells are healthy.

DNA is usually recovered by the normal sorts of extraction procedures. A complication may arise owing to the properties of yolk proteins. They are rather resistant to proteases, are heavily phosphorylated, and tend to come through the purification procedures like nucleic acids. Thus, DNA recovered from whole cells or from whole-cell extracts is contaminated with yolk proteins. Some analytical procedures are impervious to yolk proteins, but the proteins can interfere with bacterial transformation assays (Hays et al., 1990); they completely inhibit packaging of λ DNA (Carroll, 1983), and they may interfere with digestion by some restriction enzymes (D. Carroll and C. W. Lehman, unpublished observations). Centrifugation through a CsCl cushion effectively removes the contaminating proteins but is time consuming, particularly with multiple samples. As an alternative, we have routinely begun our recoveries by manually dissecting the nucleus from each injected oocyte. This adds effort to the overall experiment; however, DNAs can be recovered from GVs or GV extracts with simpler extraction procedures (Carroll, 1983), and they are active in all the assays to which we have subjected them.

Recovered DNAs are frequently analyzed directly by gel electrophoresis. Because of the large capacity of each oocyte or egg, products from just a few cells can often be visualized in ethidium bromide-stained gels (see Fig. 4). The only endogenous DNA abundant enough to cause confusion is mitochondrial DNA, which is avoided when GVs are isolated. The sensitivity and specificity of these analyses can be increased by Southern blot hybridization, by using labeled substrates, or by incorporating radioactive precursors during the repair or recombination process. Specific oligonucleotide probes can be used to obtain detailed information about sequences present in recovered intermediates and products (Maryon and Carroll, 1989; 1991a,b).

When the substrate DNA is a bacterial plasmid or phage chromosome, it can also be analyzed by transformation, transfection, or infection. In some cases, the simple measurement of transformation efficiency reports on the effectiveness of DNA repair or recombination. Introduction into bacteria also permits the cloning of individual recombination or repair products for more detailed analysis, for example, by DNA sequencing. One must be careful that the host bacteria do not contributed to the outcome, by processing intermediates that emerged from the oocyte/egg incubations or by doing some repair or recombination of their own. Some problems can be averted by use of bacterial mutants deficient, for example, in the repair of mismatches or particular types of DNA damage (Pfeiffer and Vielmetter, 1988; Hays *et al.*, 1990; Sweigert and Carroll, 1990).

III. Results and Discussion

A. Recombination

1. HOMOLOGOUS RECOMBINATION IN OOCYTES

When the injected substrate DNA meets several important criteria, recombination can be extremely efficient in oocyte nuclei (Carroll *et al.*, 1986). These criteria are as follows: (1) the substrate must have homologous sequences to support recombination; (2) the participating molecules must have ends; (3) the homologies must be near both participating molecular ends. As illustrated above, these criteria are met by our usual plasmid substrate (Fig. 1). A single oocyte can convert roughly 10^9 of these substrate molecules to covalently closed products in several hours, with very little loss (Fig. 4). Larger amounts of injected DNA are recombined more slowly, but we have not determined the ultimate capacity of the oocytes. The third requirement cited above is illustrated by the fact that leaving nonhomologous (Ad2) sequences on one end of the substrate greatly reduces the yield of recombination products (Fig. 4) (Jeong-Yu and Carroll, 1991).

The high capacity of oocytes and relatively leisurely kinetics of recombination have made it possible to identify and characterize intermediates in the process (Maryon and Carroll, 1989, 1991b). In addition, we have tested the influence of substrate alterations, and we have constructed and injected molecules with structures deduced for intermediates and shown that they are accelerated in their appearance in recombination products (Maryon and Carroll, 1991a,b). These and other studies have led to a model of homologous recombination in oocytes (Fig. 5).

Quite recently we have found that extracts of oocyte nuclei are also capable of supporting homologous recombination (Lehman and Carroll, 1991). With GVs isolated as described in Section II, the capacity for recombination is

24. DNA RECOMBINATION AND REPAIR

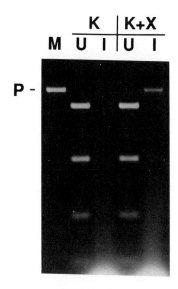

FIG. 4. Results of homologous recombination in oocytes. As diagrammed in Fig. 1, the substrate was pDC10 digested with *Kpn*I alone (K) to leave a terminal nonhomology, or with *Kpn*I plus *Xho*I (K + X) to leave only terminal homologies. The lanes marked I contain DNA that was injected into 10 oocytes, incubated overnight, recovered, and digested with *Pst*I and *Xho*I prior to electrophoresis. The lanes marked U contain uninjected substrate DNA digested with the same enzyme mixture. Lane M contains *Pst*I-digested pBR322, which marks the position diagnostic of recombination products (P). The gel was stained with ethidium bromide: products from 10 oocytes are clearly visible, and the inhibitory effect of the terminal nonhomology (K) is evident.

reduced substantially compared to injected oocytes, but completed recombination products are easily scored (Fig. 6). The reduced capacity may be due to dilution of nuclear components or to leakage during the isolation procedure. Recombination in this cell-free system is dependent on added nucleoside triphosphates (NTPs), and this has allowed us to show that

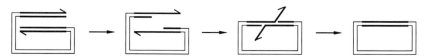

FIG. 5. Mechanism of homologous recombination postulated to occur in *Xenopus* oocytes. A double-stranded DNA with terminal direct repeats (thick lines) is shown at left; half-arrowheads denote 3' ends. In the first step, the substrate is acted on by a $5' \to 3'$ exonuclease to expose complementary sequences in single strands. These sequences anneal in the second step. The junction is completed by the continued action of the $5' \to 3'$ exonuclease, and perhaps other nuclease and polymerase activities, then sealed by DNA ligase. Although only intramolecular recombination is illustrated here, intermolecular recombination occurs by the same basic mechanism.

completion of the process requires the addition of a hydrolyzable NTP, while nuclease digestion, the first step, does not. As seen in Fig. 6, formation of the discrete 4.36-kilobase (kb) product is supported by either ATP or dATP alone. Without added NTPs or in the presence of nonhydrolyzable ATP analogs (ATPγS or AMP-PNP), smears of intermediates, but no completed products, are seen. Under some conditions in these extracts, there is a moderate level of semihomologous end joining using short sequence matches very near the

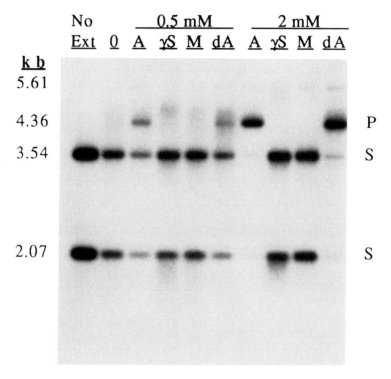

FIG. 6. Results of homologous recombination in GV extracts demonstrating nucleotide requirements. The substrate was a linear molecule with terminal homologies, like that indicated in Fig. 1 following K + X digestion. After incubation in GV extracts, the DNA was digested with PvuII (which cuts once in pBR322 sequences outside the region of homology), fractionated by electrophoresis, and blot-hybridized with labeled substrate as probe. The letters P and S at right indicate the positions of product and substrate bands, respectively. In some lanes a faint band can be seen at 5.61 kb, which represents incompletely digested substrate DNA. Discrete product bands were formed in the presence of 0.5 or 2 mM ATP (A) and dATP (dA). In the absence of any added NTP (0) or in the presence of nonhydrolyzable ATP analogs at 0.5 and 2 mM (γS, ATP γ S; M, AMP-PNP), no products were formed, but some intermediates were seen, represented by smears trailing upward from the 3.54-kb substrate band; also, there was some degradation of both substrate fragments to faster migrating smears.

24. DNA RECOMBINATION AND REPAIR

molecular ends, as had been observed previously in oocytes (Grzesiuk and Carroll, 1987).

2. NONHOMOLOGOUS RECOMBINATION IN EGGS AND EMBRYOS

When linear DNAs are injected into fertilized or unfertilized eggs by hopeful experimenters aiming to achieve stable genomic alteration, the injected molecules are joined at their ends in an apparently homology-independent fashion (Bendig, 1981; Rusconi and Schaffner, 1981). This is also the dominant fate of introduced DNAs in mammalian somatic cells (Rothand Wilson, 1988). Only rarely has chromosomal intergration been achieved in *Xenopus* (Etkin and Pearman, 1987).

The nature of the end-joining reaction has been studied in detail by Pfeiffer, Vielmetter, and colleagues (Pfeiffer and Vielmetter, 1988; Thode et al., 1990), using egg extracts (see also Bayne et al., 1984). Their observations are that ends of essentially any kind can be joined to each other. If short tails with the same polarity are present, they are joined in an overlapping register that can be set by as little as a single, fortuitously matched base pair. Blunt ends and ends of opposite polarity are also joined with reasonable efficiency. By using chain-terminating dideoxynucleotides, Thode et al., (1990) found a requirement for limited DNA synthesis when joining nonoverlapped ends. In the joining of a blunt end to a 3' protruding end, fill-in of the 3' tail is primed by the 3' terminus of the blunt end. Because of this observation and the registration of ends of like polarity by a single base pair, the authors postulated the presence of an "alignment protein" that holds the ends in proximity without base pairing. This speculation is also fueled by the observation that purified DNA ligases from *E. coli* and bacteriophage T4 will not join many of the end combinations that are efficiently joined by the egg extract.

Recently this same group has investigated the end-joining reaction during oocyte maturation induced by progesterone (W. Goedecke, W. Vielmetter, and P. Pfeiffer, personal communication). They found that the capability of efficiently making nonhomologous end joints appears several hours after germinal vesicle breakdown. It will be very interesting to see what functions are activated and/or inactivated at this time to drive the transition from homology-dependent recombination in oocytes to homology-independent recombination in eggs.

3. COMPARISONS WITH OTHER ORGANISMS

When linear DNA is introduced into mammalian somatic cells, ends are joined rapidly and efficiently, without regard for homology. When sequences are present that could support homologous recombination, this can also be detected, although the balance frequently favors end joining (Roth and Wilson, 1985). The nature of the junctions made by end joining is very similar to what has been seen in frog egg extracts: all types of ends are joined, and minimal base matches are used when they are available (Roth and Wilson, 1986).

Homologous recombination in mammalian cells appears to be largely nonconservative (Subramani and Seaton, 1988), and the data are consistent with an annealing model (Lin et al., 1984) like that derived for oocyte recombination (Fig. 5). Thus, *Xenopus* accomplishes both styles of recombination seen in mammalian cells, but separates them into different cell types. Oocytes apparently do essentially only homologous recombination, whereas eggs do primarily end joining. It should be pointed out that the ability of eggs to support homologous recombination has not been extensively tested; such a capability might be revealed by the use of substrates like pDC10 (Fig. 1). To the extent that it is real, the segregation of homologous and nonhomologous processes into different cells may make it easier to study each process without interference from the other.

B. Repair

1. UV Damage

Two groups have independently studied the processing of UV-irradiated DNA after injection into oocyte nuclei (Legerski et al., 1987; Hays et al., 1990). Using quite different methods of analysis, both found that the oocytes have a large capacity to repair UV damage: more than 10^{10} lesions are removed in a single oocyte. Interestingly, they also found that some damage was not repaired, even after many hours of incubation (Fig. 7). It may be that this is due to exhaustion of some component of the repair machinery, since the percentage of residual damage is less when fewer lesions are injected (Fig. 7; R. Legerski, personal communication). *In vitro* photoreactivation showed that the residual lesions were not inherently unrepairable and that the oocytes repair both pyrimidine dimers (the major UV photoproducts) and other types of lesions (Hays et al., 1990).

Both groups reported that recombination was not stimulated in oocytes by UV damage, and one of the substrates used by Hays et al. was designed specifically to measure such events. Some DNA synthesis accompanies repair (Legerski et al., 1987), but the lengths of these repair tracts has not been measured. Hays et al., (1990) demonstrated that there was not wholesale replication of repaired DNA. It should be mentioned in connection with these experiments that, although replicative DNA synthesis has been observed with

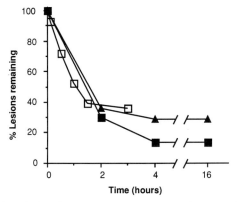

FIG. 7. Repair of UV lesions in oocytes. Data are adapted from Legerski et al. (1987) (open squares, DNA irradiated at 700 J/m^2) and from Hays et al. (1990) (solid triangles, DNA irradiated at 250 J/m^2; solid squares, DNA irradiated at 120 J/m^2). (Both excerpts by permission of the authors and the American Society for Microbiology, Journals Division.) The horizontal axis indicates time of incubation in oocytes. The DNA substrates and the methods used to assay remaining damage were quite different in the two studies, but the results are in general agreement.

DNA injected into *Xenopus* eggs, only repair-type synthesis is seen in oocytes (Gurdon and Melton, 1981). Single-stranded DNA is made double stranded in oocytes (Cortese *et al.*, 1980), but essentially no synthesis occurs on covalently closed duplex DNA. This is consistent with the interpretation of repair tracts as being lesion dependent.

2. X-RAY DAMAGE

Both biological and physical assays have been used to show that X-ray-induced lesions are repaired in oocytes (Sweigert and Carroll, 1990). Apparently strand breaks and some types of base damage are repaired, although the efficiencies were not quantitated. Unlike UV-irradiated DNAs, X-irradiated plasmids are stimulated to recombine in oocytes. It is likely that the introduction of double-strand breaks (a property of X-ray, but not UV damage) is partly responsible, but the levels of recombination observed suggest that other types of lesions may contribute as well.

3. ABASIC SITES

A number of repair pathways characterized in other organisms proceed through the generation of abasic (AP) sites. A base may be removed directly by some types of damage, or a modified base may be removed as the first step in repair. Such sites are recognized by AP endonucleases that excise the damaged strand preparatory to its accurate resynthesis. Matsumoto and Bogenhagen (1989) showed that oocyte extracts contain such an AP endonuclease, plus activities to repair and reseal a synthesic abasic site. This repair was essentially complete within about 40 minutes of incubation. Both the nature of the synthetic lesion and the location of the initial incision implicate the involvement of a type II AP endonuclease. Damage-stimulated DNA synthesis was very localized: perhaps only 1 and not more than 3–4 nucleotides (including the abasic site) were involved.

4. Mismatch Repair

DNA mismatches can result from misincorporation during DNA synthesis, pairing of strands during genetic recombination, or by chemical modification of bases. Bacteria have several mechanisms for correcting mismatches; the major, methyl-directed system has been extensively characterized (Modrich, 1987). Brooks *et al.*, (1989) found that mismatches included in circular DNAs are repaired in egg extracts. Varlet *et al.*, (1990) examined all 12 possible mismatches and found that the efficiency of repair varies among different mismatches and with context. It can be as high as 20%, but may also be less than 5% (Fig. 8). Lesion-stimulated DNA synthesis accompanies this type of repair as well.

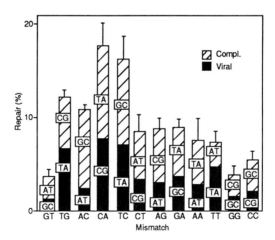

FIG. 8. Repair of various mismatches in egg extracts. (From Varlet *et al.*, 1990; by permission of the authors.) For each input mismatch (given at the bottom), the efficiency of repair to either possible matched pair is given by the bar heights (with standard deviation of the mean for total repair). Viral and Compl. indicate whether the viral or complementary strand provided the template for repair.

While localized to the region around the initial mismatch, these synthetic tracts were 150–500 bases in length, much longer than those observed for abasic sites (Matsumoto and Bogenhagen, 1989). It is possible that this reflects the different replicative capabilities of oocytes and eggs (Gurdon and Melton, 1981), rather than an inherent difference in the repair systems.

The order of repair efficiency of different mismatches does not correspond to that found in some other organisms, but context effects need further examination. It is not known whether all mismatches are repaired by the same machinery or whether several different systems might be operating. Two observations ensure that repair is dependent on mismatch recognition, not due to nonspecific processes: DNA synthesis is mismatch stimulated and very localized, and different mismatches are corrected with quite different efficiencies. Furthermore, the direct analysis of recovered DNA avoids possible complications owing to preferential replication or retention of specific sequences in assays that require propagation of repair products.

Mismatch repair also appears to operate on heteroduplexes produced during homologous recombination in oocytes (Lehman *et al.*, 1991). We have injected nicked-circular heteroduplexes containing eight mismatches spread over about 1.2 kb and found that oocytes convert them to homoduplexes, each of which carries information derived mostly from one parental strand or the other (at equal frequency) (C. W. Lehman, S. Jeong-Yu, and D. Carroll, unpublished results). This suggests that the repair tracts are quite long.

C. Prospects

One question frequently asked is, how good a "model" is the *Xenopus* oocyte (or egg) for other cells and other organisms? We should begin by asserting that the properties and capabilities of these cells are of inherent interest and denying that motivation for their investigation depends on comparisons to other systems. Nonetheless, as stated above, oocytes and eggs have DNA metabolic activities very reminiscent of other organisms, from bacteria to mammals. This reflection of the conservation of biological functions and mechanisms ensures that lessons learned from the study of frog materials will be broadly generalizable.

The chief advantages of using oocytes and eggs derive from the extraordinary size of these single cells. One oocyte has the internal volume and many components and activities equivalent to about 10^5 vertebrate somatic cells (Gurdon and Wickens, 1983). When challenged to perform recombination or repair, a single oocyte can process several nanograms (about 10^9 molecules) of substrate DNA. This substrate can be manipulated arbitrarily before introduction into cells or extracts to yield the maximum amount of information about the process.

The most promising future prospects lie in the pursuit of the cell-free extracts, described above, that accomplish complete or partial recombination or repair events. It is possible to collect many grams of oocytes or eggs to serve as starting material for purification of individual activities. Given the capacity of each cell, it should be possible to purify biochemically significant quantities of enzymes that catalyze individual steps in each of the processes described.

The tool that is not available with *Xenopus* is genetics. Therefore, proof of the involvement of specific catalysts in particular events must rely on reconstitution of those events using purified components, or interference with their completion. An attractive approach is to raise antibodies that block the activities of purified catalysts, then add the antibodies to *in vitro* reactions or inject them into oocytes where recombination (for example) normally occurs and inquire whether the antibodies interfere. Antibodies can also be used to determine when during oogenesis or embryogenesis particular catalysts are made, and the blocking approach can report on the natural functions of these enzymes in normal developmental processes (Warner *et al.,* 1984; Wright *et al.,* 1989).

Acknowledgments

We are grateful to the individuals who responded to our requests for unpublished information and who gave us permission to reproduce their findings. Particularly helpful were Petra Pfeiffer, John Hays, Randy Legerski, and Peter Brooks. We thank Isabelle Varlet, Miro Radman, Peter Brooks, and Petra Pfeiffer for allowing us to see and quote their results prior to publication. We are grateful to members of our laboratory, particularly Ed Maryon and Sunjoo Jeong-Yu, whose results are quoted. Work in our laboratory was supported by grants from the National Institutes of Health and the National Science Foundation.

References

Abastado, J.-P., Darche, S., Godeau, F., Cami, B., and Kourilsky, P. (1987). Intramolecular recombination between partially homologous sequences in *Escherichia coli* and *Xenopus laevis* oocytes. *Proc. Natl. Acad. Sci. U.S.A.* **84,** 6496–6500.

Almouzni, G., and Mechali, M. (1988). Assembly of spaced chromatin promoted by DNA synthesis in extracts from *Xenopus* eggs. *EMBO J.* **7,** 665–672.

Bayne, M. L., Alexander, R. F., and Benbow, R. M. (1984). DNA binding protein from ovaries of the frog, *Xenopus laevis*, which promotes concatenation of linear DNA. *J. Mol. Biol.* **172,** 87–108.

Benbow, R. M., Pestell, R. Q. W., and Ford, C. C. (1975). Appearance of DNA polymerase activities during early development of *Xenopus laveis. Dev. Biol.* **43,** 159–174.

Bendig, M. M. (1981). Persistence and expression of histone genes injected into *Xenopus* eggs in early development. *Nature (London)* **292,** 65–67.

Birkenmeier, E. H., Brown, D. D., and Jordan, E. (1978). A nuclear extract of *Xenopus laevis* oocytes that accurately transcribes 5S RNA genes. *Cell (Cambridge, Mass.)* **15,** 1077–1086.

Brooks, P., Dohet, C., Almouzni, G., Mechali, M., and Radman, M. (1989). Mismatch repair involving localized DNA synthesis in extracts of *Xenopus* eggs. *Proc. Natl. Acad. Sci. U.S.A.* **86**, 4425–4429.

Carroll, D. (1983). Genetic recombination of bacteriophage λ DNAs in *Xenopus* oocytes. *Proc. Natl. Acad. Sci. U.S.A.* **80**, 6902–6906.

Carroll, D., Wright, S. H., Wolff, R. K., Grzesiuk, E., and Maryon, E. B. (1986). Efficient homologous recombination of linear DNA substrates after injection into *Xenopus laevis* oocytes. *Mol. Cell. Biol.* **6**, 2053–2061.

Cortese, R., Harland, R. M., and Melton, D. A. (1980). Transcription of tRNA genes *in vivo*: Single-stranded compared to double-stranded templates. *Proc. Natl. Acad. Sci. U.S.A.* **77**, 4147–4151.

Etkin, L., and Pearman, B. (1987). Distribution, expression and germ line transmission of exogenous DNA sequences following microinjection into *Xenopus laevis* eggs. *Development* **99**, 15–23.

Friedberg, E. C. (1985). "DNA Repair." Freeman, New York.

Glikin, G. C., Ruberti, I., and Worcel, A. (1984). Chromatin assembly in *Xenopus* oocytes: *In vitro* studies. *Cell (Cambridge, Mass.)* **37**, 33–41.

Grzesiuk, E., and Carroll, D. (1987). Recombination of DNAs in *Xenopus* oocytes based on short homologous overlaps. *Nucleic Acids Res.* **15**, 971–985.

Gurdon, J. B., and Melton, D. A. (1981). Gene transfer in amphibian eggs and oocytes. *Annu. Rev. Genet.* **15**, 189–218.

Gurdon, J. B., and Wickens, M. P. (1983). The use of *Xenopus* oocytes for the expression of cloned genes. *In* "Methods in Enzymology" (R. Wu, L. Grossman, and K. Moldave, eds.), vol. 101, pp. 370–386. Academic Press, New York.

Hays, J. B., Ackerman, E. J., and Pang, Q. (1990). Rapid and apparently error-prone excision repair of nonreplicating UV-irradiated plasmids in *Xenopus laevis* oocytes. *Mol. Cell. Biol.* **10**, 3505–3511.

Jeong-Yu, S., and Carroll, D. (1991). In preparation.

Kaiserman, H. B., Ingebritsen, T. S., and Benbow, R. M. (1988). Regulation of *Xenopus laevis* DNA topoisomerase I activity by phosphorylation *in vitro*. *Biochemistry* **27**, 3216–3222.

Kressmann, A., Clarkson, S. G., Telford, J. L., and Birnstiel, M. L. (1978). Transcription of *Xenopus* tDNA$_1^{met}$ and sea urchin histone DNA injected into the *Xenopus* oocyte nucleus. *Cold Spring Harbor Symp. Quant. Biol.* **42**, 1077–1082.

Kucherlapati, R., and Smith, G. R., eds. (1988). "Genetic Recombination." American Society for Microbiology, Washington, D. C.

Legerski, R. J., Penkala, J. E., Peterson, C. A., and Wright, D. A. (1987). Repair of UV-induced lesions in *Xenopus laevis* oocytes. *Mol. Cell. Biol.* **7**, 4317–4323.

Lehman, C. W., and Carroll, D. (1991). Homologous recombination catalyzed by a nuclear extract from *Xenopus* oocytes. Submitted for publication.

Lehman, C. W., Jeong-Yu, S., Dohrmann, P., Dawson, R. J., Trautman, J. K., and Carroll, D. (1991). In preparation.

Lin, F. -L., Sperle, K., and Sternberg, N. (1984). Model for homologous recombination during transfer of DNA into mouse L cells: Role for the ends in the recombination process. *Mol. Cell. Biol.* **4**, 1020–1034.

Lohka, M. J., and Masui, Y. (1983). Formation *in vitro* of sperm pronuclei and mitotic chromosomes induced by amphibian ooplasmic components. *Science* **220**, 719–721.

Low, K. B., ed. (1988). "The Recombination of Genetic Material." Academic Press, New York.

Lund, E., and Dahlberg, J. E. (1989). *In vitro* synthesis of vertebrate U1 snRNA. *EMBO J.* **8**, 287–292.

Maryon, E., and Carroll, D. (1989). Degradation of linear DNA by a strand-specific exonuclease activity in *Xenopus laevis* oocytes. *Mol. Cell. Biol.* **9**, 4862–4871.

Maryon, E., and Carroll, D. (1991a). Involvement of single-stranded tails in homologous recombination of DNA injected into *Xenopus laevis* oocyte nuclei. *Mol. Cell. Biol.* **11,** 3268–3277.

Maryon, E., and Carroll, D. (1991b). Characterization of recombinant intermediates from DNA injected into *Xenopus laevis* oocytes: Evidence for a nonconservative mechanism of homologous recombination. *Mol. Cell. Biol.* **11,** 3278–3287.

Matsumoto, Y., and Bogenhagen, D. F. (1989). Repair of a synthetic abasic site in DNA in a *Xenopus laevis* oocyte extract. *Mol. Cell. Biol.* **9,** 3750–3757.

Modrich, P. (1987). DNA mismatch correction. *Annu. Rev. Biochem.* **56,** 435–466.

Orr-Weaver, T. L., and Szostak, J. W. (1985). Fungal recombination. *Microbiol. Rev.* **49,** 33–58.

Pfeiffer, P., and Vielmetter, W. (1988). Joining of nonhomologous DNA double-strand breaks in vitro. *Nucleic Acids Res.* **16,** 907–924.

Poll, E. H. A., and Benbow, R. M. (1988). A DNA helicase from *Xenopus laevis* ovaries. *Biochemistry* **27,** 8701–8706.

Roth, D. B., and Wilson, J. H. (1985). Relative rates of homologous and nonhomologous recombination in transfected DNA. *Proc. Natl. Acad. Sci. U.S.A.* **82,** 3355–3359.

Roth, D. B., and Wilson, J. H. (1986). Nonhomologous recombination in mammalian cells: Role for short sequence homologies in the joining reaction. *Mol. Cell. Biol.* **6,** 4295–4304.

Roth, D. B., and Wilson, J. (1988). Illegitimate recombination in mammalian cells. *In* "Genetic Recombination" (R. Kucherlapati and G. R. Smith, eds.), pp. 621–653. American Society for Microbiology, Washington, D. C.

Ruberti, I., Beccari, E., Bianchi, E., and Carnevali, F. (1989). Large scale isolation of nuclei from oocytes of *Xenopus laevis*. *Anal. Biochem.* **180,** 177–180.

Rusconi, S., and Schaffner, W. (1981). Transformation of frog embryos with a rabbit β-globin gene. *Proc. Natl. Acad. Sci. U.S.A.* **78,** 5051–5055.

Scalenghe, F., Buscaglia, M., Steinheil, C., and Crippa, M. (1978). Large scale isolation of nuclei and nucleoli from vitellogenic oocytes of *Xenopus laevis*. *Chromosoma* **66,** 299–308.

Stephens, D. L., Miller, T. J., Silver, L., Zipser, D., and Mertz, J. (1981). Easy-to-use equipment for the accurate microinjection of nanoliter volumes into the nuclei of amphibian oocytes. *Anal. Biochem.* **114,** 299–309.

Subramani, S., and Seaton, B. L. (1988). Homologous recombination in mitotically dividing mammalian somatic cells. *In* "Genetic Recombination" (R. Kucherlapati and G. R. Smith, eds.), pp. 549–573. American Society for Microbiology, Washington, D. C.

Sweigert, S. E., and Carroll, D. (1990). Repair and recombination of X-irradiated plasmids in *Xenopus laevis* oocytes. *Mol. Cell. Biol.* **10,** 5849–5856.

Thaler, D. S., and Stahl, F. W. (1988). DNA double-chain breaks in recombination of phage λ and of yeast. *Annu. Rev. Genet.* **22,** 169–197.

Thode, S., Schafer, A., Pfeiffer, P., and Vielmetter, W. (1990). A novel pathway of DNA end-to-end joining. *Cell (Cambridge, Mass.)* **60,** 921–928.

Varlet, I., Radman, M., and Brooks, P. (1990). DNA mismatch repair in *Xenopus* egg extracts: Repair efficiency and DNA repair synthesis for all single base-pair mismatches. *Proc. Natl. Acad. Sci. U.S.A.* **87,** 7883–7887.

Warner, A. E., Guthrie, S. C., and Gilula, N. B. (1984). Antibodies to gap-junctional protein selectively disrupt junctional communication in the early amphibian embryo. *Nature (London)* **311,** 127–131.

Wright, C. V. E., Cho, K. W. Y., Hardwicke, J., Collins, R. H., and DeRobertis, E. M. (1989). Interference with function of a homeobox gene in *Xenopus* embryos produces malformations of the anterior spinal cord. *Cell (Cambridge, Mass.)* **59,** 81–93.

Zierler, M. K., Marini, N. J., Stowers, D. J., and Benbow, R. M. (1985). Stockpiling of DNA polymerases during oogenesis and embryogenesis in the frog, *Xenopus laevis*. *J. Biol. Chem.* **260,** 974–981.

Chapter 25

Expression of Ion Channels by Injection of mRNA into Xenopus Oocytes

ALAN L. GOLDIN

Department of Microbiology and Molecular Genetics
University of California, Irvine
Irvine, California 92717

I. Introduction
II. Studies Using Oocytes to Express Ion Channels
III. Cautionary Notes
IV. Preparation of RNA for Injection
    A. Lithium Chloride–Urea RNA Extraction Procedure
    B. Guanidine Hydrochloride RNA Extraction Procedure
    C. Transcription of RNA *in Vitro* from Cloned Channel Genes
V. Preparation and Injection of Oocytes
VI. Biochemical Analysis of Expression
    A. Immunoprecipitation of Channel Proteins
    B. Ligand Binding Assay
VII. Electrophysiological Analysis
    A. Whole-Cell Voltage Clamping
    B. Macropatch Recording
    C. Single Channel Analysis
VIII. Conclusions
    References

I. Introduction

The *Xenopus* oocyte system is perhaps uniquely suited to the study of ion channels and neural receptors in a controlled *in vivo* environment. Miledi and co-workers first demonstrated that oocytes injected with RNA from rat brain express a variety of neurotransmitter receptors and voltage-sensitive ion channels in a functional configuration (Barnard et al., 1982; Miledi et al., 1982; Gundersen et al., 1984). Using two-microelectrode voltage clamping to analyze the oocytes for electrophysiological function, they demonstrated expression of the nicotinic acetylcholine receptor (nAChR) and voltage-sensitive sodium channel. Since that time many different channels and receptors have been successfully expressed following injection of mRNA isolated from a variety of tissue sources (for reviews, see Dascal, 1987; Snutch, 1988). In addition, the oocyte system has been used as an assay for the isolation of cDNA clones encoding receptors for which the only information available was a functional response. More recently, molecular biological techniques have been used to make structural alterations in the cDNA clones followed by electrophysiological analysis to correlate structure with function. The goal of this chapter is to briefly outline the approaches and methods used for these studies.

II. Studies Using Oocytes to Express Ion Channels

One of the earliest uses for *Xenopus* oocytes in the study of ion channel expression was to examine individual responses in the absence of other channels or receptors. Miledi et al. (1982) analyzed the properties of acetylcholine receptors following injection of cat muscle RNA into oocytes to study the current–voltage relationship, block by local anesthetics, and single channel properties by noise analysis. Gundersen et al. (1984) used similar approaches to study the characteristics of human brain receptors for serotonin and kainate as well as voltage-sensitive sodium channels, all of which were impractical to study *in vivo*. Methfessel et al. (1986) demonstrated that patch-clamp techniques could be used in oocytes to directly study the single channel properties of nAChR and voltage-sensitive sodium channels as well as endogenous stretch-activated channels. The subunit composition necessary for functional sodium channels was examined by various investigators using sucrose gradient fractionation to demonstrate that high molecular weight RNA, presumably encoding only the high molecular weight α subunit, was sufficient to encode functional channels in oocytes (Goldin et al., 1986; Hirono et al., 1985; Sumikawa et al., 1984a). Isolation of full-length cDNA clones for the α subunit of the sodium channel then confirmed these results (Auld et al., 1988; Noda et al., 1986). Taking advantage of the higher levels of expression obtainable with mRNA synthesized from cloned cDNA, Stühmer et al. (1987) demonstrated that macroscopic sodium channels could be recorded with submillisecond time resolution by using macropatches. It has now become commonplace for investigators to study the function of individual channels or receptors by expression of mRNA in oocytes.

The oocyte expression system has been particularly useful for the isolation of cDNA clones encoding ion channels for which no sequence information is available. Two basic approaches have been used: hybrid selection or depletion, and direct expression. An example of the first approach was the cloning of the 5-HT_{1C} serotonin receptor by Lübbert et al. (1987a). They constructed a cDNA library from mRNA isolated from a tumor of the rat choroid plexus that expressed the receptor at a high level, then screened the library by hybrid depletion of rat brain mRNA to remove the 5-HT_{1C} response. They isolated one clone that depleted the response to serotonin and could be used to hybrid select an mRNA that was sufficient for expression of the 5-HT_{1C} receptor in oocytes. A cDNA clone for the same receptor was isolated using the second approach by Julius et al. (1988). They constructed a cDNA library in a plasmid vector containing T7 and T3 promoters at each end, then screened the library by transcribing RNA for injection into oocytes. They isolated a full-length cDNA clone that encoded the 5-HT_{1C} receptor. The first approach is significantly more labor intensive, but it can be used to isolate clones for individual subunits even if the final response requires the presence of multiple subunits. The second approach is less time consuming and results in the isolation of a full-length clone for the receptor. It has been used to identify cDNA clones for receptors which were previously unknown, as demonstrated by the isolation of a potassium channel which shares no structural similarity with any previously identified channel or receptor (Takumi et al., 1988).

The most recent use of oocytes in the study of ion channels and receptors has been to correlate molecular structure with both biochemical and electrophysiological functions. For example, Kobilka et al. (1988) constructed chimeric channels between the β_2-adrenergic receptor and the α_2-adrenergic receptor, both of which are stimulated by epinephrine but which couple to different G proteins and so have opposite effects on the adenylate cyclase system. They then examined the effects of the different chimeras on adenylate cyclase activity to determine that the region responsible for coupling to G proteins is a proposed intracellular loop between the fifth and sixth transmembrane regions. The structural regions involved in electrophysiological functions have been studied using both the voltage-sensitive sodium and potassium channels. Stühmer et al. (1989) used site-specific mutagenesis to demonstrate that the positively charged residues in the fourth proposed membrane-spanning helix of each of two domains of the sodium channel are involved in the voltage sensitivity of the molecule. Auld et al. (1990) used comparable approaches to show that the nonpolar amino acids in the same helix were equally as important in voltage sensitivity. The region involved in inactivation of the sodium channel has been suggested to be the intracellular loop between domains III and IV by similar site-specific mutagenesis experiments (Moorman *et al.,* 1990a) and also by modulation of inactivation through the binding of antipeptide antibodies (Vassilev *et al.,* 1988, 1989). Similar studies have defined the regions involved in activation (Papazian *et al.,* 1991), inactivation (Hoshi *et al.,* 1990; Zagotta *et al.,* 1990), toxin binding (Yellen *et al.,* 1991; MacKinnon and Miller, 1989; MacKinnon and Yellen, 1990; MacKinnon *et al.,* 1990), and permeation (Yool and Schwarz, 1991; Hartmann *et al.,* 1991) in the voltage-sensitive potassium channel.

III. Cautionary Notes

There are a number of important considerations in using *Xenopus* oocytes to study ion channel and receptor functions. The first of these is that uninjected oocytes express a variety of electrophysiological responses which might interfere with the one being studied. Kusano *et al.* (1982) detected responses in uninjected oocytes to cholinergic and catecholaminergic agents, rare and very weak responses to serotonin, and no responses to aspartate, glutamate, γ-aminobutyric acid, and glycine. Voltage-sensitive channels, including those specific to sodium, potassium, and calcium, have also been detected in uninjected oocytes (reviewed by Dascal, 1987). A partial listing of endogenous oocyte responses is given in Table I. These responses usually do not cause a problem for a number of reasons. First, not all oocytes express the endogenous receptors, so that it is frequently possible to obtain oocytes which have no response to the neurotransmitter of interest. Second, the induced receptor response is usually much greater than the endogenous one. For example, endogenous voltage-sensitive sodium channels are only observed in oocytes from occasional frogs (fewer than 10%), and when present the amount of current detected is in the range of 10–20 nA. Following injection of rat brain mRNA or *in vitro* transcribed RNA, however, the currents can easily exceed 10 μA. A final reason why endogenous responses may not interfere is that some of them are a property of the follicle cells, so that they are no longer observed after defolliculation (see Table I). For example, potassium channels open in response to cyclic nucleotides because the cyclic nucleotides are acting on the follicle cells, which are electrically coupled to the oocyte through gap junctions (Miledi and Woodward, 1989). For these reasons it is frequently possible to eliminate all endogeneous background before injection of RNA.

A second consideration in the study of ion channels using *Xenopus* oocytes is whether the particular channel is functional in the oocyte. There are a variety of reasons why this might not be the case. One possibility is that the

TABLE I

ENDOGENOUS RECEPTOR RESPONSES IN *Xenopus* OOCYTES

Receptor	Response	Second messenger	Defolliculated oocytes[a]	Reference
β-Adrenergic	Potassium current	cAMP[b]	No	Sumikawa et al. (1984b)
Purinergic	Potassium current	cAMP	No	Lotan et al. (1982)
Gonadotropic	Potassium current	cAMP	No	Woodward and Miledi (1987a)
VIP[c]	Potassium current	cAMP	No	Woodward and Miledi (1987b)
Muscarinic AChR	Chloride current	IP_3[d]	No	Kusano et al. (1982)
Calcium	Chloride current	Calcium	Yes	Miledi and Parker (1984)
Voltage-gated	Sodium current	—	Yes	Parker and Miledi (1987)
Voltage-gated	Potassium current	—	Yes	Parker and Miledi (1988b)
Voltage-gated	Calcium current	—	Yes	Dascal et al. (1986)
Voltage-gated	Chloride current	—	Yes	Parker and Miledi (1988a)

[a] Response is present after follicle cells have been removed.
[b] cAMP second messenger system through induction or inhibition of adenylate cyclase.
[c] Vasoactive intestinal peptide.
[d] Inositol trisphosphate second messenger system through induction of phospholipase C.

receptor complex is composed of multiple subunits, like the nAChR. A functional nAChR is a pentameric complex with the stoichiometry $\alpha_2\beta\gamma\delta$. All four subunits are required for normal levels of electrophysiological function, although expression of the α subunit together with any two of the other subunits (β, γ, and δ) results in a low level of function (Kurosaki et al., 1987). Expression of any of the four subunits by itself results in no electrophysiological response to acetylcholine (Kurosaki et al., 1987). If the assay is ligand binding, however, expression of the α subunit by itself can be detected by bungarotoxin binding, because this subunit encodes the acetylcholine- and bungarotoxin-binding domains (Kurosaki et al., 1987). An added complication is that oocytes express low levels of mRNA encoding the α subunit of the nAChR, so that injection of the β, γ, and δ subunits can result in low levels of functional expression by combining with the endogenous α subunit (Buller and White, 1990a). Therefore the subunit composition of a particular channel or receptor is an important consideration for expression in oocytes.

A second explanation for the lack of an observed response is the absence of a second messenger system to which the particular receptor is coupled *in vivo*. For example, there are multiple subtypes of the serotonin receptor, including 5-HT_{1A}, 5-HT_{1B}, 5-HT_{1C}, 5-HT_{1D}, 5-HT_2, 5-HT_3, and 5-HT_4. The 5-HT_{1A}, 5-HT_{1B}, 5-HT_{1D}, and 5-HT_4 receptors activate or inhibit adenylate cyclase, the 5-HT_{1C} and 5-HT_2 receptors stimulate phospholipase C-catalyzed hydrolysis of phosphatidylinositol bisphosphate, and the 5-HT_3 receptor is a ligand-gated ion channel (reviewed by Julius, 1991). However, when rat brain mRNA was injected into *Xenopus* oocytes, a response to serotonin with the pharmacological profile of the 5-HT$_{1C}$ receptor was the only one observed (Lübbert *et al.*, 1987b). Evidently all of the components for the inositol trisphosphate second messenger cascade are present in the oocyte for coupling to the 5-HT$_{1C}$ receptor, but the same may not be the case for the adenylate cyclase second messenger system. The 5-HT$_{1A}$ receptor was cloned by low stringency hybridization using a β-adrenergic receptor cDNA clone as a probe (Kobilka *et al.*, 1987), but its expression has only been demonstrated in mammalian cells (Fargin *et al.*, 1989) and not in oocytes. Absence of the appropriate G protein or second messenger cascade can prevent a functional response following injection of mRNA for a ligand-gated receptor.

Finally, RNA transcribed from cDNA clones for some channels or receptors have thus far been nonfunctional in oocytes for unknown reasons. The dihydropyridine receptor is a multisubunit calcium channel consisting of α_1, α_2, β, γ, and δ subunits (Catterall, 1988). The complexes from cardiac and skeletal muscle are comparable, and in fact the amino acid sequences of the α_1 subunits from the two tissues are homologous (Tanabe *et al.*, 1987; Mikami *et al.*, 1989). This subunit is structurally very similar to the α subunit of the voltage-sensitive sodium channel (Mikami *et al.*, 1989), so it seemed likely that it might encode a functional channel when injected into oocytes. However, only the cardiac muscle α_1 subunit induced a functional channel in oocytes (Mikami *et al.*, 1989), whereas injection of the skeletal muscle α_1 subunit did not (Tanabe *et al.*, 1987). The skeletal muscle subunit is functional in other systems, since expression of the same α_1 subunit clone in skeletal muscle cells from mice with muscular dysgenesis (Tanabe *et al.*, 1988) and in L cells (Perez-Reyes *et al.*, 1989) resulted in functional calcium channels. The reason for the lack of function in oocytes is as yet unknown.

One final cautionary note which should be considered is that *Xenopus* oocytes may not provide the same processing machinery or accessory proteins that would be present in the tissue of origin. This could result in electrophysiological properties that are not identical to those normally seen *in vivo*. For example, the *Torpedo* nAChR synthesized in oocytes is missing the N-linked complex oligosaccharides that are present on the native receptor (Buller and White, 1990b). In this case the glycosylation difference does not appear to affect function. On the other hand, the *Electrophorus* sodium channel cannot be functionally expressed in oocytes because of a deficiency in processing (Thornhill and Levinson, 1987). A case in which electrophysiological differences have been observed is the rat brain sodium channel. This channel, a complex consisting of three subunits termed α, β_1, and β_2, is normally a rapidly inactivating channel *in vivo*. When only the α subunit was expressed in oocytes, however, the inactivation kinetics of the channel were markedly slower than those seen in neuronal cells (Auld *et al.*, 1988; Moorman *et al.*, 1990b; Trimmer *et al.*, 1989). This was restored to normal by the coinjection of low molecular weight RNA from rat brain [less than 2 kilobases (kb) in size], but the specific protein responsible for the restoration of normal inactivation has not yet been identified (Auld *et al.*, 1988). When the same α subunit clone was expressed in Chinese hamster ovary cells, the kinetics of inactivation were completely normal (Scheuer *et al.*, 1990). The explanation for this difference is not known, although it may be due to the absence of some low molecular weight subunits in oocytes or differential processing in mammalian cells compared to oocytes.

IV. Preparation of RNA for Injection

The best source of RNA for ion channel expression is dependent on the specific channels or receptors being studied. The most commonly used tissue source is probably the brain, because this organ expresses a large number of different types of receptors and channels (Gundersen *et al.*, 1983; 1984). The most important consideration, however, is the quality of the mRNA. Many brain mRNA species are extremely large, greater than 10 kb. In addition, some of the mRNA species are highly susceptible to degradation by heat and mechanical or chemical agents. Therefore, procedures should be followed which ensure isolation of intact, high molecular weight RNA. A variety of different procedures can be used to isolate RNA from tissues or cultured cells, but not all of them result in RNA that is functional in oocytes. Two procedures which have worked quite well are the lithium chloride–urea extraction procedure (Auffray and Rougeon, 1980) and the guanidine hydrochloride extraction procedure (Chirgwin *et al.*, 1979).

A. Lithium Chloride–Urea RNA Extraction Procedure

The lithium chloride–urea RNA extraction procedure has been quite efficacious in isolating functional RNA from brain tissue, which is neither very high in ribonuclease activity nor very fibrous. This technique does not work well on tissues that are high in ribonuclease (such as liver) or very difficult to homogenize (such as muscle), but it has the advantage of being fast and simple. In addition, the relatively small number of manipulations involved decreases the likelihood of RNA degradation by mechanical or chemical actions. The following procedure is specifically designed for the isolation of RNA from neonatal rat brains (animals under 2 weeks old), but it should be applicable to any brain tissue with adjustments in volume.

1. Dissect out the brains and immediately homogenize each one in 6 ml of LiCl–urea solution (see below) with 10 strokes of a Dounce homogenizer. Use approximately 100 ml of LiCl–urea solution per 10 to 15 brains.
2. Leave the homogenized tissue solution on ice at 4°C overnight.
3. Vortex the solution to decrease the viscosity and pour it into 50-ml polyallomer centrifuge tubes (autoclaved). Spin at 10,000 g at 4°C for 30 minutes.
4. Pour off the supernatants and dissolve the pellets in LSB (see below). Use approximately 20 ml LSB per 10 brains.
5. Extract the RNA with phenol–chloroform–isoamyl alcohol (25:24:1, by volume) until the interface is clean, generally about 3 times. Then extract 2 times with chloroform–isoamyl alcohol (24:1).
6. Add 0.1 volume of 3 M sodium acetate and 2.5 volumes of ethanol. Leave at $-20°C$ overnight in 35-ml Corex tubes (baked to inactivate ribonuclease).
7. Spin at 10,000 g for 20 minutes at 4°C. Resuspend the pellets in elution buffer for oligo (dT) column chromatography, or resuspend in water and precipitate with ethanol 2 more times for use as total RNA.

LiCl–Urea Solution
   3 M LiCl
   6 M urea
   10 mM sodium acetate buffer, pH 5.2
   0.1% sodium dodecyl sulfate (SDS)
   The SDS should not be added until immediately before use. Once added, it begins to precipitate out. This does not cause a problem as long as the solution is being stirred constantly.

LSB
   10 mM Tris-HCl, pH 7.4
   2 mM EDTA
   0.1% SDS

The amount of DNA contamination with this procedure is variable and seems to increase with increasing time the LiCl–urea homogenate is left at 4°C before centrifugation. As stated above, this procedure is not appropriate for tissues with high levels of ribonuclease because the urea does not inactivate ribonucleases completely. However, a modified version which uses guanidinium isothiocyanate as a denaturant does work well for the isolation of RNA from tissue culture cells (Cathala *et al.*, 1983).

B. Guanidine Hydrochloride RNA Extraction Procedure

For tissues that are more fibrous or higher in ribonuclease content, procedures involving more effective protein denaturation are necessary. Guanidine hydrochloride is a very effective denaturant, and it has been used to isolate active mRNA from muscle or heart. The procedure works well for soft or fibrous tissue, and results in little or no contamination by protein or genomic DNA. However, it is not efficient for the purification of low molecular weight RNA or for small quantities of RNA. In addition, closed circular DNA will copurify with the RNA to some extent. The procedure is outlined below.

1. Homogenize the tissue with either a Dounce homogenizer (soft tissue) or a Polytron (fibrous tissue) in about 8–10 ml of ice-cold 6 M guanidine solution (see below) per gram of tissue. Use about 50–60 ml per milliliter of packed tissue culture cells. If the tissue is fibrous, centrifuge at 10,000 g for 5 minutes at 4°C to remove any remaining tissue.
2. Add 0.6 volume of 95% ethanol and leave at $-20°C$ overnight.
3. Pellet the RNA by centrifuging at 10,000 g for 10 minutes at 4°C. Discard the supernatant.
4. Dissolve the pellet in ice-cold 7.5 M guanidine solution (about 1–2 ml per gram of tissue or 20 ml per milliliter of packed cells). Triturate the solution to aid dissolution but do not heat (RNA is very sensitive to heat in 7.5 M guanidine). If the material does not dissolve after a reasonable time, spin the tube briefly to remove the insoluble debris, which is not RNA.
5. Add 0.025 volume of 1 M acetic acid to lower the pH and facilitate RNA precipitation, then add 0.5 volume of 95% ethanol. Leave at $-20°C$ overnight or for at least 3 hours.
6. Pellet the RNA as in Step 3.
7. Suspend the pellet in sterile, autoclaved water, using about 0.5 ml per gram of tissue or 10 ml per milliliter of packed cells.
8. Extract with phenol–chloroform–isoamyl alcohol until the interface is clean. This generally requires about 3 extractions. Then extract 2 times with chloroform–isoamyl alcohol alone.
9. Add 1/10 volume of 2.5 M sodium acetate, pH 5.0, and 2.5 volumes of 95% ethanol. Leave at $-20°C$ for 1 hour to overnight.
10. Pellet the RNA as in Step 3. Rinse the pellet once in 70% ethanol, then dry it completely and resuspend the RNA in sterile, autoclaved water.
11. Precipitate the RNA with ethanol again as in Step 10.
12. Pellet the RNA. Rinse the pellet in 70% ethanol and suspend in sterile, autoclaved water at a concentration of 1–2 mg/ml. Store the RNA at $-70°C$.

6 M *Guanidine Solution*
  6 M guanidine hydrochloride
  200 mM sodium acetate, pH 5
  50 mM 2-mercaptoethanol
  Dissolve the guanidine hydrochloride in 150 ml of sodium acetate solution. Heat if necessary to dissolve all of the guanidine. Then add the mercaptoethanol and additional sodium acetate solution to a final volume of 200 ml.

7.5 M *Guanidine Solution*
  7.5 M guanidine hydrochloride
  25 mM sodium citrate, pH 7.0
  50 mM 2-mercaptoethanol
  Dissolve the guanidine hydrochloride in 400 ml of water. Heat if necessary to dissolve all of the guanidine. Then either add 1.76 g of sodium citrate and adjust the pH to 7.0, or add 12.5 ml of 1 M sodium citrate solution, pH 7.0. Finally add the mercaptoethanol and adjust the volume to 500 ml with water. Filter sterilize.

C. Transcription of RNA *in Vitro* from Cloned Channel Genes

Once a full-length cDNA clone for a neural receptor or ion channel has been isolated, it is easy to synthesize biologically active RNA from the clone. The coding region should be cloned following a bacteriophage promoter, either SP6, T7, or T3. All three bacteriophage polymerases work well to transcribe RNA uniquely from the appropriate promoter. However, the T7 polymerase has been cloned, and so is significantly less expensive than SP6 polymerase. For many receptors the coding region can simply be inserted after the promoter for successful transcription and translation. In some cases the resulting message will not translate well in oocytes, either because of poor stability or poor ribosome binding. In those cases the coding region can be cloned in a plasmid designed for expression in *Xenopus* oocytes, such as pSP64T (Krieg and Melton, 1984) or pBSTA (K. J. Kontis and A. L. Goldin, unpublished, 1990). These plasmids contain the *Xenopus* β-globin 5' and 3' untranslated mRNA regions, with either an SP6 promoter (pSP64T) or a T7 promoter (pBSTA) at the 5' end and poly(A) tail at the 3' end. This enhances both stability and translatability of some messages, and can result in very significant increases in level of expression.

The plasmid DNA containing the insert should be cut with a restriction enzyme which linearizes the plasmid past the 3' end of the coding region. Both pSP64T and pBSTA have polylinker regions after the poly(A) tail which can be used for this purpose. For linearization it is best to use a restriction enzyme that leaves either a 5' overhang or a blunt end, since a 3' overhang can function as a primer for synthesis in the wrong direction, making antisense RNA which has the potential to interfere with translation of the mRNA. After linearization the DNA should be extracted with phenol–chloroform–isoamyl alcohol, precipitated with ethanol, and resuspended in RNase-free water for use as a transcription template. The transcription reaction is carried out as described below to make RNA for use as either a probe (highly radioactive) or for translation (full-length).

1. The reactions are set up at room temperature as follows:

Reagent	Probe	Full-length
10 × SP6/T7 Buffer	2 μl	5 μl
10 × ATP, CTP, UTP (5 mM each)	2 μl	5 μl
10 × GTP (5 mM)	2 μl	1 μl
10 × GpppG (5 mM)	—	5 μl
1 M dithiothreitol (DTT)	0.2 μl	0.5 μl
RNasin (40 U/μl)	0.5 μl	1.25 μl
[^{32}P] CTP (100 μCi/10 μl)	10 μl	—
[^{32}P] CTP (diluted 1:10)	—	1 μl
Water	1.3	21–28
Linearized DNA (1 μg/μl)	1 μl	1–5 μl
SP6 or T7 polymerase (20 U/μl)	1 μl	2–5 μl
Total reaction volume	20 μl	50 μl

2. Incubate the reaction at 37°C for 1 hour for the probe reaction or for 2–3 hours for full-length transcripts. Additional enzyme can be added after 1 hour to the full-length reaction, which may increase the yield. Incubation at temperatures lower than 37°C (down to 4°C) may increase the yield of full-length transcript in some cases (Krieg, 1990).
3. Add RNasin to 1 unit/μl (0.5 μl or 1.25 μl). Then add 1 μl DNase (RNase-free, either 1 U/μl or 1 mg/ml), and digest at 37°C for 10 minutes to remove the template DNA.
4. Add water to 75 μl total volume. Add 1 μl of yeast carrier RNA (10 mg/ml) to the probe reaction (none to the full-length reaction), and extract once with phenol–chloroform–isoamyl alcohol.
5. Purify the transcripts on a spun column (Sambrook et al., 1989) and freeze the eluate. The spun column can be either Sephadex G-50 or Sephacryl S-400, which also removes low molecular weight RNA [less than about 400 base pairs (bp) size]. The full-length capped transcript should be precipitated with ethanol 1–2 times before injection.

10 × SP6/T7 Buffer
400 m*M* Tris-HCl, pH 7.5
60 m*M* MgCl$_2$
20 m*M* spermidine
100 m*M* NaCl

The unlabeled nucleotides should be dissolved as concentrated stocks (100 m*M*) in RNase-free 0.1 *M* Tris-HCl, pH 7.5, since the acidic nucleotides are unstable in water. The labeled CTP from Amersham (Arlington Heights, IL) and NEN (Boston, MA) at a specific activity of 400 Ci/mmol both work well. However, older labeled CTP can inhibit the reaction because of breakdown products, so it is a good idea to use fresh [^{32}P]CTP (within a few days for a probe reaction, or within 4 weeks for a full-length reaction). SP6 and T7 polymerase from New England Biolabs (Beverly, MA), U.S. Biochemicals (Cleveland, OH), Pharmacia (Piscataway, NJ), and Promega (Madison, WI) have all been satisfactory. RNasin is obtained from Promega Biotec and is used to inhibit RNase activity. GpppG can be obtained from New England Biolabs or from Pharmacia. It need not be used in the methylated form, as it will become methylated in either the cell or the oocyte, but the methylated cap analog from New England Biolabs is the same price as the nonmethylated form. It is important to set up the reaction at room temperature (not on ice), as it is possible for the DNA and spermidine to precipitate out at 4°C.

The specific activity of the probe should be approximately 6.7×10^8 dpm/ μg of RNA, assuming [^{32}P]CTP at a specific activity of 400 Ci/mmol. The yield of the full-length transcription reaction can be calculated by counting a 1-μl aliquot of the reaction mix before and after the spun column. Approximately 0.33 μg of RNA is synthesized for each 1% incorporation of the label, assuming a 50-μl reaction volume. If it is desirable not to have any radioactivity in the full-length transcript, the reaction can be run without [^{32}P]CTP. In this case the final yield of product can be quantitated by absorbance at 260 nm (be sure to use RNase-free cuvettes) or by comparing the ethidium bromide fluorescence of spots of diluted RNA with spots of diluted RNA standards (Sambrook *et al.*, 1989).

V. Preparation and Injection of Oocytes

The procedures for preparation of oocytes and injection with mRNA are no different for the study of ion channels than for other proteins and are well described elsewhere in this volume (see Chapter 9 and Appendix B). The only unique consideration is whether the follicle cells should be removed from the oocytes. Oocytes can be injected and voltage clamped with follicle cells around them. This is technically more difficult, because the follicle cells are harder to pierce with the needle and it is time-consuming to separate out individual oocytes. However, as stated earlier, some electrophysiological responses measured in oocytes either depend on the presence of follicle cells or are actually occurring in the follicle cells (Table I). If any of these responses are desired, such as the ability to use forskolin to increase cAMP levels, then the follicle cells must be left attached to the oocyte. Otherwise it is usually best to use defolliculated oocytes.

VI. Biochemical Analysis of Expression

Channels and receptors expressed in oocytes can be studied by standard biochemical techniques, such as immunoprecipitation or ligand binding assays. The major consideration before using these techniques is that they are significantly less sensitive than the electrophysiological techniques discussed below. Using whole cell voltage clamping, it is possible to detect as few as 10^5 channel molecules in a single oocyte (less than 10^{-18} mol). On the other hand, biochemical techniques are generally reliable down to the level of 10^{-12} mol, although this depends strongly on the specific activity of the reagents being used and the number of oocytes examined. Therefore, it is necessary to express the channels at a much higher level for biochemical analysis than is required for electrophysiological recording.

The other consideration is that solubilization of oocytes results in isolation of both cytoplasmic and membrane proteins. Ion channels and receptors are all located in the membrane when they are functional, but there is frequently a large intracellular pool of molecules that either have not or cannot be inserted into the membrane (Schmidt et al., 1985). Electrophysiological recording examines only the molecules that have been inserted into the membrane in a functional configuration, whereas immunoprecipitation of total oocyte proteins examines both membrane and cytoplasmic proteins. Since the properties of the cytoplasmic proteins may be different from the functional proteins inserted into the membrane, it may not be possible to correlate the results from the two types of analysis. One way to avoid this problem is to use membrane preparations for solubilization (Colman, 1984), but this technique is difficult and will not remove proteins bound to internal membranes. Ligand binding assays, on the other hand, can easily be used to measure either the membrane-bound or total cellular protein.

A. Immunoprecipitation of Channel Proteins

The procedures for lysis and immunoprecipitation of oocyte proteins do not differ markedly from those used for other cells, with the one exception that the large quantity of yolk protein can interfere and should be removed. One procedure for lysis of oocyte proteins and immunoprecipitation is described below.

1. Incubate the oocytes in medium containing [^{35}S]methionine at a concentration of 1 mCi/ml (specific activity >1,000 Ci/mmol) for 1–2 days.
2. Solubilize 10 oocytes in 250 μl solubilization buffer (see below) at 4°C. Homogenize with 10 strokes of an Eppendorf tube Dounce homogenizer.
3. Remove the yolk proteins in either of the following ways:
   a. Centrifuge the homogenate in an Eppendorf tube for 15 minutes at 4°C. This is the simplest procedure, but some proteins may also precipitate with the yolk proteins.
   b. Extract the homogenate 2 times with Freon. Centrifuge in an Eppendorf tube for 10 minutes each time to separate the phases.
4. Prepare the sample in either of the following ways, depending on whether the antibody recognizes native or denatured protein:
   a. If the antibody recognizes denatured proteins, add SDS to 0.5% and boil for 8 minutes, then cool at room temperature. Add an equal volume of 2× IP buffer (see below) with bovine serum albumin (BSA).
   b. If the antibody does not recognize denatured proteins, add an equal volume of 2× IP buffer with BSA.
5. Incubate on a rotator at 4°C for 30 minutes, then centrifuge in an Eppendorf microcentrifuge for 15 minutes and save the supernatant.
6. Add the antibody to the sample (the amount depends on the titer of the antiserum). Incubate on a rotator at 4°C for 1–2 hours.
7. Add 50 μl of 40% protein A–Sepharose beads in 1× IP buffer. Continue incubation on a rotator at 4°C for 1 hour.
8. Centrifuge in a microcentrifuge for 30 seconds to pellet the beads. Discard the supernatant (it may be helpful to save a portion to examine the composition of proteins which have not been precipitated by the antiserum).
9. Wash the beads in 1 ml of 1× IP buffer without BSA. Centrifuge in a microcentrifuge and save the pellet.
10. Repeat Step 9 for two additional times (a total of 3 washes).
11. After the last centrifugation remove all of the supernatant and resuspend the pellet in 20 μl sample preparation buffer for SDS-polyacryl-

amide gel electrophoresis. The samples can be stored at $-20°C$ and should be boiled for 3 minutes before loading.

Solubilization Buffer
  65 mM Tris-HCl, pH 7.5
  1 mM sodium vanadate
  1 mM phenylmethylsulfonyl fluoride (PMSF)
  10 mM EGTA
  1 U/ml aprotinin
  1 µg/ml leupeptin

$2 \times$ *IP Buffer*
  130 mM Tris-HCl, pH 7.5
  150 mM NaCl
  5% Triton X-100
  100 mM NaH$_2$PO$_4$
  1 mM sodium vanadate
  1 mM PMSF
  10 mM EGTA
  1 U/ml aprotinin
  1 µg/ml leupeptin
  Add 10 mg/ml BSA for all steps other than the final washes.

B. Ligand Binding Assay

Ligand binding assays are again not terribly different in oocytes than in mammalian cells. The specific buffers depend on the conditions necessary for interaction of the ligand and the receptor, but the most important parameter is the specific activity of the labeled ligand. The following procedures are representative assays for determining ^{125}I-labeled α-bungarotoxin binding to the nAChR (Buller and White, 1990b). The first procedure measures surface labeling, and the second procedure measures total cellular labeling.

1. Surface Receptors

1. After the receptors have been expressed by incubation of the oocytes for 1–2 days, incubate the oocytes for 15 minutes at room temperature in standard incubation medium containing 1 mg/ml BSA and 0.1 mg/ml cytochrome *c*.
2. Transfer the oocytes to the same solution containing 1 nM ^{125}I-labeled α-bungarotoxin (17 µCi/µg) and continue incubation with shaking at room temperature for 90 minutes.
3. Transfer the oocytes to 50 ml cold incubation solution with BSA and

cytochrome c and incubate for about 5 minutes. Repeat this wash again to remove unbound ^{125}I-labeled α-bungarotoxin.
4. Remove individual oocytes to a γ counter tube and determine the amount of ^{125}I bound. Nonspecific binding is determined by using uninjected oocytes as controls. If the levels of expression are too low, multiple oocytes can be counted in each tube.

2. Total Receptors

1. Homogenize groups of 5–10 oocytes in 0.5 ml binding buffer (see below) and incubate at room temperature for 10 minutes.
2. Centrifuge for 10 minutes to remove the yolk proteins and transfer the supernatant to a new tube.
3. Add ^{125}I-labeled α-bungarotoxin to a final concentration of 1 nM and continue incubation at room temperature for 90 minutes.
4. Filter through Whatman DE81 filter paper that has been preincubated with binding buffer containing 2 nM unlabeled α-bungarotoxin. Wash the filters with 3 washes of 10 ml binding buffer, and count in a γ counter.

Binding Buffer
50 mM sodium phosphate buffer, pH 7.2
1% Triton X-100
1 mM EGTA
1 mM EDTA
1 mM PMSF
1 mg/ml BSA
0.1 mg/ml cytochrome c
0.1 U/ml aprotinin

VII. Electrophysiological Analysis

The primary advantage of using *Xenopus* oocytes for the expression of channels and receptors is the ability to perform detailed electrophysiological recordings using an *in vivo* system. Essentially all of the standard electrophysiological techniques can be performed on oocytes, including whole-cell recording and patch-clamp recording of both macroscopic and single channel currents. It is beyond the scope of this chapter to describe detailed procedures for voltage and patch clamping of oocytes, but these procedures are described in detail in the volume in *Methods in Enzymology* on ion channels (Rudy and Iverson, 1991). The general considerations involved in the various techniques are presented here, along with a description of the representative kinds of results that can be obtained. Particular attention will be given to the differences between electrophysiological recording from oocytes compared to mammalian cells.

A. Whole-Cell Voltage Clamping

Whole-cell voltage clamping of oocytes involves using two electrodes inserted into the oocyte, rather than using one electrode to make a patch on the surface followed by rupturing the membrane as is done in mammalian cells. The large size of the oocyte (about 1 mm in diameter and 0.5–1 μl in volume for stage V oocytes) makes this feasible, and is both the major advantage and disadvantage of the system. The advantage is that it is possible to insert multiple electrodes and injection needles into the same oocyte. Therefore, modulators of channel function can be injected inside the cell while recording, so that a rapid and direct response to an intracellular signal can be observed. The disadvantage is that the large size results in an extremely large membrane capacitance (about 150–200 nF), which causes a slow clamp settling time following voltage shifts. This makes it difficult to obtain any data during the first 1–2 msec of a depolarization, the time during which rapidly activating voltage-sensitive channels such as the sodium channel open. The large capacitance is not a serious problem in examining slow responses or ligand-gated responses in the absence of voltage shifts.

Figure 1A shows representative current traces recorded from an oocyte injected with *in vitro* synthesized RNA for the α subunit of the voltage-sensitive sodium channel. The oocyte was depolarized from a holding potential of -100 mV to potentials ranging from -30 to $+30$ mV in 10-mV steps. The depolarization was initiated at 4 msec, and the capacitive settling time for the potential of the oocyte to reach the new potential is about 1 msec in this experiment. The oocyte was injected with 50 pg of RNA into the cytoplasm, which resulted in a peak inward current of about 1500 nA. Currents ranging from about 10 nA to over 100 μA can be obtained by injecting different amounts of RNA, but it is difficult to clamp the membrane potential of the oocyte accurately if the currents are larger than 5–10 μA.

B. Macropatch Recording

One method of circumventing the problems caused by the large size of the oocyte is to depolarize only a small fraction of the membrane. This procedure was developed and used extensively by Stühmer *et al.* (1987). An electrode with a relatively large opening (about 10 μm in diameter) is used to make a gigaohm seal with the oocyte membrane, and then only the channels in that

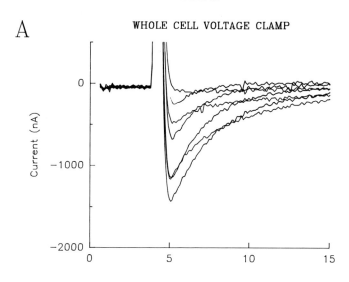
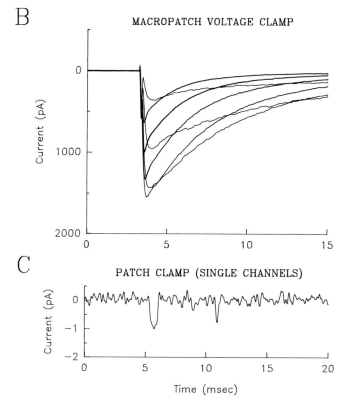

portion of the oocyte membrane are studied. Recording can be performed in the cell-attached mode or, after excising the patch, in the inside-out mode. In the cell-attached mode it is important to know the internal potential of the oocyte, which can be accomplished in either of two ways. The first method is to insert two electrodes into the oocyte and use a voltage clamp to maintain a fixed holding potential for the whole cell. This has the advantage that the potential of the patch need not be adjusted, and channels throughout the remainder of the oocyte will be held at the desired potential so that slow inactivation of voltage-dependent channels will not occur. The disadvantage is that removal of the patch electrode after recording usually leaves a hole which results in a significant increase in membrane leak, thus causing high holding currents in the whole-cell voltage clamp that can kill the oocyte. This can be avoided by turning off the clamp when removing the patch electrode and allowing time for the hole in the membrane to seal before clamping again, but this negates the major advantage of clamping the oocyte membrane potential. It is therefore generally easier to use the second method, which is to insert an electrode into the oocyte and constantly record the intracellular potential. The holding potential of the patch electrode is then adjusted to take into account the membrane potential of the oocyte so that the difference across the patch remains constant. An important consideration with using the macropatch technique is that there is no longer access to the external surface of the portion of the membrane being studied. Therefore, macropatch

---

FIG. 1. Sodium currents expressed in *Xenopus* oocytes. (A) Currents recorded with a two-microelectrode whole-cell voltage clamp. Oocytes were injected with 50 pg of *in vitro* transcribed RNA encoding the rat IIA sodium channel α subunit and incubated for 2 days at 20°C. The incubation and recording solution consisted of 96 mM NaCl, 2 mM KCl, 1.8 mM CaCl$_2$, 1 mM MgCl$_2$, 5 mM HEPES, pH 7.5. The electrodes had resistances of 0.5–1 MOhm, and a virtual ground circuit was used for recording. The oocyte was maintained at a holding potential of -100 mV and depolarized from -90 to $+50$ mV in 10-mV steps. The data have been filtered at 3 kHz. Only the traces from -30 to $+30$ mV are shown. (B) Currents recorded with a cell-attached macropatch voltage clamp. Oocytes were injected with 500 pg of *in vitro* transcribed RNA and incubated as described in (A). Currents were recorded using a macropatch electrode with a tip diameter of approximately 10 μm in the cell-attached configuration. The resting potential of the oocyte was -60 mV, as determined by an intracellular voltage electrode, and the patch electrode holding potential was $+40$ mV (-100 mV effective holding potential). The oocyte was depolarized from -90 to $+50$ mV in 10-mV steps. The data have been filtered at 3 kHz. Only the traces from -30 to $+30$ mV are shown. (C) Currents recorded with an excised outside-out patch electrode (single channels). Oocytes were injected and incubated as in (B). Currents were recorded using a patch electrode with a tip diameter of approximately 0.5 μm in diameter in the outside-out configuration. The pipette solution consisted of 90 mM KCl, 10 mM NaCl, 10 mM EGTA, 10 mM HEPES, pH 7.4 with KOH. The patch was maintained at a holding potential of -100 mV and depolarized to -30 mV between 4 and 16 msec. The data have been filtered at 2 kHz.

recording is not a suitable system for studying the interactions of toxins which directly bind to a channel from the external surface. On the other hand, it is an excellent system for studying modulation through second messenger systems.

Figure 1B shows sodium currents obtained from an oocyte injected with sodium channel RNA and recorded using a cell-attached macropatch configuration. The resting potential of the oocyte as measured with an intracellular electrode was -60 mV, so the patch electrode potential was set to $+40$ mV. This resulted in a -100 mV holding potential difference across the patch, comparable to the holding potential used in the experiment shown in Fig. 1A. As in that experiment, the patch was depolarized from -100 mV to potentials ranging from -30 to $+30$ mV in 10-mV steps. Electrode capacitance has been electronically neutralized, so there is minimal loss of resolution owing to capacitive settling following the voltage shift. The peak inward current is about 1000-fold less than in Fig. 1A (1500 pA), but again a wide range of current densities can be recorded.

C. Single Channel Analysis

The most detailed electrophysiological analysis of channel function in oocytes is to examine individual channel molecules through the use of single channel recording. A complete discussion of the approaches used for these kinds of analysis can be found in *Single-Channel Recording* by Sakmann and Neher (1983). For recording single channels, electrodes with a small tip diameter of about 1 μm are used to make a gigaohm seal with the oocyte membrane. To make the seal, the vitelline membrane must first be removed from the oocyte. This is accomplished by placing the oocyte in a hypertonic solution (normal incubation solution containing 200 mM NaCl), which causes the oocyte to shrink and the vitelline membrane to become exposed. The membrane is then carefully pulled off using a pair of fine forceps. Once the oocyte is devitellinized it must not be exposed to air, as it will explode. In addition, the devitellinized oocyte will attach tightly to plastic, so it should be transferred immediately to a dish which will be the recording chamber, or maintained in a dish to which it will not attach. The devitellinized oocyte will not attach to either plexiglass or agarose-coated petri dishes.

Oocyte patches can be analyzed in either the cell-attached or excised configuration, as in mammalian cells. The disadvantages of the cell-attached mode are that the intracellular potential must be determined and that the activity of endogenous stretch-activated channels may interfere with the signal of interest (Yang and Sachs, 1990). The patches can be excised in either of two configurations, inside-out (which allows access from the bath to the intracellular surface of the membrane) or outside-out (which allows access from the bath to the extracellular surface of the membrane). An inside-out patch is excised simply by moving the electrode away from the oocyte after a tight seal is obtained. An outside-out patch is more difficult, in that first the membrane must be ruptured, after which the electrode is moved away so that the membrane will reseal around it in the opposite configuration. Rupturing the oocyte membrane is more difficult than rupturing a mammalian cell membrane. The most effective technique is positive pressure, and this works well only with relatively small electrode tip openings. Suction does not work well. Once the patch is excised, recording single channels is comparable to recording from mammalian cell patches.

Figure 1C shows single channel records from an oocyte injected with sodium channel RNA and recorded in the outside-out configuration. The patch was held at -100 mV and depolarized to -30 mV between 4 and 16 msec. Two openings can be seen, with a maximal current of about 1 pA. For analysis a large number of records would be obtained and statistical methods would be used to determine the mean open time, single channel conductance, latency to first opening, and other properties of the channel. This type of information would then be used to construct kinetic models for the various states involved in channel gating (Sakmann and Neher, 1983).

VIII. Conclusions

In summary, the *Xenopus* oocyte system has become a widely used approach for the study of ion channels and receptors. It has the advantage that channels can be rapidly expressed and analyzed both biochemically and electrophysiologically in an *in vivo* situation. Most standard electrophysiological recording techniques can be used, so that whole cell currents and single channels can be examined. The system can be used quite effectively as an assay for the functional cloning of channels that have only been identified by their electrophysiological properties. Once cDNA clones have been isolated, oocytes are an excellent system for correlating structure with function using a combination of molecular biological and electrophysiological techniques.

Acknowledgments

The author is a Lucille P. Markey Scholar, and work in his laboratory is supported by grants from the U.S. National Institutes of Health (NS-26729), the Lucille P. Markey Charitable Trust, and the March of Dimes Basil O'Connor Starter Scholar Program.

References

Auffray, C., and Rougeon, F. (1980). *Eur. J. Biochem.* **107,** 303–314.
Auld, V. J., Goldin, A. L., Krafte, D. S., Marshall, J., Dunn, J. M., Catterall, W. A., Lester, H. A., Davidson, N., and Dunn, R. J. (1988). *Neuron* **1,** 449–461.
Auld, V. J., Goldin, A. L., Krafte, D. S., Catterall, W. A., Lester, H. A., Davidson, N., and Dunn, R. J. (1990). *Proc. Natl. Acad. Sci. U.S.A.* **87,** 323–327.
Barnard, E. A., Miledi, R., and Sumikawa, K. (1982). *Proc. R. Soc. London* **215,** 241–246.
Buller, A. L., and White, M. M. (1990a). *Mol. Pharmacol.* **37,** 423–428.
Buller, A. L., and White, M. M. (1990b). *J. Membr. Biol.* **115,** 179–189.
Cathala, G., Savouret, J.-F., Mendez, B., West, B. L., Karin, M., Martial, J. A., and Baxter, J. D. (1983). *DNA* **2,** 329–335.
Catterall, W. A. (1988). *Science* **242,** 50–61.
Chirgwin, J. M., Przybyla, A. E., MacDonald, R. J., and Rutter, W. J. (1979). *Biochemistry* **18,** 5294–5299.
Colman, A. (1984). *In* "Transcription and Translation: A Practical Approach" (B. D. Hames and S. J. Higgins, eds.), pp. 271–302. IRL Press, Oxford.
Dascal, N. (1987). *CRC Crit. Rev. Biochem.* **22,** 317–387.
Dascal, N., Snutch, T. P., Lübbert, H., Davidson, N., and Lester, H. A. (1986). *Science* **231,** 1147–1150.
Fargin, A., Raymond, J. R., Regan, J. W., Cotecchia, S., Lefkowitz, R. J., and Caron, M. G. (1989). *J. Biol. Chem.* **264,** 14848–14852.
Goldin, A. L., Snutch, T., Lübbert, H., Dowsett, A., Marshall, J., Auld, V., Downey, W., Fritz, L. C., Lester, H. A., Dunn, R., Catterall, W. A., and Davidson, N. (1986). *Proc. Natl. Acad. Sci. U.S.A.* **83,** 7503–7507.
Gundersen, C. B., Miledi, R., and Parker, I. (1983). *Proc. R. Soc. London, Ser. B* **220,** 131–140.
Gundersen, C. B., Miledi, R., and Parker, I. (1984). *Nature (London)* **308,** 421–424.
Hartmann, H. A., Kirsch, G. E., Drewe, J. A., Taglialatela, M., Joho, R. H., and Brown, A. M. (1991). *Science* **251,** 942–944.
Hirono, C., Yamagishi, S., Ohara, R., Hisanaga, Y., Nakayama, T., and Sugiyama, H. (1985). *Brain Res.* **359,** 57–64.
Hoshi, T., Zagotta, W. N., and Aldrich, R. W. (1990). *Science* **250,** 533–538.
Julius, D. (1991). *Annu. Rev. Neurosci.* **14,** 335–360.
Julius, D., MacDermott, A. B., Axel, R., and Jessell, T. M. (1988). *Science* **241,** 558–564.
Kobilka, B. K., Frielle, T., Collins, S., Yang-Feng, T., Kobilka, T. S., Francke, U., Lefkowitz, R. J., and Caron, M. G. (1987). *Nature (London)* **329,** 75–79.
Kobilka, B. K., Kobilka, T. S., Daniel, K., Regan, J. W., Caron, M. G., and Lefkowitz, R. J. (1988). *Science* **240,** 1310–1316.
Krieg, P. A. (1990). *Nucleic Acids Res.* **18,** 6463.
Krieg, P. A., and Melton, D. A. (1984). *Nucleic Acids Res.* **12,** 7057–7070.
Kurosaki, T., Fukuda, K., Konno, T., Mori, Y., Tanaka, K., Mishina, M., and Numa, S. (1987). *FEBS Lett.* **214,** 253–258.
Kusano, K., Miledi, R., and Stinnakre, J. (1982). *J. Physiol. (London)* **328,** 143–170.
Lotan, I., Dascal, N., Cohen, S., and Lass, Y. (1982). Nature *(London)* **298,** 572–574.
Lübbert, H., Hoffman, B. J., Snutch, T. P., VanDyke, T., Levine, A. J., Hartig, P. R., Lester, H. A., and Davidson, N. (1987a). *Proc. Natl. Acad. Sci. U.S.A.* **84,** 4332–4336.
Lübbert, H., Snutch, T. P., Dascal, N., Lester, H. A., and Davidson, N. (1987b). *J. Neurosci.* **7,** 1159–1165.
MacKinnon, R., and Miller, C. (1989). *Science* **245,** 1382–1385.
MacKinnon, R., and Yellen, G. (1990). *Science* **250,** 276–279.

MacKinnon, R., Heginbotham, L., and Abramson, T. (1990). *Neuron* **5**, 767–771.
Methfessel, C., Witzemann, V., Takahasi, T., Mishina, M., Numa, S., and Sakmann, B. (1986). *Pfluegers Arch* **407**, 577–588.
Mikami, A., Imoto, K., Tanabe, T., Niidome, T., Mori, Y., Takeshima, H., Narumiya, S., and Numa, S. (1989). *Nature (London)* **340**, 230–233.
Miledi, R., and Parker, I. (1984). *J. Physiol. (London)* **357**, 173–183.
Miledi, R., and Woodward, R. M. (1989). *J. Physiol. (London)* **416**, 601–621.
Miledi, R., Parker, I., and Sumikawa, K. (1982). *EMBO J.* **1**, 1307–1312.
Moorman, J. R., Kirsch, G. E., Brown, A. M., and Joho, R. H. (1990a). *Science* **250**, 688–691.
Moorman, J. R., Kirsch, G. E., VanDongen, A. M. J., Joho, R. H., and Brown, A. M. (1990b). *Neuron* **4**, 243–252.
Noda, M., Ikeda, T., Suzuki, H., Takeshima, H., Takahashi, T., Kuno, M., and Numa, S. (1986). *Nature (London)* **322**, 826–828.
Papazian, D. M., Timpe, L. C., Jan, Y. N., and Jan, L. Y. (1991). *Nature (London)* **349**, 305–310.
Parker, I., and Miledi, R. (1987). *Proc. R. Soc. London* **232**, 289–296.
Parker, I., and Miledi, R. (1988a). *Proc. R. Soc. London* **233**, 191–199.
Parker, I., and Miledi, R. (1988b). *Proc. R. Soc. London* **234**, 45–53.
Perez-Reyes, E., Kim, H. S., Lacerda, A. E., Horne, W., Wei, X., Rampe, D., Campbell, K. P., Brown, A. M., and Birnbaumer, L. (1989). *Nature (London)* **340**, 233–236.
Rudy, B., and Iverson, L. E., eds. (1992). *Methods Enzymol.* Academic Press, San Diego.
Sakmann, B., and Neher, E. (1983). "Single-Channel Recording." Plenum, New York.
Sambrook, J., Fritsch, E. F., and Maniatis, T. (1989). "Molecular Cloning: A Laboratory Manual," 2nd Ed. Cold Spring Harbor Laboratory, Cold Spring Harbor, New York.
Scheuer, T., Auld, V. J., Boyd, S., Offord, J., Dunn, R., and Catterall, W. A. (1990). *Science* **247**, 854–858.
Schmidt, J. W., Rossie, S., and Catterall, W. A. (1985). *Proc. Natl. Acad. Sci. U.S.A.* **82**, 4847–4851.
Snutch, T. P. (1988). *Trends Neurosci.* **11**, 250–256.
Stühmer, W., Methfessel, C., Sakmann, B., Noda, M., and Numa, S. (1987). *Eur. Biophys. J.* **14**, 131–138.
Stühmer, W., Conti, F., Suzuki, H., Wang, X., Noda, M., Yahagi, N., Kubo, H., and Numa, S. (1989). *Nature (London)* **339**, 597–603.
Sumikawa, K., Parker, I., and Miledi, R. (1984a). *Proc. Natl. Acad. Sci. U.S.A.* **81**, 7994–7998.
Sumikawa, K., Parker, I., and Miledi, R. (1984b). *Proc. R. Soc. London* **223**, 255–260.
Takumi, T., Ohkubo, H., and Nakanishi, S. (1988). *Science* **242**, 1042–1045.
Tanabe, T., Takeshima, H., Mikami, A., Flockerzi, V., Takahashi, H., Kangawa, K., Kojima, M., Matsuo, H., Hirose, T., and Numa, S. (1987). *Nature (London)* **328**, 313–318.
Tanabe, T., Beam, K. G., Powell, J. A., and Numa, S. (1988). *Nature (London)* **336**, 134–139.
Thornhill, W. B., and Levinson, S. R. (1987). *Biochemistry* **26**, 4381–4388.
Trimmer, J. S., Cooperman, S. S., Tomiko, S. A., Zhou, J., Crean, S. M., Boyle, M. B., Kallen, R. G., Sheng, Z., Barchi, R. L., Sigworth, F. J., Goodman, R. H., Agnew, W. S., and Mandel, G. (1989). *Neuron* **3**, 33–49.
Vassilev, P. M., Scheuer, T., and Catterall, W. A., (1988). *Science* **241**, 1658–1661.
Vassilev, P. M., Scheuer, T., and Catterall, W. A. (1989). *Proc. Natl. Acad. Sci. U.S.A.* **86**, 8147–8151.
Woodward, R. M., and Miledi, R. (1987a). *Proc. Natl. Acad. Sci. U.S.A.* **84**, 4135–4139.
Woodward, R. M., and Miledi, R. (1987b). *Proc. R. Soc. London* **231**, 489–497.
Yang, X.-C., and Sachs, F. (1990). *J. Physiol. (London)* **431**, 103–122.
Yellen, G., Jurman, M. E., Abramson, T., and MacKinnon, R. (1991). *Science* **251**, 939–942.
Yool, A. J., and Schwarz, T. L. (1991). *Nature (London)* **349**, 700–704.
Zagotta, W. N., Hoshi, T., and Aldrich, R. W. (1990). *Science* **250**, 568–571.

Chapter 26

Tissue Culture of Xenopus Neurons and Muscle Cells as a Model for Studying Synaptic Induction

H. BENJAMIN PENG, LAUREN P. BAKER, AND QIMING CHEN

Department of Cell Biology and Anatomy
University of North Carolina
Chapel Hill, North Carolina 27599

I. Introduction
II. Preparation of Neuron and Myotomal Muscle Cultures
   A. Materials and Solutions
   B. Preparation of Mixed Nerve–Muscle Culture
   C. Preparation of Separate Nerve or Muscle Cultures
   D. Preparation of Enriched Nerve–Muscle Cocultures
   E. Preparation of Neural Tube Explants
III. Characteristics of *Xenopus* Cultures and Development of Neuromuscular Junctions *in Vitro*
   A. Development of Neuromuscular Junctions
   B. Formatiom of Synaptic Specializations without Nerve
   C. Comparison with Neuromuscular Junction Development *in Vivo*
IV. Preparation of Myotube Cultures
V. Discussion
   References

I. Introduction

Synaptogenesis is a central event in the development of the nervous system. The most convenient system to study this process is the development of the neuromuscular junction (NMJ). During the formation of the NMJ, processes from motoneurons interact with muscle cells and elicit the development of synaptic specializations. This process of nerve–muscle interaction bears similarities to the embryonic induction in that the cellular interaction at the cell–cell contact signals the local differentation of the pre- and postsynaptic cells.

NMJ development can be conveniently studied in tissue culture. This was pioneered by Ross Harrison (1907), who explanted the neural tube and the myotomes from frog embryos on coverglass and observed contractions of the muscle fibers within a few days, thus witnessing the formation of the NMJ *in vitro*. Since that time, numerous investigators have placed embryonic neurons and muscle cells from a variety of animals, including amphibian, avian, and mammalian species, into tissue culture to study the innervation process (reviewed by Bloch and Pumplin, 1988). The amphibian culture is one of the easiest systems to grow and requires minimum maintenance (Cohen, 1980). It also responds to a variety of synaptogenic stimuli (Peng, 1987). This versatility is unparalleled by other culture systems. The culture system involving the neural tube and the myotomal muscle cell from the *Xenopus* embryo is described in this chapter.

II. Preparation of Neuron and Myotomal Muscle Cultures

A. Materials and Solutions

1. Culture Substratum

Cover Glass
  Gold Seal #1 cover glass (Clay Adams, Becton Dickinson, Oxnard, CA) is heat sterilized in an oven at 150°C for 1 hour.
Tissue Culture Dishes
  Polystyrene tissue culture dishes of a vaiety of brands (e.g., Falcon, Corning, Costar, Nunc) provide an adequate substratum for muscle cell attachment. For electron microscopy, cells can be cultured in Lux Permanox tissue culture dishes (Nunc, Naperville, IL). These dishes are more resistant to solvents used in dehydration and embedding procedures than conventional tissue culture plastics. After embedding, the cell block can be easily removed from the Permanox dishes.

The substrata described above work well for muscle cells. Neurons generally do not attach well to these substrata. However, if neurons are plated together with muscle cells or are plated onto a lawn of muscle cells, these substrata offer adequate support for growth. If neurons are to be cultured alone, substrata coated with collagen or poly-D-lysine should be used. Rat tail collagen (Type I) is available from several sources (e.g., Upstate Biotechnology, Lake Placid, NY). It is spread onto the cover glass or culture dish to form a thin layer and cured with ammonia vapor. Alternatively, the substratum is treated with a 100 µg/ml solution of poly-D-lysine (Sigma Chemical Co., St. Louis, MO) for 1 hour and then rinsed with distilled water. The coated substratum can be sterilized under UV light.

2. SOLUTIONS

Holtfreter's Solution (According to Hamburger, 1960)
  NaCl, 60.00 mM
  KCl, 0.60 mM
  $CaCl_2$, 0.90 mM
  $NaHCO_3$, 0.20 mM
  A 10% solution is used to raise *Xenopus* embryos. For sterilizing embryos, a 10% Holtfreter's solution plus thimerosal (1:10,000 dilution; Sigma) is used.

Steinberg's Solution (Modified from Hamburger, 1960)
  NaCl, 60.00 mM
  KCl, 0.67 mM
  $Ca(NO_3)_2$, 0.34 mM
  $MgSO_4$, 0.83 mM
  N-2-Hydroxyethylpiperazine-N'-2-Ethanesulfonic acid (HEPES), 10.00 mM pH adjusted to 7.4

Calcium, Magnesium-Free (CMF) Steinberg's Solution
  NaCl, 60.00 mM
  KCl, 0.67 mM
  EDTA, 0.40 mM
  HEPES, 10.00 mM
  pH adjusted to 7.4

Culture Medium
  Steinberg's solution, 89%
  Leibovitz L-15 medium, 10%
  Fetal bovine serum, 1%
  Gentamicin sulfate, 100 µg/ml
  Leibovitz L-15 medium (Cat. No. 430-1300) is obtained from GIBCO (Grand Island, NY). Gentamicin sulfate is purchased from Schering Corp. (Kenilworth, NJ; GRS-garamycin reagent solution) or United States Biochemical (Cleveland, OH; Cat. No. 16051).

Frog Ringer's Solution
  NaCl, 115.00 mM
  KCl, 2.50 mM
  CaCl$_2$, 1.80 mM
  HEPES, 5.00 mM
  pH adjusted to 7.4

3. Equipment

Dissection Tools
  Dissection microscope
  Watchmaker's forceps (Dumont No. 5)
  Microdissecting (iridectomy) scissors
  Disposable hypodermic needles, gauge 25 and 30 (to be used as microknives)
  Disposable 1-ml syringes (to be used as handles)

Glassware
  Pasteur pipettes, cotton plugged, 5-3/4 and 9 inches (Cat. No. 1278-10005 and 1278-10009, from Bellco Glass, Vineland, NJ). Pipettes are autoclaved in canisters prior to use.
  Pipettes for transferring embryos: Cut tips off 5-3/4 inch sterilized Pasteur pipettes with a diamond marker and fire polish the cut ends.

Specialized Culture Chambers
  Anderson–Cohen chamber (see Anderson *et al.*, 1977)
  Sykes–Moore chamber (Cat. No. 1943, Bellco Glass)
  These chambers sandwich cultures between two pieces of cover glass and offer good optics.

Plastics
  Embryo transfer net: Cut the bottom off a plastic beaker. Fix a peice of nylon mesh screen to the cut opening.
  Disposable plastic petri dishes (60 or 35 mm)
  Disposable sterile centrifuge tubes

Other Equipment
  Ultraviolet light sterilization chamber (from Carolina Biological Supply, Burlington, NC)
  Tabletop centrifuge (e.g., IEC HN-SII)
  Culture incubator: *Xenopus* cultures can be maintained at a wide range of temperatures between 12–24°C. Lower temperature lengthens the lifetime of the cultures. An under-the-counter refrigerator can be converted to a low-temperature incubator by replacing the existing thermostat with one which is sensitive to temperature change above 10°C. A circulating

fan should also be installed to maintain the uniformity in chamber temperature.
Inverted microscope for microscopy

B. Preparation of Mixed Nerve–Muscle Culture

1. Set up a series of petri dishes containing 70% ethanol, 10% Holtfreter's solution, 10% Holtfreter's solution plus thimerosal, Steinberg's solution, and CMF Steinberg's solution.
2. Embryos at stage 20 are collected into the transfer net. One to two embryos per culture is sufficient to yield adequate cell density for microscopic studies (see Step 8). Sterilize embryos in 70% ethanol for 10 seconds. Wash twice with 10% Holtfreter's solution. Sterilize again by passing through the thimerosal solution, followed by two washes in 10% Holtfreter's solution.
3. In 10% Holtfreter's solution, the jelly coat and the vitelline membrane are removed with two 25-gauge hypodermic needles (Fig. 1A,B). The denuded embryos are fragile. They are especially prone to rupturing by surface tension. Care should be taken to avoid exposing the embryos to the air–water interface.
4. The embryos are transferred to Steinberg's solution. Under the dissecting microscope, the dorsal part of the embryo, containing the skin, the neural tube, somites, and the notochord, is excised with a pair of 25-gauge hypodermic needles (Fig. 1C).
5. The dorsal pieces are treated with CMF Steinberg's solution for 10–20 minutes (Fig. 1D). The skin is then lifted off with fine forceps. Continue bathing the dorsal pieces in CMF solution for an additional 10 to 30 minutes until the pieces are dissociated. This is seen as a collapse of the cell mass (Fig. 1E).
6. The cell masses are transferred to a test tube containing 1 ml of Steinberg's solution. In this solution, the cell suspension can be gently triturated a few times with a fire-polished Pasteur pipette to complete the dissociation. *Note*: Overtrituration damages cells and increases the number of yolk granules released into the culture by damaged cells.
7. Pellet the cells in a tabletop centrifuge ($250\,g$ for 1 minute). Resuspend the pellet in culture medium by gentle agitation and trituration.
8. Plate the cells onto culture substratum. For microscopic studies, cells should be plated into a small area to increase the local cell density.
9. Cultures are maintained at $15°–22°C$. To facilitate cell spreading and attachment, cultures are maintained at $22°C$ during the first day and then kept at $15°C$.

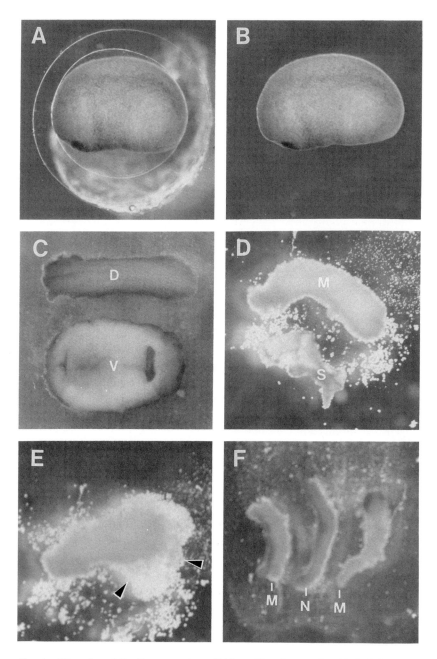

FIG. 1. Dissection of the *Xenopus* embryo. (A) Stage 22 embryo enclosed in its jelly coat and vitelline membrane. (B) The same embryo after removal of the membranes. (C) The dorsal (D) and ventral (V) parts are separated. (D) In CMF Steinberg's solution, the skin (S) is lifted off the myotomal part of the embryo. (E) After an incubation in CMF Steinberg's solution, the myotomal part is dissociated as shown by the collapsed cell mass (arrowheads). (F) The myotomes (M) and the neural tube (N) can be separated by collagenase treatment.

C. Preparation of Separate Nerve or Muscle Cultures

The inclusion of an enzymatic digestion step enables one to separate the neural tube from the myotomes. Thus, separate nerve and muscle cultures can be prepared. After Step 4 of the above procedure, the dorsal pieces of the embryos are transferred to Steinberg's solution containing 1 mg/ml collagenase. After 30 minutes, they are transferred to enzyme-free Steinberg's solution. Under the dissecting microscope, the dorsal embryo piece is positioned with hypodermic needles such that the ventral part of the embryo faces up. The two halves of the dorsal piece are separated with needles to expose the myotomes. They are then cut away from the neural tube and the notochord by a slicing action with the needles (Fig. 1F). The notochord can be separated from the neutral tube by gently teasing them apart with needles. The pieces are then collected separately with fine pipettes made by narrowing the tips of Pasteur pipettes with flame. The myotomes and the neural tubes are then dissociated with CMF Steinberg solution and plated onto appropriate culture substratum in culture medium.

D. Preparation of Enriched Nerve–Muscle Cocultures

Although nerve–muscle pairs are present in the mixed nerve–muscle cultures described above, they are not abundant in number. To enrich the NMJs, it is necessary to seed large number of neurons into each muscle culture. A ratio of five neural tubes per muscle culture yields an adequate density of nerve–muscle pairs. To make these enriched cocultures, muscle cultures are prepared according to above procedure. The neural tubes isolated during the preparation of muscle cultures are stored in culture medium at 4°C until the muscle cells are well spread (1–3 days). At this time, the neural tubes are dissociated with CMF dissection solution and seeded into muscle cultures

E. Preparation of Neural Tube Explants

The neural tube explant is an alternative to dissociated neuron culture. Although cell bodies of individual neurons cannot be seen in the explants, vigorous neuritic outgrowths can be easily achieved, and they generally survive longer than the dissociated neurons. The neural tubes are isolated from stage 20 to 22 embryos. They are cut into slices about 0.5 mm thick with hypodermic needles or with microdissecting scissors in Steinberg's solution. The slices are collected into culture medium and plated onto a suitable substratum, such as polylysine- or collagen-coated cover glass or culture dish.

After explanting, care should be taken to avoid agitating the cultures. This is best accomplished by culturing explants in a sealed cover glass chamber, such as the Anderson–Cohen chamber or the Sykes–Moore chamber described

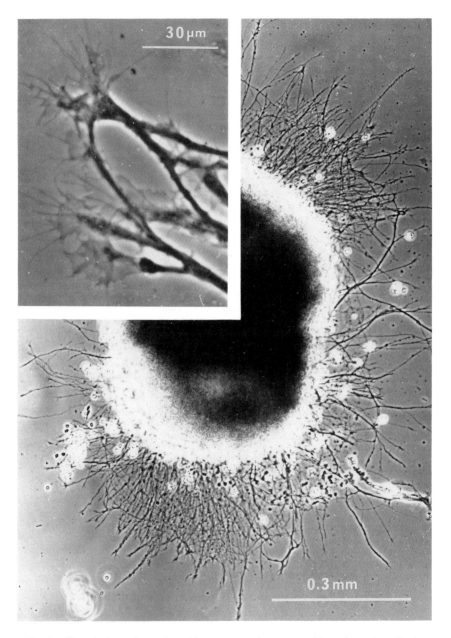

FIG. 2. Neural tube explant culture. Numerous neurites emanated from the explant. The inset shows the growth cones at the tips of the neurites.

above. When properly assembled to be free of air bubbles, explant cultures can be kept alive for a relatively long time. An example of a *Xenopus* neural tube explant is shown in Fig. 2. Numerous neurites with vigorous growth cones at their tips can be observed in this type of culture.

III. Characteristics of *Xenopus* Cultures and Development of Neuromuscular Junctions *in Vitro*

Myotomal muscle cells attach to the substratum and spread out within 12 hours after plating. During the first day, they appear as spindle-shaped cells. Yolk platelets, which appear as bright intracellular granules under phase-contrast optics, provide nutrients for cell growth. Because of this intrinsic nutrient supply, *Xenopus* cells can be cultured in an essential salt solution with minimal added nutrients. On the first day, yolk platelets are prominent in the cells. As the yolk is consumed during development, a clearing of the cytoplasm is gradually seen. Cross-striations, indicative of the development of the contractile apparatus, can first be detected in the central portion of the myotomal cell in day 1, and they gradually appear toward the ends of the cell (Peng *et al.*, 1981). By day 3, they are observed throughout the cell. Despite the rapid development of the contractile apparatus, these cells do not contract spontaneously. There is evidence that the development of the contractile apparatus can be influenced by innervation (Kidokoro and Saito, 1988).

Myotomal muscle cells do not divide and remain mononucleated in culture. In contrast, their counterparts in the tail musculature of the tadpoles do become multinucleated during development (Chen *et al.*, 1990; see Fig. 4A,B). This may be due to the fact that, under culture conditions, few myoblasts (or satellite cells) come into contact with the myotomal cells, and thus cell fusion does not take place. In comparison, satellite cells are associated with myotomal muscle fibers in the tadpole. Despite the lack of multinucleation, cellular differentiation, such as the myofibrillogenesis, the development of excitability, and the synaptic development, proceeds in the same manner as *in vivo*.

A. Development of Neuromuscular Junctions

Neuromuscular junctions are formed in culture when myotomal muscle cells are contacted by nerve processes. Such nerve–muscle contacts are best visualized in live cultures. In chemically fixed cultures, the exact position of nerve–muscle contact is often not clearly seen. However, antibodies against neural-specific proteins, such as the neural cell adhesion molecules (NCAM), can be used to aid the identification of nerve–muscle contacts in fixed cultures (Rochlin and Peng, 1990). Spontaneous and evoked synaptic potentials can be recorded from the muscle cell almost immediately after such contacts are established (Kidokoro and Yeh, 1982; Xie and Poo, 1986; Evers *et al.*, 1989).

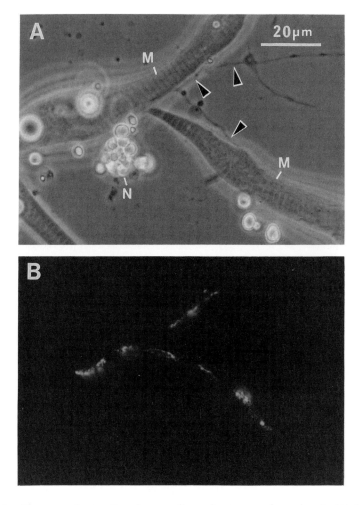

FIG. 3. Neuromuscular junctions in a coculture of neurons and muscle cells. (A) A single neuron (N) sent out neurites (arrowheads) which contacted two adjacent muscle cells (M). (B) The nerve–muscle contacts induced NMJ formation as shown by the clustering of acetylcholine receptors, which are visualized by the fluorescence of rhodamine-conjugated α-bungarotoxin. This culture was labeled with 0.3 μM fluorescent toxin and observed in the living state.

The development of synaptic specializations follows this initial contact within the first day (Takahashi *et al.,* 1987).

The postsynaptic specialization is manifested by the clustering of nicotinic acetylcholine receptors. This can be visualized by labeling the culture with fluorescently conjugated α-bungarotoxin (available from Molecular Probes, Eugene, OR) as shown in Fig. 3. The presynaptic specialization can be studied by labeling cultures with antibodies which recognize synaptic vesicle proteins (e.g., mAB 48 which recognizes a 65-kDa glycoprotein on synaptic vesicles; see Peng *et al.,* 1987; Cohen *et al.,* 1987). In addition, other components of the synaptic apparatus also develop *in vitro*. These include a concentration of cholinesterase in the synaptic cleft (Moody-Corbett and Cohen, 1982) and the accumulation of postsynaptic cytoskeletal proteins (Burden, 1985; Rochlin *et al.,* 1989). Elaboration of the synaptic specializations can also be seen at the ultrastructural level. This is manifested by the appearance of postsynaptic densities, the basal lamina, and the clustering of synaptic vesicles (Weldon and Cohen, 1979; Takahashi *et al.,* 1987). All these bear striking resemblance to the NMJ *in vivo*. Thus, the cultured NMJ is a faithful representation of the innervation process *in vivo*.

B. Formation of Synaptic Specializations without Nerve

In addition to innervation, synaptic specializations can be induced in the *Xenopus* myotomal muscle cell by other stimuli. Polycation-coated latex beads induce an extensive postsynaptic development when applied to myotomal muscle cells (Peng and Cheng, 1982). These beads also induce the presynaptic development when applied to cultured neural tube explants (Peng *et al.,* 1987). In addition, clustering of acetylcholine receptors can also be elicited by constant electric field in cultured myotomal muscle cells (Orida and Poo, 1978; Luther and Peng, 1985; Stollberg and Fraser, 1990). In the absence of exogenously applied stimuli, acetylcholine receptor clusters are formed spontaneously (Anderson *et al.,* 1977). These "hot spots" are often located on the ventral surface of the cell where the cell membrane comes into close proximity with the substratum (Peng, 1987). This versatility in response to synaptogenic signals makes the *Xenopus* culture an ideal system to elucidate the mechanism of synaptic induction.

C. Comparison with Neuromuscular Junction Development *in Vivo*

The myotomal muscle cells form the tail musculature of the larva. Since it is easy to raise *Xenopus* embryos and larvae, a comparison with the developmental processes *in vivo* can be conveniently performed. As mentioned before,

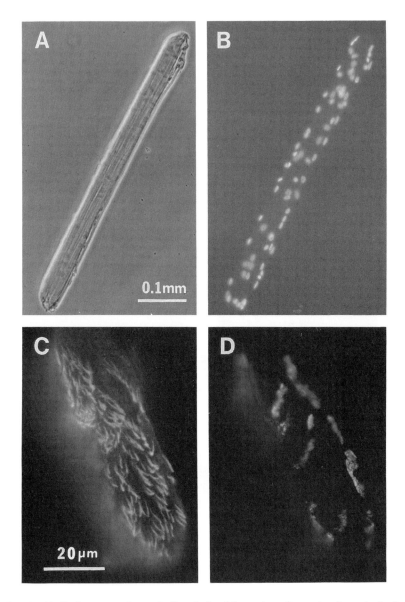

FIG. 4. (A, B) A myotomal muscle fiber isolated from the tail muscle of a tadpole. It is multinucleated as shown by DAPI (4′, 6-diamidino-2-phenylindole) staining (B). (C, D) The end of a myotomal fiber has two kinds of junctions: the myotendinous junction (C) visualized by labeling with an antibody against dystrophin and the NMJ (D) as seen by fluorescent α-bungarotoxin labeling.

myotomal muscle cells undergo fusion *in vivo* and develop into multincleated muscle fibers. However, innervation precedes the process of multinucleation. NMJs are formed at the ends of the muscle fibers *in vivo*, and their development can be followed in developing larvae with fluorescent α-bungarotoxin or with antibodies against synaptic vesicle antigens in tail whole mounts (Peng *et al.*, 1989; Cohen *et al.*, 1987). Myotomal muscle cells can also be easily dissociated from the tail musculature of tadpoles by collagenase (Fig. 4A,B). The dissociated muscle fibers have intact postsynaptic membrane and myotendinous junctions (Fig. 4C,D). In addition to providing an *in vivo* comparison with cultured cells, these dissociated muscle fibers can also be maintained in culture. With appropriate conditions, it is possible to cause a dedifferentiation in these fibers to yield multinucleated myotubes (cf. Bischoff, 1990).

IV. Preparation of Myotube Cultures

In addition to myotomal muscle cells, which remain mononucleated in culture, cultures of myoblasts and multinucleated myotubes can also be prepared from *Xenopus* larvae (Kay *et al.*, 1988). Premetamorphic tadpoles (stage 50–55) are anesthetized with MS222 (ethyl 3-aminobenzoate methanesulfonate) and sterilized with a 1:10,000 solution of thimerosal. Myoblasts can be isolated from several places in stage 50–55 tadpoles. These include limb buds (or developing limbs) and the submaxillaris muscle underneath the jaw. Tissues are excised in frog Ringer's solution. After the skin and the bones are removed, the tissues are teased into small pieces and incubated in Ca^{2+}, Mg^{2+}-free Ringer's solution containing 0.4 mM EDTA, 2.5 mg/ml trypsin, and 2 mg/ml collagenase for 1 hour at room temperature. The tissue pieces are triturated with a fire-polished Pasteur pipette, and the cells are collected by centrifugation with a tabletop centrifuge. The cells are plated onto collagen- or poly-D-lysine-coated cover glass and cultured in 50% L-15 (Leibovitz) medium containing 5% fetal bovine serum and 10 mM HEPES buffer (pH 7.4) at 22°C.

Spindle-shaped myoblasts proliferate in culture after spreading (Fig. 5A). If the cell density is sufficiently high, multinucleated myotubes can be observed in culture within 2 days (Fig. 5B,C). These myotubes express acetylcholine receptors on their surface as shown by autoradiography after treating the cultures with ^{125}I-labeled α-bungarotoxin. They can be innervated in culture with neurons from the neural tubes of *Xenopus* embryos.

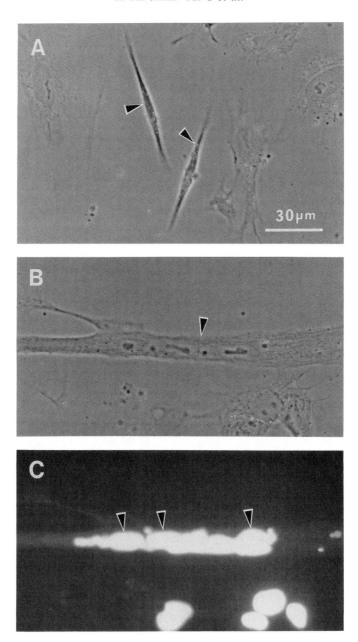

FIG. 5. Myoblast and myotube cultures from *Xenopus*. (A) Myoblasts (arrowheads) from developing limbs of a larva. (B, C) Myotubes (arrowhead) are formed due to the fusion of myoblasts when they come into contact with each other. In (C), DAPI staining shows multiple myonuclei (arrowheads) within the myotube.

V. Discussion

This chapter describes the use of *Xenopus* cell cultures in the study of neuromuscular synapse development. In addition to this application, these cultures are useful tools in understanding a wide range of problems in muscle development, such as myogenesis and the development of membrane excitability. The feasibility in manipulating the development of the embryos also adds to the versatility of this culture system. For example, one can easily inject markers or proteins into blastomeres and follow the fate of the descendent cells in culture (see Sanes and Poo, 1989). This opens the possibility to alter components of the synaptic apparatus selectively with protein or nucleic acid probes and then to study the effect in culture. With new methods of cell transfection, it may be possible to express foreign genes directly in cultured myotomal cells in the future (see Chapter 32, this volume).

In using an *in vitro* system to study development, one always has to be cautious about artifacts owing to culture conditions. Based on morphological criteria at both the light and electron microscopic levels, the myotomal neuromuscular synapses formed *in vitro* appear to be a faithful representation of those *in vivo*. Because of the small size and near transparency of the *Xenopus* larva, the myotomal NMJs can also be conveniently studied in the tail of the animal. This allows a ready comparison between the *in vivo* and the *in vitro* systems. New methods of imaging at the light microscopic level, such as confocal microscopy and the use of specific fluorescent membrane and organelle probes, should further enhance the usefulness of the *Xenopus* myotomal muscle system in developmental studies.

As described above, the development of the synapse can be considered as a special form of embryonic induction. Recent studies have shown that peptide growth factors play important roles in embryonic induction. For example, basic fibroblast growth factor (bFGF) and members of the transforming growth factor β (TGF-β) family appear to be involved in the mesoderm induction process (Smith, 1989; Ruiz i Altaba and Melton, 1990). Recently, we have shown that bFGF, when locally presented to cultured *Xenopus* muscle cells, induces the development of postsynaptic specializations (Peng et al., 1991). Therefore, there may be a mechanistic similarity between the general embryonic induction and the more specialized synaptic induction. Synaptic induction is marked by a sequence of well-defined events which lead to the development of local specializations such as the aggregation of nicotinic acetylcholine receptors and various cytoskeletal and extracellular matrix proteins in the muscle cell. Thus, in addition to being a good model for studying synaptogenesis, the nerve–muscle coculture system described in this chapter may also prove useful in understanding the cellular and molecular mechanisms of embryonic inductions.

References

Anderson, M. J., Cohen. M. W., and Zorychta, E. (1977). *J. Physiol. (London)* **268**, 731–756.
Bischoff, R. (1990). *Development (Cambridge, U.K.)* **109**, 943–952.
Bloch, R. J., and Pumplin, D. W. (1988). *Am. J. Physiol.* **254**, C345–C364.
Burden, S. J. (1985). *Proc. Natl. Acad. Sci. U.S.A.* **82**, 8270–8273.
Chen, Q., Sealock, R., and Peng, H. B. (1990). *J. Cell Biol.* **110**, 2061–2071.
Cohen, M. W. (1980). *J. Exp. Biol.* **89**, 43–56.
Cohen, M. W., Rodriquez-Marin, E., and Wilson, E. M. (1987). *J. Neurosci.* **7**, 2849–2861.
Evers, J., Laser, M., Sun, Y., Xie., Z., and Poo, M.-M. (1989). *J. Neurosci.* **9**, 1523–1539.
Hamburger, V. (1960). "A Manual of Experimental Embryology." Univ. of Chicago Press, Chicago, Illinois.
Harrison, R. G. (1907). *Proc. Soc. Exp. Biol. Med.* **4**, 140–143.
Kay, B. K., Schwartz, L. M., Rutishauser, U., Qiu, T. H., and Peng, H. B. (1988). *Development (Cambridge, U.K.)* **103**, 463–471.
Kidokoro, Y., and Saito, M. (1988). *Proc. Natl. Acad. Sci. U.S.A.* **85**, 1978–1982.
Kidokoro, Y., and Yeh, E. (1982). *Proc. Natl. Acad. Sci. U.S.A.* **79**, 6727–6731.
Luther, P. W., and Peng, H. B. (1985). *J. Cell Biol.* **100**, 235–244.
Moody-Corbett, F., and Cohen, M. W. (1982). *J. Neurosci.* **2**, 633–646.
Orida, N., and Poo. M.-M. (1978). *Nature (London)* **275**, 31–35.
Peng, H. B. (1987). *CRC Crit. Rev. Anat. Sci.* **1**, 91–131.
Peng, H. B., and Cheng, P.-C. (1982). *J. Neurosci.* **2**, 1760–1774.
Peng, H. B., Wolosewick, J. J., and Cheng, P.-C. (1981). *Dev. Biol.* **88**, 121–136.
Peng, H. B., Markey, D. R., Muhlach, W. L., and Pollack, E. D. (1987). *Synapse* **1**, 10–19.
Peng, H. B., Chen, Q., DeBiasi, S., and Zhu, D. (1989). *J. Comp. Neurol.* **290**, 533–543.
Peng, H. B., Baker, L. P., and Chen, Q. (1991), *Neuron* **6**, 237–246.
Rochlin, M. W., and Peng, H. B. (1990). *Dev. Biol.* **140**, 27–40.
Rochlin, M. W., Chen, Q., Tobler, M., Turner, C. E., Burridge, K., and Peng, H. B. (1989). *J. Cell Sci.* **92**, 461–472.
Ruiz i Altaba, A., and Melton, D. A. (1990). *Trends Genet.* **6**, 57–64.
Sanes, D. H., and Poo, M.-M. (1989). *Neuron* **2**, 1237–1244.
Smith, J. C. (1989). *Development (Cambridge, U.K.)* **105**, 665–677.
Stollberg, J., and Fraser, S. E. (1990). *J. Cell Biol.* **111**, 2029–2039.
Takahashi, T., Nakajima, Y., Hirosawa, K., Nakjima, S., and Onodera, K. (1987). *J. Neurosci.* **7**, 473–481.
Weldon, P. R., and Cohen, M. W. (1979). *J. Neurocytol.* **8**, 239–259.
Xie, Z. P., and Poo, M.-M. (1986). *Proc. Natl. Acad. Sci. U.S.A.* **83**, 7069–7073.

Chapter 27

The Xenopus Embryo as a Model System for the Study of Cell–Extracellular Matrix Interactions

DOUGLAS W. DeSIMONE

*Department of Anatomy and Cell Biology
and the Molecular Biology Institute
University of Virginia Health Sciences Center
Charlottesville, Virginia 22908*

KURT E. JOHNSON

*Department of Anatomy
George Washington University Medical Center
Washington, D.C. 20037*

I. Introduction
II. Methods
   A. Required Materials and Solutions
   B. Isolation and Purification of *Xenopus* Plasma Fibronectin
   C. Immunocytochemical Methods to Localize Extracellular Matrix Molecules
   D. Mesodermal Cell Adhesion *in Vitro*
III. Concluding Remarks
   References

I. Introduction

There is now a substantial body of evidence, obtained from several experimental systems, that the extracellular matrix (ECM) plays an important role in supporting morphogenetic cell movements during development. *Xenopus* and other amphibian embryos are well suited to investigations of cell–ECM interactions (reviewed in Johnson et al., 1991) because the system can be investigated on several experimental levels, which include (1) whole-embryo perturbation studies, (2) in vitro adhesion assays, and (3) immunohistochemical analyses. This combination of approaches makes it possible to study the structure, function, and developmental expression of ECM molecules and their receptors in a single model system.

Xenopus gastrulas synthesize substantial quantities of glycoconjugates that they secrete into the extracellular spaces surrounding cells (Johnson, 1984). Dramatic increases in the amount of fibrillar extracellular material are also observed on the basal surface of the blastocoel roof during gastrulation (Nakatsuji and Johnson, 1983). This fibrillar material was shown to contain fibronectin (FN) (Lee et al., 1984; Nakatsuji et al., 1985). Several workers have also studied the behavior of Xenopus blastomeres and tissue fragments on FN in vitro and found regional differences in cell adhesion to this ECM protein (Nakatsuji, 1986; Winklbauer, 1988, 1990; Smith et al., 1990; DeSimone et al., 1991a).

In recent years, our understanding of cell–ECM interactions has progressed rapidly owing, in large part, to the identification and detailed characterization of several ECM molecules and their cellular receptors, the integrins. Fibronectins are among the most intensively studied components of vertebrate ECMs, and a great deal is now known about their structure and function (for reviews, see Mosher, 1989; Hynes, 1990). Fibronectins are large, multifunctional glycoproteins that can bind to the surfaces of cells and to other ECM proteins. There are multiple cell binding sites on the FN molecule, and multiple integrin receptors for FNs are expressed by many cell types. Further complexity is observed in the structure of the FN molecule itself, several forms of which are derived from the alternative splicing of a common transcript. Multiple integrins (DeSimone and Hynes, 1988; Smith et al., 1990; DeSimone et al., 1991a) and alternatively spliced forms of FNs (DeSimone et al., 1991b) are expressed in early Xenopus embryos. A major challenge, therefore, is to understand how this complex array of cell adhesion molecules may function during morphogenesis.

In this chapter we describe general methods used to study the ECM during Xenopus development. We have chosen to concentrate on FN as a "prototypical" ECM molecule. This is due to the fact that FN is one of the best understood components of the ECM and has received a great deal of attention from investigators interested in the control of cell migration in vertebrate embryos. The first section deals with the isolation of Xenopus FN from adult plasma. Immunocytochemical methods are then considered and a protocol described for the preparation of the blastocoel roof in order to observe the organization of ECM proteins expressed at gastrulation. Finally, the suitability of *in vitro* adhesion assays is discussed and methods used to condition substrates with ECM molecules presented.

II. Methods

A. Required Materials and Solutions

Gelatin and Heparin Sepharose

Gelatin sepharose is available commerically (Pharmacia/LKB). Although perhaps not as convenient, higher concentrations of bound gelatin to resin can be obtained in the laboratory by directly coupling porcine gelatin to CNBr activated sepharose (Pharmacia/LKB) using standard protocols (Yamada, 1982). Gelatin sepharose is packed into siliconized (e.g., Sigmacote, Sigma) glass columns. Our standard columns are 2.5 × 10 cm (Bio-Rad "econo-columns") and contain 35-40 ml bed volume of gelatin-sepharose. Heparin sepharose CL-6B (Pharmacia/LKB) is prepared in the same size column but at half the bed volume.

Modified Barth's Saline (MBS; see Appendix A, this volume) Modified Stearn's Solution (MSS; to make 1 liter)

NaCl, 4.360 g
Na_2SO_4, 0.270 g
HEPES, 1.190 g
KCl, 0.180 g
Na_2HPO_4, 0.089 g
KH_2PO_4, 0.019 g
$MgCl_2$, 0.095 g
$CaCl_2$, 0.012 g
Phenol red, 0.008 g
Adjust the pH to 8.3 with 5 N NaOH

Tris saline (plus EDTA)

0.15 M NaCl
10 mM Tris-Cl (pH 7.4)
10 mM EDTA

B. Isolation and Purification of *Xenopus* Plasma Fibronectin

The most common source of FN used in cell biology experiments is that found circulating in the plasma (pFN). The primary reason for this is one of convenience; pFN is present in soluble form in the plasma at high concentrations (0.3 mg/ml in mammals) and is readily isolated by binding to gelatin. In contrast, the matrix, or cellular, form of the protein (CFN) is relatively insoluble and must be extracted with urea from cell surfaces (see Yamada, 1982; Hynes, 1990). Once purified, cFNs will frequently precipitate from solution unless buffer conditions are maintained at high pH (e.g., CAPS buffer, pH 11.0). The difficulties associated with the isolation and handling of cFNs have precluded detailed analyses of the functional properties of cFN isoforms. Recent studies using recombinant FNs, however, suggest that several of the biological activities of these multiple forms are similar (Guan et al., 1990).

We have determined that the FN expressed during embryogenesis in *Xenopus* is composed primarily (>95%) of cellular forms of the protein (DeSimone et al., 1991b). It should be noted, however, that we (Smith et al., 1990) and others (Winklbauer, 1988, 1990) have used mammalian pFNs to investigate the adhesive properties of *Xenopus* embryo fragments and dissociated cells *in vitro*. Future studies will require that a complete analysis of the behavior of these cells and tissues be examined using appropriate cFNs of *Xenopus* origin. One strategy used to circumvent this problem has been to utilize artificial substrates conditioned with the FN-rich ECM produced by the blastocoel roof, as discussed in Section II,D. The following protocols are useful starting points for the isolation of relatively large quantities of pFN, which are suitable for *in vitro* adhesion assays and the production of immunogen.

1. Isolation of Adult Plasma

Generally, only 2–5 ml of blood can be obtained easily from an adult frog (depending on size), which makes large-scale *Xenopus* pFN isolations an expensive undertaking. At least 50 ml of plasma is applied to a gelatin–Sepharose column, from which approximately 3–4 mg of *Xenopus* pFN may be purified as described below. This yield is severalfold less than that obtained from human plasma (human pFN concentration is 0.15–0.3 mg/ml plasma) under identical conditions. It is unclear whether this is due to a lower concentration of circulating pFN in *Xenopus* or to differences in FN gelatin-binding affinity.

We have found that the simplest and most efficient way to obtain blood from *Xenopus* adults is by exsanguination. Animals are anaesthetized in 0.03% benzocaine, the thoracic cavity opened, and the heart carefully exposed. The animal is held ventral side down so that the exposed heart can be placed over the lip and along the inside of a 50-ml polypropylene centrifuge tube. The ventricle is then opened with a pair of scissors and the blood collected directly into the tube. Clotting is extremely rapid in *Xenopus* (which may account for some FN loss in these preparations), but coagulation can be reduced by adding EDTA (10 mM final) to the collection tube. The collected blood is pooled and centrifuged at 2000 rpm for 10 minutes to pellet cells. The supernatant is then transferred to polypropylene tubes and stored at −80°C. Polypropylene tubes (or siliconized glass) should be used to prevent appreciable losses of FN during purification and storage because the protein sticks to glass and polystyrene.

2. Affinity Purification of Plasma Fibronectin

Two columns are prepared in tandem. The first is a precolumn containing 10 ml of Sepharose CL-4B, which is used to remove proteins from the serum that stick nonspecifically to Sepharose. The second column containing the gelatin–Sepharose is attached to the precolumn so that the flow-through can be applied directly to the affinity resin. Both columns are extensively prewashed with Tris–saline containing 10 mM EDTA and 2 mM PMSF (phenylmethylsulfonyl fluoride).

Before applying to the gelatin column, plasma is centrifuged at 10,000 g for 10 minutes, and 10 mM EDTA and 2 mM PMSF are added to the supernatant. The supernatant is then passed over the packed columns at room temperature. It will normally take 15–30 minutes for 50 ml of plasma to pass through both columns. The FN-depleted flow-through can be collected and saved, but, in our experience, we have not observed an increase in the final yield following reapplication to the column. The precolumn is washed with 10 ml of Tris–saline and then disconnected from the gelatin column. The gelatin column is rapidly washed with 2 bed volumes of 0.5 M NaCl, 10 mM Tris-Cl (pH 7.4) followed by 3 bed volumes of Tris–saline. The pFN is then slowly eluted from the column (approximately 1 drop every 2–3 seconds) with 4 M urea, 10 mM Tris-Cl (pH 7.5) and 3-ml fractions collected. The peak of eluted protein is determined by absorbance at 280 nm or by gel electrophoresis. We routinely analyze column fractions by sodium dodecyl sulfate (SDS)-polyacrylamide gel electrophoresis (SDS-PAGE) in order to identify peak fractions and to determine the relative concentration and purity of a given preparation (Fig. 1). After use, the gelatin column can be regenerated by washing extensively with 8 M urea, 10 mM Tris-Cl (pH 7.4), followed by Tris–saline. The column is then equilibrated and stored in 10 mM Tris-Cl (pH 7.4) containing 0.02% sodium azide.

Fractions containing FN are pooled and dialyzed exhaustively against Tris–saline at 4°C. We have found that the protein can be concentrated following dialysis without substantial loss by placing the dialysis bag in 1%

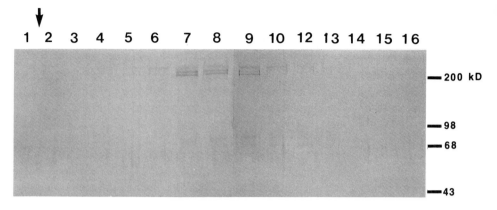

FIG. 1. Coomassie-blue stained profile of *Xenopus* plasma fibronectin (pFN) fractions eluted from a gelatin column with 4 M urea, 10 mM Tris-Cl (pH 7.5) and separated by SDS-PAGE. Each lane represents 80 μl from each 3.0-ml fraction collected from the column. Samples were suspended in Laemmli (1970) sample buffer containing 2-mercaptoethanol and analyzed on 7% acrylamide minigels. pFN runs as a doublet under these conditions with subunits of approximately 220 kDa (fractions 7–10). The arrow indicates the point of application of 4 M urea, 10mM Tris-Cl (pH 7.5) to the column.

aquacide (Calbiochem, San Diego, CA) for several hours. The concentrated FN is then stored in small aliquots at $-80°C$.

A typical preparation of pFN is shown in Fig. 1, where each lane of the gel represents 80 μl of each 3-ml fraction obtained from the gelatin column. *Xenopus* pFN runs as a doublet on reduced SDS gels at a subunit size of approximately 220 kDa. Additional minor bands represent contaminants and FN breakdown products. *Xenopus* pFN can be purified further by passing the flow-through from the gelatin column directly over a heparin–Sepharose column (bed volume approximately 1/10 that of gelatin column). This takes advantage of the fact that FN will bind to heparin in the presence of 4 M urea, whereas most contaminants will not. The pFN is then eluted from the column with high salt (0.5 M NaCl, 10 mM Tris-Cl, pH 7.4). We have purified *Xenopus* FN to virtual homogeneity (determined by silver stained gels, not shown) in this way, although such highly purified preparations are generally unnecessary for most applications. Details concerning the use and preparation of heparin affinity columns to purify FN can be found in Yamada (1982).

3. Preparation of Fibronectin Antisera

Fibronectins generally elicit strong immune responses in animals, and high-titer antisera are easily obtained following injection of purified pFNs.

We have used *Xenopus* pFNs eluted from gelatin columns to immunize mice for monoclonal antibody production with excellent results (DeSimone and Ramos, 1991). For polyclonal antibody production in rabbits, pFN can be purified further by SDS-PAGE and the gel slice containing the protein used as immunogen (see Lee *et al.*, 1984). In our experience, antisera directed against such SDS-denatured preparations of pFN will also recognize the native protein.

C. Immunocytochemical Methods to Localize Extracellular Matrix Molecules

1. General Considerations

Histological procedures often pose special difficulties with regard to the fixation and preservation of ECM structure. Shrinkage artifacts and the differential solubility of ECM molecules during the processing of tissues are among the most serious commonly encountered problems that can lead to a distorted view of how the ECM is organized *in vivo*. This is particularly true in the case of early amphibian embryos with their large, fluid-filled blastocoels, which are rich in extracellular materials.

We have made use of several immunocytochemical methods over the years to detect ECM components such as FN in *Xenopus* embryos. Cryosectioning is generally quite advantageous for retaining the immunogenicity of target ECM molecules, but the morphology is often unacceptable for early stage embryos (i.e., cleavage stages through gastrulation). Several protocols have also been reported for the sectioning of *Xenopus* embryos embedded in polyethylene glycol distearate (Heasman *et al.*, 1984), acrylamide (Hausen and Dreyer, 1981), and plastic (see Chapter 21, this volume). These procedures offer excellent preservation of tissue morphology (e.g., Wedlich *et al.*, 1989), but it is difficult to reconstruct a three-dimensional pattern of matrix organization by analyzing sectioned materials. Whole-embryo immunocytochemical procedures (see Chapter 22, this volume) also maintain tissue morphology quite nicely. In our experience, however, whole-embryo preparations are best suited to localizing molecules such as FN in later stage embryos that are producing large amounts of ECM. In order to investigate the ontogeny and organization of the ECM, it is necessary to resolve fine, fibrous materials that are initially deposited onto the surfaces of blastomeres lining the blastocoel. The following protocol describes a whole-mount preparation of the blastocoel roof that makes it possible to observe FN fibrils *en face* as they are laid down in the embryo.

2. Whole-Mount Preparations of Blastocoel Roof

Embryos of the desired stage are dejellied by treating with 2% cysteine (pH7.8) followed by extensive washing in 10% MBS. The embryos, with vitelline membranes intact, are then fixed in 10% MBS containing 2% (w/v) trichloroacetic acid (TCA) for 1 hour to overnight at room temperature. After fixation, embryos are dissected into appropriate fragments with sharpened tungsten needles or Dumont No. 5 watchmaker's forceps. Embryos can also be dissected prior to fixation; however, in this case the vitelline membranes must first be removed. When dissecting the roof of the blastocoel from blastula and gastrula stage embryos, we often find it helpful to mark the "polarity" of the fragment removed. This is done simply by cutting a small wedge of tissue from one end of the fragment (e.g., dorsal side).

Fixed fragments are transferred to phosphate-buffered saline (PBS) in Falcon (#3008) multiwell tissue culture plates and allowed to settle. Fluid is carefully removed from each well with Pasteur pipettes and the fragments washed 2 times, 10 minutes each, in changes of PBS. We find the multiwell dishes useful because several different antibodies and embryonic stages may be processed simultaneously. The PBS is replaced with "blocking" solution containing 3% bovine serum albumin (BSA) (or powdered nonfat milk) in PBS for 1 hour at room temperature. After blocking, the tissue is washed in 2 changes of 0.5% BSA in PBS, and incubated overnight at 4°C in the same solution containing appropriate dilutions of primary antibody (we routinely use IgG purified antibodies at 10–50 μg/ml). Antibody incubation is followed by three 10-minute washes in 0.5% BSA in PBS (with very gentle rocking). The specimens are then incubated for 1–2 hours at room temperature in fluorescein isothiocyanate (FITC)-conjugated secondary antibody and washed in three 10-minute changes of PBS. Finally, the fragments are rinsed briefly in 50% (v/v) glycerol and mounted in the same solution. A trace amount of p-phenylenediamine may be added to the mounting medium to reduce quenching of the fluorescent signal.

The fixed tissue is brittle, and care must be taken during transfer of the whole mounts. Specimens are mounted on standard glass slides by preparing a rectangular dam with a thin bead of petroleum jelly (dispensed from a syringe needle with the bevel filed off). The whole mounts are transferred with wide-bore Pasteur pipettes to a drop of mounting medium on the slide and then oriented so that the basal surface of the blastocoel roof faces up. A #1 glass coverslip is seated on the petroleum jelly chamber to form a seal. These preparations are carefully flattened by pressing down on the coverslip to remove excess mounting medium and to bring the specimen in contact with the coverslip. The tissue will soften somewhat as it is infiltrated with glycerol, so it is possible to compress the specimen slightly without damage to the

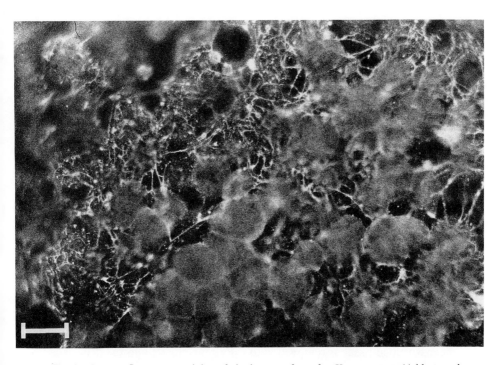

FIG. 2. Immunofluorescent staining of the inner surface of a *Xenopus* stage 11 blastocoel roof with anti-FN primary antibody and FITC-conjugated goat anti-rabbit secondary antibody. A whole mount of the blastocoel roof was prepared as described in the text. In this example, an IgG purified primary antibody directed against *Ambystoma mexicanum* pFN was used to detect *Xenopus* FN, which forms a fibrillar matrix that underlies the ectodermal cells along the roof of the blastocoel. Bar equals 25 μm.

underlying tissue layers. Whole mounts are examined with a fluorescence microscope equipped for epi-illumination. A typical result is shown in Fig. 2.

D. Mesodermal Cell Adhesion *in Vitro*

One of the primary difficulties associated with studying cell adhesion and migration in developing amphibians is the opacity of the intact embryo. Unlike many transparent invertebrate, avian, and fish embryos, amphibian blastomeres are heavily pigmented and filled with yolk platelets and other cytoplasmic inclusions. It is not generally possible, therefore, to make direct observations of cell–ECM interactions in living amphibian embryos. Most workers have instead opted to correlate static morphological observations with studies of embryonic fragments and cells *in vitro*. *In vivo* studies have relied largely on probes (e.g., antibodies and synthetic peptides) that, when injected into embryos, disrupt specific cell–ECM interactions (for review, see Johnson et al., 1991). The following sections describe two general methods for analyzing the behavior of mesodermal cells on conditioned substrates in vitro.

1. EXTRACELLULAR MATRIX-CONDITIONED SUBSTRATES

We have developed methods for conditioning plastic tissue culture dishes with the ECM from the basal surface of the blastocoel roof. Our conditioning medium (Nakatsuji and Johnson, 1982) consists of modified Stearn's solution (MSS) with 110 μM $CaCl_2$. First, the vitelline membrane is removed manually and then the entire animal cap dissected from a devitellinized stage 10.5 gastrula and placed in conditioning medium with the basal surface flattened against the surface of a tissue culture dish. The explant is then held in place with a small fragment of a #1 coverslip, cut to the approximate size of the tissue. After 2–4 hours of conditioning, the coverslip is removed and the outline of the conditioning fragment traced on the plastic dish. The conditioning medium is replaced by 3 changes of Ca^{2+}- and Mg^{2+}-free MSS containing 2 mM EDTA (CMFMSS). After 1 hour of incubation in CMFMSS, the conditioning fragment dissociates and detaches from the deposited fibrils. The dissociated tissue is removed with a Pasteur pipette and the dish rinsed 3 times with culture medium, which is the same as conditioning medium but with 100 μM $CaCl_2$, 0.5% BSA, and gentamicin sulfate (10 $\mu g/ml$) added. Culture medium is made up fresh each day and sterile filtered before use.

2. DISSOCIATION OF MESODERMAL CELLS

Fragments of the dorsal lip of the blastopore from stage 10 embryos are dissected from whole embryos in CMFMSS and incubated for 1 hour. The tissue will dissociate into a neat pile of single cells, which can be transferred to a conditioned substrate or to a culture dish precoated with matrix proteins (see below). Once mesodermal cells are seeded onto conditioned substrates the cultures are left undisturbed for 1 hour. During this time, cells attach to the conditioned area, form locomotory organelles, and begin rapid movements.

The movement of cells on conditioned substrata can be analyzed by time-lapse cinemicrography. We float a round #1 coverslip on top of the culture medium and then remove most of the fluid with a Pasteur pipette. Cultures are filmed for up to 3 hours and then discarded. During this brief filming period, there is little fluid evaporation from the culture chamber and no bacterial growth. We use a heat filter and a green interference filter between the light source and the specimen. Preparations are examined by phase-contrast microscopy. Films of cell movement are exposed at 8-second intervals and then projected at 16 frames per second for analysis.

3. Cell Adhesion to Fibronectin Substrates

The ability of a given ECM molecule to support the adhesion and spreading of cells and tissues can be tested *in vitro* by adsorbing purified ECM proteins to plastic substrates. Cells or tissue fragments are then placed in contact with these substrates and their adhesive and migratory behaviors recorded. One advantage of this procedure is that a number of different molecules can be rapidly tested for their ability to support cell adhesion.

A solution containing the ECM protein of interest (e.g., in PBS or MBS) is used to cover the bottom of a tissue culture dish. The dish is then incubated at room temperature in a humidified chamber for several hours (4–18). After rinsing with MBS, the dish is again incubated for 1 hour but with 0.5% BSA in MBS in order to "block" any uncoated regions of the plastic substrate. *Xenopus* blastomeres will not adhere to control dishes coated with BSA alone. We use concentrations of 10–100 μg/ml pFN to condition substrates. In order to conserve on matrix protein (e.g., *Xenopus* pFN or cFN) a drop of the ECM protein solution can be incubated in a defined area previously scratched into the plastic with a needle. Cells can later be added and observed within this conditioned area of the dish.

We have used this culture method to demonstrate that one of the earliest responses of induced mesodermal tissues is the ability to adhere, spread, and migrate on pFN (Fig. 3; see also Smith *et al.*, 1990; DeSimone *et al.*, 1991a). Cells derived from the marginal zone adhere to FN whereas those isolated from animal cap tissues do not (Fig. 3A). When animal cap cells are treated with mesoderm-inducing factors, however, they rapidly begin to spread and migrate on FN substrates (Fig. 3B).

III. Concluding Remarks

The study of cell–ECM interactions in amphibians is likely to yield important new information about the structure and function of ECM molecules and their cellular receptors. Among the advantages of this system is that cell–matrix adhesion can be reconstituted *in vitro* using individual cells, fragments of embryos, and conditioned substrates. The availability of *Xenopus* cDNA clones for FNs (DeSimone *et al.*, 1991b) and integrins (DeSimone and Hynes, 1988; Ransom and DeSimone, 1990) will make it possible to express

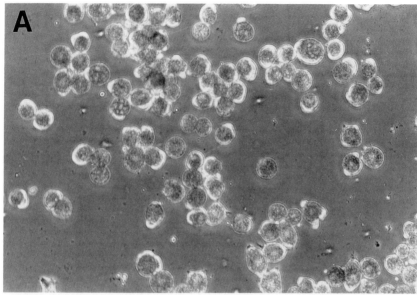

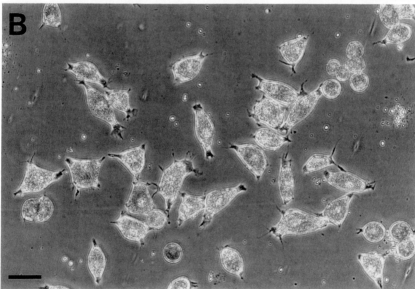

FIG. 3. Dissociated *Xenopus* animal cap cells from stage 8 embryos adhere to FN-coated substrates following induction with activin A. Substrate was prepared by treating plastic tissue culture dishes with 50 μg/ml bovine pFN in MBS followed by a "blocking" solution containing 0.5% BSA in MBS. (A) Uninduced and (B) induced cells from dissociated stage 8 animal caps cultured on FN substrates display markedly different adhesive behaviors. Induced cells (B) rapidly adhere and spread following exposure to activin, whereas uninduced cells do not. Details of these experiments are reported in Smith *et al.* (1990) and DeSimone *et al.* (1991a). (Photomicrographs courtesy of Dr. Jim Smith.) Bar in (B) equals 50 μm.

functional domains of these proteins for adhesion assays and to interfere with normal patterns of ECM expression in the embryo. This combination of *in vitro* and *in vivo* approaches will be critical in evaluating the relative importance of individual receptors and ECM molecules expressed during embryogenesis.

ACKNOWLEDGMENTS

We are grateful to Drs. Richard O. Hynes, Jim Smith, Jean-Claude Boucaut, and Chris Wylie for many helpful discussions concerning FN isolation, antibody production, and cell adhesion assays. This work is supported by grants from the National Institute of Child Health and Human Development (#R01-HD26402) and by the Pew Scholars Program in the Biomedical Sciences.

REFERENCES

DeSimone, D. W., and Hynes, R. O. (1988). *J. Biol. Chem.* **263,** 5333–5340.
DeSimone, D. W., Smith, J. C., Howard, J. E., Ransom, D. G., and Symes, K. (1991a). *In* "Gastrulation: Movements, Patterns and Molecules" (R. Keller, W. Clark, and F. Griffin, eds.). Plenum, New York.
DeSimone, D. W., Norton, and Hynes, R. O. (1991b). Submitted for publication.
Guan, J. L., Trevithick, J. E., and Hynes, R. O. (1990). *J. Cell Biol.* **110,** 833–847.
Hausen, P., and Dreyer, C. (1981). *Stain Technol.* **56,** 287–293.
Heasman, J., Wylie, C. C., Hausen, P., and Smith, J. C. (1984). *Cell (Cambridge, Mass.)* **37,** 185–194.
Hynes, R. O. (1990). "Fibronectins." Springer-Verlag, New York.
Johnson, K. E. (1984). *Am. Zool.* **24,** 605–614.
Johnson, K. E., Nakatsuji, N., and Boucaut, J. C. (1990). *In* "Cytoplasmic Organization Systems" (G. Malacinski, ed.), pp. 349–374. McGraw-Hill, New York.
Johnson, K. E., Boucaut, J. C., and DeSimone, D. W. (1991). *Curr. Top. Dev. Biol.* (in press).
Laemmli, U. K. (1970). *Nature (London)* **227,** 680–685.
Lee, G., Hynes, R. O., and Kirschner, M. (1984). *Cell (Cambridge, Mass.)* **36,** 729–740.
Mosher, D. F. (1989). "Fibronectin." Academic Press, New York.
Nakatsuji, N. (1986). *J. Cell Sci.* **86,** 109–118.
Nakatsuji, N., and Johnson, K. E. (1982). *Cell Motil.* **2,** 149–161.
Nakatsuji, N., and Johnson, K. E. (1983). *J. Cell Sci.* **59,** 61–70.
Nakatsuji, N., Smolira, M. A., and Wylie, J. C. (1985). *Dev. Biol.* **107,** 264–268.
Ransom, D. G., and DeSimone, D. W. (1990). *J. Cell Biol.* **111,** 142a.
Smith, J. C., Symes, K., Hynes, R. O., and DeSimone, D. W. (1990). *Development (Cambridge, U.K.)* **108,** 229–238.
Wedlich, D., Hacke, H., and Klein, G. (1989). *Differentiation (Berlin)* **40,** 77–83.
Winklbauer, R. (1988). *Dev. Biol.* **130,** 175–183.
Winklbauer, R. (1990). *Dev. Biol.* **142,** 155–168.
Yamada, K. M. (1982). *In* "Immunocytochemistry of the Extracellular Matrix" (H. Furthmayr, ed.), Vol. 1, pp. 111–122. CRC Press, Boca Raton, Florida.

Chapter 28

Chromatin Assembly

ALAN P. WOLFFE

Laboratory of Molecular Biology
National Institute of Diabetes, Digestive, and Kidney Diseases
National Institutes of Health
Bethesda, Maryland 20892

CAROLINE SCHILD

Institut de Biologie Animale
Université de Lausanne
CH-1015 Lausanne,
Switzerland

I. Introduction
II. Preparation of Extracts
   A. Egg Extracts
   B. An Oocyte Extract (Oocyte S150)
   C. Fractionated Systems
   D. Choosing the Best Preparation
III. Assays for Chromatin Assembly
   A. Chromatin Assembly on Double-Stranded DNA
   B. Chromatin Assembly on Replicating DNA
   C. Choosing the Best Method
IV. Unresolved Problems
   A. Special Features of Chromatin Assembled in Extracts of *Xenopus* Eggs and Oocytes
   B. Positioning and Spacing of Nucleosomes
   C. Higher-Order Structures and the Histone H1 Enigma
   D. Reconstructing Functional Chromosomes and Nuclei
References

I. Introduction

Chromatin represents an enigma for many molecular and developmental biologists. We know that the compaction of DNA into the eukaryotic nucleus is an important function for all of the proteins within the chromosome. We are aware that the structure of the nucleus changes during development, yet we are unable to relate these changes to effects on nuclear function or to understand their developmental significance. A major current challenge for molecular and developmental biologists is therefore to understand chromatin. We need to comprehend how and when chromatin structure contributes to developmentally important events such as gene activation. *Xenopus* has many advantages for these studies. Chromatin and nuclear structure change during *Xenopus* embryogenesis. These changes relate to alterations in the length of the cell cycle, the activation of transcription, and the repression of certain genes. Cell-free systems exist that are derived from *Xenopus* eggs and oocytes which allow the reconstruction of chromatin and eventually nuclei *in vitro*. The reconstruction of chromatin in these extracts is the subject of this chapter. This *in vitro* reconstruction of chromatin templates is the first step to the analysis of how transcription and replication occur within a chromosome. The solution of this problem will lead to important insights into how these processes actually occur within the nucleus *in vivo*.

The first experiments that clearly demonstrated the utility of cell-free preparations derived from *Xenopus* eggs and oocytes for assembling chromatin came from the discovery of Laskey *et al.* (1977) that cloned circular simian virus 40 (SV40) DNA would be assembled into nucleosomes by a high speed supernatant (145,000 g) of homogenized *Xenopus* eggs. Fractionation of these homogenates led to the purification of nucleoplasmin, an acidic thermostable protein which promotes nucleosome assembly from purified histones and DNA (Laskey *et al.*, 1978). More recently two other acidic proteins called N1/N2 have been purified from eggs (Kleinschmidt *et al.*, 1990). It is now clear that N1/N2 are complexed with histones H3/H4 and nucleoplasmin with histones H2A/H2B *in vivo* (Dilworth *et al.*, 1987).

The next major technical advance followed from a small change in the way *Xenopus* eggs were disrupted in order to prepare extracts. Instead of homogenizing the eggs, Lohka and Masui (1983) broke the eggs open by low speed centrifugation. The advantage of this procedure was that yolk platelets within the egg were no longer disrupted and the highly phosphorylated phosvitin and lipovitellin molecules were no longer released to inhibit chromatin assembly. Worcel and colleagues used this procedure to prepare a high speed extract of *Xenopus* oocytes (Glikin *et al.*, 1984). The properties of this oocyte S150 are perhaps the most thoroughly documented of any of the *Xenopus* chromatin assembly extracts (Shimamura *et al.*, 1988, 1989a,b; Rodriquez-Campos *et al.*, 1989; Tremethick *et al.*, 1990). It is unfortunate that, like any other process concerning *Xenopus* oocytes, the quality of the final extract depends on the quality of the oocytes used. Inexperienced investigators often have problems determining oocyte quality; hence, they may prefer to utilize similar preparations from *Xenopus* eggs.

Xenopus eggs have several advantages over oocytes for the preparation of chromatin assembly extracts. The egg is a fairly well-defined biochemical entity, whereas an oocyte may be within one of several developmental stages, each containing a distinct store of macromolecules. Aside from the physiological integrity of the *Xenopus* egg compared to an oocyte, the molecules involved in chromatin assembly may be more active when isolated from eggs. There is clear evidence that this is the case for *Xenopus* nucleoplasmin (Sealy *et al.*, 1986), which is more highly phosphorylated and more competent for depositing histones onto DNA when isolated from eggs rather than oocytes. Lohka and Masui (1985) fractionated a low speed (9000 g) extract of *Xenopus* eggs further by high speed centrifugation (150,000 g). The resulting high speed supernatant readily assembles chromatin on duplex DNA (Almouzni and Mechali, 1988a). This extract will also replicate single-stranded DNA and assemble chromatin very efficiently as relication progresses (Almouzni and Mechali, 1988b).

II. Preparation of Extracts

A. Egg Extracts

1. Egg Preparation

The crucial step in preparing *Xenopus* egg extracts is to obtain a large number of high quality, dejellied eggs. A high quality egg is defined here as having a clear demarcation between the dark (animal) and light (vegetal) hemispheres. An excellent summary of egg collection procedures has recently been described by the Chalkley laboratory (Sealy *et al.*, 1989). To obtain eggs each frog is injected into the dorsal lymph sac with 1 ml (1000 units) of human chorionic gonadotropin, 12 to 14 hours before the eggs are required. Groups of three frogs are placed onto a mesh through which eggs can fall into collection buckets. Otherwise, the frogs will eat the eggs. The frogs are covered overnight with high salt Barth's solution (110 mM NaCl, 2 mM KCl, 1 mM MgSO$_4$, 0.5 mM Na$_2$HPO$_4$, 2 mM NaHCO$_3$, 15 mM Tris-HCl, pH 7.4). It is important that the temperature of the Barth's solution should not fall below 18°–20°C, otherwise a considerably longer period of time will elapse before the frogs begin to release their eggs. The injected frogs can also be *gently* pressed on the abdomen and back to strip additional eggs from the oviduct. More than 100 ml of eggs is required to prepare a useful quantity of egg extract; more is better because of the effort expended in characterizing each individual preparation. This requires that 20–30 frogs contribute their eggs to the preparation.

The most difficult step in preparing egg extracts is to remove the jelly coat surrounding the eggs. The simplest way to do this is to dissolve the jelly in 20% modified Barth's solution [20% MBS: 18 mM NaCl, 0.2 mM KCl, 0.5 mM NaHCO$_3$, 2 mM HEPES–NaOH, pH 7.5, 0.15 mM MgSO$_4$, 50 μM Ca(NO$_3$)$_2$, 0.1 mM CaCl$_2$] containing 2% cysteine (pH 8.0). The eggs are gently swirled in a large beaker for 5 minutes or until the jelly coat dissociates from the egg. It is important that greater than 95% of the jelly coats dissociate, since contaminating jelly interferes with extract preparation. Dissociation can be easily followed by watching how closely the eggs pack together: they occupy a much smaller volume without the jelly coat. As soon as the jelly coat dissociates, the eggs should be repeatedly washed with large volumes of 20% MBS. At this point the eggs should be quickly sorted, with damaged or discolored eggs being discarded. Preparations of egg extracts should not begin with less than 90% high quality eggs.

2. Low Speed Supernatant

The simplest extract preparation capable of assembling chromatin on exogenous DNA (Wolffe and Brown, 1987) and eventually nuclei (Blow and Laskey, 1986) is made by breaking open dejellied eggs at a low centrifugal force (Lohka and Masui, 1983). The dejellied eggs are washed thoroughly in 20% MBS, followed by ice-cold extraction buffer [50 mM HEPES–KOH (pH 7.4), 50 mM KCl, 5 mM MgCl$_2$, 2 mM 2-mercaptoethanol, 3 μg/ml leupeptin (Sigma Chemical Co., St. Louis, MO), and 10 μg/ml cytochalasin B]. The eggs are packed into 15-ml Corex tubes and centrifuged in a swing-out HB-4 rotor at 10,000 g (8000 rpm) using a Sovrall RC 5B centrifuge for 30 minutes. This produces four major fractions (Fig. 1A): an insoluble plug of yolk platelets and pigment, a brown cytoplasmic layer, a pale yellow layer containing buffer and some cytoplasmic components, and a lipid pellicle. The brown cytoplasmic layer is removed using a pipette and recentrifuged as above, in order to remove residual debris. (An alternative to the HB-4 and RC5-B centrifugation is to use a Beckman ultracentrifuge and SW50Ti rotor; the corresponding volumes are 5 ml, and the spin is at 9000 rpm.) The extract

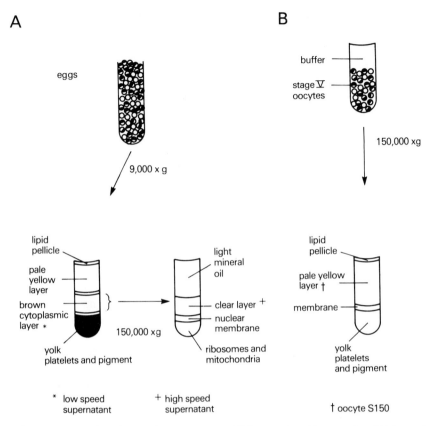

FIG. 1. Preparation of egg and oocyte extracts. (A) Preparation of low speed and high speed chromatin assembly extracts. (B) Preparation of oocyte S150 chromatin assembly extract. For details, see text.

can be frozen at $-70°C$ without affecting any chromatin assembly capacity and is stable for several months.

3. High Speed Supernatant

The high speed supernatant retains all of the chromatin assembly capability of the low speed supernatant but will not form "pseudonuclei" owing to the absence of components such as the nuclear membrane (Lohka and Masui, 1985; Wolffe and Brown, 1987). The first step in the preparation of this extract is to prepare the low speed supernatant. This extract is then centrifuged at

150,000 g for 60 minutes at 4°C, typically in a Beckman ultracentrifuge using a SW50Ti rotor at 40,000 rpm. Ribosomes and mitochondria form a golden brown pellet; the pellet is covered with a membrane fraction containing nuclear membrane and annulate lamellae. The low speed supernatant is above these components and can be taken off with a pipette (Fig. 1A). If the tube cannot be filled with eggs, it helps to cover the eggs with light mineral oil. The clear high speed supernatant can be frozen at $-70°C$ without adverse effects on DNA replication or chromatin assembly.

4. Egg Homogenates and Heat Supernatant

Extracts capable of chromatin assembly have been described which use eggs that have been disrupted by homogenization (Laskey et al., 1977). These preparations assemble chromatin, but with less reproducibility than when eggs are disrupted by centrifugation. The major problem appears to be yolk contamination following homogenization. Egg homogenates will still assemble chromatin after heating to 80°C for 10 minutes to precipitate the majority of proteins present if exogenous histones are added. This reaction depends on the heat-stable molecule nucleoplasmin, which remains in solution (Laskey et al., 1978; Sealy et al., 1989). Nucleoplasmin is capable of assembling nucleosomes in the presence of all four core histones; however, this is not a physiological reaction and is not recommended for routine use. For extensive discussion of the problems concerning this system, see Rhodes and Laskey (1989). A separate issue concerning the use of this system is that other soluble denatured proteins present in the heat-treated extracts may well interfere with the processes of interest under investigation, for example, transcription or replication.

B. An Oocyte Extract (Oocyte S150)

Extracts of stage V oocytes have been extensively used to study chromatin assembly and the subsequent effects on class III gene transcription (Glikin et al., 1984; Shimamura et al., 1988, 1989a). A detailed description of the basic protocols used in this work from Worcel's laboratory has recently been published (Shimamura et al., 1989b).

The protocol is similar to that of preparing the egg extracts with two exceptions. First, the follicle must be removed from the oocyte, and, second, the oocyte cytoplasm is diluted following centrifugation of the oocytes. Oocytes are defolliculated by digestion with collagenase (0.2% w/v) in OR2 buffer (5 mM HEPES, pH 7.6, 1 mM Na_2HPO_4, 82.5 mM NaCl, 2.5 mM KCl, 1 mM $CaCl_2$, 1 mM $MgCl_2$). Digestion is generally for 3–6 hours at room temperature with constant gentle shaking for example, in a 50-ml conical capped tube. Dispersed oocytes are transferred using a plastic transfer pipette to a beaker where they are washed thoroughly with ice-cold extraction buffer [20 mM HEPES, pH 7.5, 5 mM KCl, 1.5 mM MgCl$_2$, 1 mM EGTA, 10% glycerol, 10 mM β-glycerophosphate, 0.5 mM dithiothreitol (DTT)]. The oocytes are then transferred so as to halfway fill a 5-ml tube for the Beckman SW50.1 rotor. The tube is filled with extraction buffer, and the oocytes are disrupted by centrifuation in the Beckman SW50.1 rotor for 30 minutes at 150,000 g (40,000 rpm). The clear supernatant that forms below the lipid pellicle and above the yolk pellet is removed using a pipette (Fig. 1B). This is the oocyte S150 chromatin assembly extract, which can be stored at $-70°C$ for several months without loss of activity.

C. Fractionated Systems

Considerable effort has led to a detailed understanding of the biochemistry of chromatin assembly in *Xenopus* egg and oocyte extracts. These systems may be of use to investigators who detect effects of chromatin assembly on the biological activity of interest in the crude egg or oocyte extracts. This particular methods chapter is an inappropriate place to review this work; however, the reader is referred to papers from the Laskey (Dilworth *et al.*, 1987), Franke (Kleinschmidt *et al.*, 1990), and Worcel laboratories (Tremethick *et al.*, 1990).

D. Choosing the Best Preparation

Which preparation for chromatin assembly is appropriate for a particular experiment depends on the biological activity under investigation. A major consideration is the reproducibility of the chromatin assembly system. There is clearly variation between egg and oocyte extracts, and among individual extracts. In general, egg extracts are more reproducible simply because an egg is a more defined cell than an oocyte. The egg extracts are also prepared without the addition of exogenous proteases to separate the follicle cells.

The concentration of protein and RNA in the various extracts may influence the type of experiment that can be carried out. The low speed egg extract has a protein concentration of 40–50 mg/ml and an RNA content of 7 mg/ml. These protein and RNA concentrations correspond to those estimated to occur in *Xenopus* egg cytoplasm. The high speed egg extract has low levels of contamination with RNA but a protein content of 20–30 mg/ml. In contrast, the oocyte S150 has variable ribosomal RNA contamination owing to the low Mg^{2+} content in the extraction buffer and has a protein content of 3–5 mg/ml.

It should be emphasized that all of these preparations contain activities other than those involved in chromatin assembly. The low speed egg extract will almost do everything we might except of the living egg cytoplasm, namely, replication, transcription, and recombination. The high speed egg extract is competent for all these processes to a lesser degree (Almouzni and Mechali, 1988a; Tafuri and Wolffe, 1990). The oocyte extract will transcribe class II and class III genes (Shimamura et al., 1988; Corthesy et al., 1990). Careful controls are required to exclude endogenous activities when investigating any process in *Xenopus* extracts.

III. Assays for Chromatin Assembly

There are three simple assays for chromatin assembly that make use of the change in properties of DNA following the association of histones. The first follows from the fact that each nucleosome introduces a negative superhelical turn into DNA in the presence of topoisomerase. The second assay makes use of the fact that DNA in a nucleosome is inaccessible to nucleases relative to the DNA between nucleosomes. Finally, the association of histones with DNA generates a structure that has a greater mass and a more compact structure than DNA alone.

A. Chromatin Assembly on Double-Stranded DNA

The type of chromatin assembly reaction used depends on the mass of DNA that is to be reconstituted and the amount of extract that is available. If a large amount of DNA is to be assembled in unlimited quantities of extract, then the following protocol can be used. The same protocol is applicable irrespective of the extract preparation.

To 1 μg of DNA (1 μl) add 200 μl of extract in a 1.5-ml microcentrifuge tube plus 20 μl of 30 mM ATP and 50 mM MgCl$_2$ at room temperature for 4–6 hours. At the end of this time chromatin assembly will be complete. If small amounts of radioactive DNA can be used, the reaction can be scaled down 10- to 20-fold without any problems. A simple procedure for generating adequate quantities of radioactive plasmid DNA for the supercoiling assay is to linearize the plasmid at a unique restriction site, treat the termini with alkaline phosphatase, use T4 polynucleotide kinase to radiolabel the 5′ termini, and then religate the DNA with T4 ligase (Razvi et al., 1983). Once the chromatin assembly process is complete the next step is to assay for the presence of nucloeosomes on DNA.

1. Supercoiling Assay

The chromatin assembly reaction is stopped by the addition of 1/4 volume of a 2.5% sarkosyl, 100 mM EDTA solution. One microliter of a 10 mg/ml RNase A solution (DNase free) is added and incubation continued at 37°C for 30 minutes. This step removes any RNA in the egg extract. Removal of RNA is necessary for the clear visualization of ethidium bromide-stained DNA following resolution on agarose gels (see below). Proteins are removed by adding sodium dodecyl sulfate (SDS) to a concentration of 0.2% and proteinase K to a final concentration of 1 µg/ml. After a 1-hour incubation at 37°C, the reaction is extracted with phenol–chloroform. The aqueous supernatant is made 3 M with ammonium acetate, 2 volumes of ethanol is added, and the tube is placed on dry ice for 5 minutes before centrifugation at 10,000 g for 10 minutes. The pellet is washed with 80% ethanol, dried, and resuspended in TE buffer (10 mM Tris-HCl, 1 mM EDTA, pH 7.4). One-tenth volume of 50% glycerol containing bromophenol blue is added, and the samples are electrophoresed on 0.8% agarose gels. A typical supercoiling reaction for radiolabeled double-stranded DNA is shown in Fig. 2.

2. Micrococcal Nuclease Digestion

Chromatin is assembled onto double-stranded DNA as described above. Nuclease digestion is carried out by adding 1/10 volume of 30 mM $CaCl_2$ followed by 2.4 units of micrococcal nuclease (Boehringer Mannheim, Indianapolis, IN) per microliter of extract at room temperature. Aliquots are removed after various times (typically 2–20 minutes). Each aliquot has 1/4 volume of 2.5% sarkosyl–100 mM EDTA added, then endogenous RNA is digested by the addition of 1 µl of a 10 mg/ml RNase A solution, followed by incubation at 37°C for 30 minutes. Proteins are digested by adding SDS and proteinase K to final concentrations of 0.2% and 1 mg/ml, respectively, and incubating the reaction for 1 hour at 37°C. The samples are then extracted with phenol–chloroform. The aqueous phase is made 3 M with ammonium acetate, 2 volumes of ethanol is added, and the tube is placed on dry ice for 5 minutes. Nucleic acids are pelleted by centrifugation for 10 minutes at 10,000 g, washed with 80% ethanol, and resolved on a 2.0% agarose gel containing TAE (40 mM Tris–acetate, 1 mM EDTA). A typical result using radioactive double-stranded DNA as the starting material is shown in Fig. 3.

3. Sucrose Gradient Centrifugation

A detailed treatment of sucrose gradient centrifugation techniques in the analysis of chromatin structure has been presented recently (Noll and Noll,

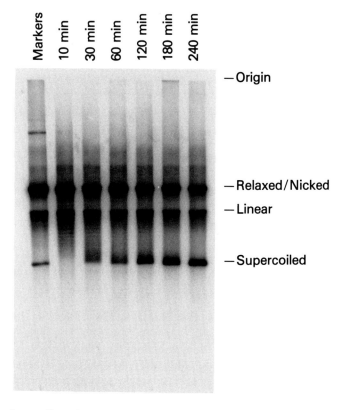

FIG. 2. Supercoiling of double-stranded DNA. Double-stranded DNA was prepared by replicating single-stranded M13 DNA in the egg extract in the presence of [α-^{32}P]dATP. Chromatin was assembled as described in the text. Aliquots were taken after 10, 30, 60, 120, 180, and 240 minutes, processed for electrophoresis on 0.8% agarose gels, and autoradiographed for 1 hour with X-ray film.

1989). In practice it is fairly simple and useful to resolve DNA molecules that have undergone varying degrees of chromatin assembly on sucrose gradients (Laskey et al., 1977; Shimamura et al., 1988). Free DNA is left at the top of the gradient.

Following chromatin assembly, samples for sedimentation are diluted to 300 µl by the addition of 10 mM EDTA, 10 mM Tris-HCl (pH 7.0), 0.25% Triton X-100. Samples are then loaded onto 5-ml 5–20% sucrose gradients containing 60 mM KCl, 20 mM Tris-HCl (pH 7.5), and 1 mM EDTA. Centrifugation is for 2.5 hours at 40,000 rpm (150,000 g) in the Beckman SW50.1Ti rotor at 4°C. Fractions (100 µl) are collected and aliquots resolved on a 1% agarose gel if nonradioactive DNA is used for the assembly reaction. If

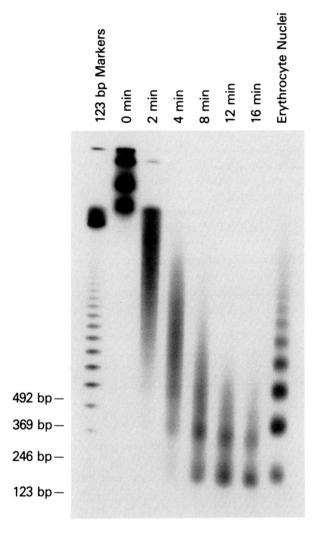

Fig. 3. Micrococcal nuclease digestion of chromatin assembled on double-stranded DNA. Chromatin was assembled for 4 hours on radioactive double-stranded DNA as described in the text. The reaction was made 3 mM with $CaCl_2$, and micrococcal nuclease (300 U/μg DNA) was added at room temperature. Aliquots were taken after 0, 2, 4, 8, 12, and 16 minutes and processed as described before resolution on a 2% agarose gel. Molecular weight markers are a 123-base pair (bp) DNA ladder (BRL, Gaithersburg, MD) and a micrococcal nuclease digestion of erythrocyte nuclei, both radiolabeled with [γ-^{32}P]ATP and T4 polynucleotide kinase.

radioactive DNA is used in the experiment, the various fractions can be assayed by determining the distribution of radioactivity in the gradient (Fig. 4).

B. Chromatin Assembly on Replicating DNA

In vivo, chromatin assembly occurs on replicating DNA (Worcel *et al.*, 1978; Senshu *et al.*, 1978). A role for the replication process itself in promoting chromatin assembly has been suggested from *in vitro* experiments using mammalian (Stillman, 1986; Stillman and Gluzman, 1985) and *Xenopus* cell-free extracts (Almouzni and Mechali, 1988a; Almouzni *et al.*, 1990b). Chromatin assembly on replicating single-stranded (M13) DNA is very efficient and rapid compared to that on double-stranded DNA. We regard it as the method of choice. However, the following protocol has only been examined in detail using the high speed egg supernatant.

To 1–3 μg of DNA (1–3 μl of a 1 μg/μl stock of single-stranded M13 DNA) add 200 μl of the high speed egg extract in a 1.5-ml microcentrifuge tube, supplementing the reaction with 20 μl of 30 mM ATP and 50 mM MgCl$_2$. The reaction is incubated for 1–2 hours at room temperature. If desired, the efficiency of replication and DNA supercoiling can be monitored

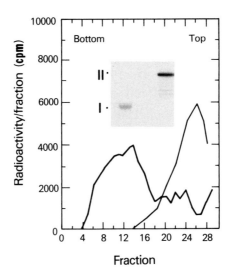

FIG. 4. Sucrose gradient sedimentation. Chromatin was assembled for 4 hours on radioactive double-stranded DNA as described in the text. Chromatin and relaxed naked DNA were resolved on sucrose gradients as described; the distributions of radioactivity are shown. The DNA in fractions 10–14 is shown at left and that in fractions 24–28 is shown at right, resolved in a 0.8% agarose gel. The positions of supercoiled (I) and nicked DNA (II) are indicated.

by the inclusion of 1–10 μCi of [α-^{32}P]dATP in the reaction mixture. Examples of DNA supercoiling owing to chromatin assembly and micrococcal nuclease digestions of the reconstitutes are shown in Fig. 5.

C. Choosing the Best Method

Which procedure is used for chromatin assembly depends once more on the biological activity under investigation. If the efficient assembly of a chromatin template is required, we strongly recommend the use of the high speed egg extract and replicating single-stranded DNA as template. The only limitation to this technique is the fact that most laboratories do not routinely prepare constructs in M13. However, many plasmids now carry appropriate bacteriophage origins of replication that can readily yield single-stranded DNA with the appropriate helper phage. Procedures for cloning into M13 and preparing single-stranded DNA are straightforward and well documented (Maniatis *et al.*, 1982). This methodology also allows simple assays for chromatin assembly following incorporation of [α-^{32}P]dATP into the reaction mixture. Other experiments using this method of chromatin assembly have suggested that it has the closest parallels to *in vivo* chromatin assembly yet described (Almouzni *et al.*, 1990a,b, 1991).

IV. Unresolved Problems

A. Special Features of Chromatin Assembled in Extracts of *Xenopus* Eggs and Oocytes

Chromatin assembly systems derived from *Xenopus* eggs and oocytes are possible to create simply because of the enormous stores of histones that accumulate during *Xenopus* oogenesis (Woodland, 1980). There is sufficient excess histone to assemble over 20,000 nuclei. The stored histones are highly modified and exist in both acetylated and phosphorylated forms (Woodland, 1979). Moreover, the histones are stored in the egg as soluble complexes by association with the acidic polypeptides nucleoplasmin (Laskey *et al.*, 1978; Earnshaw *et al.*, 1980) and N1/N2 (Kleinschmidt and Franke, 1982; Kleinschmidt *et al.*, 1990). This presents two problems: first, nucleoplasmin is not ubiquitous, for example, somatic cell nuclei exist (in *Xenopus laevis* liver, kidney, brain, heart, and pancreas) in which nucleoplasmin cannot be detected (Dreyer and Hausen, 1983), raising questions as to whether nucleoplasmin functions in somatic cells; second, the histones found complexed with nucleoplasmin and N1/N2 correspond to basal variants described by Wu

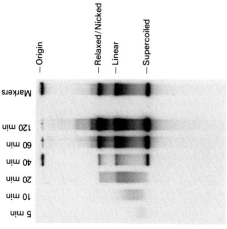

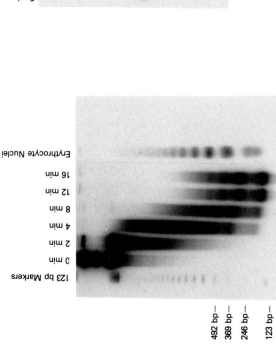

FIG. 5. Replication and chromatin assembly of single-stranded DNA incubated in the egg extract. (A) Replication of single-stranded DNA is as described in the text. Reactions were stopped after 5, 10, 20, 40, 60, and 120 minutes of incubation at 22°C. After phenol extraction and ethanol precipitation, the samples were resolved in a 0.8% agarose gel. (B) After a 2-hour incubation, the reaction was made 3 mM which $CaCl_2$ and micrococcal nuclease added (300 units/μg DNA). Digestion at room temperature was allowed to proceed for 0, 2, 4, 8, 12, and 16 minutes, before aliquots were removed and the nuclei acid extracted and resolved in a 2% agarose gel. Molecular weight markers are a 123-bp DNA ladder (BRL) and a micrococcal nuclease digestion of erythrocyte nuclei, both radiolabeled with [γ-^{32}P]ATP and T4 polynucleotide kinase. (A) and (B) are autoradiographs.

and Bonner (1981) that, unlike S-phase histones, are synthesized throughout the cell cycle independent of DNA replication (Dilworth et al., 1987). The same histones are assembled onto DNA in the chromatin assembly reaction (Dilworth et al., 1987; Shimamura et al., 1988). The *Xenopus* oocyte and egg extracts may therefore represent a specialized example of chromatin assembly that occurs outside of S phase (Dilworth and Dingwall, 1988). Chromatin assembled in these systems may have effects on biological activities distinct from those seen in normal somatic cells.

B. Positioning and Spacing of Nucleosomes

Recent progress in understanding the structure of the nucleosome has enabled experiments to be carried out which clearly demonstrate that the position of nucleosomes relative to a particular DNA sequence can have major effects on the interaction of other proteins with the sequence (Wolffe and Drew, 1989; Simpson, 1990). Other experiments using yeast have shown that the precise architecture of a gene promoter in terms of both histone–and transcription factor–DNA interactions can have significant effects on the regulation of transcription from that promoter (reviewed by Grunstein, 1990). Similar effects may occur on other biological processes assayed on chromatin templates assembled in *Xenopus* extracts. Worcel and colleagues have convincingly demonstrated that nucleosomes assembled in the oocyte S150 extract will position themselves on a *Xenopus borealis* somatic 5 S RNA gene exactly as if reconstituted from purified components on the same DNA (Rhodes, 1985; Shimamura et al., 1988; Hayes et al., 1990). However, positioning of nucleosomes on *Xenopus* 5 S DNA does not always occur in these extracts (Tremethick et al., 1990; Almouzni et al., 1990a).

Although a single nucleosome will position itself on a particular DNA segment because of special properties of that sequence (Hayes et al., 1990), a more difficult problem concerns how other nucleosomes away from such a special region of DNA are arrayed relative to each other. Many systems using purified components reconstitute nucleosomes onto DNA (see Wassarman and Kornberg, 1989); however, only crude extracts such as those prepared from *Xenopus* eggs and oocytes generate nucleosomes that have a physiological spacing (180–190 bp; Lam and Carroll, 1983). What is responsible for this difference is unkown. However, it is clear that in the *Xenopus* S150 two factors can influence the spacing between nucleosomes. The first variable is the number of nucleosomes assembled onto the template (Shimamura et al., 1988). When the template is only 50% occupied with nucleosomes, correct spacing occurs, but when more nucleosomes are assembled, they pack together until the length of DNA per nucleosome declines from 180–190 to 160 bp. The second variable is the presence of exogenous histone H1 in the extract

(Rodriquez-Campos et al., 1989). When exogenous histone H1 is added to an oocyte S150, the spacing of nucleosomes can be altered during assembly such that they do not become close packed. Why this should happen is not resolved.

C. Higher-Order Structures and the Histone H1 Enigma

A major problem with respect to chromatin assembly in egg and oocyte extracts relates to the presence or absence of histone H1. Direct analysis of the proteins associating with DNA during chromatin assembly in *Xenopus* extracts does not reveal any histone H1 (Dilworth et al., 1987; Shimamura et al., 1988). Moreover, indirect assays relating to the repression of genes and nucleosome spacing suggest that histone H1 is not present on the assembled chromatin (Wolffe, 1989a,b; Shimamura et al., 1989; Rodriquez-Campos et al., 1989b). These experiments do suggest that if exogenous histone H1 is added to *Xenopus* extracts, then that exogenous histone H1 can be assembled into chromatin in the correct way.

Developmental analysis of histone H1 gene expression and synthesis during oogenesis and embryogenesis suggests that histone H1 protein is deficient in mature oocytes and eggs relative to core histones (Van Dongen et al., 1983; Wolffe, 1989b). In contrast, large stores of histone H1 mRNA exist whose translation is activated following fertilization (Woodland, 1980). However, at each time during development there is clearly enough of the somatic form of histone H1 to assemble all of the endogenous DNA in the embryo. What is questionable is whether there is enough free histone H1 protein in oocyte and egg extracts to be detectable in the assembled minichromosomes. The situation is further complicated by the detection of a histone H1-like molecule in *Xenopus* oocytes and early embryos, whose expression ceases in later development (Smith et al., 1988).

Xenopus egg extracts clearly have the capacity to condense chromatin into chromosomes (Lohka and Masui, 1985; Newport, 1987). Our current understanding of chromosome structure suggests that histone H1 or a molecule like it must mediate this process (Felsenfeld and McGhee, 1986). It should be noted that the formation of higher-order chromatin structures such as the 30-nm fiber is probably irrelevant to most experiments on small plasmid molecules below 4–5 kilobases (kb) owing to the topological problems of folding such as small number of nucleosomes into a fiber. Clearly many interesting experiments remain to be carried out with respect to resolving this enigma.

D. Reconstructing Functional Chromosomes and Nuclei

The addition of core histones and histone H1 to DNA to assemble chromatin is only the first step in the reconstruction of chromosomes and nuclei.

All of the later processes concerning chromosome and nuclear assembly are possible in the low speed egg extract and are reviewed in detail in Chapter 31 of this volume.

ACKNOWLEDGMENTS

I thank Dr. D. Clark for review of the manuscript and much valuable discussion. I am also grateful to Ms. Thuy Vo for preparing the manuscript.

REFERENCES

Almouzni, G., and Mechali, M. (1988a). Assembly of spaced chromatin promoted by DNA synthesis in extracts from *Xenopus* eggs. *EMBO J.* **7,** 665–672.
Almouzni, G., and Mechali, M. (1988b). Assembly of spaced chromatin: Involvement of ATP and DNA topoisomerase activity. *EMBO J.* **7,** 4355–4365.
Almouzni, G., Mechali, M., and Wolffe, A. P. (1990a). Competition between transcription complex assembly and chromatin assembly on replicating DNA. *EMBO J.* **9,** 573–582.
Almouzni, G., Clark, D. J., Mechali, M., and Wolffe, A. P. (1990b). Chromatin assembly on replicating DNA *in vitro*. *Nucleic Acids Res.* **18,** 5767–5774.
Almouzni, G., Mechali, M., and Wolffe, A. P. (1991). Transcription complex disruption caused by a transition in chromatin structure. *Mol. Cell. Biol.* **11,** 655–665.
Blow, J. J., and Laskey, R. A. (1986). Initiation of DNA replication in nuclei and purified DNA by a cell-free extract of *Xenopus* eggs. *Cell (Cambridge, Mass.)* **47,** 577–587.
Corthesy, B., Leonnard, P., and Wahli, W. (1990). Transcriptional potentiation of the vitellogenin B1 promoter by a combination of both nucleosome assembly and transcription factors: An *in vitro* dissection. *Mol. Cell. Biol.* **10,** 3926–3933.
Dilworth, S. M., and Dingwall, C. (1988). Chromatin assembly *in vitro* and *in vivo*. *BioEssays* **9,** 44–49.
Dilworth, S. M., Black, S. J., and Laskey, R. A. (1987). Two complexes that contain histones are required for nucleosome assembly *in vitro*: Role of nucleoplasmin and N1 in *Xenopus* egg extracts. *Cell (Cambridge, Mass.)* **51,** 1009–1018.
Dreyer, C., and Hausen, P. (1983). Two-dimensional gel analysis of the fate of oocyte nuclear proteins in the development of *Xenopus laevis*. *Dev. Biol.* **100,** 412–425.
Earnshaw, W. C., Honda, B. M., Laskey, R. A., and Thomas, J. O. (1980). Assembly of nucleosomes: The reaction involving *X. laevis* nucleoplasmin. *Cell (Cambridge, Mass.)* **21,** 373–383.
Felsenfeld, G., and McGhee, J. D. (1986). Structure of the 30 nm chromatin fiber. *Cell (Cambridge, Mass.)* **44,** 375–377.
Glikin, G. C., Ruberti, I., and Worcel, A. (1984). Chromatin assembly in *Xenopus* oocytes: *In vitro* studies. *Cell (Cambridge, Mass.)* **37,** 33–41.
Grunstein, M. (1990). Histone function in transcription. *Annu. Rev. Cell Biol.* **6,** 643–678.
Hayes, J. J., Tullius, T. D., and Wolffe, A. P. (1990). The structure of DNA in a nucleosome. *Proc. Natl. Acad. Sci. U.S.A.* **87,** 7405–7409.
Kleinschmidt, J. A., and Franke, W. W. (1982). Soluble acidic complexes containing histones H3 and H4 in nuclei of *Xenopus laevis* oocytes. *Cell (Cambridge, Mass.)* **29,** 799–809.
Kleinschmidt, J. A., Seiter, A., and Zentgraf, H. (1990). Nucleosome assembly *in vitro*: Separate histone transfer and synergistic interaction of native histone complexes purified from nuclei of *Xenopus laevis* oocytes. *EMBO J.* **9,** 1309–1318.
Lam, B. S., and Carroll, D. (1983). Tandemly repeated DNA sequences from *Xenopus laevis*.

I. Studies on sequence organization and variation in satellite I DNA (741 base-pair repeat). *J. Mol. Biol.* **165,** 567–585.

Laskey, R. A., Mill, A. D., and Morris, N. R. (1977). Assembly of SV40 chromatin in a cell-free system from *Xenopus* eggs. *Cell (Cambridge, Mass.)* **10,** 237–243.

Laskey, R. A., Honda, B. M., Mills, A. D., and Finch, J. T. (1978). Nucleosomes are assembled by an acidic protein which binds histones and transfers them to DNA. *Nature (London)* **275,** 416–420.

Lohka, M. J., and Masui, Y. (1983). Formation *in vitro* of sperm pronuclei and mitotic chromosomes induced by amphibian ooplasmic components. *Science* **220,** 719–721.

Lohka, M. J., and Masui, Y. (1985). Roles of cytosol and cytoplasmic particles in nuclear envelope assembly and sperm pronuclear formation in cell-free preparation from amphibian eggs. *J. Cell Biol.* **98,** 1222–1230.

Maniatis, T., Fritsch, E. F., and Sambrook, J. (1982). "Molecular Cloning: A Laboratory Manual." Cold Spring Harbor Laboratory, Cold Spring Harbor, New York.

Newport, J. (1987). Nuclear reconstitution *in vitro*: Stages of assembly around protein-free DNA. *Cell (Cambridge, Mass.)* **48,** 205–217.

Noll, H., and Noll, M. (1989). Sucrose gradient techniques and applications to nucelosome structure. *In* "Methods in Enzymology" (P. M. Wassarman and R. D. Kornberg, eds., Vol. 170, pp. 55–116. Academic Press, New York.

Razvi, F., Garfiulo, G., and Worcel, A. (1983). A simple procedure for parallel sequence analysis of both strands of 5′-labelled DNA. *Gene* **23,** 175–183.

Rhodes, D. (1985). Structural analysis of a triple complex between the histone octamer. a *Xenopus* gene for 5S RNA and transcription factor IIIA. *EMBO J.* **4,** 3473–3482.

Rhodes, D., and Laskey, R. A. (1989). Assembly of nucleosomes and chromatin *in vitro*. *In* "Methods in Enzymology" (P. M. Wassarman and R. D. Kornberg, eds.) Vol. 170, pp. 575–585. Academic Press, New York.

Rodriquez-Campos, A., Shimamura, A., and Worcel, A. (1989). Assembly and properties of chromatin containing histone H1. *J. Mol. Biol.* **209,** 135–150.

Sealy, L., Cotten, M., and Chalkley, R. (1986). *Xenopus* nucleoplasmin: Egg vs oocyte. *Biochemistry* **25,** 3064–3072.

Sealy, L., Burgess, R. R., Cotten, M., and Chalkley, R. (1989). Purification of *Xenopus* egg nucleoplasmin and its use in chromatin assembly *in vitro*. *In* "Methods in Enzymology" (P. M. Wassarman and R. D. Kornberg, eds.), Vol. 170, pp. 612–630. Academic Press, New York.

Senshu, T., Fukuda, M., and Ohashi, M. (1978). Preferential association of newly synthesized H3 and H4 histones with newly replicated DNA. *J. Biochem. (Tokyo)* **84,** 985–988.

Shimamura, A., Tremethick, D., and Worcel, A. (1988). Characterization of the repressed 5S DNA minichromosomes assembled *in vitro* with a high speed supernatant of *Xenopus laevis* oocytes. *Mol. Cell. Biol.* **8,** 4257–4269.

Shimamura, A., Sapp, M., Rodriquez-Campos, A., and Worcel, A. (1989a). Histone H1 represses transcription from minichromosomes assembled *in vitro*. *Mol. Cell. Biol.* **9,** 5573–5584.

Shimamura, A., Jessee, B., and Worcel, A. (1989b). Assembly of chromatin with oocyte extracts. *In* "Methods Enzymology" (P. M. Wassermaan and R. D. Kornberg, eds.), Vol. 170, pp. 603–612. Academic Press, New York.

Simpson, R. T. (1990). Nucleosome positioning can affect the function of cis-acting DNA element *in vivo*. *Nature (London)* **343,** 387–389.

Smith, R. C., Dworkin-Rastl, E., and Dworkin, M. B. (1988). Expression of a histone H1-like protein is restricted to early *Xenopus* development. *Genes Dev.* **2,** 1284–1295.

Smith, S., and Stillman, B. (1989). Purification and characterization of CAF-1 a human cell factor required for chromatin assembly during DNA replication *in vitro*. *Cell (Cambridge, Mass.)* **58,** 12–25.

Stillman, B. (1986). Chromatin assembly during SV40 DNA replication *in vitro*. *Cell (Cambridge, Mass.)* **45,** 555–565.

Stillman, B. (1989). Initiation of eukaryotic DNA replication *in vitro*. *Annu. Rev. Cell Biol.* **5,** 197–245.

Stillman, B. W., and Gluzman, Y. (1985). Replication and supercoiling of simian virus 40 DNA in cell extracts from human cells. *Mol. Cell. Biol.* **5,** 2051–2060.

Tafuri, S., and Wolffe, A. P. (1990). The *Xenopus* Y-box transcription factors, molecular cloning, functional analysis and developmental regulation. *Proc. Natl. Acad. Sci. U.S.A.* **87,** 9028–9032.

Tremethick, D., Zucker, K., and Worcel, A. (1990). The transcription complex of the 5S RNA gene, but not transcription factor IIIA alone, prevents nucleosomal repression of transcription. *J. Biol. Chem.* **265,** 5014–5023.

Van Dongen, W. M. A. M., Moorman, A. F. M., and Destree, O. H. J. (1983). The accumulation of the maternal pool of histone H1A during oogenesis in *Xenopus laevis*. *Cell Differ.* **12,** 257–264.

Wassarman, P. M., and Kornberg, R. D., eds. (1989). "Methods in Enzymology" Vol. 170 (Nucleosomes). Academic Press, New York.

Wolffe, A. P. (1989a). Transcriptional activation of *Xenopus* class III genes in chromatin isolated from sperm and somatic nuclei. *Nucleic Acids Res.* **17,** 767–780.

Wolffe, A. P. (1989b). Dominant and specific repression of *Xenopus* oocyte 5S RNA genes and satellite I DNA by histone H1. *EMBO J.* **8,** 527–537.

Wolffe, A. P., and Brown, D. D. (1987). Differential 5S RNA gene expression *in vitro*. *Cell (Cambridge, Mass.)* **51,** 733, 740.

Wolffe, A. P., and Drew, H. R. (1989). Initiation of transcription on nucleosomal templates. *Proc. Natl. Acad. Sci. U.S.A.* **86,** 9817–9821.

Woodland, H. R. (1979). The modification of stored histones H3 and H4 during the oogenesis and early development of *Xenopus laevis*. *Dev. Biol.* **68,** 360–370.

Woodland, H. R. (1980). Histone synthesis during the development of *Xenopus*. *FEBS Lett.* **121,** 1–7.

Worcel, A., Han, S., and Wong, M. L. (1978). Assembly of newly replicated chromatin. *Cell (Cambridge, Mass.)* **15,** 969–977.

Wu, R. S., and Bonner, W. M. (1981). Separation of basal histone synthesis from S-phase histone synthesis in dividing cells. *Cell (Cambridge, Mass.)* **27,** 321–330.

Chapter 29

DNA Replication in Cell-Free Extracts from *Xenopus laevis*

GREGORY H. LENO AND RONALD A. LASKEY

Wellcome Trust and Cancer Research Campaign
Institute of Cancer and Developmental Biology
University of Cambridge
Cambridge CB2 1QR, England

I. Introduction
II. Preparation of Egg and Oocyte Extracts
   A. Buffers for Extract Preparation
   B. Preparation and Partial Fractionation of Egg Extracts
   C. Preparation and Partial Fractionation of Oocyte Extracts
III. Analysis of DNA Replication in *Xenopus* Cell-Free Systems
   A. Preparation of DNA Templates
   B. *In Vitro* Conditions for DNA Replication
   C. Assays for DNA Replication
IV. Selecting an Appropriate System
   References

I. Introduction

Eggs of *Xenopus laevis* have proved to be exceptionally favorable sources of cell-free systems for studying chromosome replication and assembly of the cell nucleus. Systems have been developed that assemble nucleosomes and nuclear envelopes on purified DNA, decondense highly compact sperm chromatin, induce interphase nuclei to enter mitosis, initiate and complete semiconservative DNA replication on either nuclei or purified DNA, and even undergo spontaneous mitotic cycles *in vitro*. Some but not all of these activities can also be performed by oocyte extracts. Oocytes, the female germ cells in the ovary, are in prolonged meiotic prophase. During meiotic maturation the oocyte nuclear envelope breaks down and the chromosomes condense. They arrest again at the second meiotic metaphase to undergo ovulation and await fertilization. This stage is the "unfertilized egg," which is the source of most of the extracts described in this chapter.

The development of cell-free systems has been facilitated by earlier microinjection studies. Thus, the behavior of DNA templates or nuclei injected into eggs or oocytes has been documented (Laskey et al., 1985), and these studies have defined a performance target for the development of cell-free systems that mimic microinjected cells.

In this chapter, we focus primarily on extracts that replicate DNA and assemble nuclei *in vitro*. Nucleosome assembly systems have been reviewed previously (Rhodes and Laskey, 1989), and the use of egg extracts to study other aspects of the cell cycle is covered in Chapter 15 of this volume. *Xenopus* egg extracts are presently unique as the only eukaryotic cell-free systems that are able to initiate and complete efficient semiconservative DNA replication *in vitro*. Not only can they replicate nuclei, but, remarkably, they can also replicate purified DNA, under strict cell cycle control, irrespective of the DNA sequence.

A crucial feature that allows such efficient cell-free systems is the extraordinarily rapid rate of cell division cycles following fertilization. The first division occurs 90 minutes after fertilization, and the next 11 cell cycles last only 30 minutes each. To sustain these remarkable rates, each egg is provided with a maternal stockpile of materials for chromosome replication and nuclear assembly. This is the key advantage of *Xenopus* eggs for the study of these processes.

In the next section, we describe the methodological details for preparation of a range of specific extracts from *Xenopus* eggs and oocytes. We then consider the methods of studying replication in the extracts in Section III. Finally, in Section IV, we consider the choice of extract for specific applications and discuss relative advantages and disadvantages.

II. Preparation of Egg and Oocyte Extracts

A. Buffers for Extract Preparation

High-Salt Barth's (Blow and Laskey, 1986)
  110 mM NaCl
  2 mM KCl

1 mM MgSO$_4$
0.5 mM Na$_2$HPO$_4$
2 mM NaHCO$_3$
15 mM Tris–HCl, pH 7.4
Barth's (Blow and Sleeman, 1990)
  88 mM NaCl
  15 mM Tris-HCl, pH 7.6
  2 mM KCl
  1 mM MgCl$_2$
  0.5 mM CaCl$_2$
Extraction Buffer (Blow and Laskey, 1986)
  50 mM HEPES–KOH, pH 7.4
  50 mM KCl
  5 mM MgCl$_2$
  2 mM 2-mercaptoethanol
  3 µg/ml leupeptin
  10 µg/ml cytochalasin B
Extraction Buffer (Newport, 1987)
  250 mM sucrose
  2.5 mM MgCl$_2$
  50 mM KCl
  100 mg/ml cycloheximide
  5 µg/ml cytochalasin B
  1 mM dithiothreitol (DTT)
Extraction Buffer (Hutchison et al., 1987)
  110 mM KCl
  5 mM MgCl$_2$
  2 mM 2-mercaptoethanol
  20 mM HEPES, pH 7.5
  3 µg/ml leupeptin
  0.3 mM phenylmethylsulfonyl fluoride (PMSF)
Extraction Buffer (XB) (Murray and Kirschner, 1989)
  100 mM KCl
  1.0 mM MgCl$_2$
  0.1 mM CaCl$_2$
  10 mM K HEPES, pH 7.7
  50 mM sucrose
  Plus protease inhibitors: 10 µg/ml leupeptin, chymostatin, and pepstatin
OR2 Buffer (Wallace et al., 1973)
  82.5 mM NaCl
  2.5 mM KCl
  1.0 mM CaCl$_2$

1.0 mM $MgCl_2$
1.0 Na_2HPO_4
5.0 mM HEPES–NaOH, pH 7.8

B. Preparation and Partial Fractionation of Egg Extracts

1. Selection and Preparation of Frogs

The selection and preparation of frogs to obtain eggs for extracts is of critical importance. Sexually mature *Xenopus laevis* females purchased from supply companies may require a period of one to several months of controlled feeding prior to egg laying. Only robust frogs should be chosen for egg production. Initially, frogs should be primed by an injection of 50–100 IU serum gonadotropin [e.g., Folligon (Intervet; Cambridge, U.K.) or pregnant mare serum gonadotropin (PMSG; Sigma Chemical Co., St. Louis, Mo)] (Lohka and Maller, 1985; Coppock *et al.*, 1989) 2–4 days prior to egg collection. Twelve to fifteen hours before the desired collection time, frogs should be injected with 350–800 IU chorionic gonadotropin (e.g., Chorulon, Intervet; chorionic gonadotropin, Sigma) into the dorsal lymph sac. The amount of hormone administered is dependent in part on the size of the frog that is, larger frogs may require more hormone to induce laying effectively.

2. Collection of Eggs

Following injection with chorionic gonadotrophin, frogs should be placed in one of a variety of routinely used salt solutions (e.g., high salt barth's, Blow and Laskey, 1986; saline tap water, Hutchison *et al.*, 1987, 1988; MMR, Murray and Kirschner, 1989; 0.3 × NKH, Coppock *et al.*, 1989) that will prevent spontaneous activation of the eggs. Note that saline tap water (Hutchison *et al.*, 1987) is not suitable for egg collection in many regions, for example, regions with high calcium ion concentrations. Frogs should be kept singly in tanks in a quiet environment maintained at 21°–22°C. Eggs laid spontaneously or those stripped from the frog are collected 12–15 hours after hormone treatment in a beaker or petri dish.

3. Removal of Jelly Coat and Necrotic Eggs

The external jelly coat must be removed from the eggs prior to processing. Eggs are dejellied by incubation in 2% cysteine hydrochloride, pH 7.8–7.9, for approximately 10 minutes at room temperature (Blow and Laskey, 1986; Newport, 1987; Murray and Kirschner, 1989; Lohka and Maller, 1985;

Coppock et al., 1989). Complete removal of the jelly coat results in the eggs "settling" together, that is, no visible space is observed between adjacent eggs. Eggs should then be rinsed repeatedly with one of a number of salt solutions (e.g., 20% modified Barth's solution, Blow and Laskey, 1986; Barth's solution, Blow and Sleeman, 1990; saline tap water, Hutchison, 1987, 1988; 2× MMR, Newport, 1987). During rinsing, any eggs that show necrotic changes or that appear discolored or misshapen must be removed. It is essential that all suspect eggs be removed prior to centrifugation to obtain functional extracts.

4. Activation of Eggs

Activation of eggs results in entry of the cell into interphase and is illustrated by contraction of the pigment in the animal hemisphere. This is routinely achieved by incubation of the eggs with the calcium ionophore A23187 for approximately 5 minutes [e.g., 0.2–0.5 μg/ml in a rinse solution (as described in Section II,B,3); Blow and Laskey, 1986; Newport, 1987] or by electric shock (e.g., two 1-second pulses of 12-V alternating current; Murray and Kirschner, 1989). Following activation, eggs should be rinsed extensively and may be incubated for 10–20 minutes in rinse solution. Prior to centrifugation, eggs should be washed several times in ice-cold extraction buffer (see Section II,A), which may be supplemented with a variety of protease inhibitors (e.g., 3 μg/ml leupeptin, Blow and Laskey, 1986; 3 μg/ml leupeptin and 0.3 mM PMSF,Hutchison et al., 1987; leupeptin, chymostatin, and pepstatin at 10 μg/ml, Murray and Kirschner, 1989). Cytochalasin B (5–10 μg/ml, Blow and Laskey, 1986; Newport, 1987), which prevents actin polymerization, can also be included.

5. Extraction of Egg Cytoplasm by Centrifugation

a. Packing Spin. After the final rinse in extraction buffer, eggs should be poured into centrifuge tubes held on ice and all excess buffer removed. To remove additional buffer and thereby avoid excessive dilution of the final extract, the eggs may be packed by a low speed centrifugation [e.g., 100 g for 30 seconds in an Eppendorf centrifuge at 4°C, (Newport, 1987), or 1500 rpm for 2 minutes in an SW50.1 rotor (Beckman Fullerton, CA), at 4°C, (Blow and Sleeman, 1990)]; all buffer can then be removed from the surface. Recently, Murray and Kirschner (1989) have demonstrated that excess buffer may be separated from the eggs by overlaying the eggs with Versiluble F-50 oil (General Electric) followed by centrifugation at 200 g for 1 minute.

b. Crushing Spin. To separate the various egg components, packed eggs should be crushed by centrifugation. In general, eggs can be centrifuged at

9000–10,000 g for 10–15 min in a SW50.1 Ti or SW60Ti rotor (Beckman) at 2°–4°C (Blow and Laskey, 1986; Hutchison et al., 1987, 1988; Blow and Sleeman, 1990). Alternatively, eggs can be crushed in an Eppendorf centrifuge by spinning for 5–15 minutes at 4°C (Newport, 1987; Coppock et al., 1989). Using either of these approaches, egg components are separated into three or four distinct fractions. The three common fractions include (1) a plug of yolk platelets and pigment, (2) a golden cytoplasmic layer, and (3) a plug or cap of lipid. The cytoplasmic layer can be collected by puncturing the side of the tube with a syringe or by using a cooled Pasteur pipette to break through the lipid cap. This cytoplasmic extract can be supplemented with cytochalasin B (10–50 μg/ml, Hutchison et al., 1987, 1988), cyclohexamide (100 μg/ml, Blow and Laskey, 1988), and a variety of protease inhibitors [e.g., leupeptin, chymostatin, and pepstatin (10 μg/ml), Murray and Kirschner, 1989].

c. *Clearing Spin.* The cytoplasmic layer, obtained following the crushing spin, is often contaminated with pigment, yolk, and lipid material. This extract can be "cleared" by a second centrifugation step similar to that described in Section II,B,5, b. Following collection of the cleared extract, often referred to as the "low speed supernatant" (LSS), it may be supplemented with potease inhibitors, if not included following the crushing step [e.g., aprotinin (80 KKU/ml), leupeptin (6 μg/ml), Hutchison et al., 1987], and an energy-regenerating system (Blow and Laskey, 1986) (e.g., 10 mM creatine phosphate, 10 μg/ml creatine phosphokinase, and 1 mM ATP, Coppock et al., 1989) if the extract is to be used fresh. If the extract is to be frozen (see Section II,B,7), it is necessary to supplement it with 2–10% glycerol (Blow and Laskey, 1986; Blow and Sleeman, 1990). The extract does not require supplementation, as described in this section, if further fractionation is to be conducted.

6. Fractionation of Egg Low Speed Supernatant

Further fractionation of the cleared extract (LSS), if desired, may be achieved by high-speed centrifugation (Lohka and Masui, 1983, 1984; Lohka and Maller, 1985; Newport, 1987; Sheehan et al., 1988). According to the method of Sheehan et al. (1988), approximately 1.5 ml of LSS is placed in precooled ultraclear centrifuge tubes and overlaid with chilled liquid paraffin. The extract is centrifuged at 100,000 g for 1 hour in an SW50.1 Ti or SW60Ti rotor (Beckman) at 4°C. This approach results in the generation of three fractions: (1) a transparent soluble fraction or high-speed supernatant (HSS), (2) a membranous or vesicular fraction, and (3) a golden, hard ribosomal pellet. In certain cases, the HSS may not be entirely separated from the vesicular fraction, and, collectively, these fractions must be recentrifuged. If initial columns of LSS are long (i.e., >1.5 ml), further centrifugation will nearly always be required. In this case, both fractions should be collected by syringe and recentrifuged at 100,000 g for 30 minutes to 1 hour. This step normally results in a "cleared" HSS and a distinct vesicular fraction. These fractions should be collected by side-puncture of the tube with a syringe. It is essential that the membranous vesicular fraction does not contaminate the HSS during recovery. In addition, to remove the contaminating HSS from the vesicular fraction, the latter should be diluted with 5 ml extraction buffer and recentrifuged for 10 min at 100,000 g. The resultant vesicular pellet should be resuspended with "extraction buffer" in an amount equivalent to one-tenth the volume of the corresponding HSS.

7. Storage of Extracts

Extracts to be used on the day prepared should be held on ice. If bulk extract is prepared, it should be stored as frozen aliquots in liquid nitrogen, which can be achieved by adding glycerol to the extract and dropping the extract as beads into liquid nitrogen (Blow and Laskey, 1986). The LSS can be frozen in 2–10% glycerol (Blow and Laskey, 1986; Blow and Sleeman, 1990), with HSS and the vesicular fraction in 7 and 2% glycerol, respectively. Frozen extracts often remain functional for several years.

8. Miscellaneous Comments

Frogs induced to lay eggs should be used on a rotated schedule, allowing a period of recovery prior to subsequent induction. The recovery period varies greatly between laboratories, namely, from 1 month up to 1 year. The length of time required for recovery probably depends on a variety of factors, such as feeding schedule, tank size, type of food, and temperature. However, it should be noted that, in general, the shorter the recovery period allowed, the fewer eggs will be obtained per frog following the next induction. This may prove to be critical in extract preparation as low egg numbers may be prohibitive depending on the method chosen.

The number of eggs needed to make an extract is dependent on the method of extract preparation as well as the type of experiments to be done. If one is preparing bulk extract according to a method similar to that of Blow and Laskey (1986), the minimum number of eggs that can be handled is about 1000. This limitation arises because fewer eggs will not give a sufficient volume, in the tubes fitting the SW50.1 or SW60 rotors, so as to allow removal of the cleared extract (LSS) without lipid or pellet contamination. In addition, if HSS is to be prepared, an even greater number of eggs is required at the outset to offset the losses resulting from the additional centrifugation steps. Preparation of bulk extract has the advantage that one can thoroughly characterize the extract and use it in several different experimental settings.

However, small-scale extract preparation, using bench-top centrifugation procedures (Newport, 1987; Coppock *et al.*, 1989), has the advantage that fewer eggs per frog are required than in the large-scale preparation procedure described in this section. In addition, bulk extracts may be stored frozen, whereas small batches of fresh extract may be used directly following preparation; the latter extracts are more efficient at "cycling" than frozen extracts.

Eggs from individual frogs must always be kept separate. "Pooling" eggs from several different frogs, to prepare bulk extract, may result in loss of replication activity in the entire extract, even if only one of the constituent egg populations was suboptimal. No advantage is gained by processing a large volume of suboptimal extract.

C. Preparation and Partial Fractionation of Oocyte Extracts

1. SELECTION AND PREPARATION OF FROGS

Female *Xenopus laevis* frogs should be primed with 50 IU serum gonadotropin (Folligon, Intervet) 1 week prior to use. Priming increases the number of stage V and VI oocytes. Typically, the entire ovary from one large frog is required for preparation of approximately 2 ml of oocyte high speed supernatant (HSS). This is, of course, dependent on the quality and number of stage V or VI oocytes (see Section II,C,2). Frogs can be sacrificed by a lethal injection of 3-aminobenzoic acid ethyl ester (MS222, Sigma) into the dorsal lymph sac.

2. COLLECTION AND PREPARATION OF OOCYTES

The ovaries should be removed into Barth's solution (Section II,A) and the oocytes separated into small clumps by manual dissection. During separation of oocytes, the Barth's solution should be changed frequently. Following the final rinse with Barth's solution, the oocytes should be incubated in 0.2% collagenase (type IV, Sigma) in OR2 buffer (Wallace *et al.*, 1973) for 4–8 hour at 23°C. During incubation in collagenase, the oocytes should be gently agitated to facilitate removal of follicle cells and to prevent anaerobiosis. Following separation of oocytes they should be washed several times in Barth's solution (Blow and Sleeman, 1990). This removes any residual collagenase and facilitates removal of follicle cells from the oocytes. In addition, repeated rinsing selectively removes much of the tissue debris and many of the unwanted early stage oocytes. Following rinsing, stage V and VI oocytes (Nieuwkoop and Faber, 1956) should be selected. Oocytes that show degenerative changes should be discarded. Healthy oocytes should be rinsed several times in extraction buffer at 4°C (Blow and Laskey, 1986) with 8–10 μg/ml cytochalasin B in the final rinse. The protease inhibitors leupeptin, pepstatin, and aprotinin, at a final concentration of 10 μg/ml, may also be included in the final rinse.

3. EXTRACTION OF OOCYTE CYTOPLASM BY CENTRIFUGATION

a. Crushing Spin. To separate the various oocyte components, oocytes can be crushed by centrifugation at 10,000 rpm for 10 minutes in an SW 50.1Ti or SW 60Ti rotor (Beckman) at 4°C (Cox and Leno, 1990). As with crushed egg cytoplasm, at least three distinct fractions are obtained: (1) a yolk and pigment pellet, (2) a biphasic soluble cytoplasmic fraction, and (3) a surface plug of lipid. The biphasic cytoplasmic layer can be collected by puncturing the side of the tube with a syringe or by using a cooled Pasteur pipette.

b. Clearing Spin. The soluble cytoplasmic fraction obtained from crushed oocytes should be transferred to a fresh, cooled centrifuge tube and recentrifuged at 15,000 rpm for 10 minutes at 4°C in an SW50.1Ti or SW60Ti rotor (Beckman). The resultant cleared soluble cytoplasmic fraction (i.e., the LSS) should be removed as described in Sections II,B,5,b and II,B,5,c. This fraction is much more viscous and difficult to collect than the egg LSS. The oocyte LSS may be used directly or fractionated further using high-speed centrifugation (see Section II,B,6). If produced in bulk, the LSS may be frozen see (Section II,C,5).

4. FRACTIONATION OF OOCYTE LOW SPEED SUPERNATANT

Oocyte LSS can be fractionated by high-speed centrifugation (Cox and Leno, 1990) following the protocol described for the egg system (Sheehan *et al.*, 1988; also see Section II,B,6). As previously mentioned, the oocyte LSS is considerably more viscous than the egg LSS. Thus, a second spin of 100,000 g for 1 hour is nearly always required for complete separation of the HSS and vesicular fractions, even when short column volumes are used (i.e., $\sim$ 1.5 ml). Once these fractions are clearly separated, the HSS may be removed with a syringe. The amount of vesicular material present following the 100,000 g spin varies between batches of oocytes. Usually a small amount of this fraction is recoverable following HSS collection. As with the vesicular fraction from egg, oocyte vesicles should be washed with 5 ml extraction buffer and pelleted by centrifugation at 100,000 g for 10 minutes. The vesicular pellet may be resuspended in extraction buffer in a volume equivalent to one-tenth the total volume of HSS recovered.

5. Storage of Oocyte Extracts

Oocyte LSS prepared in bulk should be frozen by dropping small aliquots into liquid nitrogen after addition of glycerol to 3–10%. The HSS and vesicular fractions should be supplemented with 7 and 2% glycerol, respectively.

III. Analysis of DNA Replication in *Xenopus* Cell-Free Systems

A. Preparation of DNA Templates

1. Purified DNA

Protein-free, double-stranded DNA from a variety of sources can be used as a template for *in vitro* replication. Purified *Xenopus* genomic DNA (Blow and Laskey, 1986), bacteriophage λ DNA (Newport, 1987), and DNA from a variety of different plasmids [e.g., pUCm54, pBR322 (Blow and Laskey, 1986); RK2 (Blow and Sleeman, 1990)] are replicated in the egg extract. The procedure for preparing *Xenopus* sperm DNA is described by Blow and Laskey (1986), λ DNA by Newport (1987), and plasmid DNA by Maniatis *et al.* (1982). In the case of plasmid DNA, Blow and Laskey (1986) introduced additional purification steps to reduce the possibility of contamination with RNA primers.

2. Demembranated Nuclei

a. Xenopus Sperm. Demembranated *Xenopus* sperm nuclei may be prepared following the method of Gurdon (1976) with modifications (Blow and Laskey, 1986; Murray and Kirschner, 1989; Cox and Leno, 1990) or according to the method of Lohka and Masui (1983), Hutchison *et al.* (1987), or Coppock *et al.* (1989). In all cases, isolation of nuclei involves the demembranization of the cells with L-α-lysophosphatidylcholine (lysolecithin Calbiochem, San Diego, CA). When demembranated *Xenopus* sperm nuclei are added to the egg extract, they undergo a very rapid initial phase of decondensation followed by a slower, membrane-dependent phase. The rapid first stage of decondensation is mediated by the acidic nucleosome assembly factor, nucleoplasmin (Philpott *et al.*, 1991).

b. Quiescent Somatic Cells. Nucleate erythrocytes from *Xenopus laevis* or domestic chickens are metabolically quiescent, arrested in G_0 of the cell cycle. However, nuclei from these cells can be reactivated for DNA replication in the egg extract (Coppock et al., 1989; Leno and Laskey, 1991). These nuclei are easily obtainable in large quantity and possess unique features that allow analysis of the role of nuclear structure in the regulation of DNA replication (Coppock et al., 1989; Leno and Laskey, 1991).

    c. Tissue Culture Cells. Nuclei from slow-growing mouse myeloma cells (SP2/0) and Chinese hamster ovary (CHO) cells, which can be synchronized in G1 of the cell cycle by isoleucine starvation (Schlegel and Pardee, 1986), are adequate templates for replication in the egg extract (Cox and Leno, 1990). These nuclei can be prepared according to the method of Gurdon (1976) or by using a low salt lysis buffer (Evan and Hancock, 1985). The low salt lysis buffer reduces the extent of nuclear clumping seen with lysolecithin-permeabilized nuclei.

    d. Storage of Nuclear Templates. Demembranated nuclei can be stored as aliquots at $-80°C$ or in liquid nitrogen. Prior to freezing, nuclei should be supplemented with 30–50% glycerol. With glycerol at 50%, working stocks of nuclei can be kept at $-20°C$, which prevents freeze–thaw effects.

B. *In Vitro* Conditions for DNA Replication

Fresh, unfrozen or freshly thawed–frozen extracts may be supplemented with an energy-regenerating system to improve the efficiency of replication (Blow and Laskey, 1986). From 10 to 60 mM creatine phosphate and 10–150 μg/ml creatine phosphokinase may be used (Blow and Laskey, 1986; Coppock et al., 1989; Blow and Sleeman, 1990). In addition, 1–2 mM ATP may be added as an initial energy source (Blow and Laskey, 1986; Murray and Kirschner, 1989). If inhibition of protein synthesis is desired, from 100 (Blow and Laskey, 1988; Cox and Leno, 1990) to 250 μg/ml cycloheximide (Blow and Sleeman, 1990) may be added to the extract.

The optimal size and concentration of DNA to be added to the extract for analysis of replication clearly depend on the nature of the study undertaken (Section IV), although certain general guidelines are discussed here. Blow and Laskey (1986) found that larger plasmids were replicated more efficiently than smaller plasmids in the egg extract. However, in general both plasmid DNA and *Xenopus* sperm DNA are replicated less efficiently than sperm nuclei in this system (Blow and Laskey, 1986; Newport, 1987). This may be due to the requirement for nuclear assembly in the initiation of DNA replication. Blow and Sleeman (1990) have shown that the extent of replication of plasmid DNA correlates with the efficiency of assembly of plasmid DNA into pseudonuclei, the sites of DNA replication. However, the maximum efficiency of replication occurs between 2 and 10 ng DNA/μl extract whether sperm nuclei or plasmid DNA (e.g., pUCm54) are used (Blow and Laskey, 1986). Thus, as would be expected, there appears to be an upper and lower threshold of template concentration tolerated by noncycling extracts. Recently, Dasso and Newport (1990) demonstrated that the concentration of *Xenopus* sperm nuclei determines the duration of S phase in egg extracts cycling between S phase and mitosis. They have shown that above a critical threshold level, unreplicated DNA prevents the extracts from entering mitosis. Thus, depending on the type of extract prepared and the context within which replication is studied, the template concentration may be of critical importance.

C. Assays for DNA Replication

1. INCORPORATION AND DETECTION OF RADIOLABELED DEOXYRIBONUCLEOTIDE PRECURSORS

Radiolabeled deoxynucleotides (e.g., [α-^{32}P]dATP at a specific activity of ~ 800 Ci/mmol) are routinely used to determine the mass of DNA synthesized from a known mass of DNA added to the extract. This procedure requires knowledge of the endogenous deoxynucleotide pools in the extract. Using isotope dilution analysis, the dATP (Blow and Laskey, 1986) and dCTP (Hutchison *et al.*, 1987) pool sizes in egg extracts were shown to be approximately 50 μM. In oocyte extracts the dATP pool is between 5 and 10 μM (Cox and Leno, 1990). The use of the oocyte extract in conjunction with the egg extract in analysis of DNA replication (see Section IV) requires supplementing the endogenous oocyte dATP pool with cold deoxynucleotides. This step can be conveniently done by supplementing the oocyte extract with all four deoxynucleotides, up to 50 μM, assuming a 5 μM endogenous oocyte dATP pool, prior to experimentation. Supplementation not only ensures sufficient deoxynucleotide precursors for replication, but it also allows direct comparison of radiolabeled precursor incorporation between reactions containing egg extract with those containing egg extract supplemented with oocyte extract (Cox and Leno, 1990).

a. Acid Precipitation of DNA and Detection by Liquid Scintillation. Following incubation of the radiolabeled precursor in extract, the reactions are terminated by addition of a solution containing 0.5% sodium dodecyl sulfate (SDS), 20 mM EDTA, and 20 mM Tris–HCl, pH 8.0. Proteinase K (0.5 mg/ml) may be added and the sample incubated for 1 hour at 37°C (Blow and Laskey, 1986). The DNA is then extracted with phenol, phenol–chloroform, and chloroform and applied to glass-fiber filters (GF-C, Whatman, Maidstone, Kent). The DNA is precipitated with 10% trichloroacetic acid (TCA) containing 2% sodium pyrophosphate for 30 minutes, and the filters are washed 3 times with 5% TCA and 2 times in 95% ethanol. Dried filters are counted in scintillant (e.g., Optiscint Hi-safe, LKB, Uppsala, Sweden). Alternative protocols for isolation of DNA and quantitation of incorporated precursor are described by Hutchison et al., (1987) and Coppock et al. (1989).

b. *Detection of Radiolabeled DNA by Agarose Gel Electrophoresis.* For agarose gel electrophoresis, the reactions may be terminated by addition of replication sample buffer (80 mM Tris–HCl, pH 8.0, 8 mM EDTA, 0.13% phosphoric acid, 10% Ficoll, 5% SDS, 0.02% bromophenol blue) and incubated with 0.5 mg/ml proteinase K for 1 hour at 37°C (Hutchison et al., 1987; Dasso and Newport, 1990). Samples are loaded and electrophoresed on 0.8–0.9% agarose gels. The gels are dried, and incorporated precursor is visualized by autoradiography.

c. *Detection of Radiolabeled DNA by Autoradiography of Slides.* The sites of incorporation of radiolabeled precursor within an individual nucleus can be visualized by autoradiography. Lohka and Masui (1983, 1984) described a method for detection of radiolabeled precursor into *Xenopus* sperm nuclei replicating in a cell-free extract derived from *Rana pipiens* eggs. Briefly, the incubation is fixed in cold ethanol–acetic acid (3:1), and samples are transferred to a drop of 50% acetic acid on a microscope slide and squashed. Squashed preparations are frozen in dry ice–ethanol, and coverslips should be removed prior to air-drying. Slides should then be washed in cold 5% TCA, rinsed in water for 1 hour, dried, and coated with emulsion. Slides should be exposed for several hours at 4°C and developed. Although providing insight into the spatial organization of the sites of DNA replication within the nucleus, this approach is limited by the degree of sensitivity of detection and low resolution of these sites. An approach that offers improved sensitivity and resolution of replication sites involves fluorescent detection of deoxynucleotide analogs (see Section III,B,2).

2. INCORPORATION AND DETECTION OF DEOXYRIBONUCLEOTIDE ANALOGS

The deoxynucleotide analogs of thymidine, 5-bromodeoxyuridine triphosphate (BrdUTP), and biotinylated deoxyuridine triphosphate (biotin-dUTP) can be used for analysis of DNA replication in the extract. In nascent DNA, BrdUTP is nearly entirely substituted for thymidine (Harland and Laskey, 1980), and biotin-dUTP incorporation has been shown to increase linearly with DNA content in *Xenopus* sperm nuclei replicating in egg extract (Blow and Watson, 1987). Thus, the incorporation of these analogs accurately represents the extent of DNA replication in *Xenopus* extracts. Two specific features of these analogs, namely, density and specificity of detection, can be exploited in the analysis of DNA replication.

In density subsitution experiments, BrdUTP, a dense precursor, can be used in combination with a radiolabeled precursor (as a tracer). This type of procedure allows semiconservative DNA replication to be distinguished from repair synthesis and also indicates whether replication is occurring under cell cycle control, namely, one round per S phase (Blow and Laskey, 1986, 1988; Hutchison et al., 1987, 1988; Coppock et al., 1989; Leno and Laskey, 1991). In brief, DNA is incubated in extract containing a radioactive precursor and approximately 0.25 mM BrdUTP. Following incubation, free label can be removed by gel filtration (Blow and Laskey, 1986) or the DNA can be precipitated with ethanol and loaded directly onto a cesium chloride (CsCl) gradient (Hutchison et al., 1987). Details of the conditions required for separation of substituted DNA on a CsCl equilibrium gradient have been reported by several laboratories (e.g., Blow and Laskey, 1986; Hutchison et al., 1987; Coppock et al., 1989). The end result is the separation of completely substituted DNA on one strand (heavy–light DNA) from partially substituted DNA (incomplete or repair synthesis) or complete substitution of DNA on both strands (heavy–heavy DNA), indicating a second round of replication. Care must be taken when centrifuging such high concentrations of CsCl as any of the following factors can cause crystallization and rotor failure: excessive rotor speed, excessive CsCl concentration, overfilled tubes, or very low temperature.

In addition to detecting BrdUTP incorporation on the basis of differing densities of substituted and unsubstituted DNA, incorporated BrdUTP can be detected directly by BrdU–specific monoclonal antibodies (e.g., Beckton Dickinson, Oxnard, CA). In this case, sites of incorporation can be detected either by indirect immunofluorescence (Nakamura et al., 1986) or by indirect immunocytochemistry (van Dierendonck et al., 1989). These methods of detection are much more sensitive and allow greater resolution of the sites of replication within a nucleus than that achieved by autoradiography. Replication domains have been visualized by indirect immunofluorescence, in cultured mammalian cells replicating *in vivo* (Nakamura et al., 1986) and in *Xenopus* sperm nuclei replicating in the egg extract (Mills et al., 1989). Specifically, nuclei assembled in the extract are fixed with an equal volume of 4% paraformaldehyde at 4°C for 15 minutes or with ethylene glycol–bis (Succinimidyl succinate)(EGS; Pierce, Rockford, IL) in dimethyl sulfoxide for 10 minutes and spun onto polylysine-coated coverslips (Mills et al., 1989). For access of antibodies, the DNA must be denatured which can be achieved by incubation in 4 N HCl–phosphate-buffered saline (PBS) (1:1) with 0.1% Tween-20 for 30 minutes at room temperature. Nuclei are incubated in anti-BrdU monoclonal antibody at 37°C for 30 minutes and then stained with a fluorescent conjugated polyclonal antibody against mouse immunoglobulin G (IgG). The sites of BrdUTP incorporation are then visualized by

FIG. 1. Sites of DNA replication *in vitro* visualized by incorporation of biotin-dUTP and streptavidin staining. Four consecutive confocal optical sections taken at 1-μm intervals through a single *Xenopus* sperm nucleus are seen after a 5-minute pulse with biotin-dUTP in a *Xenopus* egg LSS. Each bright focus represents a cluster of at least 300 replication forks (Mills *et al.*, 1989).

fluorescence microscopy. A protocol for immunocytochemical detection of incorporated BrdU in cultured cells is described in detail by van Dierendonck et al. (1989).

An alternate approach for the visualization of nascent DNA in replicating nuclei involves the use of another TTP analog, biotin-dUTP. Biotin-dUTP is incorporated into nascent DNA, which can be visualized by staining with fluorescent streptavidin (Fig. 1) (Blow and Watson, 1987; Hutchison et al., 1988; Nakayasu and Berezney, 1989; Mills et al., 1989; Leno and Laskey, 1991). In addition, using flow cytometry to detect the fluorescent streptavidin–biotin-dUTP complex, Blow and Watson (1987) demonstrated that biotin-dUTP incorporation is proportional to DNA synthesis. Some advantages of the biotin–streptavidin system over the BrdU antigen–antibody system are as follows: (1) There is no requirement for specific denaturation of the DNA to facilitate streptavidin binding to biotin as in the antibody–antigen system. (2) The greater affinity of streptavidin for biotin relative to the antibody–antigen complex translates into fewer and shorter incubation steps. (3) There is, in general, a lower "background" staining with the biotin–streptavidin system.

The biotin–streptavidin system has proved extremely useful for analysis of DNA replication in *Xenopus* extracts (Blow and Watson, 1987; Hutchison et al., 1988; Mills et al., 1989; Leno and Laskey, 1991). Briefly, extract is supplemented with 8–40 μM biotin-11-dUTP or biotin-19-dUTP and incubated at 23°C. The reactions are then diluted with buffer, and nuclei are fixed with ethylene glycol–bis(succinimidyl succinate) (EGS, Pierce) in dimethyl sulfoxide for 10 minutes. For flow cytometry, nuclei are spun into sucrose (Blow and Watson, 1987); for analysis by conventional or confocal microscopy, they are spun through sucrose onto underlying coverslips. Nuclei are stained with fluorescent streptavidin and a total DNA stain such as Hoechst 33258 or propidium iodide. To eliminate any possible nonspecific signal from labeled RNA, RNase A (50 $\mu g/ml$) may be included in the stain mixture. Figure 1 illustrates the high resolution that can be obtained by the biotin–streptavidin approach (Mills et al., 1989).

IV. Selecting an Appropriate System

In Section II, we described a range of different extract preparation procedures from either eggs or oocytes. These preparations have been used to study DNA replication, nucleosome assembly, chromatin decondensation, assembly and function of the nuclear membrane, and cell cycle control. In this section we discuss the choice of the most appropriate type of extract for each of these purposes and summarize the conclusions (Table I).

Three main distinctions can be drawn. First, only unactivated egg extracts can induce mitosis, by virtue of the maturation-promoting factor that they contain. Second, although oocyte extracts can assemble nucleosomes and elongate replicating intermediates, they are unable to assemble nuclei or to initiate DNA replication. Third, HSSs of eggs resemble oocyte extracts in their synthetic and assembly capacities. Unlike LSSs of eggs, they are unable to assemble nuclei or to initiate DNA replication.

Thus, low speed supernatants of eggs are the most versatile extracts (Table I). Their preparation is detailed in Section II,B,5. Low speed egg supernatants are capable of initiating and completing semiconservative DNA replication on either demembranated sperm nuclei or purified DNA. In addition, replication is under strict cell cycle control, limited to one round in S phase (Blow and Laskey, 1986). In general, there appears to be a clear requirement for complete nuclear assembly before initiation of DNA replication occurs, whether demembranated nuclei or purified DNA are used as template. Evidence that the initiation of DNA replication is dependent on nuclear structure and that the regulation of DNA replication is tightly coupled to nuclear assembly comes from the following observations:

1. Initiation of DNA replication occurs within nuclear structures (Lohka and Masui, 1983, 1984; Blow and Laskey, 1986; Newport, 1987; Blow and Sleeman, 1990).

TABLE I

SELECTING AN APPROPRIATE EXTRACT

Activity	Extract	Section
Nucleosome assembly	Egg or oocyte LSS or HSS	II,B,5, II,B,6, II,C,3, and II,C,4
Chromatin decondensation	Egg LSS or HSS (oocyte capable but more slowly)	II,B,5, II,B,6, II,C,3, and II,C,4
Nuclear assembly	Egg LSS	II,B,5
DNA replication		
Initiation	Egg LSS	II,B,5
Elongation	Egg LSS or HSS or oocyte LSS or HSS + dNTPs	II,B,5 II,B,6, II,C,3, and II,C,4
Mitosis	Unactivated egg LSS or HSS	See Chapter 30, this volume

2. If nuclear membrane formation is inhibited, by removal of membrane vesicles from the extract, initiation of DNA replication does not occur; however, readdition of membrane vesicles to a depleted extract results in reconstitution of both nuclear assembly and DNA replication (Newport, 1987; Sheehan et al., 1988; Blow and Sleeman, 1990).

3. The nuclei formed within the egg extract act as independent and integrated units of replication (Blow and Watson, 1987; Leno and Laskey, 1991).

4. Integrity of the nuclear membrane is essential for restricting DNA replication to a single round per cell cycle, because permeabilization of the membrane is sufficient for rereplication of DNA to occur without passage through mitosis (Blow and Laskey, 1988). In addition, the nuclear membrane is also the feature of nuclear structure that defines the nucleus as an individual unit and determines the timing of initiation of DNA replication in egg extracts (Leno and Laskey, 1991).

5. Synthesis of DNA occurs at clustered sites or foci in nuclei replicating in egg extract. Each of the replication foci represents hundreds of clustered replication forks, which apparently remain clustered throughout the period of DNA synthesis, suggesting an underlying association of these "fixed" sites with the nuclear matrix (Mills et al., 1989; Leno and Laskey, 1991).

When high speed supernatants can be used, they offer the advantage of being fractionated by conventional chromatographic procedures, whereas this is not possible for low speed supernatants when membrane vesicles or other particulate fractions are required.

Oocyte extracts do not offer many advantages over eggs for studies of chromosome replication or nuclear assembly. However, oocytes may be obtained directly from newly purchased frogs, which may be an advantage if facilities are not available for maintaining a frog colony. Of course, this does not offset the fact that some of the most useful activities of egg extracts (Table I) cannot be performed with oocyte extracts. Nevertheless, oocyte extracts offer another potential advantage that has not been realized yet, namely, they provide a possible novel approach to fractionating egg extracts and thus to identifying active components from the egg (Cox and Leno, 1990).

The underlying premise of this scheme is that the dependence of DNA replication on the assembly of complex nuclear architecture makes conventional fractionation too complex. Before a replication initiation factor can be purified in this system, it is necessary to separate it from all other activities required to assemble nucleosomes and nuclear envelopes as well as all activities required to elongate replicating intermediates. The oocyte extract offers a means of providing many of these activities so that it becomes more realistic to assay egg fractions to find the crucial factors that oocytes lack but that allow eggs to initiate DNA replication. This approach is the reverse of that used by Engelke *et al.* (1980) to purify TFIIIA from *Xenopus* oocytes by assaying oocyte fractions for their ability to stimulate transcriptionally inactive egg extracts.

The fact that DNA replication in the extract depends on a complex nuclear structure is clearly a disadvantage for biochemical fractionation. However, it offers compensating advantages. First, extracts can be used to study the assembly of the cell nucleus (Laskey and Leno, 1990), second, the structural basis of DNA replication can be analyzed. For both purposes, *Xenopus* egg extracts can play an important future role.

Acknowledgments

We are grateful to the Cancer Research Campaign for generous support and to Jackie Robbins for help with the manuscript.

References

Blow, J. J., and Laskey, R. A. (1986). Initiation of DNA replication in nuclei and purified DNA by a cell-free extract of *Xenopus* eggs. *Cell (Cambridge, Mass.)* **47,** 577–587.

Blow, J. J., and Laskey, R. A. (1988). A role for the nuclear membrane in controlling DNA replication within the cell cycle. *Nature (London)* **332,** 546–548.

Blow, J. J., and Sleeman, A. M. (1990). Replication of purified DNA in *Xenopus* egg extract is dependent on nuclear assembly. *J. Cell Sci.* **95,** 383–391.

Blow, J. J., and Watson, J. V. (1987). Nuclei act as independent and integrated units of replication in a *Xenopus* cell-free DNA replication system. *EMBO J.* **6,** 1997–2002.

Coppock, D. L., Lue, R. A., and Wangh, L. J. (1989). Replication of *Xenopus* erthrocyte nuclei in a homologous egg extract requires prior proteolytic treatment. *Dev. Biol.* **131,** 102–110.

Cox, L. S., and Leno, G. H. (1990). Extracts from eggs and oocytes of *Xenopus laevis* differ in their capacities for nuclear assembly and DNA replication. *J. Cell Sci.* **97,** 177–184.

Dasso, M., and Newport, J. W. (1990). Completion of DNA replication is monitored by a feedback system that controls the initiation of mitosis *in vitro*: Studies in *Xenopus*. *Cell (Cambridge, Mass.)* **61,** 811–823.

Engelke, D. R., Ng, S. Y., Shastry, B. S., and Roeder, R. G. (1980). Specific interaction of a purified transcription factor with an internal control region of 5S RNA genes. *Cell (Cambridge, Mass.)* **19,** 717–728.

Evan, G. I., and Hancock, D. C. (1985). Studies on the interaction of the human c-*myc* protein with cell nuclei: $p62^{c-myc}$ as a member of a discrete subset of nuclear proteins. *Cell (Cambridge, Mass.)* **43,** 253–261.

Gurdon, J. B. (1976). Injected nuclei in frog oocytes: Fate, enlargement and chromatin dispersal. *J. Embryol. Exp. Morphol.* **36,** 523–540.

Harland, R. M., and Laskey, R. A. (1980). Regulated replication of DNA microinjected into eggs of *Xenopus laevis*. *Cell (Cambridge, Mass.)* **21,** 761–771.

Hutchison, C. J., Cox, R., Drepaul, R. S., Gomperts, M., and Ford, C. C. (1987). Periodic DNA synthesis in cell-free extracts of *Xenopus* eggs. *EMBO J.* **6,** 2003–2010.

Hutchison, C. J., Cox, R., and Ford, C. C. (1988). The control of DNA replication in a cell-free extract that recapitulates a basic cell cycle *in vitro*. *Development (Cambridge, U.K.)* **103,** 553–566.

Laskey, R. A., and Leno, G. H. (1990). Assembly of the cell nucleus. *Trends Genet.* **72**, 406–410.
Laskey, R. A., Kearsey, S. E., and Mechali, M. (1985). Analysis of chromosome replication with eggs of *Xenopus laevis*. *In* "Genetic Engineering: Principles and Methods" (J. Setlow and A. Hollaender, eds.), Vol. 7. pp. 135–148. Plenum, New York.
Leno, G. H., and Laskey, R. A. (1991). The nuclear membrane determines the timing of DNA replication in *Xenopus* egg extracts. *J. Cell Biol.* **112**, 557–566.
Lohka, M. J., and Maller, J. L. (1985). Induction of nuclear envelope breakdown, chromosome condensation, and spindle formation in cell-free extracts. *J. Cell Biol.* **101**, 518–523.
Lohka, M. J., and Masui, Y. (1983). Formation *in vitro* of sperm pronuclei and mitotic chromosomes induced by amphibian ooplasmic components. *Science* **220**, 719–721.
Lohka, M. J., and Masui, Y. (1984). Roles of cytosol and cytoplasmic particles in nuclear membrane assembly and sperm pronuclear formation in cell-free preparations from amphibian eggs. *J. Cell Biol.* **98**, 1222–1230.
Maniatis, T., Fritsch, E. F., and Sambrook, J. (1982). "Molecular Cloning: A laboratory Manual." Cold Spring Harbor Laboratory, Cold Spring Harbor, New York.
Mills, A. D., Blow, J. J., White, J. G., Amos, W. B., Wilcock, D., and Laskey, R. A. (1989). Replication occurs at discrete foci spaced throughout nuclei replicating *in vitro*. *J. Cell Sci.* **94**, 471–477.
Murray, A. W., and Kirschner, M. W. (1989). Cyclin synthesis drives the early embryonic cell cycle. *Nature (London)* **339**, 275–280.
Nakamura, H., Morita, T., and Sato, C. (1986). Structural organizations of replicon domains during DNA synthetic phase in mammalian nucleus. *Exp. Cell Res.* **165**, 291–297.
Nakayasu, H., and Berezney, R. (1989). Mapping replicational sites in the eucaryotic cell nucleus. *J. Cell Biol.* **108**, 1–11.
Newport, J. (1987). Nuclear reconstitution *in vitro*: Stages of assembly around protein free DNA. *Cell (Cambridge, Mass.)* **48**, 205–217.
Nieuwkoop, P. D., and Faber, J. (1956). "Normal Table of *Xenopus laevis* (Daudin)" North-Holland Publ. Amsterdam.
Philpott, A., Leno, G. H., and Laskey, R. A. (1991). Sperm decondensation in *Xenopus* egg cytoplasm is mediated by nucleoplasmin. *Cell* **65**, 569–578.
Rhodes, D., and Laskey, R. A. (1989). Assembly of nucleosomes and chromatin *in vitro*. *In* "Methods in Enzymology" (P. Wasserman and R. Kornberg, eds.), Vol. 170, pp. 575–585. Academic Press, New York.
Schlegel, R., and Pardee, A. B. (1986). Caffeine-induced uncoupling of mitosis from the completion of DNA replication in mammalian cells. *Science* **232**, 1264–1266.
Sheehan, M. A., Mills, A. D., Sleeman, A. M., Laskey, R. A., and Blow, J. J. (1988). Steps in the assembly of replication competent nuclei in a cell-free system from *Xenopus* eggs. *J. Cell Biol.* **106**, 1–12.
Wallace, R. A., Jared, D. W., Dumont, J. N., and Sega, M. W. (1973). Protein incorporation into isolated amphibian oocytes. III. Optimum incubation conditions. *J. Exp. Zool.* **184**, 321–334.
van Dierendonck, J. H., Keyzer, R., van de Velde, C. J. H., and Cornelisse, C. J. (1989). Subdivision of S-phase by analysis of nuclear 5-bromodeoxyuridine staining patterns. *Cytometry* **10**, 143–150.

Chapter 30

Cell Cycle Extracts

ANDREW W. MURRAY

Department of Physiology
University of California, San Francisco
San Francisco, California 94143-0444

I. Introduction
II. Reagents and Equipment
   A. Solutions for Extract Preparation
   B. Glassware
   C. Activation Chamber
   D. Laid versus Squeezed Eggs
III. Methods
   A. Priming Frogs and Inducing Ovulation
   B. Sperm Nucleus Preparation
   C. Cycling Extract Preparation
   D. Cycling Extract Use
   E. Cytostatic Factor-Arrested Extracts
   F. Nuclease-Treated Extracts
IV. Conclusion and Prospects
   References

I. Introduction

Early embryos have been popular for studies of the cell cycle since the nineteenth century (Wilson, 1928). Their advantages include the availability of large quantities of material and rapid, synchronous cell cycles. The large size of amphibian eggs confers the additional advantages of easy microinjection and the ability to perform biochemical analysis on microinjected cells. However, microinjection allows only a limited variety of perturbations of the cell cycle, and the refractoriness of the yolk platelets has made microscopic examination of intracellular events in living frog embryos impossible. Recently both of these problems have been overcome by the development of egg extracts that carry out the events of the cell cycle *in vitro* (Blow and Laskey, 1986; Hutchison *et al.*, 1987; Lohka and Masui, 1983; Murray and Kirschner, 1989a). The development of such extracts has greatly increased the range of biochemical manipulations available to the investigator, and it has opened the way to real-time microscopy of cell cycle events in *Xenopus* (Belmont *et al.*, 1990). In this chapter I briefly review the early *Xenopus* cycle and the history of cell cycle extracts before describing in detail the techniques for preparing and using such extracts.

The early *Xenopus* cell cycles are controlled by the activation and inactivation of a protein kinase named maturation-promoting factor (MPF) (reviewed in Murray and Kirschner, 1989b). MPF is also known as the cdc2 kinase and growth-associated histone H1 kinase, and consists of two principal subunits, the catalytic $p34^{cdc2}$ subunit and cyclin. Activation of MPF induces both mitosis and meiosis, whereas its inactivation triggers the onset of anaphase and the progression into interphase. The fluctuation of MPF activity in early *Xenopus* embryos is shown in Fig. 1. Immature oocytes are arrested with intact nuclear envelopes. Stimulation with progesterone induces them to mature, that is, to break down their nucleus, complete meiosis I, and then enter meiosis II. The oocytes are arrested at metaphase of meiosis II by the action of cytostatic factor (CSF), one of whose components appears to be the product of the c-*mos* protooncogene (Sagata *et al.*, 1989). These arrested oocytes pass down the oviduct to emerge from the female frog as unfertilized eggs. On fertilization, the fusion of the sperm with the egg induces a transient increase in the cytoplasmic calcium concentration that inactivates both CSF and MPF, allowing the egg to escape from its metaphase arrest, pass through anaphase, and enter interphase. The first mitotic cell cycle lasts 75 minutes at 22°C, and subsequent cycles last 25 minutes. Unfertilized eggs can also be induced to escape from CSF arrest by electrical shock or by treatment with the calcium ionophore A23187.

The first cell cycle extracts were produced by Lohka and Masui (1983). They activated *Rana pipiens* eggs and then crushed them by centrifugal force, to separate the yolk, pigment granules, and lipid droplets from the cytoplasm. When demembranated *Xenopus* sperm nuclei were added to such an extract, they decondensed, began DNA replication, and ultimately underwent chromosome condensation and nuclear envelope breakdown, thus demonstrating all the events of a single cell cycle *in vitro*. Subsequently, Blow and Laskey (1986) and Félix *et al.* (1989) developed extracts that also performed the events of a single cell cycle and could be frozen, and both Hutchison *et al.* (1987) and Murray and Kirschner (1989a) developed extracts that could undergo multiple cell cycles *in vitro*. I refer to extracts prepared from activated

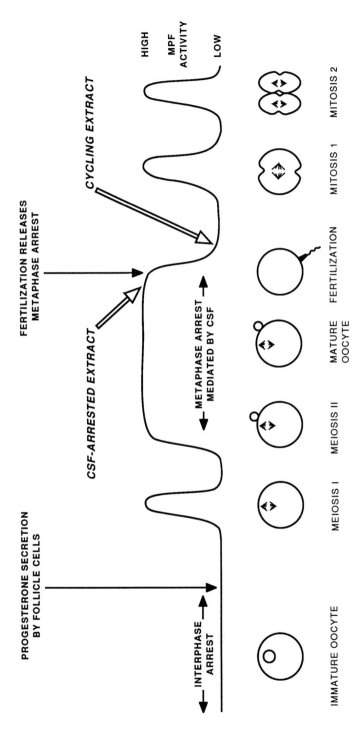

FIG. 1. Schematic representation of early stages of the *Xenopus* life cycle. The levels of MPF activity are shown during oocyte maturation and the early cleavage divisions of the *Xenopus* embryo. The points in the life cycle at which CSF-arrested and cycling extracts are prepared are indicated.

eggs as cycling extracts. These extracts have been widely used to investigate the regulation of the cell cycle and faithfully mimic many aspects of the cell cycle *in vivo*, including semiconservative DNA replication (Blow and Laskey, 1986; Hutchison *et al.*, 1987), nuclear envelope breakdown and reformation (Hutchison *et al.*, 1988; Murray and Kirschner, 1989a), cell cycle changes in microtubule dynamics (Belmont *et al.*, 1990; Verde *et al.*, 1990), and membrane vesicle fusion (Tuomikoski *et al.*, 1989).

Lohka and Maller (1985) were the first to make extracts from unfertilized eggs in the presence of EGTA, which I refer to as CSF-arrested extracts. They showed that demembranated sperm nuclei added to such extracts would form spindles. Addition of calcium to these extracts induced the formation of interphase nuclei. We subsequently developed techniques for preparing CSF-arrested extracts that could be induced to enter interphase by much lower concentrations of calcium (Murray *et al.*, 1989), and such extracts have been used to investigate the role of cyclin degradation in the exit from metaphase (Murray *et al.*, 1989) and to develop *in vitro* models for spindle assembly (Sawin and Mitchison, 1991) and anaphase (C. Shamu and A. W. Murray, unpublished results).

II. Reagents and Equipment

A. Solutions for Extract Preparation

Extract Buffer (XB)
   100 mM KCl
   0.1 mM CaCl$_2$
   1 mM MgCl$_2$
   10 mM potassium HEPES, pH 7.7
   50 mM sucrose

Prepared fresh from
   20× XB salts (2M KCl, 20 mM MgCl$_2$, 2 mM CaCl$_2$), sterile filtered and stored at 4°C.
   1.5 M sucrose, sterile filtered and stored in aliquots at -20°C.
   1 M HEPES, titrated with KOH to yield a pH of 7.7 when diluted to 10 mM, sterile filtered and stored in aliquots at -20°C.

MMR
   100 mM NaCl
   2 mM KCl
   1 mM MgCl$_2$
   2 mM CaCl$_2$

0.1 mM EDTA
5 mM HEPES, pH 7.8
A 10× stock is prepared, titrated with NaOH to pH to 7.8, autoclaved in 500-ml bottles, and stored at room temperature.

Dejellying Solution
1× XB salts (100 mM KCl, 0.1 mM $CaCl_2$, 1 mM $MgCl_2$)
2% w/v cysteine, free base (Sigma Chemical Co., St. Louis, MO, C-7755)
Made up within 1 hour of use and titrated to pH 7.8 with NaOH (4.5 ml of 1 M NaOH per 2 g of cysteine yields the desired pH).

Gelatin
5% w/v in water (Sigma G-9382), autoclaved and stored frozen at $-20°C$

Energy mix
150 mM creatine phosphate (Boehringer Mannheim, Indianapolis, IN, 127 574)
20 mM ATP (Boehringer Mannheim, 519 979), pH 7.4
2 mM EGTA, pH 7.7
20 mM $MgCl_2$
Stored in aliquots at $-20°C$.

Protease Inhibitors
Mixture of leupeptin [Chemicon (100 Lomita St., El Segundo, CA 90245), EI 8), chymostatin (Chemicon, EI 6), and pepstain (Chemicon, EI 9) dissolved to a final concentration of 10 mg/ml each in dimethyl sulfoxide (DMSO). Stored in small aliquots at $-20°C$.

Cytochalasin B
10 mg/ml cytocholasin B (Sigma C-6762) in DMSO, stored in small aliquots at $-20°C$

Pregnant mare serum gonadotropin (PMSG)
100 U/ml PMSG (Calbiochem, San Diego, CA, 367222) made up in water and stored at $-20°C$

Human chorionic gonadotropin (HCG)
250 U/ml HCG (Sigma CG-2) made up in water and stored at 4°C (good for at least 2 weeks)

Oil
Versilube F-50 is made by General Electric and may be purchased from Andpak-EMA [1890 Dobbin Drive, San Jose, CA 95133; Tel. (408) 272-8007]

B. Glassware

Rinse all glassware with glass-distilled water immediately before use. We find that glass petri dishes are the easiest vessels to use for washing and dejellying small to moderate quantities of eggs (eggs stick to polystyrene dishes). We use 60-mm dishes for 3 ml of eggs (volume after dejellying) or less, and 100-mm dishes for larger quantities. Eggs are transferred between vessels by pouring or by using Pasteur pipettes whose tips have been broken off to give an orifice about 4 mm in diameter and then fire polished. These pipettes are also used to remove damaged eggs and debris during the series of rinses before egg crushing. After dejellying and washing, eggs can stick even to glass dishes, leading to spontaneous activation or lysis. Sticking can be prevented by incubating the dishes for 10 minutes in XB containing 100 μg/ml gelatin, then washing dishes once with XB before adding eggs to them.

C. Activation Chamber

The activation chamber is an 11 cm square, 5 cm high plexiglass box and is shown in Fig. 2 and 3a. The bottom of the interior and the underside of the lid are electrodes made of sheets of perforated stainless steel. The bottom electrode is covered to a depth of 1 cm with 2% agarose (Sigma, A-0169) in $0.2 \times$ MMR. The eggs rest on the agarose and are covered with $0.2 \times$ MMR which the lid touches to make electrical contact. Activation is accomplished by two 1-second pulses of 12 V alternating current, separated by 5 seconds. A convenient power supply for this purpose is the Heathkit Model IP-17 electrophoresis power supply, which has outlets that deliver 12 V AC.

The agarose is removed and the chamber is cleaned with 1 M HCl, weekly. Before pouring new agarose the chamber is extensively washed with water, then ethanol, and finally dried.

D. Laid Versus Squeezed Eggs

Both cycling and CSF-arrested extracts can be prepared from two types of eggs: eggs collected during an overnight ovulation (laid eggs), or eggs that have been squeezed from frogs immediately prior to extract preparation (squeezed eggs). In our hands extracts prepared from squeezed eggs have more reproducible properties and are generally of higher quality. For instance, the cycle time of a cycling extract prepared from squeezed eggs is less than that of a laid egg extract prepared from eggs from the same group of frogs. However, each frog yields 5 to 10 ml of laid eggs, but only 0.5 to 1 ml of squeezed eggs.

The performance of the cell cycle extracts is strongly influenced by the quality of the eggs that they are prepared from. Unfortunately, there is no rigorous definition of egg quality. Good eggs are uniform in appearance and size, and give very low levels of spontaneous activation or lysis during the preparations for crushing the eggs. In some years we have seen seasonal variation in egg quality, with the poorest eggs being obtained in August.

III. Methods

A. Priming Frogs and Inducing Ovulation

Female frogs are primed for ovulation by injecting them with pregnant mare serum gonadotropin (PMSG) and then induced to ovulate by a subsequent injection of human chorionic gonadotropin (HCG). During and after priming frogs are kept at 16° to 22°C in tap water that has been made 30 mM in NaCl (Sigma S-9625) or rock salt (restaurant grade, obtained from a local supplier). The water should be changed twice a week. Frogs are sensitive to chlorine and should not be kept in tap water that is being constantly replenished. In most locations chlorine concentrations are sufficiently low that there is no need to dechlorinate the water before putting the frogs in fresh water. If necessary water can be dechlorinated by allowing it to stand overnight. Keeping frogs in rooms where the temperature exceeds 22°C leads to a variable, but sometimes severe, decline in the quality of the eggs. All hormone injections are made into the dorsal lymph sac using a 27-gauge needle.

Frogs are primed for ovulation by injecting 50 units of PMSG on day 1 and 25 units on day 3. Ovulation can be induced on any day from day 5 to at least

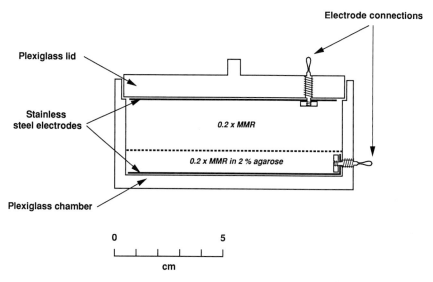

FIG. 2. Working drawing for a frog egg activation chamber, showing a cross-sectional view. From above, the chamber is square (see Fig. 3a). The electrodes are fabricated from a perforated stainless steel plate. This apparatus can be assembled with about 4 hours of labor by a lightly equipped machine shop.

day 12. Ovulation is induced by injecting 150 units of HCG, after which each frog is placed in a separate container containing 1 liter of MMR precooled to 16°C, and the containers are placed in a 16°C incubator. The time from HCG injection to the beginning of egg laying is somewhat variable from frog to frog, but averages 12 to 14 hours. At higher temperatures the onset of ovulation is much more rapid (7 hours at 23°C), but the quality of the eggs is sometimes poor.

For squeezed egg extracts the eggs are squeezed from frogs into 60-mm glass petri dishes containing 15 ml of MMR. Some frogs secrete copious quantities of a milky exudate from their skins when squeezed. If any of this secretion drips into a petri dish it will induce lysis of a fraction of the eggs, and the contents of the dish should be discarded. Individuals with allergies can become allergic to frogs and may wish to wear gloves while handling frogs. Any batch of eggs that shows more than 5% spontaneous lysis, extensive pigment mottling or variegation, or more than 10% spontaneous activation should also be discarded.

For laid eggs, the frogs are removed from the containers, most of the MMR is poured off, and the eggs are inspected visually. All batches of eggs that meet the criteria discussed above are pooled and then washed several times in MMR to remove shed frog skin, excrement, and other detritus.

Extracts should be prepared as promptly as possible. We do not allow squeezed eggs to remain in MMR for more than 10 minutes before beginning to prepare the extract. Laid eggs slowly deteriorate when kept in MMR at 16°C, and they should not be used if the interval between the beginning of ovulation and preparation of extracts exceeds 10 hours.

B. Sperm Nucleus Preparation

The protocol for the preparation of sperm nuclei is based on the method of Gurdon (1976).

1. Reagents

Nuclear Preparation Buffer (NPB)
    250 mM sucrose
    15 mM HEPES (titrated with KOH to give a pH of 7.4 at 15 mM)
    1 mM EDTA, pH 8.0
    0.5 mM spermidine trihydrochloride (Sigma S-2501)
    0.2 mM spermine tetrahydrochloride (Sigma S-1141)
    1 mM Dithiothreitol (Sigma D-0632)
    10 μg/ml leupeptin
    0.3 mM phenylmethylsulfonyl fluoride (PMSF) (Sigma P-7626)
    Made up as a fresh 2× solution from stock solutions stored at −20°C

Lysolecithin
10 mg/ml lysolecithin (Sigma L-4129) in water made up fresh
Bovine Serum Albumin (BSA)
10% (w/v) BSA (fraction V, Sigma A-7906), made up in water, titrated to pH 7.6 with KOH, and stored at $-20°C$

2. Procedure

Perform all steps on ice unless otherwise noted. All spins are performed in 14-ml round-bottomed polypropylene tubes in a swinging-bucket rotor (HB-4 or equivalent) in a high speed centrifuge (Sorvall RC-5B or equivalent).

1. Prime four male frogs with 25 U of PMSG 3 days before sperm collection and then with 125 U of HCG the day before collection.
2. Anesthetize the frogs by immersion in ice water for 20 minutes and then pith them. Cut through the skin and body musculature to expose the the body cavity. With a pair of blunt forceps carefully move the yellow fat bodies aside to expose the two testes which are attached to the base of the fat bodies. The testes are shiny, oval, white bodies 1 to 2 cm long, one on either side of the midline. Using a pair of fine scissors remove the testes and place them in a petri dish containing cold MMR, making every effort to avoid cutting major blood vessels and to minimize contamination with the fat bodies.
3. Rinse the testes 3 times in cold MMR and remove any remaining fat body.
4. Wash the testes twice in NPB, remove as much NPB as possible, and macerate the testes finely with a pair of dissecting scissors or a razor blade.
5. Add 2 ml of NPB and gently pipette the testes fragments up and down through a fire-polished truncated pasteur pipette with an opening of about 3 mm in diameter.
6. Filter the testes through eight thicknesses of cheesecloth, rinsing with 8 ml of NPB, and spin the sperm down at 3000 rpm for 10 minutes.
7. Resuspend sperm in 8 ml of NPB, then spin down at 3000 rpm for 10 minutes.
8. Resuspend sperm in 1 ml of NPB, warm to room temperature, add 50 μl of 10 mg/ml lysolecithin, mix, and incubate for 5 minutes at room temperature. The progress of demembranation can be followed by withdrawing aliquots of the reaction, adding them to NPB containing 1 μg/ml Hoechst 33342, and observing them under a fluorescence microscope. Before demembranation only a few damaged sperm stain with Hoechst, whereas after demembranation all the sperm heads stain.

9. Add 10 ml of cold NPB containing 3% BSA, mix gently, and spin down for 10 minutes at 3000 rpm.
10. Resuspend gently in 5 ml NPB containing 0.3% BSA and spin down for 10 minutes at 3000 rpm.
11. Resuspend in 1 ml of NPB, without PMSF and containing 0.3% BSA and 30% (w/v) glycerol. Count the sperm density using a hemocytometer, adjust to a final density of 10^7 sperm nuclei/ml, and quick freeze 25-μl aliquots in liquid nitrogen and store at $-70°$C. This procedure routinely yields $3-10 \times 10^7$ sperm from four frogs.

C. Cycling Extract Preparation

Preparation of successful extracts requires considerable organization and takes about 1.5 hours. All buffers and solutions should be prepared prior to starting to manipulate the eggs. Once the eggs have been squeezed, and especially once they have been dejellied, they should not be stored before beginning extract preparation. There is no point during extract preparation at which the eggs can be left for more than a few minutes.

1. Remove as much MMR as possible from the vessels containing the eggs. Rinse the eggs once with room temperature distilled or deionized water and leave for 10 minutes. This treatment ensures that the subsequent activation step is complete.
2. Remove as much water as possible from the eggs and add dejellying solution, 10 ml per 60-mm dish containing squeezed eggs, or 30 ml per frog for laid eggs. Gently swirl the eggs at intervals. For squeezed eggs we dejelly the eggs from each frog separately. Dejellying takes 3 to 8 minutes and is complete when the eggs pack as tight spheres (Fig. 3b), without any visible separation by their jelly coats. Pour off the dejellying solution and rinse briefly with half the original volume of dejellying solution.
3. Wash the eggs twice with $0.2 \times$ MMR and then transfer to an activation chamber filled with $0.2 \times$ MMR. Activation consists of two 1-second pulses of 12 V AC separated by 5 seconds.
4. After activation remove the eggs from the chamber and wash four times in extract buffer (XB). During these washes the eggs should be monitored for the success of activation. Activation causes a contraction of the pigmented region of the cortex (compared Fig. 3b and 3c), and the white spot at the center of the pigmented region becomes less distinct. With experience the pigment contraction is visible with the naked eye, but initially it is helpful to view the eggs under a dissecting microscope. If more than 20% of the eggs fail to activate, they should be subjected to an additional activation pulse.

30. CELL CYCLE EXTRACTS 591

5. Wash the eggs twice more with XB containing 10 µg/ml each leupeptin, pepstatin, and chymostatin. Using a wide-mouthed Pasteur pipette, transfer the eggs to centrifuge tubes containing XB plus protease inhibitors and 100 µg/ml cytochalasin B. Try to transfer the eggs in a

FIG. 3. Stages in the preparation of cell cycle extracts. (a) Egg activation chamber with lid inverted to show the stainless steel electrode. (b) Eggs that have been dejellied but not activated. Note that the eggs pack together tightly. (c) Eggs from the same batch after activation. Note that the pigment has contracted to occupy a smaller fraction of the egg surface, and note the increase in density of pigment within the pigmented area. (d) Eggs placed in an SW50.1 tube after washing. (e) The same tube after the low speed spin to replace the buffer between the eggs with Versilube. (f) The same tube after the high speed spin to crush the eggs. To collect the cell cycle extract a syringe needle is inserted at the site marked by the arrow, with and the region of the tube corresponding to the black bar being removed.

minimal volume of buffer, in order to keep the cytochalasin as concentrated as possible. For 2.5 ml of eggs or more we use ultraclear SW50.1 tunes (Beckman, Fullerton, CA, 344057) containing 1 ml XB plus cytochalasin B at 100 μg/ml. After transferring the eggs, remove excess XB and add 1 ml of Versilube F-50, a silicone oil that is less dense than the eggs but more dense than XB (Fig. 3d). For volumes less than 2.5 ml use 1.5-ml Eppendorf tubes. These should contain 0.2 ml of Versilube underneath 0.3 ml XB plus cytochalasin. For very large preparations of extract we have successfully used SW40.1 and SW28 tubes and rotors.

6. After the addition of Versilube the eggs are spun at low speeds to displace the interstitial XB trapped between the eggs and replace it with Versilube, thereby decreasing the dilution of egg cytoplasm in the final extract. Spin the eggs in a clinical centrifuge at room temperature for 60 seconds at 1000 rpm (150 g) and then 30 seconds at 2000 rpm (600 g) (eggs after this spin are shown in Fig. 3e). Remove all XB and Versilube from the top of the tube, using a Pasteur pipette or Pipetman. From activation to this stage usually takes 10 to 15 minutes. Continue incubation at room temperature until 15 minutes after the time of activation, then transfer the centrifuge tubes to an ice–water bath and incubate for a further 15 minutes.

7. Crush the eggs by spinning them in a swinging-bucket rotor. For eggs in SW50.1 tubes we use an SW50.1 rotor (or equivalent) in an ultracentrifuge for 10 minutes at 10,000 rpm at 2°C. [If no ultracentrifuge is available, the SW50.1 tube can be placed inside a 12-ml polypropylene tube (Sarstedt), which is then placed inside a rubber adaptor (e.g., Sorvall 00363) and spun in a swinging-bucket rotor (e.g., Sorvall HB-4 or Beckman JS-13) in a high speed centrifuge for 10 minutes at 10,000 rpm].

8. At the end of the centrifugation the eggs will have been ruptured and separated into three layers: lipid, cytoplasm, and yolk, running from top to bottom (Fig. 3f). The residual Versilube will be present as a band of droplets near the top of the cytoplasmic layer. Collect the cytoplasmic layer very slowly using a 1-ml syringe and 20-gauge needle via side puncture at the bottom of the cytoplasmic layer. Attempts to collect the cytoplasmic layer too rapidly will draw lipid and large quantities of Versilube droplets into the extract. Some contamination with Versilube is inevitable and does not affect the performance of the extracts. In general we recover about one-fifth of the original volume of packed eggs as cytoplasmic extract.

9. To the cytoplasmic fraction add leupeptin, chymostatin, pepstatin and cytochalasin B to a final concentration of 10 μg/ml each (1/1000 volume

each of the protease inhibitor and cytochalasin stocks) and 1/20 volume of energy mix. All mixing should be preformed gently, either by minimal vortexing at low speed, by gently flicking inclined tubes with a finger (small volumes), or by inversion (large volumes).
10. If the extract is straw colored, moderately turbid, and no lipid and little Versilube contamination has occurred during removal, no clarifying spin is required. If any of these conditions are not met, perform a second clarifying spin for 15 minutes at 10,000 rpm, 2°C, in a swinging-bucket rotor. For extract volumes less than 1.5 ml we perform this spin in Eppendorf tubes in a high speed centrifuge (e.g., Sorvall HB-4 rotor in an RC-5B centrifuge) with appropriate adaptors (e.g., Sorvall 00364). In our hands, the superior temperature regulation of a high speed centrifuge makes it superior to a microcentrifuge for this step. For larger volumes we use an SW50.1 rotor or equivalent in an ultracentrifuge or high speed centrifuge as described above. This second spin is not essential, but it removes stray pigment granules and other unwanted particulate material and improves the lifetime of the extracts on ice.

D. Cycling Extract Use

Cycling extracts have been used for a number of purposes. For any experiment that requires more than one cell cycle, freshly prepared extracts must be used. In our hands these have a lifetime of about 4 hours on ice. The cell cycle length shows considerable variability from extract to extract and is strongly influenced by the quality of the eggs. The best extracts have cycle lengths of 35 to 55 minutes at 23°C, whereas the cycle length of extracts made from lower quality eggs may be up to 75 minutes. We do not use extracts whose cycle length exceeds 75 minutes. It is not uncommon for the first *in vitro* cycle to be shorter than subsequent ones. This probably reflects the time period between activation and the crushing of the eggs. In addition the extent of DNA synthesis is often markedly lower in the first cycle than in subsequent cycles (A. W. Murray, unpublished results), reflecting the inability to decondense the sperm DNA and complete replication fully in the limited period before the onset of the first mitosis.

For experiments that do not require multiple cell cycles, frozen extracts may be acceptable. Our best success has been with extracts that were made 200 mM in sucrose by addition of 1.5 M sucrose and then quick frozen in liquid nitrogen in aliquots of no more than 0.1 ml contained in 0.5-ml Eppendorf tubes. This concentration of sucrose reduces the rate of protein synthesis and cell cycle progress in the initial extracts, but no further reduction is seen after freezing and thawing. Entry into mitosis in frozen and thawed extracts takes between 75 and 150 minutes. Any extract that is to be frozen must be subjected to a clarifying spin. The frozen extract can be stored at $-70°C$ for at least 2 months without significant loss of activity. Do not freeze and thaw extracts more than once.

For both fresh and frozen extracts, performing pilot experiments to test the quality of the extracts is strongly recommended. For such pilots demembranated sperm nuclei are added to the extract, the extract is incubated at 23°C, and samples are withdrawn every 10 minutes, fixed, and examined for nuclear morphology (see below). We follow such pilots until the extract has passed through the first mitosis and returned to interphase (in extracts that will be subjected to a clarifying spin, the pilot can be started with material that has been subjected only to the first spin, while the clarifying spin is in progress). In our hands less than 10% of the extracts made from high quality eggs are rejected after the pilot. The two major causes for rejection are failure to enter mitosis or arrest in first mitosis. For frozen extracts a pilot reaction is performed on the unfrozen extract and also on an aliquot of the extract after freezing and thawing.

There appears to be an effect of reaction volume or surface-to-volume ratio on extract performance. We find that the extracts perform best when incubated in small volumes (50 μl or less for 0.5-ml Eppendorf tubes, 200 μl or less for 1.5-ml tubes). To keep extracts homogeneous we mix them periodically (at least once every 15 minutes). An easy way of doing this is to gently stir the extract with a pipette tip while taking samples.

1. Monitoring the Cell Cycle in Extracts

The cycling of extracts may be monitored in a number of different ways.

a. Morphology of Added Sperm Nuclei

Sperm Dilution Buffer
  1 mM MgCl$_2$
  100 mM KCl
  150 mM sucrose
  5 mM HEPES, pH 7.7
  Stored in aliquots at $-20°C$

Fix
  0.3 volumes 37% formaldehyde
  0.6 volumes 80% (w/v) glycerol
  0.1 volumes 10 × MMR
  1 μg/ml Hoechst 33342 (Sigma B-2261, stored as a 10 mg/ml solution in water, in a light-tight vessel at $-20°C$) or 1 μg/ml diamidinophenyl-

indole (DAPI, Sigma D-1388, stored as a 1 mg/ml solution in water, in a light-tight vessel at $-20°C$)

Made up fresh from stocks.

Checking sperm morphology is the simplest assay and is the only one that can be performed while an experiment is in progress. Thus, even though the cyclic activation and inactivation of MPF occurs in the absence of added nuclei, we strongly urge the addition of sperm nuclei at a final concentration of 10^5/ml to all experiments. Sperm nuclei are added on ice before incubating the extracts at 23°C. For extract volumes of more than 100 μl we add concentrated sperm nuclei directly, but for smaller volumes we dilute the sperm nuclei to 10^6/ml in sperm dilution buffer, and then add 10% of the extract volume to the extract. Nuclear morphology is scored by withdrawing 1 μl of extract and depositing it on a slide, then adding 4 μl of fix on top of this drop and gently lowering an 18 × 18 mm coverslip on top of the combined drops. With this technique, although the liquid is spread over the entire surface of the coverslip, the vast majority of the nuclei are found near the center of the coverslip at the spot where the drop of extract was originally deposited. The slides are observed by fluorescence microscopy using a filter set suitable for visualizing Hoechst or DAPI fluorescence, initially at a total magnification of about 100× and then at 500× for a more detailed assessment of morphology. At the higher magnification phase-contrast is often a useful addition to fluorescence observation. For phase-contrast microscopy the background of particulate material decreases the further the nucleus is from the site of the original drop of extract.

The morphological behavior of an extract is shown in Fig. 4. This extract is slightly atypical, since it formed visible spindles in mitosis I. This phenomenon is rare, probably because the rate of spindle assembly in extracts is slow relative to the period for which MPF levels are high. Operationally we define mitosis as the combined events of chromosome condensation and nuclear envelope breakdown.

The synchrony of nuclei in extracts is good. The range of nuclear morphologies at a given time point is always small, for example, fully condensed chromosomes plus some nuclei in midtelophase, or midtelophase plus some interphase nuclei. All of the nuclei in the extract participate in the cycle, except for very occasional large clumps of tangled sperm nuclei. Inhibition of protein synthesis prevents entry into mitosis, but substantial chromosome condensation does eventually occur, accompanied by swelling of the nuclei to much larger sizes than are observed in extracts where protein synthesis occurs normally.

The inclusion of glycerol in the fixative means that the slides do not dry out, even if left at room temperature overnight. If they are to be preserved for any

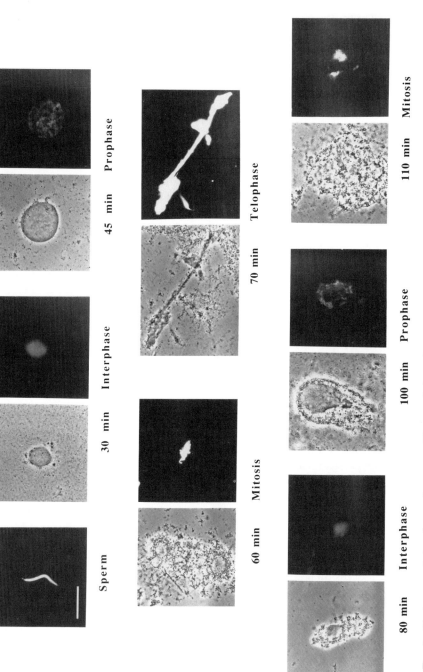

Fig. 4. Nuclear morphology in a cycling extract. The morphology of sperm nuclei is shown at the indicated times after adding them to a cycling extract. An input sperm nucleus is visualized by fluorescence with the DNA-binding dye Hoechst 33342. For samples withdrawn from extracts each time point is represented by a phase-contrast view (left) and Hoechst 33342 fluorescence (right). Bar equals 20 µm.

longer period they should be sealed with nail varnish and stored at $-20°C$ in the dark.

b. *Histone H1 Kinase Assays*

H1 Kinase Buffer
  80 mM β-glycerophosphate, pH 7.4
  15 mM $MgCl_2$
  20 mM EGTA
  0.1% Nonidet P-40 NP-40
  Components above may be stored mixed at 4°C for up to 1 month; components below added fresh:
  50 μM ATP
  15 μg/ml benzamidine (Sigma B-6506)
  10 μg/ml leupeptin
  10 μg/ml aprotinin
  10 μg/ml soybean trypsin inhibitor (Boehringer Mannheim 109 886)

Histone H1
  10 mg/ml histone H1 (Boehringer Mannheim 223 459) in water, stored in aliquots at $-20°C$

Sodium Dodecyl Sulfate (SDS) Sample Buffer
  2% (w/v) SDS
  80 mM Tris-HCl, pH 8.8
  10 mM EDTA
  2 μg/ml bromophenol blue
  10% (w/v) glycerol

With the discovery that MPF appeared to be identical with the growth associated histone H1 kinase (Langan *et al.*, 1989), it became possible to assay progress through the cell cycle by monitoring H1 kinase activity. This assay is normally performed on samples that have been removed from the reaction and then frozen. Typically, we remove a 1-μl sample from the reaction and freeze it in liquid nitrogen. At the time of assay the samples are thawed by adding 9 μl of H1 kinase buffer containing 1.25 μg of histone H1 and 0.6 μCi of [γ-^{32}P]ATP (Amersham, Arlington Heights, IL, PB10168, 3000 Ci/mmol). The reactions are incubated at 30°C for 10 minutes, then stopped by the addition of 30 μl of SDS sample buffer; 10-μl samples are run on 15% polyacrylamide gels (Anderson *et al.*, 1973), which are dried down and autoradiographed. For quantification of H1 kinase activity the amount of phosphate incorporated into H1 is determined by cutting out and counting the gel bands. H1 kinase activity is expressed as picomoles phosphate incorporated per minute per milligram of protein.

As described here, the H1 kinase assay contains a substantial amount of cold ATP (up to 100 μM final) contributed by the extract. We have never seen fluctuations in the amount of measured H1 kinase activity that could be attributed to variations in the amount of cold ATP added to the kinase assay. To minimize the possibility of such variation, H1 kinase assays can be performed using a final dilution of the extract of 80-fold in a 10-μl assay mixture containing ATP at 0.4 mM and with the other components as described above. Under these conditions the final ATP concentration in the assay is very insensitive to the ATP concentration in the extract; however, an intermediate dilution of the extract is required, and the amount of γ-^{32}P incorporated into H1 is greatly decreased.

  c. *Cyclin Synthesis and Degradation.* Cyclin abundance fluctuates during the cell cycle, increasing during interphase and then declining abruptly at the end of mitosis. Since cyclin is a major newly synthesized protein in the early embryonic frog cell cycles, this fluctuation can be followed by continuously labeling extracts with [^{35}S]methionine and monitoring the abundance of cyclin during the reaction. We routinely add [^{35}S]methionine (Amersham SJ.1515) to a final concentration of 0.4 mCi/ml, take 2-μl samples into 20 μl of sample buffer, and run 5 μl on 15% polyacrylamide gels (3–4 mm well width). In this gel system the B type cyclins run with an apparent molecular mass of 56 kDa that on well-resolved gels show as many as four distinct closely spaced bands, representing different stages of posttranslational modification. The cyclin A bands run at 60 kDa and are considerably less intense than cyclin B; therefore, cyclin A bands are not easily seen against the background of other newly synthesized proteins.

2. Troubleshooting

Below are some common problems experienced with cycling extracts and possible solutions.

  a. *Extracts Will Not Enter Mitosis.* Possible causes and solutions are as follows:

1. Extracts too dilute. Cycling is greatly slowed in extracts where buffer or other additions exceed 30% of the final reaction volume.

2. Glassware or reagents contaminated with ribonuclease. Extracts are very sensitive to ribonuclease, and care should be taken to keep all reagents and chemicals free of ribonucleases. For instance, certain preparations of aprotinin contain levels of RNase that are sufficient to reduce the rate of protein synthesis below that required for entry into mitosis. Another source of RNase contamination is the activation chamber if it is not cleaned regularly and washed extensively after each use.

3. Extracts prepared from batches of eggs heavily contaminated with necrotic or otherwise damaged eggs. We believe that damaged eggs release hy-

drolytic enzymes that can destroy mRNA and other essential components in the extract. For this reason we reject all batches of eggs that are of questionable quality.

b. Extracts Arrest at First Mitosis. A possible cause is incomplete destruction of CSF. The most common cause of incomplete CSF destruction is failure of egg activation. It is essential to perform a visual check for egg activation after attempting to activate eggs. The most common cause of activation failure is corrosion of electrodes in the egg activation chamber. We have never encountered a batch of eggs that could not be activated, and we have successfully prepared extracts from batches of eggs that have been activated on a second attempt.

E. Cytostatic Factor-Arrested Extracts

Extract Buffer for CSF Extracts (CSF-XB)
 100 mM KCl
 0.1 mM CaCl$_2$
 2 mM MgCl$_2$
 10 mM potassium HEPES, pH 7.7
 50 mM sucrose
 5 mM EGTA, pH 7.7
CSF-Energy Mix
 150 mM creatine phosphate
 20 mM ATP
 20 mM MgCl$_2$
 Stored in 0.1-ml aliquots at -20 C

The preparation of CSF-arrested extracts is broadly similar to that of cycling extracts. Only those steps that differ in detail from the cycling extract protocol are described in detail. All steps are at room temperature unless noted otherwise.

1. Collect laid eggs in or squeeze fresh eggs into 1 × MMR and then dejelly by incubation in dejellying solution (there is no incubation of the eggs in distilled water before dejellying, washes in 0.2 × MMR, or activation step).
2. The dejellied eggs are washed 4 times in standard XB and then twice in CSF-XB containing protease inhibitors.
3. Transfer the eggs to centrifuge tubes containing CSF-XB plus protease inhibitors and 100 μg/ml cytochalasin B.
4. Spin the eggs in a clinical centrifuge at room temperature for 60 seconds at 1000 rpm and then 30 seconds at 2000 rpm. Remove all CSF-XB and Versilube.

5. Perform the crushing spin at 16°C, 10,000 rpm, for 10 minutes.
6. Collect the cytoplasmic layer very slowly using a needle and syringe via side puncture. After this step the extracts should be kept on ice.
7. To the cytoplasmic fraction add leupeptin, chymostatin, pepstatin, and cytochalasin B to a final concentration of 10 μg/ml each (1/1000 volume of the protease inhibitor and cytochalasin stocks) and 1/20 volume of CSF-energy mix.
8. If a clarifying spin is needed it should be for 10 minutes at 10,000 rpm at 2°C.

1. Use of Cytostatic Factor-Arrested Extracts

CSF-arrested extracts remain arrested with high levels of MPF and cyclin, unless calcium is added to inactivate the CSF. We usually perform a pilot experiment to determine the amount of calcium addition required to release the arrest. Three identical pilot reactions are set up with sperm nuclei at 10^5/ml and are incubated for 20 minutes at 23°C. At this time 4 mM CaCl$_2$ in sperm dilution buffer is added to final concentrations of 0.2 or 0.4 mM to two of the reactions, and the third reaction receives no added calcium. By 40 minutes after calcium addition, the nuclei in one or both of the reactions that received added calcium should be in interphase, while the control extract should demonstrate no visible chromosome decondensation, although it may show the reorganization of the homogeneous sperm nucleus into individually visible condensed chromosomes. By 80 to 100 minutes after calcium addition, the released extracts will have returned to mitosis. They often arrest at this stage, suggesting that although the CSF was inactivated at the time of calcium addition, there was no CSF destruction and it had become reactivated by the time of the first mitosis. We have preliminary evidence that less fractionated extracts, made with a 5-minute crushing spin and no clarifying spin, retain the capacity to destroy CSF in response to calcium addition.

CSF-arrested extracts and CSF-arrested extracts after calcium treatment are often used to compare interphase and mitotic states. To ensure that the interphase extracts do not return to mitosis, cycloheximide is added to 100 μg/ml before calcium addition. In the absence of cyclin synthesis, these interphase extracts cannot enter mitosis.

An alternative method for producing extracts arrested in mitosis involves the use of derivatives of cyclins that cannot be degraded. Deletion of the 90 N-terminal amino acids of sea urchin cyclin B yields a molecule named cycΔ90, that can drive extracts into mitosis, but cannot be degraded, and therefore arrests them in mitosis (Murray *et al.,* 1989). We refer to these as cycΔ90-arrested extracts to distinguish them from CSF-arrested extracts. The main difference between these extracts is that the machinery which degrades full length cyclins is inactive in CSF-arrested extracts but is fully active in cycΔ90-arrested extracts. Although the initial cycΔ90-arrested extracts were prepared by adding mRNA encoding cycΔ90 to extracts (Murray *et al.*, 1989), a much more convenient preparation is to add bacterially synthesized cycΔ90 protein to interphase extracts (Glotzer *et al.*, 1991).

2. Troubleshooting

a. Extract Is Not Arrested in Metaphase. Eggs or extract may have been activated during preparation. The most common problem is spontaneous activation of eggs during dejellying and washing. The current protocol reduces the frequency of spontaneous activation, but any batch of eggs that shows more than 50% spontaneous activation is unlikely to yield a CSF-arrested extract.

F. Nuclease-Treated Extracts

The ability to use nuclease treatment to make protein synthesis in reticulocyte lysates dependent on exogenous mRNA (Pelham and Jackson, 1975) revolutionized the use of *in vitro* translation systems. We have prepared nuclease-treated *Xenopus* cell cycle extracts by treating cycling extracts or CSF-arrested extracts with low doses of pancreatic ribonuclease and then inhibiting the ribonuclease with placental RNase inhibitor. This procedure destroys the endogenous mRNA but preserves the ability to translate exogenous mRNA that is added after the nuclease treatment. These extracts have so far had two principal uses: (1) To examine the protein synthesis requirement for particular cell cycle events; we used this approach to demonstrate that the synthesis of cyclin was sufficient to meet the protein synthesis requirement for multiple cell cycles (Murray and Kirschner, 1989a). (2) Nuclease-treated extracts may be used to follow the fate of a particular protein during the cell cycle in terms of synthesis and degradation or changes in posttranslational modification, free from the background of endogenously synthesized proteins. This approach may be useful for studying the secretion and sorting of newly synthesized proteins. In order to produce cell cycles after nuclease treatment, cyclin mRNA must be added in addition to the mRNA for the protein of interest.

1. Preparation

Pancreatic RNase
  Pancreatic RNase (Sigma R5503) is dissolved at 10 mg/ml in 100 mM KCl, 1 mM EDTA, 10 mM HEPES, pH 7.4, boiled for 10 minutes, and then

diluted to 100 μg/ml in sperm dilution buffer and stored frozen in small aliquots at −20°C.

Placental RNase Inhibitor

Placental RNase inhibitor is from Promega (Madison, WI), N211

Calf Liver tRNA

Calf liver tRNA (Boehringer 647 225) is dissolved in water at 5 mg/ml and stored in small aliquots at −70°C.

Test mRNA

Any well-translated eukaryotic mRNA that is easy to prepare is suitable. We prepare tobacco mosaic virus (TMV) RNA and store it dissolved in water at 2 mg/ml in small aliquots at −70°C. Alternatively, brome mosaic virus (BMV) RNA is commercially available (Promega D1541).

2. Procedure

The basic approach is to treat the extract with ribonuclease at 10°C for 20 minutes and then add placental RNase inhibitor and incubate for a further 10 minutes at 10°C, before adding calf liver tRNA to 50 μg/ml and any desired mRNA and continuing incubation at 23°C. The total volume of additions in this protocol can approach 20% of the extract volume. For solutions with low ionic strength or osmolarity, adjust the parameters to those found in sperm dilution buffer by adding concentrated sperm dilution buffer either to the solution in question or directly to the extract.

Since different preparations of RNase and RNase inhibitor will differ in activity, it is necessary to titrate the amount of RNase and RNase inhibitor to get good destruction of endogenous mRNA without damaging the translational machinery. Once the optimal levels for the particular preparations of RNase and RNase inhibitor have been determined, they can be used for both cycling and CSF-arrested extracts.

First titrate the level of RNase required to achieve complete mRNA destruction without causing significant damage to the translational machinery:

1. Set up several small tubes containing 10 μl of extract and perform 2-fold serial dilutions of RNase from 1 to 0.06 μg/ml. Include a control without added RNase.
2. Incubate for 20 minutes at 10°C and then add RNase inhibitor to 4000 U/ml.
3. Incubate for a further 10 minutes at 10°C and then add tRNA to 50 μg/ml and [^{35}S]methionine to 0.4 mCi/ml.
4. Split each reaction into two and incubate one-half without added mRNA and the other with the test mRNA (BMV or TMV RNA) at a final concentration of 50–100 μg/ml for 60 minutes at 23°C.
5. Determine the amount of trichloroacetic acid (TCA)-precipitable radio-

activity: withdraw a 2-μl sample into a test tube containing 0.5 ml of water, vortex, add 0.5 ml of 1 M NaOH (to hydrolyze methionyl-tRNA) containing 1 mM cold methionine, vortex, wait 5 minutes, add 1 ml of 25% TCA, and collect the precipitated protein on a glass-fiber filter (Whatman GF/C) and count.

We use a concentration of RNase 2-fold higher than the lowest dose which gives more than 90% destruction of the endogenous mRNA, as judged by comparison with the control to which RNase inhibitor, but not RNase, had been added. Having determined the optimal dose of RNase we repeat the experiment described above varying the amount of RNase inhibitor added after digestion with a constant dose of RNase. We use a dose of RNase inhibitor 2-fold higher than the dose at which the level of translation of the added test mRNA plateaus. Typical values of the final concentrations for RNase and RNase inhibitor are 0.25 μg/ml of RNase and 1000 U/ml of RNase inhibitor.

3. Notes

Since DNA that is added to *Xenopus* egg extracts can be incorporated into nuclei and alter the cell cycle (Dasso and Newport, 1990), it is important to remove any DNA from synthetic mRNA preparations that are to be added to the extract. For *in vitro* transcription reactions, we do this by adding RQ1 DNase (Promega M6101) to a concentration of 1 U/μg of template DNA, NaCl to 10 mM, and incubating the reaction for 1 hour at 37°C before phenol extraction and ethanol precipitation.

Some nuclease-treated cycling extracts go through one round of mitosis even in the absence of added mRNA. We believe that this reflects the occurrence of a level of cyclin synthesis, between egg activation and the destruction of the endogenous mRNA, that permits entry into the first mitosis. Such extracts never enter a second mitosis.

The activity of the RNase inhibitor is critically dependent on keeping the sulfhydryl groups in the protein reduced. We recommend adding dithiothreitol (DTT) to 25 mM to the RNase inhibitor stock, minimizing the length of time it spends at temperatures higher than $-20°C$, and also adding 1 mM DTT to the extract to be treated with nuclease.

IV. Conclusion and Prospects

In this chapter I have described methods for preparing and utilizing frog cell cycle extracts. These extracts have so far been used mainly to investigate questions of cell cycle regulation. However, since the extracts appear competent to carry out most of the processes that occur in intact cells, it seems that they will also be useful for investigating many important cell biological and biochemical problems that are not directly related to the regulation of the cell cycle.

Finally, an obvious question concern the ability to make *in vitro* cell cycle extracts from other types of cells. Considerable success has been achieved with extracts from clam eggs (Luca and Ruderman, 1989) which can also undergo cell cycles *in vitro*. Like frog eggs, clam eggs are large cells which have short cell cycles and which can be obtained in large quantities and broken easily. Somatic cells have longer cell cycles and must increase in mass during the cell cycle. Therefore, it seems unlikely that extracts can be prepared from them that will perform the reactions of a complete cell cycle. However, it should be possible to prepare extracts from cells at one stage of the cell cycle that can be induced to make the transition to the next stage of the cell cycle *in vitro*. Thus, it should be possible to prepare extracts from cells in G_1 that can be induced to initiate DNA synthesis by adding G_1 cyclins (Cross, 1988; Hadwiger *et al.*, 1989; Nash *et al.*, 1988) and extracts from cells in G_2 that can be induced to enter mitosis after the addition of purified MPF. The primary candidates for such attempts are mammalian tissue culture cells, budding and fission yeasts, and *Drosophila*. It is hoped that in the next few years cell cycle extracts will help to unravel the detailed mechanism of cell cycle transitions and the controls that regulate them.

Acknowledgments

The methods for producing cell cycle extracts were developed while I was a postdoctoral fellow in Marc Kirschner's laboratory. I am pleased to acknowledge his support, generosity, enthusiasm, insightful advice, and constant encouragement which made this work possible. I thank Michael Glotzer, Jeremy Minshull, and Mark Solomon for helpful comments on the manuscript. I am a Lucille P. Markey Scholar, and this work was supported in part by a grant from the Lucille P. Markey Charitable Trust.

References

Anderson, C. W. W., Baum, P. R., and Gesteland, R. F. (1973). Processing of adenovirus 2-induced proteins. *J. Virol.* **12**, 241–252.

Belmont, L. D., Hyman, A. A., Sawin, K. E., and Mitchison, T. J. (1990). Real-time visualization of cell cycle-dependent changes in microtubule dynamics in cytoplasmic extracts. *Cell (Cambridge, Mass.)* **62**, 579–589.

Blow, J. J., and Laskey, R. A. (1986). Initiation of DNA replication in nuclei and purified DNA by a cell-free extract of *Xenopus* eggs. *Cell (Cambridge, Mass.)* **47**, 577–587.

Cross, F. R. (1988). DAF1, a mutant gene affecting size control, pheromone arrest, and cell cycle kinetics of *Saccharomyces cerevisiae*. *Mol. Cell. Biol.* **8**, 4675–4684.

Dasso, M., and Newport, J. W. (1990). Completion of DNA replication is monitored by a feedback system that controls the initiation of mitosis *in vitro*: Studies in *Xenopus*. *Cell (Cambridge, Mass.)* **61**, 811–823.

Félix, M.-A., Pines, J., Hunt, T., and Karsenti, E. (1989). A post-ribosomal supernatant from activated *Xenopus* eggs that displays post-translationally regulated oscillation of its $cdc2^+$ mitotic kinase activity. *EMBO J.* **8**, 3059–3069.

Glotzer, M., Murray, A. W., and Kirschner, M. W. (1991). Cyclin is degraded by the ubiquitin pathway. *Nature (London)* **349**, 132–138.

Gurdon, J. B. (1976). Injected nuclei in frog oocytes: Fate, enlargement, and chromatin dispersal. *J. Embryol. Exp. Morphol.* **36**, 523–540.

Hadwiger, J. A., Wittenberg, C., Richardson, H. E., de Barros-Lopes, M., and Reed, S. I. (1989). A novel family of cyclin homologs that control G1 in yeast. *Proc. Natl. Acad. Sci. U.S.A.* **86**, 6255–6259.

Hutchison, C. J., Cox, R., Drepaul, R. S., Gomperts, M., and Ford, C. C. (1987). Periodic DNA synthesis in cell-free extracts of *Xenopus* eggs. *EMBO J.* **6**, 2003–2010.

Hutchison, C. J., Cox, R., and Ford, C. C. (1988). The control of DNA replication in a cell-free extract that recapitulates a basic cell cycle *in vitro Development (Cambridge, U.K.)* **103**, 553–566.

Langan, T. A., Gautier, J., Lohka, M., Hollingsworth, R., Moreno, S., Nurse, P., Maller, J., and Sclafani, R. A. (1989). Mammalian growth-associated H1 histone kinase: A homolog of $cdc2^+/CDC28$ protein kinase controlling mitotic entry in yeast and frog cells. *Mol. Cell. Biol.* **9**, 3860–3868.

Lohka, M., and Maller, J. (1985). Induction of nuclear envelope breakdown, chromosome condensation, and spindle formation in cell-free extracts. *J. Cell Biol.* **101**, 518–523.

Lohka, M. J., and Masui, Y. (1983). Formation *in vitro* of sperm pronuclei and mitotic chromosomes induced by amphibian ooplasmic components. *Science* **220**, 719–721.

Luca, F. C. and Ruderman, J. V. (1989). Control of programmed cyclin destruction in a cell-free system. *J. Cell Biol.* **109**, 1895–1909.

Murray, A. W., and Kirschner, M. W. (1989a). Cyclin synthesis drives the early embryonic cell cycle. *Nature (London)* **339**, 275–280.

Murray, A. W., and Kirschner, M. W. (1989b). Dominoes and clocks: The union of two views of cell cycle regulation. *Science* **246**, 614–621.

Murray, A. W., Solomon, M. J., and Kirschner, M. W. (1989). The role of cyclin synthesis and degradation in the control of maturation promoting factor activity. *Nature (London)* **339**, 280–286.

Nash, R., Tokiwa, G., Anand, S., Erickson, K., and Futcher, A. B. (1988). The $WHI1^+$ gene of *Saccharomyces cerevisiae* tethers cell division to cell size and is a cyclin homolog. *EMBO J.* **7**, 4335–4346.

Pelham, H. R. B., and Jackson, R. J. (1975). An efficient mRNA dependent translation system from reticulocyte lysates. *Eur. J. Biochem.* **67**, 247–256.

Sagata, N., Watanabe, N., Vande Woude, G. F., and Ikawa, Y. (1989). The c-*mos* proto-oncogene product is a cytostatic factor responsible for meiotic arrest in vertebrate eggs. *Nature (London)* **342**, 512–518.

Sawin, K. E., and Mitchison, T. J. (1991). Mitotic spindle assembly by two different pathways *in vitro*. *J. Cell Biol.* **112**, 925–940.

Tuomikoski, T., Félix, M.-A., Doree, M., and Gruenberg, J. (1989). Inhibition of endocytic vesicle fusion by the cell cycle control protein kinase *cdc2*. *Nature (London)* **342**, 942–945.

Verde, F., Labbe, J.-C., Doree, M., and Karsenti, E. (1990). Regulation of microtubule dynamics by *cdc2* protein kinase in cell-free extracts of *Xenopus* eggs. *Nature (London)* **343**, 233–238.

Wilson, E. B. (1928). "The Cell in Development and Heredity," 3rd Ed. Macmillan, New York.

Chapter 31

Egg Extracts for Nuclear Import and Nuclear Assembly Reactions

DONALD D. NEWMEYER

La Jolla Cancer Research Foundation
La Jolla, California 92037

KATHERINE L. WILSON

Department of Cell Biology and Anatomy
Johns Hopkins University School of Medicine
Baltimore, Maryland 21205

I. Introduction
   A. Nuclear Envelope Components and Their Fates during Mitosis
   B. Cell-Free Extracts for Nuclear Assembly and Transport Studies
   C. Properties of Nuclear Transport
   D. Nuclear Protein Import *in Vitro*
   E. Constructing Mutant Organelles *in Vitro*
II. Preparatory Methods
   A. Buffers and Reagents
   B. Demembranated Sperm Chromatin Preparation
   C. Nuclear Assembly Extracts from Unactivated Eggs
   D. Mitotic Vesicle Preparation
   E. Embryonic Extracts Containing Embryonic Nuclei for Nuclear Transport Studies
   F. Isolation of Rat Liver Nuclei for Nuclear Transport Studies
III. Nuclear Assembly Reactions and Assays
   A. Assembly of Nuclei around Demembranated Sperm Chromatin
   B. Assaying Nuclear Assembly by Fluorescence Microscopy
IV. Nuclear Transport Substrates, Reactions, and Assays
   A. Preparation of Transport Substrates
   B. Standard Nuclear Import Assay
   C. Identification of Broken and Intact Nuclei
   D. Staining Nuclear Pore Complexes with Fluorescein Isothiocyanate—Wheat Germ Agglutinin

E. ATP Depletion of Egg and Embryo Extracts
F. Video Image Analysis System to Quantitate Nuclear Import by Fluorescence Microscopy
References

I. Introduction

The nuclear envelope is the dynamic boundary that separates the eukaryotic genome from the cell cytoplasm. In addition to the chromosomes, all the proteins and enzymes needed for replication, transcription, and RNA maturation are confined to the nucleus. Ribosomes and translational factors, on the other hand, are restricted to the cytoplasm. This segregation of nucleoplasmic and cytoplasmic constituents is established early after mitosis, when the nuclear envelope components reassemble directly onto chromosome surfaces (Lohka and Masui, 1984a). After the chromosomes are enclosed by the envelope, segregation is maintained by nuclear pore complexes in the envelope, which regulate the movement of macromolecules into and out of the nucleus (Feldherr, 1962; Stevens and Swift, 1966; Feldherr et al., 1984; Dworetzky and Feldherr, 1988). Furthermore, the nuclear envelope and associated skeletal structures bind chromatin and DNA, and may help organize chromatin structure (reviewed by Franke et al., 1981; Newport and Forbes, 1987; Scheer et al., 1988; and see, e.g., Agard and Sedat, 1983; Hochstrasser et al., 1986; Paddy et al., 1990). In this chapter, we describe extracts that can be used to study the assembly, structure, function, and cell cycle regulation of the nuclear envelope and pore complexes.

A. Nuclear Envelope Components and Their Fates during Mitosis

The nuclear envelope is sketched in Fig. 1. Also illustrated is the fate of each nuclear component during mitosis, when the nucleus, Golgi, and endoplasmic reticulum disassemble.

1. INNER AND OUTER NUCLEAR MEMBRANES

The outer nuclear membrane is continuous with (and functions as part of) the endoplasmic reticulum (ER). However, the nuclear membranes are functionally distinct from ER membranes in terms of their ability to reform the nuclear envelope after mitosis (Wilson and Newport, 1988). The inner and

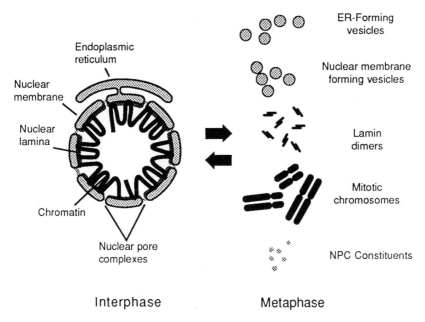

Fig. 1. The natural mitotic processes of nuclear disassembly and reassembly provide a basis for *in vitro* systems derived from *Xenopus* egg extracts.

outer membranes are joined at nuclear pore complexes. During mitosis the membranes are dispersed throughout the cytoplasm as small vesicles, approximately 200 μm in diameter (Lohka and Masui, 1984a).

2. Nuclear Lamina

The nuclear lamina, a polymeric network of lamin proteins, is located under the inner membrane. Lamins are related to intermediate filament proteins (Aebi *et al.*, 1986; Fisher *et al.*, 1986; McKeon *et al.*, 1986). The lamina provides strength to the nuclear envelope. Because the expression of different lamin proteins is regulated during development, it has been speculated that changes in the composition of the lamina may be related to changes in chromatin organization (see Gerace and Burke, 1988). Lamin proteins are substrates for the mitotic cyclin/cdc2 kinase activity (Peter *et al.*, 1990); hyperphosphorylation of lamins during mitosis appears to trigger the depolymerization of the lamina (Gerace and Blobel, 1980; Ottaviano and Gerace, 1985; Miake-Lye and Kirschner, 1985; Ward and Kirschner, 1990; Heald and McKeon, 1990).

3. NUCLEAR PORE COMPLEXES

Nuclear pore complexes (NPCs) are huge macromolecular complexes (molecular weight ~150,000,000) that span both nuclear membranes and anchor into the lamina. Pore complexes are the sites of RNA efflux (Stevens and Swift, 1966; Dworetzky and Feldherr, 1988) and protein influx (Feldherr *et al.*, 1984). Only a handful of pore components have been identified, including myosin (Berrios and Fisher, 1986), a novel family of O-linked glycoproteins (Davis and Blobel, 1986; Holt and Hart, 1986; Hanover *et al.*, 1987; Holt *et al.*, 1987; Park *et al.*, 1987; Schindler *et al.*, 1987; Snow *et al.*, 1987; Davis and Fink, 1990; Nehrbass *et al.*, 1990; Starr *et al.*, 1990), and an integral membrane protein named gp210 (formerly gp190; Gerace *et al.*, 1982; Wozniak *et al.*, 1989; Greber *et al.*, 1990). Except for gp210, it is thought that the known pore proteins become soluble during mitosis, and the pores disappear.

B. Cell-Free Extracts for Nuclear Assembly and Transport Studies

Xenopus eggs are a rich source of the protein and membrane components required to assemble a functional nuclear envelope around exogenously added DNA or chromatin (Lohka and Masui, 1983; Newport and Forbes, 1985; Newmeyer *et al.*, 1986a; Newport, 1987; reviewed in Newport and Forbes, 1987; Lohka and Maller, 1987; Newmeyer, 1990). Materials required for organelle biogenesis are stockpiled in the egg for use during early development (Laskey *et al.*, 1979). Moreover, the egg is arrested in metaphase of the second meiotic division, which means that stockpiled components (including lamins, pore proteins, and nuclear membrane vesicles) are stored in their disassembled mitotic form until the egg is fertilized or activated.

Lohka and Masui (1983) were the first to show that a cell-free extract of frog eggs could assemble nuclei around sperm chromatin, mimicking the normal process of sperm pronucleus formation after fertilization. The extract reproduced an entire embryonic cell cycle: nuclear assembly followed by DNA replication and mitotic disassembly. The presence of a functioning cell cycle regulatory system in *Xenopus* egg extracts has since been extensively exploited (Lohka and Masui, 1984b; Lohka and Maller, 1985; Miake-Lye and Kirschner, 1985; Hutchison *et al.*, 1987; Newport and Spann, 1987; Dunphy and Newport, 1988; Dunphy *et al.*, 1988; Felix *et al.*, 1989; Newport and Dasso, 1989; Dasso and Newport, 1990; reviewed by Murray and Kirschner, 1989). In this chapter, we describe how to make extracts stably arrested in either interphase or metaphase. Interphase extracts support nuclear assembly and transport, whereas mitotic extracts disassemble nuclei.

C. Properties of Nuclear Transport

The fundamental properties of nucleocytoplasmic exchange are diagrammed in Fig. 2. Pore complexes allow small macromolecules (e.g., proteins smaller than ~60 kDa) free diffusional access to the nuclear interior. Larger macromolecules are restricted to the cytoplasm unless they possess a nuclear location signal (NLS), in which case they accumulate rapidly in the nucleus. Such proteins are termed karyophilic. The NLSs for many karyophilic proteins have been identified. An NLS is usually a short peptide sequence enriched in the basic amino acids lysine and arginine. No obvious consensus sequence has emerged from various studies. However, it appears that charge alone is not sufficient for NLS function. For the simian virus 40 (SV40) large T antigen, the NLS consists of the peptide Pro-Lys-Lys-Lys-Arg-Lys-Val; these seven amino acids in reverse order do not function as an NLS (Lobl *et al.*, 1990). Although the amino acids surrounding the NLS can influence its function, in general the NLS behaves as if autonomous (Roberts *et al.*, 1987). Indeed, several NLSs have been shown to possess nuclear targeting activity as synthetic peptides covalently coupled to nonkaryophilic proteins (Goldfarb *et al.*, 1986; Lanford *et al.*, 1986; Chelsky *et al.*, 1989; Lanford *et al.*, 1990).

RNA molecules, in the form of ribonucleoprotein particles, are exported through the same pores that import proteins (Dworetzky and Feldherr, 1988). As one might expect, the functional complexity of the nuclear pore is matched by the intricacy of its structure (Unwin and Milligan, 1982; Stewart and Whytock, 1988; Akey, 1989; Akey and Goldfarb, 1989). To begin to

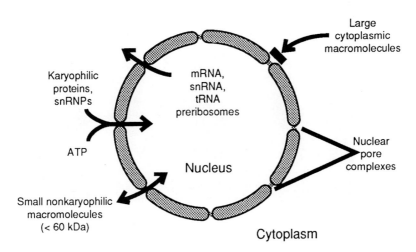

FIG. 2. Basic properties of the transport of RNA and protein molecules between nucleus and cytoplasm.

understand pore function, biochemical assays such as the cell-free system described in this chapter are essential.

D. Nuclear Protein Import *In Vitro*

Nuclei assembled around added DNA or chromatin in *Xenopus* egg extracts are fully functional in nuclear protein import (Newmeyer *et al.*, 1986a,b). In addition, nuclei isolated from exogenous tissues or cultured cells are also transport competent when incubated in the extract. To study transport *in vitro*, the import of fluorescently labeled karyophilic proteins can be assayed by light microscopy, as shown in Fig. 3. Using this *in vitro* assay, nuclear import was shown to be an active process, since it requires both ATP and an intact nuclear envelope (Newmeyer *et al.*, 1986a,b). Studies with the *Xenopus* extract also showed that nuclear import *in vitro* is inhibited by the lectin wheat germ agglutinin (WGA; Finlay *et al.*, 1987). Inhibition of import by WGA and by antibodies directed against the WGA-binding glycoproteins in the nuclear pore provided a much needed experimental handle with which to study pore constituents (Dabauvalle *et al.*, 1988a,b; Featherstone *et al.*, 1988; Yoneda *et al.*, 1987).

The inhibition of import by WGA or by ATP depletion was used to separate two steps in nuclear import: (1) binding of the protein to the pore, mediated by

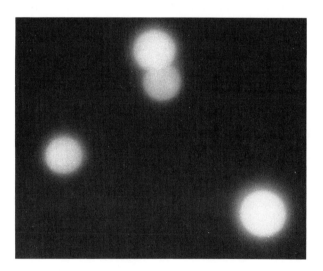

FIG. 3. Transport of a fluorescent protein into rat liver nuclei in a *Xenopus* egg extract. Rat liver nuclei were incubated for 30 minutes in the extract before addition of the transport substrate, a TRITC-labeled conjugate of human serum albumin with a nuclear localization signal peptide from SV40 T-Antigen (cys-thr-pro-pro-lys-lys-lys-arg-lys-val). After a further 30-minute incubation, an aliquot of the mixture was examined by fluorescence microscopy and photographed. × 1250.

the nuclear localization signal of the protein, and (2) translocation of the protein through the pore, which requires ATP and is inhibited by WGA (Newmeyer and Forbes, 1988; Richardson et al., 1988). The observation that WGA blocked translocation but not binding suggested that the pore complex glycoproteins are involved in the translocation step.

E. Constructing Mutant Organelles *In Vitro*

A major strength of the *Xenopus* extracts is the opportunity to remove a given component selectively to determine its role in nuclear envelope assembly or pore function. In an elegant experiment, Finlay and Forbes (1990) assembled "biochemically mutant" nuclear pores in extracts that had been depleted of pore glycoproteins by affinity chromatography. The resulting pore complexes looked fairly normal by transmission electron microscopy but were inactive for nuclear import. Interestingly, the mutant pores were defective in the binding step. It thus appears that one or more of the glycoproteins are required for the assembly of functional pores, in addition to their role during protein translocation.

Nuclei with mutant nuclear envelopes (no lamina) have also been constructed *in vitro*. Nuclei lacking a lamina were assembled by incubating sperm chromatin or naked λ phage DNA in extracts immunodepleted of the embryonic lamin, L_{iii} (Newport et al., 1990). Lamins have been shown to assemble on chromosome surfaces in a manner suggesting that lamins are important early components of the reassembling nuclear envelope (Burke, 1990; Glass and Gerace, 1990). Nevertheless, the *in vitro* mutant nuclei demonstrated that L_{iii} is not required for assembly of the nuclear membrane or pore complexes. However, the nuclear lamina is needed for DNA replication and for envelope growth beyond certain sizes; the mutant nuclei were small, did not replicate, exhibited high pore density with respect to nuclear surface area, and were easily broken. The "mutant organelle" approach will be increasingly valuable to determine the role of specific proteins in nuclear assembly and function.

Direct biochemical fractionation of the extract can be used to identify soluble factors required for nuclear assembly and import. For example, Newmeyer and Forbes (1990) identified an N-ethylmaleimide-sensitive soluble factor, NIF-1, that is required for the binding step during nuclear import. A second cytosolic factor, NIF-2, was shown to interact synergistically with NIF-1 to promote nuclear import. One proposed function for these factors is to bind proteins that have nuclear location signals and escort them to the pore, much like the signal recognition particle (SRP) that accompanies newly translated secretory proteins to the endoplasmic reticulum (reviewed by Walter et al., 1984).

II. Preparatory Methods

A. Buffers and Reagents

1. Reagents

Reagent	Final composition	Comments
Aprotinin/leupeptin	1 mg/ml Aprotinin, 1 mg/ml leupeptin	In 10 mM Tris, pH 7.4, 1000× stock; store aliquots at −20°C
Hoechst 33258 (Calbiochem, San Diego, CA)	10 mg/ml in dimethyl sulfoxide (DMSO)	Carcinogen
Creatine phosphokinase	5 mg/ml in 50% glycerol	100× stock; store at −20°C

2. Buffers

10 × Buffer A Salts
  800 mM KCl, 150 mM NaCl, 50 mM EDTA, 150 mM PIPES–NaOH (pH 7.4)

Membrane Wash Buffer (MWB)
  50 mM KCl, 250 mM sucrose, 2.5 mM $MgCl_2$, 50 mM HEPES–NaOH (pH 8.0), 1 mM dithiothreitol (DTT), 1 mM ATP, 1 µg/ml leupeptin, 1µg/ml aprotinin. Store aliquots at −20°C. Option: Add ATP and protease inhibitors just before use (ATP interferes with the BCA protein assay; Pierce Chemical Co., Rockford, IL.).

Hoechst Buffer
  200 mM sucrose, 5 mM $MgCl_2$, buffer A salts 1 ×, 10 µg/ml Hoechst 33258. Option: Add formaldehyde to 3.7%. Store in the dark at 4°C.

10 × Lysolecithin Stock
  0.5% lysolecithin, 200 mM maltose, 7 mM $MgCl_2$. Store at −20°C.

B. Demembranated Sperm Chromatin Preparation

Sperm chromatin is a convenient substrate for studying nuclear envelope assembly and function because (1) assembly is synchronous, (2) the DNA content per nucleus is uniform, (3) the number of nuclei per reaction can be controlled, and (4) nuclear assembly is faster than with naked DNA, which must be assembled into chromatin before envelope assembly begins. All steps in the preparation of demembranated sperm chromatin (Lohka and Masui,

1983) are done at room temperature unless otherwise noted. A clinical centrifuge (IEC; VWR Scientific) is used for all pelleting steps. Use 1.5-ml Eppendorf tubes, one pair of testes per tubes. The approximate buffer volumes needed per frog are as follows: 4.0 ml solution SB (1 × buffer A salts, 0.2 M sucrose, 7 mM $MgCl_2$), 1.0 ml SB with 0.05% lysolecithin, and 5.0 ml SB with 3% bovine serum-albumin (BSA).

1. THE DAY BEFORE

1. Inject each frog with 100 units of human chorionic gonadotropin (hCG); wait 1 hour.
2. Anesthetize male frogs 15–20 minutes in tricaine [3-aminobenzoic acid ethyl ester (Sigma Chemical Co., St. Louis, MO), 2.5 g per liter water]. Lay a frog on its back; no movement means the frog is ready. Pith frog; place belly-up on clean paper.
3. Make an incision into the lower abdomen to expose the intestinal area. Testes are about 1 cm long, white, roughly kidney shaped, and are located in the low central area on either side of the spine. Remove the testes. Do not puncture them!
4. Incubate testes overnight at 18°C in MMR solution (see Appendix A) with 10 units/ml hCG.

2. PROCEDURE

1. Carefully cut away contaminating fatty tissue and membranes from testes.
2. Rinse testes in solution SB. Use one Eppendorf tube per 1–2 testes.
3. Chop and squeeze testes to release sperm in 1 ml of SB. Be brutal but do not disintegrate the outer tissue (somatic cells).
4. Pellet the tissue 10 seconds at setting 3 in the clinical centrifuge.
5. Transfer the supernatant to a new Eppendorf tube.
6. Add 500 µl SB to the pelleted tissue from Step 4, mix gently, and recentrifuge. Combine the two supernatants.
7. Centrifuge to pellet the sperm, 2 minutes at setting 7 (maximum speed).
8. Remove and discard the supernatant.
9. Slowly and carefully resuspend the white part of the pellet in SB, 50–100 µl at a time, for a total of approximately 500 µl. The reason for caution is to avoid resuspending the red bottom of the pellet; this contains red blood cells and somatic cells whose nuclei will contaminate the final preparation.
10. Pellet the sperm for 2 minutes at setting 7.

11. Resuspend the sperm pellet as in Step 9 (500 μl final). Again, avoid the red part of the pellet, but do not leave behind too many sperm.
12. Pellet the sperm for 2 minutes at setting 7.
13. Resuspend the sperm in 100 μl SB, then add 1 ml SB/0.05% lysolecithin. Lysolecithin removes the plasma membrane and nuclear membranes.
14. Leave at room temperature for 5 minutes.
15. Add 3 ml of SB/3% BSA and pipette gently to mix.
16. Pellet the demembranated sperm for 8–10 minutes at setting 4.
17. Resuspend the pellets in 200 μl each of SB/3% BSA; mix gently, pool into one tube, and add 1 ml SB/3% BSA.
18. Pellet the sperm 8–10 minutes at setting 4.
19. Resuspend the pellet in 50 μl SB.
20. From now on, keep the sperm on ice.
21. Count the sperm using a hemacytometer. (Mix gently; sperm tend to settle.) Transfer 0.5 μl of sperm to 100 μl Hoechst buffer (200-fold dilution). Mix, place under the coverslip of hemacytometer (Spotlight; Baxter/Sci. Products, McGaw Park, IL). Count the total number of sperm in the central cross-gridded area. The cross-grid area is a 5×5 matrix of boxes. Each box has a 4×4 matrix inside.
22. Calculate the sperm concentration as follows: number of sperm per microliter in preparation = $N \times 10 \times D$, where N is the total number of sperm in entire 5×5 grid and D is the dilution factor.
23. Dilute sperm with SB to get a final concentration of 40,000 sperm/μl. Be conservative; count a higher dilution if necessary.
24. Recount the final adjusted sperm preparation to ensure accuracy.
25. Mix thoroughly by pipetting and distribute small aliquots (~4 μl) to small Eppendorf tubes on ice.
26. Freeze aliquots in liquid nitrogen; store at -80°C.

C. Nuclear Assembly Extracts from Unactivated Eggs

Xenopus eggs are arrested in second meiotic metaphase, which is for all practical purposes a mitotic state. "Mitotic" extracts do not assemble nuclei; rather, they are capable of disassembling exogenous nuclei added to them. The egg cytoplasm is normally converted from metaphase to interphase by the act of fertilization, which stimulates the release of internal Ca^{2+} stores. Hence, earlier protocols for nuclear assembly extracts called for mock-fertilization using a calcium ionophore and Ca^{2+}, or electric shock (Newport 1987; Lohka and Masui, 1983). However, these treatments cause many eggs to self-destruct. Experience has shown that egg activation is unnecessary for nuclear assembly extracts; apparently, disruption of the eggs is sufficient to release the mitotic block. The following protocol (modified from Newport, 1987, and Wilson and Newport, 1988) therefore omits the activation step.

1. INITIAL PREPARATIONS

1. Obtain eggs (see Chapter 30, this volume). Estimate extract volumes:
   - X = starting egg volume
   - $X \div 5$ = volume of dejellied eggs
   - $X \div 15$ = approximate volume of crude cytoplasm from 10,000 g spin
   - $X \div 50$ = approximate volume of soluble fraction from 200,000 g spin
2. Decant each batch of eggs into a separate beaker.
3. Examine eggs and discard bad batches (big, white, blotchy, or furrowed).
4. Remove debris from good eggs.
5. Measure the volume of good eggs. It is a good idea to keep batches of eggs separate, if possible. Occasionally, batches of eggs that appear good are not and can poison other batches of eggs mixed with them.
6. Make the following solutions; keep at room temperature. (Wear gloves. Cycloheximide and cytochalasin B are toxic. Cycloheximide inhibits protein synthesis and thus prevents the extract from progressing into mitosis. Cytochalasin B inhibits actin polymerization and thus prevents "gelling.")

Solution	Approximate volume needed	Purpose
Cysteine	4 × egg volume	Dejelly the eggs
NaCl or MMR	4 × egg volume	Remove cysteine
S-Lysis buffer	2 × egg volume	Lysis buffer

S-Lysis Buffer

Final concentration	Stock solution	For 50 ml
	Pure water	34.5 ml
250 mM sucrose (ultrapure)	1 M	12.5 ml
50 mM KCl	1 M	2.5 ml
2.5 mM MgCl$_2$	1 M	125 μl
1 mM DTT	1 M	50 μl
50 μg/ml cycloheximide	10 mg/ml in water	250 μl (*Toxic!*)
5 μg/ml cytochalasin B	10 mg/ml in DMSO	25 μl (*Toxic!*)

Dejellying Solution

2% cysteine, pH 7.8–7.9. Add 2 g cysteine to 100 ml of either water or 4-fold diluted MMR, then add 12–15 drops of 10 N NaOH. (If not neutralized with NaOH, the solution will cause eggs to stick and will not remove the jelly coat.)

NaCl Solution
100 mM NaCl. Or use MMR.

2. Procedure

Be gentle and thorough with all rinses; keep eggs swirling gently. Use a Pasteur pipette to remove as much buffer as possible between rinses. Avoid jarring eggs.

a. *Crude Assembly Extract*
1. Optional: rinse eggs once in NaCl or MMR.
2. Dejelly the eggs in cysteine (pH 7.8) for up to 10 minutes, until the eggs are closely packed.
3. Rinse 4 times with NaCl or MMR to remove cysteine. Transfer eggs to glass petri dishes and examine with a dissecting microscope. With a Pasteur pipette, pick out all the bad eggs. Work quickly.
4. Transfer the eggs to a clean beaker.
5. Rinse 3 times in S-lysis buffer.
6. Remove excess buffer.
7. Add protease inhibitors (aprotinin/leupeptin, 1000 × stock) to an approximate final concentration of 3 × (3 μg/ml); gently swirl to mix.
8. Transfer the eggs gently to 15-ml centrifuge tubes (Falcon 2059). Do not fill the tubes above 12 ml. Remove excess buffer.
9. Pack the eggs gently in clinical centrifuge (10 seconds at setting 3, then 10 seconds at setting 1; or 500 g).
10. Remove excess buffer.
11. Add more protease inhibitors to 2 × final concentration; do not mix.
12. Lyse the eggs in a centrifuge with swinging-bucket rotor (Sorvall HB-4 or Beckman J-21) at 10,000 rpm ($\sim$ 10,000 g) for 12 minutes at 4°C. (For small volumes, lyse at 4°C at maximum speed in a Beckman Microfuge or an Eppendorf centrifuge with a horizontal rotor.) After lysis, keep extracts on ice.
13. Insert a 21-gauge needle, bevel up, just above the pigment layer (see Fig. 4) and withdraw the crude cytoplasmic fraction.
14. Transfer the cytoplasm to a clean tube on ice.

Note: Stop here if a fresh unfractionated assembly extract is desired. Add an ATP-regenerating system. Crude extracts are good several hours on ice but do not freeze reliably.

b. *Fractionation of Crude Assembly Extract.* The ultracentrifugation step separates the membranes and soluble components. Each component can be frozen and stored for at least 6 months. *Caution*: To form nuclei around λ DNA we suggest using the crude assembly extract; λ nuclei do not assemble

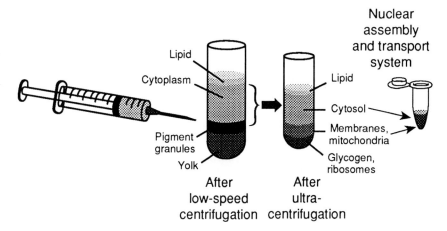

FIG. 4. Preparation of crude and fractionated *Xenopus* egg extracts.

well from ultracentrifuged components (P. Hartl and D. Forbes, personal communication).

1. Add more protease inhibitors to the extract (final concentration 2×); pipette to mix.
2. Transfer the extract into an ultraclear ultracentrifuge tube, on ice.
3. Centrifuge at 200,000 g for 1 hour at 4°C, in a swinging-bucket rotor. Ideal conditions are to use a Beckman tabletop TL-100 ultracentrifuge with a TLS-55 swinging-bucket rotor at 55,000 rpm. Other rotors may not achieve adequate separation in 1 hour. An alternate method is to spin at 45,000 rpm (~135,000 g) for 1.5 hours. One of us (D.D.N) prefers this lower speed because the membranes are easier to resuspend.
4. Aspirate carefully to remove the lipid layer from the surface.
5. Pipette the soluble fraction into a new ultraclear tube (see Fig. 4), or withdraw from the side using a syringe and needle as before. Centrifuge again at 200,000 g for 20 minutes at 4°C to remove residual membranes. Transfer the membrane-depleted soluble fraction to a fresh tube. Record the volume of soluble extract obtained. Add an ATP-regenerating system (stocks kept at −20°C): 1 mM ATP, pH 7, 10 mM phosphocreatine (Sigma, disodium salt; the di-Tris salt is not recommended), and 50 μg/ml creatine phosphokinase (Sigma). Freeze aliquots in liquid nitrogen, store at −80°C. Alternatively, the ATP-regenerating system can be added to aliquots of the extract after thawing.
6. Resuspend membranes in at least 20 volumes of cold membrane wash buffer (MWB).
   a. Pellet the membranes by centrifugation at 40,000 g for 15 minutes at 4°C (25,000 rpm in TLS-55 rotor). Optionally, underlay the

membranes with 1.3 M sucrose in MWB to form a cushion. If supernatant is cloudy, spin longer or harder.
  b. Aspirate to remove the supernatant.
  c. Recover the membranes in the smallest possible volume of MWB. Attempt to avoid the darker, denser part of the membrane fraction.
  d. The desired final volume of membranes is 1/10 of soluble fraction volume. (More concentrated membranes are fine.)
  e. Record the volume of membranes.
  f. Supplement the membranes with sucrose to at least 0.5 M final concentration prior to aliquoting for freezing.
  g. Freeze small aliquots in liquid nitrogen; store at $-80°C$.
6A. *Alternate method* (D.D.N.).
  a. Collect the membrane layer with a wide-bore pipette tip.
  b. Add DMSO to 10% and mix gently with a wide-bore pipette tip. The DMSO acts as a cryoprotectant.
  c. Freeze 60 to 80-μl aliquots directly at $-80°C$. *Do not immerse in liquid nitrogen!*
  d. Membranes must be washed once after thawing to remove the DMSO: Resuspend an aliquot of membranes in 3.6 ml S-lysis buffer. Centrifuge in three Eppendorf tubes at 4°C in a Beckman Microfuge or Eppendorf centrifuge. Resuspend and pool the pellets in 20 μl of S-lysis buffer or egg cytosol fraction.
7. Thaw and test aliquots (membrane–soluble ratios of 1:5 to 1:20) to determine the best reconstitution conditions.

c. Mitotic extracts from Xenopus eggs. The only difference between a mitotic extract (Newport and Spann, 1987) and our nuclear assembly extract is the buffer in which the eggs are lysed. M-lysis buffer has two ingredients that maintain the mitotic biochemistry of the eggs: β-glycerophosphate, a competitive inhibitor of phosphatase activity, and EGTA, which chelates Ca^{2+} and thus prevents inactivation of cytostatic factor (CSF). CSF is the *Xenopus* egg c-*mos* protooncogene product, formerly identified as a Ca^{2+}-sensitive factor that stabilizes the mitotic cyclin/cdc2 (MPF) kinase in its active form (Meyerhof and Masui, 1979; Sagata *et al.*, 1989; Watanabe *et al.*, 1989).

M-Lysis Buffer

Final concentration	Stock solution	For 100 ml
	Pure water	59.4 ml
240 mM β-glycerophosphate	1 M, pH 7.4	24 ml
60 mM EGTA	0.5 M, pH 8	12 ml
45 mM MgCl$_2$	1 M	4.5 ml
1 mM DTT	1 M	100 μl

Please refer to the nuclear assembly extracts protocol (Section II,C) for details of the procedure. As an alternative to that method, dilute the crude mitotic extract with an equal volume of 80 mM β-glycerophosphate, pH 7.4, 20 mM EGTA, pH 8, 15 mM $MgCl_2$, 1 mM DTT. Dilution will facilitate clearing of membranes from the soluble fraction at lower ultracentrifugation forces. *Caution*: Do not attempt to dilute nuclear assembly extracts, or the cytosol will become inactive for membrane fusion.

D. Mitotic Vesicle Preparation

The membrane fraction of *Xenopus* extracts is heterogeneous. In the following protocol (Wilson and Newport, 1988), vesicles active in nuclear envelope assembly are separated from other membranes. This fractionation works *only* with mitotic membrane preparations (see Wilson and Newport, 1988). Mitotic membranes are fractionated by density in a sucrose step gradient. This yields two membrane fractions, "light" aand "heavy." The light fraction contains vesicles of diameter 100–500 nm and includes the vesicles that form the nuclear envelope. The light membrane fraction also contains ER vesicles and probably Golgi vesicles. The heavy fraction contains mitochondria and other dense membranes, and it is inactive for nuclear envelope assembly. All steps are performed at 4°C. *Note*: To maintain vesicle activity, avoid unnecessary pelleting steps. Prepare sucrose gradient solutions in advance and cool to 4°C: 0.7, 1.1, and 1.3 M sucrose in 20 mM HEPES (pH 8.0), 1 mM DTT, and 1× aprotinin/leupeptin. (Optionally include 0.5 mM ATP.) Usually it is sufficient to set up one gradient for each ultracentrifugation tube used in the first 200,000 g spin.

1. Measure the volume of membranes.
2. Resuspend the membranes in MWB/1.4 M sucrose to the appropriate volume (25% of the ultracentrifuge tube volume). The final sucrose concentration of the membrane solution should be 1.38 to 1.4 M.
3. Mix the membrane/sucrose solution thoroughly. Pipette or vortex at low speed, but do not froth.
4. Place the membranes at the bottom of a clean ultraclear ultracentrifuge tube.
5. Overlay with MWB/1.3 M sucrose (~15% of tube volume).
6. Overlay with MWB/1.1 M sucrose (~30% of tube volume).
7. Overlay with MWB/0.7 M sucrose (~20% of tube volume).
8. Centrifuge at 200,000 g for 1 hour at 4°C in a swinging-bucket rotor (e.g., 55,000 rpm in a Beckman TLS-55 rotor). Light membranes should be visible as a white cloudy layer. Heavy membranes are yellowish.
9. Carefully remove most of the 0.7 M sucrose layer by aspiration or pipette.

10. Recover the light (vesicle) fraction at the 0.7/1.1 M interface.
11. Aspirate to remove the 1.1 M zone and recover the heavy membranes at the 1.1/1.3 interface.
12. Resuspend the membranes evenly (gradient solutions are satisfactory for freezing). Freeze small aliquots (in 0.5 M sucrose/MWB) in liquid nitrogen, and store at $-80°C$. (If vesicle activity is important, and if vesicles will be subjected to additional wash steps, consider pelleting onto a small 1.3 M sucrose cushion.)

E. Embryonic Extracts Containing Embryonic Nuclei for Nuclear Transport Studies

The following protocol is from J. Newport (personal communication) with slight modifications.

1. Fertilize eggs:
   a. Obtain testes from a male frog (see Section II,B). Testes can be stored for a week or so in 20% fetal bovine serum with gentamicin added. With a razor blade, chop off a small fragment of one testis. Tease gently with a pair of fine forceps to release sperm.
   b. Squeeze eggs from a female frog into a dry petri dish.
   c. Take the testis fragment with forceps and swirl it around for a minute or so in the dish of eggs.
   d. Add about 5 ml of distilled water. In the dissecting microscope, look for the appearance of a pigmented area at the sperm entry point and other signs of fertilization: contraction of the pigmented region at the animal pole, flattening of the eggs, and separation of the vitelline membrane from the egg surface.
   e. Transfer the fertilized eggs into ten-fold diluted MMR.
2. Collect the embryos at the desired stage (up through gastrula). Extracts can be made from fewer than 50 embryos. For the highest yield of nuclei, embryos should be arrested in interphase by incubation in cycloheximide (50 $\mu g/ml$) for 30–60 minutes prior to homogenization.
3. Rinse the embryos in S-lysis buffer. Allow the embryos to settle and remove excess buffer. Pack the embryos by centrifuging briefly at low speed in a clinical centrifuge or in a variable-speed Eppendorf centrifuge. Remove excess buffer. Add protease inhibitors aprotinin and leupeptin to $3 \times$ concentration.
4. Lyse by gently pipetting with an Eppendorf pipettor, first with a wide-bore yellow tip (chop off 1–2 mm with a razor blade), and then with a standard yellow tip.
5. Centrifuge for 5 minutes in a clinical centrifuge at maximum speed or in an adjustable-speed Eppendorf centrifuge at 5000 rpm.

6. Collect the supernatant layer above the pigment/yolk pellet, puncturing the side of the tube with a needle attached to a 1-ml syringe. The embryonic nuclei remain in the supernatant under these conditions.
7. Add an ATP-regenerating system as for the standard assembly extract and add a transport substrate (see below) to monitor import into the embryonic nuclei.

F. Isolation of Rat Liver Nuclei for Nuclear Transport Studies

The following protocol is from Blobel and Potter (1966), with crucial buffer modifications (Newport and Spann, 1987). Rat liver nuclei prepared using Mg^{2+}-containing buffers function poorly in nuclear transport (D. D. Newmeyer and D. J. Forbes, unpublished observations). Make all solutions in advance and cool to 4°C. Precool the ultracentrifuge, SW28 rotor, and buckets to 4°C.

Final concentration	Stock solution	For 500 ml
Solution A		
	Pure water	316 ml
250 mM sucrose (ultrapure)	1 M	125 ml
1 mM DTT	1 M	0.5 ml
1 × buffer A salts	10 ×	50 ml
0.5 mM spermidine	100 mM	2.5 ml
0.2 mM spermine	100 mM	1.0 ml
Add fresh before use:		
1 mM phenylmethylsulfonyl fluoride (PMSF)	0.1 M	5 ml
Solution B		
	Pure water	(Trace; add last)
2.3 M sucrose (ultrapure)	393.65 g	Crystals
1 mM DTT	1 M	0.5 ml
1 × buffer A salts	10 ×	50 ml
0.5 mM spermidine	100 mM	2.5 ml
0.2 mM spermine	100 mM	1.0 ml
Add fresh before use:		
1 mM PMSF	100 mM	3 ml

Procedure

1. Sacrifice rat(s) by asphyxiation with CO_2.
2. Remove the liver, keep on ice.
3. Cut away the tough connective tissue between lobes.
4. Record the weight of liver tissue here: $W =$ _____ grams.

5. Transfer the tissue to clean glass plate on ice; mince with razors.
6. Scrape the minced tissue into a chilled Dounce chamber (30 ml size).
7. Add ($2 \times W$) ml of chilled solution A.
8. Homogenize with 12–14 strokes (by hand or with motorized Dounce homogenizer). Determine the percentage of unbroken cells on a microscope slide, using Hoechst buffer and phase-contrast optics.
9. Continue homogenizing until there are less than 5% unbroken cells.
10. Filter the homogenate through four layers of cheesecloth in a funnel.
11. Add ($2 \times W$) ml of solution B; mix thoroughly by inversion.
12. Pipette into SW28 centrifuge tubes (ultraclear; 30 ml per tube).
13. Drip 5 ml of solution B into the tube to form a cushion at bottom.
14. Balance the tubes carefully.
15. Centrifuge 1.5 hours at 22,000 rpm at 4°C, with the brake on.
16. The nuclei form a white layer at the bottom of the tube. Aspirate and remove all supernatant debris. Use a lint-free paper towel to clean the inside of the tube wall.
17. Resuspend each pellet gently in 200 μl of solution A.
18. Count nuclei (see Section II,B).
19. Adjust the concentration to 160,000–200,000 nuclei/μl.
20. Freeze small (5–20 μl) aliquots in liquid nitrogen; store at $-80°$C.

III. Nuclear Assembly Reactions and Assays

A. Assembly of Nuclei Around Demembranated Sperm Chromatin

Each egg contains enough nuclear components to construct about 4000 embryonic nuclei (Newport and Forbes, 1987). In a fractionated and reconstituted system, however, some components may be damaged or lost. Nuclear growth will cease (or never begin) if any component is limiting, particularly the membranes (Wilson and Newport, 1988). Calculate the approximate number of "egg equivalents" in the reaction and try to include about 800–2000 nuclear substrates (i.e., demembranated sperm) per egg-equivalent volume. The amount of extract obtained from one egg is approximately 0.7 μl. For example, the reaction below contains 10 μl extract or 14 eggs' worth of nuclear components. A typical nuclear assembly reaction contains 10 μl soluble fraction (includes an ATP-regenerating system), 1.0 μl membranes, and 0.7 μl sperm chromatin (at 40,000 sperm/μl; final sperm concentration ~2000 per egg equivalent). Nuclear assembly reactions reconstituted from frozen components typically yield decent nuclear envelopes and decondensed chromatin within 1 hour.

B. Assaying Nuclear Assembly by Fluorescence Microscopy

To visualize DNA, use Hoechst buffer (the microscope must have the appropriate fluorescence filter set). Add a small amount of the sample to Hoechst buffer on a microscope slide and cover gently with coverslip. The total (sample + Hoechst) volume depends on the coverslip size.

To visualize membranes, use phase-contrast microscopy or include the membrane dye 3,3'-dihexyloxacarbocyanine (DHCC) in the Hoechst buffer and view in the fluorescence microscope using a filter set appropriate for fluorescein. DHCC is light sensitive; a working solution should be made fresh each day. (Dilute the stock DHCC solution, 10 mg/ml in DMSO, into approximately 50 volumes of Hoechst buffer, centrifuge 2 minutes in microcentrifuge to pellet insoluble chunks of DHCC, and remove the supernatant as the working solution.)

IV. Nuclear Transport Substrates, Reactions, and Assays

A. Preparation of Transport Substrates

1. Nucleoplasmin Purification

The following protocol is from Dingwall et al. (1982), as modified by Newmeyer et al. (1986a).

1. Collect eggs from at least 4–6 frogs. Dejelly the eggs as described above and rinse several times in MMR.
2. Rinse the eggs several times in S-lysis buffer containing 20 mM potassium HEPES. Resuspend the eggs with an equal volume of lysis buffer. Homogenize the eggs gently. Centrifuge at 10,000 rpm ($\sim$ 10,000 g) for 10 minutes at 4°C in the Sorvall centrifuge, HB-4 rotor (or equivalent). Carefully aspirate off the lipid and collect the crude cytoplasmic supernatant.
3. Centrifuge at 100,000 g or higher for 1.5 hours at 4°C. Collect the cytosol layer from the side of the tube with a syringe.
4. Heat at 80°C for 10 minutes. This precipitates most proteins; nucleoplasmin and a few other proteins are heat stable and remain soluble.
5. Centrifuge at 10,000 rpm ($\sim$ 10,000 g) for 10 minutes at 4°C in Sorvall centrifuge.
6. To the supernatant, add 1.2 volumes of saturated ammonium sulfate (i.e., bring to 55% saturated, or 2.25 M). Let sit 1 hour on ice. Nucleoplasmin remains soluble.

7. Centrifuge at 10,000 rpm for 30 minutes at 4°C in the Sorvall centrifuge. *Be careful*: Ammonium sulfate will corrode aluminum rotors.
8. Dilute the supernatant with 20 mM Tris-HCl (pH 7.6) to 1.5 M in ammonium sulfate and apply it to a 5-ml Phenyl-Sepharose column equilibrated with 1.5 M ammonium sulfate, 20 mM Tris-HCl (pH 7.6).
9. Wash the column with 20 ml of the equilibration buffer.
10. Elute with a 50-ml 1.5–0.0 M ammonium sulfate gradient in 20 mM Tris (pH 7.6).
11. Load an aliquot from every other fraction on sodium dodecyl sulfate (SDS)-7.5% polyacrylamide gels. Nucleoplasmin runs mostly as the pentamer (150K) because of the high salt. There may be more than one band.
12. Pool the fractions containing nucleoplasmin. Nucleoplasmin elutes in a broad peak in the first one-third of the gradient. If you plan to label the nucleoplasmin with tetramethylrhodamine-β-isothiocyanate (TRITC), dialyze against 0.1 M sodium carbonate or sodium borate buffer (pH 9.0), 50 mM NaCl. Otherwise, dialyze against a neutral buffer. Do not remove immediately from the dialysis bags; first concentrate the nucleoplasmin by packing the bags in Sephadex or equivalent, being careful not to allow them to dry completely.
13. The protein concentration must be greater than 1 mg/ml for TRITC labeling. The yield from the eggs of 6 frogs should be roughly 0.5 mg.

2. Rhodamine Labeling of Protein Transport Substrates

1. Dissolve TRITC (isomer R, Sigma) in DMSO to 1 mg/ml.
2. Add the TRITC solution to 100 μl of protein solution in carbonate or borate buffer (pH 9.0–9.5); the final concentration should be 100 μg TRITC per milliliter for purified nucleoplasmin (Dingwall *et al.*, 1982) and 50 μg/ml for human serum albumin (HSA). Incubate for 1 hour at room temperature.
3. Separate protein from free TRITC on a BioGel P6 column (or rough equivalent) equilibrated in 20 mM Tris, 50 mM NaCl.
4. Freeze TRITC–nucleoplasmin in aliquots at -80°C. Use TRITC–HSA for peptide coupling (see below).

3. Conjugation of Fluorescent Proteins with Signal Peptides

The following protocol is adapted from Goldfarb *et al.* (1986) and Newmeyer and Forbes (1988).

1. Dissolve HSA (Calbiochem) at 5 mg/ml in 0.1 M sodium carbonate

buffer (pH 9.5) and label with TRITC as described above. One can label the conjugates with TRITC afterward, but this reduces their activity, since some of the lysine residues in the NLS essential for nuclear targeting activity are modified by the TRITC. Note, however, that TRITC labeling lowers the number of sites in HSA available for peptide coupling.
2. Dissolve m-maleimidobenzoyl-N-hydroxysuccinimde ester (MBS; Pierce Chemical Co., Rockford, IL), at 10 mg/ml in dimethylformamide. Add the MBS solution to the TRITC–HSA, to a final concentration of 2.5 mg MBS per milliliter. Incubate 1 hour at room temperature.
3. Remove unreacted MBS by passage over a BioGel P6 column (or equivalent) equilibrated in 0.1 M sodium phosphate, pH 6.0. Save 10 μl or so of the resulting TRITC–HSA–MBS for estimating coupling efficiency (Step 6).
4. Add the peptide solution (5 mg/ml in 0.1 M sodium phosphate, pH 6.0) dropwise, while mixing, to the TRITC–HSA–MBS. To ensure that the desired coupling ratio is obtained, several coupling reactions are performed for each peptide, using a range between 0.1 and 0.4 mg of peptide per milligram of HSA. Rotate end over end for 1 hour at room temperature.
5. Remove free peptide by chromatography over a second BioGel P6 column equilibrated in 20 mM potassium HEPES, pH 7.5, 50 mM KCl. Store aliquots at $-80°$C.
6. Determine the average number of peptides coupled per molecule of HSA by SDS gel electrophoresis. The average apparent molecular weights of TRITC–HSA–peptide conjugate and TRITC–HSA–MBS are subtracted. The difference, divided by the molecular weight of the peptide, is an estimate of the coupling ratio.

B. Standard Nuclear Import Assay

The following protocol is from Newmeyer *et al.* (1986b).

1. Prepare a nuclear assembly extract, either a fresh crude assembly extract or a reconstituted extract (cytosol + ATP system + membranes).
2. Obtain nuclei (a) by adding isolated nuclei (from rat liver or cultured cells) to the extract or (b) by assembling synthetic nuclei around demembranated sperm, or DNA (in crude extract only).
3. Allow time for nuclei to form or "heal." Healing of isolated nuclei is not always necessary, especially if the nuclei are fresh; in any event, not all the nuclei will be healed. With sperm chromatin or λ DNA, incubate for

30 or 60–90 minutes, respectively, to allow nuclear assembly before adding the transport substrate.
4. Add transport substrate, for example, TRITC–nucleoplasmin or TRITC–HSA–peptide conjugates (see above).
5. At various times remove an aliquot for fluorescence microscopy. Mix 3.5 µl of the assay mixture with 0.5 µl of fixative (37% formaldehyde, 100 µg/ml Hoechst 33258) on a microscope slide, add a coverslip, and examine with the microscope.

C. Identification of Broken and Intact Nuclei

The following methods are after Newmeyer *et al.* (1986b).

1. Method 1

Add a large fluorescent macromolecule lacking a nuclear localization signal, such as phycoerythrin, fluorescein isothiocyanate–immunoglobulin G (FITC–IgG), or FITC–dextran (molecular weight 150,000), to the sample. Use a concentration just sufficient to produce a noticeable background in the fluorescence microscope. Incubate for at least 5 minutes after addition, then examine by fluorescence microscopy. Intact nuclei will exclude these large molecules and appear as dark regions in a bright fluorescence field. Damaged nuclei fluoresce at nearly the same brightness as the surrounding medium.

2. Method 2

Add FITC–concanavalin A (Con A) to a final concentration of 0.1 mg/ml. Most damaged nuclei are stained at the nuclear envelope rim by FITC–Con A. The reason is that Con A binds to proteins with glucose- or mannose-containing carbohydrate side chains. Such glycosylated proteins are abundant in the lumen of the ER and nuclear envelope. In an intact nuclear envelope, Con A has no access to the lumenal space between the inner and outer nuclear membranes. Damage to the nuclear membrane will often allow Con A access to its substrates in the lumen. However, the Con A test is not always reliable, as there are some nuclei that are permeable to 150K dextran but are not stained by Con A.

D. Staining Nuclear Pore Complexes with Fluorescein Isothiocyanate–Wheat Germ Agglutinin

Add FITC–WGA (Sigma; final concentration 0.1 mg/ml) to the sample. Examine by fluorescence microscopy (Finlay *et al.*, 1987).

E. ATP Depletion of Egg and Embryo Extracts (Newmeyer et al., 1986a,b)

1. Enzymatic Hydrolysis of ATP

a. Irreversible. Add apyrase (100 units/ml; Sigma, grade VIII) to the extract; incubate 30 minutes at room temperature. Apyrase treatment is irreversible because the enzyme hydrolyzes ADP as well as ATP, and it can deplete the extract of ATP even in the presence of an ATP-regenerating system.

b. Reversible. Add a mixture of glucose (10 mM) and hexokinase (100 units/ml) to the extract, which must not contain an ATP-regenerating system. Hexokinase catalyzes the reaction:

$$ATP + glucose \rightleftharpoons ADP + glucose\ 6\text{-phosphate}$$

The ADP so produced can be recharged by creatine kinase using phosphocreatine as the phosphate donor. Thus, to restore ATP to extracts depleted by hexokinase/glucose, add phosphocreatine to twice the concentration of glucose, that is, to 20 mM. Under these conditions the rate of ATP synthesis is greater than the rate of ATP hydrolysis, and ATP is restored to the extract.

To monitor ATP depletion and restoration qualitatively, observe the nuclear chromatin. Depletion of ATP causes chromatin condensation within the nuclei, whereas restoration of the ATP level results in decondensation of the chromatin.

2. Removal of ATP by Dialysis

1. Dialyze the egg cytosol fraction to remove small molecules, including ATP.
2. Add washed membranes to reconstitute the nuclear assembly/import system. If crude membranes are used, contaminating mitochondria must be uncoupled by adding 100 μM carbonyl cyanide *m*-chlorophenylhydrazone (CCCP; Sigma).
3. Under these conditions, nuclear transport is strictly dependent on the presence of an ATP-regenerating system. Consider adding submillimolar quantities of GTP to the system to enhance nuclear membrane stability.

F. Video Image Analysis System to Quantitate Nuclear Import by Fluorescence Microscopy

It is important to have a reliable method for quantitation of nuclear import. One way to measure nuclear accumulation of a fluorescent protein is by photographing the fluorescent nuclei, scanning the negatives with a densitometer, and calculating the ratio of absorbances inside and outside each nucleus. This ratio is then multiplied by a factor that accounts for the dilution of the extract by the fixative. Many nuclei from each sample must be averaged to obtain an accurate measurement (Newmeyer et al., 1986b). Although adequate for simple experiments, this method is too time consuming for routine use.

Video image analysis is a superior method for quantitation (Newmeyer and Forbes, 1990). A video camera is mounted on the fluorescence microscope, and its signal is fed to a digitizing board ("frame-grabber") installed in a personal computer. Image analysis software is then used to extract the luminance data. A flexible and relatively inexpensive video system consists of a CCD camera (Dage MTI model CCD-72), an 8-bit frame-grabber (Imaging Technologies Model PC Vision Plus) installed in an IBM AT-compatible computer, and a video image analysis software package (Optimas, produced by Bioscan, Inc., Edmonds, WA). Less expensive software, called JAVA (Jandel Scientific, Corte Madera, CA) can also be used. Video analysis systems can also be used to measure nuclear or cellular areas, diameters, and so on.

1. Calibrate the system.
   a. Set the brightness and contrast on the frame-grabber. With the Dage video camera, one uses the test pattern of vertical bars of different intensities generated by the camera control unit. Use the software to adjust the brightness and contrast (gain and offset) levels on the frame-grabber so that a pure black signal from the video camera corresponds to a value of zero, and pure white corresponds to 255 (the maximum value encodable in eight binary bits).
   b. Set the "black level" control on the video camera so that the background light level from the optics (i.e., that seen with a nonfluorescent sample) corresponds to a numerical value slightly above zero. This background value is subtracted from all measurements.
   c. Set the gain of the video camera such that the brightest objects to be measured do not exceed the range of the camera. If the video camera lacks a gain control, use neutral density filters to reduce the light levels that reach the camera.
2. Make measurements. First, select an appropriate field at random and capture it in the frame-grabber. Circle each nucleus with a mouse, and use the software to calculate the average luminance within the marked region. This value is stored in a disk file. When enough measurements have been taken for a given sample (at least 15–20 nuclei), go on to the next sample. Finally, calculate the mean and standard error for each sample.

References

Aebi, U., Cohn, J., Buhle, L., and Gerace, L. (1986). The nuclear lamina is a meshwork of intermediate-type filaments. *Nature (London)* **323,** 560–564.

Agard, D. A., and Sedat, J. W. (1983). Three-dimensional architecture of a polytene nucleus. *Nature (London)* **302,** 676–681.

Akey, C. W. (1989). Interactions and structure of the nuclear pore complex revealed by cryoelectron microscopy. *J. Cell Biol.* **109,** 955–970.

Akey, C. W., and Goldfarb, D. S. (1989). Protein import through the nuclear pore complex is a multistep process. *J. Cell Biol.* **109,** 971–982.

Berrios, M., and Fisher P. (1986). A myosin heavy chain-like polypeptide is associated with the nuclear envelope in higher eukaryotic cells. *J. Cell Biol.* **103,** 711–724.

Blobel, G., and Potter, V. R. (1966). Nuclei from rat liver: Isolation method that combines purity with high yield. *Science* **154,** 1662–1665.

Burke, B. (1990). On the cell-free association of lamins A and C with metaphase chromosomes. *Exp. Cell Res.* **186,** 169–176.

Chelsky, D., Ralph, R., and Jonak, G. (1989). Sequence requirements for synthetic peptide mediated translocation to the nucleus. *Mol. Cell. Biol.* **9,** 2487–2492.

Dabauvalle, M.-C., Schulz, B., Scheer, U., and Peters, R. (1988a). Inhibition of nuclear accumulation of karyophilic proteins in living cells by microinjection of the lectin wheat germ agglutinin. *Exp. Cell Res.* **174,** 291–296.

Dabauvalle, M.-C., Benavente, R., and Chaly, N. (1988b). Monoclonal antibodies to a M_r 68,000 pore complex glycoprotein interfere with nuclear protein uptake in *Xenopus* oocytes. *Chromosoma* **97,** 193–197.

Dasso, M., and Newport, J. W. (1990). Completion of DNA replication is monitored by a feedback system that controls the initiation of mitosis *in vitro*: Studies in *Xenopus*. *Cell (Cambridge, Mass.)* **61,** 811–823.

Davis, L. I., and Blobel, G. (1986). Identification and characterization of a nuclear pore protein. *Cell (Cambridge, Mass.)* **45,** 699–709.

Davis, L. I., and Fink, G. R. (1990). The NUP1 gene encodes an essential component of the yeast nuclear pore complex. *Cell (Cambridge, Mass.)* **61,** 965–978.

Dingwall, C., Sharnick, S. V., and Laskey, R. A. (1982). A polypeptide domain that specifies migration of nucleoplasmin into the nucleus. *Cell (Cambridge, Mass.)* **30,** 449–458.

Dunphy, W. G., and Newport, J. W. (1988). Mitosis-inducing factors are present in a latent form during interphase in the *Xenopus* embryo. *J. Cell Biol.* **106,** 2047–2056.

Dunphy, W. G., Brizuela, L., Beach, D., and Newport, J. (1988). The *Xenopus* cdc2 protein is a component of MPF, a cytoplasmic regulator of mitosis. *Cell (Cambridge, Mass.)* **54,** 423–431.

Dworetzky, S. I., and Feldherr, C. M. (1988). Translocation of RNA-coated gold particles through the nuclear pores of oocytes. *J. Cell Biol.* **106,** 575–584.

Featherstone, C., Darby, M. K., and Gerace, L. (1988). A monoclonal antibody against the nuclear pore complex inhibits nucleocytoplasmic transport of protein and RNA *in vivo*. *J. Cell Biol.* **107,** 1289–1297.

Feldherr, C. M. (1962). The nuclear annuli as pathways for nucleo-cytoplasmic exchanges. *J. Cell Biol.* **14,** 65–72.

Feldherr, C. M., Kallenbach, E., and Schultz, N. (1984). Movement of a karyophilic protein through the nuclear pores of oocytes. *J. Cell Biol.* **99,** 2216–2222.

Felix, M.-A., Pines, J., Hunt, T., and Karsenti, E. (1989). A post-ribosomal supernatant from activated *Xenopus* eggs that displays post-translationally regulated oscillation of its $cdc2^+$ mitotic kinase activity. *EMBO J.* **8,** 3059–3069.

Finlay, D. R., and Forbes, D. J. (1990). Reconstitution of biochemically altered nuclear pores: Transport can be eliminated and restored. *Cell (Cambridge, Mass.)* **60**, 17–29.

Finlay, D. R., Newmeyer, D. D., Price, T. M., and Forbes, D. J. (1987). Inhibition of *in vitro* nuclear transport by a lectin that binds to nuclear pores. *J. Cell Biol.* **104**, 189–200.

Fisher, D. Z., Chaudhary, N., and Blobel, G. (1986). cDNA sequencing of nuclear lamins A and C reveals primary and secondary structural homology to intermediate filament proteins. *Proc. Natl. Acad. Sci. U.S.A.* **83**, 6450–6454.

Franke, W. W., Scheer, U., Krohne, G., and Jarasch, E.-D. (1981). The nuclear envelope and the architecture of the nuclear periphery. *J. Cell Biol.* **91**, 39s–50s.

Gerace, L., and Blobel, G. (1980). The nuclear envelope lamina is reversibly depolymerized during mitosis. *Cell (Cambrige, Mass.)* **9**, 277–287.

Gerace, L., and Burke, B. (1988). Functional organization of the nuclear envelope. *Ann. Dev. Cell Biol.* **4**, 335–374.

Gerace, L., Ottaviano, Y., and Kondor-Koch, C. (1982). Identification of a major polypeptide of the nuclear pore complex. *J. Cell Biol.* **95**, 826–837.

Glass, J. R., and Gerace, L. (1990). Lamins A and C bind and assemble at the surface of mitotic chromosomes. *J. Cell Biol.* **111**, 1047–1057.

Goldfarb, D. S., Gariepy, J., Schoolnik, G., and Kornberg, R. D. (1986). Synthetic peptides as nuclear localization signals. *Nature (London)* **322**, 641–644.

Greber, U. F., Senior, A., and Gerace, L. (1990). A major glycoprotein of the nuclear pore complex is a membrane-spanning polypeptide with a large lumenal domain and a small cytoplasmic tail. *EMBO J.* **9**, 1495–1502.

Hanover, J. A., Cohen, C. K., Willingham, M. C., and Park, M. K. (1987). O-Linked N-acetylgucosamine is attached to proteins of the nuclear pore: Evidence for cytoplasmic glycosylation. *J. Biol. Chem.* **262**, 9887–9894.

Heald, R., and McKeon, F. (1990). Mutations of phosphorylation sites in lamin A that prevent nuclear lamina disassembly in mitosis. *Cell (Cambridge, Mass.)* **61**, 579–589.

Hochstrasser, M., Mathog, D., Gruenbaum, Y., Saumweber, H., and Sedat, J. W. (1986). Spatial organization of chromosomes in the salivary gland nuclei of *Drosophila melanogaster*. *J. Cell Biol.* **102**, 112–123.

Holt, G. D., and Hart, G. W. (1986). The subcellular distribution of terminal N-acetylglucosamine moieties: Localization of a novel protein-saccharide linkage, O-linked GlcNAc. *J. Biol. Chem.* **261**, 8049–8057.

Holt, G. D., Snow, C. M., Senior, A., Haltiwanger, R. S., Gerace, L., and Hart, G. W. (1987). Nuclear pore complex glycoproteins contain cytoplasmically disposed O-linked N-acetylglucosamine. *J. Cell Biol.* **104**, 1157–1164.

Hutchison, C. J., Cox, R., Drepaul, R. S., Gomperts, M., and Ford, C. C. (1987). Periodic DNA synthesis in cell-free extracts of *Xenopus* eggs. *EMBO J.* **6**, 2003–2010.

Hutchison, N., and Weintraub, H. (1985). Localization of DNase I-sensitive sequences to specific regions of interphase nuclei. *Cell (Cambridge, Mass.)* **43**, 471–482.

Lanford, R. E., Kanda, P., and Kennedy, R. C. (1986). Induction of nuclear transport with a synthetic peptide homologous to the SV40 T antigen nuclear transport signal. *Cell* **46**, 575–582.

Lanford, R. E., Feldherr, C. M., White, R. G., Dunham, R. G., and Kanda, P. (1990). Comparison of diverse transport signals in synthetic peptide-induced nuclear transport. *Exp. Cell Res.* **186**, 32–38.

Laskey, R. A., Gurdon, J. B., and Trendelenburg, M. (1979). Accumulation of materials involved in rapid chromosomal replication in early amphibian development. *Br. Soc. Dev. Biol. Symp.* **4**, 65–80.

Lobl, T. J., Mitchell, M. A., and Maggiora, L. L. (1990). SV40 Large T-antigen nuclear signal analogues: Successful nuclear targeting with bovine serum albumin but not low molecular weight fluorescent conjugates. *Biopolymers* **29,** 197–203.

Lohka, M. J., and Maller, J. L. (1985). Induction of nuclear envelope breakdown, chromosome condensation, and spindle formation in cell-free extracts. *J. Cell Biol.* **101,** 518–523.

Lohka, M. J., and Maller, J. L. (1987). Regulation of nuclear formation and breakdown in cell-free extracts of amphibian eggs. *In* "Molecular Regulation of Nuclear Events in Mitosis and Meiosis" (R. A. Schlegel, M. S. Halleck, and P. N. Rao, eds.), pp. 67–109. Academic Press, New York.

Lohka, M. J., and Masui, Y. (1983). Formation *in vitro* of sperm pronuclei and mitotic chromosomes by amphibian ooplasmic components. *Science* **220,** 719–721.

Lohka, M. J., and Masui, Y. (1984a). Roles of cytosol and cytoplasmic particles in nuclear envelope assembly and sperm pronuclear formation in cell-free preparations from amphibian eggs. *J. Cell Biol.* **98,** 1222–1230.

Lohka, M. J., and Masui, Y. (1984b). Effects of Ca^{2+} ions on the formation of metaphase chromosomes and sperm pronuclei in cell-free preparations from unactivated *Rana pipiens* eggs. *Dev. Biol.* **103,** 434–442.

McKeon, F. D., Kirschner, M. W., and Caput, D. (1986). Homologies in both primary and secondary structure between nuclear envelope and intermediate filament proteins. *Nature (London)* **319,** 463–468.

Meyerhof, P. G., and Masui, Y. (1979). Properties of a cytostatic factor from *Xenopus laevis* eggs. *Dev. Biol.* **72,** 182–187.

Miake-Lye, R., and Kirschner, M. W. (1985). Induction of early mitotic events in a cell-free system. *Cell (Cambridge, Mass.)* **41,** 165–175.

Murray, A. W., and Kirschner, M. W. (1989). Dominoes and clocks: The union of two views of the cell cycles. *Science* **246,** 614–621.

Nehrbass, U., Kern, H., Mutvel, A., Horstmann, H., Marshallsay, B., and Hurt, E. C. (1990). NSP1: A yeast nuclear envelope protein localized at the nuclear pores exerts its essential function by its carboxy-terminal domain. *Cell (Cambridge, Mass.)* **61,** 979–989.

Newmeyer, D. D. (1990). Nuclear protein import *in vitro*. *Prog. Mol. Subcell. Biol.* **11,** 12–50.

Newmeyer, D. D., and Forbes, D. J. (1988). Nuclear import can be separated into distinct steps *in vitro*: Nuclear pore binding and translocation. *Cell (Cambridge, Mass.)* **52,** 641–653.

Newmeyer, D. D., and Forbes, D. J. (1990). An *N*-ethylmaleimide-sensitive cytosolic factor necessary for nuclear protein import: Requirement in signal-mediated binding to the nuclear pore. *J. Cell Biol.* **110,** 547–557.

Newmeyer, D. D., Lucocq, J. M., Buerglin, T. R., and De Robertis, E. M. (1986a). Assembly *in vitro* of nuclei active in nuclear protein transport: ATP is required for nucleoplasmin accumulation. *EMBO J.* **5,** 501–510.

Newmeyer, D. D., Finlay, D. R., and Forbes, D. J. (1986b). *In Vitro* transport of a fluorescent nuclear protein and exclusion of non-nuclear proteins. *J. Cell Biol.* **103,** 2091–2102.

Newport, J. (1987). Nuclear reconstitution *in vitro*: Stages of assembly around protein-free DNA. *Cell (Cambridge, Mass.)* **48,** 205–217.

Newport, J., and Dasso, M. (1989). On the coupling between DNA replication and mitosis. *J. Cell Sci.* **12,** (Suppl.) 149–160.

Newport, J., and Forbes, D. J. (1985). Fate of DNA injected into *Xenopus* eggs and in egg extracts: Assembly into nuclei. *Banbury Rep.* **20,** 243–250.

Newport, J., and Forbes, D. J. (1987). The nucleus: Structure, function and dynamics. *Annu. Rev. Biochem.* **56,** 535–565.

Newport, J., and Spann, T. (1987). Disassembly of the nucleus in mitotic extracts: Membrane vesicularization, lamin disassembly, and chromosome condensation are independent processes. *Cell (Cambridge, Mass.)* **48,** 219–230.

Newport, J. W., Wilson, K. L., and Dunphy, W. G. (1990). A lamin-independent pathway for nuclear envelope assembly. *J. Cell Biol.* **111,** 2247–2259.

Ottaviano, Y., and Gerace, L. (1985). Phosphorylation of the nuclear lamins during interphase and mitosis. *J. Biol. Chem.* **260,** 624–632.

Paddy, M. R., Belmont, A. S., Saumweber, H., Agard, D. A., and Sedat, J. W. (1990). Interphase nuclear envelope lamins form a discontinuous network that interacts with only a fraction of the chromatin in the nuclear periphery. *Cell (Cambridge, Mass.)* **62,** 89–106.

Park, M. K., D'Onofrio, M., Willingham, M. C., and Hanover, J. A. (1987). A monoclonal antibody against a family of nuclear pore proteins (nucleoporins) recognizes a shared determinant: O-linked *N*-acetylglucosamine. *Proc. Natl. Acad. Sci. U.S.A.* **84,** 6462–6466.

Peter. M., Nakagawa, J., Dorée, M., Labbé, J. C., and Nigg, E. A. (1990). *In vitro* disassembly of the nuclear lamina and M phase-specific phosphorylation of lamins by cdc2 kinase. *Cell (Cambridge, Mass.)* **61,** 591–602.

Richardson, W. D., Mills, A. D., Dilworth, S. M., Laskey, R. A., and Dingwall, C. (1988). Nuclear protein migration involves two steps: Rapid binding at the nuclear envelope followed by slower translocation through nuclear pores. *Cell (Cambridge, Mass.)* **52,** 665–664.

Roberts, B. L., Richardson, W. D., and Smith, A. E. (1987). The effect of protein context on nuclear location signal function. *Cell (Cambridge, Mass.)* **50,** 465–475.

Sagata, N., Watanabe, N., Vande Woude, G. F., and Ikawa, Y. (1989). The c-*mos* proto-oncogene product is a cytostatic factor responsible for meiotic arrest in vertebrate eggs. *Nature (London)* **342,** 512–518.

Scheer, U., Dabauvalle, M.-C., Merkert, H., and Benavente, R. (1988). The nuclear envelope and the organization of the pore complexes. *Cell Biol. Int. Rep.* **12,** 669–689.

Schindler, M., Hogan, M., Miller, R., and Gaetano, D. (1987). A nuclear specific glycoprotein representative of a unique pattern of glycosylation. *J. Biol. Chem.* **262,** 1254–1260.

Snow, C. M., Senior, A., and Gerace, L. (1987). Monoclonal antibodies identify a group of nuclear pore complex glycoproteins. *J. Cell Biol.* **104,** 1143–1156.

Starr, C. M., D'Onofrio, M., Park, M. K., and Hanover, J. A. (1990). Primary sequence and heterologous expression of nuclear pore glycoprotein p62. *J. Cell Biol.* **110,** 1861–1871.

Stevens, B. J., and Swift, H. (1966). RNA transport from nucleus to cytoplasm in *Chironomus* salivary glands. *J. Cell Biol.* **31,** 55–77.

Stewart, M., and Whytock, S. (1988). The structure and interactions of components of nuclear envelopes from *Xenopus* oocyte germinal vesicles observed by heavy metal shadowing. *J. Cell Sci.* **80,** 409–423.

Unwin, N. T., and Milligan, R. A. (1982). A large particle associated with the perimeter of the nuclear pore complex. *J. Cell Biol.* **93,** 63–75.

Walter, P., Gilmore, R., and Blobel, G. (1984). Protein translocation across the endoplasmic reticulum. *Cell (Cambridge, Mass.)* **38,** 5–8.

Ward, G. E., and Kirschner, M. W. (1990). Identification of cell cycle-regulated phosphorylation sites on nuclear lamin C. *Cell (Cambridge, Mass.)* **61,** 561–577.

Watanabe, N., Vande Woude, G. F., Ikawa, Y., and Sagata, N. (1989). Specific proteolysis of the c-*mos* proto-oncogene product by calpain on fertilization of *Xenopus* eggs. *Nature (London)* **342,** 505–511.

Wilson, K. L., and Newport, J. (1988). A trypsin-sensitive receptor on membrane vesicles is required for nuclear envelope formation *in vitro*. *J. Cell Biol.* **107,** 57–68.

Wozniak, R. W., Bartnik, E., and Blobel, G. (1989). Primary structure analysis of an integral membrane glycoprotein of the nuclear pore. *J. Cell Biol.* **108,** 2083–2092.

Yoneda, Y., Imamoto-Sonobe, N., Yamaizumi, M., and Uchida, T. (1987). Reversible inhibition of protein import into the nucleus by wheat germ agglutinin injected into cultured cells. *Exp. Cell Res.* **173,** 586–595.

Chapter 32

Xenopus Cell Lines

J. C. SMITH AND J. R. TATA

National Institute for Medical Research
London NW7 1AA, England

I. Introduction
II. Uses of *Xenopus* Cell Lines
III. Solutions
IV. Preparation of *Xenopus* Cell Cultures
    A. Blastula and Gastrula Stages
    B. Neurula Stage and Tadpoles
    C. Adult Tissues
V. Culture Methods
    A. Freezing *Xenopus* Cell Lines
    B. Shipping *Xenopus* Cell Lines
    C. Cloning *Xenopus* Cell Lines
VI. Transfection of *Xenopus* Cell Lines
    A. Methods for Transient Gene Expression
    B. Results from Transfection of *Xenopus* Cell Lines
    C. Future Prospects of Transfection
VII. XTC Mesoderm-Inducing Factor
    A. Preparation of XTC-Conditioned Medium
    B. Purification of XTC Mesoderm-Inducing Factor
    C. Other Factors from XTC-Conditioned Medium
VIII. Conclusions
    References

I. Introduction

Xenopus laevis is a popular experimental animal for many reasons. One is that *Xenopus* is easy and relatively inexpensive to keep. In addition, *Xenopus* has the advantages that the embryos are easy to obtain, they are plentiful, they develop quickly, they are accessible to manipulation at all developmental stages, and they are large. These advantages also apply to the other uses to which *Xenopus* embryos are put: studies of egg maturation and of metamorphosis. Finally, for the cell biologist, the *Xenopus* egg and oocyte provide useful "living test-tubes" into which one can microinject mRNAs, antisense oligonucleotides, or antibodies in order to study the function of the molecules encoded by, or recognized by, these reagents. What, then, is the use of *Xenopus* cell lines when it would appear that the organism itself has most of the advantages that cell lines are traditionally supposed to provide? In this chapter we suggest that *Xenopus* cell lines have many uses to the developmental biologist and the endocrinologist and that every laboratory studying *Xenopus* should maintain one or two of the several cell lines that are available. We go on to describe how *Xenopus* cell lines can be prepared and how they can transfected, and then we describe the purification of a mesoderm-inducing factor (MIF) that is secreted by the XTC-2 cell line (Pudney, *et al.*, 1973; see Smith, 1987; see also Chapter 17, this volume).

II. Uses of *Xenopus* Cell Lines

Cells from amphibian species have been grown *in vitro* from the earliest days of cell culture (see Freed and Mezger-Freed, 1970). Despite this long history, amphibian cultures have not been as widely used as those from mammalian species. The main reason for this, of course, is that most cell biologists, being mammals, study mammalian cells. Nevertheless, as we shall describe, amphibian cell cultures have not been without their uses.

Over the last 20 years *Xenopus* cultures have offered the opportunity to study ribosomal RNA synthesis, using the nucleolar mutants of *Xenopus* (Miller and Daniel, 1977), and they have been employed to analyze the nutritional requirements of amphibian cells together with the isolation of drug-resistant variants (Chinchar, and Sinclair, 1978a,b). In addition, amphibian cell lines (especially XTC-2) have been used to study the growth of arboviruses such as bunyavirus (Leake *et al.*, 1977; Watret *et al.*, 1985, and references therein), and cell lines from haploid embryos allow one to perform somatic cell genetics with relative ease, because mutant phenotypes are not obscured by normal alleles (Freed and Mezger-Freed, 1970). It is also convenient to use *Xenopus* cell lines to discover whether antibodies or other reagents will recognize *Xenopus* proteins (see, e.g., Evans *et al.*, 1990). Finally, since *Xenopus* cell lines grow at room temperature, usually without the need for a 5% CO_2 atmosphere, they are a valuable teaching aid. More recently, however, amphibian cultures have been used for two main purposes: for transfection studies and as a source of mesoderm-inducing factors. These uses are described below, following a general discussion of how to establish and maintain *Xenopus* cell lines.

III. Solutions

All culture media should be prepared using sterile components. If in doubt solutions can be sterilized by passage through a 0.2-μm filter such as those made by Nalgene, Rochester, NY (Cat. No. 120-0020).

Modified L-15 Medium Containing 10% Fetal Calf Serum (for 1 liter)
  Leibovitz L-15 medium (with L-glutamine), 610 ml
  Sterile distilled water, 280 ml
  Fetal calf serum, 100 ml
  Penicillin–streptomycin solution, 10 ml (e.g., Gibco/BRL, Grand Island, NY, Cat. No. 043-05070D; other antibiotics may be used if desired)

Calcium- and Magnesium-Free Medium (CMFM) (Sargent et al., 1986)
  NaCl, 88 mM
  KCl, 1 mM
  NaHCO$_3$, 2.4 mM
  Tris, pH 7.6, 7.5 mM
  This solution should be prepared with water that has been deionized with, for example, a MilliQ system (Millipore, Bedford, MA).

Trypsin/EDTA
  Add to 1 liter of CMFM:
  0.5 g trypsin
  0.2 g EDTA
  1.0 g glucose

Serum-Free Modified L-15 Medium (for 1 liter)
  Leibovitz L-15 medium (with L-glutamine), 650 ml
  Sterile distilled water, 325 ml
  Penicillin–streptomycin solution, 10 ml
  1 M HEPES (H 7.0), 10 ml

Wolf and Quimby Medium (for 1 liter)
  Eagle's minimum essential medium, 750 ml
  Earle's balanced salt solution, without NaCl, 250 ml
  Adjust pH 7.6–7.8

Normal Amphibian Medium (Slack, 1984)
  NaCl, 110 mM
  KCl, 2 mM
  Ca(NO$_3$)$_2$·4H$_2$O, 1 mM

MgSO$_4$·7H$_2$O, 1 mM
Na$_2$EDTA, 0.1 mM
Sodium phosphate, pH 7.5, 2 mM
NaHCO$_3$, 1 mM
Gentamycin, 25 μg/ml

IV. Preparation of *Xenopus* Cell Cultures

Preparation of cell lines from anuran amphibian species has been covered in some detail by Freed and Mezger-Freed (1970), and this remains the most detailed reference. The following methods are based on their paper, with additional comments, where appropriate, from us. Primary cultures can be prepared from any stage of embryo or tadpole and from many adult tissues. Cultures from blastula and gastrula stage embryos do not give rise to permanent lines, but they are of great interest in elucidating the mechanisms by which blastomeres become determined to form particular differentiated cell types (see Jones and Elsdale, 1963; Godsave and Slack, 1989). For the later stages of development, it is advantageous to prepare cultures from tadpoles or embryos before feeding commences to minimize the problem of contamination of cultures from intestinal fauna.

A. Blastula and Gastrula Stages

The method for preparing cultures from blastula and gastrula stage embryos depends on the use to which they are to be put.

1. General Observation

For general observation, Freed and Mezger-Freed (1970) recommend that dejellied embryos be treated for 15 minutes with 100 mg/ml Merthiolate and then washed several times in an amphibian balanced salt solution such as normal amphibian medium (NAM, Slack, 1984). The vitelline membranes of about 10 embryos are removed with a pair of sterile forceps using a dissecting microscope in a tissue-culture hood, and the yolky vegetal cells of the embryo are then removed and discarded; these do not survive well in culture. The animal halves are then transferred to a 3.5-cm petri dish containing 3 ml of modified L15 medium containing 10% fetal calf serum (see above) and cultured in a humidified environment at 25°C. Preparations of this sort can be subcultured (see below), and the cells will survive for about 2 months until cell division slows and the cultures apparently become senescent.

2. CELL DETERMINATION

For studies which aim to discover and study the factors involved in cell determination, it is important that the cultures do not contain serum and that the medium is as simple as possible. Godsave and Slack (1989) have dissected animal pole regions from early blastulas and dissociated the cells in "PhoNak" (50 mM sodium phosphate, pH 7.0, 35 mM NaCl, 1 mM KCl). Individual cells are then cultured in 15 μl of 50% NAM in humidified Terasaki plates which have been pretreated with laminin and fibronectin. The medium also contains 2 mg/ml γ-globulins and, where appropriate, the mesoderm-inducing factor acidic fibroblast growth factor (aFGF; see Slack et al., 1987). The cells are cultured for 68 hours at 25°C, and the differentiation of the cells is monitored. It is not necessary to include nutrients in the medium for this short culture period; the cells survive on their yolk reserves. Identification of cell types in this sort of experiment is best done with specific antibodies, if they are available (see Godsave and Slack, 1989). In the absence of such antibodies it is possible to use morphological criteria, which is satisfactory for melanocytes, neural cells with long processes, and muscle but not for cells with less dramatic phenotypes, such as epidermal cells or fibroblasts.

B. Neurula Stage and Tadpoles

Cells obtained from later-staged embryos and tadpoles can be used to establish permanent cell lines. These cultures are set up in a similar manner as described in Section IV,A,1 except that surface sterilization of the embryos is carried out after removal of the vitelline membrane. Furthermore, it is important then to remove the epidermis from the embryos by incubating them for 30 minutes in a dissociating medium such as the calcium- and magnesium-free medium (CMFM) of Sargent et al. (1986) containing 1 mM EDTA (see above). Once the epidermis is removed it is possible to dissect out the tissues of interest, if desired, and then to mince the tissue with a pair of tungsten or glass needles. The tissue is then transferred to a 3.5-cm petri dish containing modified L15 medium and 10% fetal calf serum as above.

C. Adult Tissues

Cultures can also be prepared from adult tissue, the most popular of which is the kidney. A detailed protocol is provided by Freed and Mezger-Freed (1970).

V. Culture Methods

Cell lines from *Xenopus* and other amphibian species can be established by subculture of primary cultures obtained as described above. The simplest method of subculture, which does not require centrifugation of cells, is to rinse the culture briefly with sterile CMFM lacking EDTA. This removes serum proteins which inhibit the action of trypsin. The cultures are incubated for 1 minute in a trypsin/EDTA solution (see below), which is then removed and allowed to continue its action as a thin film covering the cells. Within a few minutes the cells become rounded and can be seen to move around the dish. For some lines, however, such as XL2 (Anizet *et al.*, 1981), it is necessary to wait longer for the cells to detach. It is then possible to suspend the cells in modified L15 medium containing 10% fetal calf serum and to replate the cells at the desired density in another flask(s). The density of replating depends on the cells being studied. For primary cultures Freed and Mezger-Freed (1970) recommend pooling three small petri dish cultures into one 25-cm^2 flask; for vigorous and hardy cell lines like XTC-2, a replating at one-twentieth of the confluent density is quite reasonable.

Cultures should be inspected every day, and the cells should be subcultured when they reach confluence. For established lines, with known growth rates, it is convenient to establish a routine in which cells are fed once a week (say, Wednesday) and subcultured once (say, Friday). This frequency of cell maintenance might seem rather leisurely to a mammalian cell culturist, but the slower growth rate (see Fig. 1), and therefore metabolic rate, of amphibian cells makes this quite acceptable. For newly established cultures, where growth rates might vary, more vigilance is required.

Table I lists some of the *Xenopus* cell lines that have been prepared, and Fig. 2 shows the morphologies of two of them: XTC-2 (Pudney *et al.*, 1973) and XL2 (Anizet *et al.*, 1981).

A. Freezing *Xenopus* Cell Lines

It is possible to freeze *Xenopus* cell lines using techniques similar to those used for mammalian cells. The technique used in our laboratories is as follows. A confluent 75-cm^2 flask is trypsinized as described above, and the cells are suspended in 10 ml of modified L15 medium containing 10% fetal calf serum. The cells are then pelleted using a bench-top centrifuge for about 5 minutes at 1000 g, and all the supernatent is removed. The cells are suspended in 1 ml of 10% dimethyl sulfoxide (DMSO) in modified L15 medium containing 20% fetal calf serum, and the suspension is then transferred to a freezing vial. Vials are placed in a sandwich box surrounded by cotton-wool which is placed in

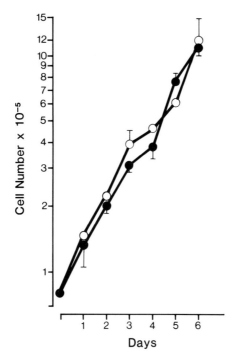

Fig. 1. Growth rates of (●) XTC-2 (Pudney et al., 1973) and (○) XL2 cells (Anizet et al., 1981). Each cell line shows a doubling time of approximately 40 hours.

TABLE I

Xenopus CELL LINES

Cell line	Derivation	Comments	Ref.
A6	Kidney	Used in transfection studies	Rafferty (1969)
B3.2	Kidney	Non-estrogen-responsive line used in transfection studies	Seiler-Tuyns et al. (1988)
KR	Clone of A6	Used in transfection studies	Rafferty (1969); Ellison et al. (1985)
XF	Fibroblast	Used in transfection studies	Chang and Shapiro (1990)
XL2	Stage 35 tadpoles	Used in transfection studies[a]	Anizet et al. (1981)
XL110	Hepatocytes	Used in transfection studies	Chang and Shapiro (1990)
XL-177	Tadpoles	—[b]	Miller and Daniel (1977)
XTC-2	Tadpole/toad carcass	Source of XTC-MIF; used for growth of arboviruses, transfection[b]	Pudney et al. (1973)

[a] estrogen responsive
[b] T_3 responsive

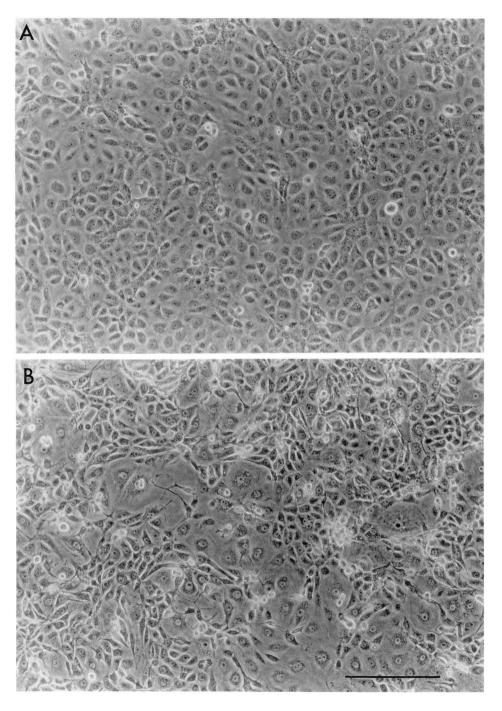

FIG. 2. Morphologies of (A) XTC-2 (Pudney *et al.*, 1973) and (B) XL2 (Anizet *et al.*, 1981) cells at confluence. Bar equals 100 μm.

a $-80°C$ freezer for 24 hours. They are then transferred to liquid nitrogen. Alternatively, the initial freezing can be done in the head of a liquid nitrogen canister.

It is wise to freeze several vials at once and to test one of them for viability before transferring all stocks to liquid nitrogen. Cells can be recovered by holding a frozen vial in water at about 30°C until the solution thaws and then immediately transferring the contents of the vials to a centrifuge tube containing 9 ml of modified L15 medium with 10% fetal calf serum. The cells are pelleted as described above, resuspended in growth medium, and plated in a 75-cm² flask. The medium should be changed the following day to remove "floaters."

B. Shipping *Xenopus* Cell Lines

Although it is possible to ship *Xenopus* cell lines to other laboratories in a frozen state, it is more convenient, and less expensive, to send growing cells in a flask filled to the neck with medium and carefully sealed with Parafilm. Naturally, Post Office and Customs regulations should be adhered to. As long as they are not subjected to extremes of temperature (above 27°C or below 4°C) the cells should survive perfectly well. Indeed, cells that have been lost in the recesses of the postal system for some months have been known to be perfectly healthy on arrival in the recipient laboratory.

C. Cloning *Xenopus* Cell Lines

Cell lines derived from whole embryos or from dissociated tissues are likely to be very heterogeneous. For many purposes, this may not be a problem, but for others it may be necessary to derive clonal derivatives of the original line. One good example of this is in the study of the mesoderm-inducing factor produced by the XTC-2 cell line (Smith, 1987; see chapter 17, this volume). Snoek *et al.* (1990) have shown that a clonal derivative of the XTC-2 cell line called XTC-GTX-11 produces very high levels of a mesoderm-inducing factor and, interestingly, that the specific activity of this factor is not increased by heat treatment. In this, the GTX-11 inducing activity differs from that derived from the parent cell line, and it may be that the parent line contains cells which secrete inhibitors of induction.

Xenopus cell lines have been cloned by at least two techniques: by limiting dilution and, more recently, by growth in soft agar (Snoek *et al.*, 1990). In our laboratory we have had success with the former technique (J. B. A. Green and J. C. Smith, unpublished work, 1989). XTC-2 cells are trypsinized, counted using a hemocytometer, and diluted to a density of approximately 3 cells/ml. The dilution medium consists of modified L15 medium containing 10% fetal calf serum mixed with an equal volume of similar medium which had been "conditioned" for 24 hours by XTC-2 cells. Aliquots of 0.15 ml of this suspension are placed in the wells of microtiter plates (culture area 0.28 cm^2), and the plates are placed in an incubator at 25°C for 3 weeks. In our experiments, 7 of 96 wells contained visible colonies, each assumed to have derived from a single cell. The colonies are trypsinized by the technique described above and replated in the wells of multiwell plates (culture area of each well 2 cm^2). When these cultures reach confluence the cells are again trypsinized and then transferred to 35-mm petri dishes. In this way the sizes of the clones are gradually increased until there are sufficient cells to freeze and, if desired, to repeat the cloning procedure to ensure that the colonies are true clones.

If it proves difficult to derive clones by this procedure, it is possible to use irradiated or mitomycin C-treated cells to provide a feeder layer for low density cells. To prepare irradiated feeder layers, confluent nonclonal cultures in 6-cm petri dishes can be treated with 50 Gy X-irradiation, or cells in suspension can be treated with a similar does of γ-radiation in a ^{60}Co unit before being plated out at confluent density. Alternatively, if irradiation facilities are not available, confluent cells can be treated with 4 μg/ml mitomycin C for 2 hours before being washed with mitomycin-free medium. After treatment of cells, they should be left for 24–48 hours to condition the medium, and then the parent cell line can be seeded onto the plates at a density of about 100–200 cells per culture. Some plates should be left without added cells to confirm that irradiation or mitomycin C treatment was adequate.

After 2 to 3 weeks, colonies of healthy dividing cells should be visible on the plates while the irradiated cells appear abnormally large and unhealthy. The positions of colonies are marked on the dish, and all the medium is removed. Sterile stainless steel cylinders, each about 6 mm in diameter and 12 mm high (a "cloning ring"), are then dipped in sterile vacuum grease and placed over the colonies of interest. The cells thus isolated are trypsinized and transferred to larger vessels as described above.

These techniques should be adequate to clone cells from most *Xenopus* cell lines. However, if a particular line proves refractory it is worth trying the alternative of cloning in soft agar as described by Snoek *et al.* (1990).

VI. Transfection of *Xenopus* Cell Lines

The availability of techniques for introducing cloned, precisely defined genetic material into animal cells in culture has led to an extraordinary advancement in our understanding of the regulation of gene expression.

Methods for directly injecting DNA (or RNA) into somatic or germ line cell nuclei and cytoplasm are described elsewhere in this volume (see Appendix B). This section therefore deals only with indirect methods of somatic cell transfection. Although these have been extensively employed for analyzing the structure–function relationships of homologous and heterologous genes in mammalian cells, there is relatively little published information yet on the transfection of amphibian cell lines.

Transfection of cells with DNA is carried out for two major purposes: to study transient gene expression or promoter activity and to permanently transform cell lines by stable integration of the DNA introduction. No successful stable transfection studies have so far been reported for *Xenopus* cell lines.

A. Methods for Transient Gene Expression

Methods most commonly used for transient transfection are based on (a) formation of precipitates or aggregates of DNA with polymeric or insoluble complexes followed by active uptake by the cells, (b) electroporation or electric field-mediated DNA transfer, and (c) the use of RNA viral vectors to introduce the DNA. Several excellent laboratory manuals describing the principles and different methods of transfection of DNA into mammalian cells are now available. The reader's attention is particularly drawn to two of them (Ausubel *et al.*, 1987; Keown *et al.*, 1990). Of the above methods, only the calcium phosphate precipitation technique has been successfully used for *Xenopus* cells.

1. CALCIUM PHOSPHATE METHOD

The calcium phosphate method is the most commonly used approach for introducing DNA into animal cells for both transient and stable transfection. It is based on exposing the cells in culture to a fine precipitate of DNA and calcium phosphate and facilitating the uptake of the complex by osmotic shock. The following is a brief description of the coprecipitation method used in one of our laboratories for *Xenopus* cell lines (Q.-L. Xu and J. R. Tata, unpublished).

Xenopus cells are plated to a density of approximately 10^4 cells/cm^2 in serum-free Wolf and Quimby medium supplemented with insulin to 5 µg/ml (0.12 U/ml). Three hours before the DNA is to be introduced the culture medium is replaced with one containing 10% fetal calf serum. Then 5–50 µg of sterile DNA in 0.5 ml of 0.25 M CaCl$_2$ is mixed dropwise into an equal volume of 2 × HEPES-buffered phosphate–saline. The precipitate is usually allowed to sit for about 30 minutes. The cells are covered evenly with the DNA–calcium phosphate precipitate, and the dish is gently tilted and shaken. The cells are incubated for about 4 hours at 25°C, after which they are washed 2 times with serum-free medium. (It is useful to monitor the fineness of the precipitate with a microscope during the incubation.) For some cells, a brief osmotic shock will improve transfection efficiency.

Numerous variables determine the reproducibility and efficiency of transfection. These include the type of cells to be transfected, the amount, nature, and purity of DNA, pH, and the degree of fineness of the DNA–calcium phosphate precipitate. For each type of cell and DNA, it is important to work out the optimum conditions. Osmotic shock, usually produced by exposing the cells for 0.5–3.0 minutes to 10% glycerol or 20% dimethyl sulfoxide (DMSO), which has often been found to be beneficial for transfection of mammalian cells, has not turned out to be consistently advantageous in our hands for *Xenopus* cells. Replacing HEPES buffer with BBS [N,N-bis(2-hydroxyethyl)-2-aminoethanesulfonic acid], which results in the slow formation of a finer DNA–calcium phosphate precipitate than with HEPES, and lowering the pH to 6.95 have been found to improve stable transfection of a variety of mammalian cell lines (Chen and Okayama, 1988). In our hands, these changes do not improve transient efficiency of *Xenopus* cell lines substantially.

2. DEAE-Dextran-Mediated DNA Transfection

Many variants of the simple and efficient DEAE-dextran method of transient DNA transfection have been described for mammalian cells (Keown *et al.*, 1990). The method can be used with cultured cells in monolayer or in a large-scale batch procedure, in both cases the cells being osmotically shocked when exposed to DNA–DEAE-dextran complex. No work has been reported with *Xenopus* cell lines using this procedure.

3. Lipid-Mediated Transfection (Lipofection)

A number of techniques of transfection which exploit the fusion of DNA-containing liposomes with the plasma membrane to introduce DNA into recipient cells have been described. In general these methods are more cumbersome and difficult to work with than the calcium phosphate or DEAE-dextran procedures. A variant of this technique, described as lipofection, in which DNA is trapped in unilamellar vesicles with a synthetic cationic lipid, commercially available as DOTMA (BRL, Bethesda, MD) and TRANSFECTAN (Promega, Madison, WI), has recently been introduced. The first reports claimed higher efficiencies of transfection with DOTMA, particularly for stable transformations, than with the calcium phosphate coprecipitation procedures (Felgner *et al.*, 1987). In our laboratory this procedure has not proved to be more effective than the calcium phosphate method for *Xenopus* XTC-2 and XL2 cells. However, two other laboratories (D. J. Shapiro and W. Wahli, personal communication) have found DOTMA to give high, reproducible efficiencies of transfection of *Xenopus* fibroblasts, providing that Tris and EDTA are left out of the culture medium.

4. Electroporation

Electric field-mediated DNA transfection was initially used for cells that could not be readily transfected by the calcium phosphate or DEAE-dextran procedures. Since the mid-1980s it has gained increasing acceptance for other cell types as well, owing to several advantages, such as ease of operation, reproducibility, transfection of cells grown in suspension, and better control of DNA copy number introduced. Usually $1-20 \times 10^6$ cells are subjected to an electric field in a cuvette, the strength, pulse height, and duration of the electric shock varying according to the cells used. The optimum conditions for mammalian cells are such that about 50% of cells are killed by this procedure. In our hands, it is difficult to limit cell loss to this figure when using *Xenopus* cells, thus resulting in poor reproducibility.

B. Results from Transfection of *Xenopus* Cell Lines

Successful transfection of wild-type and mutant gene constructs into *Xenopus* cell lines have been reported from the laboratories of Wahli, Shapiro, and our own. These have included *Xenopus* kidney cell lines (B3.2 and A6), fibroblasts (XF), a liver-derived cell line (XL110), and two embryo-derived lines (XTC-2 and XL2). All the published results are based on the calcium phosphate coprecipitation method, although lipofection has also been found to be successful (D. J. Shapiro, personal communication).

Particularly worth noting are the studies carried out by Wahli's group in which they compared in parallel transfection of human tumor-derived cell lines, namely, HeLa cells and the breast cancer cell line MCF-7, with the *Xenopus* kidney cell line B3.2 (Wahli *et al.*, 1989; Martinez and Wahli, 1989). They thus analyzed the expression of wild-type and mutant promoter constructs of the *Xenopus* vitellogenin gene linked to the CAT (chloramphenicol acetyltransferase) reporter gene, cotransfected or not with the human estrogen receptor (ER) in an expression vector. These cotransfection studies with *Xenopus* cells were particularly useful in determining the important role played by sequences neighboring the estrogen response element (ERE) in activation of the dormant vitellogenin genes by estrogen. Their studies also demonstrated that *Xenopus* cell cultures can be transfected with high efficiencies, and emphasized the importance of working with homologous host cells for understanding the regulation of *Xenopus* gene expression.

Mohun et al. (1987, 1989) have used the calcium phosphate technique to transfect *Xenopus* A6 kidney cells with CAT constructs of *Xenopus* cytoskeletal actin and human and *Xenopus* c-*fos* promoters. By demonstrating that the *Xenopus* c-*fos* gene is serum inducible when introduced in a *Xenopus* cell line they were able to reinforce their conclusions of developmental and tissue-specific expression of the protooncogene in *Xenopus*. Shapiro's laboratory have successfully exploited the *Xenopus* cell transfection approach to investigate the as yet poorly understood problem of how the stability of a given induced mRNA is regulated by the inducer itself in animal cells. These workers cotransfected XF and XL110 cells with the *Xenopus* ER gene in an expression vector and various *Xenopus* vitellogenin promoter constructs, on the one hand (Chang and Shapiro, 1990), and a "mini-vitellogenin" gene containing essentially the 5' and 3' untranslated regions of the mRNA (Nielsen and Shapiro, 1990), on the other. The results obtained allowed them to confirm that not only does the receptor–hormone complex regulate transcription of the gene, but it also controls the stability of the mRNA that is transcribed. They have clearly identified sequences near the ERE in the promoter and untranslated 3' region of the mRNA that are directly or indirectly the targets of ER in the nucleus and cytoplasm. More recently, the same group has shown by comparison of XF and XL110 cell lines, transfected with the *Xenopus* ER gene, that the endogenous vitellogenin gene of the liver-derived cell line was several times more readily activated than those in fibroblasts (M. C. Barton, A. M. Nardulli, T.-C. Chang and D. J. Shapiro, unpublished). Thus, transfection of *Xenopus* cell lines has yielded information not only about trans-acting factors, but also about tissue-specific acquisition of differential chromatin conformation of genes.

In one of our laboratories, the transfection of *Xenopus* and human gene constructs was compared in three *Xenopus* cell lines (Q.-L. Xu and J. R. Tata, unpublished). These were the embryo- or tadpole-derived XTC-2 and XL2 cell lines and an adult kidney-derived KR cell line (see above), all of which are unresponsive to estrogen by virtue of the absence of a functional ER. Transfection efficiencies were measured with both a viral–reporter construct [cytomegalovirus (CMV)–CAT] as well as a functional assay which involved the acquisition of responsiveness to estrogen by these cells following cotransfection with human or *Xenopus* ER genes and various ER target promoter–CAT constructs. XL2 cells were found to be more efficiently transfected than the other two cell lines, as determined by this test. More recent investigations have revealed the reason for this higher responsiveness: it is due to the selective expression at high levels in XL2 cells of a basal transcription factor which may function in a cooperative fashion with the ERE. The transient transfection approach in *Xenopus* cell lines, as shown earlier in mammalian cell lines, can thus also provide valuable information on interaction between trans-activating factors.

C. Future Prospects of Transfection

Xenopus genes have been extensively studied by developmental biologists and, more recently, by cell and molecular biologists. Compared with mammalian cell lines, the use of *Xenopus* cell lines for DNA transfection is still in its infancy. Nevertheless, the few examples given above quite clearly indicate the potential of this powerful technique. It is still necessary to carry out a more thorough analysis of the behavior of different *Xenopus* cell lines that are available under different transfection procedures, and to work out optimum conditions for them. We have yet to see successful stable transfections to give rise to new *Xenopus* cell lines. Once these technical problems are overcome, increasing use of *Xenopus* cells for DNA transfection will undoubtedly enhance our understanding of developmental, spatial, and hormonal regulation of gene expression.

VII. XTC Mesoderm-Inducing Factor

One of the most widely-used *Xenopus* cell lines in recent years has been the XTC-2 line of Pudney *et al.* (1973). This is because the line was found to produce a factor which induced isolated *Xenopus* blastula animal pole tissue to form mesoderm rather that its usual fate of epidermis (Smith, 1987; see Chapter 17, this volume). Here, we describe the method used to obtain serum-free conditioned medium from the XTC-2 cell line and a method for the purification of XTC-MIF, the mesoderm-inducing factor know known to be a *Xenopus* homolog of activin A (Smith *et al.*, 1990).

A. Preparation of XTC-Conditioned Medium

XTC-2 cells are routinely cultured in modified L15 medium containing 10% fetal calf serum, as described above. Conditioned medium is prepared when the cells reach confluence. Then, they are washed 3 times with serum-free modified L15 medium before being cultured in the same medium. This modified medium is diluted with water to a slightly lesser extent, to allow for the higher osmolarity of fetal calf serum, and also contains 10 mM HEPES as a buffer; without HEPES the serum-free medium rapidly becomes rather acidic. The volume of serum-free medium should be the minimum that the culture vessel will allow. For example, we use 6 ml in a 100-cm^2 flask. The XTC-2 cells are allowed to condition the medium for 24–48 hours, and the medium is then removed and replaced by serum-containing medium for 1 or 2 days. The cells can then be used for the preparation of more conditioned medium. This cycle can be repeated many times; we have obtained batches of conditioned medium from cultures 3 times a week over periods of up to 3 months.

For preparation of large volumes of XTC-2-conditioned medium one can use roller bottles, Cytodex beads, or the "Cell Factories" manufactured by Nunc (Naperville, IN). We have had success with roller bottles but particularly with Cell Factories, used according to the manufacturer's instructions. With three Cell Factories it is possible to obtain 10 liters of conditioned medium a week. We routinely concentrate this medium 10-fold using a Minitan apparatus (Millipore).

B. Purification of XTC-Mesoderm-Inducing Factor

The purification of XTC-MIF, which has recently been shown to be a *Xenopus* homolog of activin A (Smith *et al.,* 1990; van den Eijnden-Van-Raaij *et al.,* 1990), was made possible by designing a purification protocol which allowed active column fractions to be applied immediately to the next column without need for buffer change, and by use of a rapid assay. The assay we employed is a bioassay that takes advantages of one of the earliest responses to induction that is made by animal pole tissue: the onset of gastrulation-like movements (Symes and Smith, 1987). Animal pole regions are dissected from midblastula (stage 8) embryos and cultured in wells containing 50% NAM, 1 mg/ml bovine serum albumin (BSA), and an appropriate dilution of the column fraction to be tested. This dilution can be estimated from a knowledge of the amount of mesoderm-inducing activity loaded onto the column; we define 1 unit of mesoderm-inducing activity as that amount which must be present in 1 ml of medium in order to cause induction. Thus, if a sample of 1 liter of medium is active at a dilution of 1/1000 but not at 1/2000, there are a total of 10^6 units present (Cooke *et al.,* 1987). Column fractions containing mesoderm-inducing activity can be recognized because the test animal pole tissue undergoes a dramatic elongation, which resembles the gastrulation movements of the dorsal marginal zone. This behavior is visible within a few hours of doing the assay (Symes and Smith, 1987; see also Smith *et al.,* 1990).

The first step in the purification of XTC-MIF involves hydrophobic interaction chromatography. Concentrated XTC-conditioned medium is first heated to 100°C for 10 minutes and then rapidly cooled to room temperature. This increases the specific activity of the medium 3- to 10-fold (Smith, 1987). The medium is then adjusted to 1 M NaCl and applied to a column of Phenyl-Sepharose CL-4B equilibrated in 20 mM Tris, pH 8.0, 1 M NaCl. The volume of sample applied to the column should be no more than 5 times the volume of the column. Thus, for a 5 × 30 cm column, a suitable sample volume would be approximately 3 liters. After applying the sample to the column, it should be washed with at least 1 column volume of 20 mM Tris, pH 8.0, 1 M NaCl and then with a similar volume of 20 mM Tris, pH 8.0. This step removes significant quantities of contaminating protein. Inducing activity can then be eluted with a gradient of 0–80% ethylene glycol in 20 mM Tris, pH 8.0. The volume of the gradient should be about 5 times the column volume; thus for the 5 × 30 cm column above a suitable volume would be 3 liters.

Under these conditions, inducing activity elutes at approximately 40% ethylene glycol. Active fractions from this step can be pooled and then applied to an anion-exchange column such as DEAE-Sepharose CL-4B equilibrated in 20 mM Tris, pH 8.0. It is again convenient to use a column whose volume is approximately one-fifth that of the sample to be loaded, and a suitable size is approximately 2.6 × 60 cm. After loading the sample, the column is washed with approximately 1 column volume of 29 mM Tris, pH 8.0, and the column is developed with a 2 liter gradient of 0–1 M NaCl; mesoderm-inducing activity elutes at approximately 0.6 M NaCl.

Active fractions from this step are acidified to pH 2.2 with trifluoroacetic acid (TFA) and pumped directly onto a 4.6 × 100 mm C_8 reversed-phase high-performance liquid chromatography (HPLC) column such as the Brownlee RP300 equilibrated in 0.1% TFA. Mesoderm-inducing activity is relatively stable under acidic conditions, and all the activity is retained by the column. The column is developed at a flow rate of 1 ml/minute with a steep gradient of acetonitrile (1.1%/minute), and under these conditions mesoderm-inducing activity elutes at approximately 32% acetonitrile. This material can be further purified and concentrated by a second round of reversed-phase HPLC on a microbore RP300 column of dimensions 2.1 × 100 mm using a shallower acetonitrile gradient of 0.18%/minute and a flow rate of 0.2 ml/minute. Material from this step is electrophoretically homogeneous (Fig. 3) and has been used for determining the N-terminal amino acid sequence of XTC-MIF. It is capable of inducing muscle-specific actin expression from animal caps at a concentration of 0.2 ng/ml (Smith *et al.*, 1990).

C. Other Factors from XTC-Conditioned Medium

The mesoderm-incuding factor derived from XTC-conditioned medium is a homolog of activin A. It is clear from the work of Roberts *et al.* (1990), however, that other growth factor-like molecules are also produced by XTC cells. These include transforming growth factor (TGF) types $\beta2$ and the newly discovered $\beta5$. These results suggest that *Xenopus* cell lines are likely to be of great use in providing sources of novel growth factors, but they also emphasize that for many experiments on mesoderm induction it is necessary to purify the active principle rather than simply use conditioned medium. Rosa *et al.* (1988) have shown, for example, that TGF$\beta2$ has quite strong mesoderm-inducing

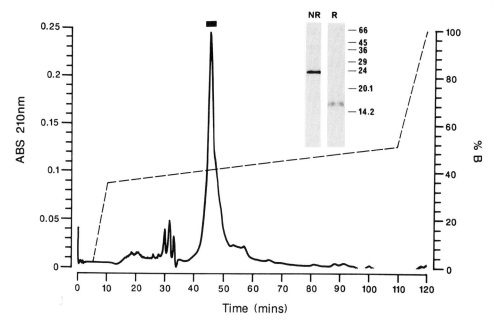

FIG. 3. Final step in the purification of XTC-MIF. Active fractions from a preliminary reversed-phase HPLC step (see text) were pooled and applied to a 2.1 × 100 mm Brownlee RP300 column which was run at 0.2 ml/minutes. The elution profile from the column is shown here. Buffer A was 0.1% trifluoracetic acid (TFA), and buffer B was 80% acetonitrile in 0.1% TFA. Mesoderm-inducing activity eluted at the position shown by the bar, coincident with a single ultraviolet-absorbing peak. When aliquots of this peak were analyzed by polyacrylamide gel electrophoresis (inset) a single major peak was observed under both nonreducing (NR) and reducing (R) conditions. Positions of relative molecular mass markers are shown. (Reproduced, with permission, from Smith et al., 1990.)

activity, and it is quite possible that in addition it acts synergistically with activin. Attempts to make deductions about the effects of activin from experiments using XTC-conditioned medium are therefore to be viewed with caution.

VIII. Conclusions

In this chapter we have described the uses of *Xenopus* cell lines for two main purposes: transfection studies and in the purification of XTC-MIF, a potent mesoderm-inducing factor and a homolog of activin A. There are, of course, other uses of amphibian cultured cells that are beyond the scope of this chapter, such as their use as hosts for arboviruses and in studies of nutrition and somatic cell genetics. We hope that the techniques described above will also provide a starting point for the use of *Xenopus* cell lines in these and other fields.

REFERENCES

Anizet, M. P., Huwe, B., Pays, A., and Picard, J. J. (1981). Characterization of a new cell line, XL2, obtained from *Xenopus laevis* and determination of optimal culture conditions. *In Vitro* **17**, 267–274.

Ausubel, F. M., Brent, R., Kingston, R. E., Moore, D. D., Seidman, J. G., Smith, J. A., and Struhl, K. (1987). Introduction of DNA into mammalian cells. *In* "Current Protocols in Molecular Biology" (F. M. Ausubel, ed.), Vol. 1, Chapter 9. Greene Publishing Associates and Wiley (Interscience), New York.

Chang, T.-C., and Shapiro, D. J. (1990). An NF1-related vitellogenin activator element mediates transcription from the estrogen-regulated *Xenopus laevis* vitellogenin promoter. *J. Biol. Chem.* **265**, 8176–8182.

Chen, C. A., and Okayama, H. (1988). Calcium phosphate-mediated gene transfer: A highly efficient transfection system for stably transforming cells with plasmid DNA. *BioTechniques* **6**, 632–638.

Chinchar, G. D., and Sinclair, J. H. (1978a). Amphibian cells in culture. I. Nutritional studies. *J. Cell. Physiol.* **96**, 333–342.

Chinchar, G. D., and Sinclair, J. H. (1978b). Amphibian cells in culture. II. Isolation of drug-resistant variants and an asparagine-independent variant. *J. Cell. Physiol.* **96**, 343–354.

Cooke J., Smith, J. C., Smith, E. J., and Yaqoob, M. (1987). The organization of mesodermal pattern in *Xenopus laevis:* Experiments using a *Xenopus* mesoderm-inducing factor. *Development (Cambridge, U.K.)* **101**, 893–908.

Ellison, T. R., Mathisen, P. M., and Miller, L. (1985). Developmental changes in keratin patterns during epidermal maturation. *Dev. Biol.* **112**, 329–337.

Evans, J. P., Page, B. D., and Kay, B. K. (1990). Talin and vinculin in the oocytes, eggs, and early embryos of *Xenopus laevis:* A developmentally regulated change in distribution. *Dev. Biol.* **137**, 403–413.

Felgner, P. L., Gadek, T. R., Holm, M., Roman, R., Chan, H. W., Northrop, J. P., Ringold, G. M., and Danielsen, M. (1987). Lipofection: A highly efficient, lipid-mediated DNA-transfection procedure. *Proc. Natl. Acad. Sci. U.S.A.* **84**, 7413–7414.

Freed, J. J., and Mezger-Freed, L. (1970). Culture methods for anuran cells. *Methods Cell Physiol.* **4**, 19–47.

Godsave S. F., and Slack, J. M. W. (1989). Clonal analysis of mesoderm induction in *Xenopus laevis*. *Dev. Biol.* **134**, 486–490.

Jones, K. W., and Elsdale, T. R. (1963). The culture of small aggregates of amphibian embryonic cells *in vitro*. *J. Embryol. Exp. Morphol.* **11**, 135–154.

Keown, W. A., Campbell, C. R., and Kucherlapati, R. S. (1990). Methods for introducing DNA into mammalian cells. *In* "Methods in Enzymology" (D. V. Goeddel, ed.), Vol. 185, pp. 527–537. Academic Press, New York.

Leake, C. J., Varma, M. G. R., and Pudney, M. (1977). Cytopathic effect and plaque formation by arboviruses in a continuous cell line (XTC-2) from the toad *Xenopus laevis*. *J. Gen. Virol.* **35**, 335–339.

Martinez, E., and Wahli, W. (1989). Cooperative binding of estrogen receptor to imperfect estrogen-responsive DNA elements correlates with their synergistic hormone-dependent enhancer activity. *EMBO J.* **8**, 3781–3791.

Miller, L., and Daniel, J. C. (1977). Comparison of *in vivo* and *in vitro* ribosomal RNA synthesis in nucleolar mutants of *Xenopus laevis*. *In Vitro* **13**, 557–567.

Mohun, T. J., Garrett, N., and Treisman, R. (1987). *Xenopus* cytoskeletal actin and human c-*fos* gene promoters share a conserved protein-binding site. *EMBO J.* **6**, 667–673.

Mohun, T. J., Garrett, N., and Taylor, M. V. (1989). Temporal and tissue-specific expression of the proto-oncogene c-*fos* during development in *Xenopus laevis*. *Development (Cambridge, U.K.)* **107**, 835–846.

Nielsen, D. A., and Shapiro, D. J. (1990). Estradiol and estrogen receptor-dependent stabilization of a mini-vitellogenin mRNA lacking 5,100 nucleotides of coding sequence. *Mol. Cell. Biol.* **10**, 371–376.

Pudney, M., Varma, M. G. R., and Leake, C. J. (1973). Establishment of a cell line (XTC-2) from the South African clawed toad, *Xenopus laevis*. *Experientia* **29**, 466–467.

Rafferty, K. A. (1969). Mass culture of amphibian cells: Methods and observations concerning stability of cell type. *In* "Biology of Amphibian Tumors" (M. Mizell, ed.), pp. 52–81. Springer-Verlag, Berlin.

Roberts, A. B., Rosa, F., Roche, N. S., Coligan, J. E., Garfield, M., Rebbert, M. L., Kondaiah, P., Danielpour, D., Kehrl, J. H., Wahl, S. M., Dawid, I. B., and Sporn, M. B. (1990). Isolation and characterization of TGFβ2 and TGFβ5 from medium conditioned by *Xenopus* XTC cells. *Growth Factors* **2**, 135–147.

Rosa, F., Roberts, A. B., Danielpour, D., Dart, L. L., Sporn, M. B., and Dawid, I. B. (1988). Mesoderm induction in amphibians: The role of TGF-β2-like factors. *Science* **239**, 783–785.

Sargent, T. D., Jamrich, M., and Dawid, I. B. (1986). Cell interactions and the control of gene activity during early development of *Xenopus laevis*. *Dev. Biol.* **114**, 238–246.

Seiler-Tuyns, A., Mérillat, A.-M., Haefliger, D. N., and Wahli, W. (1988). The human estrogen receptor can regulate exogenous but not endogenous vitellogenin gene promoters in a *Xenopus* cell line. *Nuclei Acids Res.* **16**, 8291–8305.

Slack, J. M. W. (1984). Regional biosynthetic markers in the early amphibian embryo. *J. Embryol. Exp. Morphol.* **80**, 289–319.

Slack J. M. W., Darlington, B. G., Heath, J. K., and Godsave, S. F. (1987). Mesoderm induction in early *Xenopus* embryos by heparin-binding growth factors. *Nature (London)* **326**, 197–200.

Smith, J. C. (1987). A mesoderm inducing factor is produced by a *Xenopus* cell line. *Development (Cambridge, U.K.)* **99**, 3–14.

Smith J. C., Price, B. M. J., van Nimmen, K., and Huylebroeck, D. (1990). Identification of a potent *Xenopus* mesoderm-inducing factor as a homologue of activin A. *Nature (London)* **345**, 729–731.

Snoek, G. T., Koster, C. H., de Laat, S. W., Heideveld, M., Durston, A. J., and van Zoelen, E. J. J. (1990). Effects of cell heterogeneity on production of polypeptide growth factors and mesoderm-inducing activity by *Xenopus laevis* XTC cells. *Exp. Cell Res.* **187**, 203–210.

Symes, K., and Smith, J. C. (1987). Gastrulation movements provide an early marker of mesoderm induction in *Xenopus laevis*. *Development (Cambridge, U.K.)* **101**, 339–349.

van de Eijinden-Van-Raaij, A. J. M., van Zoelen, E. J. J., van Nimmen, K., Koster, C. H., Snoek, G. T., Durston, A. J., and Huylbroeck, D. (1990). Activin-like factor from a *Xenopus laevis* cell line responsible for mesoderm induction. *Nature (London)* **345**, 32–734.

Wahli, W., Martinez, E., Corthésy, B., and Cardinaux, J.-R. (1989). Cis- and trans-acting elements of the estrogen-regulated vitellogenin gene B1 of *Xenopus laevis*. *J. Steroid Biochem.* **34**, 17–32.

Watret, G. E., Pringle, C. R., and Elliott, R. M. (1985). Synthesis of bunyavirus-specific proteins in a continuous cell line (XTC-2) derived from *Xenopus laevis*. *J. Gen. Virol.* **66**, 473–482.

Part V. Appendixes

Appendix A

Solutions and Protocols

H. BENJAMIN PENG

Department of Cell Biology and Anatomy
University of North Carolina at Chapel Hill
Chapel Hill, North Carolina 27599

Antibiotics (according to R. Keller)
Stock solution
  Penicillin, 10,000 units/ml
  Streptomycin, 10 mg/ml
  Amphotericin B, 25 μg/ml
Add 5 ml stock per liter of solution.

Calcium- and Magnesium-Free Medium (CMFM; according to T. Sargent)
NaCl, 88.00 mM
KCl, 1.00 mM
NaHCO$_3$, 2.40 mM
Tris, 7.50 mM
pH 7.6

Clearing Solution for Whole Mounts (according to A. Murray)
  1. Mix 2 parts benzyl benzoate with 1 part benzyl alcohol. *Caution*: This clearing solution is toxic.
  2. Dehydrate specimen in methanol and treat with clearing solution.

Danilchik Solution (Original)
NaCl, 53.00 mM
NaHCO$_3$, 15.00 mM
Potassium gluconate, 4.50 mM
MgSO$_4$, 1.00 mM
CaCl$_2$, 1.00 mM
Bicine, 5.00 mM

Dissolve above in 90% of final volume in glass-distilled water. Adjust the pH to 8.3 with a measured volume of 1.0 M Na_2CO_3. Calculate the total sodium concentration and bring to 95 mM with sodium isethionate. Bring to final volume with distilled water.

Danilchik's Solution (modified by Shih)
NaCl, 53.00 mM
Na_2CO_3, 10.00 mM
Potassium gluconate, 4.25 mM
$MgSO_4$, 1.00 mM
$CaCl_2$, 1.00 mM
Bicine, 6.00 mM
Bovine serum albumin (BSA), 1.00%
The pH is adjusted to 8.3 with bicine. Aliquot in 50-ml containers and stored at $-20°C$.

DeBoer's Solution (5%)
NaCl, 110.00 mM
KCl, 1.30 mM
$CaCl_2$, 0.44 mM
Adjust to pH 7.3 with 0.1 M $NaHCO_3$.

Dejellying Eggs
1. Mix up 2% cysteine hydrochloride solution with double-distilled water and pH to 8.1 by adding 10 M NaOH.
2. Place eggs in the cysteine solution for about 15 to 20 minutes; swirl the eggs gently at intervals and notice that when the jelly coats are removed, the eggs will lie close to one another, whereas before they are separated by the thickness of the jelly.
3. Rinse 4 times in one-third strength modified Barth's solution (MBS) with gentle swirling.
4. Transfer to fresh one-third strength MBS and hold until needed at temperatures between 16° and 25°C.

Dent's Fixative
20% Dimethyl sulfoxide (DMSO) in methanol.

Holtfreter's Solution (full strength)
NaCl, 60.00 mM
KCl, 0.60 mM
$CaCl_2$, 0.90 mM
$NaHCO_3$, 0.20 mM

I Buffer (according to Dana Carroll)
KCl, 75.00 mM

MgCl$_2$, 7.00 mM
Tris-HCl (pH 7.5), 20.00 mM
Dithiothreitol (DTT), 2.00 mM
Polyvinylpyrrolidone 360, 2%
EDTA 0.1 mM

Marc's Modified Ringer's Solution (1× MMR)
NaCl, 100.00 mM
KCl, 2.00 mM
CaCl$_2$, 2.00 mM
MgCl$_2$, 1.00 mM
HEPES, 5.00 mM
pH 7.4

Modified Barth's Solution (MBS)
NaCl, 88.00 mM
KCl, 1.00 mM
CaCl$_2$, 0.41 mM
Ca(NO$_3$)$_2$, 0.33 mM
MgSO$_4$, 0.82 mM
NaHCO$_3$, 2.4 mM
HEPES, 10.00 mM
pH 7.4

Modified Barth's Solution–High Salt (high salt MBS)
NaCl, 110.00 mM
KCl, 2.00 mM
MgSO$_4$, 1.00 mM
NaHCO$_3$, 2.00 mM
Sodium phosphate, 0.50 mM
Tris base, 15.00 mM
pH 7.6

Modified L-15 Medium for Xenopus Tissue Culture (according to J. C. Smith)
For 1 liter
    Leibovitz L-15 medium (with L-glutamine), 610 ml
    Sterile distilled water, 280 ml
    Fetal calf serum, 100 ml
    Penicillin–streptomycin solution (e.g., Gibco/BRL Cat. No. 043-05070D; other antibiotics may be used if desired), 10 ml

Normal Amphibian Medium (NAM; according to J. M. W. Slack)
NaCl, 110.00 mM
KCl, 2.00 mM

Ca $(NO_3)_2$, 1.00 mM
MgSO$_4$, 1.00 mM
Disodium EDTA, 0.10 mM
Sodium phosphate, pH 7.5, 2.00 mM
NaHCO$_3$, 1.00 mM
Gentamicin 25 μg/ml

Niu–Twitty Solution (modified)
NaCl, 58.00 mM
KCl, 0.70 mM
MgSO$_4$, 0.40 mM
Ca $(NO_3)_2$, 0.30 mM
HEPES, 5.00 mM
pH 7.6

OR2 Solution for Oocyte Storage (according to R. A. Wallace)
NaCl, 82.50 mM
KCl, 2.50 mM
CaCl$_2$, 1.00 mM
MgCl$_2$, 1.00 mM
Na$_2$HPO$_4$, 1.00 mM
HEPES, 5.00 mM
pH 7.8

Ron Stewart's Medium (low CMR/R/5)
NaCl, 43.00 mM
KCl, 0.78 mM
CaCl$_2$, 0.37 mM
MgCl$_2$, 0.18 mM
HEPES, 3.00 mM
pH 7.2

Smith's Fixative for General Histology
Solution A
    2.5 ml glacial acetic acid
    0.5 g potassium dichromate
    100 ml double-distilled water
To make working solution, mix 8–10 parts of solution A with 1 part of 37% formalin.

Somite Staining Protocol Using Monoclonal Antibody 12–101 (according to P. A. Wilson)
1. Fix overnight at room temperature in 4% formaldehyde solution [4% formaldehyde diluted in phsophate-buffered saline (PBS) with 2.5% DMSO].

2. Wash in large volume of PBS for 4 hours minimum at room temperature.
3. Extract in 4% Triton X-100 in PBS for 4 hours at room temperature.
4. Incubate in primary antibody, 12-101 [see Kintner and Brockes, *Nature (London)* **308,** 67–69 (1984)], at 4°C overnight. Wash in a large volume of PBS for 4 hours minimum at room temperature.
5. Incubate in secondary antibody solution at 4°C overnight. Wash in a large volume of PBS for 4 hours at room temperature.
6. Transfer specimen to 100% methanol and then into clearing solution (this appendix).
7. Protocol can be used with Dent's fixative, in which case the Triton extraction is not done. Antibody incubations are at room temperature for 2 hours each, and washings are 30 minutes each.

Steinberg's Solution (modified)
NaCl, 60.00 mM
KCl, 0.67 mM
Ca$(NO_3)_2$, 0.34 mM
MgSO$_4$, 0.83 mM
HEPES, 10.00 mM
pH 7.4

Tipping and Marking the Dorsal Side of Xenopus Eggs
Ringer's Solution
   NaCl, 100.00 mM
   KCl, 2.00 mM
   CaCl$_2$, 2.00 mM
   MgCl$_2$, 1.00 mM
   HEPES, 5.00 mM
   pH 7.4
Sorenson's Solution
   4 parts 0.25 M NaH$_2$PO$_4 \cdot$H$_2$O
   46 parts 0.25 M Na$_2$HPO$_4$
   Dilute this solution 1:3.75, adjust pH to 7.8

Ficoll: 7% type 400DL in 1:1 mix of Sorenson's solution and 1/3 Ringer's
Nile blue sulfate: 1% in water, spin 5 minutes at 5000 rpm or filter
NaCO$_3$: 100 mM

1. Put newly fertilized and dejellied eggs in Ficoll until the space under the membrane disappears and eggs stick to the dish.
2. Tip the eggs 90° with a pair of forceps. This should be done before 0.4 of first cell cycle (about 36 minutes). It is probably best to do it on a cold (18°C) table to slow development.

3. In separate dish, place a drop of Nile blue near a drop of carbonate. Pull a stream of one into the other. Red crystals should form.
4. Pull a thin glass filament with a rounded end. Collect the crystals on the end.
5. Touch eggs just below pigment boundary until a blue spot forms.
6. After first cleavage, the eggs can be untipped and transferred out of Ficoll.
7. The dorsal side will form on the side of the blue spot.

Winklbauer's Cell Dissociation Solution
NaCl, 50.3 mM
KCl, 0.70 mM
Na_2HPO_4, 9.20 mM
KH_2PO_4, 0.90 mM
$NaHCO_3$, 2.40 mM
EDTA (pH 7.3), 1.00 mM

Explants of embryos are treated with this solution for 50 minutes followed by mechanical dispersion by rinsing with the same solution. The cells are then washed in MBS. However, optimum treatment time will vary. Do not dissociate any longer than necessary to separate cells.

Wolfe and Quimby Medium
For 1 liter:
Eagles's minimum essential medium, 750 ml
Earle's balanced salt solution, without NaCl, 250 ml
Adjust pH to 7.6–7.8

Appendix B

Injection of Oocytes and Embryos

BRIAN K. KAY

Department of Biology
The University of North Carolina at Chapel Hill
Chapel Hill, North Carolina 27599

I. General Information
    A. Pipettes
    B. Samples
    C. Injection System
    D. Pipette Calibration
    E. References
II. How to Inject Oocytes
    A. Oocyte Preparation
    B. Injections into the Cytoplasm
    C. Injections into the Nucleus
    D. Controls
III. How to Inject Embryos

I. General Information

Because of an enormous flexibility in the choice of equipment and methods many techniques give satisfactory results, only a few of the many options are described here.

A. Pipettes

Prepare micropipettes from something like R6 glass capillary tubes, with an outside diameter of 0.63 mm and an inside diameter of 0.2 mm. These are supplied in 4-inch lengths by Drummond Scientific Co. [Broomall, PA; (215) 353-0200]. The pipettes can be pulled on a commerical vertical micropipette puller [i.e., Model 720, David Kopf Instruments, Tujunga CA, (818) 352-3274; Narashige Model PB-7, Medical Systems Corp., Greenvale NY, (516) 621-9190]. The pipettes can also be pulled horizontally (Model MI, Industrial Science Associates Inc., Ridgewood, NY). The pipette tip should have an outer diameter of 5–15 μm. To give the micropipette an ellipsoid opening, the pipette can be gently broken at an angle with a pair of Dumont No. 5 forceps, or the edge can be bevelled at a 20°–30° angle with a Narishige EG-3 grinder and 0.3 mm abrasive film (Thomas Scientific Co., Swedesboro, NJ, Cat. No. 6775-E54). Pulled pipette can be stored on their side by sticking them into two rows of art modeling clay or Silly Putty inside a large petri plate. If gloves are worn during the preparations of the pipettes, they should remain RNase free; otherwise the pipettes can be baked briefly. To clean a pulled pipette of abrasive material or glass chips which may clog the tip, pull distilled deionized water up the tip, rinse with absolute alcohol, and then air-dry.

B. Samples

DNA or mRNA is dissolved in injected buffer (88 mM NaCl, 10 mM HEPES, pH 6.8, or 88 mM NaCl, 15 mM Tris-HCl, pH 7.5) at concentrations of 100 to 1000 μg/ml. Sometimes it is convenient to dissolve the nucleic acid sample in water and then add 2× injection buffer to an aliquot; this is important if an mRNA sample is to be translated *in vitro*, since this reaction is sensitive to sodium and magnesium ions. More concentrated solutions, particularly of high molecular weight, double-stranded DNA can be quite viscous and difficult to draw through the narrow pippette tip. The injection solution can be spun 1–2 minutes at top speed in a microcentrifuge to remove particles that might clog the injection pipette.

C. Injection System

A number of injectors can be used. First, there is the Eppendorf Microinjector, Fremont, CA, Model 5242 (∼ $6100), which controls the volume of material injected. To facilitate micropipette filling, this apparatus can be equipped with an external vacuum connector and switch [Baltimore Instruments, Inc., (301) 426-3656]. Second, another injector that works well, and which is well suited to a low budget, is the Drummond "Nanoinject" (Cat. No. 3-00-203-X; ∼ $500). An electronic chip controls the incremental movement of a metal piston inserted into the pipette; chips are available which dispense volumes in increments of 4.7 nl (i.e., 4.7 to 47 nl). On the control box, there are also buttons which are used to fill and empty the pipette. For both calibration as well as for the actual injections, the pipette should first be filled with light mineral oil (Sigma Chemical Co., St. Louis, MO; Cat. No. M3516), before filling the pipette with the liquid sample to be dispensed. A third possibility is the Picoinjector (PLI-100; ~$2500) from Medical Systems Corporation [Greenvale, NY; (516) 621-9190]. This device uses compressed gas (N_2 or He) to deliver precise, reproducible volumes. The digital system has fill/hold, balance, and clear functions as well. Some individuals consider the pneumatic (air-driven) system to be preferable to the hydraulic (fluid-driven) system, because it avoids the mess of dealing with oil. Foot pedals are convenient.

A modest amount of desk or laboratory bench space needs to be dedicated for the injection system. The area should be level and free of most vibrations. To visualize the specimens properly, a dissecting microscope (i.e., Nikon SMZU) and a non-heat-transferring illumination system (i.e., dual fiber light pipes) are required. For maximal working distance, the microscope can be mounted on a swing-out or boom arm. The micropipettes are held in a micromanipulator (i.e., Narashige, Model M152; Brinkmann Instruments, Westbury, NY, Model MM33; Singer Instrument Co., Watchet, England, FAX: 011-44-984-41166, Model Mark 1) at a 45° angle, and the micromanipulator can be secured via a magnetic base to a steel plate beneath the microscope.

D. Pipette Calibration

In theory, each pipette should be calibrated individually. Calibration can be done in one of several ways. First, expel liquid for a given unit of time (milliseconds), pipettor device setting, or felt pen mark on the capillary wall and measure the diameter of the spherical droplet at the pipette tip (or as the aqueous droplet sits in mineral oil). With a micrometer and optical reticle present in a microscope eyepiece, calculate the volume ($V = \frac{4}{3}\pi r^3$) based on the measured diameter ($2r$); for droplets of 10 and 50 nl, the diameters should be 0.27 and 0.46 mm respectively. Second, the injection volume can be determined by using the known internal diameter of the capillary to calculate the volume delivered for each division on a eyepiece reticle, after viewing the distance traversed by the meniscus. Third, expel the liquid for 100 units of time (or setting) onto a piece of Parafilm, and measure the actual volume (1–5 μl) with a graduated glass capillary; ultimately, calculate the amount dispensed for 1 volume unit. Fourth, take a solution of known specific activity [i.e., radioactive amino acid or nucleotide, or an assayable enzyme such as chloramphenicol acetyltransferase (CAT) or luciferase], inject into a liquid droplet (~5 μl), and then measure (i.e., by liquid scintillation counting or with an enzyme assay) the actual amount of material dispensed by the micropipette.

In practice and depending on the experiment, calibration only needs to be done occasionally. If the pipette puller settings remain constant, there will be only slight variability ($\pm 10\%$) between needles. What may be more important for experimental reproducibility is that the same needle is used to inject the same set of experimental oocytes (or embryos).

E. References

Colman, A. (1984). Expression of exogenous DNA in *Xenopus* oocytes. *In* "Transcription and Translation: A Practical Approach" (B. D. Hames and S. J. Higgins, eds.), pp. 51–68. IRL Press, London.

Colman, A. (1984). Translation of eukaryotic messenger RNA in *Xenopus* oocytes. *In* "Transcription and Translation: A Practical Approach" (B. D. Hames and S. J. Higgins, eds.), pp. 271–302. IRL Press, London.

Gurdon, J. B., and Wickens, M. P. (1983). The use of *Xenopus* oocytes for the expression of cloned genes. *In* "Methods in Enzymology" (R. Wu, L. Grossman, and K. Moldave, eds.), Vol. 101, pp. 370–386. Academic Press, New York.

Melton, D. A. (1987). Translation of messenger RNA in injected frog oocytes. *In* "Methods in Enzymology" (S. L. Berger and A. R. Kimmel, eds.), Vol. 152, pp. 228–296. Academic Press, New York.

Moon, R. T., and Christian, J. L. (1989). Microinjection and expression of synthetic mRNAs in *Xenopus* embryos. *Technique* **1**, 76–89.

Stephens, D. L., Miller, T. J., Silver, L., Zipser, D., and Mertz, J. E. (1981). Easy-to-use equipment for the accurate microinjection of nanoliter volumes into the nuclei of amphibian oocytes. *Anal. Biochem.* **114**, 299–309.

II. How to Inject Oocytes

A. Oocyte Preparation

Oocytes can be injected either as they sit within their ovarian follicle or after release from the follicle. A segment of ovary can be teased into a small piece, containing about 10 oocytes, and then injected directly. This is fine for cytoplasmic injections but awkward for nuclear injections. In addition, sometimes the follicular material is an insurmountable barrier to the sharpened micropipettes. Alternatively, oocytes can be easily injected, once they are freed of their accessory cells; see Chapter 4 (this volume) which describes the isolation of oocytes through manual dissection or by treatment of the segment of ovary tissue with collagenase. The oocytes should be kept in MBS or OR2 buffer. After the oocytes have been injected, they can be maintained at 18°C in the presence of antibiotics (i.e., 10 μg/ml penicillin, 10 μg/ml streptomycin sulfate) for 1–5 days, with modest levels of morbidity.

B. Injections into the Cytoplasm

An injection vessel can be prepared from a 60-mm plastic petri plate that has a piece of polyethylene Nitex mesh (Fisher Scientific, Pittsburgh, PA, Cat No. 8-670-176) on the bottom. The grids wells of the mesh are similar in dimension to the diameter of the oocyte, and the sheet of Nitex can be affixed permanently to the bottom half of the petri dish with a few drops of chloroform. Transfer oocytes to the dish in MBS or OR2 buffer and turn the nonpigmented, vegetal hemisphere upward with a blunt instrument, such as a toothpick, so that the nucleus will not be injected. (The buffer level can be lowered so that the meniscus is just below the surface of the oocyte; this minimizes back-flow into the capillary. Usually, the needle is preloaded before exposing the superficial surface of the oocytes to air, as the membranes eventually harden owing to dehydration.) After injection of 10–50 nl, culture the oocytes in buffer (MBS or OR2) with antibiotics (i.e., 10 μg/ml gentamycin). With practice, an injector should be able to inject about 10 oocytes in a few minutes.

C. Injections into the Nucleus

When the injections are to be made in the nucleus, the oocytes should be oriented in the Nitex mesh with the brown (or black) pigmented, animal hemisphere pointing up. The oocytes can be gingerly oriented with a toothpick or a pair of blunt forceps. When injecting, insert the micropipette near the apex (center) of the pigmented hemisphere and drive the tip in approximately one fifth the diameter of the oocyte. Since the nucleus (germinal vesicle) has a volume of approximately 50 nl, only about 10 nl should be injected.

The success rate of hitting the nucleus in this "blind" method is 50–80%. Generally, this is tolerable, because the experimentalist is looking for a qualitative and not a quantitative result. However, to increase the likelihood of injecting the nucleus, the oocytes can be lightly centrifuged ($\sim$650 g, 8–10 minutes) with the animal hemisphere oriented upward in the Nitex mesh. (The oocytes can also be centrifuged in depressions formed in pink dental wax that sits at the bottom of a petri dish.) In the centrifuged oocyte, the nucleus rises to the surface of the cell, displaces some cortical pigment, and generates a bull's-eye. This presents a clear target, although some caution should be taken not to inject *through* the nucleus because it is flatter in the centrifuged state. One other method of positioning the nucleus is to treat oocytes with 50 μg/ml nocadazole (Sigma Chemicals, Co., St. Louis, MO) for 5-6 hours. The nucleus will rise to the oocyte surface creating a cleared spot.

D. Controls

There are several convenient controls for proper injection of the cytoplasm. Native rabbit β-globin mRNA (Bethesda Research Laboratories, Gaithersburg, MD, Cat. No. 8103) or Brome mosaic virus RNAs (Promega, Madison, WI, Cat. No. D1541) can be obtained commercially. When oocytes are injected with these mRNAs and incubated overnight with [^{35}S]methionine in the culture medium (100 μCi/ml), the proteins encoded by the exogenous message should be evident against the background of endogenously synthesized oocyte proteins, after sodium dodecyl sulfate-polyacrylamide gel electrophoresis (SDS-PAGE) and X-ray film autoradiography. Controls for proper translation of synthetic mRNA are also available. *In vitro* transcripts can be prepared for secretory forms of alkaline phosphatase [Tate *et al.*, (1990). *FASEB J.* **4,** 227–2274] and amylase [Urnes and Carroll, (1990). *Gene* **95,** 267–274], injected into oocyte, and the medium can be monitored, without destroying the oocyte. Either synthetic RNA can be mixed with the mRNA of interest, and monitoring for secretion of alkaline phosphatase or amylase permits rapid identification of those oocytes that efficiently express injected mRNAs.

Reporter plasmids that are transcribed well in oocyte nuclei are pSV$_2$CAT or pT109luciferase, available from the American Type Culture Collection [Cat. No. 37155 and 37583, respectively; 12301 Parklawn Drive, Rockville, MD, (800) 638-6597]. RNA transcripts from these plasmids can be measured directly using primer-extension or S1 protection assays, or indirectly by assaying the activity of their protein products. These plasmids can serve as internal controls for other injected plasmids.

As a visible control for injection of nuclei, the injector should practice injecting dye (i.e., Coomassie blue, trypan blue) into the cell, and then dissect out the nuclei for visual confirmation that the nucleus was targeted appropriately. In experiments where quantitative injection of the nucleus is important, colloidal gold particles, coated with polyvinylpyrolidine, can be included in the injection sample, and thus only oocytes that have red nuclei upon nuclear dissection can be identified for further biochemical analysis [Dworetzsky and Feldherr, (1988). *J. Cell Biol.* **106,** 575–584; Bataillé *et al.*, (1990). *J. Cell Biol.***111,** 1571–1582].

III. How to Inject Embryos

DNA, dissolved in injection buffer, should be injected deep into the animal hemisphere during either the first 30 minutes of the first cell cycle or anytime during the second cell cycle. Zygotes should be prepared by artifical fertilization (swabbing minced testis over ovulated eggs, see Chapter 31, this volume) and dejellied in 2% cysteine (Sigma Chemicals Co., St. Louis, MO). Generally, the zygote will cleave into a two-cell embryo within 90 minutes of fertilization; subsequent cell divisions occur on a 30-minute schedule. (If necessary, they can be delayed by lowering the temperature to 16°C.) The amount of DNA that must be injected will depend on the activity of the construct being studied. However, injection of greater than 200 pg frequently results in abnormal gastrulation and embryonic death. During the injection process, the embryos should be positioned in the Nitex mesh or in a special slide chamber [Moon and Christian, (1989) *Technique* **1,** 76–89], in 1× MMR, 5% Ficoll 400. After several hours transfer the embryos to 0.1 × MMR for culturing at 16°–18°C until they reach the desired stage; exogastrulation will occur if they remain in 1x MMR.

Appendix C

Mutants of Xenopus laevis

ANNE DROIN

Département de Zoologie et Biologie Animale
Université de Genève
CH-1224 Chêne-Bougeries, Switzerland

Mutation	Stage of expression	Phenotype	Reference
a^p, periodic albinism	Egg–adult	Viable, maternal effect; lack of melanin pigmentation	Hoperskaya (1975). *J. Embryol. Exp. Morphol.* **34,** 253–264.
abj, abnormal joints	58	Lethal; abnormal limb development	Droin (1988). *Alytes* **7,** 45–51.
abl, abnormal limbs	54–adult	Semilethal; syndactyly, polydactyly, brachydactyly	Droin and Fischberg (1980). *Experientia* **36,** 1286–1287.
ar, arrest	49	Lethal; arrest of development	Krotoski *et al.* (1985). *J. Exp. Zool.* **233,** 443–449.
bh, bubblehead	42–45	Lethal; fluid accumulation in the head	Krotoski *et al.* (1985). *J. Exp. Zool.* **233,** 443–449.
b1-3, bloated-3	33–46	Lethal; edema of trunk; heart and gut abnormal	Droin (1988). *Acta Embryol. Morphol. Exp.* **9,** 115–127.
bn, balloon	43–63	Viable; fluid accumulation in coelom and under skin	Krotoski *et al.* (1985). *J. Exp. Zool.* **233,** 443–449.
bt, bent tail	42–48	Lethal; edemata, internal haemorrhages, tail bent up	Droin *et al.* (1970). *Rev. Suisse Zool.* **77,** 596–603.
dcg, degenerative cement gland	30–45	Lethal; cement gland degeneration, and of entire embryo	Droin (1981). *Acta Embryol. Morphol. Exp.* **2,** 101–113.

(continued)

Mutation	Stage of expression	Phenotype	Reference
dl, distended lungs	49–66	Semilethal; delay of development, distension of lungs	Droin and Chavane (1979). *Acta Embryol. Morphol. Exp.* **3**, 273–289.
dm, delayed metamorphosis	49–58	Subvital; delay of development until metamorphic climax	Droin and Buscaglia (1978). *Acta Embryol. Exp.* **1**, 95–111.
dp, dilated pupil	48–metamorphosis	Lethal; transient disorganization of the neural retina	Krotoski *et al.* (1985). *J. Exp. Zool.* **233**, 443–449.
dtp, droopy tailtip	39–47	Lethal; microcephaly, microphthalmy, edema, drooping of the tail	Droin (1976). *Rev. Suisse Zool.* **83**, 853–858.
dw-1, dwarf-1	48	Lethal; arrest of development	Droin (1974). *Ann. Embryol. Morphol.* **7**, 141–150.
dw-2, dwarf-2	47	Lethal; arrest of development	Droin (1974). *Ann. Embryol. Morphol.* **7**, 141–150.
dw-3, dwarf-3	47	Lethal; arrest of development	Droin (1988). *Acta Embryol. Morphol. Exp.* **9**, 115–127.
dw-4, dwarf-4	46	Lethal; arrest of development	Droin (1988). *Acta Embryol. Morphol. Exp.* **9**, 115–127.
ee, enlarged eye	51–adult	Viable in females, lethal in males (?); large eyes	Krotoski *et al.* (1985). *J. Exp. Zool.* **133**, 443–449.
fj, folded jaw	42–48	Lethal; deformation of lower jaw	Droin *et al.* (1968). *Rev. Suisse Zool.* **75**, 531–538.
g, goitre	Metamorphosis	Semilethal (?); hyperplastic goiter, delayed metamorphosis	Uehlinger (1965). *Experientia* **21**, 271.
im, immobile	23–46	Lethal; lack of muscular contraction	Droin and Beauchemin (1975). *J. Embryol. Exp. Morphol.* **34**, 435–449.
jk, jerky	46–metamorphosis	Viable; phasic loss of mobility	Krotoski *et al.* (1985). *J. Exp. Zool.* **233**, 443–449.
kt, kinky tailtip	42–66	Subvital; edema of abdomen, kink in the tail	Uehlinger and Reynaud (1965). *Rev. Suisse Zool.* **34**, 680–685.
lb, laidback	47–49	Lethal; balance dysfunction, spastic	Krotoski *et al.* (1985). *J. Exp. Zool.* **233**, 443–449.
ll, light lethal	39–54	Lethal; deficient melanophore and iridophore	Kaye *et al.* (1983). *Am. Zool.* **23**, 969.
ml, Muizenberg lethal	38–42	Lethal; degeneration of ectoderm and mesoderm	Krotoski and Tompkins (1982). *Am. Zool.* **22**, 917.
mo, muscle opacity	47–48	Lethal, linked to *ry*; opacity of ventral muscles in the head	Droin (1991). *Genet. Res.* **56**, 279–282.
oe, oedema	52–post metamorphosis	Semilethal; edemata of lymph sacs	Uehlinger and Beauchemin (1968). *Rev. Suisse Zool.* **75**, 697–706.

APPENDIX C. MUTANTS OF XENOPUS LAEVIS

Mutation	Stage of expression	Phenotype	Reference
O-nu, anucleolate		*O-nu*/+ viable, deficiency of NOR	Elsdale et al. (1958). *Exp. Cell Res.* **14**, 642–643;
		O-nu/*O-nu* lethal at stages 37–45	
		Several other complete or (*p-nu*) NOR deficiencies	Miller and Gurdon (1970). *Nature (London)* **7**, 1108–1110. Krotoski et al. (1985). *J. Exp. Zool.* **233**, 443–449.
otl, otolithless	36–48	Lethal; absence of otoliths, ears swollen	Droin (1967). *Rev. Suisse Zool.* **74**, 628–636.
pc, partial cleavage	3–9	Lethal, maternal effect; abnormal cleavage	Droin and Fischberg (1984). *Roux's Arch. Dev. Biol.* **193**, 86–89.
pd, polydactyly	54–adult	Subvital; polydactyly, mainly of posterior limbs	Uehlinger (1969). *J. Embryol. Exp. Morphol.* **21**, 207–218.
pe, pale eggs	1–41	Viable, maternal effect; pale color of eggs, small amount of melanosomes	Droin and Fischberg (1984). *Roux's Arch. Dev. Biol.* **193**, 86–89.
p.oe, precocious oedema	39–47	Lethal; generalized edema	Droin (1974). *Ann. Embryol. Morphol.* **7**, 141–150.
ry, rusty	43–49	Viable, linked to *mo*; persistence of egg pigment	Uehlinger and Droin (1969). *Arch. Jul. Klaus Stift.* **43–44**, 48–54.
S, Screwy	S/+ 52–66	Subvital; spiral torsion of tail	Uehlinger (1966). *Rev. Suisse Zool.* **73**, 527–534.
	S/S 32–40	Lethal; absence of cephalic mesenchyme	
sg, short gut	47–49	Lethal; shortened gut tube	Krotoski et al. (1985). *J. Exp. Zool.* **233**, 443–449.
tb, tubby	39–54	Semilethal, maternal effect; partial albinism	Krotoski et al. (1985). *J. Exp. Zool.* **233**, 443–449.
te, tiny eye	47–49	Lethal; small eyes, development arrest at stage 49	Krotoski et al. (1985). *J. Exp. Zool.* **233**, 443–449.
tr, turner	47–53	Semilethal; abnormal morphology and position of otoliths	Droin (1971). *Rev. Suisse Zool.* **78**, 559–568.
ur, unresponsive	22–49	Viable; delayed motility, abnormal swimming	Reinschmidt and Tompkins (1984). *Differentiation* **26**, 189–193.
yr, yolky rectum	31–42	Lethal; microcephaly, microphthalmy, absence of endoderm differentiation	Reynaud and Uehlinger (1965). *Rev. Suisse Zool.* **72**, 675–680.

Appendix D

Codon Usage Table for Xenopus laevis

J. MICHAEL CHERRY

Massachusetts General Hospital
Department of Molecular Biology
Boston, Massachusetts 02114

The codon usage table was produced with the GCG program CodonFrequency. It is based on 135 genes found in GenBank (Release 63). Duplicates, pseudogenes, mutant, and synthetic genes were not included. Coding regions were specified using the Feature Table of each entry, then checked for accuracy. If more than one stop codon was found the sequence was not included.

Amino acid	Amino acid	Codon	Fraction
A	Ala	GCG	0.05
	Ala	GCA	0.32
	Ala	GCT	0.35
	Ala	GCC	0.28
C	Cys	TGT	0.48
	Cys	TGC	0.52
D	Asp	GAT	0.57
	Asp	GAC	0.43
E	Glu	GAG	0.46
	Glu	GAA	0.54
F	Phe	TTT	0.53
	Phe	TTC	0.47
G	Gly	GGG	0.18
	Gly	GGA	0.36
	Gly	GGT	0.23
	Gly	GGC	0.23

(continued)

Amino acid	Amino acid	Codon	Fraction
H	His	CAT	0.49
	His	CAC	0.51
I	Ile	ATA	0.21
	Ile	ATT	0.41
	Ile	ATC	0.38
K	Lys	AAG	0.51
	Lys	AAA	0.49
L	Leu	CTG	0.32
	Leu	CTA	0.09
	Leu	CTT	0.19
	Leu	CTC	0.14
	Leu	TTG	0.16
	Leu	TTA	0.09
M	Met	ATG	1.00
N	Asn	AAT	0.51
	Asn	AAC	0.49
P	Pro	CCG	0.07
	Pro	CCA	0.36
	Pro	CCT	0.33
	Pro	CCC	0.24
Q	Gln	CAG	0.64
	Gln	CAA	0.36
R	Arg	AGG	0.22
	Arg	AGA	0.28
	Arg	CGG	0.09
	Arg	CGA	0.10
	Arg	CGT	0.16
	Arg	CGC	0.15
S	Ser	AGT	0.16
	Ser	AGC	0.23
	Ser	TCG	0.04
	Ser	TCA	0.17
	Ser	TCT	0.22
	Ser	TCC	0.18
T	Thr	ACG	0.06
	Thr	ACA	0.34
	Thr	ACT	0.32
	Thr	ACC	0.28
V	Val	GTG	0.34
	Val	GTA	0.18
	Val	GTT	0.28
	Val	GTC	0.20

APPENDIX D. CODON USAGE FOR XENOPUS LAEVIS

Amino acid	Amino acid	Codon	Fraction
W	Trp	TGG	1.00
Y	Tyr	TAT	0.52
	Tyr	TAC	0.48
*	End	TAG	0.17
	End	TAA	0.36
	End	TGA	0.47

Appendix E

Pictorial Collage of Embryonic Stages

MIKE DANILCHICK

Department of Biology
Hall-Atwater and Shankin Laboratories
Wesleyan University
Middletown, Connecticut 06457

H. BENJAMIN PENG

Department of Cell Biology and Anatomy
The University of North Carolina at Chapel Hill
Chapel Hill, North Carolina 27599

BRIAN K. KAY

Department of Biology
The University of North Carolina at Chapel Hill
Chapel Hill, North Carolina 27599

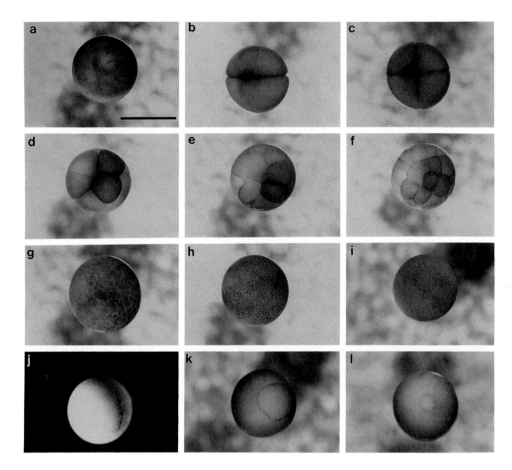

(a) zygote, animal pole, 0.8 units into the first cell cycle, bar = 1 mm (b) 2 cell, animal pole (c) 4 cell, animal pole (d) 8 cell, animal pole (e) 16 cell, animal pole (f) 32 cell, animal pole (g) stage 8, animal pole (h) stage 9, animal pole (i) stage 10, animal pole (j) stage 10, vegetal pole (k) stage 10.5, vegetal pole (l) stage 12, vegetal pole.

APPENDIX E. PICTORIAL COLLAGE OF EMBRYONIC STAGES 681

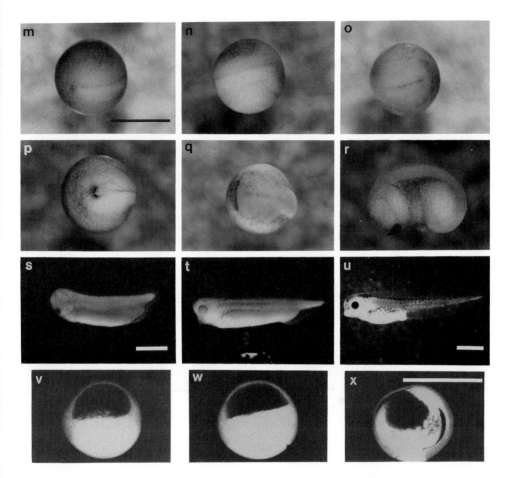

(m) stage 13, dorsal view, bar = 1 mm (m–r) (n) stage 15, dorsal view (o) stage 17, anterior view (p) stage 18, anterior view (q) stage 19, anterior view (r) stage 22, lateral view (s) stage 31, bar = 1 mm (s–t) (t) stage 35 (u) stage 40, bar = 1 mm (v) stage 9, internal view with confocal microscope (w) stage 10, internal view with confocal microscope (x) stage 11.5, internal view with confocal microscope, bar = 1 mm (v–x). All stages are according to Nieuwkoop and Faber, 1967, *In* "Normal Tables of *Xenopus laevis (Daudin)*." North-Holland, Amsterdam.

Appendix F

Xenopus Suppliers in the United States

Carolina Biological Supply Company
2700 York Road
Burlington, North Carolina
Tel: (800) 334-5551

Nasco
901 Janesville Avenue
Fort Atkinson, Wisconsin 53538
Tel: (800) 558-9595

Xenopus I
716 Northside
Ann Arbor, Michigan 48105
Tel: (313) 426-2083

Appendix G

In Situ Hybridization: An Improved Whole-Mount Method for Xenopus Embryos

RICHARD M. HARLAND

Department of Molecular and Cell Biology
Division of Biochemistry and Molecular Biology
University of California, Berkeley
Berkeley, California 94720

Introduction

The whole-mount *in situ* hybridization method introduced by Tautz and Pfiefle (1989) has great potential for application to *Xenopus* embryos. The ability to stain a large number of embryos at one time allows a thorough analysis of gene expression in different embryonic stages. Furthermore, batch staining of embryos is applicable to manipulated embryos, or embryo explants, so that a more detailed picture of gene expression can be obtained than from analysis of extracted RNA. To date, the method has been limited by lack of signal and excessive background, so that only relatively abundantly expressed genes, like muscle actin, are easy to assay (Hemmati-Brivanlou *et al.*, 1990). Here I describe modifications to the method which both increase the signal and reduce the background, so that tissue-specific mRNAs such as that for the neural cell adhesion molecule (NCAM) (Kintner and Melton, 1987) or regionally expressed homeobox-containing mRNAs such as En-2 (Hemmati-Brivanlou *et al.*, 1991) can be assayed routinely, reproducibly, and unambiguously. Below I describe the procedure currently in use. Methods not explicitly described here or by Hemmati-Brivanlou and co-workers (Hemmati-Brivanlou and Harland, 1989; Hemmati-Brivanlou *et al.*, 1990) can be found in Sambrook *et al.* (1989).

No unusual precautions are taken against RNase contamination. Where practical, solutions are treated with diethyl pyrocarbonate (DEPC). Many solutions (e.g., PBS–Tween) are autoclaved, but autoclaving would be dangerous or damaging for others (e.g., formaldehyde or hybridization solution).

Synthesis of Probe

1. A standard RNA synthesis (Melton et al., 1985) using SP6, T7, or T3 RNA polymerase is carried out incorporating a digoxigenin-substituted ribonucleotide. Although sensitive components should be kept on ice, the reaction should be mixed at room temperature to prevent precipitation of DNA. The amount of nucleotide is reduced relative to the original procedure, because nucleotide is seldom the limiting component. The following mixture is allowed to react at 37°C for 2 hours:

5 × SP6 Buffer	10 μl
1 M Dithiothreitol (DTT)	0.5 μl
2.5 mN Nucleotide triphosphate (NTP) mix	10 μl
Water	25 μl
DNA (1 μg/μl)	2.5 μl
[^{32}P]CTP	1 μl
RNasin (20 U/μl)	0.5 μl
Enzyme	90 U

(5 × SP6 buffer contains 200 mM Tris. HCl pH 7.5, 30 mM $MgCl_2$, 20 mM spermidine.)

The 2.5 mM NTP mix with digoxigenin-11 UTP (Boehringer Mannheim, Indianapolis, IN, Cat. No. 1209 256; 25 nmol) contains 10 μl of 10 mM CTP, 10 μl of 10 mM GTP, 10 μl of 10 mM ATP, 6.5 μl of 10 mM UTP, and 3.5 μl of 10 mM digoxigenin-11 UTP.

2. To the RNA synthesis mixture add 1 μl of 1 mg/ml RNase-free DNase I and incubate at 37°C for 10 minutes.

3. Dilute the mixture to 100 μl with 1% sodium dodecyl sulfate (SDS), 20 mM EDTA, 20 mM Tris (pH 7.5), 100 mM NaCl or similar buffer.

4. Take a 1-μl aliquot for determination of total counts per minute (cpm) in the reaction.

5. Remove most of the unincorporated nucleotide on a 1-ml Sephadex G-50 spin column. (Sephadex should be treated with DEPC, washed, and autoclaved in 0.3 M sodium acetate, pH 5.5, 0.1% SDS.)

6. Add 2 volumes of 100% ethanol to the eluate to precipitate RNA. The yield should be sufficient so that no carrier is needed. Monitor the pellet and supernatant to ensure that most of the RNA has precipitated. Carrier (*Torula* RNA or glycogen, 10 μg) can be added if desired.

7. Resuspend the pellet in 50 μl of 40 mM sodium bicarbonate–60 mM sodium carbonate. Count the Cerenkov radiation in the resuspended RNA

and total sample (from step 4) to estimate yield. One hundred percent incorporation would correspond to 33 μg RNA so the following equation may be used:

Incorporated cpm/total available cpm × 33 μg = yield (μg)

Heat for 35–50 minutes at 60°C (this hydrolyzes the probe to a length of ~300 nucleotides). Add 200 μl water, 25 μl of 3 M sodium acetate, and 600 μl ethanol to precipitate. Almost all the counts should precipitate.

8. Resuspend the pellet at 10 μg/ml in hybridization buffer (see below, usually 0.5–1 ml for yields of 5–10 μg) as a probe stock. Store at −20°C. Probes over 1 year old have been used satisfactorily. For hybridization, the probe is diluted to a final concentration of 1 μg/ml, though I have used a concentration as low as 0.1 μg/ml.

Preparation of Embryos

Although pigmented embryos can be used, albino embryos are preferred. To assist staging they are stained with Nile blue after dejellying (0.01% Nile blue in 50 mM phosphate buffer, pH 7.8). For penetration of reagents it is essential to remove the vitelline membrane. The membrane is particularly difficult to remove during gastrulation, so embryos are pretreated with proteinase K (5 μg/ml) at the blastula stage for 5–10 minutes to loosen the membrane. Monitor the digestion under a microscope; when the membrane is just lifting off the embryos, or if any embryo breaks out of the membrane, wash immediately in one-third strength Ringer's or similar buffer. All embryos can be treated with proteinase K at the blastula stage. Some batches of embryos display difficulty in closing the blastopore in a loose membrane, so place in 2% Ficoll in one-third strength Ringer's during gastrulation. Embryos that develop in a loose membrane also develop more detailed surface morphology during neurula and tailbud stages, so staging must be performed with this in mind.

Just before fixation remove the vitelline envelope with forceps. Blastula through tailbud stages develop background staining in the blastocoel and more intensely in the archenteron. With a needle or eyebrow knife, poke a hole in the blastocoel (of blastula stages) and the archenteron of neurula stages to allow the contents to wash out.

Select embryos and transfer them to a 5-ml screw cap glass vial (Fisher, Fairlawn, NJ, 3338B) filled with distilled water (to wash). The vial should be marked with a diamond pencil. After the embryos settle, remove most of the water and add MEMFA buffer (0.1 M MOPS, pH 7.4, 2 mM EGTA, 1 mM $MgSO_4$, and 3.7% formaldehyde) to the brim. An autoclaved stock of salts in a 10 × concentrated solution can be stored. Formaldehyde should be added freshly. Paraformaldehyde can also be used (MEMPFA), especially if fresh formaldehyde is not available. Rotate the vial end over end for 1 to 2 hours at room temperature (e.g., 8 rpm, Labquake rotator; Labindustries, Inc. Berkeley, CA). If many embryos are being prepared, use a larger vessel (e.g., a large scintillation vial) to prevent clumping. Early-stage embryos will flatten considerably unless kept in suspension during the early part of the fixation.

Replace MEMFA with methanol. Replace the buffer with methanol. After a few minutes freeze the embryos and store at −20°C (embryos are stable for months).

Hybridization Procedure

For the hybridization procedure use enough embryos to accommodate two time points in the chromogenic reaction (e.g., 4 hours and overnight) and to include at least 2 embryos of each stage per time point. At least 30 embryos (and probably many more) can be processed in a single vial. A sense probe control is necessary, particularly to judge the rate of appearance of background staining in the chromogenic reaction.

Hybridization

1. Set up the vials with the various stages that will be hybridized. Do not try to transfer them directly from the vial with a pipette; this tends to lead to breakage. To manipulate embryos (both now and later) fill the vial with liquid, make sure embryos are not stuck to the glass, and invert the tube into a dish of liquid. Under the dissecting microscope use a large-diameter Pasteur pipette to transfer embryos to the vial.

2. Rehydrate embryos by successive 5-minute washes in methanol, 75% methanol–25% water, 50% methanol–50% water, 25% methanol–75% PBS–Tween [containing 1× phosphate-buffered saline (PBS) and 0.1% Tween-20], and 100% PBS–Tween, then wash 3 times for 5 minutes each in PBS–Tween.

3. If proteinase K is to be omitted skip to Step 4; otherwise, incubate embryos at room temperature in 1 ml of 10 μg/ml proteinase K in a rack placed on a nutator (Becton Dickinson, Sparks, MD). Observe the embryos carefully and stop if signs of damage occur. The reaction time must be adjusted for different batches of protease, but 30 minutes is a good guideline for gastrula stages. Proteinase K is not absolutely required for staining, but it does increase the sensitivity. Embryos should still be strong enough that almost all

the buffer can be removed during wash steps without surface tension destroying them.

4. Rinse the embryos twice for 5 minutes in 0.1 M triethanolamine, pH 7–8 (Sigma, St. Louis, MO, T-1502; 5 ml per wash, vials set horizontally on a nutator).

5. To the rinsed embryos in 5 ml of triethanolamine add 12.5 μl acetic anhydride (Sigma). Rock the tubes horizontally on a nutator and after 5 minutes add another 12.5 μl acetic anhydride. Rock for 5 minutes.

6. Wash the embryos twice for 5 minutes in 5 ml PBS–Tween on a nutator. (If no proteinase K was used, skip to Step 8.)

7. Refix for 20 minutes with 4% paraformaldehyde (4 ml PBS–Tween plus 1 ml of 20% paraformaldehyde: to make 20% paraformaldehyde in water, neutralize the cloudy solution with NaOH and heat at 65°C with shaking until clear; store cold). Wash 5 times for 5 minutes each time in PBS–Tween.

8. Remove all but about 1 ml of the PBS–Tween and add 250 μl hybridization buffer. Hybridization buffer has the following composition:

Final concentration	For 100 ml
50% Formamide	50 ml
5× SSC	25 ml of 20× stock
1 mg/ml *Torula* RNA	2 ml of 50 mg/ml stock
100 μg/ml Heparin (Sigma H-3125)	10 mg
1× Denhart's	1 ml of 100× stock
0.1% Tween 20 (Sigma P-1379)	0.1 g
0.1% CHAPS (Sigma C3023)	0.1 g
5 mM EDTA	2.5 ml of 0.2 M EDTA

The formamide is redistilled (EM Sciences, Gibbstown, NJ, FX0421-3 or BRL, Gaithersburg, MD, 5515UB). *Torula* RNA (Calbiochem, San Diego, CA, 55711) is made up in DEP-treated water and frozen in 1-ml aliquots. 100× Denhart's contains 2% bovine serum albumin (BSA) (ICN, Costa Mesa, CA, 810661), 2% polyvinyl pyrrolidone (PVP-40, Sigma), 2% Ficoll 400 (Pharmacia, Piscataway, NJ). To make 100×, add a small amount of water to the powder, make a slurry, then dilute.

9. Once the embryos have settled through the dense buffer, remove all the buffer and replace with 0.5 ml hybridization buffer. Place the embryos at 60°C in a shaking water bath for 10 minutes.

10. Replace hybridization buffer. Prehybridize for at least 6 hours at 60°C.

11. Replace with 0.5 ml probe solution (hybridization solution with 1 μg/ml probe). Hybridize overnight at 60°C. Remove the probe and keep. Probe can be recycled up to 2 times.

Washing

12. Replace with hybridization buffer at 60°C for 10 minutes, then 50% hybridization buffer–50% 2× SSC containing 0.3% CHAPS at 60°C for 10 minutes, then 25% hybridization buffer–75% 2× SSC containing 0.3% CHAPS for 10 minutes (start cooling in a water bath to 37°). Wash the embryos 2 times for 20 minutes with 2× SSC containing 0.3% CHAPS.

13. Next wash in 2× SSC, 0.3% CHAPS with RNase A (20 μg/ml), RNase T1 (10 units/ml) at 37°C for 30 minutes. To prepare RNase A (Sigma R-5000), dissolve at 10 mg/ml in Tris–EDTA buffer (TE) and boil for 10 minutes. To prepare RNase T1 (Sigma R-8251), dissolve at 10,000 units/ml in 0.1 M sodium acetate, pH 5.5, and boil. Store the RNases frozen in aliquots. Keep a working stock at 4°C; avoid repeated freezing and thawing.

14. Wash the embryos once in 2× SSC containing 0.3% CHAPS for 10 minutes at room temperature (all wash volumes are 5 ml).

15. Wash twice in 0.2× SSC containing 0.3% CHAPS for 30 minutes at 60°C.

16. Transfer the embryos back to PBS–Tween containing 0.3% CHAPS twice for 10 minutes, each at 60°C.

17. Wash 1 time for 10 minutes in PBS–Tween (without CHAPS) at room temperature.

Antibody Incubation

18. Replace the PBS–Tween with PBT (PBS plus 2 mg/ml BSA and 0.1% Triton X-100). Fill the glass vial and rock horizontally at room temperature for 15 minutes.

19. Remove the PBT and replace with 500 μl of fresh PBT plus 20% heat-treated lamb serum [lamb serum (Gibco, Grand Island, NY) pretreated at 55°C for 30 minutes]. Rock the vial vertically at room temperature for 1 hour.

20. Replace this solution with a fresh solution of PBT plus 20% serum containing a 1:2000 dilution of the affinity-purified sheep anti digoxigenin antibody coupled to alkaline phosphatase (Boehringer 1093 274). Rock vertically overnight at 4°C.

21. To remove excess antibody, wash the embryos at least 5 times, 1 hour each at room temperature, with PBT. One of the washes can be overnight at 4°C.

22. For the chromogenic reaction with alkaline phosphatase, wash the embryos twice, 5 minutes each at room temperature, with alkaline phosphatase buffer 100 mM Tris, pH 9.5, 50 mM MgCl$_2$, 100 mM NaCl, 0.1% Tween 20, 5 mM levamisol (an inhibitor of endogenous phosphatase, add freshly just before use)].

23. Replace the last wash with the same solution containing 4.5 μl of NBT (Nitro blue tetrazolium; 75 mg/ml in 70% dimethyl formamide) and 3.5 μl of BCIP (5-bromo-4-chloro-3-indolylphosphate; 50 mg/ml in 100% dimethyl formamide) per milliliter of buffer. Use 0.5–2 ml/tube and incubate at room temperature. If a slight precipitate of $Mg(OH)_2$ forms, do not worry. There is no need to keep the reaction dark. The color reaction is visible from 5 minutes to 1 day. The embryos can be examined periodically by tipping the vial and placing it under a dissecting microscope.

24. After staining is apparent, the embryos can be tipped into a dish of alkaline phosphatase buffer. Lightly stained embryos can be removed and fixed, and the rest of the embryos replaced in fresh NBT/BCIP solution for more intense staining.

25. Stop the chromogenic reaction when satisfied with the signal and background by replacing the alkaline phosphatase buffer with MEMFA (see above). After a wash with MEMFA (which stabilizes the stain), the embryos can be refixed for an extended time (e.g., overnight). I have also used Bouin's to refix, which gives the embryos an interesting yellow counterstain but also makes them less transparent. Judging by the amount of stain that leaches out of the embryos over time, Bouin's stabilized the stain better. (Bouin's fix contains 1 g picric acid in 70 ml water, 25 ml of 37% formaldehyde, and 5 ml glacial acetic acid.)

For mounting in benzyl benzoate/benzyl alcohol it is essential that the blue stain be reasonably intense, or the embryos should be refixed. Faint stains will dissolve on clearing.

If pigmented embryos are used, the embryos can be bleached at this point with only a small loss of signal. Embryos should be incubated in 70% methanol, 30% H_2O_2 solution (10% final H_2O_2) under strong fluorescent light for several hours.

26. Dehydrate the embryos for 5 minutes in methanol and mount in 2:1 benzyl benzoate/benzyl alcohol (BB/BA). Transfer the embryos to a depression slide (thick hanging drop slide/concavity slide; 3–5 mm depth of well) which accommodates about 0.5 ml BB/BA. Remove most of the methanol and add the BB/BA. After embryos settle, change the BB/BA. Embryos can be observed and pushed around with a hair loop, but the stain will dissolve gradually if embryos are left exposed to the air. The stain is stable for months if the embryos are sealed.

Suitable mounting chambers can be made using Sylgard 184 elastomer (0.5 kg size is available from Dow Corning distributors; call 800 248 2481 for information). Make a foil boat containing 16 microscope slides pushed together, preferably on a microscope slide warmer at 60°C. Fold up the sides of the foil to prevent loss of elastomer. Prepare about 40 ml of the elastomer, pour it onto the slides, and cure overnight. Cut the slides apart and cut a rectangular well with a razor blade, as well as an end for marking the slide. The elastomer is more easily removed while fairly fresh. By adjusting the volume of elastomer, different thicknesses of slide can be made to accommodate different thicknesses of specimen. Use a diamond pencil for marking the slide since BB/BA removes ink. Transfer the embryos from BB/BA to the well and remove excess BB/BA to leave only a small convex meniscus. Add a coverslip. Tip the slide and blot excess BB/BA from the edge of the coverslip. The coverslip should stay in place, but after several days the BB/BA may need to be replaced. Take care with BB/BA. It is an irritant and is said to dissolve microscope glue. It can be cleaned up with 95% ethanol.

Observations

There are a number of changes from the procedure reported by Hemmati-Brivanlou et al. (1990) that have greatly increased the signal-to-noise ratio. The changes have not only reduced the background, allowing long staining reactions, but have also increased the rate of appearance of signal. There are more blocking reagents in the hybridization buffer, and the prehybridization is carried out for a longer time. The posthybridization washes are also more extensive.

The concentration of the probe was varied over a 25-fold range (below the 5-10 μg/ml originally recommended). This has a small effect on background. The concentration of the probe could be reduced even further than in the present protocol without appreciable loss of signal.

A rationale for using CHAPS detergent in the buffers is to prevent nonspecific sticking of the probe via the digoxigenin moiety. The cardiac glycosides are all quite insoluble, and the CHAPS molecule, which resembles part of the digoxigenin, may help maintain the digoxigenin in solution, as well as blocking hydrophobic binding sites. Deoxycholic acid was also tried, and, although this appeared to work, it precipitates some component of the hybridization buffer. Digitoxin (which resembles digoxigenin in structure but has a slight cross-reaction with the antibody) was also tried, and it eliminated the hybridization signal.

Acetic anhydride was incorporated to block positively charged groups (Hayashi et al., 1978). Nonetheless, neurula stages often stain intensely in the archenteron. Blastula/gastrula stages sometimes stain a little in the blastocoel. Staining does not always occur, and if the archenteron is punctured prior to fixation, the problem is reduced. This staining is not dependent on having probe in the reaction, so presumably it is due to nonspecific sticking of antibody to archenteron glycoproteins, or to an extremely tough alkaline phosphatase.

I have tested various reagents to inhibit the endogenous phosphatases. Acetic anhydride (at pH 8, 1 hour), N-ethylmaleimide, and hydrogen peroxide were all ineffective. Both diethyl pyrocarbonate and 5 mM EDTA (65°C, 1 hour) were effective. EDTA is included in the hybridization mix. DEPC treatment can be carried out at step 15, but gives little further improvement.

For superior morphology in sectioned embryos, it may be better to fix the embryos in paraformaldehyde (MEMPFA) and avoid the proteinase K treatment. Fixation in paraformaldehyde has little effect on the intensity of staining, but omitting the proteinase K does decrease the sensitivity; En-2 transcripts in the brain are still easily detected without proteinase K, but the branchial arch transcripts (see Hemmati-Brivanlou et al., 1991) are only detected well with proteinase K. Embryos can be embedded in Paraplast or plastic and sectioned without losing the stain.

An alternative to NBT/BCIP is the Vectastain II detection kit (see Dent et al., 1989), although the contrast is not so good as with NBT/BCIP. If the signal is too intense, consider using the horseradish peroxidase (HRP)-linked antidigoxygenin antibody. This is inherently much less sensitive than alkaline phosphatase-linked antibody, but it gives a good brown product that is very stable to clearing agents. In principle, HRP products could also be intensified for electron microscopy (Tomlinson et al., 1987).

Short Protocol

Rehydrate embryos with PBS-Tween (1 × PBS plus 0.1% Tween-20) by conducting successive 5-minute washes in methanol, 75% methanol–25% water, 50% methanol–50% water, 25% methanol–75% PBS-Tween, and 100% PBS-Tween. Then wash 3 times, for 5 minutes each, in PBS-Tween.

Incubate embryos at room temperature in 1 ml of 10 μg/ml proteinase K for about 30 minutes. Rinse 2 times, 5 minutes each, in 0.1 M triethanolamine at pH 7–8. Add 12.5 μl acetic anhydride. After 5 minutes repeat acetic anhydride addition. Wash 2 times, 5 minutes each, in PBS-Tween.

Refix for 20 minutes with 4% paraformaldehyde (4 ml PBS-Tween plus 1 ml of 20% paraformaldehyde). Wash 5 times, 5 minutes each, in PBS-Tween. Remove all but about 1 ml PBS-Tween.

Add 250 μl hybridization buffer. Once embryos have settled, replace with 0.5 ml hybridization buffer. Place embryos at 60°C in a shaking water bath for 10 minutes. Replace hybridization buffer. Prehybridize for 6 hours at 60°C.

Replace with 0.5 ml probe solution (1 µg/ml probe). Hybridize overnight at 60°C. Remove probe and keep.

Replace probe with hybridization buffer at 60°C for 10 minutes, then, with 50% hybridization buffer–50% 2× SSC with 0.3% CHAPS at 60°C for 10 minutes, then 25% hybridization buffer–75% 2× SSC with 0.3% CHAPS for 10 minutes (water bath can be allowed to cool to 37°C during this phase). Wash 2 times, 20 minutes each, with 2× SSC with 0.3% CHAPS.

Wash in 2× SSC with RNase A (20 µg/ml) or RNase T1 (10 units/ml) at 37°C, 30 minutes. Wash once in 2× SSC with 0.3% CHAPS for 10 minutes (to 1 hour) room temperature, then wash twice in 0.2× SSC with 0.3% CHAPS for 30 minutes at 60°C. Transfer back to PBS–Tween with 0.3% CHAPS and wash 2 times, 10 minutes each, at 60°C. Wash once for 10 minutes in PBS–Tween (without CHAPS) at room temperature.

Transfer to PBT (PBS plus 2 mg/ml BSA and 0.1% Triton X-100) by replacing the PBS–Tween with PBT, 15 minutes. Replace with 500 µl of fresh PBT plus 20% lamb serum. "Nutate" vertically at room temperature for 1 hr.

Replace with a fresh solution of PBT plus 20% serum containing a 1:2000 dilution of the antidigoxigenin alkaline phosphatase antibody. Nutate vertically overnight at 4°C. To remove excess antibody, wash embryos at least 5 times, 1 hour each, with PBT (fill up vials and nutate horizontally). Wash twice, 5 minutes each, at room temperature with alkaline phosphate buffer.

Replace the last wash with the same solution containing 4.5 µl of NBT (75 mg/ml in 70% dimethylformamide) and 3.5 µl of BCIP (50 mg/ml in 100% dimethylformamide) per milliliter of buffer. Use 0.5–2 ml/tube and let nutate at room temperature. The color reaction is visible from 5 minutes to 1 day.

Stop the chromogenic reaction when satisfied with signal and background by replacing the solution with MEMFA. After a wash with MEMFA, refix with MEMFA overnight.

Dehydrate for 5 minutes in methanol and amount in 2:1 benzyl benzoate/benzyl alcohol.

Acknowledgments

Many people made suggestions, but I am especially grateful to Ali Hemmati-Brivanlou. B. G. Herrmann (Tübingen) and R. Conlon (Toronto) kindly provided alternate protocols prior to publication. This work was supported by the National Institutes of Health.

References

Dent, J. A., Polson A. G., and Klymkowsky M. W. (1989). A whole-mount immunocytochemical analysis of the expression of the intermediate filament protein vimentin in *Xenopus*. *Development* **105,** 61–74.

Hayashi, S., Gillam, I. C., Delaney, A. D., and Tener, G. M. (1978). Acetylation of chromosome squashes of *Drosophila melanogaster* decreases the background in autoradiographs from hybridization with ^{125}I-labelled RNA. *J. Histochem. Cytochem.* **36,** 677–679.

Hemmati-Brivanlou, A., and Harland, R. M. (1989). Expression of an *engrailed*-related protein is induced in the anterior neural ectoderm of early *Xenopus* embryos. *Development* **106,** 611–617.

Hemmati-Brivanlou, A., Frank, D., Bolce, M. B., Brown, B. D., Sive, H. L., and Harland, R. M. (1990). Localization of specific mRNAs in *Xenopus* embryos by whole mount *in situ* hybridization. *Development* **110,** 325–330.

Hemmati-Brivanlou, A., de la Torré, J. R., Holt, C., and Harland, R. M. (1991). Cephalic expression and molecular characterization of *Xenopus En-2*. *Development* **111,** 715–724.

Kintner, C. R., and Melton, D. A. (1987). Expression of *Xenopus N-CAM* RNA in ectoderm is an early response to neural induction. *Development* **99,** 311–325.

Melton, D. A., Kreig, P. A., Rebagliati, M. R., Maniatis, T., Zinn, K., and Green, M. R. (1985). Efficient *in vitro* synthesis of biologically active RNA and RNA hybridization probes from plasmids containing a bacteriophage SP6 promoter. *Nucleic Acids Res.* **12,** 7035–7056.

Sambrook, J., Fritsch, E. F., and Maniatis, T. (1989). "Molecular Cloning: A Laboratory Manual," 2nd ed. Cold Spring Harbor Laboratory Press, Cold Spring Harbor, New York.

Tautz, D., and Pfeifle, C. (1989). A non-radioactive *in situ* hybridization method for the localization of specific RNAs in *Drosophila* embryos reveals translational control of the segmentation gene *hunchback*. *Chromosoma* **98,** 81–85.

Tomlinson, A., Bowtell, D. D. L., Hafen, E., and Rubin, G. M. (1987). Localization of the *sevenless* protein, a putative receptor for positional information, in the eye imaginal disc of *Drosophila*. *Cell*, **51,** 143–150.

INDEX

A

Actins, 26–27
Activins, 320
Adult frogs:
 breeding, 16
 commercial sources, 15
 diseases, 11–12
 feeding, 9–10
 fungal infections, 13
 genetics, 17, 19–34
 genome organization, 21–24
 laboratory sources, 15–16
 maintenance, 15
 marking, 10–11
 nematodes, 12
 polyploidy, 20
 population density, 10
 recycling and disposal, 16
 space requirements, 15–16
Aeromonas hydrophila, 12
Albumin, 26–27
Allotriploidy, 27–28
Ambystoma mexicanum, embryos, 102
Ambystoma spp., 118
Amphibian eggs:
 ECM, composition, 232
 egg envelope, 232
 jelly coat, 232
 perivitelline space, 232
Aneuploidy, 39
Animal cap explants:
 elongation, 316
 size, 312–314
 time of exposure, 315
Anterior neural plate, specification stage, 334
Antibiotics, 12
 for instrument sterilization, 107–108
Antibodies:
 binding, visualization, 426–429
 injection, 376, 381

 preparation, 369–370
Autoradiography, tissue sections, 445–456, 454–455
Axes, embryo, 62–64
Axis determination, and cell differentiation, 322–323
Axolotl, 334

B

Baker's yeast, 8
Barth's medium, 118
Beads, 10
Beef liver, 9
Bisbenzimide, stain, 398–400
Bivalents, number, 21
Blastocoel roof, whole-mount preparation, for fibronectin, 533–535
Blastopore lip:
 dorsal, source of neural induction signals, 332–333, 336
 transplant, 329–331
Blastula:
 cleavage and structure, 65–66
 gastrulation, 275–277
 mesoderm formation, 273–275
 microsurgery tools, 102
 stages, fate maps, 67
Bleaching, in whole-mount imaging, 421–422
Bombina orientalis, embryos, 102
Bombinator spp., 333
Bone, clearing agents, 421
Bone cartilage staining:
 bleaching, 422
 fixation, 426
Bone and cartilage staining:
 whole-mount, 430–433
 double staining, 437
Bone meal, 8

698   INDEX

Bottle cells:
  behavior, 82–84
  formation, 70–71, 76, 80
  misconceptions, 78
  pigmentation, 84
Bouin's fixative, 392
Branding, 10
Bright-field microscopy, tissue preparation, 413
Bufo japonicus, egg envelope, 232

C

Caenorhabditis elegans, nematode, 285–286
Calcium, microinjection, 256
Calmodulin, 27
Capillaria xenopodis, nematode, 13
Carnoy's fixative, 392–393
Cartilage, clearing agents, 421
cDNA:
  in vitro transcription, by RNA polymerases, 173
  restriction endonuclease sites, 169
  subcloning, 169
  transcription template, 173, 175
Cell behavior, investigation, 82–101
Cell development, microscopy, 108–109
Cell differentiation, and axis determination, 322–323
Cell intercalations, 84–96
Cell labeling, for tracing deep-cell intercalation, 93–94
Cell-lineage specificity, of labeling media, 294
Cell-matrix adhesion, in *in vitro* reconstitution, 537–539
Cell proliferation, dilution of clone markers, 295
Cellular signals, interactions, 332
Cephalin, for neural induction, 331
Ceratophrys ornata, embryos, 102
Chiasmata, number, 21
Chromosomes, 20
  histone genes, 24
  homeobox genes, 24
  lampbrush, 22
  mapping, 27–28
  mitotic, 22
  morphology, 22
  number, 21

sex, 22
Citrobacter spp., 12
Class II gene regulation, analysis, 347–364
Clearing agents, in whole-mount imaging, 420–421
Cleavage, blastula, 65–66
Clones:
  analysis, lineage tracer microinjection, 285–287
  markers,
    dilution by cell proliferation, 295
    fluorescent dextran, 285–295
Coelomic egg, 232
Coelomic envelope, from coelomic eggs, 238–241
Collagenase, oocyte treatment, 49–50
Colloidal gold, for visualization of bound antibodies, 426
Column chromatography, fluorescent dextran, 291
Commercial chow diets, 9
Confocal microscope, for visualization of bound antibodies, 426
Conserved duplications, estimation, 24–27
Convergence and extension, in sandwich explants, 85–88
Copper-phenanthroline, for DNA digestion, 353, 356, 359–361
Cyanogen bromide activation, of dextrans, 290
Cysteine, for vitelline membrane removal, 221
Cytokeratin, 27
Cytological localization, in situ hybridization, 28
Cytoskeletal proteins, heat extraction, 140–142
Cytoskeleton, detergent-resistant, preparation, 265–268
Cytosolic proteins, from oocytes, 142–143
Cytotoxicity, of fluorescent dextrans, 293–294

D

Deadenylation, mRNAs, 170–171
Deep cell motility, in morphogenesis, analysis, 89
Defolliculation:
  manual, 50
  microinjection, 50

Density gradient centrifugation, for organelle isolation, 146
Deuterium oxide, dorsoanterior embryo effects, 273, 280
Dextrans, fluorescent, 288
Dichlorotriazinyl aminofluorescein dihydrochloride, 289
Diet, adaptation, 9–10
DiI labeling, in microscopy, 108
Dimethyl sulfate, for DNA digestion, 361–362
Diseases, 11–13
Dispersal media, for lampbrush chromosomes, 157–158, 163
Dissection, for fractionation, 135–137, 260–265
Dissociated embryonic cells, in mesoderm induction, 323–324
DNA:
  analysis, 475–476
  homologous recombination, in oocytes, 476–479
  incubation, 474–475
  injection, 469
    distribution and expression, 377–379
    early experiments, 350–353
  linear, 375
  nonhomologous recombination, in eggs and embryos, 479
  recombination,
    comparisons with other organisms, 479–480
    homologous, 468
    homology-independent, 468
    substrates, 471–472
  recombination and repair, 467–484
  recovery, 475
  repair,
    abasic sites, 481–482
    mismatching, 482–483
    ultraviolet damage, 480–481
    X-ray damage, 481
  repair substrates, 472–474
  synthesis, 369
Dominant negative proteins, expression, 382
Dorsal ectoderm:
  induction signals, 336
  neural bias, 335–336
Dorsal mesoderm, invaginated, induction signals, 337–338
Dorsoanterior index, for phenotypes, 278

Double staining, for embryos, 437
Drosophila spp., 379
Dumont watchmaker's forceps, 103

E

Early diplotene, meiotic stage, 150
ECM:
  embryo protection, 232
  isolation, 232–246
  localization, immunocytochemical methods, 533–535
  role in morphogenetic cell movements, 527
Ectoderm:
  induction, development stage, 333–334
  neural bias, 335–337
  neural competence, 332–333
  and induction, 335
  ventral, induction, 330
Egg activation, calcium ion, 251–252
Egg envelope:
  conversion, molecular mechanisms, 245
  glycoprotein composition, 237
  isolation,
    by sieving, 235–236
    radioiodination, 242
  macromolecular composition, 232
  permeability, 232
  structure and function, 243–246
Egg extracts, whole-cell, preparation, 469–470
Eggs:
  cytoskeletons, 265–268
  DNA recombination and repair, 467–484
  harvesting, 14
  nuclear transplantation, 299–309
  oviposited, 233
  polarity, 62–64
  preparation,
    for nuclear transplantation, 306
    for signal agonists and antagonists, 254
  spatial fractionation, 260–265
  tissue sections, in situ hybridization, preparation, 446–448
Electron microscopy, 414
  tissue preparation, 404–407
Electrophoresis:
  enzyme loci, 25
  isoenzymes, 24–26
Electrophorus spp., ion channels, 492

Electrophysiological analysis, ion channels, 503
Embedding:
    in paraffin, 393–394
    in plastic, 395–396
Embryo:
    axes, 62–64
    axis-deficient, from ultraviolet irradiation, 278–279
    cell experimental design and context-dependent function, 101
    class II gene transfer, 348–349
    coinjected mRNA, 375
    culture, 216–217
    cytoskeleton, 265–268
    death, 383
    dejelling, 391
    development, maternal control, 214–215
    DNA injection, 348–349
    dorsal side, 65
    dorsoanterior deficient, agents, 281–282
    dorsoanterior enhanced, agents, 280
    double staining, 437
    early development, 61–109
    gene function, assays, 367–384
    grafting, 104–105
    histological preparation, 389–414
    host transfer fertilization, 216
    in vitro copper-phenanthroline digestion, 360–361
    in vitro dimethyl sulfate treatment, 361–362
    in vivo copper-phenanthroline digestion, 359–360
    in vivo dimethyl sulfate treatment, 361
    isolation, 216–217
    kinked, 382
    mesoderm induction, conditions, 314–315
    microsurgical methodology and techniques, 102–108
    "mutant", production, 215
    oocyte germinal vesicle insertion effects, 277
    orientation, 65
    pigmentation, 63
    polarity, 62–64
    removal of jelly coats and vitelline envelopes, 103–104
    retinoic acid effects, 276–277
    solutions and healing, 106–107
    spatial fractionation, 260–265
    sperm penetration, 215
    stages of formation, 80–82
    staining, 286, 374–375
    supernumary axis, from lithium, 280
    tissue sections,
        polyacrylamide embedding, 448–449
        in situ hybridization, preparation, 448–449
    transfer to host, 215
    vitelline membrane removal, 215
    whole-mount immunostained, sectioning, 435
    whole-mount tissue,
        antibody preabsorption, 458
        immunohistochemistry, 458–459
        mounting, 459
    Xenopus, model for cell-ECM interaction study, 527–539
Embryogenesis, cell lineage, 285–287
Endoderm induction, 324
Enriched nerve-muscle cocultures, preparation, 517
Enucleation, under ultraviolet light, 303, 307
Enzyme-based secondary reagents, disadvantages, 427
Enzyme gene expression, 25–26
Eosin, stain, 396–397
Epiboly, animal cap, 100–101
Epi-illumination, of explants, 108
Epithelial cells, superficial, 79–80
Equipment, for *Xenopus* culture, 5–11
Escherichia coli, 369
Ethanol, in dissection, 262–263
Eukaryotes, nuclear transplantation, 300
Explants, 105–106
    development diagram, 86
External application, signal agonists and antagonists, 253
Extracellular matrix, see: ECM
Eyebrow hair knife, 102

F

Fate maps:
    embryos, 312
    fluorescent dextrans, 288
    gastrulation and neurulation, 68–74
    for prospective tissues, 67–68
Female host, oocyte transfer, 217
Fertilization envelope, from fertilized eggs, 236–238

Fertilized egg:
  axis perturbations, sensitive periods, 272–277
  dorsal-ventral polarity, 273
  effects on development, 273
  mesoderm formation, 273–275
  modified egg envelope, 232
Fibroblast growth factors, 27, 321
Fibronectins:
  composition, 530
  in ECMs, 528
  from whole-mount preparation of blastocoel roof, 533–535
  in plasma, 529–530
Fixation, in whole-mount imaging, 422–426
Fixatives:
  for light microscopy, 392–393
  in whole-mount imaging,
    alcohol-based, 422–423
    aldehyde-based, 423
Fluorescein dextran amine, in microscopy, 108
Fluorescence microscope, 295
  tissue preparation, 413–414
Fluorescent dextran:
  clonal marker, 285–295
  column chromatography, 291
  conjugation, 289
  free amino groups, assay, 291
  free-dye contamination, assay, 291
  lysinated, fixation, 293
  lysine substitution, assay, 291–292
  preparation, 288–289
  tracer microinjection, 292–293
Fluorescent markers, 287–288
Fluorescent secondary reagents, disadvantages, 426
Fluorophore isothiocyanate, 289
Fluorophores:
  degree of substitution, 291
  reaction with lysine-dextran, 290–291
Freon, protein extraction, 138–139
Frog culture:
  aeration, 8
  in brackish water, 3
  feeding, 8–10
  laboratory, 3–4
  light requirements, 7–8
  media, 51
  pH requirements, 4
  tadpoles, 3–4

temperature requirements, 7–8
water salinity, 7
water tanks, 5–6
Fungal infections, 13

G

ß-Galactosidase, detection, 371–372
Gastrula:
  cleavage and blastula stages, 62–68
  microsurgery tools, 102
  origins, 62–68
  staging, 80–82
  structure, 75–78
Gastrulation, 68–82, 133, 334, 349
  cell behavior, 82–101
  cell intercalations, 84–96
  misconceptions, 78–79
  morphogenesis, 275–277
  movements, 68–74
  neural induction, 275–277
Gene expression, using *in situ* hybridization, 444
Gene function:
  antibodies, 369–370
  control injections, 370–371
  in developing embryos, assays, 367–384
  microinjections, 371
  synthetic mRNA injection, 368–369
Gene mapping, 27–30
  by polymerase chain reaction footprinting, 353–363
Gene mutagenesis, for function, 363–364
Genes:
  analysis, 20
  conserved duplications, estimation, 24–27
  structure, molecular analysis, 26–27
Gene selection, for transfer, 349–350
Genetic markers:
  in nuclear transplantation, 305–306
  in *Xenopus*, 300
Gene transcripts, density change, from *in situ* hybridization, 444
Genome:
  footprinting, using Taq reactions, 363
  gene clusters, 23–24
  manipulations, 36
  nuclear, 20
  organization, 21–24

Genome *(continued)*
  replication, 21
  sequences, 23
  tetraploidy, 214
Germinal vesicle, 154–157
  failure to disperse, 159
  removal, procedure, 261–262
Germinal vesicle sap, for macrocephalic embryo, 282–283
Glass needles, 102
Globins, 26–27
Glycoproteins, in jelly coat layer, 232–233
Gonadotropins:
  in oocyte maturation, 56
  in ovulation, 13–14
Grafts, healing, 105
Gynogenesis, 36, 40–41

H

Hair loop, 102–103
Hatching, 232
Heat extraction, cytoskeletal proteins, 140–142
Hematoxylin, stain, 396–397
Hemoglobin, 26–27
Heparin:
  for dorsoanterior deficient embryo, 281–282
  effect on gastrulation, 276
Heterozygosity, 40–41
Homeobox gene, 27, 378
  as antibody, 318–319
Homozygosity:
  error probability, 40–43
  experimental, 35–36
  methodology, 36–43
  primary, 38–39
  statistics, 40–43
Homozygous diploids, 37–39
  cloning, 39–40
  defining, 40–43
  genetics, 39
  primary, 38–40
Horseradish peroxidase, 286–287
Hybridization, *in situ,* 444–462
Hyperdiploid hybrids, in mapping, 27–28
Hypothermia, as anesthetic, 48

I

Immunocytochemistry:
  combined with Alcian blue staining, whole-mount, 433
  cortical whole-mount, 427–428
  total whole-mount, 428
  whole-mount, for antibodies, 425
Immunoelectron microscopy, tissue preparation, 407–412
Immunofluorescence:
  lampbrush chromosomes, 160–164
  mounting medium, 164
Immunogold conjugates, for visualization of bound antibodies, 426
Immunohistochemistry:
  light microscope level, 401–403
  staining, 402–403
Immunoperoxidase staining:
  effectiveness, 437
  whole-mount, 431
Inbreeding, 35–36
Infusoria, 8
Injection artifacts, 382–383
In situ hybridization, 444–462
  protocol evaluation, 459–462
Instruments, sterilization, 107–108
Integrins, 27
Intracellular signaling:
  analysis of events, 259–260
  in oocytes, 252
  role of free calcium, 256–258
  role of pH, 258
  sequence determination, 258
Invaginated dorsal mesoderm, neural induction, 337–338
Iodination, protein oxidation reactions, 126
Ion channels:
  electrophysiological analysis, 502–507
  expression,
    biochemical analysis, 499–502
    by mRNA injection into oocytes, 487–507
    electrophysiological analysis, 502–507
    source of RNA, 493
    from cDNA clone, for RNA transcription, 496–498
    immunoprecipitation of proteins, 500–501

ligand binding assay, 501–502
Iontophoresis, using fluorescent dextrans, 293
Isoenzymes, electrophoresis, 24–26
Isolation media, for lampbrush chromosomes, 157–158, 163

J

Jelly coat layer:
  function, 243
  labeling, 235
  macromolecular composition, 232, 243
  preparation, 234–235
  removal, 103–104
  solubilization, 235

K

Karnovsky fixative, modified, 390
Karyotype, 21–22

L

Lampbrush chromosomes, 22, 150–165
  dispersal, 157–158, 164–165
  genomic content, 150
  germinal vesicle, 154–157
  Giemsa staining, 162
  gonadotropin, 152–153
  history, 151
  immunofluorescence, 160–161
  influence of magnesium, 159
  isolation media, 157–158
  materials preparation, 164–165
  microscopy, 153
  morphology, 160
  number in *Xenopus*, 150
  preparation, 152–157
  protocols, 151–152
  in situ hybridization, 161–162
  slides, 164
  solution preparation, 162–164
  staining, 160
  tungsten needles, 165
  use, 159–162
Leptotene, meiotic stage, 150

Leupeptin, 141, 142
Ligand binding assay:
  of ion channels,
    surface receptors, 501–502
    total receptors, 502
Light energy, photoablation by fluorochromes, 294–295
Lineage tracing, fluorescent dextrans, 288
Linear sucrose density gradients, in mRNA isolation, 180–181
Linkage mapping, 28–30
Lithium:
  for dorsoanterior enhanced embryo, 280–281
  effect on blastula, 275
Liver powder, 9
Lysine-dextran, fluorophore reaction, 290–291

M

Macrocephalic embryo, from germinal vesicle sap, 282–283
Macropatch recording, in ion channel analysis, 503–506
Mapping:
  gene-centromere, 30
  hyperdiploid hybrids, 27–28
  linkage, 28–30
Marginal zones:
  convergence and extension, 84–96
  involuting and noninvoluting, 84–96
Meat, problems, 9
Mediolateral intercalation, 93–94
Meiosis, stages, 150
Mesodermal cells:
  adhesion to fibronectin substrates, 537
  adhesion in *in vitro*, 535–537
  conditioned substrates in *in vitro*, 536–537
  conditioning substrata, 100
  culture, 98
  dissociation, 536–537
  ECM-conditioned substrates, 536
  identification, 98–100
  isolation, 99
  migration, analysis, 96–100
Mesoderm induction, 311–325
  age, 312–314

704 INDEX

Mesoderm induction *(continued)*
  agents, 320
  antibody detection, 317
  assay, 312–315, 316–319
  axis determination, 322–324
  cell fate, 322–324
  dissociated embryonic cells, 323–324
  factors, 319–321
  gene regulation, 317–319
  markers, 316–319
  morphological criteria, 316–317
  time of exposure, 315
  tissue organization, 322–324
Microinjection:
  for oocyte labeling, 54–55
  signal agonists and antagonists, 252–253
Microscopy, in embryo development, 109
Microsurgery, tools, 102–103
Milk, 9
Mimea spp., 12
Mitosis, suppression, by pressure, 36–38
Mixed nerve-muscle culture, preparation, 515–516
Molecular signals, interactions, 332
Mosaic expression, 378–379
mRNA:
  analysis, 176–178
  "cold", synthesis, 174
  extraction, 175
  "hot", synthesis, 174–175
  injected,
    distribution, 374–375
    effects, 376
    overexpression, 372–375
    protein generation, 373
  labeled, 174
  nonpolysomal, isolation, 180
  poly(A) tail, 54
  polysomal,
    isolation, 180
    in microinjected oocytes, 178–182
    release, by EDTA, 181–182
  synthesis, 368–369
    from cDNA, 167–168, 173
  synthetic,
    capping, 170–171
    in vitro transcription, 173–175
    isolation, 173–175
    microinjection, 176–178
    polyadenylation, 170–171

  posttranscriptional polyadenylation, 175–176
  transcription, solutions and buffer preparation, 171–173
  transcription products, analysis, 176–178
  translation, in microinjected oocytes, 169
Muscle cells, tissue culture, for synaptic induction study, 511–525
Myoblast cultures, preparation, 523–524
MyoD, 27
  as antibody, 318–319
Myotomal muscle cell, synaptic specializations without nerves, 521
Myotomal muscle cultures:
  characteristics, 519–523
  preparation, 512–519
  solutions and equipment, 513–515
  substratum, 512–513
Myotube cultures, multinucleated, preparation, 523–524

N

Nematodes, 13
Nettle powder, 8
Neural cell adhesion molecule, 27
Neural cell adhesion molecules, 520
Neural competence, and transplantation, 332
Neural induction, 329–344
  aging of tissues in culture, 341
  by exogastrula, 340–341
  dissociation and reaggregation, 342
  ectoderm response to signals, 331–332
  edge-on grafting, 341
  einsteck method, 338–339
  initiation, 335–338
  Keller sandwiches, 342
  sandwich method, 339–340
  signals, 331, 343
Neural plate formation, signals, 335
Neural tube explants, preparation, 517–519
Neurectoderm, regional specification stage, 334
Neuromuscular junctions:
  in vitro development, 519–521
  in in vivo, comparison with cultures, 521–523
  postsynaptic specialization, 521
  synaptic potentials, 520

Neuron culture:
    preparation, 512–519
    solutions and equipment, 513–515
    substratum, 512–513
Neurons, tissue culture, for synaptic induction study, 511–525
Neurula:
    embryo, microsurgery tools, 102
    structure, 75–78
Neurulation, 68–82
    cell intercalations, 84–96
    intercalation, 94–96
    misconceptions, 78–79
    movements, 68–74, 94–96
Newt, lampbrush chromosomes, 151
Niu and Twitty solution, 118
Nonepithelial cells, deep, 79–80
Northern blot analysis, 197–198
Notophthalmus spp., germinal vesicle, 158
Nuclear transfer, in eggs, 306–309
Nuclear transplantation:
    1-nucleolate nuclear marker, 305
    albino mutant marker, 305–306
    donor cell preparation, 304
    equipment, 302–304
    media, 301–302
    methodology, 301–309
    in *Xenopus* eggs, 299–309
Nuclei, lampbrush chromosomes, 156
Nucleic acids, preparation from oocytes, 197
N. viridescens (newt), lampbrush chromosomes, 151

O

Oligodeoxynucleotides, see: Oligos
Oligos:
    antisense,
        controls, 371
        injection, 379–380
        for mRNA removal, 226–227
        in *Xenopus* oocytes, 186–195
    cleavage of snRNA, 200–204
    conversion to nucleotide form, 198–200
    deleterious effects, 194–195
    double-stranded, 198–200
    effects on snRNA degradation, 196
    function in *Xenopus* oocytes, 185–186
    influence on amphibian development, 188

inhibition of protein synthesis, 195
instability, 198–200
and membrane proteins, 190
microinjection into oocytes, 197
mixed-linkage, 189–190
modified, 189
mRNA specific cleavage induction, 186–187
nonspecific effects, 194–195
and pre-mRNA splicing, 196
in RNA degradation efficiency, 187
snRNA function, 192–194, 196–208
splicing inhibition in RNA, 204–208
synthesis, 185–186, 196–197, 370
in targeted degradation, 187–188
Oocytes:
    amphibian, growth characteristics, 150–151
    antibody injection, 381
    antisense experiments, 190
    antisense oligos injection, 379–380
    antisense RNA injection, 379
    biochemical fractionation, 133–147
    calcium microinjection, 256
    culture, 120–122
    cytoskeletons, 265–268
    defolliculation, 49
    "determinants", 214
    DNA recombination and repair, 467–484
    dominant negative proteins, 382
    effect of anesthetic, 48
    expression of ion channels, 488–490
    fertilization, 214–219, 222
    fractionation, 146–147
    freon extraction of proteins, 138–139
    functional channels, 490–492
    germinal vesicle, size, 150
    histological preparation, 389–414
    incubation, 51
    injection artifacts, 382–383
    intracellular signaling, 252
    in vitro culture, 117–130
    in vitro fertilization, 219–221
        isolation, 220–221
        materials, 220
        preparation, 220–221
    iodinated proteins, 125
    isolations, 45–57, 120, 135–137
    labeling, 54–56
    maintenance, media, 130
    manipulation, 217, 219

Oocytes (continued)
   maturation, 56–57, 190–192, 217–219
   meiotic resumption, 250–251
   membrane proteins, 190
   microinjection,
      foreign DNA, 223–226
      for intracellular signaling, 252–253
      mRNA, 168, 176–178, 180
      polysomal association of mRNAs, 178–182
      stability, 178–182
      total RNA isolation, 179
   mRNA content, 51–54
   mRNA identification, 46
   mRNA injection, for ion channel expression, 498–499
   and neurotransmitters, 190
   nonfunctional channels, from cDNA clones, 492
   nuclear extracts, preparation, 470–471
   nuclear injection, substrate design, 471–474
   nucleus breakdown, 56
   for ovarian envelope preparation, 241–242
   permeability, 55
   preparation, 196
      for signal agonists and antagonists, 253–254
   protein binding, 128
   protein fractionation, 142–143
   protein synthesis, 134
   protein uptake, 118, 126–129
   removal of theca, 119–120
   ribozyme injection, 380–381
   RNA extraction, 143–145
   separation, 48
   spatial fractionation, 260–265
   sperm penetration, 251–252
   stages, protein synthesis, 53–54
   tissue sections, in situ hybridization, preparation, 446–448
   ultraviolet irradiation, 222–223
   uninjected, electrophysiological responses, 490
   vitelline membrane, removal, 221
   vitellogenin uptake, 117–130
   Xenopus, as model organism, 483–484
Oogenesis, 45–57, 133–135
   asynchronous, 46
   overview, 46–47
   protein synthesis, 51–54

Open-faced explants:
   cell behavior analysis, 89–92
   development, 86
Organelles, isolation, by density gradient centrifugation, 146
Ovarian envelope, of oocyte, isolation, 137
Ovarian tissue:
   collagenase treatment, 48
   handling, 119–120
Oviposited egg, 232
Ovulation, induction, 233, 240–241

P

Pachytene, meiotic stage, 150
Paraffin, for embedding tissue, 393–395
Paraformaldehyde, in dissection, 263
Paraformaldehyde-glutaraldehyde in phosphate buffer, fixative, 393
Pepsin, for vitelline membrane removal, 221
Phenotypes:
   generation, 272–283
   perturbed, 274
   scoring, 277–278
Phorbol esters, role in intracellular signaling, 255
Photomicrography, tissue preparation, 412–413
Plasma, from adult frogs, isolation, 530–531
Plasma fibronectin:
   affinity purification, 531–532
   antisera preparation, 532–533
Plastic, for embedding tissue, 395–396
Platanna, 20
Pleurodeles spp., 381
   germinal vesicle, 158
Polarity, animal-vegetal, 63
Polyadenylated RNA, extraction, 176
Polyadenylation, mRNAs, 170–171, 175–176
Poly(A) RNAs, 54
Polymerase chain reaction:
   footprinting analysis, 353–363
   genomic footprinting, 363
   in vitro copper-phenanthroline digestion, 360–361
   in vivo copper-phenanthroline digestion, 359–360
   kinasing of primers, 362
   linker ligation reaction, 362

primer 1 extension reaction, 362
recipes, 358–359
Taq reactions, 363
Polyploidy, origin, 20
Pregastrular movements, 67
Progesterone, intracellular signaling, 249–250
Proopiomelanocortin, 27
Protein kinases, role in intracellular signaling, 254–256
Protein synthesis, rates in oocytes, 52–54
Protein uptake, by oocytes, 118, 126–129
Protooncogenes, 27
    in oocyte maturation, 190–192
Pseudocapillaroides xenopi, nematode, 13

R

Radioactive precursors, for oocyte labeling, 54–55
Rana pipiens, 25, 45
    disease treatment, 12
    egg envelope, 232
    oocyte culture media, 51
    oocytes, 136
Rana spp., 118
Raw meat, 9
Recombinant assay, 312–314
Red leg, 11–12
Repetitive sequences, chromosomal location, 23
Retinoic acid, dorsoanterior deficient embryo, 276–277, 281–282
Ribosomal proteins, 27
Ribozymes, 380–381
RNA:
    antisense, 379
    experiments, 378
    for mRNA removal, 226–227
    guanidine hydrochloride extraction, 494–496
    for ion channel expression, sources, 493
    lithium chloride-urea extraction, 493–494
    synthesis from cDNA clone, for ion channels, 494–496
RNA extraction, from oocytes, 143–145
RNA polymerases, transcription efficiencies, 168–169
Rotation, effect on egg microtubules, 273

S

Salamanders, lampbrush chromosomes, 150
Salinity, in embryo, 106
Sandwich explants:
    development, 86
    gastrula, 85–89
    tissue interactions, 88–89
Scanning electron microscopy, tissue preparation, 407
Separate nerve or muscle cultures, preparation, 517
Shaved open-faced explants:
    for cell behavior analysis, 92–93
    development, 86
Signal agonists and antagonists:
    external application, 253
    microinjection, 252–253
Signal mimetics, in oocytes, 252–260
Single channel analysis, in ion channel analysis, 506–507
SI nuclease mapping, of spliced RNA, 198
Skin disease, nematode, 13
Skin transplants, 10
Somatic cell nuclei, transplantation to oocytes, 300
Spatial fractionation:
    analysis of fractions, 263–265
    dissection, 260–265
Spemann pipette, 103
Sperm suspension, preparation, 219
Sphingosine, role in intracellular signaling, 256
Staining:
    histological preparation, 396–400
    whole-mount, 420–439
Staphylococcus spp., 12
Stereomicroscopy, in gastrulation and neurulation, 75–78
Sucrose step gradients, in mRNA isolation, 180
Supernumary axis, from microinjected lithium, 280
Suramin:
    for dorsoanterior deficient embryo, 281–282
    effect on gastrulation, 276
Surface cells, time-lapse records, 93
Synaptogenesis, and neuromuscular junction development, 511–512

T

Tadpoles:
  feeding, 8–9
  metamorphosis, 14
  RNA, 144
  sex reversal, 14–15
  sources, 13, 15
Tap water, filters, 6–7
Taq reactions, for genomic footprinting, 363
Taricha torosa, embryo, 102
Tetraploidy, 20, 39
Time-lapse records, of surface cells, 93
Tissue extracts, DNA recombination and repair, 467–484
Tissue identification, antibodies, 317
Tissue processing:
  for electron microscopy, 404–407
  embedding and sectioning, 393–396
  fixation, 391
  fixation and embedding, 404–405
  for light microscopy, 391–400
  sectioning, 405–406
  staining, 396–400, 406–407
Tissue sections:
  autoradiography, 454–455
  drying prohibition, 449
  hybridization,
  conditions, 452
  preliminary procedure, 452–453
  probe addition, 453
  probe synthesis, 450–452
  microscopy, 455–456
  prehybridization, 449–450
  acid and acetic anhydride treatment, 450
  proteinase K treatment, 450
  in situ hybridization,
  autoradiography, 445–456
  preparation, 446–449
  staining and mounting, 455
  washing, 453–454
Toenail clipping, 10
Toluidine Blue O, stain, 398
Torpedo californica, 190
Torpedo spp., ion channels, 492
Transforming growth factors, 320–321
Triton alpestris, 333
Triton spp., 332
Triturus spp., 118
Trypan blue:
  for dorsoanterior deficient embryo, 281–282
  effect on gastrulation, 276
Tungsten needles, 102

U

Ultraviolet light:
  for axis-deficient embryo, 278–279
  dose-dependent effects, 222
  effect on egg polarization, 273
  effect on gastrulation, 276
  in nuclear transplantation, 306
Urodele spp., embryo, 330

V

Vegetal cells, 314
Vegetalizing factor, 320–321, 324
Vegetal microtubular network, disruption, by ultraviolet irradiation, 223
Ventral ectoderm, epidermal bias, 335–336
Vesicle proteins, from oocytes, 142–143
Videomicroscopy, for cell behavior analysis, 92
Vimentin, 27
Vinculin, 146
Vitelline envelope:
  from oviposited eggs, 236
  removal, 103–104, 221
Vitellogenin, 26–27
  isolation, 122–124
  labeling, 124–126
  oocyte surface bound, 128
  purification, 118–119
  synthesis, 134
  titer measurement, 123–124
  uptake, by oocytes, 117–130

W

Water:
  changing, 5–6
  quality, 6–7
  salinity, 7
Water tanks:
  capacity, 5–6
  sizes, 5–6
  water quality, 5–7
Whole-cell voltage clamping, in ion channel analysis, 503

Whole-mount imaging, 420
  bleaching, 421–422
  clearing agents, 420–421
  fixation, 422–426
  staining, 420–439
Whole-mount staining:
  formulations, 437–438
  importance of sectioning, 433–434
  labeling in practice, 434–437
Whole-mount tissue:
  blastocoel roof, for fibronectin, 533–535
  hybridization, probe synthesis, 457–458
  prehybridization, 456–457
  postfix, 457
  protease digestion, 457
  in situ hybridization, 456–459

X

Xenopus borealis hybrids, 286
Xenopus gilli, chromosome mapping, 28
Xtwi, as antibody, 318

Y

Yolk proteins, freon extraction, 138–139

Z

Zygotene, meiotic stage, 150
Zygotes, spatial fractionation, 260–265

CONTENTS OF RECENT VOLUMES

Volume 31

Vesicular Transport

Part A.

Part I. *Gaining Access to the Cytoplasm*

1. LYSED CHROMATOPHORES: A MODEL SYSTEM FOR THE STUDY OF BIDIRECTIONAL ORGANELLE TRANSPORT
   Leah T. Haimo and Moshe M. Rozdzial
2. DIGITONIN PERMEABILIZATION PROCEDURES FOR THE STUDY OF ENDOSOME ACIDIFICATION AND FUNCTION
   Ruben Diaz and Philip D. Stahl
3. ATP PERMEABILIZATION OF THE PLASMA MEMBRANE
   Thomas H. Steinberg and Samuel C. Silverstein
4. PORATION BY α-TOXIN AND STREPTOLYSIN O: AN APPROACH TO ANALYZE INTRACELLULAR PROCESSES
   Gudrun Ahnert-Hilger, Wolfgang Mach, Karl Josef Fohr, and Manfred Gratzl
5. PREPARATION OF SEMIINTACT CHINESE HAMSTER OVARY CELLS FOR RECONSTITUTION OF ENDOPLASMIC RETICULUM-TO-GOLGI TRANSPORT IN A CELL-FREE SYSTEM
   C. J. M. Beckers, D. S. Keller, and W. B. Balch
6. PERFORATED CELLS FOR STUDYING INTRACELLULAR MEMBRANE TRANSPORT
   Mark K. Bennett, Angela Wandinger-Ness, Ivan deCurtis, Claude Antony, Kai Simons, and Jürgen Kartenbeck
7. RECONSTITUTION OF PROTEIN TRANSPORT USING BROKEN YEAST SPHEROPLASTS
   David Baker and Randy Schekman
8. RECONSTITUTION OF TRANSPORT FROM THE ER TO THE GOLGI COMPLEX IN YEAST USING MICROSOMES AND PERMEABILIZED YEAST CELLS
   Hannele Ruohola, Alisa Kastan Kabcenell, and Susan Ferro-Novick
9. DELIVERY OF MACROMOLECULES INTO CELLS EXPRESSING A VIRAL MEMBRANE FUSION PROTEIN
   Harma Ellens, Stephen Doxsey, Jeffrey S. Glenn, and Judith M. White

Part II. *Cell-Free Reconstitution of Transport; Proteins That Interact with the Endodomain of Membranes*

10. RECONSTITUTION OF INTRACELLULAR VESICLE FUSION IN A CELL-FREE SYSTEM AFTER RECEPTOR-MEDIATED ENDOCYTOSIS
    Luis S. Mayorga, Ruben Diaz, and Philip D. Stahl
11. FUSION OF ENDOCYTIC VESICLES IN A CELL-FREE SYSTEM
    Philip G. Woodman and Graham Warren
12. PURIFICATION AND BIOCHEMICAL ASSAY OF SYNEXIN AND OF THE HOMOLOGOUS CALCIUM-DEPENDENT MEMBRANE-BINDING PROTEINS, ENDO-NEXIN II AND LIPOCORTIN I
    Harvey B. Polland, A. Lee Burns, Eduardo Rojas, D. D. Schlaepfer, Harry Haigler, and Keith Broklehurst
13. CHARACTERIZATION OF COATED-VESICLE ADAPTORS: THEIR REASSEMBLY WITH CLATHRIN AND WITH RECYCLING RECEPTORS
    Barbara M. F. Pearse

Part III. *Subcellular Fractionation Procedures*

14. Lectin-Colloidal Gold-Induced Density Perturbation of Membranes: Application to Affinity Elimination of the Plasma Membrane
    Dwijendra Gupta and Alan Tartakoff
15. Immunoisolation Using Magnetic Solid Supports: Subcellular Fractionation for Cell-Free Functional Studies
    Kathryn E. Howell, Ruth Schmid, John Ugelstad, and Jean Gruenberg
16. Flow Cytometric Analysis of Endocytic Compartments
    Russell B. Wilson and Robert F. Murphy
17. Endosome and Lysosome Purification by Free-Flow Electrophoresis
    Mark Marsh
18. Fractionation of Yeast Organelles
    Nancy C. Walworth, Bruno Goud, Hannele Ruohola, and Peter J. Novick

Part IV. *Morphological Procedures*

19. Fluorescence Microscopy Methods for Yeast
    John R. Pringle, Robert A. Preston, Alison E. M. Adams, Tim Stearns, David G. Drubin, Brian K. Haarer, and Elizabeth W. Jones
20. Preservation of Biological Specimens for Observation in a Confocal Fluorescence Microscope and Operational Principles of Confocal Fluorescence Microscopy
    Robert Bacallao and Ernst H. K. Stelzer
21. Organic-Anion Transport Inhibitors to Facilitate Measurement of Cytosolic Free Ca^{2+} with Fura-2
    Francesco Di Virgilio, Thomas H. Steinberg, and Samuel C. Silverstein
22. Postembedding Detection of Acidic Compartments
    Richard G. W. Anderson
23. Transmission Electron Microscopy and Immunocytochemical Studies of Yeast: Analysis of HMG-CoA Reductase Overproduction by Electron Microscopy
    Robin Wright and Jasper Rine
24. Postembedding Labeling on Lowicryl K4M Tissue Sections: Detection and Modification of Cellular Components
    J. Roth
25. Immunoperoxidase Methods for the Localization of Antigens in Cultured Cells and Tissue Sections by Electron Microscopy
    William J. Brown and Marilyn G. Farquhar

Index

Volume 32

Vescular Transport
Part B.
Part I. *Monitoring and Regulating the Progress of Transport–Secretory Path*

1. A Hitchhiker's Guide to Analysis of the Secretory Pathway in Yeast
   Jonathan Rothblatt and Randy Schekman
2. Methods to Estimate the Polarized Distribution of Surface Antigens in Cultured Epithelial Cells
   Enrique Rodriguez-Boulan, Pedro J. Salas, Massimo Sargiacomo, Michael Lisanti, Andre LeBivic, Yula Sambuy, Dora Vega-Salas, Lutz Graeve
3. Analysis of the Synthesis, Intracellular Sorting, and Function of Glycoproteins Using a Mammalian Cell Mutant with Reversible Glycosylation Defects
   Monty Krieger, Pranhitha Reddy, Karen Kozarsky, David Kingsley, Lawrence Hobbie, and Marsha Penman
4. Using Temperature-Sensitive Mutants of VSV to Study Membrane Protein Biogenesis
   John E. Bergmann

CONTENTS OF RECENT VOLUMES

5. Enzymatic Approaches for Studying the Structure, Synthesis, and Processing of Glycoproteins
   Anthony L. Tarentino, Robert B. Trimble, and Thomas H. Plummer, Jr.

6. Separation and Analysis of Glycoprotein Oligosaccharides
   Richard D. Cummings, Roberta K. Merkle, and Nancy L. Stults

7. Protein Folding and Intracellular Transport: Evaluation of Conformational Changes in Nascent Exocytotic Proteins
   Mary-Jane Gething, Karen McCammon, and Joe Sambrook

8. Glycosaminoglycan Modifications of Membrane Proteins
   Karen S. Giacoletto, Tara Rumbarger, and Benjamin D. Schwartz

9. Identification and Analysis of Glycoinositol Phospholipid Anchors in Membrane Proteins
   Terrone L. Rosenberry, Jean-Pierre Toutant, Robert Haas, and William L. Roberts

10. Low Temperature-Induced Transport Blocks as Tools to Manipulate Membrane Traffic
    Esa Kuismanen and Jaakko Saraste

Part II. *Monitoring and Regulating the Progress of Transport—Endocytic and Transcytotic Path*

11. Affinity Labeling of Binding Proteins for the Study of Endocytic Pathways
    Tae H. Ji, Ryuichiro Nishimura, and Inhae Ji

12. Quantitative Evaluation of Receptor-Mediated Endocytosis
    Dwain A. Owensby, Phillip A. Morton, and Alan L. Schwartz

13. Expression and Analysis of the Polymeric Immunoglobulin Receptor in Madin–Darby Canine Kidney Cells Using Retroviral Vectors
    Philip P. Breitfeld, James E. Casanova, Jeanne M. Harris, Neil E. Simister, and Keith E. Mostov

14. Remodeling of Glycoprotein Oligosaccharides after Endocytosis: A Measure of Transport into Compartments of the Secretory Apparatus
    Martin D. Snider

15. A Flow-Cytometric Method for the Quantitative Analysis of Intracellular and Surface Membrane Antigens
    Jerrold R. Turner, Alan M. Tartakoff, and Melvin Berger

16. Control of Coated-Pit Function by Cytoplasmic pH
    Kirsten Sandvig, Sjur Olsnes, Ole W. Peterson, and Bo van Duers

Part III. *Selection and Screening of Vesicular-Transport Mutants of Animal Cells*

17. Replica Plating of Animal Cells
    Jeffrey D. Esko

18. Analysis, Selection, and Sorting of Anchorage-Dependent Cells under Growth Conditions
    Melvin Schindler, Lian-Wei Jiang, Mark Swaisgood, and Margaret H. Wade

19. Positive and Negative Lipsome-Based Immunoselection Techniques
    Lee Leserman, Claire Langlet, Anne-Marie Schmitt-Verhulst, and Patrick Machy

Index

Volume 33

Flow Cytometry

1. Dissociation of Cells from Solid Tumors
   Robert Cerra, Richard J. Zarbo, and John D. Crissman

2. Flow Cytometric Measurement of Cell Viability Using DNase Exclusion
   Oskar S. Frankfurt
3. Cell Cycle Phase-Specific Analysis of Cell Viability Using Hoechst 33342 and Propidium Iodide after Ethanol Preservation
   Alan Pollack and Gaetano Ciancio
4. Cell Membrane Potential Analysis
   Howard M. Shapiro
5. Flow Cytometric Measurement of Intracellular Ionized Calcium in Single Cells with Indo-1 and Fluo-3
   Carl H. June and Peter S. Rabinovitch
6. Measurement of Intracellular pH
   Elizabeth A. Musgrove and David W. Hedley
7. Flow Cytometric Techniques for Measurement of Cytochrome P-450 Activity in Viable Cells
   A. Dusty Miller
8. Fluorescent Staining of Enzymes for Flow Cytometry
   Frank Dolbeare
9. Supravital Cell Staining with Hoechst 33342pnm and DiOC$_5$(3)
   Harry A. Crissman, Marianne H. Hofland, Anita P. Stevenson, Mark E. Wilder, and Robert A. Tobey
10. Specific Staining of DNA with the Fluorescent Antibiotics, Mithramycin, Chromomycin, and Olivomycin
    Harry A. Crissman and Robert A. Tobey
11. DAPI Staining of Fixed Cells for High-Resolution Flow Cytometry of Nuclear DNA
    Friedrich Otto
12. High-Resolution DNA Measurements Using the Nuclear Isolation Medium, DAPI, with the RATCOM Flow Cytometer
    Jerry T. Thornthwaite and Richard A. Thomas
13. Rapid DNA Content Analysis by the Porpidium Iodide–Hypotonic Citrate Method
14. An Integrated Set of Methods for Routine Flow Cytometric DNA Analysis
    Lars Vindeløv and Ib Jarle Christensen
15. DNA Analysis from Paraffin-Embedded Blocks
    David W. Hedley
16. Detection of M and Early-G$_1$ Phase Cells by Scattering Signals Combined with Identification of G$_1$, S, and G$_2$ Phase Cells
    Elio Geido, Walter Giaretti, and Michael Nüsse
17. Controls, Standards, and Histogram Interpretation in DNA Flow Cytometry
    Lynn G. Dressler
18. Cell Division Analysis Using Bromodeoxyuridine-Induced Suppression of Hoechst 33258 Fluorescence
    Ralph M. Böhmer
19. Cell Cycle Analysis Using Continuous Bromodeoxyuridine Labeling and Hoechst 33258-Ethidium Bromide Bivariate Flow Cytometry
    Martin Poot, Manfred Kubbies, Holger Hoehn, Angelika Grossmann, Yuhchyau Chen, and Peter S. Rabinovitch
20. Detection of Bromodeoxyuridine-Labeled Cells by Differential Fluorescence Analysis of DNA Fluorochromes
    Harry A. Crossman and John A. Steinkamp
21. Using Monoclonal Antibodies in Bromodeoxyuride–DNA Analysis
    Frank Dolbeare, Wen-Lin Kuo, Wolfgang Beisker, Martin Vanderlaan, and Joe W. Gray

22. IDENTIFICATION OF PROLIFERATING CELLS BY KI-67 ANTIBODY
    Heinz Baisch and Johannes Gerdes
23. WASHLESS DOUBLE STAINING OF A NUCLEAR ANTIGEN (KI-67 OR BROMODEOXYURIDE) AND DNA IN UNFIXED NUCLEI
    Jørgen K. Larsen
24. ANALYSIS OF PROLIFERATION-ASSOCIATED ANTIGENS
    Kenneth D. Bauer
25. THE STATHMOKINETIC EXPERIMENT: A SINGLE-PARAMETER AND MULTI-PARAMETER FLOW CYTOMETRIC ANALYSIS
    Frank Traganos and Marek Kimmel
26. DETECTION OF INTRACELLULAR VIRUS AND VIRAL PRODUCTS
    Judith Laffin and John M. Lehman
27. DIFFERENTIAL STAINING OF DNA AND RNA IN INTACT CELLS AND ISOLATED CELL NUCLEI WITH ACRIDINE ORANGE
    Zbigniew Darzynkiewicz
28. FLOW CYTOMETRIC ANALYSIS OF DOUBLE-STRANDED RNA CONTENT DISTRIBUTIONS
    Oskar S. Frankfurt
29. SIMULTANEOUS FLUORESCENT LABELING OF DNA, RNA, AND PROTEIN
    Harry A. Crissman, Zbigniew Darzynkiewicz, John A. Steinkamp, and Robert A. Tobey
30. FLOW CYTOMETRIC CELL-KINETIC ANALYSIS BY SIMULTANEOUSLY STAINING NUCLEI WITH PROPIDIUM IODINE AND FLUORESCEIN ISOTHIOCYANATE
    Alan Pollack
31. FLOW CYTOMETRIC METHODS FOR STUDYING ISOLATED NUCLEI: DNA ACCESSIBILITY TO DNASE I AND PROTEIN–DNA CONTENT
    Ryuji Higashikubo, William D. Wright, and Joseph L. Roti Roti
32. ACID-INDUCED DENATURATION OF DNA IN SITU AS A PROBE OF CHROMATIN STRUCTURE
    Zbigniew Darzynkiewicz
33. FLUORESCENT METHODS FOR STUDYING SUBNUCLEAR PARTICLES
    William D. Wright, Ryuji Higashikubo, and Joseph L. Roti Roti
34. CHROMOSOME AND NUCLEI ISOLATION WITH THE $MgSO_4$ PROCEDURE
    Barbara Trask and Ger van den Engh
35. UNIVARIATE ANALYSIS OF METAPHASE CHROMOSOMES USING THE HYPOTONIC POTASSIUM CHLORIDE–PROPIDIUM IODIDE PROTOCOL
    L. Scott Cram, Frank A. Ray, and Marty F. Bartholdi
36. POLYAMINE BUFFER FOR BIVARIATE HUMAN FLOW CYTOGENETIC ANALYSIS AND SORTING
    L. Scott Cram, Mary Campbell, John J. Fawcett, and Larry L. Deaven
37. FLUORESCENCE IN SITU HYBRIDIZATION WITH DNA PROBES
    Barbara Trask and Dan Pinkel
38. FLOW CYTOMETRIC ANALYSIS OF MALE GERM CELL QUALITY
    Donald P. Evenson
39. CELL PREPARATION FOR THE INDENTIFICATION OF LEUKOCYTES
    Carleton C. Stewart
40. MULTIPARAMETER ANALYSIS OF LEUKOCYTES BY FLOW CYTOMETRY
    Carleton C. Stewart
41. IDENTIFICATION AND PURIFICATION OF MURINE HEMATOPOIETIC STEM CELLS BY FLOW CYTOMETRY
    Jan W. M. Visser and Peter de Vries
42. FLUORESCENT CELL LABELING FOR IN VIVO AND IN VITRO CELL TRACKING
    Paul Karl Horan, Meryle J. Melnicoff, Bruce D. Jensen, and Sue E. Slezak
43. RAPID DETERMINATION OF CELLULAR RESISTANCE-RELATED DRUG EFFLUX IN TUMOR CELLS
    Awtar Krishan

44. SLIT-SCAN FLOW ANALYSIS OF CYTOLOGIC SPECIMENS FROM THE FEMALE GENITAL TRACT
Leon L. Wheeless, Jr., Jay E. Reeder, and Mary J. O'Connell

45. CELL SORTING WITH HOESCHST OR CARBOCYANINE DYES AS PERFUSION PROBES IN SPHEROIDS AND TUMORS
Ralph E. Durand, David J. Chaplin, and Peggy L. Olive

46. DNA MEASUREMENTS OF BACTERIA
Harald B. Steen, Kirsten Skarstad, and Erik Boye

47. ISOLATION AND FLOW CYTOMETRIC CHARACTERIZATION OF PLANT PROTOPLASTS
David W. Galbraith

48. FLOW CYTOMETRIC ANALYSIS OF PLANT GENOMES
David W. Galbraith

49. PLANT CELL CYCLE ANALYSIS WITH ISOLATED NUCLEI
Catherine Bergounioux and Spencer C. Brown

50. ENVIRONMENTAL HEALTH: FLOW CYTOMETRIC METHODS TO ASSESS OUR WATER WORLD
Clarice M. Yentsch

51. FLOW MICROSPHERE IMMUNOASSAY FOR THE QUANTITATIVE AND SIMULTANEOUS DETECTION OF MULTIPLE SOLUBLE ANALYTES
Mack J. Fulwyler and Thomas M. McHugh

52. ON-LINE FLOW CYTOMETRY: A VERSATILE METHOD FOR KINETIC MEASUREMENTS
Kees Nooter, Hans Herweijer, Richard Jonker, and Ger van den Engh

53. CALIBRATION OF FLOW CYTOMETER DETECTOR SYSTEMS
Ralph E. Durand

54. SPECTRAL PROPERTIES OF FLUOROCHROMES USED IN FLOW CYTOMETRY
Jan Kapuscinski and Zbigniew Darynkiewicz

INDEX

Volume 34

Vectorial Transport of Proteins into and Across Membranes

1. THE USE OF ANTIIDIOTYPE ANTIBODIES FOR THE CHARACTERIZATION OF PROTEIN–PROTEIN INTERACTIONS
David Vaux and Stephen D. Fuller

2. IN VIVO PROTEIN TRANSLOCATION INTO OR ACROSS THE BACTERIAL PLASMA MEMBRANE
Ross E. Dalbey

3. ANALYSIS OF MEMBRANE PROTEIN TOPOLOGY USING ALKALINE PHOSPHATASE AND β-GALACTOSIDASE GENE FUSIONS
Colin Manoil

4. EXPRESSION OF FOREIGN POLYPEPTIDES AT THE *ESCHERICHIA COLI* CELL SURFACE
Maurice Hofnung

5. IN VITRO BIOCHEMICAL STUDIES ON TRANSLOCATION OF PRESECRETORY PROTEINS ACROSS THE CYTOPLASMIC MEMBRANE OF *ESCHERICHIA COLI*
Shoji Mizushima, Hajime Toduka, and Shin-ichi Matsuyama

6. PREPROTEIN TRANSLOCASE OF *ESCHERICHIA COLI:* SOLUBILIZATION, PURIFICATION, AND RECONSTITUTION OF THE INTEGRAL MEMBRANE SUBUNITS SECY/E
Arnold J. M. Driessen, Lorna Brundage, Joseph P. Hendrick, Elmar Schiebel, and William Wickner

7. IN VITRO PROTEIN TRANSLOCATION INTO *ESCHERICHIA COLI* INVERTED MEMBRANE VESICLES
Phang C. Tai, Guoling Tian, Haoda Xu, Jian P. Lian, and Jack N. Yu

8. MEMBRANE COMPONENTS OF THE PROTEIN SECRETORY MACHINERY
Koreaki Ito and Yoshinori Akiyama

9. SIGNAL SEQUENCE-INDEPENDENT PROTEIN SECRETION IN GRAM-NEGATIVE BACTERIA: COLICIN V AND MICROCIN B17
Rachel C. Skvirsky, Lynne Gilson, and Roberto Kolter

10. TRANSCRIPTION OF FULL-LENGTH AND TRUNCATED mRNA TRANSCRIPTS TO STUDY PROTEIN TRANSLOCATION ACROSS THE ENDOPLASMIC RETICULUM
    Reid Gilmore, Paula Collins, Julie Johnson, Kennan Kellaris, and Peter Rapiejko

11. PROBING THE MOLECULAR ENVIRONMENT OF TRANSLOCATING POLYPEPTIDE CHAINS BY CROSS-LINKING
    Dirk Görlich, Teymuras V. Kurzchalia, Martin Wiedmann, and Tom A. Rapoport

12. RECONSTITUTION OF SECRETORY PROTEIN TRANSLOCATION FROM DETERGENT-SOLUBILIZED ROUGH MICROSOMES
    Christopher Nicchitta, Giovanni Migliaccio, and Günter Blobel

13. ANALYSIS OF PROTEIN TOPOLOGY IN THE ENDOPLASMIC RETICULUM
    Hans Peter Wessels, James P. Beltzer, and Martin Spiess

14. PROTEIN IMPORT INTO PEROXISOMES *IN VITRO*
    Paul B. Lazarov, Rolf Thieringer, Gerald Cohen, Tsuneo Imanaka, and Gillian Small

15. *IN VITRO* RECONSTITUTION OF PROTEIN TRANSPORT INTO CHLOROPLASTS
    Sharyn E. Perry, Hsou-min Li, and Kenneth Keegstra

16. ANALYSIS OF MITOCHONDRIAL PROTEIN IMPORT USING TRANSLOCATION INTERMEDIATES AND SPECIFIC ANTIBODIES
    Thomas Söllner, Joachim Rassow, and Nikolaus Pfanner

17. IMPORT OF PRECURSOR PROTEINS INTO YEAST SUBMITOCHONDRIAL PARTICLES
    Thomas Jascur

18. PULSE LABELING OF YEAST CELLS AS A TOOL TO STUDY MITOCHONDRIAL PROTEIN IMPORT
    Anders Brandt

19. THE PROTEIN IMPORT MACHINERY OF YEAST MITOCHONDRIA
    Victoria Hines and Kevin P. Baker

20. PROTEIN IMPORT INTO ISOLATED YEAST MITOCHONDRIA
    Benjamin S. Glick

21. MITOCHONDRIAL INNER MEMBRANE PROTEASE I OF *SACCHAROMYCES CEREVISIAE*
    André Schneider

22. PURIFIED PRECURSOR PROTEINS FOR STUDYING PROTEIN IMPORT INTO YEAST MITOCHONDRIA
    Ute C. Krieg and Philipp E. Scherer

23. CROSS-LINKING REAGENTS AS TOOLS FOR IDENTIFYING COMPONENTS OF THE YEAST MITOCHONDRIAL PROTEIN IMPORT MACHINERY
    Philipp E. Scherer and Ute C. Krieg

INDEX

Volume 35

Functional Organization of the Nucleus: A Laboratory Guide

Part I. *Visualization of Nucleic Acids*

1. DNA SEQUENCE LOCALIZATION IN METAPHASE AND INTERPHASE CELLS BY FLUORESCENCE *IN SITU* HYBRIDIZATION
   Barbara J. Trask

2. LOCALIZATION OF mRNAs BY *IN SITU* HYBRIDIZATION
   Lynne M. Angerer and Robert C. Angerer

3. FLUORESCENT DETECTION OF NUCLEAR RNA AND DNA: IMPLICATIONS FOR GENOME ORGANIZATION
   Carol Villnave Johnson, Robert H. Singer, and Jeanne Bentley Lawrence

4. VISUALIZATION OF DNA SEQUENCES IN MEIOTIC CHROMOSOMES
   Peter B. Moens and Ronald E. Pearlman

5. NUCLEIC ACID SEQUENCE LOCALIZATION BY ELECTRON MICROSCOPIC *IN SITU* HYBRIDIZATION
   Sandya Narayanswami, Nadja Dvorkin, and Barbara A. Hamkalo

Part II. *Visualization of Proteins*

6. THE USE OF AUTOANTIBODIES IN THE STUDY OF NUCLEAR AND CHROMOSOMAL ORGANIZATION
   W. C. Earnshaw and J. B. Rattner

7. MEIOTIC CHROMOSOME PREPARATION AND PROTEIN LABELING
   C. Heyting and A. J. J. Dietrich

8. DISTRIBUTION OF CHROMOSOMAL PROTEINS IN POLYTENE CHROMOSOMES OF DROSOPHILA
   Robert F. Clark, Cynthia R. Wagner, Carolyn A. Craig, and Sarah C. R. Elgin

9. THE USE OF MONOCLONAL ANTIBODY LIBRARIES
   H. Saumweber

10. OPTICAL SECTIONING AND THREE-DIMENSIONAL RECONSTRUCTION OF DIPLOID AND POLYTENE NUCLEI
    Mary C. Rykowski

Part III. *Identifying Specific Macromolecular Interactions*

11. YEAST MINICHROMOSOMES
    Sharon Y. Roth and Robert T. Simpson

12. NUCLEOSOMES OF TRANSCRIPTIONALLY ACTIVE CHROMATIN: ISOLATION OF TEMPLATE-ACTIVE NUCLEOSOMES BY AFFINITY CHROMATOGRAPHY
    Vincent G. Allfrey and Thelma A. Chen

13. THE NUCLEOPROTEIN HYBRIDIZATION METHOD FOR ISOLATING ACTIVE AND INACTIVE GENES AS CHROMATIN
    Claudius Vincenz, Jan Fronk, Graeme A. Tank, Karen Findling, Susan Klein, and John P. Langmore

14. PROTEIN—DNA CROSS-LINKING AS A MEANS TO DETERMINE THE DISTRIBUTION OF PROTEINS ON DNA *IN VIVO*
    David S. Gilmour, Ann E. Rougvie, and John T. Lis

15. PROTEIN—DNA INTERACTIONS *IN VIVO*—EXAMINING GENES IN *SACCHAROMYCES CEREVUSUAE* AND *DROSOPHILA MELANOGASTER* BY CHROMATIN FOOTPRINTING
    Melissa W. Hull, Graham Thomas, Jon M. Huibregtse, and David R. Engelke

Part IV. *Reconstitution of Functional Complexes*

16. CONTROL OF CLASS II GENE TRANSCRIPTION DURING *IN VITRO* NUCLEOSOME ASSEMBLY
    Jerry L. Workman, Ian C. A. Taylor, Robert E. Kingston, and Robert G. Roeder

17. SYSTEMS FOR THE STUDY OF NUCLEAR ASSEMBLY, DNA REPLICATION, AND NUCLEAR BREAKDOWN IN *XENOPUS LAEVIS* EGG EXTRACTS
    Carl Smythe and John W. Newport

18. *IN VITRO* NUCLEAR PROTEIN IMPORT USING PERMEABILIZED MAMMALIAN CELLS
    Stephen A. Adam, Rachel Sterne-Marr, and Larry Gerace

Part V. *Genetic Approaches*

19. MUTATIONS THAT AFFECT CHROMOSOMAL PROTEINS IN YEAST
    M. Mitchell Smith

20. MUTATIONS THAT AFFECT NUCLEAR ORGANIZATION IN YEAST
    Ann O. Sperry, Barbara R. Fishel, and W. T. Garrard

21. MUTATIONS AFFECTING CELL DIVISION IN *DROSOPHILA*
    Maurizio Gatti and Michael L. Goldberg

22. POSITION EFFECT VARIEGATION: AN ASSAY FOR MUTATIONS IN CHROMOSOMAL PROTEINS AND CHROMATIN ASSEMBLY AND MODIFYING FACTORS
    T. Grigliatti

INDEX
CONTENTS OF RECENT VOLUMES

RANDALL LIBRARY-UNCW

3 0490 0441854 %